D1268423

Scientific Style and Format

EIGHTH EDITION

The mission of the Council of Science Editors is to serve editorial professionals in the sciences by creating a supportive network for career development, providing educational opportunities, and developing resources for identifying and implementing high-quality editorial practices; our purpose is to serve as an authoritative resource on current and emerging issues in the communication of scientific information.

Style Manual Subcommittee: Lindsey S Buscher (project manager); Heather Goodell, Russell Harper, Jody Hundley (advisory group); Louise R Adam; Emmanuel A Ameh; Dana M Compton; Jo Ann M Eliason; Tina L Fleischer; Kelly Gerrity; Beth E Hazen; Devora Krischer; Raymond Lambert; Tom Lang; Kate Mason; Beva Nall-Langdon; Mary Warner

Copy editor: Mary B Corrado

The Council of Science Editors is grateful to the University of Chicago Press for their contributions to the production of this manual.

Other publications of the Council of Science Editors:

CSE's White Paper on Promoting Integrity in Scientific Journal Publications, 2012 Update
Global Theme Issue on Poverty and Human Development (2007)
Science Editor (bimonthly publication of the CSE)

Direct inquiries to:

Council of Science Editors
10200 W 44th Avenue, Suite 304
Wheat Ridge, CO 80033
Email: CSE@CouncilScienceEditors.org
www.councilscienceeditors.org

Scientific Style and Format

The CSE Manual for Authors, Editors, and Publishers

EIGHTH EDITION

Style Manual Subcommittee
Council of Science Editors

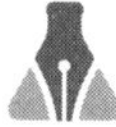
Published by the Council of Science Editors in cooperation with the University of Chicago Press

The University of Chicago Press Chicago and London

The University of Chicago Press, Chicago 60637

The University of Chicago Press, Ltd., London

Printed in the United States of America

23 22 21 20 19 18 17 16 15 14   1 2 3 4 5

ISBN-13: 978-0-226-11649-5 (cloth)

Library of Congress Cataloging-in-Publication Data

Scientific style and format : the CSE manual for authors, editors, and publishers. — Eighth edition.
pages cm
Includes bibliographical references and indexes.
ISBN 978-0-226-11649-5 (cloth : alk. paper) 1. Technical writing—Handbooks, manuals, etc.
I. Council of Science Editors. Style Manual Committee.
T11.S386 2014
808.06′6—dc23

2013033867

♾ This paper meets the requirements of ANSI/NISO Z39.48-1992 (Permanence of Paper).

TABLE OF CONTENTS

PREFACE

ABOUT THIS EDITION

Previous editions[1–7] of the Council of Science Editors (CSE) style manual have contained major expansions, reorganizations, and updates. In 1994, the sixth edition[6] aimed to expand the scope of the manual to include recommendations on scientific style for publications about not only plant sciences, zoology, microbiology, and medical sciences, but also other experimental and observational sciences. CSE officially changed the name of the organization in 2000 from the Council of Biology Editors to the Council of Science Editors, and the seventh edition,[7] published in 2006, more clearly defined and implemented the intended expansion from a society dedicated to editors of publications in the field of biology to a more wide-ranging and inclusive group of disciplines consisting of all physical and life sciences. Now, with the eighth edition, the changes to overall makeup and structure are not as drastic as in the previous several editions, but content updates include some important revisions to language, particularly relating to the perpetually evolving electronic and online environments in scientific writing and publishing; updated references throughout; and a shift in the previously recommended citation–name reference style to citation–sequence (see Chapter 29). We are also thrilled that for the first time in the history of *Scientific Style and Format* and its previous iterations, the CSE flagship style manual is now available in a fully searchable online version, as well as an optimized version for mobile platforms. This much-anticipated online version is available for subscription at www.scientificstyleandformat.org.

The search for a project manager for the eighth edition began in 2010. Once the project manager was appointed, the SSF8 Subcommittee was formed in early 2011 under the auspices of the Publications Committee. CSE members were recruited to review the seventh edition material, make suggestions for updates and changes, and write drafts that were then peer-reviewed by experts in the specific field for each chapter. An Advisory Group consisting of the project manager, a CSE Board member, and two others was also formed to further assist the subcommittee and provide guidance throughout the life of the project.

The *Scientific Style and Format* Eighth Edition Subcommittee worked to ensure the continued integrity of the CSE style and to provide a progressively up-to-date resource for our valued users, which will be adjusted as needed on the website. We hope that this new edition will prove to be an authoritative tool used to help keep the language and writings of the scientific community alive and thriving, whether the research is printed on paper or published online.

SCIENTIFIC SCOPE

Because the traditional boundaries between scientific disciplines are dissolving, Part 3 of this manual, which covers special scientific conventions, is organized not by the various sciences but rather by their subjects, as was the case in the seventh edition. Genetics and biochemistry, for example, were for many years quite separate in the subjects they covered. Now, the molecular structure of genes and how they chemically produce their effects are integrated with the more traditional subjects of genetic inquiry. A report on an inherited characteristic of a particular organism might thus make use of recommendations in the chapters on genetic nomenclature (Chapter 21), chemical formulas and names (Chapter 17), organism structure and function (Chapter 23), and even taxonomy (Chapter 22). This manual's content on special scientific conventions is generally organized according to a rising scale of dimensions, starting with the fundamental units of matter and proceeding up through chemical and cellular components, microorganisms and more complex organisms, to the planet Earth and the rest of the universe.

This manual's recommendations on scientific style focus primarily on nomenclature and symbolization. The principles governing these topics are presented in sufficient detail to make clear the basis and rationale for appropriate style, but the manual does not include comprehensive lists of scientific terms, symbols, and abbreviations for all disciplines. Authoritative documents are available in many disciplines to guide authors who are responsible for coining new terms and symbols and those who need to use them. References for these documents, many of them now available online, are given at the end of the relevant chapters.

ORGANIZATION

The text is divided into chapters grouped in 4 parts. Part 1 provides an overview of important issues related to scientific publication, including editorial practices and copyright. Part 2 covers conventions widely applied in both general and scientific scholarly publishing. Part 3 covers details of style mainly applicable in the sciences. Part 4 recommends formats applicable in scientific journals, books, and other media, including formats for bibliographic citations and references.

Within each chapter, the section numbering is hierarchical, and section numbers are linked to headings.

In a departure from the seventh edition, references and their citations are presented in the citation–sequence format. This choice was based on certain advantages of that system (see Section 29.2.3), in particular, ease of finding references as the user reads the chapter from beginning to end. Authoritative online resources, where such are available, are provided in chapter reference lists.

CHANGES FROM THE PREVIOUS EDITION

Overview

The most significant changes in the eighth edition include closing up "email" and "website" and making them lowercase. The subcommittee recognizes that these moves are widely debated and may still be considered controversial by some, but we ultimately decided that dropping the hyphen in "email" and closing the space in "website" are the progressive and responsible ways to go and will keep CSE style on the front end of linguistic evolution. Other noteworthy changes include shifting the reference citation format within each chapter to citation–sequence, as previously mentioned, so that now each chapter's references section is numbered in the order in which each reference is cited instead of numbered alphabetically (as in the seventh edition); and adding new recommendations and examples to Chapter 29 for citing e-books, online images and infographics, podcasts and webcasts, online videos, blogs, and social networking sites.

We have also updated the language throughout the manual to be in line with a more integrated online and electronic publishing environment: we changed terms such as "World Wide Web" to "Internet" or rephrased to say "available/published online" or similar, and we removed most mentions of antiquated media such as magnetic tapes, floppy disks, typewriters, handwritten manuscripts, videocassettes, and audiocassettes, except where noted in historical contexts.

In an effort to reduce confusion for authors and editors using this manual, we removed most typesetting- and publishing-specific details from each chapter as appropriate and instead provide that information as an online-only publishing appendix that is available at www.scientificstyleandformat.org.

Significant Chapter-Specific Updates

In Chapter 1, "Elements of a Scientific Publication", Section 1.2.1, information was added about abstracts, which are now considered an essential part of a scientific publication, and how abstracting and indexing databases have become increasingly important in a publication's visibility in scientific and academic communities.

Chapter 2, "Publication Policies and Practices", provides new details about online databases and repositories (Section 2.2) and includes a new section titled "Responsibilities of Publishers, Journal Owners, and Sponsoring Societies" (Section 2.9).

Recent changes to copyright application instructions are outlined in Chapter 3, "The Basics of Copyright", Section 3.2.2.2. There is also a new section in Chapter 3, "Creative Commons Licenses" (Section 3.3.4), which provides details to help users interested in making their work available to others and the various levels of licensing possible through the Creative Commons licensing office.

Chapter 4, "Alphabets, Symbols, and Signs", contains updated information about STIX fonts (Section 4.3), including Version 1.1.0, which was released in 2012.

In response to requests from users of the seventh edition, a new section was added to Chapter 7: "Active vs. Passive Voice" (Section 7.9). Though other style manuals may have similar sections on this topic, it is as important in scientific writing as it is in other prose writing to use the correct voice. The authors of this new section provide recommendations on the appropriate place within a scientific manuscript to use each type of voice.

Chapters 12 ("Numbers, Units, Mathematical Expressions, and Statistics") and 30 ("Accessories to Text: Tables, Figures, and Indexes") were both significantly revised and reorganized for ease of use and to provide the user with recent updates in their respective areas. Also specified in Chapter 12 (and mentioned in Chapter 5) is a change in recommendation to now use a comma in whole numbers with 4 or more digits. The seventh edition recommended using thin spaces to separate 5 or more digits, which created more keystrokes and led to differences in the way various typesetting systems interpreted the amount of space a thin space should have, so the authors decided to simplify this and avoid discrepancies and limitations by using a comma (Section 12.1.3.1).

The title of Chapter 13 was changed from "Time and Dates" in the seventh edition to "Time, Dates, and Age Measurements", which better encompasses the contents of the chapter. Clarifications were made to more accurately describe the use of time units and their approved abbreviations (Section 13.2.1) and how best to refer to clock time (Section 13.2.2).

In Chapter 14, "Geographic Designations", outdated references to the USSR were removed and other country names were updated, as well as URLs for online databases of country and geographic names and postal codes.

Within Chapter 15, "The Electromagnetic Spectrum", Section 15.2, "Longer Wavelengths", was simplified for better understanding.

Updates were made to Chapter 18, "Chemical Kinetics and Thermodynamics", to be in line with the most recent edition of the International Union of Pure and Applied Chemistry (IUPAC) Green Book, such as the fact that the symbol for "apparent" is now "app" and the abbreviation for "fluid" is now "f" instead of "fl".

In Chapter 19, "Analytical Chemistry", Section 19.2.1, "Types of Techniques", 2 new techniques were added: electron microscopy and lab-on-a-chip.

Chapter 21, "Genes, Chromosomes, and Related Molecules", contains new information in Section 21.2, "Genetic Units, Measures, and Tools", to address new technologies and the naming of single-nucleotide polymorphisms. There are also updated sections on cotton nomenclature (Section 21.14.7.1), soybean genetics (Section 21.14.5.3), and human genetic nomenclature (Section 21.15.7).

In Chapter 23, "Structure and Function", Section 23.5.10, "Thermal Physiology", was reorganized and modified for clarity.

Several significant updates were made in Chapter 26, "Astronomical Objects and Time Systems". In Section 26.4.2, the title was changed from "Minor Planets (Asteroids) and Comets" to "Dwarf Planets and Small Solar System Bodies (Asteroids and Comets)" and the use of "minor planets" was removed throughout the chapter in accordance with the 2006 definitions from the International Astronomical Union, which famously downgraded Pluto from a planet to a dwarf planet.

Chapter 27, "Journal Style and Format", contains some significant updates to language, including removing references to videocassettes as a modern, viable option for publication (Section 27.2, "Format"), acknowledging the increasingly prevalent use of online-only publications and parts of publications (Section 27.3.1.2), and updating references to browsers (e.g., removed mention of AltaVista; Section 27.3.1.2.2, "Production Issues").

The title of Chapter 28 was changed from "Books, Technical Reports, Conference Proceedings, and Other Monographs" to "Published Media", which is inclusive of all forms of publications, which are constantly changing in this age of rapid advancement in technology. The chapter was revised accordingly to allow for this in the coming years until the next edition of this manual is published.

There are also some significant updates in Chapter 29, "References". In Section 29.3.7.11, "Audiovisuals", we have changed our recommendation from including a physical description in a reference listing to not including it. The title of Section 29.3.7.13 was changed from "Homepages and Other Internet Material" to "Websites and Other Online Formats" and, as previously mentioned, recommendations and examples were added for reference citations to e-books, online databases, online images and infographics, podcasts and webcasts, online videos, blogs, and social networking sites. This chapter will likely need to be updated again in the next edition of this manual.

Distinctions were added in Chapter 31, "Typography and Manuscript Preparation", to clarify the differences between "typeface", "text attributes", and "font" (Sections 31.1 and 31.2.1). Language was also updated throughout the chapter to accommodate electronic software systems as opposed to more print-centric tools, and Section 31.3.2, "Traditional and Electronic Mark-Up", includes additional information about XML, HTML, and electronic devices.

Finally, Chapter 32, "Proof Correction", was drastically updated to feature the specification in Section 32.4, "Marking Paper Proofs", indicating mark-up recommendations to hard-copy proofs, and the addition of a new section, "Annotating PDF Proofs" (Section 32.5). The new section does not lay out its own best practices, as many are available, but it does provide helpful information covering the basics of PDF annotation as well as a few new figures depicting tools an author or editor could use, depending on the version of Adobe Acrobat that is available.

RECOMMENDATIONS FOR FUTURE EDITIONS

Monitoring the scientific literature for documents establishing or recommending new nomenclature, notations, or formats is a huge task. We urge scientific societies, committees, working groups, individual scientists, and others to send to the Council of Science Editors both formal and informal documents with recommendations on nomenclature, symbols, and style, whether they are new or simply not represented in this edition. The committee responsible for the style manual will review suggestions in consultation with subject experts and decide on the recommendations to be represented in future editions.

Send suggestions to the Council of Science Editors:

Attention: Council of Science Editors
10200 W 44th Avenue, Suite 304
Wheat Ridge, CO 80033
Email: CSE@CouncilScienceEditors.org
Website: www.councilscienceeditors.org

CITED REFERENCES

1. Conference of Biological Editors, Committee on Form and Style. Style manual for biological journals. Washington (DC): American Institute of Biological Sciences; 1960.

2. Conference of Biological Editors, Committee on Form and Style. Style manual for biological journals. 2nd ed. Washington (DC): American Institute of Biological Sciences; 1964.

3. Council of Biology Editors, Committee on Form and Style. CBE style manual. 3rd ed. Washington (DC): American Institute of Biological Sciences; 1972.

4. Council of Biology Editors, CBE Style Manual Committee. Council of Biology Editors style manual: a guide for authors, editors, and publishers in the biological sciences. 4th ed. Bethesda (MD): Council of Biology Editors; 1978.

5. Council of Biology Editors, CBE Style Manual Committee. CBE style manual: a guide for authors, editors, and publishers in the biological sciences. 5th ed., revised and expanded. Bethesda (MD): Council of Biology Editors; 1983.

6. Council of Biology Editors, Style Manual Committee. Scientific style and format: the CBE manual for authors, editors, and publishers. 6th ed. New York (NY): Cambridge University Press; 1994.

7. Council of Science Editors, Style Manual Committee. Scientific style and format: the CSE manual for authors, editors, and publishers. 7th ed. Reston (VA): The Council; 2006.

ACKNOWLEDGMENTS

Language is a living thing. We can feel it changing. Parts of it become old: they drop off and are forgotten. New pieces bud out, spread into leaves, and become big branches, proliferating.

Gilbert Highet
Professor, Columbia University
1906–1978

In an episode of the podcast Lexicon Valley,[1] the hosts discuss the importance of documenting language even though it will, in all likelihood, eventually become extinct or change so drastically over time that it takes on a completely new form or characteristic, leaving the original version unrecognizable. They conclude that even if extinction is inevitable, it is "still worth the effort to document [language] because it honors . . . the effort of the people who spent centuries or more developing it."[1] Though the Council of Science Editors has published editions of its style manual only since 1960, not for centuries, it is with this deep appreciation that the Eighth Edition Style Manual Subcommittee gives profound thanks and much credit to the seventh edition committee members, as well as those of all previous editions, for their devotion to documenting the style and format guidelines and recommendations found in these pages, as the eighth edition is very largely based on the content and organization of all previous editions, especially that of the seventh.

To go one step further, it could also be said that documentation of the evolution of language is equally important for future lexicographers, language historians, and researchers of cultural and scientific knowledge.[1] As Gilbert Highet suggests, "language is a living thing", so with this eighth edition, we have massaged and nurtured the verbiage of previous editions to help the manual evolve and remain alive and relevant as an authoritative reference text in the writing and editing of physical and life sciences publications.

Many individuals contributed to the updates in this edition of *Scientific Style and Format*. Without the help of the following people, this edition would not be what it is today. An enormous amount of gratitude goes to those who were involved, in ways both great and small. If we have omitted the names of anyone who helped with the preparation of this manual, it is unintentional and we sincerely apologize for the error.

PROJECT MANAGER

Lindsey S Buscher, ELS
Allen Press, Inc
Chapters 1, 28, 29, 31, 32

ADVISORY GROUP

Heather Goodell, MIS, American Heart Association
Russell Harper, freelance editor
Jody Hundley, American Heart Association

STYLE MANUAL SUBCOMMITTEE

Louise R Adam, ELS
Federation of Animal Science Societies
Chapters 21, 22, 24

Emmanuel A Ameh, MBBS, FWACS, FACS
Annals of African Medicine and Ahmadu Bello University
Chapters 2, 27

Dana M Compton
National Academy of Sciences
Chapters 4, 10

Jo Ann M Eliason, MA, ELS
JNS Publishing Group
Chapters 2, 3, 8, 13, 14, 28

Tina L Fleischer
Dartmouth Journal Services
Chapters 19, 23, 29

Kelly Gerrity
National Academy of Sciences
Chapters 25, 26

Beth E Hazen, PhD
Willows End Scientific Editing and Writing
Chapters 15, 16, 17

Devora Krischer, ELS
CVS Caremark (retired)
Chapters 5, 6, 9, 11, 29

Raymond Lambert
Duke University Press
Chapter 18

Thomas A Lang, MA
Tom Lang Communications and Training International
Chapters 12, 30

Kate Mason, ELS
JNS Publishing Group
Chapters 5, 7, 27

Beva Nall-Langdon
Biotext
Chapter 20

Mary Warner, MS
American Geophysical Union
Chapters 16, 17

REVIEWERS

Jean Baldwin, Allen Press, Inc (Chapter 14)
Carolyn Brown, freelance editor (Chapters 2, 30, 31)
Monica L Helton, *The Journal of Pediatrics* (Chapter 13)
Lynn Jaluvka, MFA, ELS, GlaxoSmithKline (Chapter 8)
Kathryn Kadash-Edmondson, Write Science Right (Chapter 15)
Russell Kohel, US Department of Agriculture (Chapter 21, Cotton)
Raymond Lambert, Duke University Press (Chapter 12)
Beva Nall-Langdon, Biotext (Chapter 7)
Teresa A Oblak, The JB Ashtin Group, Inc (Chapters 16, 17, 18)
Nicola Parry, Massachusetts Institute of Technology (Chapters 19, 20, 21, 24, 25, 26)
R Michelle Sauer, University of Texas Health Science Center at Houston (Chapters 6, 11)
Rupinder Sayal, Michigan State University (Chapters 3, 4, 5)
Jonathan Schultz, American Heart Association (Chapters 1, 22, 23, 27, 28, 32)
Ann Tennier, Stanford University (Chapters 9, 10, 22)

PROOFREADERS

Tracy Candelaria, Allen Press, Inc (Chapters 25, 26)
Lindsey Christopher, Allen Press, Inc (Chapters 5, 6)
Myron Grotta, Allen Press, Inc (Chapters 11, 12)
Leslie Hunter, Allen Press, Inc (Chapters 7, 8)
Raymond Lambert, Duke University Press (Chapters 19, 20)
Joel T Luber, Duke University Press (Chapters 2, 3)
Suzanne Meyers, ELS, Suzanne Meyers Editing (Chapter 13)
David Nadziejka, Van Andel Research Institute (Chapters 15, 16, 17, 18)
Danielle R Smith, Allen Press, Inc (Chapters 9, 30)
Linda Tamblyn, Allen Press, Inc (Chapters 23, 24, 27)
Ann Tennier, Stanford University (Chapters 14, 22)

Unending thanks also go to David Morrow, senior editor, University of Chicago Press, for his constant support and guidance throughout this adventure, and without whom I could not have managed this project. Much gratitude also goes to Jenny Gavacs, assistant editor, University of Chicago Press, for her knowledgeable assistance and infinite patience during the beginning stages of the print production process.

Lindsey S Buscher, ELS
Project Manager

Cited Reference

1. Vuolo M. Our dying words [podcast, episode 16]. Lexicon Valley. Slate. 2012 July 9, 21:10 minutes. [accessed 2012 Oct 23]. http://www.slate.com/articles/podcasts/lexicon_valley/2012/07/lexicon_valley_why_should_we_care_if_a_language_goes_extinct_.html.

1

Publishing Fundamentals

1 Elements of a Scientific Publication

1.1 TYPES OF SCIENTIFIC PUBLICATIONS

Scientific publications take myriad forms, typically that of a scholarly journal or book. The basis for the development of knowledge in the sciences is the original research report, describing observational or experimental investigations of some phenomenon of interest. A review article attempts to answer a particular question by analyzing in aggregate the results of previously published research. A systematic approach to the review helps to ensure that the conclusions reached are supported by the data; hence, systematic reviews may include meta-analysis, "a set of statistical techniques that combine quantitative results from independent studies".[1] Methodological articles describe new methodological approaches or modifications of existing ones but do not report original data.[2] Journal editorials offer a forum for expert opinion, usually on a narrow topic (e.g., providing additional context or interpretation for an article appearing in the same issue). The letters to the editor section of a journal affords readers the opportunity to respond to articles that have been published. Like journal articles, scholarly books may report original research or review the literature,[3] but the subject of a book will be broader or treated in more depth than is possible in an article. See Section 27.7.5 for additional description of research articles, reviews, and editorials.

The editor, associate or assistant editors, and/or editorial board of a journal determine the scope of the journal's content and the categories of articles that will be considered for publication. This information is usually made available to prospective authors in the instructions for authors section of the journal, on the journal's website, or in the publisher's style manual.[1–4] Authors should use such statements of scope to select the target journal for their publications and thereby avoid delay that may result from submission to an inappropriate journal.

1.2 COMPONENTS OF SCIENTIFIC PUBLICATIONS

The basic components of an original research article (introduction, methods, results, and discussion [IMRAD]) have become well established, but just as journal editors specify the scope of content and types of articles that will appear in their journals, they may also outline how each type of article should be organized. This information forms an important part of a journal's instructions for authors. Therefore, once the target journal has been identified, authors should consult and follow the journal's instructions, again to avoid unnecessary delay. Such guidelines impose structure within a document, and this structure has important benefits for readers, including other researchers.

1.2.1 Abstract

As noted in Section 27.7.1.4, both the National Information Standards Organization (NISO)[5] and the International Organization for Standardization (ISO)[6] stipulate that an abstract should be published with every journal article, essay, and discussion. An abstract helps readers to decide whether the article is of interest; as such, its content must reflect the content of the article as closely as possible, within the length available. Most journals specify in their instructions for authors how an abstract is to be organized. The use of a structured format (with formal headings) is one way to ensure that all pertinent information from the article appears in the abstract. The *American Medical Association Manual of Style*[1] provides detailed information on abstracts for medical articles, including structured abstracts.

The presence and quality of an abstract has become increasingly important in scientific publications. Abstracting and indexing databases such as Thomson Reuters' BIOSIS, Elsevier's Scopus (http://www.scopus.com), PubMed (http://www.ncbi.nlm.nih.gov/pubmed/), and EBSCO (http://www.ebsco.com) are a few of the common services available to publishers wanting to expand the reach of a publication within the scientific community. Registration with an abstracting service can increase a publication's exposure and citation by making abstracts easily searchable and available to the public.

1.2.2 Text

The text of an article should follow the outline specified in the journal's instructions for authors. For research articles, this will usually be some variation on the IMRAD form along with a references section, but authors should note any variations requested by the journal. The organization of other types of articles is much more variable.

The methods section is particularly important for allowing a reader to assess whether the study results are pertinent to his or her own work, either research or application. As such, it must include sufficient detail to allow another researcher to repeat the experiment. The following is a general checklist of the types of information that should be included:

1) for biological studies, unambiguous identification of genus, species, and strain; the source of any organisms (e.g., cell line, animal stock); and the age, sex, weight, and condition of organisms, as appropriate

2) for clinical studies, pertinent details about human subjects, including methods of recruitment and relevant physical characteristics
3) for studies involving fieldwork, the specific location of study sites, with appropriately detailed maps[4,7]
4) unambiguous identification of nonbiological materials used (e.g., chemicals), including the source of such materials[3]
5) the types of apparatus used, including model number and manufacturer for specialized equipment
6) the experimental procedure (by reference to a previous report using the same procedure or by detailed description), including potential hazards, if applicable[3]
7) the types of tests performed, including statistical tests[8]

More detailed checklists may be provided in the target journal's instructions for authors. For medical research, many medical and health care journals and editorial groups, including the Council of Science Editors, support the Consolidated Standards of Reporting Trials (CONSORT).[9] Other checklists include that of the *BMC Medical Research Methodology*[10] and the National Institutes of Health.[11]

Consideration should also be given to the presentation of the data and other information; tables and figures are often clear and concise ways to do this. If space is limited, as in a printed publication, supplemental tables and figures that are available only online may be a better option. Depending on a journal's online capabilities, videos and other data formats may also be possible. See Chapter 30 for details about accessories to text.

1.2.3 Other Requirements

Depending on the subject of the article and the journal in which it will be published, there may be other required elements. For example, the medical journals represented by the International Committee of Medical Journal Editors now require that clinical trials be registered in an acceptable trial registry before patient enrollment begins.[12] In experimental animal studies, a statement confirming adherence to ethical guidelines for the care and treatment of animals (e.g., that of the American Psychological Association[13]) is usually required. Ethics approval by an institutional review board and consent for participation are required for research involving human subjects. Acknowledgment of the sources of financial and other support and statements of potential conflict of interest help editors and readers to assess whether there might be any bias (see Section 2.5). Signed copyright release forms or license agreements are also common requirements, typically stating that authors agree to transfer all or some rights to the publisher on acceptance of a manuscript (see Section 3.1.3.3).

1.3 STRUCTURE AND XML TAGGING

Ensuring that scientific articles follow an established format appropriate to the article type and that they include all necessary information, as outlined above, is integral to the comparison of data across studies. Such comparisons and other analyses of the literature usually begin with searches of electronic databases, a process made easier by the use of structured tagging.

The Extensible Markup Language (XML),[14] based on the Standardized General Markup Language (SGML),[15] is a metalanguage that is used to "make up and define markup systems"[16] for documents intended for electronic delivery. Like SGML, XML tagging specifies the structure of a document, rather than its appearance. User-defined tags identify various components; these may be parts of the article as a whole, such as the title and author byline, or parts of the content, such as the organism under study or the statistical tests performed. Although not required in XML, the role of each element can be defined in a formal model called the document type definition (DTD).[17] Documents that have been tagged with XML can be searched and analyzed for various types of content, regardless of the terminology used.

The scientific community is taking advantage of the potential offered by XML in a variety of ways. Discipline-specific tagging systems have been or are being developed (e.g., the Mathematical Markup Language[18]). Research groups are defining DTDs to "[facilitate] viewing, sharing and merging . . . data from different laboratories".[19] Publishers and publisher groups have developed DTDs and XML tag sets. The National Center for Biotechnology Information (NCBI) has developed a DTD that is recommended for journal articles submitted to PubMed Central (a "free full-text archive of biomedical and life sciences journal literature at the U.S. National Institutes of Health's National Library of Medicine" [http://www.ncbi.nlm.nih.gov/pmc/]).[20] Since its inception in 2002, NCBI has expanded the original DTD intended only for article archiving and interchange into a 3-part tag suite (http://jats.nlm.nih.gov/index.html). The full benefit of these tools depends on well-structured scientific publications.

See Section 28.6.3 for a discussion of metadata.

CITED REFERENCES

1. Iverson C, Christiansen S, Flanagin A, Fontanarosa PB, Glass RM, Gregoline B, Lurie SJ, Meyer HS, Winker MA, Young RK. American Medical Association manual of style: a guide for authors and editors. 10th ed. New York (NY): Oxford University Press; 2007. Chapter 1, Types of articles; p. 3–6.

2. Publication manual of the American Psychological Association. 6th ed. Washington (DC): The Association; 2009.

3. Dodd JS, editor. The ACS style guide: a manual for authors and editors. 2nd ed. Washington (DC): American Chemical Society; 1997.

4. Hansen WR, editor. Suggestions to authors of the reports of the United States Geological Survey. 7th ed. Reston (VA): United States Geological Survey; 1991 [modified 2005 Jan 25; accessed 2011 Nov 16]. 289 p. http://www.nwrc.usgs.gov/lib/lib_sta.htm.

5. National Information Standards Organization. Guidelines for abstracts. Bethesda (MD): NISO Press; 2010. (ANSI/NISO Z39.14-1997 (R2009)). http://www.niso.org/.

6. International Organization for Standardization. Documentation—abstracts for publications and documentation. Geneva (Switzerland): ISO; 1976. (ISO 214). http://www.iso.org/iso/home/store/catalogue_tc/catalogue_detail.htm?csnumber=4084.

7. Bates RL, Adkins-Heljeson MD, Buchanan RC, editors. Geowriting: a guide to writing, editing, and printing in earth science. 5th rev. ed. Alexandria (VA): American Geological Institute; 1995.

8. Lang TA, Secic M. How to report statistics in medicine: annotated guidelines for authors, editors, and reviewers. 2nd ed. Philadelphia (PA): American College of Physicians; 2006.

9. Moher D, Schulz KF, Altman DG. CONSORT 2010 checklist of information to include when reporting a randomized trial. [place unknown]: CONSORT Group; 2010 [accessed 2011 Nov 17]. http://www.consort-statement.org/consort-statement/overview0/#checklist.

10. Bornhöft G, Maxion-Bergemann S, Wolf U, Kienle GS, Michalsen A, Vollmar HC, Gilbertson S, Matthiessen PF. Checklist for the qualitative evaluation of clinical studies with particular focus on external validity and model validity. BMC Med Res Methodol. 2006 [accessed 2011 Nov 17];(6):56. http://dx.doi.org/10.1186/1471-2288-6-56.

11. National Institutes of Health, National Institute of Neurological Disorders and Stroke. Clinical trial checklist. Bethesda (MD): NINDS; [accessed 2011 Nov 17]. http://www.ninds.nih.gov/research/clinical_research/toolkit/clinical_trial_checklist.pdf.

12. International Committee of Medical Journal Editors. Uniform requirements for manuscripts submitted to biomedical journals: obligation to register clinical trials. Philadelphia (PA): ICMJE; [updated 2010 Apr; accessed 2011 Nov 16]. http://www.icmje.org/publishing_10register.html.

13. APA Board of Scientific Affairs, Committee on Animal Research and Ethics. Guidelines for ethical conduct in the care and use of animals. Washington (DC): American Psychological Association; 2011 [accessed 2011 Nov 16]. http://www.apa.org/science/leadership/care/guidelines.aspx.

14. W3C XML Core Working Group. Extensible Markup Language (XML) 1.0. 5th ed. [place unknown]: W3C Consortium; 2008 [accessed 2011 Nov 16]. http://www.w3.org/TR/REC-xml/. W3C Recommendation.

15. International Organization for Standardization. Information processing—text and office systems—Standard Generalized Markup Language (SGML). Geneva (Switzerland): ISO; 1989. (ISO 8879). ISO 8879:1986/Amd 1:1988; ISO 8879:1986/Cor 1:1996; ISO 8879:1986/Cor 2:1999; ISO/IEC TR 9573-11:2004; ISO 11748-3:2002; ISO/IEC TR 19758:2003; ISO/IEC 19757-3:2006. http://www.iso.org/iso/home/search.htm?qt=8879&sort=rel&type=simple&published=on.

16. Kasdorf B. XML and PDF: why we need both. An introduction to the two core technologies for publishing. JP News J Publ. 2003;(2):1,3–14. Also available at http://www1.allenpress.com/newsletters/pdf/JP-2003-02.pdf.

17. Bryan M. An introduction to the Extensible Markup Language (XML). Bull Am Soc Inf Sci. 1998 Oct–Nov [accessed 2011 Nov 16];25(1). http://www.asis.org/Bulletin/Oct-98/bryanart.html.

18. W3C XML Core Working Group. Mathematical Markup Language (MathML). Ver. 2.0. 2nd ed. [place unknown]: W3C Consortium; 2003 [accessed 2011 Nov 16]. http://www.w3.org/TR/MathML2/. W3C Recommendation.

19. Nohle DG, Ayers LW. The tissue microarray data exchange specification: a document type definition to validate and enhance XML data. BMC Med Inform Decis Mak. 2005;5(1):12. http://www.biomedcentral.com/1472-6947/5/12.

20. Journal publishing DTD. Bethesda (MD): National Library of Medicine (US), National Center for Biotechnology Information; [updated 2012 Aug; accessed 2013 Jan 14]. http://jats.nlm.nih.gov/publishing/1.0/.

2 Publication Policies and Practices

2.1 SOURCES AND SCOPE

This chapter is drawn largely from policy statements prepared by the Editorial Policy Committee of the Council of Science Editors,[1] *Ethics and Policy in Scientific Publication*,[2] and the *CSE's White Paper on Promoting Integrity in Scientific Journal Publications, 2012 Update*.[3] Further valuable detailed treatment of editorial policy is available in statements by the International Committee of Medical Journal Editors (ICMJE),[4] the World Association of Medical Editors (WAME),[5] the Committee on Publication Ethics (COPE),[6] and the EQUATOR Network (Enhancing the QUAlity and Transparency Of health Research),[7] as well as in the *American Medical Association Manual of Style*.[8] Because most professional societies have produced guidelines on ethical publication, discipline-specific guidance may also be available; a sample of these guides is included in the Additional References at the end of this chapter. We suggest that readers seek out information specific to their own subject of study.

Situations involving unethical behavior, such as scientific or publishing misconduct, may arise or be revealed during the editorial review of a manuscript or after its publication. Thus journal editorial offices, editors, and editorial boards often become involved in resolving misconduct issues. For specific situations involving misconduct, most professional societies will rely on their internal ethical guidelines or codes of conduct regarding publication issues; these guidelines also inform authors of proper conduct. In some countries, independent bodies exist to arbitrate or investigate certain situations of misconduct—for example, in the United States, the Office of Research Integrity of the Public Health Service[9] and the Office of Inspector General of the National Science Foundation.[10]

2.2 RESPONSIBILITY TO PUBLISH RESEARCH RESULTS

Historically, publication has been the standard method for formally presenting new findings to fellow scientists and the public at large.[11] In recent years, however, concerns have been expressed about the appropriate avenue for publication ("primary" versus "gray" literature), about the "cloistering" of data whereby study findings and materials are purposefully made unavailable to other scientists,[12] and about the suppression of findings that could be used for illicit purposes.[13]

Primary literature is an account of original research that is readily available via library or Internet searches because it appears in print or electronic books or journals. Gray literature is work that is not as readily available because it may be an internal government document, white paper, or in-house literature. Gray literature is an appropriate venue for a wide variety of scientific findings (e.g., data reports, interim progress reports), but sometimes authors and their employers choose to release information in the form of gray literature for reasons of convenience rather than appropriateness: publishing gray literature is easier and faster because, unlike primary literature, it does not require peer review and editorial oversight.[14] This practice impedes ready accessibility to important findings, undermines the credibility of such findings, increases the likelihood that fallacious findings will obfuscate and thereby slow scientific advances, and constrains the credit that authors receive for their contributions.[11,14]

Suppression of research findings may be related to protecting intellectual property and/or proprietary interests. In some instances, it may be legitimate; however, in its most egregious form, research findings are suppressed that would, if published, act against the financial interests of authors or their employers. Although a certain delay is sometimes warranted, extended delay or total suppression prevents science from advancing because it allows redundant research, which wastes time and limited resources.[15]

Attention has increasingly been focused on sharing of existing data (as opposed to development of new data). This has been at least partially facilitated by improved technologies that enable large data sets to be readily accessed and shared. Recognizing publication and data sharing as wheels moving science forward, in 2001, the US National Research Council appointed an interdisciplinary group, the Committee on Responsibilities of Authorship in the Biological Sciences, to examine publication practices and determine how the upswing of conflicting forces (e.g., time constraints, financial interests, even bioterrorism) might be addressed and incorporated into the traditional fabric of publishing and sharing data. The committee concluded that all scientific investigators, whether from academia, government, or commercial enterprises, are obligated "not only [to] release data and materials to enable others to verify or replicate published findings (as journals already implicitly or explicitly require) but also to provide them in a form on which other scientists can build with further research."[16] Underpinning this conclusion is the understanding that all members of the scientific community derive benefits from an open communication system and, therefore, information should flow in multiple directions. It follows that because all scientists are beneficiaries of this open system, they are likewise obligated to support it with their own information and data. From this basic tenet, the committee developed guiding principles and made explicit recommendations related to making data and materials available.

Online public databases have been created where researchers can place large data sets that can be shared with others. Examples include clinical trial registration websites (e.g., ClinicalTrials.gov[17] and the EudraCT [European Union Drug Regulating Authorities Clinical Trials] database[18]) and databases offering high-resolution structural data and nucleotide sequences (such as the Research Collaboratory for Structural Bioinformatics (RCSB) Protein Data Bank[19] and database members of the International Nucleotide Database Collaboration[20]). Specialties have identified areas in which sharing of data is essential to drive science forward. A journal's instructions for authors should list these areas and, if appropriate, website addresses. They also should specify whether authors are encouraged or mandated to place their information on these sites.

"Negative findings", which show that a hypothesis is incorrect or anticipated results are not forthcoming, provide valuable information for researchers and may keep them from traveling a false path of experimentation and repeating experiments that are unlikely to succeed. Unfortunately, negative findings are often viewed as uninteresting and unlikely to be cited. Because of this, traditionally journals rarely publish them. To alert other researchers that a research plan will not work, authors should be encouraged to write and submit papers on their negative findings, and editors *should* include these papers in their journals. Journals should also be encouraged to print negative results. To date they do so rarely, but in at least one case negative results are the focus of a journal. In 2002, the *Journal of Negative Results in Biomedicine*[21] began publication.

To ensure broad dissemination of scientific findings, in recent years many funding organizations have mandated deposition of funded publications in public databases that are accessible to scientists and lay persons alike. One such database is PubMed Central,[22] sponsored by the National Institutes of Health's (NIH) National Library of Medicine. The NIH and the Wellcome Trust, for example, require that published papers be deposited within 6 to 12 months after publication in scientific journals. The authors are responsible for seeing that is done; in many cases the journals will automatically deposit the papers as an aid to the authors.

Many educational institutions, such as universities and research centers, also require their faculty and staff to post research papers in institutional repositories. This allows increased accessibility to important research papers that may not be available if published in non–open access journals. While we recognize that many publishers wrestle with the problem of balancing free access and adequate compensation for the publication process, we urge editors, societies, and publishers to make scientific material available in free-access repositories, such as PubMed and institutional databases, as soon as possible to further the progression of scientific knowledge.

The heightened threat of global terrorism has led to a relatively new type of publication suppression and controversy: When and how should publication of scientific findings be constrained to prevent their use by terrorists? Experience with terrorism has given rise to US recommendations that provide at least initial guidance on this problematic question pitting scientific advancement against national security.[13,23] These sometimes become very public and controversial. In late 2011, the United States requested suppression of scientific data that would show how influenza A virus subtype H5N1 (commonly known as "avian influenza") was made more contagious in the laboratory; papers on this subject, one from a research group in Wisconsin[24] and the other from a group in the Netherlands,[25] were scheduled to appear in *Nature* and *Science*, respectively. Editors of both journals agreed to redact the articles; however, in February 2012, the World Health Organization publicly disagreed with suppression of the data. Despite much controversy, which spread across scientific and lay communities, the two articles were published shortly thereafter.

2.2.1 Author's Responsibilities

Authors should understand their obligation to publish results and should comply with the ethical guidelines of their institutions *before* submitting a manuscript to a journal. Authors should also become familiar with the editorial policies and instructions for authors of the journal to which their manuscript will be submitted and should attempt to comply fully with those policies and disclosure requirements. For example, if full compliance is not possible, authors should explain in a cover letter why they are unable to comply with certain requirements; this enables the editor to decide whether to allow an exception. In all ethical situations, full transparency is the best guide. Authors should follow the journal's instructions for authors for manuscript preparation. In addition, authors must not submit a manuscript to more than one journal at the same time. However, if this does occur in error and 2 or more editors accept the same manuscript for publication at about the same time, the author should immediately inform the editors of the situation to forestall problems that otherwise could be caused by redundant publication and copyright infringement (see Section 2.7).

2.2.2 Editor's Responsibilities

Journal editors, sometimes in conjunction with their societies/associations, should develop internal and external editorial policies to guide authors, reviewers, editors, and journal staff through the wide range of ethical issues and pragmatic details related to journal publishing. Editors must treat the review of all manuscripts consistently, fairly, and with equitable timelines; they must also be protected from and be able to resist external pressures that could otherwise infringe on the integrity of the review process. If they feel they cannot meet this requirement for a particular manuscript, they should recuse themselves and have a process in place to handle the situation.

Publication-related misconduct, whether attributable to authors, reviewers, editors, or journal staff, can erode a journal's credibility. Editors should therefore take precautionary measures to ensure that misconduct is constrained and should publish notices to inform readers when correction of an honest error is required or when published articles have been tainted by misconduct.

2.2.3 Reviewer's Responsibilities

Reviewers should not agree to review a manuscript unless they believe they are qualified to conduct the review; can meet the requested timelines; and can conduct the review diligently, without prejudice, and without compromising the confidential nature of the process. If reviewers suspect or have knowledge of unusual circumstances regarding a manuscript they have been given to review (e.g., author-related misconduct or redundant publication), they should report such information to the editor but to no one else. In such cases, it is insufficient for reviewers to simply decline to review the manuscript.

Reviewers should remember that confidentiality means not communicating anything about the manuscript under review to others and not using any knowledge gained from the review for their own personal research.

2.3 DATA FABRICATION AND FALSIFICATION, DATA FILTERING, AND PLAGIARISM

Fabrication or falsification of data (intentional creation or synthesis of data or fraudulent manipulation of data, respectively, as distinguished from errors in collecting or processing data) and plagiarism are serious breaches of ethics with legal implications.[26,27] Allegations are often difficult to substantiate because clear intent and corroborating evidence may be lacking. Accusations must be pursued with diplomacy, discretion, and scrupulous attention to due process because fallacious or contrived charges can seriously affect a researcher's professional credibility and reputation.

Data falsification should not be confused with data filtering (sometimes also referred to as "data dredging"[2]), in which a researcher elects to report certain data in full and summarize other data. In most cases, this sort of discretionary practice is acceptable because peripheral findings are condensed but not hidden, which simplifies reading and interpretation. In some cases, however, researchers may overextend their latitude and omit findings that are pertinent or even critical. Data and information that could be even remotely relevant to the findings must not be omitted simply to avoid complicating the explanations or weakening the conclusions. Data filtering must be explained (and, if data are available elsewhere, the original source referenced), and it should be used only with great care and when fully justified. When the ultimate meaning and import of the findings are hidden, disguised, distorted, or misrepresented, then data filtering becomes publication falsification.

Plagiarism—misrepresenting ideas or words taken from the intellectual efforts of another as one's own or without crediting the source—is a serious ethical breach because it involves a deception that places personal interests ahead of giving credit where it is actually due. The *American Medical Association Manual of Style*[8] lists 4 types of plagiarism (page 158):

1. Direct plagiarism: Verbatim lifting of passages without enclosing the borrowed material in quotation marks and crediting the original author.
2. Mosaic: Borrowing the ideas and opinions from an original source and a few verbatim words or phrases without crediting the original author.
3. Paraphrase: Restating a phrase or passage, providing the same meaning but in a different form, without attribution to the original author.
4. Insufficient acknowledgment: Noting the original source of only part of what is borrowed or failing to cite the source material in such a way that a reader will know what is original and what is borrowed.

Questions of plagiarism also can arise when similar words or ideas are arrived at independently or by mental processes in which the source may have been lost from memory and only the product of the prior communication remains; that is, the author may believe his or her idea is original when, in fact, it is not. In such instances, the lack of intent to deceive eliminates any ethical breach, but the situation is nonetheless regrettable because the appearance of possible plagiarism could taint the findings and the author's reputation.

Plagiarism, which is an ethical issue, can become a legal issue when copyright infringement, an actionable violation, is involved (see Section 3.3.5). Intellectual material in the public domain (i.e., material not protected by copyright; see Section 3.1.4) can be used without the author's permission, but authors using such materials are still ethically bound to include appropriate citations.

2.3.1 Author's Responsibilities

It is impossible for authors to intentionally misrepresent data or findings without knowing they have violated the core principle of science: to advance scientific knowledge. When scientists place job pressures, financial gain, or enhancement of their reputation above that core principle, they destroy the value of their past and future research, as well as that of their coauthors, and they may undermine the value of related research performed by other scientists who assume the findings to be valid. Because fabrication and falsification of data can waste time, misdirect future research, squander research funding, produce damage stemming from practical applications, and lead to public distrust and disillusionment, offenders should expect the most severe disciplinary consequences.

In the case of data filtering (see Section 2.3), authors should at least summarize the procedures and reasons for omitting any data that are not presented in full. This ensures full disclosure of research and thwarts problems of perception that might otherwise cast doubt on valid findings.

Authors should keep detailed notes in the course of their research and should carefully examine their manuscripts before publication to be sure they have properly credited other investigators' findings, ideas, and intellectual products. Authors should also carefully examine their own ideas and words to be certain that the origins were not subconsciously derived from past contact with the product of their colleagues' intellectual activity.

2.3.2 Editor's Responsibilities

Journals' instructions to authors should state that authors may be asked to provide journal editors with all relevant data should a fact-finding process arise from an allegation of misconduct. However, these instructions to authors may also note that editors are not responsible for proving or disproving such charges.

Editors and publishers receiving an allegation of plagiarism or of fabricated or falsified research should initially evaluate the allegation to determine whether the charge seems credible. This step may include querying all authors of the work. When convinced that the charge is not simply a matter of oversight, misunderstanding, or error, editors should notify the researcher's employer and any related oversight committee or review board and provide evidentiary information. When the employer is also suspect, editors may have to rely on oversight committees or review boards; in the absence of such bodies, editors may have to do what they can to substantiate or refute the claim. Independent bodies are available to assist in resolving situations of misconduct; for example, in the United States, the Office of Research Integrity will assist in instances in which funding from the Department of Health and Human Services supported the research described in the paper.

Before publication (and as early as submission) many journals make use of software programs that can be used to detect plagiarism, such as CrossCheck[28] and eTBLAST.[29] This software can be used by journal staff, reviewers, and sometimes the authors themselves. Use of this software should drastically reduce the publication of plagiarized texts. When plagiarism, fabrication, or falsification is substantiated after publication of a paper, however, an editor's primary responsibility is to correct the written record.[2] A retraction from the author, his or her employer, or coauthors—or if that is not forthcoming, a note from the editor—should be published in a prominent location within the publication so that it can be incorporated by indexing services (e.g., MEDLINE in the health sciences) and so that readers who access the paper through such services can be made aware of the retraction. Correcting the written record also serves a disciplinary role by notifying the author's professional peers of the offense.

Online-only publications and those with enhanced online versions of their print journals should be aware that, although research has been retracted, it should not be simply deleted; it is part of the historical record and may be of value to subsequent researchers (e.g., those studying misconduct). Rather, the notice of retraction should be posted with the research article. As a precursor to a retraction, some online journals post "expressions of concern", which alert readers that the editors are concerned about some aspect of the data or the analytical integrity. See the ICMJE's discussion of corrections, retractions, and "expressions of concern"[30] and COPE's guidelines.[31] If the paper is indexed, the indexing service should be notified. MEDLINE and other indexing services may have specific requirements about how such information is presented so that it gets properly linked to the original article; editors should request such requirements from the indexing service to ensure proper linkage.

2.3.3 Reviewer's Responsibilities

Reviewers should be alert to the possibility of finding fraud both before and after publication. They should report any suspicions to the editor along with any corroborating evidence. They should not directly contact the author(s), the author's professional peers, or the author's employer about their suspicions but instead should allow the editor to do so. However, reviewers may independently investigate the published record or assemble other evidence that could help resolve the matter.

Because of their familiarity with the manuscript's subject and the related literature, reviewers should watch for possible citation omissions during the peer-review process and note these as part of their review. If an omission appears to be intentional, the reviewer should note this and any plausible reasons for it in confidential comments to the editor.

2.4 REDUNDANT PUBLICATION

Redundant publication, in its simplest form, is the description of essentially the same study in 2 or more publications. "Redundant publication" is the term originally posited by the Council of Biology Editors Editorial Policy Committee[1] and is used by COPE[6]; it has a number of synonyms, including "duplicate publication" (*Journal of the American Medical Association*), "prior publication" (American Chemical Society, *Journal of General Virology*), "dual publication" (American Fisheries Society), and "overlapping publication" (ICMJE). A redundant manuscript may include new data (the raw information to be analyzed) or even new findings (intellectual products of analyzed data), but if the practical relevance and import of the new material are at best modest or the findings remain basically unaltered, then the publication is considered redundant.

Redundant publication is widely proscribed for several reasons: 1) it needlessly expands the burgeoning scientific literature and, hence, needlessly extends the time required for literature searches; 2) it squanders limited scientific research funds because of the expense associated with the review and publication processes; and 3) it can mislead other researchers into counting the same finding twice, which would give that finding more weight than is due.

In its most egregious form, redundant publication represents an attempt to pad publication lists and thereby enhance reputation, prominence, and professional advancement. Authors resorting to this practice unfairly disadvantage their colleagues who adhere to a higher professional ethic. More often, though, redundant manuscripts are submitted by unseasoned authors who may not fully understand the definitions and ramifications of their actions or by authors who erroneously conclude that their manuscript falls into one of the special situations in which redundant publication may be acceptable.

Instances in which redundant publication may be justifiable vary among journals but generally include the following:

1) The initial publication is substantially enhanced by new data and new findings, in which case the original data and findings must be clearly distinguished from the new.
2) Significant findings published in one country and language warrant republication in another country and language to reach an audience that otherwise may not readily have access to the findings; in such a case, the journal editors and publishers of both publications should be informed of, and agree to, the redundancy (because there may be copyright issues and both publishers should be made fully aware of the situation).
3) Significant findings published in the journal of one specialty or profession warrant republication in the journal of another specialty or profession to reach an audience that otherwise might not readily have access to the findings; here, too, the journal editors and publishers of both publications should be informed of, and agree to, the redundancy.
4) The publication is intended to supplant published abstracts (generally, but not always, 250 to 300 words, sometimes with tables or figures), preliminary or in-house reports, unreviewed symposium proceedings, or other such material generally labeled as gray literature[11]: in all such cases, the journal editor should be informed of the earlier offering (however, many journals have policies proscribing republication of gray literature).

Journals should include guidelines about redundant publication and any special situations, as listed earlier, in their instructions for authors.

Fragmented publication ("salami science"), a variant of redundant publication, covers situations in which an author divides what should be reported as a single complete study into 2 or more publications. This often occurs through a lack of awareness but may also be attributable to the author's desire to pad his or her list of publications. In either case, fragmented publication should be discouraged.

2.4.1 Author's Responsibilities

Most journals have editorial policies that prohibit redundant publication and describe special situations and exceptions. Authors should become familiar with those policies, especially as they relate to disclosure. The letter accompanying a submission should state that the manuscript has not been submitted for publication elsewhere (if that is the case). Authors submitting a manuscript that includes previously published results, whether in the primary or gray literature, or a manuscript that has been submitted elsewhere, should explain this to the editor, as well as why they believe the new manuscript is not redundant. They should also enclose a copy of the manuscript or previous publication (or, if that is not possible, the source information).

2.4.2 Editor's Responsibilities

Scientific journals will serve their authors with greater equity if they develop editorial policies on redundant publication; such policies simplify editorial decisions in this area, saving time and ensuring a consistent approach. The policies should be published regularly and prominently within the journal, on the journal or publisher's website, and in the instructions to reviewers. In cases in which republication is deemed appropriate, editors should require authors to briefly note the initial publication prominently—within the first or second paragraph of text or in a footnote to the title of the new publication—and include the reason for duplication.

Policies on redundant publication should require that authors disclose, in the cover letter accompanying a manuscript, any related or previous publications of results from the research that is currently reported. On learning that an author has not disclosed duplication, the editor should gather information about the initial publication and then obtain the author's explanation. Situations judged to be oversights may warrant informing authors that their future submissions will be scrutinized with greater care and that reviewers will be informed of previous situations. When dealing with a flagrant or repeat offender, the prudent editor may decide to inform the author's employer and/or an oversight committee or board, if any, of the misconduct.

When it is determined after investigation that a published paper reiterates an earlier publication of which the editor had no prior knowledge, the editor should publish a notice—an expression of concern or a notice of retraction—explaining the violation, which may also include an explanation by the author or the author's institution. Online publications have the option of placing such notices prominently within the paper itself. The notice will inform other researchers, protect the credibility of the journal's editorial policies, and thwart professional benefits that might otherwise accrue to the authors.

2.4.3 Reviewer's Responsibilities

Many universities and some government agencies have manuscript review committees through which authors work before submitting manuscripts to journals. These committees should be alert to potentially redundant manuscripts and prevent their submission.

Likewise, journal reviewers should be familiar with the relevant journal policy and notify the editors of any prior publications that are similar to the manuscript under review.

2.5 CONFLICTS OF INTEREST

Conflicts of interest can be defined as "sets of conditions in which an author, editor, or reviewer holds conflicting or competing interests that could result in bias or improper decisions" (CSE statement "Conflicts of Interest and the Peer Review Process"[2]). Conflicts of interest can arise when researchers publish studies in which the findings could have been influenced by a profit motive. For example, an opportunity for conflict exists when an employer or company can influence the form or content of the reported research or can determine which research material is published and which is withheld from publication. A similar conflict of interest can occur when research is performed under contract, either with a company that has a financial motive in the research or with advocacy groups hoping that research they sponsor will advance their interests. Of particular pertinence here is whether the independent researchers receive only enough to cover the costs of the research or receive funds beyond their costs, which could raise the possibility of a conflict. Similarly, questions can arise when authors own patent rights or stock in a company whose product receives favorable findings in the reported research.

Another form of conflict of interest may occur in the review process when reviewers or editors possess a positive or negative bias toward one or more authors of a manuscript, are involved in a competitive situation with authors, or have prejudices concerning an author's institution or country of origin.

Substantiated misconduct involving a conflict of interest is rarely discovered and validated, because journals "don't research possible conflicts of interest and are not expected to 'police' authors" (CSE statement "Conflicts of Interest and the Peer Review Process"[2]). However, conflicts of interest may occur fairly often, and, although they may be apparent and readily confirmed, often they are not. Authors, editors, and reviewers should each be aware of this likelihood and should attempt to avert the problems that can arise when such conflicts are ignored or overlooked.

2.5.1 Author's Responsibilities

Authors should fully disclose any real, potential, or potentially perceived conflicts of interest (often called "competing interests") that could be related to any manuscript submitted for publication, even if the authors themselves are certain that no conflict affects the work. This can be done in a submission letter, in a space provided in an online submission site, in a form provided by the journal, and/or within the paper itself, depending on the journal's preference. Failure to inform the editor of any potential conflict often violates journal policies on disclosure and could appear incriminating should a question of motive arise during the review process or after publication.

Authors may request that editors bypass certain candidate reviewers of their manuscript when they believe those reviewers would be incapable of rendering a fair review. Authors should give precise reasons for such requests, and journal editors should carefully consider the requests and grant them if appropriate.

2.5.2 Editor's Responsibilities

Most journals routinely publish policies that explain conflicts of interest to potential authors and stipulate the circumstances that require disclosure. Nevertheless, editors should be alert for situations in which conflicts could exist that authors, through lack of understanding or awareness or by intent, have failed to disclose.

When a potential for conflict is disclosed or discovered, the editor must weigh the likelihood of a conflict. Even if the editor has no reason to suspect that the results have been tainted by the potential conflict, it may be advisable to disclose the potential conflict to readers. Editors may require that authors agree to publication of such a statement before initiating the review process. Some journals do this for all published articles, either publishing a disclosure section within the article or providing an Internet address in a print article (or a hyperlink in an online article) that can lead the reader to an online website that provides information on the authors' possible conflicts of interest. This disclosure permits readers to decide for themselves how much confidence to place in the results.

If there are reasons to suspect that a conflict of interest has tainted the reported findings, the editor will have to rely on good judgment and may need to investigate discreetly. Two outcomes are possible. The editor may 1) find that the evidence is scant and decide to publish the paper with a note, as described earlier, or 2) return the manuscript to the author without initiating its review. Some editors will not publish manuscripts in which the investigator has a potential profit motive tied to the findings. Other publishers use different strategies; the American Fisheries Society, for example, will publish such papers only if they take the form of an informational advertisement.[32]

When an undisclosed conflict of interest is discovered, the editor must determine whether the failure to disclose was unintended or intended and then take appropriate action. If the paper has not been published, the editor may choose either of the 2 courses described earlier. If publication precedes discovery, the editor may elect to publish a notice explaining the circumstances and possible conflict in a prominent place in the journal; online publications can place the notice prominently within the paper itself.

Editors at all levels must also be sensitive to their own conflicts of interest and recuse themselves from editorial decisions when their objectivity could be compromised, for example, by friendship, collaboration, or competition with authors or when the reported research could positively or negatively affect their own financial interests. In those situations, editors who believe they can remain fully objective need not step aside, but they should consider whether the possible perception of impropriety could cast doubt on the efficacy and robustness of the author's findings and the journal's reputation. The editor should have an associate or guest editor assume editorial duties for a particular manuscript. Publishers may find it helpful to establish a conflict of interest policy for their editors; some publish annual disclosures of instances of editorial recusal.[33]

Editors should be aware that reviewers may have conflicts of interest that could prejudice their reviews. To minimize bias in reviews, editors may provide guidelines to help reviewers evaluate possible conflicts of interest and decide when to disclose these to the editor.

2.5.3 Reviewer's Responsibilities

Reviewers should inform the editor of any personal reasons or biases that might prejudice their review of a manuscript (e.g., having financial interests that could be affected by the manuscript or an acrimonious relationship with an author). They should be prepared to recuse themselves if they have a conflict. Many journals ask reviewers to indicate any conflicts of interest when they submit their reviews. Some journals also ask potential reviewers to submit a list of financial dealings with manufacturers or firms related to their field. This information is taken into account when assigning reviewers and when making a decision based on a particular reviewer's comments.

Reviewers should also inform editors if they have any knowledge regarding conflicts of interest related to the manuscript under review, especially conflicts that might have influenced the findings (e.g., if data falsification, as described in Section 2.3, might be attributable to an author's conflict of interest).

2.6 REVIEW PROCESS AND PRIVILEGED INFORMATION

Matters of confidentiality in the manuscript review process vary widely. Some journals allow reviewers to identify themselves to the authors, some allow reviewers to know the identities of authors (single-blinding), and others conceal the identities of both authors and reviewers (double-blinding). Under any of these variations, however, it is customary to treat the materials under review with confidentiality.[34] The work represents a substantial investment of time, effort, and creativity and is usually an instrument for professional advancement and perhaps monetary return. Those participating in the process should therefore understand that the work is proprietary to the author and should respect the need for confidentiality until publication.[1]

Reviewers should treat a manuscript they are reviewing confidentially and refrain from sharing it with anyone, even close colleagues or trusted students. If they provide copies or information to sources outside the review process, a variety of problems can arise: 1) errors and deficiencies in manuscripts could become widely disseminated and mislead the public or misdirect related research, 2) unsolicited commentary from colleagues and competitors could be generated that might complicate and delay the publication process or sway the author(s) into making unwarranted changes in the manuscript,[35] and 3) colleagues might use information in the as-yet-unpublished manuscript for their own personal advantage.[34]

2.6.1 Author's Responsibilities

Authors must become familiar with a selected journal's policies regarding confidentiality during the review process. After submitting a manuscript to a journal that requires confidentiality during the review process, authors should not solicit secondary reviews from colleagues or provide copies of the manuscript to anyone else. Making changes on the basis of such secondary reviews and submitting revisions while the manuscript is under formal review could slow and complicate the journal's review process, as well as increase the burden on journal reviewers. Authors should resist the temptation to discuss their manuscript with anyone whom they suspect or know to be a reviewer of the manuscript.

Authors should check whether the journal has an embargo on prepublication release of information. Under an embargo agreement, the journal makes forthcoming publications available to news media in advance of the formal publication date, and the news media agree not to release the information until a specified date and time (usually the day of or the day before the formal publication date).[8] If an embargo is in place, the editor may expect authors to refrain from distributing materials related to, or discussing the details of, forthcoming publications, especially in public forums. The publisher may also ask authors to make themselves available to reporters during the embargo period.

Some authors are now posting full copies of their draft manuscripts on personal websites, often in conjunction with their published papers. Other authors post draft manuscripts on online prepublication archives (e.g., *Nature Precedings*,[36] 2007–2012, hosted by the Nature Publishing Group), where the public can view or download the paper, comment on the paper, and cite it. The author should be aware of the standards and expectations of journals to which they submit their papers. In some disciplines, use of preprint servers is encouraged (e.g., arXiv[37] in physics, mathematics, computer science, and other fields); in others, use of preprint servers may be discouraged.

2.6.2 Editor's Responsibilities

Editors must protect authors' interests throughout the review process, and they should clearly delineate their confidentiality requirements to authors and reviewers. For example, if a journal does not conceal reviewers' identities from the authors, it should explain whether all contacts regarding the review should be made through the editor or be conducted independently. Many journals do not condone the latter action because it increases the risk that personalities and professional eminence will affect the outcome of the review.

When developing confidentiality policies, editors must remember that their highest priority in publishing a journal is to advance science.[1] Hence, policies should ensure that the assessment of manuscripts is "unbiased, independent, [and] critical".[38] That priority should be paramount in the decision to use an open or anonymous (single- or double-blind) peer review process. Editors should carefully consider whether an open process could create situations in which the professional prominence of a reviewer or author could lead to undue influence or enable a reviewer to curry the favor of an author (or vice versa) and thereby provoke inappropriate decisions regarding the manuscript. On the other hand, open peer review may discourage a reviewer from giving a biased or unnecessarily harsh criticism of a manuscript. Single-blind review may be open to bias because the authors' reputations (good or bad) precede them and may be taken into account by the reviewer (even subconsciously) during paper evaluation. Some editors and authors argue that double-blind peer review is better because neither the authors' nor reviewers' identities are known. However, in small fields, these identities may easily be guessed.

Editors may sometimes be called on to provide recommendations regarding professional advancement of authors known to them solely on the basis of works submitted for review and publication. If the journal's review process is confidential, editors should avoid making a recommendation based on anything other than the published record because it would compromise the journal's confidentiality requirement. Even if the review process is open, there are reasons for editors to refrain from providing recommendations based on information gathered during peer review; for example, the review process may provide only a limited and therefore distorted picture of a researcher's capabilities.[1]

2.6.3 Reviewer's Responsibilities

Reviewers should fully understand and accept the journal's confidentiality requirements. If they find those stipulations unacceptable, they must decline the request to serve. Reviewers should understand that they are not only accepting a professional service responsibility but also being entrusted by the authors and editors to protect proprietary and valued material. If a reviewer is approached by an author of a manuscript that he or she is reviewing, the reviewer should neither confirm nor deny being a participant in the review process, unless the particular journal policy allows the reviewer to reveal him- or herself to the author.

2.7 WITHDRAWAL OF MANUSCRIPTS AND DUPLICATE SUBMISSIONS

Authors generally want to publish their manuscripts as quickly as possible and in the journal that will best serve their interests (e.g., reaching the greatest number of their colleagues or best able to advance their personal credit). Consequently, an author may be tempted to submit a manuscript to more than one journal at the same time, hoping to shorten the publication process should one journal accept and the other reject. However, duplicate submission is considered improper within the conventions of scientific publication. Publishers and editors of journals expend considerable resources in conducting reviews, and reviewers donate their time to prepare those reviews, so duplicate submissions squander valuable time and resources. If a manuscript is accepted by 2 journals, the author will be compelled to withdraw one of the manuscripts because most journals have strict policies prohibiting redundant publication (see Section 2.4).

2.7.1 Author's Responsibilities

Authors should be aware of the problems created by duplicate submissions and the negative effects that such disregard for established conventions of scientific publication can have on limited scientific resources. Authors must therefore submit each manuscript to only one journal at a time and must not submit that manuscript elsewhere, unless it is rejected by the first journal or has been formally withdrawn from the first journal.

Authors who discover errors or deficiencies in a submitted manuscript are encouraged to withdraw the manuscript from consideration if the problems warrant. Such withdrawals, however, should not be contrived to facilitate submitting the manuscript to another journal.

2.7.2 Editor's Responsibilities

Editors should state in the journal's instructions for authors the requirement that any manuscript be submitted to only one journal at a time. The instructions may also specify that the author's letter of submission include a guarantee that the manuscript is not under consideration by any other journal. If editors become aware of duplicate submissions made before or during their review of a manuscript, they should explain to the authors the timelines involved with review and publication and allow the authors to withdraw the paper before the review process begins.

If an author asks to withdraw a manuscript because it has been submitted to another journal at the same time, the editor may find that not granting the request is more problematic than it is worth, even if copyright has already been transferred. However, the editor should identify and notify the other journal editor of the infraction and perhaps discuss appropriate corrective measures with that editor.

2.7.3 Reviewer's Responsibilities

If, in the course of a review, a reviewer becomes aware that the author has submitted the paper to another journal (e.g., if the reviewer is asked to review the same paper by 2 or more journals), the reviewer should immediately notify the editor of the journal for which the reviewer is reviewing the paper, who may elect to suspend the review until the situation has been clarified.

2.8 AUTHORSHIP DISPUTES

Disputes over authorship (see criteria in Section 2.8.2) can arise in numerous ways. Many individuals contribute to modern scientific research; some qualify as authors but others do not. Contributors judged not to be authors may disagree about whether they should be a listed author or otherwise acknowledged, generally in an acknowledgments section. Colleagues have sometimes been listed as contributors without their knowledge, which can cause considerable dissension. Authors may also disagree about who should be the primary or senior author or, less often, about the authorship order. Disputes have frequently resulted because authors are misinformed about authorship guidelines or have been conditioned by experience to follow outdated authorship practices. In some situations, individual interpretations of authorship guidelines have produced acrimony.

In situations of flagrant authorship omissions or inclusions, special boards dealing with such disputes may be available—for example, a journal's editorial board or a panel assembled by the author's employer. Some editors may be willing and able to serve as arbiters, but if not or if the situation is not readily resolved, an editor may elect to delay review or publication of the manuscript until authorship is resolved.

2.8.1 Author's Responsibilities

Authorship should be determined when the original research plan is developed. All authors should be aware that they will be designated authors and should agree to their participation in the project and approve all paper submissions. As the project evolves, the research plan and authorship roles may shift. Any such changes should be recorded and agreed to by all authors; journals often demand that changes in authorship be signed off by all authors prior to publication.

Before preparing a manuscript for publication, authors should find out what policies or procedures the intended journal, society, or publisher may have for authorship and then attempt to comply fully with those requirements, as well as seeking clarification where ambiguity exists.

The authors should contact all people who contributed to the article in some way but do not qualify for authorship, requesting permission to include their names in the acknowledgments section.

2.8.2 Editor's Responsibilities

Editors should develop authorship guidelines for their journals. Many of the guidelines now in use (American Physical Society,[39] American Statistical Association,[40] National Academy of Sciences,[41] Society for Neuroscience[42]) embody, with some variation, the criteria recommended by the ICMJE[43]:

> Authorship credit should be based on (1) substantial contributions to conception and design, acquisition of data, or analysis and interpretation of data; (2) drafting the article or revising it critically for important intellectual content; and (3) final approval of the version to be published. Authors should meet conditions 1, 2, and 3.
>
> When a large, multicenter group has conducted the work, the group should identify the individuals who accept direct responsibility for the manuscript. These individuals should fully meet the criteria for authorship defined above.
>
> Acquisition of funding, collection of data, or general supervision of the research group, alone, does not justify authorship.
>
> All persons designated as authors should qualify for authorship, and all those who qualify should be listed.
>
> Each author should have participated sufficiently in the work to take public responsibility for appropriate portions of the content.

These authorship requirements discourage the common practice of gratuitous authorship, such as the inclusion of laboratory directors who play only a funding or supervisory role. The requirements also exclude "ghost authors" whose names appear as authors on a paper written by someone else. Professional scientific writers, sometimes referred to as "ghost writers", who draft a manuscript based on data and contributions from the researchers, may provide a valuable contribution to the work but do not qualify for authorship because they are not responsible for the research that they report. They should be credited in the acknowledgments section. The authorship requirements also attempt to reduce the relatively common practice of including as authors any employees who worked solely on data collection. The ICMJE recommends that contributors who do not meet the criteria for authorship be listed in an acknowledgments section with a clear statement about their participation[43]; editors should ensure that all those acknowledged in this way have provided their permission to be named.

Many editors require that all authors sign a statement indicating that they have met the journal's authorship criteria. Many journals also require that authors list their contributions to the study and paper; these are often published within the article. These practices not only support fairness in authorship listings across all papers published, but also ensure that all authors have agreed to authorship on the submitted manuscript. Likewise, editors may require that people who are acknowledged in the acknowledgments section grant written permission to be so listed; this provides an opportunity for them to excuse themselves from mention or to seek a position as an author.

2.8.3 Reviewer's Responsibilities

Reviewers should be familiar with the journal's authorship policies, but they are generally not responsible for commenting on the qualifications for authorship of authors listed or of those otherwise acknowledged. However, a reviewer aware of a possible impropriety in authorship should inform the editor that a problem might exist.

2.9 RESPONSIBILITIES OF PUBLISHERS, JOURNAL OWNERS, AND SPONSORING SOCIETIES

"Editorial independence" describes the responsibility, authority, and accountability that editors have with respect to the content of their journals.[3] It also describes the freedom that editors must have from any editorial influence from publishers, journal owners, or sponsoring societies. The editor decides which peer-reviewed papers are published in journal issues as well as which other materials should be included, such as editorials, news items, and appropriate types of advertising.[3] These decisions should be based on the scientific importance of the journal's content and not on the journal's financial success.[44]

Publishers, journal owners, and/or sponsoring societies are usually responsible for financial and management issues, as well as business models and policies.

A contract or terms-of-reference document should be established and signed by all parties. This document should clearly state the responsibilities of the editor and those of the publishers, journal owners, and sponsoring societies. It may clarify circumstances under which the editor's contract may be terminated by the journal publisher or owner—for example, if the editor does not meet his or her responsibilities. All parties should work together with trust and respect.[3] Institutions should be responsible for the conduct of their researchers and should create and encourage a healthy research environment. Journals, on the other hand, should be responsible for the conduct of their editors, for safeguarding the research record, and for ensuring the reliability of what they publish. Institutions and journals (including publishers and editors) should collaborate in an effective and transparent manner in order to safeguard the quality and integrity of scientific research. Excellent information on the relationships between these parties can be found in the CSE's White Paper[3] and on the websites of the ICMJE[4] and COPE.[6,45]

CITED REFERENCES

1. Council of Science Editors. Editorial policy statements approved by the CSE Board of Directors. Wheat Ridge (CO): CSE; 2005 [accessed 2012 Nov 9]. http://www.councilscienceeditors.org/i4a/pages/index.cfm?pageid=3413.

2. Bailar JC III, Council of Biology Editors Editorial Policy Committee. Ethics and policy in scientific publication. Bethesda (MD): CBE; 1990.

3. CSE's White Paper on Promoting Integrity in Scientific Journal Publications, 2012 Update. Wheat Ridge (CO): CSE; 2012 [accessed 2012 Nov 4]. http://www.councilscienceeditors.org/i4a/pages/index.cfm?pageid=3331.

4. International Committee of Medical Journal Editors [accessed 2012 Nov 5]. http://www.icmje.org.

5. World Association of Medical Editors [accessed 2012 Nov 5]. http://www.wame.org.

6. Committee on Publication Ethics [accessed 2012 Nov 5]. http://www.publicationethics.org.

7. EQUATOR Network [accessed 2012 Nov 5]. http://www.equator-network.org/home/.

8. Iverson C, Christiansen S, Flanagin A, Fontanarosa PB, Glass RM, Gregoline B, Lurie SJ, Meyer HS, Winker MA, Young RK. American Medical Association manual of style: a guide for authors and editors. 10th ed. New York (NY): Oxford University Press; 2007. Also available at http://www.amamanualofstyle.com/oso/public/index.html.

9. Office of Research Integrity, Public Health Service [accessed 2012 Nov 5]. http://ori.dhhs.gov.

10. Office of Inspector General of the National Science Foundation [accessed 2012 Nov 5]. http://www.nsf.gov/oig/.

11. Collette BB. Problems with gray literature in fishery science. In: Hunter J, editor. Writing for fishery journals. Bethesda (MD): American Fisheries Society; 1990. p. 27–32.

12. The "file drawer" phenomenon: suppressing clinical evidence [editorial]. CMAJ. 2004;170(4):437. Also available at http://www.cmaj.ca/content/170/4/437.full?sid=0149d242-5d32-4b69-8ebb-d2e624373094.

13. Atlas R, Campbell P, Cozzarelli NR, Curfman G, Enquist L, Fink G, Flanagin A, Fletcher J, George E, Hammes G, et al. Statement on scientific publication and security [editorial]. Science. 2003;299(5610):1149. Also available at http://www.sciencemag.org/content/299/5610/1149.full.

14. Wilbur RL. Gray literature: a professional dilemma. Fisheries. 1990;15(5):2–6.

15. Wager E, Field EA, Grossman L. Good publication practice for pharmaceutical companies. Curr Med Res Opin. 2003;19(3):149–154. Also available at http://www.controlled-trials.com/news/Good_Publication_Practice_for_pharmaceutical_companies_2355-Article.pdf.

16. National Research Council (US), Division on Earth and Life Studies, Board on Life Sciences, Committee on Responsibilities of Authorship in the Biological Sciences. Sharing publication-related data and materials: responsibilities of authorship in the life sciences. Washington (DC): National Academies Press; 2003. Also available at http://www.nap.edu/books/0309088593/html/.

17. ClinicalTrials.gov [accessed 2012 Nov 7]. http://clinicaltrials.gov.

18. European Union Drug Regulating Authorities Clinical Trials database [accessed 2012 Nov 7]. Available at https://eudract.ema.europa.eu.

19. Research Collaboratory for Structural Bioinformatics (RCSB) Protein Data Bank [accessed 2012 Nov 7]. http://www.rcsb.org/pdb/home/home.do.

20. International Nucleotide Database Collaboration [accessed 2012 Nov 7]. http://www.insdc.org.

21. Journal of Negative Results in Biomedicine [accessed 2012 Nov 5]. http://www.jnrbm.com.

22. PubMed Central [accessed 2012 Nov 5]. http://www.ncbi.nlm.nih.gov/pmc/.

23. Kennedy D. Two cultures [editorial]. Science. 2003;299(5610):1148.

24. Imai M, Watanabe T, Hatta M, Das SC, Ozawa M, Shinya K, Zhong G, Hanson A, Katsura H, Watanabe S, et al. Experimental adaptation of an influenza H5 HA confers respiratory droplet transmission to a reassortant H5 HA/H1N1 virus in ferrets. Nature. 2012;486(7403):420–428.

25. Herfst S, Schrauwen EJ, Linster M, Chutinimitkul S, de Wit E, Munster VJ, Sorrell EM, Bestebroer TM, Burke DF, Smith DJ, Rimmelzwaan GF, Osterhaus AD, Fouchier RA. Airborne transmission of influenza A/H5N1 virus between ferrets. Science. 2012;336(6088):1534–1541.

26. Kennedy D. Next steps in the Schön Affair [editorial]. Science. 2002;298(5593):495.

27. Service RF. Bell Labs fires star physicist found guilty of forging data. Science. 2002;298(5591):30–31.

28. CrossCheck powered by iThenticate, product of CrossRef [accessed 2012 Nov 7]. http://www.crossref.org/crosscheck/index.html.

29. eTBLAST, product of Virginia Polytechnic Institute and State University [accessed 2012 Nov 7]. http://etest.vbi.vt.edu/etblast3/.

30. International Committee of Medical Journal Editors. Publishing and editorial issues related to publication in biomedical journals: corrections, retractions and "expressions of concern". Uniform Require-

ments for Manuscripts Submitted to Biomedical Journals. Philadelphia (PA): ICMJE; 2009 [accessed 2012 Nov 4]. http://www.icmje.org/publishing_2corrections.html.

31. Wager E, Barbour V, Yentis S, Kleinert S, on behalf of COPE Council. Retraction guidelines. [place unknown]: Council on Publication Ethics; 2009 [accessed 2012 Nov 4]. http://www.publicationethics.org/files/retraction%20guidelines.pdf.

32. American Fisheries Society. Conflict of interest policy. Bethesda (MD): AFS; 1998.

33. Angell M. Acceptance address at Council of Science Editors meeting. Sci Ed. 2003;26(5):148–149.

34. Marshall E. Trial set to focus on peer review. Science. 1996;273(5279):1162–1164.

35. Alaska Department of Fish and Game. ADF&G writer's guide. 3rd ed. Juneau (AK): ADFG; 1999. Also available at http://www.adfg.alaska.gov/static/home/library/PDFs/writersguide_full.pdf.

36. Nature Precedings [accessed 2012 Nov 4]. http://precedings.nature.com.

37. arXiv [accessed 2012 Nov 4]. http://arxiv.org.

38. International Committee of Medical Journal Editors. Uniform requirements for manuscripts submitted to biomedical journals: writing and editing for biomedical publication. Philadelphia (PA): ICMJE; 2012 [accessed 2012 Nov 5]. http://www.icmje.org.

39. American Physical Society. APS guidelines for professional conduct. College Park (MD): APS; 2002 Nov 10 [accessed 2012 Nov 5]. http://www.aps.org/policy/statements/02_2.cfm.

40. American Statistical Association, Committee on Professional Ethics. Ethical guidelines for statistical practice. Alexandria (VA): ASA; 1999 [accessed 2012 Nov 9]. http://www.amstat.org/about/ethicalguidelines.cfm.

41. National Academy of Sciences (US), Committee on Science, Engineering, and Public Policy. On being a scientist: responsible conduct in research. 2nd ed. Washington (DC): National Academies Press; 1995 [accessed 2012 Nov 8]. http://www.nap.edu/openbook.php?record_id=4917.

42. Society for Neuroscience. Responsible conduct regarding scientific communication. Washington (DC): Society for Neuroscience; 2012 [accessed 2012 Nov 9]. http://www.sfn.org/index.aspx?pagename=responsibleConduct.

43. International Committee of Medical Journal Editors. Ethical considerations in the conduct and reporting of research: authorship and contributorship. Uniform requirements for manuscripts submitted to biomedical journals. Philadelphia (PA): ICMJE; 2009 [accessed 2012 Nov 4]. http://www.icmje.org/ethical_1author.html.

44. International Committee of Medical Journal Editors. Ethical considerations in the conduct and reporting of research: editorship. Uniform requirements for manuscripts submitted to biomedical journals. Philadelphia (PA): ICMJE; 2009 [accessed 2012 Nov 7]. http://www.icmje.org/ethical_2editor.html.

45. Wager E, Kleinert S on behalf of COPE Council. Cooperation between research institutions and journals on research integrity cases: guidance from the Committee on Publication Ethics (COPE). COPE; 2012 Mar [accessed 2012 Nov 11]. http://www.publicationethics.org/files/Research_institutions_guidelines_final.pdf.

ADDITIONAL REFERENCES

American Chemical Society, Publications. Ethical guidelines to publication of chemical research. Washington (DC): ACS; 2012 June [accessed 2012 Nov 7]. http://pubs.acs.org/userimages/ContentEditor/1218054468605/ethics.pdf.

American Mathematical Society. Ethical guidelines. Providence (RI): The Society; 2005 [accessed 2012 Nov 7]. http://www.ams.org/about-us/governance/policy-statements/sec-ethics.

American Society for Biochemistry and Molecular Biology. Code of ethics. http://www.asbmb.org/Page.aspx?id=70&terms=ethics.

Bulger RE, Heitman ÊE, Reiser SJ, editors. The ethical dimensions of the biological and health sciences. 2nd ed. New York (NY): Cambridge University Press; 2002. Part 4, The ethics of authorship and publication.

Jones AH, McLellan F, editors. Ethical issues in biomedical publication. Baltimore (MD): Johns Hopkins University Press; 2000.

Lafollette MC. Stealing into print: fraud, plagiarism, and misconduct in scientific publishing. Berkeley (CA): University of California Press; 1992.

National Academy of Sciences (US), Committee on Science, Engineering, and Public Policy, Panel on Scientific Responsibility and the Conduct of Research. Responsible science. Vol. 1, Ensuring the integrity of the research process. Washington (DC): National Academies Press; 1992. Also available at http://books.nap.edu/books/0309047315/html/index.html.

Online Ethics Center: the online ethics center for engineering and science. Washington (DC): National Academy of Engineering; 2003–2013 [accessed 2013 Jan 31]. http://www.onlineethics.org.

Publication manual of the American Psychological Association. 5th ed. Washington (DC): The Association; 2001. Appendix C, Ethical standards for the reporting and publishing of scientific information; p. 387–396.

The responsible researcher: paths and pitfalls. Research Triangle Park (NC): Sigma Xi, The Scientific Research Society; 1999 [accessed 2012 Nov 7]. http://www.sigmaxi.org/programs/ethics/ResResearcher.pdf.

The role and activities of scientific societies in promoting research integrity. Report of a conference; 2000 Apr 10; Washington, DC. Washington (DC): American Association for the Advancement of Science; 2000 Sep [accessed 2013 Oct 20]. Available at http://ori.hhs.gov/documents/role_scientific_societies.pdf.

Teich AH, Frankel MS. Good science and responsible scientists: meeting the challenge of fraud and misconduct in science. Washington (DC): AAAS; 1992. (AAAS publication; no. 92-13S).

3 The Basics of Copyright

3.1 COPYRIGHT

Copyright is a set of intellectual property rights granted to the creators of original works of authorship. The economic rights enable the creators to control the reproduction, performance, and broadcast of their intellectual property. The moral rights honor a creator's right to be identified as the creator of the work and object to its distortion or mutilation. Copyright encourages future creativity by protecting the investment in past efforts. Patents and trademarks, which are comparable types of intellectual property that protect the creativity invested in inventions, slogans, and other distinctive goods and services, are not within the scope of this chapter. Similarly, designs for architectural works, semiconductor chips, integrated circuits, and vessel hulls (although eligible for copyright) are not within the scope of this manual.

Although historically most important for artistic and literary creations with commercial value, copyrights have become increasingly important for intellectual property, such as computer software, databases, and compilations of research results. Copyrights allow authors to have legal recourse in the event of unauthorized exploitation of their creations and to recover damages or profits from any unauthorized use. In addition, copyrights help ensure proper credit for copyrighted materials copied or republished by others and offer the authors some control over the integrity of the works. Copyrights do not always belong to the creator of a work, however. They may, and in some cases must, be reassigned to the creator's institution or to the publisher of the work to allow publication or dissemination of the work.

Copyright law in the United States is recommended by the US Constitution[1] and current US law is embodied in the US Code, Title 17, which includes the Copyright Act of 1976,[2] amendments to it, the Berne Convention Implementation Act of 1988,[3] the Uruguay Round Agreements Act of 1994,[4] and the Digital Millennium Copyright Act of 1998.[5] The US Copyright Office of the Library of Congress administers the Copyright Act.

The following summary is based on the US law, with differences noted for the laws of Canada,[6] the United Kingdom,[7,8] Australia,[9] and New Zealand.[10] Although laws in other countries differ in some respects, those of all 164 signatories to the Berne Convention of 1971,[11] the Universal Copyright Convention of 1971,[12] or later revisions of either one should offer some protection. The goal of the Copyright Directive of the European Parliament of 2001[13] was to harmonize the laws in European Union member states. The copyright law in the country in which the work was written, created, and/or published applies to each work; the language of the work is immaterial.

Consult the agency in charge of intellectual property in your country of interest for detailed instructions on how to publish notices of copyright and register the works being protected. Seek legal counsel from someone familiar with the local laws if you are uncertain about a planned use of copyrighted materials or if you believe your own rights have been violated.

3.1.1 Eligibility

Copyright laws assign to authors of original works exclusive rights to determine how and when such works are copied, performed, reproduced, or republished. Ideas, news, facts, data, and other materials that are not the result of independent intellectual effort do not meet the minimum requirements of authorship, are not copyrightable, and are therefore in the public domain (see Section 3.1.4). Table 3.1 lists types of work and their eligibility for copyright.

Copyright does not extend "to any idea, procedure, process, system, method of operation, concept, principle, or discovery regardless of the form in which it is described, explained, illustrated, or embodied" in a work.[2] Although ideas may be the result of independent intellectual effort, copyright is granted for the expression of those ideas, not for the ideas themselves. It is also granted for the description of the news, not for the news itself, and for the compilation of facts and data, not for the facts and data themselves. Impromptu speeches, performances, and other works that are not yet fixed in a tangible form are not copyrightable, although they would be as soon as they are notated or recorded. Some works may be protected by patents (e.g., original inventions) or by trademarks (e.g., words, phrases, symbols, and designs) and some by either patent or copyright (e.g., computer software).

Table 3.1 Eligibility for copyright[a,b,c,d,e]

Eligible: original works of authorship, including
Literary works, including books, periodicals, manuscripts, poetry, computer software, and databases
Music, including any accompanying words
Drama, including any accompanying music
Pantomime and choreography
Visual arts, including photographic, pictorial, graphic, and sculptural works
Motion pictures, radio and television broadcasts, and other audiovisual works
Published editions
Sound recordings
Not eligible: works that usually do not meet the minimum requirements of authorship,[f] including
Ideas, procedures, processes, systems, methods of operation, concepts, principles, discoveries
Data, facts, news
Standard lists, calendars, and tables from public documents
Blank forms intended to be filled out
Computing and measuring devices
Names, titles, short phrases, no matter how distinctive [Seek trademark protection instead.]
Inventions [Seek patent protection instead.]
Symbols and designs [Seek trademark protection instead.]
Choreographic works, impromptu speeches, and other works that are not yet fixed in a tangible form [These works become eligible once they have been fixed by notation, sound recording, or the like.]

[a] Eligibility status based on United States Code, Title 17, Copyright Act of 1976,[2] Sections 102(a) and (b) and circulars issued by the US Copyright Office.
[b] Eligibility status based on [Canadian] Copyright Act,[6] R.S.C. 1985, c. C-42, Section 5.
[c] Eligibility status based on [United Kingdom] Copyright, Designs and Patents Act 1988[7](c. 48).
[d] Eligibility status based on [Australian] Copyright Act 1968, Act No. 63 of 1968[9] as amended.
[e] Eligibility status based on [New Zealand] Copyright Act 1994[10] as amended.
[f] Compilations or descriptions of some of these works may be eligible if they meet the test for originality.

3.1.2 Authorship and Duration of Rights

Authorship determines the duration of the copyright (see Tables 3.2 and 3.3), so the identity of the author is very important. If only one self-employed author is involved, there is no question who the author is. If there are 2 or more writers, the authorship is shared unless there is agreement to the contrary. If the work is a film, a musical drama, or other work involving different kinds of authorship, the determination of authorship becomes more difficult. In the United States, the creator of an original work is the author, no matter what kind of work, unless the work was made for hire. In Canada, the United Kingdom, Australia, and New Zealand, authorship depends on the type of material (e.g., literary work, photograph, film, sound recording).

In the United States, if the writers are employees and the work is the result of their employment, then the work was made for hire and the employer is considered the author (see Section 3.1.5). In Canada, the United Kingdom, Australia, and New Zealand, those who created the work are the authors even if the work was made during their employment and if the employer owns the copyright.

Unlike other personal property, copyright ends at a legally defined time. In the United States, its duration depends on whether the work was published, when and where it was

Table 3.2 Duration of copyright protection in the United States[a,b]

Date of creation	Description of copyright notice and registration	Duration of protection from date of original copyright[c]
Before 1 Jan 1950	Work published without a copyright notice and not registered	None
	Work published with a copyright notice and not renewed before 1 Jan 1978	28 years
	Work published with a copyright notice and renewed in 1978	95 years
Between 1 Jan 1950 and 31 Dec 1963	Work published with a copyright notice and not renewed during the 28th year	28 years
	Work published with a copyright notice and renewed during the 28th year	95 years
Between 1 Jan 1964 and 31 Dec 1977	Work published with a copyright notice, whether or not a renewal was registered during the 28th year	95 years
Before 1 Jan 1978	Works in existence but not published or registered:	
	• works created by individuals (but not for hire)	Life of the author (or of last surviving author, if more than one) + 70 years
	• works created for hire and anonymous or pseudonymous works	At least 120 years from creation, with 25 years of statutory protection
On or after 1 Jan 1978	Whether or not they include a copyright notice or are registered:	
	• works created by individuals (but not for hire)	Life of the author (or of last surviving author, if more than one) + 70 years
	• works created for hire and anonymous or pseudonymous works (unless the identity is known to the Copyright Office)	95 years from first publication or 120 years from creation, whichever is shorter

[a] Based on *Duration of Copyright: Provisions of the Law Dealing with the Length of Copyright Protection*[27] (Circular 15a) (2011).
[b] The duration in other countries varies by country and type of work (e.g., see Table 3.3).
[c] All terms begin at the end of the year in which the creation or death occurred.

published, and whether it contained a copyright notice. Before the US Copyright Act of 1976[2] went into effect on 1 January 1978, copyright in the United States was secured on the date the work was published (if it was published) or on the date it was registered in unpublished form. In either case, the copyright lasted for 28 years from the date it was secured and could be renewed for another 28 years. If the published work did not include a copyright notice or the unpublished work was not registered, it was not copyrighted. The US Copyright Act of 1976 (as amended in 1992 and 1998) has changed the rules considerably to conform to the law in other countries.

In countries that are signatories to the Berne Convention of 1971[11] and its revisions (including the United States, Canada, the United Kingdom, Australia, and New Zealand), a copyright is in effect as soon as the work is in fixed, tangible form (which includes a saved electronic form), whether or not it has been printed. With a few exceptions, new works are protected for the life of the author plus 50 years in Canada and New Zealand

and for the life of the author plus 70 years in the United States, Australia, and the United Kingdom. Works made for hire in the United States are protected for 95 years from the date of publication or for 120 years from the date of creation, whichever expires first; the employers for whom the works were prepared are considered to be the authors. In Canada, the United Kingdom, Australia, and New Zealand, the creators retain authorship (but not ownership) even when the work was part of their employment. Duration of copyright in countries other than the United States depends on the type of work. In contrast, duration of copyright in the United States depends on the date of creation—not on the type of work—whether the work contains a copyright notice, and whether it was made for hire. For all countries described here, all terms of duration begin at the end of the year in which the creation or death occurred; see Tables 3.2 and 3.3 for details.

3.1.3 Ownership

Immediately on creating an original work of authorship in a fixed, tangible form, the creator of that material automatically owns the copyright to it, unless the work was made for an employer (see Section 3.1.5). Coauthors of a work are co-owners of the copyright in the work, unless there is an agreement to the contrary. Copyright in each separate contribution to a collective work is distinct from copyright in the compilation of the work as a whole and is owned by the author of the contribution until or unless transferred by written agreement. Each contribution and the compilation can be registered separately in the United States (see Section 3.2.2). In the United Kingdom, Australia, and New Zealand, there is also a separate copyright in the layout of the work of published editions owned by the publisher; the duration of that right is 25 years from publication.

3.1.3.1 Rights and Responsibilities of Copyright Owners

Copyright conveys exclusive economic and moral rights to the copyright owner. Economic rights allow the owner to reproduce the work, to prepare derivative works from it, to distribute copies, and to perform, display, or broadcast the work. As property, economic rights can be—in whole or in part—sold, transferred, assigned, or licensed. For example, the owner may sell or assign print distribution rights but retain all electronic distribution rights. The rights may be conveyed to one's heirs by will or probate or may be confiscated by act of law.

Moral rights are those of attribution and integrity. The right of attribution is the right to be acknowledged as the creator of a particular work. The right of integrity forbids anyone to distort, mutilate, or modify a work in a way that is prejudicial to the creator's honor or reputation without permission from the creator. Moral rights may be waived in some countries (e.g., the United Kingdom) but not in others; they cannot be sold or transferred. In the United States, moral rights are restricted to certain types of visual art. In Canada, moral rights apply to all works eligible for copyright. In the United Kingdom and Australia, they extend to authors of literary, dramatic, musical, and artistic works and to screenwriters, directors, and producers of films. In New Zealand, they extend to authors of literary, dramatic, musical, and artistic works and to screenwriters and directors of films.

Table 3.3 Duration of copyright protection in Canada, Australia, United Kingdom, and New Zealand

Type of work or right	In Canada[a,b]	In Australia[b,c]	In the United Kingdom[b,d]	In New Zealand[b,e]
Unpublished literary, dramatic, musical, or artistic works (i.e., unpublished at the time of the author's death)	Life of the author + 50 years; transitional provisions apply to works unpublished prior to 31 Dec 1998	If published posthumously, 70 years after first publication; if never published, indefinitely	Until 31 Dec 2054 or life of the author + 70 years, whichever is later	Life of the author + 50 years
Published literary, dramatic, musical, or artistic works (not including photographs)	If published before 1 Jan 1994, then life of the author + 50 years from the exact date of death	Life of the author + 70 years	Life of the author + 70 years	Life of the author + 50 years
	If published on or after 1 Jan 1994, then life of the author + 50 years			
Cinematographic films	If scripted and created, then life of the author + 50 years	If created after 1 May 1969, 70 years after publication; otherwise indefinitely	Life of the last surviving author + 70 years [In this instance, the author is defined as the principal director, the author(s) of the screenplay and dialogue, and the composer of any music specially created for the film.]	50 years from creation or release to the public, whichever is later
	If not scripted but created before 1 Jan 1994, then 50 years after the creation, whether or not the film was published			
	If not scripted but created on or after 1 Jan 1994, then 50 years after publication			
Anonymous and pseudonymous works	50 years after publication or 75 years after creation, whichever is shortest	70 years after publication; otherwise indefinitely	70 years from date of creation or publication, whichever is later	50 years after publication; otherwise indefinitely

Photographs	If author is an individual, life of the author (end of calendar year) + 50 years	Life of the author + 70 years	If created by a known creator since 1989, then 70 years from death of the creator [For unpublished works, see the Copyright, Designs and Patents Act.]	Life of the author + 50 years
	If author is a corporation but natural person does not hold majority of shares, year initial negative was taken (end of calendar year) + 50 years			
Sound recordings	50 years from the first fixation of the sound	If created before 1 May 1969, then 70 years after creation	50 years from creation or release	50 years from creation or release to the public, whichever is later
		If created on or after 1 May 1969, then 70 years after first publication; otherwise indefinitely		
Broadcasts and cable programs	[not specified]	50 years from date of broadcast	50 years after creation	50 years from date of broadcast
Published editions	[not specified]	25 years from publication	25 years from publication	25 years from publication
Computer-generated works	[not specified]	Life of creator + 70 years	50 years after creation	50 years after creation
Right of integrity of authorship in cinematographic film	Same as other copyrights for cinematographic films	Life of author	Same as other copyrights for cinematographic films	Same as other copyrights for cinematographic films
Right of integrity of authorship in other works and other moral rights	Same as other copyrights for the respective type of work	Same as other copyrights	Same as other copyrights for the respective type of work	Same as other copyrights for the respective type of work
Right of attribution as author or director	[not specified]	Same as other copyrights	Same as other copyrights	Life of the author + 20 years

[a] Based on [Canadian] Copyright Act,[6] R.S.C. 1985, c. C-42, as amended, ss. 6-12.
[b] Unless otherwise noted, all terms begin at the end of the year in which the creation or death occurred.
[c] Based on [Australian] Copyright Act 1968, Act No. 63 of 1968[9] as amended, s. 34, s. 95, and s. 96.
[d] Based on [United Kingdom] Copyright, Designs and Patents Act 1988[7] (c. 48), s. 86 and Duration of Copyright and Rights in Performances Regulations 1995,[8] s. 5, 6, 13B, and 14 [Statutory Instrument 1995 No. 3297].
[e] Based on [New Zealand] Copyright Act 1994[10] as amended.

Reproduction of works by anyone other than the copyright owner requires written permission from the copyright owner, except as allowed under fair use (see Section 3.3.1). Owners of copyrighted material thereby control reproduction of that material in 4 ways:

1) license the use with a royalty or fee and impose appropriate conditions
2) license the use and impose conditions on the request without seeking a royalty or fee
3) deny the request
4) fail to reply, *which must be interpreted as equivalent to denying the request*

Copyright owners are responsible for monitoring unauthorized use of their materials by others. Copyright collectives facilitate permissions and royalties (see Section 3.3.3).

3.1.3.2 Limitations on Rights of Owners

Rights of owners are somewhat limited by exceptions to copyright that are defenses against an infringement action. The US Copyright Act[2] allows some uses of copyrighted materials without infringement or the requirement to obtain permission from the copyright owner. Users may

1) reproduce parts of works if the principle of fair use is observed (see Section 3.3.1 for more detail)
2) sell or give away legally obtained copies (other than fair-use copies made for personal use)
3) perform or display works in classrooms and certain other nonprofit settings
4) broadcast over a licensed station if no charge is made to the public to see or hear the transmission
5) make a limited number of ephemeral recordings of audiovisual works
6) reproduce and distribute copies of published, nondramatic literary works in specialized formats exclusively for use by people with visual or other disabilities
7) obtain compulsory or negotiated licenses (with payment of a royalty) to make and distribute phonorecords of nondramatic musical compositions, to broadcast the performance or display of works on television, radio, or digital audio transmission, or to introduce new recording devices that allow consumers to make copies of digital recordings

Copyrights do not prevent the use of copyrighted materials in ways that would alter their original form and content such that the restructured item no longer mirrors, in full or in part, the copyrighted item. For example, a book on the birds of Wyoming can be used by another author to develop a different but very similar book on the same subject, provided none of the graphic material or verbatim text is used without the original author's permission. Copyright, therefore, protects only against copying or other unauthorized uses and that protection is of limited duration.

In Canada, limitations on the rights of owners include fair dealing (see Section 3.3.1), parody, public recitation of extracts, the right to make a single copy of a computer program either for backup or adaptation to make it compatible with a particular computer, and reproduction of works permanently situated in a public location. Exceptions also apply for educational institutions, libraries, archives, and museums.

In the United Kingdom, Australia, and New Zealand, there are defenses of fair dealing for the purposes of research and study, criticism and review, and reporting of news. There are also free-use provisions to cover photographing, filming, or including a publicly displayed sculpture or work of artistic craftsmanship in a broadcast and to copy a computer program for specified purposes.

3.1.3.3 Transfer of Ownership

The author(s) may transfer all or part of their copyright in a work to someone else by written agreement, by will or intestate succession as personal property, or by enforcement of the law. Transfers of exclusive rights may be recorded in the Copyright Office of the United States or Canada. Many journals and other publishers require authors to transfer at least some rights on acceptance of a manuscript for publication.

3.1.4 Public Domain

In the United States, works created before 1950 that were never registered, works on which the copyright has expired, works that are considered to be common property (e.g., standard calendars, tape measures and rulers, facts), and works created by US federal employees in the course of their duties (see Section 3.1.5.3) are all in the public domain and are not protected by copyright in the United States, although the US government may seek protection in other countries. A handy listing on what works remain under copyright and those in the public domain in the United States was produced by the Cornell University Copyright Information Center.[14]

In Canada and Australia, the concept of public domain is limited to works for which copyright has expired and to works that did not meet the test for copyright protection at all. Works created for these governments by independent contractors are owned by the creators, unless there is agreement to the contrary, and are not in the public domain. Works created and published by the Canadian, British, Australian, and New Zealand governments are subject to Crown copyright and are also not automatically in the public domain.

Works in the public domain should be used carefully, cited appropriately, and not plagiarized. Public domain is the end point in the continuum of ownership of the copyright in a work. For example, a US or British author who writes a manuscript owns the copyright on it from the time the words flow onto paper or into a word-processing file. Then, when the manuscript is accepted by a journal for publication, the author may be required to transfer the copyright to the journal. The article is published and remains covered by copyright for the life of the author plus 70 years (50 years if the author is a Canadian or a New Zealander). After copyright expires, the article moves into the public domain and can be reprinted freely without infringement of the copyright. Thus, a modern publisher could now reprint the works of Charles Darwin (who died in 1882) or Louis Pasteur (who died in 1895) without permission but not those of Carl Sagan (who died in 1996) or Oliver Sacks (who is alive and flourishing in 2012).

3.1.5 Works Made for Employers

3.1.5.1 Employees

Copyrightable materials created by an employee during daily work become the intellectual property of the employer, so that the employer—not the employee—becomes the owner of record (see Section 3.1.3) and the author of record (see Section 3.1.2). In the United States, these materials are called "works made for hire", a term apparently not used elsewhere, even though the same concept may apply.

In the United Kingdom, Australia, and New Zealand, the employer owns the copyright for works created during employment. One exception is that moral rights still belong to the author.

In other countries, the creator remains the author of record and retains all rights as author, including moral rights.

3.1.5.2 Contractors

In the United States, materials created by an independent contractor, hired to create a product that is copyrightable, become works made for hire only if the parties expressly agree in writing that the work is a work made for hire. Section 101 of the US Copyright Act[2] lists 9 categories of works that can become the employer's intellectual (copyright) property when specially ordered or commissioned for use:

1) contribution to a collective work
2) part of a motion picture or other audiovisual work
3) translation
4) supplementary work (as secondary adjunct to work by another author)
5) compilation
6) instructional text
7) test
8) answer material for a test
9) atlas

If the work belongs in one of these categories, the contract should specify whether the products to be prepared are "works made for hire", thus ensuring that there is no future question about who owns the copyright to the work done. In addition, copyrights to contractual works that fall outside these categories can also be transferred to the employer, in which case contracts should expressly address the question of whether copyright ownership will be transferred. Contracts may also describe any limitations on use of copyrighted materials.

In Canada, works prepared by contractors usually belong to the contractor in the absence of an agreement to the contrary. Commissioners of engravings, photographs, and portraits who pay valuable consideration for the works own the copyright to them in the absence of an agreement stating otherwise. Copyright can be assigned; however, moral rights can only be waived.

In Australia, most works prepared by contractors belong to the contractor in the absence of an agreement to the contrary. However, copyrights to photographs, paintings, drawings, engravings, cinematographic films, and sound recordings belong to the commissioners; newspaper and periodical publishers own the right to reproduce commissioned works in books and facsimile editions of the newspaper or periodical.

In New Zealand, the copyright to any of the following commissioned works created by a contractor belongs to the commissioner unless there is agreement to the contrary:

photographs, paintings, drawings, diagrams, maps, charts, plans, engravings, models, sculptures (all of which are classified as artistic works), computer programs, films, and sound recordings.

3.1.5.3 Government Employees and Contractors

A work produced exclusively by US federal employees within their official duties is not protected under US copyright law and, if there is no other statutory basis for the government to restrict its access, is in the public domain (see Section 3.1.4). However, a work produced by an independent contractor for the federal government is not automatically in the public domain; contractors and grantees are not considered government employees for purposes of copyright. Thus, unlike works of the US government, works produced by contractors under government contracts are protected under the US Copyright Act.[2] The ownership of the copyright depends on the terms of the contract.

Works of joint authorship created by US federal employees and one or more nonfederal colleagues are protected by copyright in the United States (i.e., are not in the public domain). When the US government is joint author with a nongovernment entity, the law on how much of the work is protected by copyright is unsettled and open to differing interpretations. For example, if a joint work is interdependent, contributions are generally created independently by separate coauthors with the intention to merge them into a unitary whole, and therefore they comprise separable parts. One should be able to isolate the contributions of a government employee from the contributions of a nongovernment employee. If, on the other hand, coauthors collaborated on much or all of a joint work, it is considered inseparable, and it may be impossible to determine where the contributions of one author end and the other author or authors begin. Therefore, for an inseparable joint work, it is difficult or impossible to isolate the contribution of government employees from contributions of nongovernment employees.

Works created by Canadian, British, Australian, or New Zealand government employees or for their governments are subject to Crown copyright and are not in the public domain. Works created by contractors for the Canadian, British, and Australian governments belong to the contractors, in the absence of agreements to the contrary, and are copyrightable. Works created by contractors for the New Zealand government are subject to Crown copyright.

3.1.6 Public Disclosure

Although materials copyrighted by state and local governmental bodies may not be reproduced without their consent, this should not be confused with disclosure of public records as set forth in local, state, and federal laws. A variety of laws ensure that public records are available to the public for inspection and that reasonable numbers of copies will be provided on request. This is quite different from an individual or company making unauthorized and unlimited copies of copyrighted materials.

3.2 COPYRIGHT PROCEDURES

Copyright protection begins automatically from the moment the work is created in fixed form and begins without any notice, formality, process, or application.

3.2.1 Notice

The standard copyright notice (e.g., "© 2011 Council of Science Editors") is not required to establish copyright in the United States, Canada, the United Kingdom, Australia, New Zealand, or any other signatory to the Berne Convention of 1971.[15] Nevertheless, copyright notice should be included if the copyright owner wants to ensure that users are aware that the material is copyright protected. Notice also provides additional benefits in the United States (and protection in some non–Berne member countries), as described in Section 3.2.2.1. The notice consists of all of the following 3 parts:

1) on visually perceptible copies, the copyright symbol ©, the word "copyright", or the abbreviation "copr" (The letter "c" in parentheses, "(c)", and other variations are not acceptable substitutes for the copyright symbol. On sound recordings, the symbol ℗, the "P in a circle", is used instead.)
2) the year of first publication of the work
3) the name of the owner of the copyright

Set the copyright notice in roman type with a space between the copyright symbol, abbreviation, or word and the year of first publication. The notice may also include the phrase "all rights reserved", which is necessary to satisfy requirements in some countries; see Section 29.3.6.7 for copyright symbols in references.

Affix the 3 elements of the notice together in such a way and position that most users will find them readily. The customary position on printed publications is on the title page or the verso of the title page, at the end of the text (especially in short documents), as part of the masthead of serials, or in a similar prominent location. The position on electronic copies should be with or near the title, at the end of the work on printouts, on the sign-on screen, and/or on a label on the jacket or binder for the copy. See *Copyright Notice*,[16] Circular 3 from the US Copyright Office, for more details and examples of acceptable locations.

3.2.2 Registration

Registration of the owner's copyright is beneficial and encouraged in the United States and Canada, although not required. There is no registration system in the United Kingdom, Australia, or New Zealand.

3.2.2.1 Benefits

Although registration in the United States is not mandatory, there are benefits. Registration creates a public record of the copyright claim and establishes prima facie evidence of the validity of the copyright and the facts stated on the certificate. An infringement lawsuit cannot be filed in a US court until the copyright is registered for works of US origin. Depending on the timing of the registration, statutory damages and attorney's fees may be available to the copyright owner in court actions; otherwise, the owner will receive only actual damages and profits. Registration of the copyright also allows registration with the US Customs Service and protection from the importation of infringing copies. See *Copyright Basics*,[17] Circular 1 from the US Copyright Office, for more details.

In Canada, one of the benefits of registration is that it creates a presumption of validity—that is, a presumption that copyright subsists in the work and that the person or entity registering the work is the copyright owner.

3.2.2.2 Submission

There are two possible ways to register a work in the United States: by mail or online. The online registration is faster and less expensive. Three elements are required to register a work by mail:

1) a completed copy of the appropriate registration form (e.g., Form TX for most printed materials, Form SE for serials)
2) the required number of copies of the materials being registered
3) the nonrefundable filing fee

To submit online, you will need the following:

1) a completed application form (which can be filled out online)
2) an uploaded file or files of the material(s) being registered
3) the nonrefundable filing fee

See the pamphlet "Registering a copyright with the U.S. Copyright Office"[18] (http://www.copyright.gov/fls/sl35.pdf) for instructions on hardcopy and online applications, different types of material, copies of forms, and fee schedules.

To register a work in Canada, submit the application and filing fee to the Canadian Intellectual Property Office (CIPO). Application forms and instructions for hardcopy and online registration are available from the CIPO website[19] (http://www.cipo.ic.gc.ca/eic/site/cipointernet-internetopic.nsf/eng/Home).

There is no registration procedure in the United Kingdom, Australia, or New Zealand.

3.2.3 Deposit

If materials published in the United States include a copyright notice, the copyright owner must, within 3 months of the date of publication, send 2 complete hard copies of the material to the Register of Copyrights for filing with the Library of Congress. This must be done regardless of the manner in which registration was made (by mail or online). Copies sent via mail with application forms and filing fees for registration (as described in Section 3.2.2.2) will satisfy this requirement. Failure to comply, on written demand by the Register, can result in fines.

The Canadian Copyright Office does not accept copies of the work being registered. However, the Library and Archives of Canada Act[20] requires copies of every work published in Canada to be sent to the Library and Archives of Canada. The Australian Copyright Act[9] has a similar provision requiring the deposit of printed materials in specified libraries. The effect is the same as in the United States in that copies become part of a national library but, in Canada and Australia, the acts of registration of copyright and deposit of materials in the library are not linked the way they are in the United States.

There is no formal deposit procedure in the United Kingdom or New Zealand. However, to establish that the work existed on a certain date, copyright owners may choose to deposit a copy with a bank or solicitor or to send copies to themselves by special delivery and leave the envelopes unopened. Such evidence may be needed later in case of infringement.

3.3 USE OF COPYRIGHTED MATERIAL

3.3.1 Fair Use and Fair Dealing

Although copyright laws forbid unauthorized reproduction of protected works, US law does allow for fair use of copyrighted work for criticism, comment, news reporting, teaching, scholarship, or research without infringement. The distinction between fair use and infringement may be unclear and cannot be easily defined or described. Although there are guidelines for deciding between fair and unfair uses, there are no hard rules, no specific numbers of words, lines to be copied, or percentage of content that are fair, when one more of any would be unfair. Consider the following factors when determining whether a particular copy is fair or unfair (Copyright Act of 1976[2]):

1) the purpose and character of the intended use, including whether such use is for personal research, commercial enterprise, or nonprofit educational purposes
2) the nature of the copyrighted work
3) the amount and substantiality of the portion used in relation to the copyrighted work as a whole
4) the effect of the use on the potential market for, or value of, the copyrighted work

In light of these factors, legal copies of copyrighted materials can ordinarily be made without the copyright owner's permission under the following situations:

1) A single copy may be made for personal use in education or noncommercial research.
2) A single copy of a small portion of a work (e.g., a chapter, article, poem, or chart) may be made for each member of a group assembled for educational purposes in the United States under certain conditions. Although some copying for educational purposes is allowed in Canada, the United Kingdom, Australia, and New Zealand, some restrictions and royalties apply.

By these same factors, copies made without the owner's permission or payment of royalty under the following situations would not be fair:

1) Copies of several articles from the same volume of a scholarly journal may not be made for any use, even for personal or educational use; commercial use would also not be allowed.
2) Extensive quotations of one work in another should not be made, even with attribution.
3) Any substantial copying from a work in lieu of purchase of the original is not fair.
4) Posting portions of someone else's work on a website from which it can be printed at no charge, thereby reducing the market for the original work, is unfair. Note: some publishers/copyright holders readily allow posting of abstracts; however, one should always obtain permission to do so beforehand.

The concept of "fair dealing" (which is used in Canada, the United Kingdom, Australia, and New Zealand) is somewhat narrower than "fair use". Fair dealing is a defense that can be used if allegations of copyright violations are made. In determining whether a particular reproduction is fair, one looks at both the quality of the reproduction (i.e., the value of the part of the work being reproduced in relation to the rest of the work) and the quantity being copied. In addition, the copying must fit within one of the purposes permitted: research, private study, criticism, review, or news reporting. Even if a use fits one of these purposes, it may be fair use in the United States, but not fair dealing in the United Kingdom, or it may be fair dealing in Australia, but not fair use in the United States. Note: it is never inappropriate to check with the copyright holder if you are unsure whether copying or reuse is permitted.

3.3.2 Internet and Copyrights

Original materials posted on the Internet are usually copyrighted and subject to all protections afforded any other copyrighted materials. Likewise, fair-use provisions of the law also apply, allowing limited copying. Posting copyrighted materials owned by others online could be an infringement of copyright unless the copyright owner has licensed such use. Similarly, unauthorized copying and posting of material contained on one web page to another may constitute infringement. This is different from downloading materials for fair-use purposes, such as making a paper copy for subsequent reading and study.

It is important to note that the absence of a copyright statement or symbol should not be taken to mean that material on a website is not protected by copyright. Unless there is a statement allowing free reuse of material, one should always ask permission before copying or reusing material outside a strict notion of the fair-use law.

3.3.3 Permission to Republish

Republishing scientific articles or parts of articles (e.g., a figure or table) requires permission from the copyright owner. Some journals allow authors to retain copyright. Many other journals, however, require authors to transfer their copyright to the publisher or society as a condition of publication of their work. If all authors of a paper were US federal employees, the paper was developed within the scope of their official duties, and the publisher has not added copyrightable contributions to the paper (e.g., peer review comments), then no copyright exists within the United States. (Outside the United States, the US government can claim copyright on the work.) Within the United States, you may use the material without permission, but you should cite its source. On the other hand, if the publisher has added a copyrightable contribution to the paper as published, the paper may be protected by copyright and may not be in the public domain.[21]

If you wish to request permission from a copyright owner or designated agent to copy or republish material, first look at the journal/book publisher's website to see whether there is a standard procedure for requesting permission. Many publishers provide permission request forms on their websites that can be completed online. Some publishers deal with permissions in house; others outsource this responsibility to entities such as Copyright Clearance Center. Regardless of the procedure, you should have the following information ready when requesting permission: 1) your intended use (e.g., educational use, reuse in commercial enterprise) and its duration; 2) pertinent source information (e.g., year, volume, number, edition, pages, or portions of a page) that clearly identifies the material that you would like to reuse; and 3) information about the journal/book/ website, etc., in which you plan to reuse the material (e.g., name of book/journal, publisher, expected number of print copies, whether translations will be made, and whether the online version will be on a site that is password protected).

You may also be able to request permission to copy from a collective such as the Copyright Clearance Center in the United States (http://www.copyright.com),[22] Access Copyright in Canada (http://www.accesscopyright.ca),[15] the Copyright Licensing Agency in the United Kingdom (http://www.cla.co.uk),[23] the Copyright Agency Limited in Australia (http://www.copyright.com.au/),[24] or Copyright Licensing Limited in New Zealand (http://www.copyright.co.nz).[25] These centers can often provide prompt authorization to make copies of many copyrighted materials and bill for and receive any attendant royalty payments. They may also assist in resolving questions regarding whether fair use would allow legal copying of a needed item without permission or royalty payments.

3.3.4 Creative Commons Licenses

Creative Commons copyright licenses were first established in 2002 to allow individual and institutional copyright owners to license their work to others in a manner simpler than traditional copyright licensing. Copyright owners decide how their material should be licensed throughout the duration of the copyright. Then they simply insert a statement (and/or marks) in their work showing that the work is licensed through a particular Creative Commons license (see below). People who wish to reuse the material do not have to request permission from the author or publisher. Instead, they can immediately reuse the material in the manner specified in the work's particular Creative Commons license. More detailed information on Creative Commons licenses and how they can be applied and used can be found on the Creative Commons website[26]: http://creativecommons.org.

The copyright holder can select from 6 different Creative Commons copyright licenses*:

Attribution CC BY License

Most accommodating license. Allows others the right to distribute, adapt, and build on the original work, as long as the original work and copyright holder are acknowledged.

Attribution-NoDerivs CC BY-ND License

Allows others the right to redistribute the original work commercially or noncommercially, as long as it remains unchanged and whole, and the original work and copyright holder are acknowledged.

Attribution-NonCommercial-ShareAlike CC BY-NC-SA License

Allows others the right to redistribute, adapt, and build on the original work for noncommercial purposes, as long as the original work and copyright holder are acknowledged. Derivative works based on the original must be licensed under the same terms.

Attribution-ShareAlike CC BY-SA License

Allows others the right to distribute, adapt, and build on the original work for commercial purposes, as long as the original work and copyright holder are acknowledged. Derivative works based on the original must carry the same license.

Attribution-NonCommercial CC BY-NC License

Allows others the right to distribute, adapt, and build on the original work for noncommercial purposes, as long as the original work and copyright holder are acknowledged. Derivative works based on the original do not have to carry the same license.

Attribution-NonCommercial-NoDerivs CC BY-NC-ND License

The most restrictive license. Allows others the right to use and distribute the original work for noncommercial purposes, as long as it and copyright holder are acknowledged. No derivative works are allowed

* BY = attribution; CC = Creative Commons; NC = noncommercial; ND = no derivatives; SA = share alike.

Every Creative Commons license ensures that the creator of the original work receives credit for that work. Based on copyright laws, these licenses are honored worldwide and last as long as the copyright.

Creative Commons also provides a tool by which creators of a work can waive all rights and place their work in the public domain. This tool is called CC0.

3.3.5 Infringement

Infringement is any unauthorized use of copyrighted materials. It is illegal and subject to either civil remedies or criminal prosecution and court action (including fines and/or imprisonment), depending on the severity of the offense. Examples include

1) photocopying too much of a book or journal
2) selling what was once a fair-use copy of an article
3) reprinting a recent best seller
4) showing a film rented for home use to a paying audience
5) making copies for sale or rent
6) distributing copies for the purpose of trade
7) exhibiting a work in public without permission
8) importing illegal copies of a book or sound recording for sale or rent
9) authorizing or facilitating the infringing act of another party

Plagiarism is the intentional act of copying someone else's work and claiming it to be one's own intellectual effort. The work in question may or may not be under current copyright protection. Whether the original work is by Charles Darwin, Stephen Jay Gould, or a US federal employee, falsely claiming authorship is plagiarism, as well as a breach of moral rights in Canada, the United Kingdom, Australia, and New Zealand. Whether it is infringement depends on the copyright status of the work (see Section 2.3).

Fair use is not infringement, but the line between them is fine and sometimes difficult to draw. There are no simple rules to follow other than moderation and good judgment. When in doubt, ask for permission and/or pay a royalty.

The owners of copyrighted materials are responsible for monitoring their use and abuse by the public and for seeking legal recourse if they discover infringement. Negotiation between the parties may be sufficient to settle a dispute over rights. If not, lawsuits may follow. Although infringement of copyright may be a criminal offense, law enforcement officers will not take action unless the affected owner files a complaint. To file a lawsuit in the United States for works of US origin, the work in question must be registered with the US Copyright Office (see Section 3.2.2).

3.4 INTERNATIONAL LAW

There is no such thing as an "international copyright" that will automatically protect works throughout the world. National laws differ, but most countries offer protection to foreign works under certain circumstances. The Berne Convention for the Protection of Literary and Artistic Works of 1971[11] and the Universal Copyright Convention of 1971[12] are the major treaties involved. In addition to these treaties, bilateral agreements exist between individual pairs of countries. If you wish to protect a work in a particular country, determine the extent of protection available in that country and consult someone familiar with the laws there.

The Universal Copyright Convention calls for the use of the copyright notice consisting of the copyright symbol, the date of first publication, and the name of the copyright owner—for example, © 2014 Council of Science Editors. Neither the word "copyright" nor its abbreviation "copr" without the symbol is valid outside the United States.[27] Some Latin American countries do not recognize the symbol or the word "copyright" and insist on the phrase "All rights reserved" instead. Strong (p. 95)[28] advises using both terms when full protection is needed.

CITED REFERENCES

1. U.S. Const. art. I, sect. 8, cl. 8.

2. Copyright Act of 1976, Pub. L. No. 94-553, 90 Stat. 2541 (Oct 19, 1976).

3. Berne Convention Implementation Act of 1988, Pub. L. No. 100-568, 102 Stat. 2853 (Oct 31, 1988).

4. Uruguay Round Agreements Act of 1994, Pub. L. No. 103-465, 108 Stat. 4809, 4973 (Dec 8, 1994).

5. Digital Millennium Copyright Act of 1998, Pub. L. No. 105-304, 112 Stat. 2860, 2887 (Oct 28, 1998) [accessed 2012 Apr 2]. http://www.copyright.gov/legislation/dmca.pdf.

6. Copyright Act. R.S.C. 1985, Chapter C-42. Ottawa (ON): Department of Justice Canada; [updated 2011 Oct 31; accessed 2012 Mar 29]. http://www.wipo.int/wipolex/en/details.jsp?id=8415.

7. Copyright, Designs and Patents Act 1988 (c. 48). London (UK): Queen's Printer of Acts of Parliament; 1988 [accessed 2012 Mar 30]. http://www.legislation.gov.uk/ukpga/1988/48/contents.

8. The Duration of Copyright and Rights in Performances Regulations 1995. London (UK): Queen's Printer of Acts of Parliament; 1995 [prepared 2000 Sep 20; accessed 2012 Apr 2]. http://www.opsi.gov.uk/si/si1995/Uksi_19953297_en_1.htm.

9. Copyright Act 1968, Act No. 63 of 1968 as amended. Canberra (Australia): Australian Government, Attorney-General's Department; [prepared 2011 Jun 9; accessed 2012 Apr 2]. http://www.comlaw.gov.au/Details/C2013C00145. Takes into account amendments up to Act No. 46 of 2011.

10. Copyright Act 1994, No 143. Wellington (New Zealand): Parliamentary Counsel Office; 1994, reprinted 7 October 2011 [accessed 2012 Mar 31]. http://www.legislation.govt.nz/act/public/1994/0143/latest/DLM345634.html.

11. Berne Convention for the Protection of Literary and Artistic Works, Paris Act of July 24, 1971 as Amended on September 28, 1979. http://www.wipo.int/treaties/en/ip/berne/pdf/trtdocs_wo001.pdf.

A list of the signatories can be found at http://www.copyrightaid.co.uk/copyright_information/berne_convention_signatories.

12. Universal Copyright Convention as revised at Paris on 24 July 1971, with Appendix Declaration relating to Article XVII and Resolution concerning Article XI 1971. Paris (France): United Nations Educational, Cultural and Scientific Organization; [accessed 2012 Apr 2]. http://portal.unesco.org/en/ev.php-URL_ID=15241&URL_DO=DO_TOPIC&URL_SECTION=201.html.

13. Directive 2001/29/EC of the European Parliament and of the Council of 22 May 2001 on the harmonization of certain aspects of copyright and related rights in the information society. Off J Eur Communities (Engl Ed). 2001 Jun 22 [accessed 2012 Apr 1];44(L167):10–19. http://eur-lex.europa.eu/LexUriServ/LexUriServ.do?uri=OJ:L:2001:167:0010:0019:EN:PDF.

14. Hirtle P. Copyright term and the public domain in the United States. [Ithaca (NY)]: Cornell University Library; 2011 [accessed 2012 April 1]. http://copyright.cornell.edu/resources/publicdomain.cfm.

15. access ©™: The Canadian Copyright Licensing Agency. Toronto (ON): Access Copyright; [accessed 2011 Dec 1]. http://www.accesscopyright.ca.

16. Copyright notice. Washington (DC): US Copyright Office; 2004 [accessed 2012 Mar 30]. (Circular 3). http://www.copyright.gov/circs/circ03.pdf.

17. Copyright basics. Washington (DC): US Copyright Office; 2004 [accessed 2012 Mar 30]. (Circular 1). http://www.copyright.gov/circs/circ01.pdf.

18. Copyright: United States Copyright Office. Washington (DC): The Office; [revised 2012 Mar 12; accessed 2012 Mar 30]. http://www.copyright.gov. This site includes registration forms, "Copyright Basics", and a variety of circulars and fact sheets on many issues relating to copyright.

19. Canadian Intellectual Property Office. Gatineau (QC): The Office; [modified 2011 Dec 2; accessed 2012 Mar 29]. http://www.cipo.ic.gc.ca/eic/site/cipointernet-internetopic.nsf/eng/Home. This site includes "A Guide to Copyrights" and a variety of circulars on many issues relating to copyrights.

20. Library and Archives of Canada Act. 2004, Chapter 11. Ottawa (ON): Department of Justice Canada; [updated 2006 Dec 12; accessed 2005 Oct 12]. http://canlii.ca/en/ca/laws/stat/sc-2004-c-11/latest/sc-2004-c-11.html.

21. CENDI Copyright Working Group. Frequently asked questions about copyright: a template for the promotion of awareness among CENDI agency staff. Klein B, Hodge G, editors. Oak Ridge (TN): CENDI Secretariat; 2004 Aug [accessed 2012 Mar 30]. http://cendi.dtic.mil/publications/00-3copyright.html.

22. Copyright Clearance Center. Danvers (MA): Copyright Clearance Center; 2012 [accessed 2012 Mar 30]. http://www.copyright.com.

23. CLA: the Copyright Licensing Agency Limited. London (UK): The Agency; [accessed 2012 Mar 30]. http://www.cla.co.uk.

24. Copyright Agency Limited. Sydney (Australia): The Agency; 2010 [updated 2005 Oct 13; accessed 2012 Apr 1]. http://www.copyright.com.au.

25. Copyright Licensing Limited. Auckland (New Zealand): Copyright Licensing Limited; 2008 [accessed 2012 Apr 2]. http://www.copyright.co.nz.

26. Creative Commons. Mountain View (CA); [accessed 2012 Mar 29]. http://creativecommons.org.

27. Duration of copyright: provisions of the law dealing with the length of copyright protection. Washington (DC): US Copyright Office; 2004 [accessed 2012 April 1]. (Circular 15a). http://www.copyright.gov/circs/circ15a.pdf.

28. Strong WS. The copyright book: a practical guide. 5th ed. Cambridge (MA): MIT Press; 1999.

2

General Style Conventions

4 Alphabets, Symbols, and Signs

4.1 ALPHABETS

Most European languages are written and printed with letters of the Roman or Latin alphabet. English uses, in general, unmodified letters from the Roman alphabet. For special needs, scientific English also uses some special characters formed from these letters, as well as Greek letters and some special typefaces.

4.1.1 Roman or Latin Alphabet

The classical Roman alphabet had 23 letters. During the Middle Ages, 3 additional letters were developed: the consonant J (as distinct from the vowel I), the vowel U, and the consonant W (as distinct from the consonant V). Therefore, the present-day Roman alphabet has 26 letters, each of which can be represented as a capital (uppercase) letter or as a small (lowercase) letter (see Table 4.1). In some non-English European languages, these letters are modified by diacritics or combined in ligatures to form additional characters (see Table 4.2 and Section 5.6). These additional characters are not generally needed in English scientific texts, except for proper names of people and places and for direct quotations of non-English words.

The end references for a scientific publication may include titles of articles, serials, and books originally published in a non-English language. In the case of a non-English title in the Roman alphabet, present the original title, followed by a translation, placing the translated title within square brackets. For a title in Cyrillic or Greek characters, romanize the letters and present a translation within square brackets after the original title. For a title in a character-based alphabet (e.g., Chinese or Japanese) present only a translation, enclosed within square brackets (see Section 29.3.6.2.1).

For more detailed information about special characters in non-English languages, see *The Chicago Manual of Style*.[1]

Table 4.1 Roman and Greek alphabets

Roman letters		Greek letters[a]		
Capital (uppercase)	Small (lowercase)	Name of letter	Capital (uppercase)	Small (lowercase)
A	a	alpha	Α	α
B	b	beta	Β	β
C	c			
D	d	delta	Δ	δ
E	e	epsilon	Ε	ε
F	f			
G	g	gamma	Γ	γ
H	h	eta	Η	η
		theta	Θ	θ
I	i	iota	Ι	ι
J	j			
K	k	kappa	Κ	κ
L	l	lambda	Λ	λ
M	m	mu	Μ	μ
N	n	nu	Ν	ν
O	o	omicron	Ο	ο
P	p	pi	Π	π
Q	q			
R	r	rho	Ρ	ρ
S	s	sigma	Σ	σ
T	t	tau	Τ	τ
U	u	upsilon	Υ	υ
V	v			
W	w			
X	x	xi	Ξ	ξ
Y	y			
Z	z	zeta	Ζ	ζ
		phi	Φ	φ
		chi	Χ	χ
		psi	Ψ	ψ
		omega	Ω	ω

[a] In the normal order of the Greek alphabet, the letter gamma follows beta, zeta follows epsilon, and xi follows nu. The order here shows the relation of the Greek alphabet to the Roman alphabet.

4.1.2 Greek Alphabet

Most of the capital letters of the classical Greek alphabet can be related to Roman capital letters (see Table 4.1). Greek letters, both uppercase and lowercase, are used in many nomenclatures and notations for scientific disciplines, notably astronomy, biochemistry, chemistry, mathematics, pharmacology, physics, and statistics. These uses are discussed in Part 3 of this manual.

Lowercase Greek letters are themselves typically cursive, but italic (sloping) versions of lowercase Greek letters are widely used in mathematics, physics, and statistics. In

Table 4.2 Common diacritics and special formations for the Roman alphabet used primarily in languages other than English

Name	Mark	Example
Acute accent	´	é
Double acute accent	˝	ő
Grave accent	`	è
Breve	˘	Ğ
Caron, wedge (háček [Czech for "little hook"])	ˇ	ž
Cedilla	¸	Ç
Circumflex	ˆ	Ô
Dot	˙	İ
Ligature	[no mark]	æ
Macron	¯	Ē
Ogonek (Polish for "little tail")	˛	ę
Over-ring (kroužek [Czech for "little circle"])	˚	å
Bar, stroke	/	ø
Tilde	˜	ñ
Umlaut (diaeresis mark)	¨	ü

other disciplines, Greek letters are typically set in roman font; however, the Particle Data Group uses the italic versions of lowercase Greek letters as symbols for subatomic particles (see Section 16.2).[2]

4.1.3 Hebraic Alphabet

Hebraic letters are rarely used in English-language scientific literature except for the aleph (א), which is occasionally needed in mathematics as a full-size symbol or a superscript character.

4.2 SYMBOLS AND SIGNS

Notations representing quantities, objects, and actions have been in use through all of recorded history (e.g., numerals for numbers of objects). As the pace of scientific discovery and description accelerated in the Renaissance period and in succeeding centuries, new notations were needed to efficiently represent the much more complex knowledge that was being developed. These needs led to the present extensive systems of symbolic notation described in Part 3 of this manual for chemicals, genes, animal functions, mathematical operations, and other subjects of scientific inquiry.

Many symbolic notations in science have been developed in a logical and coherent scheme for specific functional needs. They may serve either to represent what cannot be as economically expressed by a term or to represent functional relations among various items, as in mathematical equations. In contrast, abbreviations have been developed mainly to eliminate the effort that would go into writing out what they represent or to save space; they usually represent simply the shortening of a term (see Chapter 11).

Graphic notations that stand for relations and operations, especially in mathematics, are often called signs; examples are the plus sign (+) and the minus sign (−). The International Organization for Standardization (ISO) standards handbook on quantities and units[3] distinguishes between the terms "sign" and "symbol" but does not define the distinction. In this manual, mathematical signs are generally referred to as symbols; this usage is consistent with the style manuals of the American Institute of Physics[4] and the American Mathematical Society[5] and with the National Information Standards Organization standard for electronic manuscript markup.[6] The term "sign" is used for notations that convey a direction for the reader (e.g., footnote signs, in the form of lowercase letters or other characters, are used to direct the reader to footnotes).

4.2.1 Types of Symbols

Symbols are characters or other graphic units representing a quantity, unit, element, unit structure, relation, or function. They are of 3 general kinds.

1) Single alphabetic or numeric characters or an unspaced combination of 2 or more alphabetic, numeric, or alphanumeric characters. Such a symbol may be an abbreviated form of the term it represents (e.g., V for volume or cos for cosine), but it need not be (e.g., φ for electric flux). A few symbols are graphic units derived from an alphabetic origin, for example, Ꞓ for Cambrian geologic period.
2) Punctuation marks, particularly in mathematics (e.g., the exclamation mark for factorial, the colon for a ratio).
3) Graphic units designated as specific representations (e.g., the equals symbol [=], which represents equality; the symbols for greater than [>] and less than [<]).

A symbol may stand alone or be incorporated into a larger symbol. For examples of various scientific symbols, see Table 4.3.

An abbreviation can function as a symbol when it is part of a group of symbols that together are used for the same conceptual purpose. For example, abbreviations of names for geologic periods (e.g., Pz for Paleozoic and J for Jurassic) serve as symbols when they are used with other alphabet-derived symbols (e.g., Ꞓ for Cambrian) in representing geologic systems. Chemical symbols may be abbreviated versions of the element names (e.g., C for carbon and O for oxygen), but they also serve as symbols in indicating relations among chemical elements in a unit structure (e.g., CO_3^-, the symbolic representation for the carbonate anion). When an abbreviation does not serve any of these purposes but is used solely as the short form of a term, it remains designated as an abbreviation (see Chapter 11).

In this manual, symbols for specific scientific fields are presented in the chapters on style in those fields (see Table 4.4).

4.2.2 Type Styles for Symbols

The appropriate uses of roman and italic type for symbols and the appropriate styling are specified in the international standard *Quantities and Units, Part 0: General Principles*,[7] which is also available in the ISO standards handbook *Quantities and Units* (see Table 4.5).[3] Use boldface type for symbols for vectors and tensors (see Section 12.3.1.3). Fraktur or black-letter script is used for some conventions in mathematics and physics.[4,5] These conventions are also summarized in Chapter 10.

Table 4.3 Examples of scientific and common symbols

Symbol	Name of representation	Type of subject
1	one	Unit
x	unknown quantity	Quantity
n	number	Quantity
m	mass	Quantity
P	probability	Quantity
V	volume	Quantity
CO	cardiac output	Rate
3	1 + 1 + 1	Unit structure
♂	male (e.g., in a family tree)	Unit structure
Al	aluminum, aluminium	Unit, chemical
$AlBr_3$	aluminum bromide	Unit structure, chemical
sis	oncogene related to simian sarcoma	Unit structure, genetic
Pz	Paleozoic Era	Unit, geologic time
m	meter, metre	Unit, measurement
$	dollar	Unit, monetary
€	euro	Unit, monetary
cos	cosine	Mathematical function
+	plus (for addition)	Mathematical function

Table 4.4 Scientific and common symbols presented in this manual

Subject	Section	Table
Algebra	12.3.1	12.10
Allergens	23.5.6.5	
Amino acids	17.13.2	17.13
Astronomy	26.3	26.3
Atomic and molecular states	16.6, 16.7	
Biothermodynamics	18.2	18.1
Blood groups	23.4.3, 23.5.2.1	
Bone histomorphometry	23.5.3	
Calculus	12.3.1.2	12.11
Carbohydrates	17.13.4	
Chemical elements	16.4	16.4
Chemical formulas	16.5, 17.2	
Chemical kinetics	18.3.1, 18.3.3	
Chromosomes	21.8–21.15	21.2, 21.32, 21.34
Chromosomes: components and units	21.2	
Complement	23.5.6.2	
Crystallography	25.6	25.1, 25.2
Density of water	25.11.2.4	
Drug receptors	20.3	
Electrocardiographic recordings	23.5.4.2	
Elementary particles	16.2	16.1
Enzyme kinetics	18.7	18.4
Enzymes	17.13.7	

Table 4.4 Scientific and common symbols presented in this manual (*continued*)

Subject	Section	Table
Genes and phenotypes	21.8–21.15	21.4–21.11, 21.14–21.22, 21.25–21.31, 21.33
Hemodynamic functions	23.5.4.1	23.3
Hemoglobins	23.5.2.2	
Immunoglobulins	23.5.6.1	
Immunologic systems	23.4.4, 23.5.6	
Isotopic modifications	17.10	
Lipids	17.13.9	
Mathematical functions	12.3.1, 12.4	12.9
Mathematical operators	12.3.1.1	12.10
Mathematical symbols	12.3.1	12.10
Matrices	12.3.1.4	12.13
Monetary units	12.1.6	12.2
Non-SI units	12.2.2, 12.2.3, 25.13	12.5, 12.7, 12.8
Nuclear reactions	16.3	16.3
Nucleic acids and related compounds	17.13.10	
Nuclides	16.4	
Pedigrees	23.5.11	23.6
Pharmacokinetics	20.5	20.2
Plant physiology	23.2	23.1, 23.2
Prenols	17.13.13	
Prostaglandins	17.13.14	
Proteins	17.13.15	
Purines, pyrimidines	17.13.10.1	17.16
Renal function	23.5.8	
Respiratory functions	23.5.9	23.4
Retinoids	17.13.16	17.17
Rocks and minerals	25.5	
Roman numerals	12.1.1	12.1
Set theory notations	12.3.1.2	12.12
SI base units, derived units, and prefixes	12.2.1	12.3, 12.4, 12.6
Soil horizons	25.7.1	25.3
Stereochemistry	17.11	17.11
Steroids	17.13.17	17.18
Statistics	12.5	12.14
Thermal regulation	23.5.10	23.5
Vitamins	17.13.19	

4.3 STIX FONTS FOR SCIENTIFIC USE

In an electronic workflow, problems can arise if special characters and symbols in more than one font set are imported into a document. Characters taken from different font sets may vary in style, positioning, or size. Beyond problems related to the appearance of such variable characters in print, difficulties may arise when documents are posted online; for example, some characters may not be rendered correctly on screen if the same fonts are not installed on the user's computer system.

Table 4.5 Style conventions for alphanumeric symbols

Symbol category	Typeface	Examples and notes
Quantity	Italic	Examples: *y*, *V*
		Usually a single Roman or Greek letter with no following punctuation mark except as needed for normal punctuation within a sentence; may be modified by a subscript. For a frequent exception to use of italic, see footnote a. If the text must be in italic type, words, phrases, or symbols that would normally be set in italics are set in roman type.
Subscript modifiers		
For a quantity symbol	Italic	Example: P_x, for the probability (*P*) of an unknown quantity (*x*)
For other symbols	Roman	Example: V_{L}, for volume (*V*) of the lung (L)
Unit of measurement	Roman	Examples: m (meter), kg (kilogram)
		In roman type regardless of the type used for the rest of the text. Used in conjunction with numeric values; the unit follows the numeric value and is separated from it by one space (by a hyphen when used as a compound modifier); use the singular form for plurals; no following punctuation mark except as needed for normal punctuation within a sentence. Usually lowercase letters except if the unit name is derived from a proper name (see Section 12.1.4.1).
Number	Roman	Examples: 11, 64 (arabic numerals); XVIII, xviii (roman numerals)
Mathematical operator	Roman	Examples: +, ×
		There are some rare exceptions to the roman convention.
Subatomic particles	Italic or roman	Examples: π^-, μ^+
		Italic type is widely used, but there is no accepted standard regarding use of italic or roman type for these symbols (see Section 16.2).
Chemical element	Roman	Examples: K, Rb,Sn, $\mathrm{CaCl_2}$
		No following punctuation mark except as needed for normal punctuation within a sentence; modifying subscripts and superscripts (as in symbols for compounds; see Sections 16.4 and 16.5) are also in roman type.
Chromosome	Roman	Example: 46, Yt(Xq+; 16p−) (see Chapter 21)
Gene	Italic	Examples: *sis*, *HRCT* (see Chapter 21)
Phenotype	Roman	Example: HRCT (see Chapter 21)

[a] In general, multiletter symbols are set in roman type rather than italic, even when they represent quantities. An example is CO for cardiac output, a variable quantity presented with the unit liters per minute. Use of roman characters for multiletter variables avoids potential misinterpretation of the symbol as a combination of 2 single-letter variables, in this case *C* and *O*. Nonetheless, the American Chemical Society[8] recommends italic for 2-letter variables defining chemical transport properties (e.g., *Bi* for Biot number); with this convention, special care is needed in representing multiplication involving such variables to avoid potential misinterpretation.

In response to this problem, the Scientific and Technical Information Exchange (STIX), a group of 6 major scientific publishers, has developed a comprehensive set of fonts for mathematics and other characters used in scientific, medical, and mathematical publishing. The STIX fonts have been designed for use in all stages of the publishing process, from manuscript creation through final print or electronic publication.[9] More information about this project is available at the project website (http://www.stixfonts.org).

CITED REFERENCES

1. The Chicago manual of style: the essential guide for writers, editors, and publishers. 16th ed. Chicago (IL): University of Chicago Press; 2010. Also available at http://www.chicagomanualofstyle.org.

2. Beringer J, Arguin JF, Barnett RM, Copic K, Dahl O, Groom DE, Lin CJ, Lys J, Murayama H, Wohl CG, et al. Particle Data Group. The review of particle physics. Berkeley (CA): Regents of the University of California; 2012. http://pdg.lbl.gov/.

3. International Organization for Standardization. Quantities and units. Geneva (Switzerland): ISO; 2009. (ISO 80000).

4. AIP style manual. 4th ed. New York (NY): American Institute of Physics; 1990 [accessed 2012 February 18]. http://www.aip.org/pubservs/style/4thed/toc.html.

5. Swanson E, O'Sean A, Schleyer A. Mathematics into type. Updated ed. Providence (RI): American Mathematical Society; 1999.

6. National Information Standards Organization. Electronic manuscript preparation and markup. Bethesda (MD): NISO Press; 1995. (ANSI/NISO/ISO 12083 - 1995 (R2002); updated 2013 Jan 16).

7. International Organization for Standardization. Quantities and units. Part 0: general principles. Geneva (Switzerland): ISO; 2009. (ISO 80000-1:2009).

8. Dodd JS, editor. The ACS style guide: a manual for authors and editors. 2nd ed. Washington (DC): American Chemical Society; 1997.

9. STIX fonts project. Ver. 1.1.0. [place unknown]: STI Pub Companies; [accessed 2013 Jan 17]. http://www.stixfonts.org/.

ADDITIONAL REFERENCES

Bringhurst R. The elements of typographic style. 3rd ed. London (UK): Frances Lincoln Limited; 2004.

Diringer D, Minns E. The alphabet: a key to the history of mankind. Whitefish (MT): Kessinger Publishing, LLC; 2007.

McArthur T, editor. The Oxford companion to the English language. Oxford (UK): Oxford University Press; 1992.

5 Punctuation and Related Marks

5.1 OVERVIEW

For many centuries, punctuation in English was applied mainly as a guide for readers to stops or pauses, which is its rhetorical use. Beginning in the late 17th century, its logical, or grammatical, use began to replace the rhetorical use. Today, punctuation serves mainly to clarify the relation of prose elements to one another, as with periods demarcating sentences, and to support the meaning of statements, as with the question mark and the exclamation mark. With the growth of scientific literature, punctuation has increasingly been needed in conventions for numeric and mathematical expression and for specialized nomenclature and symbolization. Especially clear discussions of the logic and development of punctuation can be found in Partridge's *You Have a Point There*[1] and Fowler and Fowler's *The King's English*.[2] Bringhurst[3] presents a complete catalog of analphabetic characters (punctuation marks and related characters) and a discussion of them from a typographer's point of view.

This chapter covers several groups of punctuation marks: first, those designated "stops" and applied mainly to indicate the close of a sentence or its equivalent; second, the marks that occur within sentences or words; and finally, marks indicating relations between 2 or more lines of text (see Table 5.1). Quotation marks and ellipses, as well as the relations between quotation marks and other punctuation marks, are discussed in Chapter 10 (see Section 10.2 and its subsections). The hyphen and apostrophe are also considered in Sections 6.2 and 6.5, respectively.

Each section of this chapter presents general uses and then, as appropriate, specialized uses in science. Many of the conventional punctuation marks serve together in applications of the various markup languages (e.g., Standard Generalized Markup Language or SGML) for marking electronic manuscripts. Chapter 31 discusses these uses in detail.

Three general principles can be applied in reaching decisions on punctuation:

1) Use specific marks of punctuation when called for by well-established and internationally accepted conventions of style, bearing in mind that some well-established uses change over time.
2) When the need for a mark (or its omission) is unclear, use the mark if it will reduce or eliminate possible ambiguity.
3) If a format or text structure logically arranges text elements and thereby makes the meaning clear, omit punctuation that might otherwise be used (e.g., in vertical lists or after introductory phrases).

5.2 STOPS

5.2.1 Period or Full Stop .

5.2.1.1 General Uses

1) To close a declarative sentence (within or outside parentheses), a short equivalent of an implied complete declarative sentence, or an imperative sentence (see also Section 5.2.3).

    Four of the soluble vitamins of the B complex have precise roles in the citric acid cycle.
    The first rat died. (It refused to eat.)
    What is the worst ethical sin in science? Fraud. What next? Plagiarism.
    Filter the solution while it is hot.

2) To separate elements of some kinds of abbreviations, especially the abbreviations for Latin phrases that are widely used in scientific publication and those specified by nomenclatural codes. (See Section 11.2 for recommendations to omit unnecessary periods in abbreviations.)

   e.g. i.e. sp., spp.

3) In Uniform Resource Locators (URLs) for websites.

   http://www.councilscienceeditors.org

4) To separate an introductory element (usually a number or letter) from text in an ordered list; a closing parenthesis mark may also be used for this purpose (see Section 5.3.6).

   1. Biochemistry a. New Jersey
   2. Chemistry b. New York
   3. Physics c. Pennsylvania

Other functions for the period include the ellipses (see Section 10.3), dot-line connectors (see Section 5.5.3), and the raised period (see below in this section).

Table 5.1 Alphabetical list of punctuation marks

Name and synonyms	Mark	Sections in this chapter
Ampersand	&	5.4.8
Angle brackets	< >	5.3.9
Apostrophe	’	5.4.4
Asterisk	*	5.4.7
“At” symbol	@	5.4.9
Braces (single, paired)	{ {}	5.3.8, 5.5.1
Comma	,	5.3.3
Diacritics		5.4.6
Ditto marks	"	5.5.2
Dots	.....	5.5.3
Em dash (and 2-em dash)	—	5.3.5
En dash	–	5.3.5
Exclamation point, exclamation mark	!	5.2.3
Guillemets, chevrons (quotation marks in French, Spanish)	« »	5.6
Hyphen	-	5.4.1
Interrogation mark in Spanish	¿	5.2.2
Octothorp, numeral symbol, pound symbol, space symbol, hash mark	#	5.4.10
Paragraph mark, pilcrow	¶	5.5.4
Parentheses, parenthesis marks, round brackets	( )	5.3.6
Period, full stop, full point	.	5.2.1
Prime symbol (single, double)	′ ″	5.4.5
Question mark, interrogation mark, interrogation point	?	5.2.2
Quotation marks (single, double)	‘ ’ “ ”	5.3.4
Semicolon	;	5.3.2
Slash, oblique bar, oblique mark, oblique stroke, shilling mark, slant line, virgule, forward slash	/	5.4.2
Square brackets, brackets, square parentheses	[ ]	5.3.7
Vertical bar, vertical line	|	5.4.3
Vinculum	‾	5.3.10

5.2.1.2 Specialized Uses: On-the-Line Period

1) As the decimal point, especially in US, British, and Canadian usage. The CSE recommends this form (see also Section 12.1.3.1).

   The infant weighed 10.7 kg at birth.

2) To indicate hierarchical divisions represented numerically (as for the numbering of section headings in this manual).
3) In end references, to indicate the end of a component (e.g., author name, article title) or the end of the reference itself (see Section 29.3.5).

   Martonyi CL. Photographic slit-lamp biomicroscopes. Ophthalmology. 1989 Sep;(Pt 2):6–19.

4) To indicate ring size in chemical nomenclature.

   spiro[2.3]hexane   bicyclo[2.2.2]octane

5.2.1.3 Specialized Uses: Raised Period (Centered Dot)

1) As a multiplication symbol in equations and other mathematical expressions and in compound units (see Section 12.3.1.1 and Table 12.10).

   $k \cdot g(a + 2)$ 1 C = 1 A · s

2) As an ellipsis symbol in a mathematical expression (see Section 12.4.3).

   $x_1 + x_2 + \cdots + x_n$

3) To indicate associated base pairs of nucleotides (see Section 17.13.10.5).

   G · C *for* guanine and cytosine

4) To indicate connection with adducts (e.g., water of hydration) in a chemical formula.

   $Na_2B_4O_7 \cdot 10H_2O$

5) As a group of 3 raised periods, to indicate a chemical association of unspecified type.[4]

   F · · · H–NH3   C · · · Pt

5.2.2 Question Mark ?

In general, the question mark indicates interrogation or uncertainty, regardless of context. In addition, it is also used as a character in URLs for electronic addresses on the Internet.

1) To close a freestanding question whether or not it is a complete sentence.

   What is the government's policy on pollution of river headwaters?
   The policy was overturned. Why? In view of the number of species lost, this change appears . . .

2) To close a declarative sentence ending with a complete question set off by a comma, even if it is not a direct quotation. In this situation, the question need not be capitalized.

   The committee asked, why were so many species killed?
   but The committee asked why so many species were killed.

3) To indicate uncertainty about an element within a sentence or another notation, e.g., chromosome or chromosomal structure (see Tables 21.21 and 21.34).

   Girolamo Fracastoro (1483?–1553) was in effect the father of the concept of infectious disease.
   45,XXY,–?8 [45 chromosomes, XXY sex chromosomes, missing chromosome that is probably chromosome 8]

Try to position the question mark to clearly indicate the queried element.

1172?–1221 *not* ?1172–1221
1546–?1627 *not* 1546–1627?
?1546–?1627

5.2.3 Exclamation Mark !

In general, the exclamation mark serves to emphasize something or to indicate surprise. This character also has some specialized uses in science.

1) To close a declarative, imperative, or interrogative sentence to give it rhetorical emphasis. Such uses are usually not justified in reports of scientific findings but may appear in commentaries on findings or in documents about science.

   Freud's "science" was pure metaphysics!

2) As a factorial symbol in mathematics (see Table 12.10).

   4! *to represent* 1 × 2 × 3 × 4

3) In botanical writing, to indicate specimens examined by the author.

   Lectotype here designated: P!; isolectotypes: K!, NY!.

5.3 INTRASENTENCE AND PHRASE MARKS

5.3.1 Colon :

The colon indicates a stronger pause than that indicated by a semicolon or a comma and clearly directs the reader to the material that follows it.

5.3.1.1 General Uses

1) To separate 2 independent clauses, specifically if the second clause amplifies or clarifies the first, a use akin to that of the colon preceding an explanatory list. The first word after the colon may be capitalized if the second clause is a formal statement or long quotation.

   We had second thoughts about the first run: the data lacked the needed precision.
   This is the rule: On closing a file, back it up on a separate disk.

   If the second independent clause is related to the first but is not directly explanatory, a semicolon is sufficient (see Section 5.3.2).

2) To introduce a list (series). The part of the sentence preceding the colon must be a complete independent clause; do not insert a colon between a verb or preposition and its object, even if the list is presented in vertical format.

   The lectures covered 3 topics: carbohydrates, lipids, and proteins.
   or The sequence of the 3 lectures was as follows: carbohydrates, lipids, proteins.
   not The lectures covered: carbohydrates, lipids, and proteins.
   His tastes ran to acorn squash, carrots, and spinach.
   not His tastes ran to: acorn squash, carrots, and spinach.

   Omit closing punctuation for items in a vertical list, except if each item forms a complete sentence.[5]

   The survey was carried out with 3 purposes in mind:
   - to estimate the size of the rodent populations
   - to estimate environmental influences on population sizes
   - to identify previously unrecognized species

3) To introduce a long quotation, especially when the quotation clarifies or amplifies the preceding independent clause.

   Lewin captured the essence of the history of genetics in a few sentences: "The gene is the unit of genetic information. The crucial feature of Mendel's work . . ."

5.3.1.2 Specialized Uses

1) To couple elements of titles such as book titles, chapter titles, and table titles.

   Scientific Style and Format: The CSE Manual for Authors, Editors, and Publishers
   Chapter 3: Pulmonary Functions

   Note that the elements in the second example could be distinguished by spacing. A period is sometimes used instead of a colon in chapter, table, and figure designations.

2) To demarcate certain elements in end references—for example, between title and subtitle of an article or book if punctuation is not already included, between issue number and page range (see Section 29.3.5).

   Trigg PI, Kondrachine AV. Commentary: malaria control in the 1990s. Bull World Health Organ. 1998;76(1):11–16.

3) To separate parts of a ratio or proportion (see Section 12.3.4).

   The ratio of females to males was 3:1. We recommended a 5:10:5 fertilizer.

4) To couple components of a time datum when the units are hours, minutes, or seconds. When the coupled element is a decimal fraction of the initial unit, use a decimal point rather than a colon.

   12:38 AM
   2345:15 *or* 23:45:15 *for* 23 h, 45 min, 15 s *but* 12.25 [as decimal notation for "12 hours 15 minutes"]

5) To symbolize a chromosomal break in detailed descriptions (see Table 21.34). A double colon symbolizes a break and reunion, the disruption of one gene by another, or the integration of one gene beside another without disruption (see various tables in Chapter 21).

   46,XX,del(1)(pter→q21::q31→qter) [breakage and reunion of bands 1q21 and 1q31 in the long arm of chromosome 1]
   ade6::ura4 [disruption of *ade6* by *ura4*]

6) To group the locants in chemical names.

   2,3:4,5-bis-*O*-(phenylmethylene)-*altro*-hexodialdose
   (3,4:3,9:5,6:6,7:7,8-penta-μH)-(3-*endo*-H)-*nido*-nonaborane
   pyrido[1′,2′:1,2]imidazo[4,5-*b*]quinoxaline

7) To separate indications of originating and sanctioning authors to protect fungal names designated as "sanctioned". Note that in this case, the colon is preceded and followed by a space (see Section 22.3.2.1.2).

   Boletus piperatus Bull : Fr.

5.3.2 Semicolon ;

As a mark of linkage, the semicolon is weaker than the colon but stronger than the comma. It is a mark of coordination.

5.3.2.1 General Uses

1) To separate 2 or more independent clauses whether or not they are joined by a conjunctive adverb such as "however", "besides", or "therefore", but generally not clauses joined

by a simple conjunction such as "and" (which can be separated with a comma). Short clauses that do not need separation with rhetorical force can also be separated with commas.

The patient was first treated in the emergency department; however, she soon had to be transferred to the intensive care unit.
Slow-stop mutants complete the round of replication; they cannot start another.
The nurse removed the sutures, and the patient was discharged.
not The nurse removed the sutures; and the patient was discharged.

2) To separate 2 or more independent clauses when at least one has internal punctuation (e.g., comma or dash), even if connected by a simple conjunction. In the latter situation, a comma is also acceptable.

The uppermost formation, first identified by Smith, is sandstone; the next lower, identified by Jones, is a shale.
The nurse, acting on an order from the surgeon, removed the sutures; and the patient, disregarding the surgeon's advice, insisted on leaving the hospital.

3) To separate 2 clauses when the second is structurally parallel, even if grammatically elliptical.

In men, the most important causative factor is a high-fat diet; in women, an estrogen deficiency.

4) To separate the elements of a complex series, specifically if at least one of the elements has internal punctuation.

His ethnographic studies concentrated on 3 groups: Chinese, Japanese, and Taiwanese; French, Germans, and Austrians; and Inuit, Mexicans, and Peruvians.

5.3.2.2 Specialized Uses

1) To demarcate certain elements of end references—for example, to separate year from volume in a journal article (see Section 29.3.5).

Raines CA. The Calvin cycle revisited. Photosynth Res. 2003;75(1):1–10.

2) To symbolize chromosomal rearrangements (see various tables in Chapter 21).

46,XX,t(12;?)(q15;?) [rearranged chromosome 12; the segment of the long arm (q) distal to band 12q15 could not be identified]

3) Followed by a space, to separate symbols for genes on different chromosomes or to separate linked genes (see various tables in Chapter 21).

bw; e; ey
Sh2/sh2; Bt2/bt2

4) To separate subscripts in selectively labeled compounds.

$[1\text{-}^{2}H_{1;2}]SiH_3OSiH_2OSiH_3$

5) To separate sets of locants in which colons have already been used.

benzo[1″,2″:3,4;5″,4″:3′,4′]dicyclobuta[1,2-*a*:1′,2′-*a*′]diindene

5.3.3 Comma ,

5.3.3.1 General Uses

The comma serves mainly to clarify grammatical structures by setting off introductory material, by separating certain elements of a sentence or an expression, or by bracketing material.

1) To set off an introductory clause beginning with a subordinating conjunction (e.g., "if", "although", "because", "since", "when", "where", "while").

   When 2 equally efficacious and safe treatments are available, the patient's preference should be taken into account.

2) To set off a transitional or parenthetic word or phrase.

   Of course, another geological survey should be done.
   In the end, 8 samples may not be enough.

3) To set off a short word of address or emphasis.

   Chemists, you cannot be held responsible for all environmental pollution.
   No, state insurance is not the answer.

4) To separate 2 independent clauses joined by a coordinating conjunction ("and", "but", "neither", "nor", "or") unless the clauses are short and the absence of a comma would not cause ambiguity.

   The survey was completed in June 1983, and the map was published in July 1984.
   but We finished in June and the map appeared in December.

5) To separate elements to clarify meaning.

   In all, 8 experiments were performed.
   not In all 8 experiments were performed.
   Keys provide, except in the most specialized works, a useful means of identification.

6) To separate a nonrestrictive (i.e., nonrequired) clause from the rest of the sentence.

   The cells, which had been sent to us by the institute's depository, were infected.

7) To separate a nonrestrictive (i.e., nonrequired) appositive from the rest of the sentence (in this situation, a pair of commas is needed).

   Raymond Turner, a mammalogist, described the 2 species.

8) To separate a parenthetic statement, term of address, or interjection from the rest of the sentence.

   Remember, students, you have come here to learn and will go forth to work.
   Penicillin, alas, was in short supply in the region.

9) To separate the elements (words, phrases, clauses) of a simple series of more than 2 elements, including a comma before the closing "and" or "or" (the so-called serial comma). Routine use of the serial comma helps to prevent ambiguity.

   Seeds of tomato, bean, and pepper were planted in April.
   The patient can be given penicillin, ampicillin, or erythromycin.
   The tumor was bloody, necrotic, and malodorous.
   He used a cheap, effective, and readily obtained remedy.

   However, an adjective forming a noun phrase with the noun it modifies is not set off from any immediately preceding adjectives. Test whether a comma is needed by substituting the word "and" for the comma and checking the sense.

   He waved a large, torn, faded US flag at commencement.
   not He waved a large, torn, faded, US flag at commencement.

10) To separate contrasting expressions and interdependent clauses.

    The greater the risks, the greater will be the probable gain from treatment.

11) To indicate an elliptical construction.

    The fetuses exposed to the drug in the third month have spinal defects; the others, no defects.

12) To set off adjacent and unrelated numbers.

    By the end of 1935, 900 experiments had been completed.

13) To separate different elements of an address or geographic designation.

    The specimens of newly identified species were deposited in the museum in Cairo, Egypt.

14) To set off the year in a date expressed as month day year; omit the comma when only the month and year are stated.

    The first celebration was set for October 12, 1991, before the facts of his birth were known.
    but. . . set for October 1991 before the facts . . .

15) To separate every 3 digits of numbers (see Section 12.1.3.1).

    12,578,896

5.3.3.2 Unnecessary and Incorrect Uses

A comma is not required in the following instances.

1) To set off a restrictive appositive (a defining word or phrase required for the desired meaning).

    The species *Bombyx mori* is distinguished from others in the genus by . . .
    not The species, *Bombyx mori*, is distinguished from . . .

2) To separate 2 relatively simple and short independent clauses connected by a coordinating conjunction if the lack of a comma would not produce ambiguity.

    The survey was completed and we went home.
    not The survey was completed, and we went home.

3) To set off a short introductory phrase or clause if the comma would not contribute to clarity or ease of reading.

    Despite the pause the run was successful.
    not Despite the pause, the run was successful.

4) To separate digits in a page number, an address number, a ZIP code, a year number, or an accession number.

    p. 6984 *not* page 6,984
    10779 Glenwood Avenue *not* 10,779 Glenwood Avenue
    ZIP code 64114 *not* 64,114
    AD 1066 *not* AD 1,066

5) To separate a noun clause from the predicate of the sentence.

    He said that the bird was a flicker.
    not He said, that the bird was a flicker.

    Where the plant grew was fertile ground.
    not Where the plant grew, was fertile ground.

6) To separate name modifiers from the stem name.

    Franklin D Roosevelt Jr *not* Franklin D Roosevelt, Jr
    Homer Smith III *not* Homer Smith, III

7) To set off the year in a date expressed as day month year or as year month day.

    The first celebration was set for 12 October 1991 before the facts of his birth were known.

5.3.3.3 Specialized Uses

1) To separate subelements of groups in bibliographic references, e.g., author names and hierarchical organization names (see Section 29.3.5).

    Smith TH, Robbit Q, Tannen M. The molecular biology of . . .

2) Unspaced, to separate elements in symbolic representation of human chromosome aberrations (see Table 21.34).

    46,XX,t(4;13)(p21;q32)

3) Followed by a space, to separate linked genes in some organisms (see Table 21.11).

   CHX1, EST1/CHX1, EST1

4) Unspaced, to separate locants in chemical formulas (i.e., letters and numbers identifying the location of an atom or group in a molecule; see Section 17.3.4).

   4,5-difluoro-2-nitroaniline   N,N,N′,N′-tetramethyldiaminomethane

5) In patent numbers.

   US patent 5,973,257

6) In European conventions for numbers, as the decimal point. This convention is recommended by the International Organization for Standardization (ISO) standard on quantities and units[6] but is generally not followed in the United States, Canada, and the United Kingdom for either general or scientific literature. The CSE recommends an on-the-line period as the decimal marker (see Section 12.1.3).

5.3.4 Quotation Marks

5.3.4.1 Double Quotation Marks “ ”

Quotation marks are used mainly to delineate quoted words, terms, or longer elements of text. These uses are discussed in detail in Section 10.2. A few more specialized uses are noted here.

1) To indicate the title of a journal article, a book chapter, or a series title in running text, but not in end references.

   The chapter “Geographic Designations” appears at the end of Part 2 of this manual.

2) To suggest a nonstandard or ironic use of a word or term.

   The politicians told us that they “cared” about big science.

3) To indicate a word or phrase used as such (i.e., as an example or explanation); this use is applied in this manual to avoid confusion with italicization required for scientific conventions.

   Avoid the sloppy use of “impact”, “case”, and “individual”.

5.3.4.2 Single Quotation Marks ‘ ’

1) To enclose the names of plant cultivars (see Section 22.3.1.6).
2) In US usage, to enclose a quotation occurring within another quotation (see Section 10.2).

5.3.5 Dashes

Four lengths of dashes are available for printed text: the en dash, the em dash, the 2-em dash, and the 3-em dash. The hyphen serves as a connector rather than as an indicator of interruption or omission and thus does not belong to the family of functions generally served by dashes. The en dash can also serve as a connector and thus is akin to the hyphen. The 3-em dash has been used in bibliographic references to represent names of authors when the authors are the same as those in the preceding reference; this use is not recommended and the 3-em dash is not described further here.

The different lengths of dashes were generally not available in typewriter typefaces but are usually available in word-processing programs.

5.3.5.1 Em Dash —

The length of the em dash is equal to the size in points of the typeface; thus, the em dash in 10-point type is 10 points long (wide). As implied by the name, this length approximates the width of the capital letter "M". The correct character is available in word-processing software, but if necessary the em dash may be represented by 2 adjacent hyphens with no space on either side.

The em dash has 4 main types of uses.

1) To set off elements within a sentence that express a parenthetic break in the line of meaning. The set-off statement usually defines, elaborates, emphasizes, explains, or summarizes and typically is a sharp break, tangential and not vital to the central message of the sentence. This function of the em dash is akin to uses of the comma and parentheses, but the 3 marks usually represent different types and strengths of interruption, em dashes being the strongest and commas the least forceful.

   Osteoporosis—perhaps the most common disorder of postmenopausal women—cannot be diagnosed without access to specialized imaging equipment.

2) To set off introductory elements in a sentence that explains their significance.

   The Japanese beetle, the starling, the gypsy moth—these pests all came from abroad.

3) To indicate the source of a quotation or editorial statement.

   What is a doctor? A licensed executioner.—Mazarinade
   The opinions expressed in this letter are those of the author and do not represent journal policy.—The Editor

5.3.5.2 Two-Em Dash ——

The 2-em dash (4 hyphens in typed text) can be used to represent unknown or missing letters in words.

1) In printed versions of written documents with illegible letters in the original text.

   My dear Mr Darwin, Your st—— about our descent is . . .

2) In quotations of incomplete spoken statements.

   His chronic harping on ecology was a steady pain in my ——.

3) In statements preserving anonymity (but it may be preferable to use wording that avoids the need to imply a name).

   Patient M—— was the first in this series to develop the complication of . . .

5.3.5.3 En Dash –

The length of the en dash approximates the width of the capital letter "N". It has several linking functions.

1) To link 2 words or terms representing items of equal rank, including compound modifiers. In this construction, the en dash is interpreted to mean "and" or "to".[7]

   north–south avenues
   the Río San Juan–Lake Nicaragua route
   gas–liquid chromatography
   cost–benefit analysis
   hexane–benzene solvent

author–editor relationship
Italian–Canadian relations *but* the Italian-Canadian immigrant population

2) To connect names in eponymous terms attributed to 2 people; however, do not add an en dash to a nonhyphenated proper name.

Michaelis–Menten kinetics
Mann–Whitney *U* test
but Bence Jones protein *from* Henry Bence Jones

3) As a coordinate connector within a term that includes hyphenated elements.

the Winston-Salem–Raleigh group of scientists [from 2 cities, Winston-Salem and Raleigh]
sugar-maple–dominated forest *but* maple-dominated forest

4) Where space is limited (e.g., tables), to link numbers representing a range of values (in reference lists and related materials, hyphens may be used instead; see Section 29.3.5). Do not use an en dash if either number in the range is negative, and do not use the word "from" in combination with an en dash.

temperatures of −5 to 25 °C *not* temperatures of −5–27 °C
weight changes of −3.5 to +4.7 kg
−4 to −6 °C *not* −4 – −6 °C
from page 6 to 10 *not* from page 6–10

5) To represent chemical bonds.

C6H5CO–O–COCH3

6) As a minus symbol if a separate minus symbol is not available.
7) For page ranges in end references (see Section 29.3.6.9.1).

J Athl Train. 2013;48(5):716–720.

5.3.6 Parentheses (Parenthesis Marks, Round Brackets) ( )

5.3.6.1 General Uses

1) To enclose a parenthetic word, phrase, or sentence.

The most common use of parenthesis marks (parentheses) is . . .
Urinary incontinence (and disorders discussed in other chapters) is a frequent problem . . .

This function is sometimes better carried out with commas or em dashes when different strengths of interruption are needed (see Sections 5.3.3 and 5.3.5.1). The text within parentheses must be grammatically independent of the sentence carrying it. In the second example above, the subject "urinary incontinence" and the verb "is" are singular; the plural verb "are" would be incorrect.

2) To enclose a symbol or abbreviation to be used later in the text.

. . . and arteriovenous malformation (AVM) was diagnosed . . .

3) To enclose directive text.

As noted recently (Smith 1992), the . . . [in-text reference in name–year system]
As discussed elsewhere (see Section 17.3), we . . . [cross reference to other text]

5.3.6.2 Specialized Uses

1) To enclose certain components of end references (see Section 29.3.5).

Mastri AR. Neuropathy of diabetic neurogenic bladder. Ann Intern Med. 1980;92(2 Pt 2):316–318.

2) To enclose mathematical elements that must be grouped for certain functions (see Section 12.3.2.1).

$z = k(a + b + c)$ $\quad$ $y = 3.47(x + z)3$

3) To group molecular components to indicate the application of a subscript number (see Section 17.3.3).

   $K_4Fe(CN)_6 \cdot 3H_2$

4) To enclose oxidation numbers in text. In formulas they are placed superscript (see Section 16.4).

   Pb(IV)

5) To enclose side chains substituted in amino acids (see Section 17.13.2.5).

   Cys(Et)

6) To enclose the stereochemical descriptors R, S, E, and Z (see Section 17.11).

   (2S)-alanine (E)-2-butene

7) For immunoglobulin notations (see Section 23.5.6.1).

   $F(ab')_2$ IgG(Pr)

8) For notation of specificities of histocompatibility antigens.

   HLA-Bw56(w22)

9) To enclose symbols indicating structural alteration of a chromosome (see Tables 21.32 and 21.34).

   46,XX,t(4;13)(p21;q32)

10) To enclose the name of the author of the original taxonomic description when a species is transferred to another genus (see Sections 22.3.1.2.1 and 22.3.3.3).

   Tetraneuris herbacea Greene, renamed as *Hymenoxys herbacea* (Greene) Cronquist
   Taenia dunubyta Rudolphi, renamed as *Hymenolepis diminuta* (Rudolphi)

5.3.6.3 Single (Right) Parenthesis)

The second mark of parentheses can be used alone after numerals or letters introducing elements of a list or series, usually in a vertical presentation.

Three projects were funded.
1) Philadelphia: Archaeology of the Late 17th Century
2) New York: Sites of Black Cemeteries
3) San Francisco: Pre-Gold-Rush Buildings

Inclusion criteria were as follows: 1) minimum patient age of 18 years; 2) histologically proven gliobastoma; and 3) complete data sets.

5.3.7 Square Brackets (Brackets) [ ]

The term "brackets" is often used for this pair of marks, but it may also be applied to other kinds of punctuation marks: parentheses (round brackets), the bracket pair (often called braces or curly brackets), and angle brackets. Hence, the term "square brackets" is preferred.

5.3.7.1 General Uses

1) To demarcate text or letters added to quoted text to amplify or clarify the original text.

   Cushing commented, "When Osler moved [to Baltimore], he was not risking his future."
   The last entry in his journal was "I took lunch with D[arwin] and spotted P[otter] with his mistress E[laine Smythe]."
   The results of the analysis by Neyman, Scott, and Smith [Science. 1969; 163:1445–1449] were released to the press.

2) To demarcate an editorial comment.

His diary included this note: "When I was in London I briefly met Darwynne [sic], that horrible chap who claims we are descended from monkeys."

This use of sic ("thus" in Latin) presumes that the reader knows the correct form. A more helpful notation is a correction of the error, also enclosed in square brackets.

"I briefly met Darwynne [Charles Darwin], that horrible chap who . . ."

3) To enclose a parenthetic statement within a parenthetic statement enclosed by parentheses.

. . . and vaccination was carried out (according to current [2004] recommendations).

5.3.7.2 Specialized Uses

1) For multiple bracketing in mathematical expressions and chemical names (see Sections 12.3.2.1 and 17.3.3).

$z = k[(a + b) - y(c + d)]$
bis(bicyclo[2.2.2]octadiene)platinum

The general sequence for multiple bracketing in mathematics is {[( )]}, which differs from the usual sequence in text (i.e., braces inside square brackets inside parentheses; see Section 5.3.7.1).

For the sequence of enclosures in chemistry, including square brackets, see Section 17.2.3. In some chemical usages, internal brackets must be used and parentheses may surround the brackets, as ([ ]); see the example above.

2) To enclose chemical concentrations.

$[Na^+]$ $[HCO_3^-]$

3) To enclose isotopic prefixes (see Section 17.10.2).

$[^{14}C_2]$glycolic acid $[^{32}P]$AMP

4) To enclose complex substituent prefixes containing internal parentheses.

N,N-bis[(phenylmethoxy)carbonyl]alanine

5) To enclose connecting points in fusion nomenclature (see Section 17.2.3).

benz[*a*]anthracene

6) To enclose numbers used as ring-size indicators in spiro and bicyclo names (see Section 17.6.3.2).

spiro[2.3]hexane bicyclo[2.2.2]octane

5.3.8 Braces {}

In mathematics and chemistry, paired left and right braces are used to signify aggregation, enclosing elements already enclosed by parentheses and square brackets (see Sections 12.3.2.1 and 17.3.3). A single multiline brace can be used to aggregate elements in adjacent lines, but this use is now uncommon outside mathematics (see Section 5.5.1).

5.3.9 Angle Brackets <>

Angle brackets are used in specific contexts to signify instructions. For example, they may be used to enclose the name of a computer key to be pressed in program instructions.

Press <Enter> to start searching.

5.3.10 Vinculum ‾

The vinculum (meaning "a bond", from vincire, Latin "to bind") can be used in mathematics to indicate a fraction or to perform the same function as parentheses; however, parentheses are preferred.

$$a - \overline{b - c}$$

$$a - (b - c)$$

$$\frac{a - b}{c + d}$$

5.4 TERM AND WORD MARKS

5.4.1 Hyphen -

5.4.1.1 General Uses

Partridge[1] summed up the 2 main uses of the hyphen as "dividing and compounding". The dividing function is exemplified by hyphens used to mark word divisions for line breaks and to indicate syllabification. The compounding use of the hyphen in word and term formation is discussed below and in Sections 6.2.1.2 and 6.2.2.2. There is a strong tendency in scientific English either to form new single-word terms by combining 2 stem words or to use a space (not a hyphen) between 2 or more modifiers. However, terms that are "in transition" from a 2-word form to a single-word form are often hyphenated. For such terms, different publishers may be using different forms simultaneously (e.g., health care, health-care, and healthcare as both a noun and an adjective).

The hyphen has numerous general uses.

1) To connect prefixes and suffixes to stem words in certain situations: if the stem word must be capitalized (pre-Colombian); with the prefixes "ex", "quasi", and "self" (ex-husband, self-inflicted); and with the suffix "-elect" (chairperson-elect). A hyphen may also be used if there is potential for visual confusion because of repeated letters (e.g., meta-analysis, shell-like); however, in US usage, the hyphen is often omitted in such terms (preeclampsia, coordinates). (See Section 6.2 for other prefixes and suffixes that are closed up to their stem words.) Consult a standard dictionary to resolve any uncertainties about hyphenation of prefixes and suffixes.
2) For some compound terms, especially if a space instead of a hyphen would leave one of the stem words with a meaning inappropriate to the context.

    light-year [a unit of length (the distance traveled by light in 1 year); without the hyphen, "light year" could appear to mean a year that is not heavy]
    cure-all [the verb "cure" is converted to a noun]
    has-been [the progressive verb "has been" is converted to a noun]

3) For compound modifiers in which the second element is a past or present participle.

seizure-inducing drugs	ill-advised procedure
all-encompassing strategy	well-established rules
well-known physicist	better-represented type

Omit the hyphen if the modifier follows the verb (predicate modifier) and the first element is an adverb. This rule follows from the lack of ambiguity in such constructions as to whether the adverb (such as "well") might modify a noun it precedes rather than the adjective.

The rules are well established. He was well known for his accomplishments in this field.

However, a hyphen should be used in terms that are listed as hyphenated entries in a standard dictionary, regardless of where the terms appear in a sentence.

The theory was well-founded. Their study design is ill-defined.

Do not hyphenate adverbial elements in compound modifiers if the adverb ends in "ly". This exception appears to be based on the identifiability of the adverb by its "ly" ending; in contrast, the adverb "well" may not be readily identified as an adverb when it is not in a predicate construction.

the widely applauded plan for conservation a clearly described new species

4) For compound modifiers when the nonhyphenated form could be ambiguous.

low-frequency amplitudes a large-bowel obstruction

Two or more modifiers of the same adjectival noun retain the hyphen.

low- and high-frequency amplitudes sodium- and potassium-conserving drugs

5) For modifiers with numeric values and units. Age terms take a double hyphen.

a 5-g dose 50-km radius a 10-woman team
a 3-year-old child a 50-year-old patient

6) For a spelled-out fraction, unless either component is already hyphenated; the hyphen appears between the numerator and denominator.

one-third of the population thirty-two hundredths

7) For compound cardinal and ordinal numbers from 21 through 99 when spelled out.

Eighty-five samples were collected. He pointed to the Sixty-sixth Congress.

8) For verbs needing hyphens for correct meaning.

He re-covered the explored well. *but* He recovered quickly from the operation.
Such patients are usually re-treated. *but* The water retreated from the structures.

9) For page ranges in end references if preferred instead of an en dash (see Section 29.3.6.9.1).

Indian J Med Sci. 1999;53(10):454-455.

10) For so-called e-terms, where "e" stands for "electronic", except for the term "email", since it is so widely used.

e-book e-commerce e-business e-alert email

Additional guidance on the formation of compounds is available in *The Chicago Manual of Style*,[5] *Editing Canadian English*,[8] the *United States Government Printing Office Style Manual*,[9] and the *American Medical Association Manual of Style*,[10] among others.

5.4.1.2 Unnecessary Uses

A hyphen is not required in the following instances.

1) For well-established compound terms. Standard scientific dictionaries are reliable guides.

freezing point determination amino acid residues
potassium chloride absorption Sertoli cell analysis

2) For Latin phrases used adjectivally.

a post hoc hypothesis in vitro testing a quid pro quo arrangement

3) For letters used as modifiers in scientific terms; hyphens are used in adjectival phrases.

   LE cells *but* LE-cell rosettes   T lymphocytes *but* T-cell lymphocyte functions

4) In a compound with a comparative or superlative adjective.

   better adjusted children   least favorable outcome

5.4.1.3 Specialized Uses

1) To represent single bonds in chemical or molecular formulas or names; for double bonds (represented by a double line), see Section 17.11.1.1.

   $(CH_3)_2$-CH-CH_2-CH(NH_2)-COOH

2) Between element symbols (e.g., C for "carbon") and the numeral designating a particular atom; also between the element name and the mass number of an isotope in spelled-out form.

   C-3   iodine-131

3) Between amino acid symbols in amino acid residues of known sequence; each hyphen represents a peptide bond (see Section 17.13.2).

   Gly-Lys-Ala-His

4) Between a prefix that specifies molecular configuration, or that serves as a locant, and the name of the chemical compound (see Section 17.3.4).

   S-benzyl-*N*-phthaloylcysteine   9-(1,3-dihydroxy-2-propoxymethyl)guanine

5) To show linkage of nucleotides in polynucleotides (see Section 17.13.10).

   pG-A-C-C-T-T-A-G-C-A-A-T-Gp

5.4.2 Slash (Virgule, Oblique Bar, Oblique Mark, Oblique Stroke, Slant Line, Shilling Mark, Forward Slash) /

Specialists in typography distinguish between the slash (virgule) and the solidus (shilling mark).[3] The solidus (⁄) has a greater slant than the slash (/) and has been used to separate units of traditional British currency (pounds, shillings, and pence) and the numerator and denominator of fractions. However, the slash has tended to displace the solidus in the representation of fractions on the line.

5.4.2.1 General Uses

The main use of the slash is as a symbol for the mathematical operation of division (meaning "divided by"). It has also come into general use to indicate alternatives. A few terms that incorporate the slash, such as "and/or" (meaning "either or both"), are used widely enough that they appear in standard dictionaries and are unlikely to be misunderstood. For the construction "he/she" (meaning a person of unspecified sex), it may be preferable to rephrase as a plural construction (using "they").[10]

Other, temporary uses of the slash may result in ambiguity, and rephrasing for clarity is advised. For example, a series can be punctuated with commas (see Section 5.3.3), and coordinate modifiers can be punctuated with en dashes or hyphens (see Sections 5.3.5.3 and 5.4.1).

   The route of the geology tour was New York, Pittsburgh, Chicago, and Salt Lake City.
   not The route of the geology tour was New York/Pittsburgh/Chicago/Salt Lake City.

The physician–patient relationship is exemplified by . . .

not The physician/patient relationship is exemplified by . . . [Here, the expression may be understood to mean the physician as patient.[10]]

The medical school has started a new hematology–oncology program.

not The medical school has started a new hematology/oncology program.

5.4.2.2 Specialized Uses

1) A mathematical symbol for division, either as an operation or implied by a unit (see Sections 12.2.1 and 12.3.1.1 and Table 12.10).

   $1/4\ y = 3.5x/(a + b)$ $\mathrm{kg \cdot m/s^2}$ for kilogram-meter per second squared

2) In expressions of rate or concentration (see Section 12.1.4.3).

   5 m/s   20 mol/L

   To avoid mathematical ambiguity, use no more than a single slash in an expression of rate or concentration.

   $1.5\ \mathrm{pCi \cdot km^{-2} \cdot y^{-1}}$ *or* $1.5\ \mathrm{pCi/(km^2 \cdot yr)}$ *not* $1.5\ \mathrm{pCi/km^2/yr}$

3) To separate symbols for mutant genes on homologous chromosomes (see various tables in Chapter 21).

   y w f/B

4) To separate clones in human chromosome nomenclature (see Table 21.33).
5) To separate symbols for homologous genes or alleles (see various tables in Chapter 21).

   ac17/AC17   *dhfrts-3/DHFRTS*   *BTU2/BTU2*

6) As a character in URLs for Internet addresses.

5.4.3 Vertical Bar (Vertical Line) |

The vertical bar, usually presented in pairs, has specialized uses in chemical and electrochemical notation and in mathematics. For example, a single vertical bar indicates a phase boundary in an electrochemical cell[11]; double vertical lines are used to enclose a matrix in running text (see Section 12.3.1.4).

$\mathrm{Zn|Zn^{2+}} \,\vdots\, \mathrm{Cu^{2+}|Cu}$ $\| a_i b_i c_i \|$

5.4.4 Apostrophe ’

The apostrophe is used to form possessives and, occasionally, plurals. For information on the use of an apostrophe for possessives, see Section 6.5, and for plurals (of abbreviations), see Section 11.1.4. The apostrophe also indicates contractions in informal writing (“I’d prefer” for “I would prefer”), but these forms are usually inappropriate in scientific material.

5.4.5 Prime Symbol, Single and Double ′ ″

The single prime symbol is not interchangeable with the single quotation mark or apostrophe, nor is the double prime symbol interchangeable with a double quotation mark. The correct characters are available in most word-processing programs, but if a substitute must be used for either character, the substitution should be indicated for the publisher (with an electronic submission) or marked with a marginal note (in a hard-copy submission) and the symbol changed during typesetting.

The prime sign has 2 main scientific uses.

1) With locants in chemical names (see Section 17.6.3.2).

   N,N′-dimethylurea

2) To indicate "minutes" and, doubled, "seconds" in geographic coordinates (see Section 14.2.1).

   The meteorite was found at latitude 52°33′05″N, longitude 33°21′10″E.

5.4.6 Diacritics

For some words of non-English origin, diacritics are retained in English usage; for others, the diacritics have been dropped and the original pronunciation has been retained; for yet others, the diacritics have been dropped and the spelling has been altered to reflect the original pronunciation.

garcon　　résumé *or* resume　　facade *from* façade　　canyon *from* cañón

Consult a standard English-language dictionary for current usage. See Table 4.2 for the most widely used diacritics in European languages. Retain diacritics in personal names and place names if the names have not been anglicized.

Word-processing programs now offer a wide variety of characters combining letters and the applicable diacritics, but such characters must be checked after typesetting to ensure that the desired characters appear. If a diacritic is needed that is not available in the character set of the word-processing program, the author must alert the publisher to the need for a special character (for an electronic submission).

Many bibliographic databases use the ASCII (American Standard Code for Information Interchange) character set, which does not include diacritics. Hence, a diacritic may be altered for presentation in the limited character set, either by presenting the letter as if it had no diacritic or by changing the spelling slightly to indicate pronunciation (e.g., Rontgen or Roentgen for Röntgen).

5.4.7 Asterisk *

The asterisk is used in gene symbolization (see Tables 21.13, 21.26, 21.28, 21.29, 21.30, and 21.32).

*MDH-B*1*　　*Ho*hl/Ho*hl*

In addition, the asterisk is one of the graphic footnote signs, but the CSE discourages this system of designating footnotes (see Section 30.1.1.8). Do not use asterisks to indicate omitted text; instead, use ellipsis points for this function (see Section 10.3).

5.4.8 Ampersand &

Retain the ampersand when it appears as part of a proper name, such as a publisher or law firm. Do not use this symbol to represent the word "and" in the title, text, or other components of a scientific publication.

5.4.9 "At" Symbol @

The "at" symbol has 2 uses in scientific publishing.

1) Placed after a gene symbol, to indicate a gene cluster (see Table 21.32).
2) In email addresses, to separate the specific component of the address from the domain (see also Section 5.7).

    headquarters@societyname.org

Do not use this symbol to represent the word "at" in the title, text, or other components of a scientific publication.

5.4.10 Octothorp (Numeral, Pound, or Space Symbol, Hash Mark)

The octothorp ("8 fields") is used in URLs for Internet addresses. Other uses for the octothorp are not applicable to scientific publishing (e.g., in cartography as a symbol for "village", in the avoirdupois system of weights as a symbol for "pound", and in general contexts as a symbol for "number").

5.5 MARKS FOR LINE RELATIONS

5.5.1 Brace

A multiline brace can be used as a fence for aggregation in mathematics.

$$\left\{ \begin{array}{c} x=a+b \\ y=c+d \\ a=e+f \end{array} \right\}$$

To simplify manuscript preparation, avoid using braces to cluster related text elements; instead, use a tabular list or a simple sentence.

Typical differences in spelling are represented by the US spellings color, leukemia, and theater and the British colour, leukaemia, and theatre.

Typical Differences in Spelling

US: color, leukemia, theater

British: colour, leukaemia, theatre

not Typical Differences in Spelling

$$\text{U.S.}\left\{ \begin{array}{c} \text{color} \\ \text{leukemia} \\ \text{theater} \end{array} \right\} \quad \text{British}\left\{ \begin{array}{c} \text{colour} \\ \text{leukaemia} \\ \text{theatre} \end{array} \right\}$$

5.5.2 Ditto Marks "

Ditto marks have been widely used in lists and tables to indicate repetition of the word or term appearing directly above the location of the ditto marks. To ensure unambiguous meaning and to avoid an unattractive appearance, avoid using this device in scientific publications (even though it might save some time in typing).

5.5.3 Dots (Leaders) . . . . .

A line of closely spaced period marks is sometimes used as a formatting device to lead the reader's eye from a text element at or near the left-hand margin to a text element at or near the right-hand margin, as in a table of contents. The decision on use of this device is best left to the designer of the publication.

Punctuation Marks. 12
Period 13
Colon. 14

See Section 10.3 for the use of spaced dots (ellipsis) to indicate an omission in a quotation.

5.5.4 Paragraph Mark (Pilcrow) ¶

The paragraph mark is one of the graphic footnote signs, but the CSE discourages this system of designating footnotes (see Section 30.1.1.8).

5.6 PUNCTUATION MARKS IN NON-ENGLISH LANGUAGES

Some punctuation marks not used in English may have to be preserved within English scientific texts (e.g., for block quotations from a non-English document). Examples are the guillemets (chevrons) used in French and Spanish to mark quotations and the inverted question mark that opens an interrogative statement in Spanish.

N'oubliez pas ces mots: «Guérir quelquefois, soulager souvent, consoler toujours».
¿Quién o qué estaba «interpretando»?

Consult *The Chicago Manual of Style*[5] for details of punctuation in non-English languages.

5.7 PUNCTUATION MARKS IN ELECTRONIC ADDRESSES

Various punctuation marks and other symbols appear as characters in electronic addresses.

Email addresses use the "at" symbol (@) to separate the specific addressee component from the domain name. One or more periods, underscore characters, or hyphens may appear within each component, but the final punctuation mark (before the code designating type of institution or country) is always a period.

a-smith@companyA.com c.a.jones@organization-CA.org
b_jones@universityB.edu mainoffice@society-name.ca

The URLs that designate Internet addresses use a variety of punctuation marks (question mark, slash, and tilde) and other symbols (equals symbol, underscore, and percent symbol).

To prevent misinterpretation of a period closing a sentence as part of an email address or URL, avoid placing either form as the last element of a sentence if possible. A

URL may also be enclosed within parentheses or angle brackets, preceded and followed by a nonbreaking space.

For URLs appearing in running text or end references, do not allow a hyphen to be inserted at a line break. Instead, break the URL at an existing slash.

Complete URL: http://www.councilscienceeditors.org/events/annualmeeting06/index.cfm

http://www.councilscienceeditors.org/
events/annualmeeting06/index.cfm
not
http://www.councilscience-
editors.org/events/annualmeeting06/index.cfm

CITED REFERENCES

1. Partridge E. You have a point there: a guide to punctuation and its allies. London (UK): Taylor & Francis; 1978.

2. Fowler HW, Fowler FG. The King's English. 3rd ed. New York (NY): Oxford University Press; 2003.

3. Bringhurst R. The elements of typographic style. 4th ed. Vancouver (BC): Hartley and Marks Publishers; 2013.

4. Coghill AM, Garson LR, editors. The ACS style guide: effective communication of scientific information. 3rd ed. Washington (DC): American Chemical Society; 2006.

5. The Chicago manual of style: the essential guide for writers, editors, and publishers. 16th ed. Chicago (IL): University of Chicago Press; 2010. Also available at http://www.chicagomanualofstyle.org.

6. International Organization for Standardization. Quantities and units. Part 1: General. 4th ed. Geneva (Switzerland): ISO; 1992. (ISO 80000-1:2009).

7. New Hart's rules: the handbook of style for writers and editors. New York (NY): Oxford University Press; 2005.

8. Editing Canadian English. 2nd ed. Toronto (ON): Macfarlane Walter & Ross; 2000. Prepared for the Editors' Association of Canada.

9. United States Government Printing Office style manual: main page. 30th ed. Washington (DC): The Office; 2008 [updated 2008 Sep 16; accessed 2013 Jul 19]. http://www.gpo.gov/fdsys/pkg/GPO-STYLE MANUAL-2008/pdf/GPO-STYLEMANUAL-2008.pdf.

10. Iverson C, Christiansen S, Flanagin A, Fontanarosa PB, Glass RM, Gregoline B, Lurie SJ, Meyer HS, Winker MA, Young RK. American Medical Association manual of style: a guide for authors and editors. 10th ed. New York (NY): Oxford University Press; 2007.

11. International Union of Pure and Applied Chemistry, Analytical Chemistry Division. Compendium of analytical nomenclature, definitive rules 1997. Oxford (UK): Blackwell Scientific Publications; 1998. Prepared for publication by J Inczédy, T Lengyel, AM Ure. Also available at http://www.iupac.org/publications/analytical_compendium/.

ADDITIONAL REFERENCES

Seely J. Oxford A–Z of grammar and punctuation. 2nd ed. New York (NY): Oxford University Press; 2010.

Truss L. Eats, shoots and leaves: the zero tolerance approach to punctuation. London (UK): Profile Books Limited; 2009.

6 Spelling, Word Formation and Division, Plurals, and Possessives

6.1 SPELLING: AMERICAN AND BRITISH DIFFERENCES

Although scientific English is almost uniform throughout the world, there are differences in some of its details, notably spelling. No national or international authority governs the spelling of words in English, as the Académie française specifies the spelling of words in French. Hence, many words in English are spelled in variant forms that can be characterized as American or British preferences. British forms tend to be preferred in the Commonwealth countries, notably Australia, Canada, India, and New Zealand. Some of the preferences can be rationalized; others relate to historical influences and cultural trends. In general, the British forms tend to reflect the continuing influence of French (stemming from the Norman Conquest) and of education stressing Latin and Greek, whereas the American forms tend to reflect trends toward clearer representation of the pronunciation of individual words and simplified spelling. Both groups of variant forms are usually represented in the major dictionaries of English published in the United States and the Commonwealth countries. An extensive summary of differences can be found in the entry "American English and British English" of *The Oxford Companion to the English Language*.[1]

Choices for American or British spellings in a publication should be governed by what forms most of its readers can be expected to prefer, and authors should be prepared to follow the preferences selected by the publication. Journal offices can specify whether an American, Australian, British, or Canadian dictionary should be the standard for spelling in papers to be submitted, and authors can more easily adhere to the expected spellings by preparing their papers with a word-processing program that includes a spelling checker that adheres to a national preference.

The generally preferred variant forms are described below with examples.

6.1.1 Nouns with Variant Endings

American and British forms of English include many nouns with variant endings (see Table 6.1).

Table 6.1 Nouns with variant spellings in American and British English

	Examples		
Category of noun	American	British	Notes
Ending in "ction" or "xion"	connection	connexion	
	deflection	deflexion	
Ending in "er" or "re"[a]	center	centre	Some words have the same form in both (e.g., manager, interpreter, mediocre)
	fiber	fibre	
	liter	litre	
	maneuver	manoeuvre	
	meter	metre [SI unit], meter [instrument]	
Ending in "or" or "our"[a,b]	behavior	behaviour	For some words, the "or" form is preferred in both (e.g., stupor, terror, director, monitor, pallor, tremor)
	color	colour	
	humor	humour	
Ending in "ense" or "ence"	defense, defence	defence	"defensible" for both
	license, licence	licence	"license" as a verb in British English
	offense	offence	"offensive" for both
	practice	practice	"practise" as a verb in British English
			Adjectives "immense" and "intense" in both
Ending in "log" or "logue"[a]	analog	analogue	British forms also accepted in American English
	catalog	catalogue	
	dialog	dialogue	"ideologue", "secretagogue" in both

Table 6.1 Nouns with variant spellings in American and British English (*continued*)

Category of noun	Examples		Notes
	American	British	
With digraphs "ae" and "oe"	anesthesia	anaesthesia	British preferences reflect Latin or Greek origin
	cesium	caesium	
	diarrhea	diarrhoea	
	edema	oedema	
	esophagus	oesophagus	
	estrogen	oestrogen	
	etiology	aetiology	
	fetus	foetus	
	hematology	haematology	
	leukemia	leukaemia	
With simpler forms	aluminum	aluminium[c]	
	artifact	artefact	
	check	cheque	
	draft	draught	
	mold	mould	
	program	programme	
	sulfur	sulphur	
Derived from verbs ending in "e"	acknowledgment	acknowledgement	
	aging	ageing	
	judgment	judgement	
	likable	likeable	

[a] American ending generally reflects a phonetically based form, and British ending reflects French derivation (e.g., "honour" from "honneur").
[b] For a word derived from a noun ending in "or" or "our", the simpler spelling is generally preferred in both American and British English (e.g., coloration, honorary, laborious).
[c] Reflects the chemist Humphrey Davy's original term, alumium.

6.1.2 Verbs Ending in "ize" and "yze" or "ise" and "yse" and Their Noun Derivatives

Preferences depend in part on etymology and pronunciation. The "ize" ending is widely used in both British and especially American English, but the "ise" form is also common in British usage.[2] For verbs derived from "lysis", British usage prefers the "yse" ending and American usage the "yze" form (e.g., "analyse" or "analyze" from "analysis"). For some words, the "ise" form is universal, even if the word is pronounced with a "z" rather than an "s" sound (e.g., advertise, advise, compromise). Consult a standard dictionary to verify the preferred form of a verb and its noun derivative.

American	British
catalyze	catalyse
civilize, civilization	civilise, civilisation
compromise	compromise
organize, organization	organise, organisation
promise	promise
rationalize, rationalization	rationalise, rationalisation

6.1.3 Verbs Ending in "l" or "ll"

The terminal letter "l" may be single or double in verbs ending with a stressed syllable in both American and British usage, although the British tendency is toward the single "l". In both usages, a single "l" is doubled when a suffix ("able", "ant", "ed", "ing") is added to verbs ending in a stressed syllable; derivatives formed by adding "ment" retain the double "l" in American usage, but not in British usage.

American	British
compel, compelling	compel, compelling
distill	distil, distilled
enroll	enrol, enrolled
fulfill	fulfil, fulfilling
propel, propellant, propelled	propel, propellant, propelled
forestall, forestallment	forestall, forestalment
install, installment	install, instalment

6.1.4 Words Derived from Non-English Words and Roots

Words transliterated or derived from other languages can have variant spellings. In general, the variant that is phonetically unambiguous is preferred. For example, "leukemia" (or "leukaemia"), derived from the Greek *leukos*, is preferred to "leucemia" or "leucaemia".

When the spelling of a word in English is identical to that in the language of origin except for one or more diacritics, and the word appears in standard English dictionaries, the diacritics are generally omitted. With anglicization, the pronunciation of these words usually changes (e.g., the French words "brassière" and "émotion" have become "brassiere" and "emotion" in English and are pronounced differently).

In a foreign-language phrase pronounced in the language of origin but less widely used as an English phrase, retain the original spelling, including diacritics.

aide-mémoire   chargé d'affaires   pièce de résistance

6.1.5 Verbs Beginning with "in" or "en"

Some verbs and their noun derivatives beginning with "en" or "in" are spelled with either opening syllable, but the resulting words may have different meanings.

same meaning:	enclose, inclose	endorse, indorse	endue, indue
different meanings:	ensure, insure		
no variant:	enact *not* inact	insurance *not* ensurance	

6.1.6 Doubled Consonants before Suffixes

Current practices for doubling terminal consonants in the formation of derivative words ending in "ed", "er", "est", and "ing" are similar for American and British usage. The following points are based on the summary presented in *Hart's Rules for Compositors and Readers at the University Press, Oxford*.[3]

1) Double a single terminal consonant when it follows a single vowel in a monosyllable and the suffix begins with a vowel.

   sag, sagging   rot, rotted   dig, digging

2) Double a single terminal consonant for a multisyllabic word with a final accented syllable or an unaccented final syllable with a short vowel sound and a suffix beginning with a vowel.

control, controlling   format, formatting   input, inputting   occur, occurring

3) Do not double a single terminal consonant if the terminal accent of the stem moves forward to the last syllable of the derived form.

infer, inference   refer, reference

4) A single terminal consonant may or may not be doubled if the final syllable is unaccented in the stem.

travel, traveler, traveller   cancel, canceled, cancelled

5) Do not double a terminal consonant "h", "w", or "x".

watch, watched   flaw, flawed   sex, sexed

6.2 WORD FORMATION

The rapid growth of knowledge in the sciences has generated many new terms and words. Some may be single words coined from Latin or Greek roots, but more often they are compound terms created from existing stem words or stem words combined into single words. Some originally compound terms have shifted to a single-word form, but many compound terms retain their original compound foım.

bloodstream *but* blood vessel   database *but* data flow   headphone *but* head shield

Some scientific fields have specific rules for vernacular terms. For recommendations in the plant sciences, see Section 22.3.1.7, and in the animal sciences, see Section 22.3.3.7. Concise summaries of general usage can be found in *The Chicago Manual of Style*,[4] the *United States Government Printing Office Style Manual*,[5] *New Hart's Rules*,[2] *Editing Canadian English*,[6] and *The Canadian Style*.[7]

The formation of new words in English by adding prefixes or suffixes to stem words is not governed by firm and unequivocal rules, and usage tends to shift over time.

6.2.1 Formations with Prefixes

6.2.1.1 Nonhyphenated Prefixes

Many scientific terms are formed by the nonhyphenated addition of prefixes to stem words. In general, these prefixes do not stand alone as words; some are derivatives in an adjectival form. They indicate action, character, location, number, state, or time.

Prefix	Example	Prefix	Example
aero	aerostatics	counter	counterimmunoelectrophoresis
after	aftershock	de	denitrification
ante	antepartum	di	diketone
anti	anticodon	electro	electrosurgery
astro	astrophysics	exo	exopathogen
auto	autoimmunity	extra	extrasystole
bi	bivalve	geo	geochemistry
bio	biomechanics	hemi	hemianesthesia
chemo	chemotherapy	hemo	hemodialysis
co	coenzyme	hyper	hyperventilation

Prefix	Example	Prefix	Example
hypo	hypomenorrhea	phyto	phytopathology
in	incoordination	poly	polyarthritis
infra	infrared	post	postprecipitation
inter	interface	pre	prediabetes
iso	isohexane	pro	proinsulin
macro	macroflora	pseudo	pseudopod
meta	metagenesis	re	recombination
micro	microfossil	semi	semiconductor
mid	midbrain	stereo	stereochemistry
milli	millisecond	sub	subspecies
mini	minicomputer	super	supersaturation
multi	multiprocessor	supra	suprascapula
non	nonconductor	trans	transpolarizer
over	overtone	tri	tribromoethanol
para	paramyxovirus	ultra	ultracentrifuge
photo	photochemistry	un	unconformity
physio	physiotherapy	under	undernutrition

When there is uncertainty about whether to hyphenate new terms formed with such suffixes, the preference should be for the closed-up form, not the hyphenated form. Some computer programs for searching text may not recognize the hyphenated form as a term. Some readers may object to the doubled vowels that can result from the general rule not to hyphenate, as in "semiindependent" and "metaanalysis". However, such formations often become acceptable to the eye over time, as for "coordination", which is now generally not hyphenated in American usage.

Authors may confuse homonymic or near-homonymic prefixes. Most such errors will be identified by spelling checkers of word-processing programs.

ante, anti:	antediluvian *not* antidiluvian	antifreeze *not* antefreeze
for, fore:	forward *not* foreward	foreword *not* forword

6.2.1.2 Hyphenated Prefixes

Prefixes must be joined to the stem word with a hyphen in some formations:

1) when the stem is capitalized

   pre-Columbian civilization   post-Copernican astronomy   sub-Saharan Africa

2) when omitting the hyphen changes the meaning

   Re-cover the flask after adding the reagent. *but* You may not recover enough precipitate.

For recommendations on hyphenation in compound nouns and adjectives, see Section 5.4.1.

6.2.2 Formations with Suffixes

6.2.2.1 Suffixes That Are Not Words

Many nouns and adjectives are formed by adding suffixes that are not complete words; these formations are not hyphenated. Some of these suffixes are similar to one another, and the words they form may be easily misspelled. Most misspellings that arise from confusion about correct suffixes will be caught by spelling checkers of word-processing programs (see also Section 7.3.1).

able	ible	eous	ous
ance	ence	erous	orous
ant	ent	ful	full
ative	ive	ified	yfied
cede	ceed	efy	ify

6.2.2.2 Words as Suffixes

Many nouns have been formed by adding a suffix that is itself a complete word to a stem verb or noun. In many such formations, the hyphen is often omitted, but it is retained in certain situations.

Suffix	Example
away	runaway reactor
down	breakdown of the varieties of species
in	cave-in at the excavation site [hyphen generally desirable]
like	wormlike lower vertebrate [hyphen preferred when stem word ends in a single or double "l" or another ascender, as in "a mold-like excrescence", "a shell-like carapace"]
off	heavy runoff causing a flood
out	turnout greater than expected
over	turnover in the faculty
up	breakup of the USSR

The form of the word may depend on its use as a noun, verb, or adjective.

They shut down the reactor. ["shut down", a phrasal verb]
The shut-down reactor needed extensive repairs. ["shut-down", an adjective]
The announcement of the shutdown was delayed several days. ["shutdown", a noun]

6.2.2.3 Variant Suffixes ("ic" and "ical")

Some word pairs ending in the suffixes "ic" and "ical" convey the same meaning. The selected spelling should be consistent within a given document. Consult a dictionary to verify whether the words in a particular pair are synonymous, such as the pairs below.

etiologic, etiological   histologic, histological   microscopic, microscopical

Some potential variants are not idiomatic (e.g., "chemic" is not accepted as a shorter form of "chemical"). Other variant forms carry different meanings. Consult a dictionary to verify currently accepted meanings for these variants.

economic theory, an economical method of collecting samples
a demanding work ethic, a violation of ethical principles
national historic sites, an outmoded historical theory
the statistic thus calculated, a statistical analysis

Use adjectives formed from stem words ending with "ology", such as "histologic" from "histology" and "pathologic" from "pathology", rather than the stem words themselves, to describe particular entities.[8]

histologic diagnosis *not* histology [to refer to a lesion diagnosed by histologic examination]
pathologic lesion *not* pathology [to refer to an abnormality]

6.3 WORD AND TERM DIVISION

For consistency in dividing words at the ends of lines and to maintain satisfactory spacing within lines, apply the divisions indicated in a standard dictionary. American divisions are generally based on syllable structure; British, on etymologic elements. See *The Chicago Manual of Style*[4] and *New Hart's Rules*[2] for more detailed guidance on word division.

1) Break hyphenated words and terms after the hyphen, if possible, to avoid the need for another hyphen.
2) Divide long chemical names by syllables, by etymologic units, or according to the hyphen rule, but at least 4 characters should appear on each line. Do not break at a hyphen connecting a locant or descriptor prefix.

   2-acetyl- *not* 2-
   aminofluorene acetylaminofluorene

3) The part carried over to the next line should not look like a separate word.

   path- *not* patho-
   ologic logic

4) Do not separate numbers from associated unit designations.

   a concentration of *not* a concentration of 250
   250 g/L g/L

5) Do not divide single-syllable and very short multisyllabic words (e.g., "also").

Word-processing programs and computer-driven composition have reduced the need for word divisions, especially in manuscripts with left-margin justification (ragged right). Manuscripts to be marked for electronic methods of publication should not include hyphens to indicate end-of-line word divisions; instead, hyphens should be used only in compound words needing hyphenation.

For recommendations about line breaks within Internet addresses, see Section 5.7.

6.4 PLURALS

6.4.1 General Principles

For most common nouns, form the plural by adding "s" (or "es", for nouns that end in sibilants: soft "ch", "s", "sh", "x", or "z").

chemicals	monarchs	axes	branches	foxes	losses
expressions	organisms	beaches	buses	grasses	washes

The plural of nouns ending in "i" is usually formed by adding "s", but some exceptions use "es".

alibis rabbis skis alkalis *or* alkalies

For nouns ending in "o" preceded by another vowel, form the plural by adding "s" to the singular. For nouns ending in "o" preceded by a consonant, there is no strict rule; some words take only "s", others take "es", and yet others use either form.

cameos	echoes	scenarios	tattoos	mosquitos *or* mosquitoes
infernos	ratios	tomatoes	halos *or* haloes	zeros *or* zeroes
duos	potatoes			

Form the plural of nouns ending in "y" preceded by a consonant or a consonant sound by changing the "y" to "i" and adding "es".

berry, berries colloquy, colloquies equity, equities
city, cities duty, duties fly, flies

6.4.2 Proper Nouns

Form the plural of most proper nouns in the same manner as for common nouns. Do not alter the spelling for proper nouns ending in "i", "o", or "y", except for the shortened forms of geographic terms.

the Charnyshes both Germanys *not* Germanies the Rockies
the Joneses the Perrys *not* the Perries the Sicilies
the Churchills the Alleghenies

If adding "s" or "es" to a proper noun would result in a false pronunciation, the plural is the same as the singular; this situation arises mainly with French names that end in an unpronounced "s", "z", or "x"; in the plural form the "s" is pronounced.

the 16 King Louis of France *not* the 16 King Louises of France
the Gervais *not* Gervaises

6.4.3 Irregular Forms

Many nouns have irregular plural forms, notably English nouns originating before the Norman Conquest. There are no formal rules for these plurals.

child, children ox, oxen die, dice louse, lice

For some single-syllable nouns the plural is formed by changing vowels within the word. The plurals of compound words that end in these words are formed in the same way.

foot, feet clubfoot, clubfeet
man, men workman, workmen
mouse, mice dormouse, dormice

The plurals of words ending in "man" that are not compounds are formed simply by adding "s".

human, humans German, Germans

Although the plurals of most nouns that end in "f", "ff", or "fe" are formed in the regular way, a few change the "f", "ff", or "fe" to "ves" in the plural.

calf, calves hoof, hoofs, hooves self, selves
dwarf, dwarfs, dwarves life, lives sheaf, sheaves
half, halves loaf, loaves staff, staffs, staves

See Section 6.4.6 for nouns with more than one plural form; some variant plural forms have different meanings.

6.4.4 Nouns with a Single Form

Some nouns have only one form, which can indicate singular or plural depending on the context.

aircraft pains [meaning "effort"] spacecraft
corps progeny species
forceps remains sperm
goods [meaning "material things"] series

The names of many nationalities have only one form, which is used to indicate both singular and plural.

Chinese Japanese Portuguese

Some nouns cannot represent a plural concept and therefore have no plural form.

equipment horsepower information

For some nouns, the sole form appears to be a plural but is construed as a singular.

hives
measles
mumps
shingles [the skin eruption, herpes zoster]
rickets
whereabouts
news
premises [meaning "property"]

For the names of certain types of animals, the singular is used to denote both one and more than one individual.

deer
fish [and individual species: trout, cod, haddock, herring]
moose
sheep
swine

For some of these nouns, as well as others that in normal scientific usage have a singular form (e.g., "oat", "wheat"), the regularly formed plural is used to indicate more than one species, strain, or variety.

6 experimental wheats 10 mutant oats 3 fishes of interest

For the names of other large mammals and some other organisms, either the singular or the regularly formed plural may be used to indicate the plural. When in doubt as to whether the singular form is acceptable as the plural, consult a dictionary.

antelope, antelopes
buffalo, buffalos, buffaloes
caribou, caribous
crab, crabs
elk, elks
giraffe, giraffes
lobster, lobsters
walrus, walruses

6.4.5 Plural Endings from Non-English Languages

English scientific and medical vocabularies have many terms that are taken directly from other languages. The plurals of such words are formed by the rules of the original language. In some cases, an anglicized form that follows the general rules for plural formation is also acceptable. See Section 6.4.6, which discusses nouns with more than one plural form. The acceptability of the English plural forms may differ with the type of publication and its audience. Editorial offices should create their own lists of preferred plural forms (see Table 6.2).

6.4.6 Other Nouns with 2 Plural Forms

For some nouns with more than 1 plural form, the 2 forms have different meanings. Use the form correct for the context.

brothers [males with the same parents], brethren [members of a society]
dies [devices for stamping], dice [gambling pieces]
indexes [alphabetic lists of topics or names], indices [numeric expressions]
ossa [bones], ora [mouths]
staffs [groups of employees or assistants], staves [poles]

For other nouns, the 2 forms of the plural are identical in meaning. Regular and specialized dictionaries may suggest a preferred form, may indicate that either form is

Table 6.2 Plural endings from non-English languages

Endings		Examples	
Singular	Plural[a]	Singular	Plural
a (Latin)	ae	abscissa	abscissae *or* abscissas
		alga	algae *or* algas
		formula	formulae *or* formulas
eu, eau (French)	eux, eaux	milieu	milieux *or* milieus
		rouleau	rouleaux *or* rouleaus
en (Latin)	ina	foramen	foramina *or* foramens
		rumen	rumina *or* rumens
ex, ix (Latin)	ices	appendix	appendices [for anatomic structures] *or* appendixes [for closing sections of books]
		fornix	fornices *or* fornixes
		index, indices	indices [in economics and mathematics] *or* indexes [for closing sections of books]
is, es (Greek or Latin)	es	analysis	analyses
		hypothesis	hypotheses
itis (Greek or Latin)	itides	arthritis	arthritides
		meningitis	meningitides
o (Italian)	i	virtuoso	virtuosi *or* virtuosos
on, oan (Greek)	a	criterion	criteria *or* criterions
		mitochondrion	mitochondria
		protozoon	protozoa
		protozoan	protozoa *or* protozoans
um (Latin)	a	addendum	addenda *or* addendums
		bacterium	bacteria
		datum	data *or* datums [for tidal benchmarks; see Section 25.11.4.1]
		medium	media *or* mediums
		phylum	phyla
		symposium	symposia *or* symposiums
us (Latin)[b]	i	bacillus	bacilli
		focus	foci *or* focuses
		fungus	fungi *or* funguses
	uses	apparatus	apparatuses
		prospectus	prospectuses

[a] For many terms, the plural may also be formed by adding "s" or "es".
[b] The plural of "genus" is "genera".

acceptable, or may list only one of the variants. In the list that follows, the first plural form is the variant recommended by this manual. Authors should create their own lists of preferred forms (see Section 6.4.4).

Singular	Plural
biceps	biceps, bicepses
femur	femora, femurs
gladiolus	gladioli, gladioluses
hoof	hooves, hoofs
thorax	thoraxes, thoraces

6.4.7 Numbers, Letters, and Abbreviations

Form the plural of a number presented as a numeral by adding "s" only.

expressed in 100s *not* expressed in 100's
the 1990s *not* the 1990's
patients in their 50s *not* patients in their 50's

Use the general rules for plurals (see Section 6.4.1) to form the plural of a number expressed as a word or of an expression indicating a number.

ones	in her teens	hundreds of specimens
at sixes and sevens	counting by tens	thousands of species

Form the plural of a single letter by adding an apostrophe and "s". This formation avoids confusion with true words ("as", "is") and abbreviations (e.g., "Ps" for photosynthesis) and is consistent for capital and lowercase letters.

x's and *y*'s   P's and Q's

For general guidance on plurals of abbreviations and initialisms, see Section 11.1.4.

6.4.8 SI and Other Units of Measurement

The symbols for most units are the same in singular and plural; see Section 12.2.1.1.

1 kg, 5 kg   1 mL/min, 18 mL/min

6.4.9 Plurals of Compound Words and Terms

For compounds that are spelled as a single word, the plural is formed regularly by adding the appropriate ending to the end of the word.

dormouse, dormice   jumpsuit, jumpsuits   spoonful, spoonfuls

Hyphenated and open compounds take the plural form of the noun that is the basis of the term.

aide-de-camp, aides-de-camp	man-of-war, men-of-war
governor general, governors general	right-of-way, rights-of-way
brother-in-law, brothers-in-law	coup d'état, coups d'état
chargé d'affaires, chargés d'affaires	surgeon general, surgeons general

If the compound contains no nouns or if none of the nouns is significant in the context, "s" is added to the last component.

forget-me-not, forget-me-nots   go-between, go-betweens

If the components of a compound term are more or less equivalent, the plurals of both are used.

woman scientist, women scientists

6.4.10 Common Names Taken from Scientific Names

For some plants, the vernacular English name is the genus name set in roman type without an initial capital letter (also see Section 22.3.1.7), and the plural is usually formed according to the general rules for plurals.

camellia, camellias	iris, irises
crocus, crocuses	gladiolus, gladioluses *or* gladioli [preferred]

Some microorganisms have a vernacular plural designation based on the name of the genus.

bacilli [for organisms in the genus *Bacillus*]
chlamydiae [for organisms in the genus *Chlamydia*]
mycobacteria [for organisms in the genus *Mycobacterium*]
pseudomonads [for organisms in the genus *Pseudomonas*]
salmonellae [for organisms in the genus *Salmonella*]
staphylococci [for organisms in the genus *Staphylococcus*]
streptococci [for organisms in the genus *Streptococcus*]
treponemes [for organisms in the genus *Treponema*]

If a form of this kind is not in general use, add a descriptive term to the italicized genus name (e.g., "*Escherichia* bacteria" or "*Candida* yeasts").

6.5 POSSESSIVES

6.5.1 General Principles

Form the possessive of most singular common and proper nouns and of some indefinite pronouns by adding an apostrophe and "s". Some indefinite pronouns, including "any", "few", "many", "none", and "such", do not have a possessive form.

the patient's condition
Canada's agreement with the United States
the wolf's territory
Pettigrew's study
everyone's attendance
one's own view
someone's responsibility

6.5.2 Singular Nouns That End in "s"

The general principle of adding an apostrophe and "s" holds for most nouns, including proper nouns, that end in "s". Pronunciation can serve as a guide: if the possessive "s" would be pronounced, it should appear in the written form.

the bus's route
the grass's texture
the moss's reproductive capacity
the lens's properties
Dr Jones's interpretation
Charles's suggestion
Professor Harris's viewpoint
Dickens's works of fiction

If the double sibilant sounds awkward, recast the expression to avoid the possessive form.

the texture of the grass
the properties of the lens

By tradition, the possessive forms of Greek and hellenized names of more than one syllable ending in "s" (which often have an unaccented ending pronounced "eez"), as well as those of "Jesus" and "Moses", are formed by adding an apostrophe only.

Archimedes' screw
Ulysses' adventures
Hippocrates' teachings
Ramses' tomb
Achilles' heel [an expression meaning "weakness", not an anatomic part]

If the final "s" (or "x" or "z") is silent, an apostrophe and "s" must be used to yield the correct pronunciation.

Agassiz's theories of glaciation
Arkansas's geography
Descartes's essays
Lemieux's reasons for proceeding

6.5.3 Proper Names Set in Italic Type

In the possessive form of names of books and journals, set the name in italic type and the apostrophe and the "s" in roman type.

The Lancet's reputation   *Surgery*'s 20th volume

6.5.4 Pronouns

Form possessive pronouns by adding only "s" to the pronoun, without an apostrophe.

His [a modification of "he" plus "s"] is the most visible display.
Here is mine; have you found hers yet?
This laboratory is ours; theirs is across the hall, and yours is in the other building.
I collected the specimen; its measurements have already been recorded. [Do not confuse the possessive pronoun "its" with the contraction "it's" for "it is".]

6.5.5 Plural Nouns

Form the possessive of common and proper plural nouns that end in "s" by adding only an apostrophe.

the patients' histories
the lenses' characteristics
the animals' behavior
the Harrises' family tree
the doctors' privileges
the workers' contract
the United States' relationship with its allies [a plural construction for the possessive is used, although "United States" is usually construed as singular]

Form the possessive of plural nouns that do not end in "s" by adding an apostrophe and "s".

the bacteria's growth patterns
the men's pulmonary capacity tests
the children's eating habits

When the noun ends in a sibilant sound, the resulting phrase may sound awkward; in such cases, recast the sentence to avoid the possessive form.

the mice's nesting material *becomes* the nesting material used by the mice
the geese's migratory formation *becomes* the migratory formation of the geese

6.5.6 Eponymic Terms

The scientific and medical vocabularies contain many eponymous terms, compound terms that incorporate a proper name referring to a theoretician, a researcher, a physician, a patient, or a place. Such terms refer to a wide variety of laws, theories, methods, anatomic parts, conditions, diseases, reagents, syndromes, and tests, among other entities. Traditionally, many of the terms that incorporate the name of a researcher have used the possessive form, whereas those representing a place or a patient have used the nonpossessive form. CSE recommends that the possessive form be eliminated from all eponymic terms, to allow clear differentiation from true possessives.

Crohn disease *not* Crohn's disease
the Carvallo sign *not* Carvallo's sign

the Korotkoff test *not* Korotkoff's test
McCune–Albright syndrome *not* McCune-Albright's syndrome

For discussion of capitalization of eponymic terms, see Section 9.7.2.

6.5.7 Compound Expressions

Form the possessive of a compound expression by adding an apostrophe and "s" to the final element.

someone else's proposal
the editor-in-chief's guidelines
everybody else's preferences
the surgeon general's warning
the deputy chief of staff's response

For nouns in a series, the form of the possessive is determined by whether joint or individual ownership is intended; for joint ownership, add an apostrophe and "s" only to the second element; for individual ownership, add an apostrophe and "s" to both elements.

Watson and Crick's landmark paper
the student and her tutor's appointment
Watson's and Crick's memoirs
the student's and her tutor's telephone numbers

In a possessive expression that incorporates one or more possessive pronouns, all nouns and pronouns take the possessive form.

Dr Denmore's and your grant proposal
my student's and my assessment of the data

6.5.8 Organization Names

Form the possessive of an organization's name by adding an apostrophe and "s" to the last element of the name (or an apostrophe only if the last element is a plural that ends in "s").

the American Psychological Association's author guidelines
the Council of Science Editors' annual meeting

The names of many organizations and institutions incorporate a possessive or a plural; the possessive form may be used to indicate "for", or the plural may be used as an adjective. Correct usage is determined by the official name of the organization itself, so the form must be verified. For names that include the possessive of a plural noun that does not end in "s" (e.g., "children"), the apostrophe is essential.

American Medical Writers Association [plural as adjective]
Editors' Association of Canada [possessive]
Children's Hospital of Eastern Ontario [possessive]

6.5.9 Expressions of Duration

Expressions of duration based on the genitive case are analogous to possessives and are formed in the same way, with an apostrophe. However, in many instances, another form that does not require the possessive is available.

a week's vacation, a week of vacation, a 1-week vacation
in 3 days' time, in 3 days
after many years' experience, after many years of experience

If the word "of" cannot be inserted into the expression, the possessive is incorrect.

6 months pregnant *not* 6 months' pregnant
8 days late *not* was 8 days' late

CITED REFERENCES

1. McArthur T, editor. The Oxford companion to the English language. Oxford (UK): Oxford University Press; 1992.

2. New Hart's rules: the handbook of style for writers and editors. Oxford (UK): Oxford University Press; 2005.

3. Hart's rules for compositors and readers at the University Press, Oxford. 39th ed. Oxford (UK): Oxford University Press; 1983.

4. The Chicago manual of style: the essential guide for writers, editors, and publishers. 16th ed. Chicago (IL): University of Chicago Press; 2010. Also available at http://www.chicagomanualofstyle.org.

5. United States Government Printing Office style manual: main page. 29th ed. Washington (DC): The Office; 2008 [issued 2008 Sep 16; accessed 2012 Nov 7]. http://www.gpoaccess.gov/stylemanual/index.html.

6. Editing Canadian English. 2nd ed. Toronto (ON): Macfarlane Walter & Ross; 2000. Prepared for the Editors' Association of Canada.

7. Public Works and Government Services Canada, Translation Bureau. The Canadian style. A guide to writing and editing. 2nd ed. Toronto (ON): Dundurn Press; 1997.

8. Iverson C, Flanagin A, Fontanarosa PB, Glass RM, Glitman P, Lantz JC, Meyer HS, Smith JM, Winker MA, Young RK. American Medical Association manual of style: a guide for authors and editors. 10th ed. New York (NY): Oxford Press; 2007.

7 Prose Style and Word Choice

7.1 SCIENTIFIC PROSE

Effective scientific prose is precise, clear, economical, fluent, and graceful. These qualities depend on myriad details: the choice of words, the length and flow of individual sentences, how sentences relate to each other, and how paragraphs are linked. Few editorial offices can afford to rewrite extensively; therefore, only details that can be economically improved during redaction are considered here. Authors seeking detailed guidance in writing clear, efficient scientific prose will find help in books listed in the sections "Style Manuals and Other Writing Guides" and "Guides to Usage and Prose Style" in the Bibliography of this manual.

7.2 GRAMMATICAL ERRORS

7.2.1 Subject and Predicate Agreement in Number

7.2.1.1 Latin Nouns

Writers frequently fail to distinguish between singular and plural forms of Latin-derived nouns, which results in a mismatch of subject and predicate. This fault may not affect understanding in popular, nonscientific writing, but scientific prose should maintain the distinction.

The best culture medium for *Legionella* is . . .
The most widely used culture media are . . .
These data support the view that . . .
A faulty datum in the survey shows that . . .

For plural forms of Latin-based nouns widely used in science and their anglicized equivalents, see Table 6.2. The English form of the plural may be required for some uses. The US Geological Survey[1] specifies "datums" (rather than "data") for benchmarks and time markers (see also Section 25.11.4.1).

7.2.1.2 Collective Nouns

Some collective nouns, such as "committee", may take a singular or plural verb in the related predicate. The choice depends on the meaning as defined by the action represented by the verb.

The committee announces decisions every Friday.
[A single announcement comes from the committee; its members do not each issue an announcement.]
The committee argue all points with care.
[The members of the committee argue among themselves.]

The plurality of a subject may be concealed by an abbreviation, but such a plural may be treated as a collective noun.

The NIH was headed by Dr Healy. [NIH = National Institutes of Health]
The CDC issues advisories about viruses. [CDC = Centers for Disease Control and Prevention]

7.2.2 Dangling Participles

Present and past participles frequently begin an adjectival phrase that is intended to modify the noun or noun phrase that is the subject of the sentence or clause. The noun modified in this way must be explicitly stated to indicate the subject responsible for the action represented by the participle. If no subject noun is apparent or if placement of the participle within the sentence appears to attach the modifier to the wrong subject noun, the participle is said to be "dangling".

Reviewing the available data, the cause of the accident was mechanical, not chemical.
[As written, the sentence implies that "the cause" is the subject of the verb "reviewing", which is clearly not the intention. An explicit statement of the agent changes the sentence to "Reviewing the available data, the committee concluded that the cause of the accident was mechanical, not chemical." This sentence can be improved further by putting the agent first: "The committee concluded from its review of the data that . . ."]

Such sentences can often be revised by using the present participle as a gerund (verb form used as a noun).

Reviewing the data led to the judgment that the cause of the accident was mechanical, not chemical.

["Reviewing the data" is now a noun phrase serving as the subject of the sentence; "the committee" as agent may be clear from the context.]

The dangling participle "based" appears frequently in scientific writing.

Based on the evidence, the accident was caused by a mechanical, not a chemical, failure.

[The construction of the sentence indicates that "the accident" is based on the evidence, but the intention is more likely that a conclusion or judgment or decision is based on the evidence.]

Potential revisions:

The conclusion, based on the evidence, was that the accident was caused by a mechanical, not a chemical, failure.

Basing its decision on the evidence, the committee concluded that the accident was caused by a mechanical, not a chemical, failure.

On the basis of the evidence, the committee concluded that the accident was caused by a mechanical, not a chemical, failure.

One way to test for a dangling participle is to move the opening phrase to the interior of the sentence so that it follows the noun it should modify, according to the original structure of the sentence.

The accident, based on the evidence, was caused by . . .

[According to this test, the participle is dangling because the accident was not in fact based on the evidence.]

That a participle is dangling may not be apparent when it does not appear at the beginning of the sentence.

The county was surveyed using a Wehrtopf pocket altimeter.

[The agent using the altimeter is unclear. Possible revision: "The workers used a Wehrtopf pocket altimeter to survey the county."]

7.3 CONFUSED AND MISUSED TERMS

7.3.1 Homophones and Near-Homophones

Homophones are pairs of words with the same sound but different meanings and spellings. Authors uncertain of the correct spelling of the intended word may misuse homophones.

albumen, albumin	discrete, discreet	principle, principal
complement, compliment	here, hear	sheer, shear

Spelling checkers in word-processing programs may not detect such misuses.

A related kind of error is the misspelling of words ending in able/ible, ance/ence, and ant/ent. Here are the correct spellings of some frequently misspelled words.

abundance	dispensable	intransigence	prominent
acceptable	divisible	occurrence	relevant
accessible	existence	permanence	resemblance
compatible	feasible	preferable	resilient
consistent	independent	preference	resistant
credible	inevitable	prevalence	solvent
disappearance			

7.3.2 Imprecisely Applied Words

Careful writers and editors in science strive to select and use the word that most accurately and precisely conveys the intended meaning. This section presents pairs or groups of words that may be misused. The emphasis of the definitions here is on usage in science and scientific contexts and on distinctions that can clarify meaning. Some of these words may have a more specific meaning in some scientific disciplines; these meanings can be found in scientific dictionaries such as the *McGraw-Hill Dictionary of Scientific and Technical Terms*.[2] Nuances of, and preferences for, many common words are thoroughly discussed in the synonym and usage notes in *The American Heritage Dictionary of the English Language*.[3]

a, an, the: The article "a" (or "an" preceding a vowel sound) is correct when the subject of the noun it precedes can exist in more than one form or as more than one case ("A new species of *Escherichia* was identified"; other species exist or other new species may be identified in the future). When no more than one instance exists or is likely to exist in the future, "the" is appropriate ("The organism identified as responsible for this outbreak was *Escherichia coli*").

absorbance, absorptance, absorptivity: "Absorbance" is the logarithm of the ratio of the intensity of light entering a solution to the intensity it transmits; "absorptance" and "absorptivity" refer to the ratio of energy absorbed by a body to the energy striking it.

absorption, adsorption: "Absorption" is the process, ongoing or completed, of taking up by capillary, osmotic, chemical, or solvent action; "adsorption" is the holding of something by the surface of a solid or liquid through physical or chemical forces.

accuracy, precision: "Accuracy" is the degree of correctness of a measurement or a statement; "precision" is the degree of refinement with which a measurement is made or stated; for example, the number 3.43 is more precise than 3.4, but it is not necessarily more accurate. When applied to a statement, "precision" implies the qualities of definiteness, terseness, and specificity.

adduce, deduce, induce: To "adduce" is to bring forward as an example or as evidence for proof in argument; to "deduce" is to reason to a conclusion or infer it from a principle; to "induce" is to reach a conclusion through inductive reasoning (going from particular facts to a principle). To "induce" also means to bring about an effect, as in "to induce labor".

affect, effect, impact: "Affect" as a verb means to influence ("His budget cut affected all members of the staff"), and as a noun it means the impression of feeling or emotion conveyed by a person's demeanor, action, or speech ("The diagnostic feature in this case was the patient's flat affect"). "Effect" as a verb means to bring about a change ("He effected a budgetary change"), and as a noun it means the result of some action ("The effect was a cut in the budget"). "Impact" as a verb should be reserved for the striking of one body against another and should not be used to mean simply "affect".

after, following: "After" can mean simply later than a particular time or event (as in "He died of anaphylactic shock soon after you saw him"). "After" can also imply cause and effect (as in "He died of anaphylactic shock after swallowing a capsule of the wrong antibiotic"). "Following" may be used synonymously with "after" but is best reserved as an indicator of position not related to time (as in "The following authors are frequently cited in genetics textbooks: Mendel, Morgan, and Wright").

aggravate, irritate: To "aggravate" means to worsen an existing condition; to "irritate" means to evoke a reaction (e.g., cause inflammation) and thus refers to a condition that is new.

aliquot, aliquant, sample: An "aliquot" is a part of a total amount of a gas, liquid, or solid that divides evenly into the whole; for example, 10 mL is an aliquot of 100 mL. An "aliquant" is a part that does not divide evenly into the total; for example, 8 mL is an aliquant of 20 mL. A "sample" is a part taken as representative of its source for analysis or study.

alternate, alternative: The verb "alternate" describes the successive passage back and forth from one state, action, or place to another ("The water alternates between liquid and solid states in this temperature range"); the corresponding adjective conveys the same sense ("The alternate states were liquid and solid"). The noun and adjective "alternative" represent a choice between 2 or more mutually exclusive states or places ("The alternative to the current method is to use a catalyst").

although: See *while*.

among, between: The preposition "among" indicates a relation involving more than 2 units of the same kind, with an emphasis on distribution across the group ("Among the antibiotics of this class, azithromycin is the best choice"); the preposition "between" indicates a relation involving 2 (or more) units of the same kind, with an emphasis on the one-to-one relation(s) ("The study examined the interaction between predator and prey"; "The choice was between penicillin, ampicillin, and erythromycin").

an: See *a*.

analog, analogous, homolog, homologous, homoeolog, homoeologous: The noun "analog" and adjective "analogous" refer to organs or other structures that are similar in function but dissimilar in structure and origin; chemical analogs are compounds that have similar structure but differ in a certain component. "Homolog" and "homologous", in biology, refer to organs or other structures that correspond in structure, position, origin, or other characteristics but not necessarily in function; these terms are similarly applied in other sciences (e.g., homologous chromosomes, homologous proteins). "Homoeolog" and "homoeologous" refer to partially homologous chromosomes.

anatomy, morphology, structure: "Anatomy" is the study of structure, especially of living things, and is also becoming accepted to mean structural makeup. "Morphology" is the study of shape, structure, or formation of living things, especially external shape and form, and is also becoming accepted to mean structural characteristics. "Structure" refers to the parts of living or nonliving things as they relate to each other.

ante, anti: The prefix "ante" indicates before in a sequence or location ("antedate", "antebrachial artery"); the prefix "anti" indicates against, opposed to, in contrast to ("antibody", "antiparticle", "anti-inflammatory").

as, because, since: To avoid ambiguity, the conjunction "as" is preferably used only in the temporal sense ("As we were completing the paper, new evidence came to light") and not in the causal sense (not "As we were away in Africa, he thought he could pilfer our data"). Use the conjunction "because" to show a causal relation. Use the conjunction "since" to show a temporal relation ("Since then, there have been 3 additional outbreaks"); it may also be used in the causal sense, but this usage may be ambiguous ("Since the initial finding was validated, the larger question has been widely examined"; here, the temporal sense would make sense, even if the causal meaning was intended).

assess, determine, evaluate, examine, measure: These terms are frequently used synonymously or almost so, but maintaining distinct meanings for them can make for clearer expression. "Assess" is best reserved to mean estimating a value, as for taxation, or setting the amount of a payment, as for a fine. The verb "determine" is commonly used as a synonym for "measure" but is best reserved to mean set a limit on or establish conclusively. "Evaluate" means to ascertain or fix a value on the object of the action, whereas "examine" is to look at, or inquire into, or test closely. "Measure" is to examine an object quantitatively, as in measuring blood glucose.

assure, ensure, insure: "Assure" preferably connotes providing surety (a promise, pledge, or guarantee); "insure" should be used when the intended meaning is related to a financial guarantee or pledge related to property or life; "ensure" should be used to mean making sure or removing doubt.

average, characteristic, typical: In scientific usage, "average" should be reserved, if possible, to mean the statistical mean; "characteristic" and "typical", conveying a sense of representativeness, are often suitable in place of "average" used in a nonstatistical sense.

axenic, gnotobiotic: The adjective "axenic" refers to organisms kept in isolation from other living things or an environment for such isolation; the adjective "gnotobiotic" refers to laboratory animals reared to be free of infectious agents except for any agent deliberately introduced.

based on, on the basis of: "Based on" should be used as a verb ("They based their conclusion on the evidence from multiple trials"), not as an adverbial form. In the latter situation, the expression "on the basis of" should be used ("The treatment rate was decreased on the basis of effective results in field trials").

because: See ***as***.

because of, due to: "Due to" (adjective with preposition) means attributable to ("The problem was due to a mechanical failure"). "Because of" (conjunction with preposition) means as a result of or owing to ("The problem occurred because of a mechanical failure"). The expressions are not interchangeable; substituting the expression "attributable to" or "as a result of" should point to the correct choice.

believe, feel, think: These subjective verbs connote an author's convictions or persuasions with different strengths of basis; "feel" implies an intuitive or not fully reasoned conviction; "think", a view based on evidence or logic; "believe", a definite conviction on the view regardless of the strength of evidence.

benchmark, criterion, standard: "Benchmark" is used in different contexts as a synonym for "criterion" (a benchmark without an assigned value) or for "standard" (a benchmark that is assigned a specific value). "Criterion" may be synonymous with "standard" but is often restricted to be synonymous with "measure", a specific type of basis for a judgment but with possibly different specifications for favorable or unfavorable judgment. "Standard" is often reserved for a criterion assigned a specific value against which a judgment is made.

between: See ***among***.

bi, semi: The prefix "bi" can indicate 2 (as in "bicorn", the adjective for an animal with 2 horns) or at intervals of 2 units (as in "bimonthly publication", every 2 months, and "bicentennial celebration"). The prefix "semi" indicates half (semicircle) or partial (semiretirement). "Semimonthly publication" could be less ambiguously phrased as "twice-a-month publication" or "biweekly publication". Also see ***quasi***.

carry out: See ***perform***.

case, patient: A "case" is an instance, example, or episode (as of disease: "She had a case of measles" or "The worst case would be 2 earthquakes within 2 days"). A "patient" is a person

("We saw 12 patients in the clinic"). In the medical literature, a case is "reported" (as the description of presentation, diagnosis, treatment, and outcome for a patient with a particular disease or condition), but the patient with the condition is "described".

cause: See ***etiology***.

characteristic: See ***average***.

circadian, diurnal: The adjective "circadian" refers to an occurrence approximately every 24 h. The adjective "diurnal" refers, in meteorology, to processes that are completed within 24 h and recur every 24 h; in biology, to processes that occur daily or every day in the daytime; in botany, to the habit of being open in the daytime and closed at night (as a flower).

common, frequent, regular: "Common" means appearing frequently ("a common finding in this type of study"); it also refers to vernacular (as opposed to scientific) names of organisms. "Frequent" means occurring often or at relatively short intervals. "Regular" indicates ordered, consistent, or at fixed times or points.

comparable, similar: Reserve "comparable" for use as an adjective indicating an item lending itself to comparison with a similar item ("Because the methods of ascertainment differed, the mortality statistics of Sweden and Chile are not comparable"—i.e., cannot be compared); "similar" is better as the adjective indicating likeness ("The mortality rates in Sweden and Chile are similar").

compared to, compared with: "Compared with" implies looking for similarities or differences in compared objects ("In the clinical trial, the low-fat diet was compared with a high-fat diet"); "compared to" focuses on similarities, often metaphoric or fanciful ones ("He compared the bird's flight to running through the air"). Also see ***versus***.

complement, compliment: The verb "complement" carries the sense of completing something, as indicated by its first 6 letters; the noun "complement" may be used in this sense (as in the full number of people needed to staff a facility), but it also refers to a group of proteins occurring in the serum that have activity in the immune system. The verb and noun "compliment" both refer to a favorable comment.

compose, comprise, constitute: "Compose" as an active verb after a plural subject means to form, to make up a single object, to go together ("Forty-eight states compose the contiguous United States of America"), but as a passive verb it is synonymous with "comprise" ("The contiguous United States of America is composed of 48 states"); "comprise" is a verb meaning to include, to contain, to be made up of ("The United States comprises 50 states"); "constitute" is in many contexts synonymous with "compose" ("Forty-eight states constitute the contiguous United States of America"), but it also can mean to amount to, to equal, to set up, to establish. Never use the phrase "is comprised of".

congenital, genetic: "Congenital" means born with, present at birth; "genetic" means having to do with genes, chromosomes, or their effects in producing phenotypes or determining heritable characteristics. A disease or abnormality caused by genetic effects is not necessarily congenital (apparent at birth), and a congenital defect is not necessarily genetic in origin.

conjecture: See ***law***.

connote, denote: "Connote" implies a meaning beyond the usual specific, exact meaning; "denote" indicates the presence or existence of a thing.

constitute: See ***compose***.

contagious, infectious, infective: "Contagious" means capable of being transmitted from one organism to another; "infectious" means harboring a potentially infecting agent or caused by an infecting agent ("an infectious disease"); "infective" describes an agent that can cause infection ("Not all the bacteria found in this environment are infective").

continual, continuous: "Continual" means a prolonged succession or recurrence over time with or without interruption ("continual state of arousal", "continual encounters within the enclosure"); "continuous" implies without interruption, never ceasing even briefly ("Illumination was continuous during the second trial").

conventional: See ***customary***.

criterion: See ***benchmark***.

currently, presently, at present: "Currently" (preferred) and "at present" mean now; "presently" means soon, shortly, in the near future.

customary, conventional, traditional, normal, norm: "Customary" means long used, commonly practiced for a long period, used as a habit; "conventional" means an established and generally agreed-on practice or characteristic; "traditional" means long used or applied and tends to connote a long-standing and general acceptance in a social or professional group or community; "normal" is an adjective indicating that the noun it modifies has the characteristic(s) of a satisfactory or desirable majority or lacks any abnormality (it also has a specific statistical meaning); "norm" implies not simply a normal characteristic but a desired characteristic.

data, database, data set: A "database" is a formal structure (such as a computer file or printed document) containing data organized for retrieval and analysis and representing a conceptually coherent subject; a "data set" is a particular coherent body of data maintained in a database; the plural noun "data" refers to individual items of information.

deduce: See ***adduce***.

definite, definitive: "Definite" means clearly limited, firmly established; "definitive" means conclusive, defining.

demonstrate, exhibit, reveal, show: "Demonstrate" and "exhibit" are often used as inflated versions of "show"; "demonstrate" should be reserved for a deliberate action intended to illustrate a procedure (as in "The technician demonstrated how to operate the pH meter") and "exhibit" for a deliberate action to make visible (as in "He exhibited the mineral specimens at the last congress"). These 2 terms should not be used to mean passively carrying something apparent; use "The patient had a rash" rather than "The patient demonstrated or exhibited a rash." "Reveal" represents an action to make visible what has been hidden or unapparent and should not be used as a synonym for "report". Note that an inanimate agent cannot demonstrate anything; therefore, avoid such constructions as "The data demonstrated an increase in the blood pressure when the dose was lowered."

denote: See ***connote***.

desirable: See ***significant***.

determine: See ***assess***.

different, diverse, disparate: "Different" means having at least some dissimilar characteristics ("The fossil evidence establishes that this outcropping is a different formation"); "diverse" means having a notable range of differences ("The staff members have diverse interests and skills"); "disparate" means distinctly different ("They reached disparate conclusions from the same evidence"). Also see ***varying***.

different from, different than: The "different from" phrase is usually the better choice because of its parallel with "differs from" ("The flora of the Quaternary is distinctly different from that of the Cretaceous"; "The flora of the Quaternary differs distinctly from that of the Cretaceous").

differing: See ***varying.***

digit: See ***number.***

disinterested, uninterested: "Disinterested" means to have no stake in the outcome; "uninterested" means unconcerned or indifferent.

disparate: See ***different.***

diurnal: See ***circadian.***

diverse: See ***different.***

dosage, dose: In medicine, "dosage" implies not only an amount but also frequency of administration ("The dosage of analgesic was changed to 50 mg every 4 h"); it is not synonymous with "dose", which means the amount of a drug administered at one time ("A 50-mg dose of analgesic was to be administered every 4 h"). "Dose" is also used to refer to the amount of radiation administered in radiotherapy and for the application of agricultural chemicals.

due to: See ***because of.***

effect: See ***affect.***

employ, use, utilize: These verbs are often used synonymously, but in most cases "use" is adequate to mean applying or drawing on for a purpose. When consumption through a use is implied, either "use up" or "consume" is more specific. To "employ" can mean to put a person to work or an object to use.

ensure: See ***assure.***

epidemic, epiphytic, epizootic: All of these terms refer to outbreaks of disease (or associated conditions), usually widespread in a population of hosts or occurring at a frequency much higher than normal. "Epidemic" refers to a disease in a human population, "epiphytic" to an outbreak of infectious disease in plants, and "epizootic" to a disease in an animal population.

etiology, cause: "Etiology" should be reserved to mean the study and description of cause; it should not be used as a synonym for "cause", the reason for an occurrence.

evaluate: See ***assess.***

examine: See ***assess.***

execute: See ***perform.***

exhibit: See ***demonstrate.***

farther, further: "Farther" as an adverb means to a more distant or advanced point, either in a physical or nonphysical dimension (as in "concept"), and as an adjective, at a more distant or advanced point. "Further" is best reserved for use as a transitive verb meaning to move along or to develop ("His theory did little to further our knowledge of the oldest galaxies").

feel: See ***believe.***

few, fewer, less: "Few" and "fewer" are adjectives indicating small and smaller in number; they are used with counted or countable items; "less", a cognate of "little", is used with uncounted or uncountable quantities and means a smaller total amount ("We have fewer astrologers per capita now than in 1900"; "We have less information on the genome of the horse than on that of man").

flammable, inflammable, nonflammable: Both "flammable" and "inflammable" are adjectives meaning capable of combustion; "nonflammable" means noncombustible.

following: See ***after.***

frequent: See ***common.***

fungus, fungal, fungous, fungoid: The noun "fungus" (plural "fungi") refers to an organism with nucleated cells having rigid walls but no chlorophyll. The adjectives "fungal" and "fungous" mean relating to or caused by a fungus (as in "a fungous or fungal infection"), but "fungal" is preferred because its use avoids confusion with the noun. The adjective "fungoid" means resembling a fungus.

further: See ***farther.***

general, generally, generic, generically, usual, usually: The adjective "general" and the adverb "generally" describe a broad or group-typical application, relevance, or characteristic. "Generic" and "generically" are first cousins, but they imply items that are of the same category, as in generic drugs (a group of drugs of a particular type, in contrast to specifically trademarked individual drugs of that type); "generic" also means relating to a genus. "Usual" and "usually" connote situations that are likely, expected, or relatively frequent.

genetic: See ***congenital.***

gnotobiotic: See ***axenic.***

great: See ***significant.***

haphazard, random: "Haphazard" means without plan or direction; "random" refers to a method of sampling in which members of a population have a known chance of being selected (e.g., by random number table).

healthy, healthful: "Healthy" denotes good health in a living thing; "healthful" means conducive or supportive of good health; thus, a healthful lifestyle can help to keep a person healthy.

heterogenous, heterogeneous: See ***homogenous.***

homoeolog, homoeologous: See ***analog.***

homogenous, homogeneous: Both terms mean having closely similar or identical characteristics, such as components, structure, or origin. "Homogeneous" also means having a uniform quality throughout. "Homogenous", as an adjective derived from "homogeny", is used in biology to mean having similar structures derived from common origins. Similar distinctions can be drawn between "heterogenous" and "heterogeneous" (from "hetero", meaning "different").

homolog, homologous: See ***analog.***

hypothesis: See ***law.***

hypothesize, hypothecate: To "hypothesize" means to form a hypothesis; to "hypothecate" means to pledge property as security without transfer of rights.

identical to, identical with: The 2 forms are regarded as equally acceptable in some usage guides, but the "with" form should be preferred because of its parallel with the noun "identity" ("A German botanist clearly showed the specimen's identity with the plant described 10 years earlier in the Canadian survey").

immunize: See ***vaccinate.***

impact: See ***affect.***

imply, infer: "Imply" is to indicate or suggest that the subject can be interpreted to a following but not necessarily correct conclusion; "infer" is to draw a conclusion from evidence ("These data imply that the sensors are defective"; "I infer from the data that the sensors are defective").

important: See ***significant.***

incidence, prevalence, point prevalence, period prevalence: In epidemiology, "incidence" is the number of new cases occurring in a population of stated size during a stated period; "prevalence" is the number of cases existing in a population of stated size at a particular time; "point prevalence" refers to the number of cases registered on a particular date and "period prevalence" to the number of cases registered during a stated period. For more detailed definitions, see the relevant entries in *A Dictionary of Epidemiology*.[4] In botany and agriculture, "incidence" refers to the frequency of a disease or condition (as opposed to its severity).

individual, person: "Individual", as noun or adjective, indicates a unit considered specifically and distinguished from a group (e.g., an individual woman or man); "person", with its connotation of a particular human being having his or her own personality, is preferred to the simpler and dehumanized connotation of "individual". Also see ***people, persons***.

induce: See ***adduce***.

infectious, infective: See ***contagious***.

infer: See ***imply***.

infested, infected with: In medicine, "infested" means harboring or carrying visible lower organisms, notably worms or insects, that do not cause inflammation or other immunologic consequences by their presence; in plant pathology, "infested" refers to microorganisms present in or on a nonliving substrate, such as dead plant material, tools, soil, or containers; "infected with" means harboring or carrying viruses or bacteria, which do cause inflammation and other immunologic consequences.

inflammable: See ***flammable***.

influential: See ***significant***.

inherent, intrinsic: Synonymous adjectives in general used to mean a characteristic necessarily in and of the noun modified; "intrinsic" is also used in anatomy to mean entirely within a structure or organism.

insure: See ***assure***.

irritate: See ***aggravate***.

kind: See ***type***.

law, theory, hypothesis, conjecture: Terms for concepts with decreasing degrees of certitude; a "law", in science, is a concept with a high degree of certitude, sufficient for high confidence in its use for predicting phenomena; a "theory" is a broad concept based on extensive observation, experimentation, or reasoning and expected to account for a wide range of phenomena covered by its scope; a "hypothesis" is a narrow concept, generally postulated as a potential explanation for phenomena and to be tested by experiment or observation; a "conjecture" is a speculation, a concept akin to a hypothesis but proposed for testing.

less: See ***few***.

localize, locate: "Localize" means to confine, restrict, or attribute to a particular place, to have the characteristic resulting from such action ("The infection localized in the antecubital space"); "locate" means to specify, place, or find in a particular place ("We finally located the infection in the right pleural cavity"); "localize" should not be used to mean "find" (not "We localized the primary site of the disseminated cancer in the pancreas").

major: See ***significant***.

majority, most: "Majority" is frequently used as a synonym for "most", but "most" is a simpler and preferred term when a quantitative expression is not needed and a preponderance needs to be implied ("Most physicians are licensed in only one state" rather than "A majority of physicians are licensed in only one state").

mean, median: The "mean" is the arithmetic average of a set of measurements; the "median" is the midpoint of the distribution of values.

measure: See ***assess***.

meiosis, mitosis, miosis: "Meiosis" is cellular division resulting in production of cells with a haploid number of chromosomes; "mitosis" is cellular division producing the diploid number of chromosomes; "miosis" is excessive smallness of the ocular pupil.

method, methodology, technique: "Method" and "technique" are both widely used to mean an analytic, quantitative, observational, or other kind of procedure, but a valuable distinction is to reserve "method" for "procedure" and "technique" for the skill, good or bad, applied in carrying out a procedure ("This bioassay is a reliable method when the analyst applies careful technique"; "Horowitz always showed impeccable technique at the keyboard"). A "methodology" is a system or combination of methods or techniques.

morphology: See ***anatomy***.

most: See ***majority***.

mucus, mucous, mucoid: "Mucus" is a thick, slimy secretion produced by body membranes and glands; "mucous" is the adjectival form indicating that something has the character of mucus or produces mucus, as in "mucous membranes"; "mucoid" means resembling mucus.

mutant, mutation: "Mutant" as a noun refers to an organism carrying or expressing one or more genetic mutations, but it is also applied as an adjective as in a "mutant gene"; "mutation" is a stable and heritable change in a nucleotide sequence in DNA or RNA.

need: See ***require***.

nonflammable: See ***flammable***.

norm, normal: See ***customary***.

number, numeral, digit: "Number" is the count of some class of objects; a "numeral" is a single character in the group of numbers from zero (0) to nine (9). For example, the number 345 is represented by the arabic numerals 3, 4, and 5 in that sequence. "Digit" can be used as a synonym for numeral but is sometimes used to refer to the number of numerals and their representation of magnitudes in the decimal system ("He reported his data with 3 digits"); it also refers to the fingers and toes.

nutrition, nutritional, nutritious: "Nutrition" is the discipline concerned with foodstuffs and feeding, including the study of deficiencies and toxic effects associated with food; it may also refer to a desirable diet. "Nutritional" means having to do with nutrition. "Nutritious" describes substances that yield desirable nutrition.

outbreak: An imprecise term often applied to mean a sudden appearance of a disease, especially of an infectious disease, or of some other kind of social phenomenon; alternative terms could be "episode", "sudden occurrence", "epidemic", or "epizootic".

parameter: A "parameter" is a variable to which a particular value can be assigned to determine the value of other variables; it should not be used as a synonym for "variable" (a quantity that can assume any of a set of values), "index" (something leading to a particular conclusion), "indicator" (something that indicates a set of conditions), or "guideline" (a recommendation).

pathology: Reserve "pathology" to mean the discipline that studies diseases, disorders, and other abnormalities in plants, animals, or humans; it is not a synonym for "abnormality", "disease", "disorder", or "lesion".

patient: See ***case***.

people, persons: "People" means a group of persons with some characteristic in common, such as nationality ("the French people") or location ("the various peoples living east of the Volga"); use "persons" to emphasize individuality ("Persons with visual impairment can use a variety of software tools to access the Internet") or when referring to a numbered group ("An ambulance took 3 injured persons to the same hospital").

percent, percentage: "Percent" represents units per 100 units (45% means "45 units per 100 units"); "percentage" refers to a quantity or rate expressed as the unit percent (45% is a percentage). The difference between 2 percentages should be stated as a difference in percentage points, not as a percent; the difference between 25% and 50% is 25 percentage points, not a 25% difference. Also see Section 12.3.4.

perform, carry out, execute: Such verbs can often be replaced by a more specific verb such as "analyze" or "operate" (not "He performed an analysis of the possible orbits in a gravitational field" but "He analyzed the possible orbits in a gravitational field").

person: See ***individual, people***.

precision: See ***accuracy***.

presently, at present: See ***currently***.

prevalence: See ***incidence***.

proven, proved: "Proven" as a verb (irregular past participle) should generally be replaced by "proved" as the past participle with regular form ("It has been proved that a retrovirus is responsible for the recently described syndrome" not "It has been proven that a . . .").

quasi, semi: The prefix "quasi" modifies a stem term to indicate to a degree or to some extent (quasiscientific); the prefix "semi" has the same meanings as "quasi" but may also carry the more specific meaning of half (semicircle). Also see ***bi***.

random: See ***haphazard***.

rare: See ***unique***.

regime, regimen: "Regime" means a regular pattern, as in seasonal weather patterns; it also refers to a government in power. "Regimen" refers to a stipulated program for treatment or activity, as in a dietary or therapeutic regimen.

regular: See ***common***.

relationship, relation: Reserve "relationship" to mean relations between 2 or more persons; "relation" is usually adequate to describe a connection between inanimate objects ("the cause-and-effect relation of HIV to AIDS").

require, need: Reserve "require" for use as a transitive verb with the stronger meaning of actively setting obligatory or compelling expectations ("The journal requires submission of 2 copies of a manuscript"); "need" is appropriate for a passive agent ("Green-leaved plants need sunlight").

reveal: See ***demonstrate***.

sample: See ***aliquot***.

semi: See ***bi, quasi***.

sensitivity, specificity: When applied to diagnostic methods, "sensitivity" is the ability to correctly identify patients who have the condition or disease; "specificity" is the capability of the method to correctly identify patients who do not have the condition or disease. For more detailed explanations, see *A Dictionary of Epidemiology*.[4]

show: See ***demonstrate***.

significant, great, important, influential, major, valuable, useful, desirable: "Significant" should be used to mean serving as a sign of or pointing to; this precise use is especially needed for the statistical sense of reaching a predefined numeric threshold and hence pointing to a specific statistical conclusion ("The mean blood pressure was significantly lowered, with a *P* value of less than 0.05"). The other adjectives are preferred when "pointing to" or "indicating" is not intended.

similar: See ***comparable.***

since: See ***as.***

specificity: See ***sensitivity.***

standard: See ***benchmark.***

structure: See ***anatomy.***

technique: See ***method.***

that, which: Use "that" to introduce a restrictive (i.e., required or essential) clause ("This is the house that Jack built") and "which" to introduce a nonrestrictive (i.e., nonrequired or nonessential) clause ("This house, which Jack built, was of shoddy construction").

the: See ***a.***

theory: See ***law.***

think: See ***believe.***

traditional: See ***customary.***

trophic, tropic: The suffix "trophic" indicates a stimulating, nourishing, or supporting function in the growth or development of an agent's target (as in "adrenocorticotrophic hormone", a hormone stimulating increased activity of the adrenal cortex) or indicating growth or development itself; in ecology, it refers to obtaining nutrition (e.g., phagotrophic). The suffix "tropic" indicates capability to respond to a modifying or changing agent (as in "chemotropic algae", algae that respond to chemical stimuli). The "trophic" form is becoming less common in endocrinologic usage.

type, kind: "Type" is often used synonymously with "kind" but should be reserved in science to mean a kind of inanimate object or a specific animal, plant, or microorganism characteristic of or standing for a larger group of closely related items; a "type specimen" is a specimen used to establish a species name.

typical: See ***average.***

unique, unusual, rare: "Unique" means one of a kind; it is not a synonym for "unusual" (uncommon) or "rare" (occurring seldom).

uninterested: See ***disinterested.***

use, utilize: See ***employ.***

useful: See ***significant.***

usual, usually: See ***general.***

vaccinate, immunize: These 2 terms are often used as synonyms, but to "vaccinate" means to purposely expose a person or an animal to an antigen in hopes of eliciting protective antibody; to "immunize" implies that exposure to an antigen through infection or vaccination successfully elicits protective antibody.

valuable: See ***significant.***

varying, differing, different, diverse: Use "varying" to mean changing, and "differing", "different", or "diverse" to mean having unlike characteristics; "Philadelphia and New York have varying mean annual temperatures in a long-term cycle" means each city has a mean temperature that changes in the cycle; "Toronto and Santa Fe have differing mean annual temperatures" means that their mean annual temperatures are not the same; "various" can be a near-synonym for "differing" ("Various racial and ethnic groups make up the population of Los Angeles").

versus, vs., v.: "Versus" means against; it is correct in legal titles (*Roe vs. Wade* or *Roe v. Wade* [legal style]) and some expressions in which it represents this meaning ("graft-versus-host disease"). Use "compared with" for comparisons or contrasts, as in a clinical trial of 2 drugs ("penicillin compared with ampicillin for the treatment of pneumococcal pneumonia").

which: See ***that***.

while, although: "While" should be used to indicate a period under consideration ("While he was waiting for surgery, his angina pectoris became steadily more frequent") and "although" for a conditional state ("Although he was being treated for the staphylococcal infection while waiting for surgery, he succumbed to rupture of his aortic valve").

7.4 EXCESSIVELY LONG COMPOUND TERMS

Scientific writing frequently includes strings of adjectival nouns modifying a noun that is the subject or object of a sentence. Often it is unclear whether some of the nouns are compound modifiers or single modifiers and hence whether particular nouns modify an adjacent modifier or the main noun.

> a new type motor skills college performance test
> a percentage transmission recording ultraviolet light absorption meter

Such constructions are especially burdensome for readers when they are repeated subsequently in the text where the main noun could be readily understood by itself, or with a single modifier, through its context.

> They designed a new type of test to measure the motor skills of college students . . .
> Users of the new test will find . . .

Rewrite excessively long compound terms to clarify meaning by phrasing some of the relationships and hyphenating directly related modifiers.

> a new kind of motor skills test used in colleges
> an ultraviolet light–absorption meter for recording percentage transmission

7.5 BIAS-FREE USAGE

Careful attention to avoiding terms that reflect stereotypic biases or habitual vocabulary is often justified mainly by the motive of not offending or insulting the persons who are the subjects of discourse. Along with this motive, which helps in promoting social harmony and the sense of community, consider also the value of scientifically accurate statement. Stereotypic or habitual statement is not scientifically acceptable because it ignores, or even simply obscures, the complexity of scientific questions. Entire books have been written on biased usage with recommendations for more accurate modes of statement[5]; this section of this manual provides just a few examples. Relevant advice is also available in some style manuals.[6–9]

7.5.1 Gender and Sex

When both men and women are the subject of the text, make this clear by referring to both; do not assume that a male referent is adequate.

> His analysis ignored the economic problems of ordinary men and women.
> *not* His analysis ignored the economic problems of the man in the street.
> Scientific discoveries in the last century have advanced the knowledge of humankind.
> Scientific discoveries in the last century have advanced the knowledge of all men and women.

"Gender" was long applied mainly in reference to the grammatical categories of masculine, feminine, and neuter. In recent years its use has been extended to refer to the social, economic, and historical categories man and woman, which are based mainly, though not entirely, on the sex of individuals, with "sex" referring to the biological categories.

7.5.2 Ethnicity and Race, Nationality and Citizenship, Religion

The term "race" does not have a precise definition in biological terms. When applied to human beings, its use depends mainly on judgments about physical characteristics that can differ widely among members of a so-called race. "Race" is also an infraspecific category that is applied to other organisms; it is defined on the basis of geographic range, physiologic traits, and other factors. "Ethnicity" is based on such factors as culture, history, nationality, and religion. Both "nationality" and "citizenship" describe a person's relationship to a specific country. Nationality may be defined by a variety of factors, including citizenship and place of birth; it should be distinguished from citizenship, which has a narrower connotation of legal status in relation to a sovereign state.

Whenever possible, descriptions of human populations or large social groups should draw on sharply definable criteria (e.g., country of birth or habitation or self-description). If ethnicity, race, citizenship, or religion is pertinent to a biomedical study (e.g., if the condition under investigation has a higher prevalence in a genetically related population than in the population at large, as for sickle cell anemia, or the study examines attitudes that can be traced to religious beliefs), this relevance should be justified explicitly in the study report and the method of measuring the variable stated (e.g., self-identification).[10] Kaplan and Bennett have presented more detailed guidelines on using race and ethnicity in biomedical publications.[11] Ethnic terms should not be used with pseudoprecision: "Caucasian" is no more scientifically precise than "white" and is considered archaic by some.

7.5.3 Disabilities and Health-Determined Categorization

The term "disability" refers to a condition that limits the ability of a person to carry out satisfactorily some common activity or function. "Handicap" is a judgmental term referring to an environmental or attitudinal barrier to usual functioning.

In referring to individuals, strive to retain their identity as persons and avoid such depersonalizing terms as "an alcoholic", "a diabetic", "an epileptic", "a schizophrenic". Instead, prefer "a patient with diabetes", "a child with epilepsy", "a boy with hemophilia".

7.6 EXCESSIVE ABBREVIATION

Much of the contemporary and growing difficulty in reading and understanding scientific literature arises from the ad hoc coining of abbreviations for noun phrases and heavy use of them. Their use where they are not needed in the context slows the reader and may even force the reader to return to the head of the text again and again for definitions of abbreviations. For example, this paragraph is problematic:

> Over a course of weeks to months, Amadori products undergo further rearrangement reactions to form fluorescent, cross-linking moieties called advanced glycosylated end products or AGEs. These products remain irreversibly bound to long-lived proteins such as collagen and accumulate as a function of age. Hyperglycemia accelerates the formation of protein-bound AGEs. Consequently, tissue AGE amounts increase rapidly in patients with diabetes mellitus.

Consider this revised version:

> Through weeks and months, Amadori products undergo rearrangement reactions to form fluorescent, cross-linking moieties called advanced glycosylated end products (AGEs). These new moieties accumulate and remain irreversibly bound to long-lived proteins such as collagen. Hyperglycemia accelerates their formation, and consequently their amounts in tissues increase rapidly in diabetic patients.

> [The coinage AGEs may be justified by some subsequent needs for textual or tabular economy, but many of the uses are not needed in the context of the entire article. Note also the potential ambiguity created by use of the abbreviation AGE in conjunction with the word "age".]

Such coinages clog current scientific literature. Unfortunately, many of them become implicitly sanctioned usages. See Chapter 11 for a detailed discussion of abbreviations.

7.7 UNNECESSARY WORDS AND PHRASES

Unnecessary words and phrases slow the reader and should be eliminated. Some expressions, such as "it is interesting to note that", add no information and can be omitted. Some constructions, such as the passive "It is reported by Smith", can be converted to the active form ("Smith reported"). Wordy phrases, such as those listed below, can be shortened.

Wordy	Concise
a majority of	most
a number of	many, several, some, numerous
absolutely essential	essential
accounted for the fact that	explained
along the lines of	like
an innumerable number of	innumerable, countless, many
an order of magnitude	10 times
are of the same opinion	agree
as a consequence of	because of
ascertain the location of	find
at a rapid rate	rapidly
at no time	never
at the conclusion of	after
at the present moment, at this point in time	now, currently
based on the fact that	because
bright green in color	bright green
by means of	by, with, using

Wordy	Concise
caused injuries to	injured
completely filled	filled
conducted inoculation experiments on	inoculated
definitely proved	proved
despite the fact that	although
due to the fact that	because
during the course of	during, while
during the time that	while, when
end result	result
fewer in number	fewer
for the purpose of examining	to examine
for the reason that	because
foreseeable future	future
future plans	plans
give rise to	cause
has the capability of	can, is able to
if conditions are such that	if, when
in a satisfactory manner	satisfactorily
in an adequate manner	adequately
in all cases	always, invariably
in case	if
in close proximity to	near
in connection with	about, concerning
in order to	to
in the course of	during, while
in the event that	if
in the near future	soon
in the vicinity of	near
in view of the fact that	because
is in a position to	can, may
it has been reported by Jones	Jones reported
it is believed that	[I, we] believe [or omit altogether]
it is often the case that	often
it is possible that the cause is	the cause may be
it is this that	this
it is worth pointing out that	note that
it is worth noting that	note that
it would thus appear that	apparently
lacked the ability to	could not
large amounts of	much
large in size	large
large numbers of	many
lenticular in character	lenticular
located in, located near	in, near
necessitates the inclusion of	necessitates, needs, requires
of such hardness that	so hard that
on account of	because
on behalf of	for
on the basis of	from, by, because
on the grounds that	because
original source	source
oval in shape, oval-shaped	oval
owing to the fact that	because
past history	history
period of time, time period	period
prior to [in time]	before
referred to as	called

Wordy	Concise
results so far achieved	results so far, results to date
round in shape	round
serves the function of being	is
smaller in size	smaller
subsequent to	after
take into consideration	consider
the fish in question	this fish, these fish
the question as to whether	whether
the treatment having been performed	after treatment
there can be little doubt that this is	this probably is
through the use of	by, with
throughout the entire area	throughout the area
throughout the whole of the experiment	throughout the experiment
two equal halves	halves
was of the opinion that	thought
with a view to getting	to get
with reference to	about [or omit]
with regard to	about, concerning [or omit]
with the result that	so that
within the realm of possibility	possible

Similarly, many modifiers can be omitted, especially if they convey a concept inherent in the unmodified term.

Careful hemodynamic monitoring is needed to prevent . . .
[Omit "careful" given that it is unlikely the author would suggest careless monitoring.]

Unnecessary modifiers are frequent in popular speech where they serve to emphasize, but they are not needed for clear meaning.

your belongings *not* your personal belongings
destination *not* final destination
fact *not* actual fact

7.8 ABSTRACT NOUNS

The frequent use of nouns formed from verbs and ending in "ion" produces unnecessarily long sentences and dull, static prose. Examples of such abstract nouns are "production" (from "produce") and "interpretation" from "interpret". Sentences become long because of the length of "ion" nouns and the need to use unnecessary prepositions and verbs with them. The dullness comes from the lack of action that would be stated by the "ion" noun's verb equivalent and the presence of the passive verbs that are needed. Writers can often remedy this fault and shorten the sentences by reworking them to introduce verb equivalents of abstract nouns. Such changes may also yield a clearer sequence.

Original: A direct correlation between serum vitamin B12 concentration and mean nerve conduction velocity was seen.

Revised version: The mean velocity of nerve conduction correlated directly with the vitamin B12 concentration in serum.

Replacing abstract nouns with their equivalent verbs may bring the subject into the sentence and make it more specific and vivid.

Original: Following termination of exposure to pigeons and resolution of the pulmonary infiltrates, there was a substantial increase in lung volume, some improvement in diffusing capacity, and partial resolution of the hypoxemia.

Revised version: After the patient stopped keeping pigeons, her pulmonary infiltrates partly resolved, lung volume greatly increased, diffusing capacity improved, and hypoxemia lessened.

7.9 ACTIVE VS. PASSIVE VOICE

In a sentence, voice refers to whether the subject of the sentence performs the action of the verb or receives the action of the verb.

In a sentence constructed with the active voice, it is clear that the subject acts.

Active voice: The chemist prepared the HPLC samples.

In this example, the phrase "the chemist" is the subject and the phrase "the HPLC samples" is the object.

In contrast, in a sentence constructed with the passive voice, the subject does not act but rather is acted upon by the verb.

Passive voice: The HPLC samples were prepared by the chemist.

In this example, the phrase "the HPLC samples" is the grammatical subject of the passive verb "were prepared".

Passive voice: The HPLC samples were prepared.

The passive voice always includes a form of the verb "to be" (e.g., is, are, was) and a past participle (e.g., "prepared", as in these examples). The actor may appear in a prepositional phrase beginning with the preposition "by" (as in the first example of passive voice above) or is not even mentioned (as in the second example of passive voice above).

The active voice is preferred because the sentence structure is more direct and the meaning of the sentence is easier for the reader to understand. The passive voice is best used when the emphasis is on a process taking place (as in the materials and methods section of a report) rather than on who is performing the action.

7.10 JARGON

The technical vocabulary of or the informal idiom used in a scientific field is jargon. If it has good etymologic bases, it may be acceptable in formal reports. However, when a word or phrase represents slang or obscures meaning for readers not familiar with the jargon of the field, it should be revised to yield clearer meaning. Nonanglophone readers may have a great deal of difficulty with English-language jargon.

Jargon can be of several types:

1) shortened forms of words that arise in conversation and may enter formal communication.

   The lab data confirmed the diagnosis. *for* The laboratory data confirmed the diagnosis.

2) verb-object relations that are ignored.

   We stocked trout in the stream. *for* We stocked the stream with trout.

3) nouns that are not used with their proper formal meaning.

   No pathology was found in the lung. *for* No abnormalities were found in the lung.

4) euphemisms that are used to soften harsh realities.

The patient expired. *for* The patient died.
The rats were sacrificed. *for* The rats were exsanguinated.

7.11 DIFFICULTIES FOR AUTHORS FOR WHOM ENGLISH IS A SECOND LANGUAGE

All languages have their own characteristic grammar and syntax, and their cultures have rhetorical conventions. An author who is not thoroughly familiar with English and whose culture has a different rhetorical style may have difficulty writing scientific papers in idiomatically acceptable English.

7.11.1 Grammatical and Syntactical Problems

7.11.1.1 Prepositions

The meanings of generally corresponding prepositions in English and other languages do not always coincide. The use of a preposition in a non-English language may not be idiomatic in English.

de [in Spanish]
of [in English]
soy de Barcelona *means* I am from Barcelona *not* I am of Barcelona

7.11.1.2 Position of Verb

In non-English languages the predicate (the verb indicating an action or state) may be properly placed at the beginning or end of a sentence, whereas in English it is idiomatically in a different position.

It was found by Smith (1974) important activity against some dimorphic fungi.
for Smith (1974) found important activity against some dimorphic fungi.

7.11.1.3 Progressive Tenses of Verbs

Some languages rarely use a progressive tense, whereas some variants of English, notably those in the Indian subcontinent and neighboring countries, may use them excessively.

We are finding fossil evidence that this formation belongs in the Cretaceous Period.
for We have found fossil evidence that this formation belongs in the Cretaceous Period.
or We have fossil evidence that this formation belongs in the Cretaceous Period.

7.11.1.4 Phrasal Verbs

Many nonanglophone authors are not familiar with English phrasal verbs, verbs usually of action or movement followed by an adverbial or prepositional particle. Phrasal verbs are sometimes used in scientific writing for a figurative or metaphoric meaning. Omitting the particle usually changes the meaning of the verb.

Ethnic tensions made it difficult to patch up community spirit. ["patch up" is used as a metaphorical equivalent of "repair"]
not Ethnic tensions made it difficult to patch community spirit. [omission of "up" distorts the metaphorical meaning of "patch up" in this context]

Many phrasal verbs cannot be found in standard English dictionaries but are included in *Oxford Phrasal Verbs*[12] and *The Oxford Dictionary of Idioms*.[13] Phrasal verbs tend to be used more in informal prose than in formal scientific writing. In general, nonanglophone authors should avoid their use and strive for more direct and less metaphoric statement.

7.11.1.5 Incorrect Use or Omission of Articles

Articles are sometimes omitted when needed in English.

> Physician should be able to communicate readily with patient.
> *for* The physician should be able to communicate readily with the patient.
> *or* A physician should be able to communicate readily with a patient.

The choice of "the" or "a" may depend on the rhetorical strength needed (see Section 7.3.2). The correct choice in this nuance in style results more from long experience in English idiom than from knowledge of syntactical rules. When a plural form implies that the noun represents a category rather than individuals, the article is not needed.

> Physicians should be able to communicate with patients.

7.11.1.6 Inappropriate Carryover of Gender from Other Languages

English nouns are generally assigned feminine or masculine gender only when they represent persons or animals of female or male sex, the type of linguistic gender known as natural gender. Many other languages assign grammatical gender to nouns without sex reference. Authors whose native languages use grammatical gender may erroneously carry it over into English.

> This new product was naftifine and his potency was similar to clotrimazole against strains of *Candida albicans*.
> *for* This new product was naftifine, and its potency was similar to that of clotrimazole against strains of *Candida albicans*.

Gender is assigned in popular speech and its print equivalent to some English nouns lacking sex reference (e.g., ships and countries), but this idiom is rare in scientific writing.

> The *Queen Mary* sailed from Southampton on Friday; by Monday, it was clear she would break the transatlantic speed record.

7.11.2 Vocabulary Problems

Many terms cannot be carried into English by apparently legitimate translation.

7.11.2.1 "False Cousins"

Many nouns in non-English languages resemble, or are cognates of, English nouns but carry different meanings. For example, the French "actuelle" means current (as in the title of a review of existing knowledge or practice), whereas in English "actual" can mean current but more frequently means existing or real, in contrast to potential, and hence would be unidiomatic in the title of a review of existing knowledge.

7.11.2.2 Transfer Coinages

Nonanglophone authors, especially those whose native tongue is a Romance language, may attempt to "translate" a noun into an apparently English equivalent, which, in fact, is not an English idiom.

> His work for many years was on the causalism of infectious diseases.
>
> ["causalisme" is a French term meaning "theory of causation"; "etiology" is an idiomatic English term in medicine for the study of causation]

7.11.3 Solutions for These Problems

Nonanglophone authors who are inexperienced in writing in English should seek review of their papers by one or more readers with a strong knowledge of English idiom before submitting them to an English-language journal; such review should take place late in the writing process, preferably after the scientific content of the paper seems satisfactory and in the right sequence.

Multilanguage scientific dictionaries, too numerous to cite here, are available for many fields.

CITED REFERENCES

1. Hansen WR, editor. Suggestions to authors of the reports of the United States Geological Survey. 7th ed. Reston (VA): United States Geological Survey; 1991 [modified 2005 Jan 25; accessed 2012 Nov 4]. 289 p. http://www.nwrc.usgs.gov/lib/lib_sta.htm.

2. McGraw-Hill dictionary of scientific and technical terms. 6th ed. New York (NY): McGraw-Hill; 2003.

3. The American heritage dictionary of the English language. 5th ed. Boston (MA): Houghton Mifflin; 2000.

4. Porta M, editor. A dictionary of epidemiology. 5th ed. New York (NY): Oxford University Press; 2008.

5. Maggio R. The dictionary of bias-free usage: a guide to nondiscriminatory language. Phoenix (AZ): Oryx Press; 1991.

6. The Chicago manual of style: the essential guide for writers, editors, and publishers. 16th ed. Chicago (IL): University of Chicago Press; 2010. Also available at http://www.chicagomanualofstyle.org.

7. Editing Canadian English. 2nd ed. Toronto (ON): Macfarlane Walter & Ross; 2000. Prepared for the Editors' Association of Canada.

8. Iverson C, Christiansen S, Flanagin A, Fontanarosa PB, Glass RM, Gregoline B, Lurie SJ, Meyer HS, Winker MA, Young RK. American Medical Association manual of style: a guide for authors and editors. 10th ed. New York (NY): Oxford University Press; 2007. Also available at http://www.amamanualofstyle.com/oso/public/index.html.

9. Publication manual of the American Psychological Association. 6th ed. Washington (DC): American Psychological Association; c2009.

10. International Committee of Medical Journal Editors. Uniform requirements for manuscripts submitted to biomedical journals: writing and editing for biomedical publication. Philadelphia (PA): ICMJE; [updated 2010 Apr; accessed 2012 Nov 4]. http://www.icmje.org.

11. Kaplan JB, Bennett T. Use of race and ethnicity in biomedical publication. JAMA. 2003;289(20):2709–2716. Erratum in: JAMA. 2004;292(9):1022.

12. Parkinson D, editor. Oxford phrasal verbs: dictionary for learners of English. Oxford (UK): Oxford University Press; 2001.

13. Siefring J. The Oxford dictionary of idioms. 2nd ed. Oxford (UK): Oxford University Press; 2004.

8 Names and Personal Designations

8.1 PERSONAL NAMES

It is simple courtesy to present personal names, titles, and honors in their correct forms. In the context of scientific publication, accurate presentation of names also helps ensure that author information is correctly tagged for electronic indexing; this in turn facilitates retrieval of documents and ensures that authors are appropriately credited for their publications.

For the formatting of bibliographic references, indexing, and other alphabetic arrangements, it is necessary to identify the surname (family name) and may be important to determine the full formal name. For personal references in text, it may be necessary to know the name, or the shortest sequence of names, that can be used after a title such as Doctor (Dr) or Professor (Prof).

In many countries, including the United States, Canada, the United Kingdom, and other countries in the European Union, the convention is to present a person's given name(s) before the family or patronymic name.

For family names consisting of a single element, it is relatively simple to identify the name to be used for formal purposes; for example, in the name Alena Rosemary Bird, the last element, "Bird", is the family name, and for indexing purposes the name would appear under the letter B.

For a hyphenated surname, the compound is used for formal purposes (e.g., John H Collins-Smith would be indexed under the letter C). However, some compound surnames are unhyphenated, in which case it may be difficult to correctly identify the surname.

For surnames that incorporate prefixes (also known as particles; e.g., Wernher von Braun), the convention of including or excluding the prefix in the surname varies by country or cultural tradition, which presents another difficulty in identifying the surname.

In some cultures, family names are positioned differently or the formal name may be determined in some other way; however, for professional purposes authors from these backgrounds may adopt the convention of placing the family name last. For example, some Chinese and Japanese authors present the family name last in their publications, even though the convention in China and Japan is to present the family name first.[1,2]

The ideal is to follow the preference of the named person. Therefore, in an author byline or similar context, clearly indicate the name or names that should be treated as the formal name, either by typography (e.g., all capital letters or initial capital and small capitals for the surname) or, for a journal article, by including the complete bibliographic citation, showing surnames and initials, on the first page of the article (see Section 27.7.1.5). To that end, authors should indicate clearly their preference in manuscripts submitted for publication and should follow that form consistently in all of their submissions. Editors dealing directly with authors should request this information if it is not provided.

In many other contexts, such as preparing or editing reference lists and indexes, the preferred form may not be obvious and it may not be possible to consult the person directly. Knowing how the person's name has appeared in the byline of previous publications may help but is not reliable. *The Chicago Manual of Style*[1] and other sources provide comprehensive guidelines for determining, on the basis of national origin, the name to be used as an entry element in bibliographic systems or indexes, but these sources may disagree on some recommendations. Furthermore, the treatment of certain parts of names, such as the prefixes "von", "de", and "los", differs according to the country or region of origin; for example, an English-speaking author with the name Robert De La Salle should be listed under D (De La Salle, Robert), but for a French-speaking person, the same name would appear under L (La Salle, Robert de). Because the author's country or region of origin is often not evident in a published document, it may be difficult to apply such conventions. See Table 8.1 for a selection of these conventions.

For the presentation of authors' surnames, the following simplified guidelines are suggested:

1) For author bylines, determine and follow the author's preference regarding the presentation of his or her surname, and indicate that preference clearly (e.g., by using all capital letters for the surname).
2) For bibliographic references and index entries prepared from published documents in which the author's surname (including or excluding any prefixes) is clearly indicated, follow the indicated form. See Section 29.3.6.1 for details on the presentation of author names in bibliographic references.
3) If the author's preference regarding presentation of the surname is unknown and cannot readily be determined, treat prefixes as part of the surname.

Designations such as Junior (Jr), Senior (Sr), the Second (II or 2nd), and the Third (III or 3rd) are part of the person's name; place them immediately after the rest of the name without a comma. A designation of this type precedes any academic degrees or honors.

Table 8.1 Selected guidelines for determining formal surnames for author attribution and indexing, according to country of origin[a,b]

Origin of name	Convention
Europe, except Hungary, Portugal, Spain	
Single family name (surname)	Use the last element of the name Bird, Rosemary Davidson, Jeff
Compound family name (surname)	Use both parts of the compound, whether or not they are hyphenated Carson-Peters, Henriette Bonham Carter, Mark
Compound family name containing prefixes	English-speaking countries and Italy: include the prefix De La Salle, Robert France: include the prefix, except "de" Beauvoir, Simone de La Salle, Marie Germany, Austria, The Netherlands: include the following prefixes as part of the surname: am, de, del, della, delle, des, di, du, l', la, las, le, les, li, los, ver, vom, zum, zur Di Giacomo, Roberto Ver Boven, Aja Omit the following prefixes from the surname: den, op de, ten, ter, van, van den, van der, von, von der Beek, Leo op de Denmark, Norway, Sweden: do not include the prefix if it is of Germanic origin (von, der); otherwise, include
Hungary	The family name precedes the given name, so the names need not be transposed, unless it can be determined that they have been inverted for use in an English context Bartók Béla (often presented as "Béla Bartók" in English contexts) *becomes* Bartók, Béla
Portugal	Use the last element of the family name Silva, Ovidio Saraiva de Carvalho e If the last element is a qualifier indicating family relationship (Filho, Neto, Sobrinho), include the next-to-last element as well Vidal Neto, Victor
Spain	For a compound family name consisting of the father's name followed by the mother's maiden name (or, if the person is a married woman, the preposition "de" and her husband's name), use all elements of the family name Perez y Fernandez, Juan If the surname consists of one name with a prefix that is an article (la, el, las, los), include the prefix and capitalize it Las Heras, Manuel If the prefix is a preposition or a preposition and an article, do not include the prefix Vega, José de la
India	In modern usage the family name is the last element; if "Sen" or "Das" precedes an Indian name, include it with the family name Sen Gupta, Bimal C

Table 8.1 **Selected guidelines for determining formal surnames for author attribution and indexing, according to country of origin[a,b] *(continued)***

Origin of name	Convention
China, Korea, Vietnam, Japan	Use the family name, which is usually the first element of the person's name; however, if it can be determined that the person has adopted the Western convention of putting the family name last, treat it as an English name Lee, Hon-Ling *for* Lee Hon-Ling Hu, J Donald *for* J Donald Hu
Arabic names	For a name without a prefix, use the family name (the last element) Khalil, Hassan Fahmy For the prefixes al- and el-, use the element that follows the prefix Hakim, Tawfiq al- Keep other prefixes (abu, abd, ibn) with the family name Ibn Auda, D

[a] Based on the *American Journal of Neuroradiology*,[5] *The Chicago Manual of Style*,[1] *The MIBIS Manual*,[6] and *The New York Public Library Writer's Guide to Style and Usage*.[7]
[b] See text for a simplified approach to be used if country of origin is not known.

Use such designations only with the full personal name (given names and family name) and in bibliographic references.

Patrick Elliott II MD PhD — Dr James Kelly Jr *not* Dr Kelly Jr
David Garrison Jr MB FRCPC — Kelly J Jr [in a bibliographic reference]

If a person's initials are needed to represent his or her given name(s), as in bibliographic references, use only letters that are capitalized when those given names are spelled out. Do not represent prefixes such as "von" or "da" in the person's initials.

William John Smith *becomes* Smith WJ *not* Smith Wm J
Maria Anna da Fonseca *becomes* Fonseca MA *or* da Fonseca MA *not* Fonseca MAD

8.2 ACADEMIC DEGREES, TITLES, AND HONORS

A wide range of degrees, honors, and other designations may be appended to personal names. These modifiers, usually abbreviated, indicate academic achievement, the recognition of peers in certain organizations, licensure, credentials, and other distinctions. See Table 8.2 for some examples of degrees and honors. More comprehensive lists can be found in the *Acronyms, Initialisms and Abbreviations Dictionary*[3] and the *American Medical Association Manual of Style*[4] (for medical degrees and honors). Many standard dictionaries include academic degrees and their abbreviated forms.

When degrees and other designations are presented after a personal name, use the abbreviated form without periods, capitalized according to Table 8.2 or another appropriate source. If in doubt about an abbreviation, check with the author or another appropriate source. Omit commas after the person's name and between consecutive designations.

For author bylines in research publications, omit such designations or include only academic degrees (note, however, that some publications may include designations for

Table 8.2 Abbreviations of selected academic degrees, honors, and other professional designations

Abbreviation	Term
BA	Bachelor of Arts
BC, BCh, BChir	Baccalaureus Chirurgia (Bachelor of Surgery)
BM	Bachelor of Medicine
BPharm	Bachelor of Pharmacy
BS, BSc	Bachelor of Science
BVSc	Bachelor of Veterinary Science
CB, ChB	Chirurgia Baccalaureus (Bachelor of Surgery)
ChD	Chirurgia Doctor (Doctor of Surgery)
CM, ChM	Chirurgia Magister (Master of Surgery)
DCh	Doctor of Surgery
DDS	Doctor of Dental Surgery
DHyg	Doctor of Hygiene
DM	Doctor of Medicine
DMSc	Doctor of Medical Science
DP, DPharm	Doctor of Pharmacy
DPhil	Doctor of Philosophy
DSc	Doctor of Science
DSW	Doctor of Social Work
DVM	Doctor of Veterinary Medicine
DVSc	Doctor of Veterinary Science
EdD or DEd	Doctor of Education
ELS	Editor in the Life Sciences
ELS(D)	Diplomate Editor in the Life Sciences
ELS(H)	Honored Editor in the Life Sciences
FAFPHM	Fellow of the Australasian Faculty of Public Health Medicine
FANZCA	Fellow of the Australian and New Zealand College of Anaesthetists
FCSSA	Fellow of the College of Surgeons, South Africa
FRACP	Fellow of the Royal Australasian College of Physicians
FRACS	Fellow of the Royal Australasian College of Surgeons
FRCP	Fellow of the Royal College of Physicians
FRCPC	Fellow of the Royal College of Physicians of Canada
FRCP(Edin)	Fellow of the Royal College of Physicians of Edinburgh
FRCPI	Fellow of the Royal College of Physicians of Ireland
FRCS	Fellow of the Royal College of Surgeons of England
FRCSC	Fellow of the Royal College of Surgeons of Canada
FRCS(Irel)	Fellow of the Royal College of Surgeons of Ireland
FRCVS	Fellow of the Royal College of Veterinary Science
FRS	Fellow of the Royal Society
FRSC	Fellow of the Royal Society of Canada
JD, JuD	Juris Doctor (Doctor of Jurisprudence *or* Doctor of Law)
LLB	Legum Baccalaureus (Bachelor of Laws)
LLD	Legum Doctor (Doctor of Laws)
MA	Master of Arts
MB	Medicinae Baccalaureus (Bachelor of Medicine)
MC, MCh, MChir	Magister Chirurgia (Master of Surgery)
MD	Medicinae Doctor (Doctor of Medicine)
MPH	Master of Public Health
MS, MSc	Magister Scientia (Master of Science)
PharmD	Doctor of Pharmacy
PhD	Philosophiae Doctor (Doctor of Philosophy)
RN	Registered Nurse
ScD	Scientia Doctor (Doctor of Science)
SM	Scientia Magister (Master of Science, Science Maître)
VMD	Veterinary Medical Doctor

licensure and credentials). Professional honors are appropriate in other contexts, such as biographical essays and obituaries. Titles of distinction awarded by governments to honor national service or especially meritorious accomplishments (e.g., OBE for Officer of the Order of the British Empire) may also be appropriate for news articles, biographical material, historical articles, and obituaries. If designations other than academic degrees are used, place abbreviations for academic achievement first, and present abbreviations for honors in increasing order of distinction.

Sequence academic degrees from lowest to highest academic level, but do not list a lower degree if it is a prerequisite for a higher degree (unless the degrees are in different subject areas or specialties). In the anglophone world, the following sequence is usual: bachelor's degree, master's degree, professional degree (e.g., DDS, MD, or PharmD), other doctoral degrees, and honorary degrees. Some publications drop the bachelor's degree if a master's degree or higher is obtained. If more than one degree at one level must be given, present them in the order in which they were earned, if known, and otherwise in alphabetic order.

Geraldine Parker PhD *not* Geraldine Parker BSc MSc PhD   Terrence Rolf MEd PhD

Use only one form to represent an academic degree: either a term of address before the personal name or the abbreviation representing the same degree after the name, but not both.

Dr James Wingler *or* James Wingler MD *not* Dr James Wingler MD
Dr Patricia R Cole *or* Patricia R Cole PhD *not* Dr Patricia R Cole PhD

Use the unabbreviated form when referring to a degree, honor, or other designation in more general terms.

A master of science degree is not necessarily a prerequisite for entry into a doctoral program.

For bibliographic and indexing purposes, omit academic titles that appear before a personal name (e.g., "Professor" or "Doctor") and academic degrees and other designations that appear after the name.

8.3 MILITARY TITLES

In the text of nonmilitary documents, it is preferable to spell out military titles denoting rank or position. Capitalize the unabbreviated forms of military ranks that precede a personal name. The Canadian and UK services use a hyphen for most compound ranks (e.g., Brigadier-General), whereas the US services do not (e.g., Brigadier General). See *Army Dictionary and Desk Reference*[8] for US military ranks and *The Canadian Press Stylebook*[9] for ranks in the Canadian services.

The award went to Captain Desmond Grover.
The researchers were greeted by Lieutenant Colonel Jennifer Ford.

Use abbreviations of military titles or ranks in tables, addresses, or military text or when many such titles must appear and space is limited. The abbreviation of a title or rank precedes the personal name, and that of a military service (if appropriate) follows the personal name. Use a comma to separate the name and the abbreviation representing military service.

COL Adrian Locke, MC, USA   LCDR Jeffrey Malone, USCG
CPT Susanna Fort, USA   Brig Gen Colleen Drysdale, USAF

The official abbreviations of ranks used by the military services may be difficult for a nonmilitary reader to understand; therefore, it may be appropriate to use a longer but more easily comprehended form. Suggested unofficial forms are in *The Chicago Manual of Style*.[1]

2nd Lt *instead of* 2LT *for* Second Lieutenant   Lt Gen *instead of* LG *for* Lieutenant General

8.4 NAMES OF ORGANIZATIONS

The complete name of an organization should be given accurately the first time it is mentioned in a document; it may subsequently be abbreviated to an initialism (e.g., USGS for the United States Geological Survey) or shortened by inversion, colloquialization, or truncation. With the exception of initialisms, these shortened forms need not be capitalized in subsequent references. However, for some audiences and types of publication—notably, documents of the named organization—capitalization may be retained in the shortened form.

On 1 July, the Survey will release the *Geological Atlas of Mars and Venus*. [in a publication of the US Geological Survey]
After mapping Mars, the survey issued the first planetary atlas. [in a science newspaper]

8.5 HUMAN GROUPS

Capitalize the names of ethnic groups, which are distinguished on the basis of cultural characteristics. When the name refers to a tribe or group as a whole, use the singular form; add "s" to the name when referring to a specific number of people (an exception is "Inuk", singular, but "Inuit", plural and adjectival). The adjectival form of a tribal name is the same as the name itself. Adding "n" or "an" at the end of the noun usually denotes the language group (e.g., the Mayan language group or the Piman language).

Group as a whole:
The Maya inhabited Tikal in AD 800.
The Chinook are one of the tribes that make up the Quinault Indian Nation.

Specific number of people:
Awaiting Cortés were 25 Mayas.
Smallpox killed hundreds of Chinooks in 1770.

Adjectival use:
The Maya calendar begins at the year equivalent to 3114 BC.
Chinook canoes were seaworthy.

Language:
The Mayan languages are numerous.
Speakers of Chinookan lived along rivers.

Capitalize the names of groups of humans if they are derived from proper names of geographic entities, if they are names for ethnically or culturally homogeneous groups living within a specific region, or if they are names for adherents to organized bodies.

Derived from geographic entity:
Nordic
Asian
New Yorker
Hispanic
Latino, Latina

Ethnically or culturally homogeneous group:
Kiowa
Eagle clan
Pygmy

Organized body:
Odd Fellow
Boy Scout
Allied Powers

Capitalize terms referring to indigenous peoples.

American Indians, Native Americans (United States)
Aboriginals, First Nations, Inuit, and Métis (Canada)
Aboriginals and Aborigines (Australia)

Use lowercase for designations based only on color or local usage, as well as for names representing scientifically ill-defined groups of wide distribution. Use such terms only if the characteristic is relevant to the topic under discussion (see also Section 7.5.2).

black   white   highlander

Do not hyphenate "American" in a compound proper noun or when "American" is an integral part of a 2-word term.

an African American   10 Irish Americans   North American species
the Mexican American   Latin American studies

Do hyphenate "American" or "Caribbean" in a compound proper adjective when the term reflects a geographic name rather than a general cultural entity and when it is combined with part of a word.

African-American students   Japanese-American art   Anglo-American exchanges
Polish-American influence   Afro-Caribbean

See Section 7.5.2 for a discussion of bias-free usage in references to ethnicity, race, nationality, citizenship, and religion.

8.6 SCIENTIFIC NAMES

Many authoritative scientific bodies in a variety of disciplines have developed style conventions to govern the formation of scientific names and the details of their proper usage. Table 8.3 lists chapters of this manual discussing nomenclature in various scientific disciplines.

8.7 GEOGRAPHIC NAMES

In general, a geographic entity should be referred to by the name accepted in the country where it is located. Online databases of official names are maintained by some government agencies (e.g., Geographic Names Information System[10] at http://geonames.usgs.gov/ for the United States and Geographical Names of Canada[11] at http://www.nrcan.gc.ca/earth-sciences/search/index_e.php). Alternatively, consult an authority such as *Merriam-Webster's Geographical Dictionary*[12] or *The Cambridge Gazetteer of the United States and Canada*.[13] Sometimes, a major feature or populated place has an anglicized name that has become accepted through long use. Unless there is a risk of misinterpretation, these anglicized versions are preferable in English text and in reference lists (see Section 29.3.6.5).

Florence *for* Firenze   Munich *for* München   Cologne *for* Köln

Table 8.3 Scientific names: relevant chapters in this manual

Subject	Chapter	Chapter title
Animals	22	Taxonomy and Nomenclature
Astronomical objects	26	Astronomical Objects and Time Systems
Bacteria	22	Taxonomy and Nomenclature
Chemical compounds	17	Chemical Formulas and Names
Chemical elements and their components	16	Subatomic Particles, Chemical Elements, and Related Notations
Chromosomes	21	Genes, Chromosomes, and Related Molecules
Diseases	24	Disease Names
Drugs	20	Drugs and Pharmacokinetics
Fossils	22	Taxonomy and Nomenclature
Genes	21	Genes, Chromosomes, and Related Molecules
Geographic entities	14	Geographic Designations
Particle physics	16	Subatomic Particles, Chemical Elements, and Related Notations
Plants	22	Taxonomy and Nomenclature
Rock formations	25	The Earth
Soils	25	The Earth
Viruses	22	Taxonomy and Nomenclature
Units of measure	12	Numbers, Units, Mathematical Expressions, and Statistics

In most scientific contexts, short forms of country names and of the names of constituent parts are sufficient to identify sites of observation or research. However, in some contexts (e.g., references to a government or its agencies), the full formal name of a country may be needed.

Islamic Republic of Afghanistan *for* Afghanistan
Hong Kong Special Administrative Region of the People's Republic of China *for* Hong Kong
Gabonese Republic *for* Gabon
Hashemite Kingdom of Jordan *for* Jordan

Many of these formal names can be found in the *Oxford English Reference Dictionary*.[14] Alternatively, the official government website of a country usually provides the complete name and should provide the most current name.

See Chapter 14 for more information about geographic names and Section 9.7.3 for information on capitalization of geographic terms.

8.8 TRADE NAMES

See Section 9.7.7 for a discussion of trade names and trademarks.

CITED REFERENCES

1. The Chicago manual of style: the essential guide for writers, editors, and publishers. 16th ed. Chicago (IL): University of Chicago Press; 2010. Also available at http://www.chicagomanualofstyle.org.

2. Sun XL, Zhou J. English versions of Chinese authors' names in biomedical journals: observations and recommendations. Sci Ed. 2002;25(1):3–4.

3. Romaniuk BR, editor. Acronyms, initialisms and abbreviations dictionary: a guide to acronyms, abbreviations, contractions, alphabetic symbols, and similar condensed appellations. 39th ed. 4 vols. Detroit (MI): Thomson/Gale; 2008.

4. Iverson C, Christiansen S, Flanagin A, Fontanarosa PB, Glass RM, Gregoline B, Lurie SJ, Meyer HS, Winker MA, Young RK. American Medical Association manual of style: a guide for authors and editors. 10th ed. New York (NY): Oxford University Press; 2007. Also available from http://www.amamanualofstyle.com/oso/public/index.html.

5. Castillo M. Authors' names in a globalized American Journal of Neuroradiology. Am J Neuroradiol. 2009;30(3):441–442, 2009. http://dx.doi.org/10.3174/ajnr.A1367.

6. Di Lauro A, Brandon E. The MIBIS manual: preparing records in microcomputer-based bibliographic information systems. 2nd ed. Ottawa (ON): International Development Research Centre; 1995. http://idl-bnc.idrc.ca/dspace/bitstream/10625/16224/1/101974.pdf.

7. The New York Public Library writer's guide to style and usage. New York (NY): HarperCollins Publishers; 1994.

8. Zurick T. Army dictionary and desk reference. 2nd ed. Harrisburg (PA): Stackpole Books; 1999.

9. Tasko P, editor. The Canadian Press stylebook: a guide for writers and editors. 13th ed. Toronto (ON): Canadian Press; 2004.

10. Geographic Names Information System (GNIS). Reston (VA): US Geological Survey; [accessed 2011 Nov 7]. http://geonames.usgs.gov/.

11. Geographical names of Canada. Ottawa (ON): Natural Resources Canada, Earth Sciences Sector; [accessed 2011 Dec 7]. http://www.nrcan.gc.ca/earth-sciences/search/index_e.php.

12. Merriam-Webster's geographical dictionary. 3rd ed. Springfield (MA): Merriam-Webster; 1997.

13. Hobson A, editor. The Cambridge gazetteer of the United States and Canada. Cambridge (UK): Cambridge University Press; 1995.

14. Pearsall J, Trumble B, editors. Oxford English reference dictionary. 2nd ed. Oxford (UK): Oxford University Press; 2002.

9 Capitalization

9.1 GENERAL GUIDANCE

Some capitalization conventions, such as those for capitalizing the names of people and cities, are widely known and accepted; others, such as those for position titles, are debated, and presentation depends on the preferences of individual authors and publishers. This manual recommends a spare style of capitalization, the use of an initial capital letter being restricted to instances in which it is clearly warranted. The present chapter includes some general recommendations for capitalization in English writing and a summary of the recommendations in particular scientific fields. More detailed guidance on English-language conventions is available in *The Chicago Manual of Style*.[1] Detailed recommendations on capitalization for specific scientific disciplines appear in the relevant chapters of Part 3 in this manual.

9.2 SYNTACTIC CAPITALIZATION

9.2.1 First Word of a Sentence

Capitalize the first word of every complete sentence, except when the sentence follows a semicolon or is within parentheses or brackets in another complete sentence.

> We delineated 6 study plots measuring 3 m × 3 m; no plots were adjoining.
>
> Subjects were assigned to groups on the basis of age and symptoms (sex was recorded for each person but was not a factor in determining group placement).
>
> Are there enough data to support the author's conclusions?

If the sentence begins with an accepted symbol that has an initial lowercase letter, restructure the sentence, if possible, to reposition the symbol. If that is not possible, retain the lowercase form, in which case it might be necessary to capitalize the first appropriate word.

The Student *t* test was applied to each set of data. *preferred to t* Tests were used in the analyses.
The pH must be carefully controlled. *preferred to* pH must be carefully controlled.

When starting a sentence with a scientific term or symbol that does not use a capital letter, the term or symbol should not be changed.

p53 control of CDK2 kinase activation is clear during apoptosis.
3T3 cells were originally obtained from Swiss mouse tissue.
c-*fms* is a protooncogene that encodes the M-CSF receptor.
rad3.d was transformed with the *nmt1* expression vector.

If the sentence begins with a chemical name that has a locant or other prefix, consider restructuring the sentence to reposition the chemical name, if this can be done without creating wordiness or changing the meaning; otherwise, capitalize the root term, not the prefix.

The solvent α-toluene can be used for this purpose.
For this purpose, α-toluene can be used.
α-Toluene can be used for this purpose.

Multiplying prefixes (uni, bi, tri, etc.) are treated as integral parts of the words they modify; capitalize them if they appear at the beginning of a sentence.

Univalent molecules are created in this reaction.
Bicarbonate concentrations in the blood were abnormally low.

If the first word of the sentence is a proper noun that usually begins with a lowercase letter, capitalize it.

We agree with the conclusion reached by du Pont.
but Du Pont argued this point in an earlier paper.

The recommendations outlined in this section also apply to footnotes in tables, even if the content of the footnote is not a grammatically complete sentence.

9.2.2 First Word after a Colon

Capitalize the first word after a colon if it begins a direct quotation.

The lead investigator addressed the nomenclature committee: "On the basis of new evidence, we propose that these 2 taxa be merged." [capital required]

The first word after a colon may be capitalized if the words that follow the colon form a complete sentence or an independent clause that does not logically depend on the preceding clause.

The evidence was enough to answer our question: Did these 2 fossil remnants represent the same organism? [capital optional]

If the words that follow the colon do not form a complete sentence, do not capitalize the first word.

Two alternatives for treatment were proposed: surgical resection or relief of pain through medication. [lowercase required]

9.2.3 Quotations and Excerpts

Capitalize the first word of a direct quotation that does not form part of the structure of a sentence, is complete in itself, and is formally introduced.

> The program director said, "We are now working to incorporate this new information into our budget."

However, do not capitalize the first word if the quotation is a syntactic part of the sentence.

> One professor stated her concern that "the quality of research is no longer the deciding factor".

Follow the same guidelines for excerpts from printed material. If the first word of the excerpt is capitalized in the original, retain the initial capital if the excerpt is not grammatically joined to what comes before. If the excerpt is a syntactic part of the sentence, use lowercase; in this situation, it is not necessary to enclose the lowercase letter in square brackets.[1]

> **Complete sentence excerpted, with formal introduction:**
> The review summarized the requirements: "To remain certified, the farm must undergo 2 inspections and laboratory testing of fish each year."

> **Complete sentence excerpted, forming integral part of sentence:**
> The review noted that "to remain certified, the farm must undergo 2 inspections and laboratory testing of fish each year."

Different guidelines regarding change of case are usually applied for legal and other works that require frequent reference to quoted material; these are described in detail in *The Chicago Manual of Style*.[1] For additional information about setting excerpts, see Section 10.2 in this manual.

9.2.4 Prefixes

Retain lowercase for such prefixes as anti, ante, ex, inter, mid, post, pre, semi, and un when connected to a proper noun that does not appear at the beginning of a sentence.

> mid-July   anti-American   post-Christmas

However, in some cases variants on this principle have become established. The prefix may be capitalized (e.g., Pan-American) or incorporated into a single capitalized word (e.g., Precambrian) or a single lowercase word (e.g., transatlantic). There are no rules for the words that have evolved in these ways; consult a dictionary when the proper form is in doubt.

9.3 TITLES OF WORKS AND OF PARTS OF WORKS

The following rules for capitalization of the title of a work (e.g., books, pamphlets, newspapers, magazines, poems, articles, and lectures) or a part of a work (e.g., chapters and sections) are referred to as "headline style" or "title case" and apply primarily when the title is used within running text. This form of capitalization is recommended to help the reader more readily distinguish a title from the adjacent text. This style may also be used for presenting a title at the beginning of a work or for headings within the text, although sentence-style capitalization is acceptable in these situations. More detail on headline-style capitalization is provided in *The Chicago Manual of Style*.[1]

For titles of works in reference lists, use sentence-style capitalization, in accordance with the spare style of capitalization recommended in this manual (see Sections 9.1, 9.4, and 29.3.6.2).

9.3.1 Titles of Works

9.3.1.1 Multiterm Titles

Capitalize the first and last words of a title, regardless of the part of speech. Capitalize all nouns, pronouns, verbs, adjectives, adverbs, and some conjunctions. Do not capitalize articles (a, an, the), coordinate conjunctions (and, or, nor, but), or prepositions unless they appear at the beginning of the title or are part of a phrasal verb.

The Beginnings of Life
Knowing Where to Look
To Tread Lightly on the Earth
Weight Loss with Exercise but without Dieting
Doing Without the Extras: An Approach to Fiscal Management

In titles with a marked break indicated by punctuation, capitalize the word that immediately follows the break, regardless of part of speech.

Hibernation in Canadian Mammals: A Review   Saturated Fats—Out of the Frying Pan

Do not capitalize the word "to" in infinitives within a title.

Learning to Read   To Move or to Stay: The Migratory Dilemma

Do not alter the form of locants and other similar prefixes of chemical names in titles, but capitalize the root terms.

β-Agonists in Respiratory Medicine   L-Erythrose and Related Sugars

In book reviews and similar settings, follow the capitalization of a title as it appears on the verso of the title page in the original of a printed work, on the opening screen of a work published as a video, or on the first page of a website. However, if the original title appears entirely in capital letters, do not use all capital letters; instead, follow the rules given in this section.

Capitalize both components of a 2-word hyphenated term if the term is a temporary compound (as in a modifier that would not otherwise be hyphenated) or a coordinate term.

Behavior of Well-Adjusted Children in the Classroom
Determining Acid–Base Status [where "Acid–Base" is a coordinate term with en dash linkage]
Nitrogen-Fixing Bacteria

Do not capitalize the second and subsequent components of a term that would normally be hyphenated, except for proper nouns or adjectives.

A Helter-skelter Approach
Frequency of Pre-eclampsia in an Urban Population
Attitudes toward Abortion among Non-Christian Women in New York

Capitalize scientific names as they would appear in running text (see Section 22.2.3.1);

hence, do not capitalize specific epithets (in botany and microbiology) or specific names (in zoology).

The Metabolism of *Escherichia coli* *Homo sapiens* and Predecessors

9.3.1.2 Single-Word Titles

In general, capitalize a single word serving as a title. For chemical names and hyphenated terms, follow the conventions in Section 9.3.1.1.

9.3.2 Titles of Parts of Works

For cross-references, capitalize the titles of parts of works according to the principles in Sections 9.3.1.1 and 9.3.1.2.

. . . according to criteria outlined in Materials and Methods . . .
. . . , who wrote the Preface for this volume, . . .
Temperature was the controlling factor (Table 1, Figure 3).

However, if the part is not referred to in its original form (e.g., if a modified form is used or the original form is used as an adjective), capitalization is not required.

A complete description can be found in the methods section [*or* in Materials and Methods].
This discrepancy is analyzed further in the discussion section [*or* in the Discussion].
. . . in the chapter on living with the chimpanzees . . . [cross-reference to a chapter entitled "Cohabitation with Chimpanzees"]

In running text, capitalize the title of a book chapter or journal article (according to the principles in Sections 9.3.1.1 and 9.3.1.2), and enclose it in quotation marks.

The last chapter, "Future Prospects", was originally published in *New Scientist*.

Capitalize nouns (or their abbreviations) used with letters or numerals representing their position in a series of similar parts of a book or chapter, including accessories to text.

Volumes 2–4, Vol. 5 Part A, Parts 3 and 4 Plate II, Plates I and II
Figure 6, Figs. 9–12 Table 3, Tables 6 and 7

Do not capitalize the word "line" or "page", and do not abbreviate these words in running text.

The photograph on page 678 of that article is a prime example.

9.4 REFERENCE LISTS

In some respects, the guidelines for capitalization in the list of references appearing at the end of a scientific publication differ from those used for other parts of such works. In particular, titles in end references are set in sentence case instead of title case. See Chapter 29 (especially Section 29.3.6.2) for recommendations on capitalization in end references and bibliographies.

9.5 INDEXES

Reserve capitalization in indexes for terms that would have an initial capital in running text, such as proper nouns. This should help the user of the index to distinguish common nouns from proper nouns—for example, in an index of scientific names for organisms—and nonproprietary drug or equipment names from trade names.

hexylresorcinol	[US Pharmacopeia nonproprietary drug name]
Hibiclens	[Mölnlycke RM Ltd. trademark for a brand of chlorhexidine gluconate]
Histalog	[Eli Lilly and Company trademark for a brand of betazole hydrochloride]
histidine	[International Nonproprietary Name (generic) for an amino acid]

Even if the book or journal editor chooses a style that calls for initial capitalization of every term in an index, the following terms should retain an initial lowercase letter:

1) compound surnames in which the prefix begins with a lowercase letter (du Pont, von Willebrand disease)
2) the italicized prefix of a chemical compound (*p*-Aminobenzoic acid, *o*-Toluic acid)
3) a standard symbol or abbreviation that begins with a lowercase letter (pH, pK′, mRNA)

See also Section 30.6 and its subsections.

9.6 TABLES

See Section 30.1 and its subsections for recommendations on capitalization in tables.

9.7 NAMES

9.7.1 Proper Nouns

Capitalize all proper nouns—including the names of people and places and the official names of organizations, institutions, and political entities—and some adjectives based on proper nouns. In running text, do not capitalize the article "the" preceding an organizational name, even if the article is part of the official name.

Albert Einstein
Vancouver, British Columbia
Council of Science Editors
Annual Meeting and Scientific Assembly of the Radiological Society of North America
Canadian weather patterns
The patient was admitted to the Hospital for Sick Children.

Use lowercase for truncated terms that eliminate the specifying element or elements and contain only a generic term, unless lack of capitalization would produce ambiguity in the particular context.

the hospital   the state   the institute

In a plural construction for organizational entities, use lowercase for the generic part of the names if it follows the specific names; capitalize the generic part if it precedes the specific names.

Harvard and Princeton universities
Toronto Western and Mount Sinai hospitals
the Departments of Botany, Forestry, and Zoology

Do not capitalize the name of a discipline or specialty except when it is part of a proper noun, such as the name of a department.

Students interested in pursuing a career in astrophysics . . .
The Department of Astrophysics will host a seminar . . .

Some organization names have, in addition to the conventional initial capital letter, one or more internal capital letters (e.g., GlaxoSmithKline). Observe these conventions, but in general convert names that use all capital letters to conventional capitalization.

9.7.2 Eponymic Terms

Capitalize the first letter of an eponymic term for a virus, disease, syndrome, named chemical reaction, or named equation that incorporates a proper noun referring to a theoretician, researcher, physician, patient, or place. Do not capitalize derivative or adjectival forms.

Bareggi reaction
Hodgkin disease
cesarean section
Hunter sore but hunterian chancre
Down syndrome
linnaean system of classification
fallopian tube
duct of Müller *but* müllerian duct
Gasser ganglion *or* gasserian ganglion
Noble agar
graafian follicle
Parkinson disease *but* parkinsonian tremor
Gram stain *but* gram-negative bacteria
Wilms tumor

For a discussion of the possessive forms of eponyms (and the recommendation to avoid this form), see Section 6.5.6.

Several kinds of eponymic terms should not be capitalized: terms established by decision of an authoritative body (such as the names of units); terms derived from proper nouns that relate to objects, especially apparatuses; and terms in well-established common usage. If in doubt, consult a dictionary or a resource appropriate to the subject matter.

ampere
angstrom
benday process
bessemer steel
bunsen burner
burley tobacco
congo red
coulomb
curie
draconian
gauss
india ink
joule
lambert
merino sheep
paris green
roman type
rutherford
timothy grass
petri dish
plaster of paris
venturi tube

Use lowercase for verbs derived from proper nouns.

italicize pasteurize

The terms "Anglophile" and "Francophile" may be capitalized for some readerships, but lowercase is preferred because of their adjectival form.

English written by francophone authors may incorporate confused translations of "false cousin" terms.

9.7.3 Geographic Designations

Capitalize the names of the great divisions (zones) of the earth's surface and the names of distinct regions or districts of the earth, but lowercase adjectives and some nouns based on these terms.

the Antipodes [in reference to Australia and New Zealand]; antipodean
the Great Divide
the Tropic of Cancer, the tropics
North Temperate Zone, the temperate zone
the Equator, equatorial

In most cases, do not capitalize the names of the points of the compass and derived adjectives. However, capitalize such a term if it is part of a name designating a specific area.

northern Saskatchewan
an eastern exposure
eastern hemisphere
traveled south
western songbirds
visible to the north of . . .
East Africa, North Africa, central Africa
the Midwest, midwesterner
Sherman's march to the sea drove a wedge through the South.
Southeast Asia, southeastern Asia, central Asia

Capitalize generic geographic terms (such as "lake", "river", "ocean", and "mountain") that form part of a proper noun, unless the generic noun and the word "the" precede the proper noun.

Atlantic Ocean | Fraser River | Mount Everest
Canadian Shield | Grouse Mountain | Nile River *but* the river Nile
Canoe Lake

Do not capitalize a generic term that follows a capitalized generic term.

the Fraser River valley, the Fraser Valley | the Rio Grande valley

In a plural construction, use lowercase for the generic noun that would be capitalized in the singular, unless the common noun precedes a group of proper nouns.

Great Slave and Lesser Great Slave lakes | Lakes Ontario and Huron
Vancouver and Saltspring islands

Capitalize the article "the" only when it is a formal part of a geographic name. Among countries, only The Bahamas and The Gambia use the article as part of the formal name.[2] For names of other geographic entities, consult a comprehensive atlas or geographic dictionary, such as *The Columbia Gazetteer of the World*[3] or *Merriam-Webster's Geographical Dictionary*.[4]

For further recommendations on geographic names, see Chapter 14 in this manual.

9.7.4 Taxonomic Names

Capitalize the Latin (scientific) name of a phylum, class, order, family, or genus or any subdivision thereof. Use lowercase for specific epithets (botany and microbiology) and specific names (zoology) and for designations of infraspecific taxa (see Section 22.2.3.1). In general, use lowercase for common or vernacular names of organisms, but see Section 22.2.4 for a discussion of this topic.

Lepidoptera | *Homo sapiens* | *Alnus viridis* ssp. *crispa*

Neither capitalize nor italicize adjectives and English nouns derived from scientific names.

salmonid | orthopteran | pneumocystis pneumonia

A few terms for organisms have more than one meaning, which may create difficulties in determining whether they should be capitalized. Capitalize "Metazoa" and "Protozoa" when they are used as the names of divisions, but use lowercase for the common-noun designations "metazoan" (plural, "metazoan") and "protozoon" (plural, "protozoa").

When the plural forms of these words are used with the article "the" (to represent the members of the division and not the division itself), use lowercase. A similar problem can arise with "Primates", the name of an order, and "primates", the members of the order.

Gorillas, chimpanzees, and baboons are representatives of the order Primates.
the adaptations of primates to communal living . . .

For further details on capitalization of the names of organisms and viruses, see Chapter 22. For use of capitalization in the representation of chromosomes and genes, see Chapter 21.

9.7.5 Geologic Names

In general, capitalize the accepted names for geologic and stratigraphic time units in formal usage, but the second element may appear in lowercase in running text (or may be omitted altogether when the first term is preceded by "the").

Archean Eon
Cenozoic Era
Pleistocene Epoch
disappeared during the Pleistocene and was not . . .

Capitalize the names of geologic formations only if they have been formally published according to the rules set out in the *North American Stratigraphic Code.*[5] Lists of names that have been published acceptably can be found in the *Lexicon of Canadian Stratigraphy.*[6–8]

See Chapter 25 for further information about capitalization of geologic names and related terms.

9.7.6 Astronomical Names

Capitalize the names of constellations, planets and their satellites, asteroids, stars, and other celestial objects.

the Milky Way
Mars
Solar System
the North Star (Polaris)
Halley's comet

Capitalize the words "sun", "earth", and "moon" when used in connection with other astronomical terms. In this context, do not use the definite article ("the") with "Earth". Use lowercase for more general contexts.

The Moon was the first destination for space travelers, and the feasibility of travel to Mars from Earth is now under investigation.
The sun provides energy for photosynthesis.

A rule of thumb for the word "earth" is to lowercase the word when preceded by the definite article ("the").

the 4 corners of the earth
the salt of the earth
The sequence of planets is Mercury, Venus, Earth, Mars . . .

Use lowercase for the adjectival form of nouns that are usually capitalized.

Jupiter, jovian
Moon, lunar

See Section 26.2 for more detail on capitalization of astronomical terms.

9.7.7 Trademarks

A trademark is a proprietary name for a product legally registered by the firm or the person making the product. Unless the particular brand is of interest, use the generic term. If the brand name must be used, capitalize the word.

> clear plastic, Plexiglas
> synthetic fiber or polyester, Dacron
> petroleum jelly, Vaseline

The trademark Aspirin is protected in some countries, including Germany, but is no longer protected in others, including Australia, Britain, New Zealand, and the United States. For publications produced in countries where the name is protected, it should be treated as a brand name for the drug acetylsalicylic acid (i.e., used only if specific mention of the brand is warranted and capitalized if used) (see Section 20.2).

Various databases are available to help in determining whether a name is a trademark.[9,10] The legitimacy of a trademark in the United States can be ascertained by using the Trademark Electronic Search System of the US Patent and Trademark Office.[10]

Designation of trademarks (® for marks that have been officially registered, ™ for marks that have not been registered but that the manufacturer wishes to identify as its own), although appropriate in marketing materials, is usually unnecessary in formal scholarly writing,[11] provided the name is presented correctly and the item to which it is applied is identified.

In an experimental study using a drug, identify the specific drug used by the proprietary name (capitalized) and manufacturer, as well as by generic name. To determine if a drug name is proprietary or to find the nonproprietary name, consult the *United States Pharmacopeia*[12] or the *Compendium of Pharmaceuticals and Specialties*.[13] See Chapter 20 in this manual for additional information about drug names.

Some trademarks have, in addition to the conventional initial capital letter, one or more internal capital letters (e.g., WordPerfect, MedEdit). Observe these conventions, but in general convert titles in all-capital letters to conventional capitalization.

9.7.8 Names for Abstractions

The names of abstract ideas or objects (including the seasons) may be capitalized when personified. Thus, "Mother Nature" is always capitalized, but "Nature" is capitalized only when the personification is clear.

> Three varieties are found in nature, but more have been developed in the laboratory.
> Living near the Arctic Circle, he learned that Nature can be a harsh taskmistress.

CITED REFERENCES

1. The Chicago manual of style: the essential guide for writers, editors, and publishers. 16th ed. Chicago (IL): University of Chicago Press; 2010. Also available at http://www.chicagomanualofstyle.org.

2. The Permanent Committee on Geographical Names for British official use. [London (UK)]: Permanent Committee on Geographical Names. Country names; 2012 Jan [accessed 2012 Feb 13]. http://www.pcgn.org.uk/Country_names.htm.

3. Cohen SB, editor. The Columbia gazetteer of the world. New York (NY): Columbia University Press; 1998. 3 vols.

4. Merriam-Webster's geographical dictionary. 3rd ed. Springfield (MA): Merriam-Webster; 2007.

5. North American Commission on Stratigraphic Nomenclature. North American stratigraphic code. Am Assoc Pet Geol Bull. 2005;(89):1547–1591. Also available at http://www.agiweb.org/nacsn/code2.html#anchor514748.

6. Glass DJ, editor. Lexicon of Canadian stratigraphy [CD-ROM]. Vol. 4, Western Canada. Calgary (AB): Canadian Society of Petroleum Geologists; 1981–1990.

7. Lexicon of Canadian stratigraphy [CD-ROM]. Calgary (AB): Canadian Society of Petroleum Geologists; 1981–1990. Vol. 1, Arctic Archipelago; Vol. 2, Yukon-MacKenzie; Vol. 6, Atlantic Region.

8. Lexique stratigraphique canadien [Lexicon of Canadian stratigraphy]. Québec (QC): Service Géologique du Québec; 1993. Vol. 5B, Région des Appalaches, des Basses-Terres du Saint-Laurent et des Îles de la Madeleine.

9. Office de las propriété intellectuelle du Canada [Canadian Intellectual Property Office]. Gatineau (QB): Industry Canada; [modified 2012 Oct 31; accessed 2012 Nov 2] http://www.cipo.ic.gc.ca/eic/site/cipointernet-internetopic.nsf/eng/h_wr00002.html.

10. Trademark Electronic Search System. Alexandria (VA): United States Patent and Trademark Office; [modified 2012 Feb 13; accessed 2012 Feb 13]. http://www.uspto.gov/trademarks.

11. Iverson C, Christiansen S, Flanagin A, Fontanarosa PB, Glass RM, Gregoline B, Lurie SJ, Meyer HS, Winker MA, Young RK. American Medical Association manual of style: a guide for authors and editors. 10th ed. Chicago (IL): American Medical Association; 2010.

12. The United States Pharmacopeia 36th and The National Formulary 31st (USP36-NF31). Rockville (MD): United States Pharmacopeial Convention; 2012.

13. Repchinsky C, editor. Compendium of pharmaceuticals and specialties. Ottawa (ON): Canadian Pharmacists Association; 2012. Annual.

10 Type Styles, Excerpts, Quotations, and Ellipses

10.1 TYPE STYLES FOR STYLE CONVENTIONS

Virtually all scientific literature in English is printed with a roman typeface, either a serif face, like that used for the text of this manual, or a sans serif face, such as Helvetica. Variants of the basic faces are used for specific conventions in prose style or for specific symbolic meaning. The 2 main variants are italics and boldface. Other conventions call for capital (uppercase) letters, small capitals, superscript characters (raised above the line of type), or subscript characters (positioned below the line); subscript and superscript characters are generally set smaller than the main text. See Chapter 31 for more detail on characteristics of type.

10.1.1 Italic Type

Italic type can be represented directly in manuscripts prepared with word processing programs. Some of the so-called italic typefaces in these programs are not true italic faces by typographers' criteria but represent "slanted type", a computer-produced slanted version of an upright serif or sans serif typeface. Nevertheless, the "italic typefaces" thus produced are acceptable representations for editorial offices and publishers.

10.1.1.1 General Uses

Seven uses of italic type are widely applied in published texts, whether general or scientific. Note that conventions 4 and 5 in the list below are not used in this manual to avoid any confusion with use of italics for special scientific conventions (see Section 10.1.1.2). Instead, terms representing words, illustrative terms, and some explanatory terms (see item 6 below) are enclosed in quotation marks.

1) In running text for the title of a book, journal, or other complete document. Some publishers also use italics for titles in bibliographic references, to make them more visible to the reader, but this practice has not been used in this manual (see Section 29.3.4). Quotation marks can be used to distinguish titles of book sections (such as chapters) from titles of whole books (italicized) and titles of journal articles from journal titles (see Section 5.3.4).

   His report, "The Gene Responsible for Multiple Sclerosis", was published in *The New England Journal of Medicine*.
   The chapter "Quotations and Dialogue" in *The Chicago Manual of Style* covers the topic in detail.

2) For many non-English words and phrases and their abbreviations, but not for non-English proper names. This rule does not apply to widely used phrases (such as "a priori" and "in vitro"), amputated phrases (such as "post hoc", representing *post hoc ergo propter hoc*, as in "a post hoc argument"), or abbreviations (such as "i.e." for *id est* [that is] and "et al." for *et alii* [and others]) that have been taken from Latin and are now considered standard English in science.
3) For a letter or number presented in text or a legend that refers to the corresponding character in an illustration, whether the label in the illustration is set in italic or roman font. The in-text letter should match the letter in the illustration with regard to capitalization.

   We could not account for the outlier *h* in Figure 3.

4) For a word or phrase when introduced in text for definition, explanation, or discussion.

   The concept of *stratum* took a long time to develop in the evolution of geology.

5) For a word or phrase that must be represented as such.

   The term *typeface* has given way in desktop publishing to *font*.

6) For explanatory words used within text that is in a roman typeface. This convention is often used for "see" and "see also" references in indexes.

   NIH *for* National Institutes of Health
   AIDS *for* acquired immunodeficiency syndrome

   [excerpt from an index]
   coronal suture 236
   coronary heart disease
     see ischemic heart disease
   corticotrophin 479

7) For a short preface or an explanatory note by a publisher or editor that must be clearly distinguished from the text of the author.

   [*The following short paper was submitted anonymously, but the editors felt it merited publication nonetheless.—Editor-in-Chief*]

10.1.1.2 Scientific Uses

The scientific uses of italics itemized below should be applied in all scientific publications, but this list may not be exhaustive.

1) For single-letter symbols standing for a quantity or a variable, known or unknown (see Section 4.2.2 and Table 4.5). This rule applies throughout science. Multiletter symbols for quantities may have to be in a roman typeface (see Table 4.5). For applications in mathematics and statistics, see Section 12.3.1 and Tables 12.9, 12.11, and 12.14; in chemical kinetics, Tables 18.1 to 18.4; in pharmacokinetics, Section 20.5 and Table 20.2; in cardiovascular and respiratory physiology, Sections 23.5.4 and 23.5.9 and Tables 23.3 and 23.4.

2) For single-letter modifiers (usually subscript) of symbols when the modifier itself is a symbol for a quantity (see Table 4.5).
3) For Greek letters symbolizing atomic particles (see Section 16.2 and Tables 16.1, 16.2, and 16.3. See also Section 4.1.2).
4) For the scientific name of a genus, species, subspecies, or variety. The name of a higher taxon is usually not italicized (see Sections 22.2.3 [general], 22.3.1 [plants], 22.3.2 [fungi, algae, lichens], 22.3.3 [animals], 22.3.4 [bacteria, archaea], and 22.3.5 [viruses]).
5) For letters and numerals used as a symbol for a gene or allele in most systems of gene symbolization (see Chapter 21).
6) For prefixes of chemical names designating structural and configurational relations (see Section 17.3).
7) For some crystallographic symbols (see Section 25.6).
8) For satellite names after christening (see Section 25.12.4).

Scientific text that should be in italic type (see item 4 above) may be changed to roman type when the surrounding text must be in italic type for some reason of design.

> Sablefish, *Anoplopoma fimbria*, is sensitive to handling but . . .
> [changes to] *Sablefish,* Anoplopoma fimbria, *is sensitive to handling but. . .*

In general, design requirements of this kind should be avoided in scientific publishing but may be needed for a level of text heading when many levels are required.

10.1.2 Boldface Type

There are no standard uses of boldface in general literature. Some publications use boldface for some elements of indexes (such as page numbers of the main text on a subject) and in bibliographic references (such as volume or year numbers, but see Section 29.3.4). Boldface can be used to indicate letters in terms that are the basis for an abbreviation that is an initialism, especially when one (or more) of the letters used is not the initial letter in a word.

> The abbreviation AIDS represents "**a**cquired **i**mmuno**d**eficiency **s**yndrome".

In a scientific context, bold is used for symbols for vectors in mathematical expressions (see Section 12.3.1).

10.1.3 Capital (Uppercase) Letters

Capital letters are usually typed as such in manuscripts and need not be specifically marked.

10.1.3.1 General Uses

The uses of initial capital letters are described in more detail in Chapter 9. In general, use initial capital letters in the following situations.

1) For the first word of a sentence unless the sentence is within parentheses or square brackets within an enclosing sentence (see Section 9.2.1).
2) For the first word after a colon if the following words form a complete independent clause not logically dependent on the preceding clause (see Section 9.2.2).
3) For trade names (trademarks; see Section 9.7.7).

4) For certain words in the titles of articles, journals, books, and other documents (see Section 9.3 and its subsections).
5) For the initial letter of the first word and for all proper nouns and proper adjectives in titles of articles and books in bibliographic references, whether abbreviated or not (see Section 29.3.6.2.1).
6) For a professional, civil, military (see Section 8.3), or religious title that immediately precedes a personal name.
7) For the official name of a private or government organization or institution (see Section 8.4).
8) For a generic geographic name that is part of a proper name (see Section 14.1.1.2).

In an author byline, use all capital letters (or initial capital and small capital letters) for surnames of Chinese and Japanese authors, to clearly distinguish the surname from the personal name, regardless of whether the surname appears first (see Section 8.1).[1]

Capital letters are widely used for contraction abbreviations whether the words represented are capitalized or not (see Section 11.1.1).

NIH *for* National Institutes of Health   DNA *for* deoxyribonucleic acid

Nouns, adjectives, and verbs derived from proper names are usually not capitalized when they have entered common usage (see Section 9.7.2).

sousaphone   petri dish   bunsen burner   pasteurize

10.1.3.2 Scientific Uses

In a scientific context, initial capital letters (or all capital letters) are used for the following purposes.

1) For the scientific (taxonomic) name of a phylum, class, order, family, or genus and their subdivisions, but not of a specific or subspecific taxon except where permitted by international codes (see Chapter 22).
2) For many gene, chromosome, blood group, cardiovascular, and respiratory physiology symbols and other symbols (see Chapters 21 and 23).
3) For the name of a formal historical epoch, geologic age or stratum, or zoogeographic zone, or other term used for convenience in classification (see Section 9.7.5 and Chapter 25).
4) For the proper name of a star or other astronomical body, but capitalize "earth", "moon", and "sun" only when used with the names of other bodies in the Solar System (see Sections 9.7.6 and 26.2).
5) For proper-name components of the vernacular or common names of organisms. See Section 22.2.4 for a detailed discussion of this topic, especially with regard to the recommendations of a few organizations to capitalize all elements of vernacular names of specific groups of organisms.

American black currant   Arizona white oak
Hawaiian bud moth   Virginia pine sawfly

10.1.4 Small Capital Letters

Small capital letters (often called "small caps") were generally designed to be the height of the letter "x" in a particular typeface and font. For typefaces that lack small capital fonts, they are often produced by an arbitrary reduction in the size of capital letters.

10.1.4.1 General Uses

Small capital letters have often been used for typographic variation, in an attempt to aid the reader in distinguishing some parts of text from each other. Examples include the use of small capitals for abbreviations for academic degrees (MD), eras (AD, BC), or time of day (AM, PM). This convention typically requires its own font type but does not convey any particular meaning beyond that of regular capital letters; it is therefore not recommended for scientific style.

10.1.4.2 Scientific Uses

Small capital letters have been designated in several conventions to indicate particular kinds of meaning.

1) Configurational prefixes (D, L) for amino acids and carbohydrates (see Sections 17.13.2.1 and 17.13.4, respectively).
2) Modifiers indicating the anatomic site of a gas phase for symbols of respiratory function (see Table 23.4).

10.1.5 Superscript and Subscript Characters

Alphabetic, numeric, and graphic characters in a superscript position (above the main line of type) are widely used in scientific notation as qualitative and quantitative modifiers of symbols on the line. These characters should be in a smaller font than that used for the main line of type. Some superscripts may themselves have superscripts. Table 10.1 illustrates some examples of superscripts and cites the sections of this manual that describe their use in detail.

Alphabetic and numeric characters in a subscript position (below the main line of type) are also widely used as qualitative and quantitative modifiers of symbols on the line. As for superscript characters, these characters should be in a smaller font than

Table 10.1 Examples of superscript symbols

Item represented	Symbol	Section or table
Celestial coordinate, right ascension	$14^h17^m13^s$	Section 26.3
Citation by reference number (citation–sequence, citation–name system)	. . . as previously reported[12]. . .	Section 29.2.4
Degrees, geographic	75°00′15″W	Section 14.2.1
Degrees, temperature	50°C	Table 12.4
Electric charge of a particle	p^+	Section 16.3
Electron spin multiplicity	$^2P_{½}$	Section 16.6
Integration limit, upper	$\int^\infty$	Table 12.11
Ionic charge	Na^+, Al^{3+}	Section 16.4
Mass number	^{14}C	Section 16.4
Oxidation number	Mn^{VII}	Section 16.4
Phenotype, wild type	Tol^+	Table 21.1
Power function	e^x	Table 12.9
Thermodynamic state	ΔG^o	Section 18.3.2

Table 10.2 Examples of subscript symbols

Item represented	Symbol	Section or table
Atomic number	${}_7\text{N}$	Section 16.4
Constant designation	K_i	Table 18.4
Derivative specified	$D_x y$	Table 12.11
Gas phase specified	$P\text{A}_{\text{CO}_2}$	Section 23.5.9, Table 23.4
Integration limit, lower	$\int_x$	Table 12.11
Member of chemical group	vitamin D_3	Section 17.13.19
Number of atoms in a chemical compound	H_2O	Sections 16.5, 17.2.1
Number of molecular units	$\alpha_2\lambda_2$	Section 23.5.6.1
Ring systems, chemical	$1_2,2_2{:}2_5,3_2$-terpyridine	Section 17.6.3.2
Scalar designation	F_x,F_y	Section 12.3.1
Thermodynamic quantities	C_p	Section 18.3.3, Table 18.1

that used for the main line of type. Table 10.2 illustrates some examples of subscripts and cites the sections of this manual that describe their use in detail.

Use italic type for superscript and subscript characters that serve as symbols for quantities when used alone. Use roman type for superscript and subscript characters that serve in these positions as qualitative (descriptive) modifiers.

10.2 EXCERPTS AND QUOTATIONS

Short passages taken from another text can usually be presented as quotations with relevant punctuation in the line of the text quoting them; such quotations are sometimes called "run-in quotations". Long passages may be more effectively presented as excerpts distinguished from the text into which they are inserted by special typographic presentation; these are often called "block quotations" or "set-off quotations".

10.2.1 Excerpts (Block Quotations, Extracts)

Various typographic devices can be applied to identify a block of text as quoted text. One pair of devices is indenting the quoted text from the left margin of the quoting text and using a smaller type font.

> Physicians in some fields may wonder why they should understand the fundamental principles of the management of fluid and electrolyte disturbances. Elkinton and Danowski made clear in their classic monograph, *The Body Fluids*[12], that any physician, no matter his or her specialty, will sooner or later have to deal with such problems.
>
>> Many disease states, which are otherwise unrelated, have certain features in common such as starvation, dehydration, vomiting, diarrhea, sweating, and renal dysfunction. These symptoms can exert profound influences upon the volume and composition of body fluids and solutes.
>
> But despite this point, many physicians still feel that . . .

If the text preceding the excerpt does not make clear its source, the excerpt should end with a parenthetical indication of source placed after the closing punctuation mark of the excerpt.

> . . . and composition of body fluids and solutes. (Elkinton and Danowski, *The Body Fluids*[12])

The selection of such devices is usually best made with the advice of a typographer or designer. A journal can have a standing style for excerpts, but a particular style may have to be developed for a particular book or report.

The excerpt should maintain the structure, typographic style, and content of the passage in its original form, according to the following rules.

1) If the original opened as an indented first sentence of a paragraph, keep the indent.

   The second paragraph of Lewin's text[17] goes directly to the heart of the matter.

   > The gene is the unit of genetic information. The crucial feature of Mendel's work, a century ago, was the realization that the gene is a distinct entity. The era of molecular biology began in 1945 when Schrödinger developed the view that the laws of physics might be inadequate to account for . . . its stability during innumerable generations of inheritance.

2) If the excerpt is from within a paragraph and opens with a complete sentence, do not indent the excerpt.
3) If the excerpt begins with a sentence fragment, do not add an indent or initial capital letter; ellipsis marks are not necessary (see Section 10.3).
4) The initial letter of the excerpt should follow the style of the original: capitalize the first word at the beginning of a sentence, use lowercase for the word following an ellipsis (unless it is properly capitalized, as for a proper noun).
5) Indicate original content that has been omitted from within the excerpt by an ellipsis. If the original content has been omitted and replaced with substituted equivalent text, enclose the replacement within square brackets.

   These symptoms can exert profound influences upon . . . body fluids and solutes.
   These [disorders] can [disturb] body fluids and solutes.

6) Excerpts in a non-English language should preserve the conventions of the original, including diacritics and similar devices, such as the guillemets ("chevrons") used in French as quotation marks. A translation within the excerpt of a word, phrase, or longer element of text should appear within square brackets after the translated matter.

Some exceptions to the general rule of excerpting the original exactly are reasonable.

1) Reference citations in the original may be omitted, but indicate the omission by an ellipsis.
2) An error that is obviously a typographic error in the original may be corrected, but idiosyncratic spelling and phrasing in historical sources should be preserved. If such idiosyncrasies might be interpreted by a reader as an error made by the quoting author rather than an accurate reflection of the original, add the term "sic" and the correct form, both within square brackets.

   and the physiological analyses of Barnard [sic, Bernard] are often pointed to as . . .

3) Preserve typographic devices used for specific scientific meaning, such as boldface for vectors in mathematical and physics texts and italics for species names, but words or terms appearing in bold (for emphasis) in the original may be presented in the normal weight of the typeface. If an author italicizes some of the quoted text for emphasis, that change should be explained by a phrase within square brackets.

   . . . and Schmidt *proved* [italics added] that Cohen's conjecture was . . .

10.2.2 Quotations

Mark run-in quotations with quotation marks and separate them from the other text by appropriate punctuation, to make clear the quoted matter and its relation to the rest of the text. Styles for run-in quotations differ between American and British usage. There are 2 main differences.

1) American style uses double quotation marks for a primary quotation and single quotation marks for a quotation within a quotation. It places the closing quotation mark at the end of the quoting sentence after the closing stop (period, exclamation mark).
2) British style uses single quotation marks for a primary quotation and double quotation marks for a quotation within a quotation. Closing quotation marks are positioned before the closing stop of the quoting sentence, unless the original quotation included the stop (i.e., placement according to sense).

In general, US and Canadian publications[2–6] follow the American style, whereas UK and Australian publications[7–9] follow the British style. The British style is more logical (and similar to the French style of quotation), but the American style is well established in North American publishing practices. Both styles are described below.

This manual recommends a hybrid form: use the American system for quotation marks (double quotation marks as the primary marks and single marks for a quotation within a quotation) and the British style for placement of quotation marks in relation to other punctuation (i.e., position other punctuation in accordance with sense).

10.2.3 Quotations within Sentences

Enclose the quoted text within primary quotation marks. Note that the 3 examples below all assume that the quoted sentence continues beyond the portion used in the quotation.

1) American style: Nininger[7] described "tiny droplets of melted country rock," and he felt he had tangible proof of an explosion.
2) British style: Nininger[7] described 'tiny droplets of melted country rock', and he felt he had tangible proof of an explosion.
3) CSE style: Nininger[7] described "tiny droplets of melted country rock", and he felt he had tangible proof of an explosion.

Enclose an internal quotation within secondary quotation marks. Note that the 3 examples below all assume that the quoted sentence continues beyond the portion used in the quotation.

1) American style: It is worth noting that "Nininger noticed tiny 'bombs' of what he described as yellow-green-brown slag," and we must recall that . . .
2) British style: It is worth noting that 'Nininger noticed tiny "bombs" of what he described as yellow-green-brown slag', and we must recall that . . .
3) CSE style: It is worth noting that "Nininger noticed tiny 'bombs' of what he described as yellow-green-brown slag", and we must recall that . . .

Treat an internal quotation within a block quotation as it would be in a within-sentence quotation.

Enclose a tertiary quotation (a quotation within a secondary quotation) within the primary quotation marks of either style; tertiary quotations are rarely needed in scientific writing.

If the quotation is a syntactical part of the quoting sentence, lowercase the first letter even if the quoted text begins with a capital in the original document.[2] If the quoted text begins with an initial capital and is not a syntactical part of the sentence, retain the initial capital (see also Section 9.2.3). Note that the 2 examples below both assume that the sentence is quoted in its entirety.

Peabody pointed out that "one of the essential qualities of the clinician is interest in humanity," and he went on with his famous aphorism about the care of the patient.
Peabody began by saying, "One of the essential qualities of the clinician is interest in humanity," and he then stated his famous aphorism.

10.2.4 Quotation Marks in Relation to Other Punctuation Marks

This manual recommends the British style for positioning of quotation marks in relation to other punctuation marks, whereby punctuation marks are positioned according to sense. If an extract ends with a period, comma, or other stop, the stop appears before the closing quotation mark; otherwise the stop appears after the closing quotation mark. *Hart's Rules for Compositors and Readers at the University Press, Oxford*[9] summarized this approach as follows: "All signs of punctuation used with words in quotation marks must be placed according to the sense."

He saw the ecological effects as "an utter disaster".
He commented that the ecological effects were "an utter disaster", and he left immediately.
His brief note read, "A disaster!"

If the quoted passage is a complete sentence ending with a period or exclamation mark and it appears at the end of the quoting sentence, that punctuation mark also serves as the closing punctuation for the quoting sentence; do not place another period after the closing quotation mark. If the quotation does not appear at the end of the quoting sentence the full stop (period) is omitted even if it was in the original.

At his retirement dinner he said, "My work is finished."
He ended his address with the statement "I am finished with this project" and sat down.

Although this manual does not recommend use of the American style in scientific literature, the following rules apply to the placement of closing quotation marks in relation to other punctuation marks in the American style.

1) Place closing quotation marks after a comma or period, even if the punctuation mark is not part of the quotation.

   He saw the ecological effects as "an utter disaster."
   He commented that the ecological effects were "an utter disaster," and he left immediately.

2) Place closing quotation marks after any other punctuation mark if it is part of the quotation.

   "A disaster!" was his brief note.

Closing quotation marks should be placed before a semicolon or colon; this rule applies to both the British and American styles.

He listed elements of "an adequate study design": statement of question, definition of . . .

Additional details on the American style can be found in *The Chicago Manual of Style*[2] and on the British style, in *New Hart's Rules*.[7]

10.3 ELLIPSES

In a run-in or block quotation, replace any material omitted from the quoted text with ellipsis points, on-the-line dots (periods) separated from each other and (usually) from adjacent characters by single spaces. Three systems for the application of ellipses are in general use; these are described in detail in *The Chicago Manual of Style*.[2]

The simplest method (which is practiced in this manual) is the 3-dot method, whereby no more than 3 ellipsis dots are used, regardless of whether material is omitted in the middle of a sentence or between sentences. In the 3- or 4-dot method (and its variant, the rigorous method that indicates adjustments in capitalization with brackets), 3 dots are used for an omission within a sentence and 4 dots for an omission of one or more sentences. In the latter situation, the first dot represents a closing period and is closed up to the last word before the ellipsis.

> He reported, "Laboratory findings . . . were similar for 3 patients."
>
> There are mutations that affect the ability of *E. coli* cells to engage in DNA repair. . . . The major known pathways are the *uvr* excision-repair system and the *dam* replication mismatch-repair system.

Retain punctuation preceding or following an omission if it will make the sense of the passage clearer.

> This is what he concluded, but only after much field work: the crater was probably not meteoric.
>
> This is what he concluded . . . : the crater was probably not meteoric.

Do not use dashes or asterisks to indicate omissions.

CITED REFERENCES

1. Sun XL, Zhou J. English versions of Chinese authors' names in biomedical journals: observations and recommendations. Sci Ed. 2002;25(1):3–4.

2. The Chicago manual of style: the essential guide for writers, editors, and publishers. 16th ed. Chicago (IL): University of Chicago Press; 2010. Also available at http://www.chicagomanualofstyle.org.

3. Editing Canadian English. 2nd ed. Toronto (ON): Macfarlane Walter & Ross; 2000. Prepared for the Editors' Association of Canada.

4. Iverson C, Christiansen S, Flanagin A, Fontanarosa PB, Glass RM, Gregoline B, Lurie SJ, Meyer HS, Winker MA, Young RK. American Medical Association manual of style: a guide for authors and editors. 10th ed. New York (NY): Oxford University Press; 2007.

5. Public Works and Government Services Canada, Translation Bureau. The Canadian style: a guide to writing and editing. Rev. and exp. Toronto (ON): Dundurn Press; 1997.

6. United States Government Printing Office style manual: main page. 30th ed. Washington (DC): The Office; 2008 [updated 2009 Apr 23; accessed 2012 Feb 21]. http://www.gpoaccess.gov/stylemanual/index.html.

7. New Hart's rules: the handbook of style for writers and editors. Oxford (UK): Oxford University Press; 2005.

8. Style manual for authors, editors and printers. 6th ed. Brisbane (Australia): John Wiley & Sons, Australia Ltd; 2002.

9. Hart's rules for compositors and readers at the University Press, Oxford. 39th ed. Oxford (UK): Oxford University Press; 1983.

ADDITIONAL REFERENCE

Bringhurst R. The elements of typographic style. 3rd ed. Point Roberts (WA): Hartley & Marks; 2004.

11 Abbreviations

11.1 FORMATION OF ABBREVIATIONS

English does not have formal and consistent rules for forming abbreviations, and many variations of abbreviations for the same terms can be found throughout the history of English prose. The punctuation of abbreviations is also subject to wide variation (see Section 11.2). The history and current practice of abbreviating are summarized in *The Oxford Companion to the English Language*[1] and *The Barnhart Abbreviations Dictionary*.[2]

The terms "abbreviation" and "symbol" are not mutually exclusive. A symbol is a notation used to represent something else, such as a variable, a concept, or a mathematical relation. Symbols may be formed by abbreviation, as with chemical symbols (e.g., Na, the chemical symbol for sodium, from the Latin *natrium*) and units of measure (e.g., kg for kilogram), or they may take graphic or numeric forms. However, many abbreviations have no symbolic function other than the representation of longer words or terms in a shorter form (see Section 4.2).

English-language abbreviations are generally of 3 types, with minor variations within each: contraction, suspension, and hybrid.

11.1.1 Contraction Abbreviations

In its simplest form, a contraction abbreviation consists of the initial letter of a word with or without one or more letters from within the word.

Celsius *becomes* C	Alberta *becomes* AB
species *becomes* sp.	isoleucine *becomes* Ile

A second form of contraction abbreviation uses the initial letters of 2 or more words, occasionally combined with internal letters from the words; such abbreviations are usually capitalized.

acquired immunodeficiency syndrome *becomes* AIDS
National Institutes of Health *becomes* NIH
quantitative trait locus or loci *becomes* QTL [Some writers use QTLs for the plural form.]
ethylenediaminetetraacetic acid *becomes* EDTA
also known as *becomes* aka
soluble ribonucleic acid *becomes* sRNA

An abbreviation of this type (an initialism) is referred to as an acronym (in viral nomenclature, the term "sigla" is used for certain initialisms; see Section 22.3.5.4) if it is a pronounceable word. Some of these terms eventually cease to be capitalized (e.g., laser for **l**ight **a**mplification by **s**timulated **e**mission of **r**adiation), but that is not the case for organization names (e.g., NASA for National Aeronautical and Space Administration).

11.1.2 Suspension Abbreviations

A suspension abbreviation consists of the initial and terminal letters of the word.

numéro [number] *becomes* no.
Doctor *becomes* Dr
Pennsylvania *becomes* PA

In the abbreviation "no." for "numéro", which is frequently used in English to mean "number", the closing period is required to avoid ambiguity with the word "no", the negative statement.

11.1.3 Hybrid Abbreviations

Hybrid abbreviations consist of the initial and terminal letters of the word plus one or more internal letters, with or without punctuation.

Mistress *becomes* Mrs.
continued *becomes* cont'd

11.1.4 Plurals of Abbreviations

To form the plural of abbreviations that do not contain periods, add only "s". If the abbreviated term is itself a plural, do not add the "s". In scientific writing do not add "s" to a symbol for a unit of measure.

MDs	CDC *not* CDCs *for* Centers for Disease Control and Prevention
PCBs	mm *not* mms
PhDs	kg *not* kgs
ELISAs	lb *not* lbs

Although the use of periods in abbreviations should be limited as much as possible (see Section 11.2), periods are sometimes needed to prevent misreading (e.g., "c.o.d." not "cod" for "collect on delivery"). In such cases, or if the abbreviation without periods forms a word that has another meaning and adding "s" alone would create confusion, use an apostrophe and "s" for the plural.

In some instances, the abbreviation of a multiword term conceals the fact that the term is a plural; in most such cases, the abbreviation may be treated grammatically as a collective noun.

The NIH [*for* National Institutes of Health] was one of the first organizations to comment on the situation.

11.2 PUNCTUATION AND TYPOGRAPHY

In recent years, American practice has generally been to place a period (full stop) after an abbreviation of a single word (e.g., Dr. for Doctor) and to omit periods in abbreviations of multiword terms (e.g., NAS for National Academy of Sciences). British practice has tended to be the reverse (e.g., Dr for Doctor, I.U.P.A.P. for International Union of Pure and Applied Physics). CSE recommends eliminating unnecessary periods in abbreviations to bring about convergence between these 2 styles, to save authors' and editors' time, and to eliminate keystrokes. This approach has been recommended by various authorities, including *The Chicago Manual of Style*.[3]

The following guidelines are recommended.

1) In general, avoid periods (full stops) within or at the end of an abbreviation.

   Dr *not* Dr. PM *not* P.M. PhD *not* Ph.D. IUPAP *not* I.U.P.A.P.

2) Retain periods if they are required by the nomenclatural recommendations of a scientific organization, as for the scientific names of organisms (see Chapter 22).

   species *becomes* sp. [singular], spp. [plural], *not* sp, spp
   Staphylococcus aureus becomes *S. aureus*

3) Omit periods after initials standing for personal names, as in the formats for end references (see Section 29.3.6.1).

   Mazan MR, Hoffman AM. Effects of aerosolized albuterol on physiologic responses to exercise in standardbreds. Am J Vet Res. 2001;62(11):1812–1817.

4) Omit commas before and after abbreviations representing parts of personal names (e.g., Jr, 3rd). Likewise, omit commas before and after abbreviations for academic degrees and honorific titles that are readily recognized as such by readers (see Section 8.2).

   The members included Russell Tomar PhD and Thomas Smith Jr MD to represent . . .

5) In general, capitalize all letters of abbreviations of multiword terms (acronyms or initialisms), even if the unabbreviated term is not usually capitalized in running text. Avoid the use of small capitals.

   AM for ante meridiem [before noon]
   AIDS for acquired immunodeficiency syndrome
   DNA for deoxyribonucleic acid

6) Set abbreviations in roman (upright) type, unless the abbreviation must follow a formal nomenclatural convention that is different. Specifically, use italics for abbreviations of terms that would be italicized in unabbreviated form.

   The dominant bacteria, *S. aureus* and *E. coli*, were identified by microscopy.
   not
   The dominant bacteria, S. *aureus* and E. *coli*, were identified by microscopy.

11.3 USAGE

11.3.1 General Rules

Avoid abbreviations in running text except for those that are widely used in the particular context of the document. However, some commonly used abbreviations are not always appropriate in running text; for example, avoid lowercase abbreviations at the start of a sentence and do not use symbols for units of measure if not preceded by a number.

The nomenclatural commission was headed by Ralph Schowitz PhD and his colleague.
Observations began at 12:01 AM and lasted 3 d.
Morning changes [*not* AM changes] in eating habits were observed each day [*not* each d].
We isolated mtDNA. *not* mtDNA was isolated.

For an unabbreviated term that is long or complex, use of an abbreviation may be helpful. Use standardized and widely recognized forms whenever possible, rather than nonstandard variants.

Texas A&M University *not* Texas AM University

Define most such abbreviations at first use in the text of a document and in tables and figures (such that the tables and figures may be understood without reference to the text); alternatively, a list of abbreviations used may be given at the beginning of each article in a journal. Some abbreviations are so well known in science (e.g., DNA) or in general use (e.g., AIDS) that an editorial office may allow them to appear without definition. However, because the same abbreviation may be used to mean different things in different fields or even within the same field,[4] the intended meaning must be unambiguous from the context.

LSD *for* lysergic acid diethylamide *or* least significant difference
SARS *for* severe acute respiratory syndrome *or* SARs *for* structure activity relationships *or* SAR *for* systemic acquired resistance in plants
ALS *for* acetolactate synthase *or* amyotrophic lateral sclerosis

Consult a comprehensive dictionary of abbreviations, either general or discipline specific, to determine accepted forms.

11.3.2 Scientific Usage

The enormous, accelerated growth of scientific knowledge in recent decades has generated great numbers of new multiword terms. Their length and the pressures for economy in using journal space have led to the widespread use of abbreviations. Some of these abbreviations are fresh coinages and do not represent established usage; others are well known, but only in relatively narrow disciplines. Abbreviations are usually justified on the grounds that they make the text easier to read, yet some long unabbreviated terms may be more readily recognized than a newly coined abbreviation. As such, the use of less widely known abbreviations can impede reading for all but those thoroughly familiar with the topic. Furthermore, nonstandard forms and newly coined abbreviations may present problems for users searching for a particular term in documents presented in electronic format (i.e., the article may not be captured). For guidance, the following recommendations are offered (see also Section 7.6).

1) Abbreviations widely known throughout science, such as DNA for deoxyribonucleic acid, may be used in titles, abstracts, and text without definition. A helpful criterion for decisions on such use is whether the abbreviation has been accepted into thesauruses and indexes widely used for searching major bibliographic databases (e.g., the medical subject headings [MeSH] thesaurus of the National Library of Medicine[5]).
2) Abbreviations not widely known throughout science and not carried in the index to a major bibliographic database relevant to the subject and scope of the journal should not appear in article titles. If such abbreviations are well established in the discipline

represented by a specialized journal, they may be used in its abstracts and text without explanation.

3) Abbreviations not acceptable by criteria 1 or 2 above must not be used in titles and abstracts but may be used in text, tables, and illustrations if they are parenthetically defined at first use. However, if an abbreviation such as this is not needed more than a few (perhaps 3 to 5) times in the following text, it should not be used; the term it represents should be spelled out, or the text in which it appears should be rewritten to eliminate the need for abbreviation.
4) Abbreviations, especially those allowed by criterion 3 above, should be avoided if the meaning would be clear without the abbreviation. Consider the following original and revised version of text; the revised version defines the abbreviation EST but uses it only where it is essential to understanding.

Original: The publication of expressed sequence tags (ESTs) derived from randomly selected clones from commercially obtained brain complementary DNA libraries[1,2] has received much attention . . . Initially we examined one EST of interest to us (EST01828, similar to the *Drosophila* homeobox gene *otd*) and found that it appears to contain an intron. Subsequently, it seemed that there were several ESTs reported by . . .

Revised version: The publication of expressed sequence tags (ESTs) derived from randomly selected clones from commercially obtained brain complementary DNA libraries[1,2] has received much attention . . . Initially we examined one tag of interest to us (EST01828, similar to the *Drosophila* homeobox gene *otd*) and found that it appears to contain an intron. Subsequently, it seemed that there were several tags reported by . . .

The broad scope of this manual precludes the presentation of comprehensive lists of definitions for abbreviations relevant to all of the fields covered in Part 3. Nonetheless, acceptable and recommended abbreviations are discussed in each chapter.

11.4 ABBREVIATIONS OF LATIN TERMS

Many Latin terms have been retained in the vocabulary of scholarly publishing, reflecting the time when all scholarly literature was in Latin. Many of these terms are typically used in abbreviated form, especially in footnotes, endnotes, and parenthetical statements in text. Several of these abbreviations are well established in the scientific literature, especially "i.e." for *id est* (that is), "e.g." for *ex genera* (for example), and "et al." for *et alii* (and others), and are used in this manual. If these abbreviations are used, ensure that the intended meaning is conveyed by using the correct form. In particular, some authors confuse "i.e." and "e.g." Alternatively, these may be replaced in text by their English equivalents.

Extensive compilations of scholarly abbreviations (both English and Latin) can be found in *The Chicago Manual of Style*[3] and the *MLA Handbook for Writers of Research Papers*.[6]

Some Latin terms are used in scholarly contexts without abbreviation: infra (below), supra (above), and sic (as written). Such usage is particularly frequent in the medical and biological literature: in utero (within the uterus), in vitro (performed or occurring outside the living organism), and in vivo (performed or occurring within the living organism). Do not italicize Latin terms that have become accepted in English and that can be used in scientific text without translation. Inclusion of a term in an English-language dictionary, either general or specific to the discipline, is one indication that a term meets these criteria.

CITED REFERENCES

1. McArthur T. The Oxford companion to the English language. Oxford (UK): Oxford University Press; 1995.

2. Barnhart RK, editor. The Barnhart abbreviations dictionary. New York (NY): John Wiley & Sons; 1995.

3. The Chicago manual of style: the essential guide for writers, editors, and publishers. 16th ed. Chicago (IL): University of Chicago Press; 2010. Also available at http://www.chicagomanualofstyle.org.

4. Davis NM. Medical abbreviations: 32,000 conveniences at the expense of communications and safety. Huntingdon Valley (PA): Neil M Davis Associates; 2011.

5. Medical subject headings. Bethesda (MD): National Library of Medicine (US); 1999 [updated 2011 Oct 28; accessed 2012 Feb 25]. http://www.nlm.nih.gov/mesh/.

6. Gibaldi J, Franklin P. MLA handbook for writers of research papers. 7th ed. New York (NY): Modern Language Association of America; 2009.

ADDITIONAL REFERENCES

Acronyms, initialisms & abbreviations dictionary. 46th ed. Detroit (MI): Thomson Gale; 2012.

The American heritage abbreviations dictionary. 3rd ed. Boston (MA): Houghton Mifflin; 2005.

Pitman LM, editor. Buttress's world guide to abbreviations of organizations. Rev. ed. London (UK): Blackie Academic & Professional; 1997.

International encyclopedia of abbreviations in science and technology yearbook 2005/2006. Berlin (Germany): De Gruyter Saur; [date unknown].

Jablonski S. Jablonski's dictionary of medical acronyms and abbreviations. 6th ed. Philadelphia (PA): Saunders/Elsevier; 2009.

Kotyk A. Quantities, symbols, units, and abbreviations in the life sciences: a guide for authors and editors. Totowa (NJ): Humana Press; 1999.

The Oxford essential dictionary of abbreviations. New York (NY): Berkeley Books; 2004.

Stedman's abbrev: abbreviations, acronyms & symbols. 3rd ed. Baltimore (MD): Lippincott Williams & Wilkins; 2003.

12 Numbers, Units, Mathematical Expressions, and Statistics

12.1 NUMBERS

A number is "an arithmetical value, expressed by a word, symbol, or figure, representing a particular quantity and used in counting, calculating, and showing order in a series or for identification".[1] The variety of ways numbers can be presented (e.g., 547.2, 6 million, 1.54×10^6, 1/4; $7\ \mathrm{g \cdot m^{-1} \cdot d^{-1}}$) creates the need for formatting guidelines. The guidelines given here are widely applicable. In instances where they are not, common sense and editorial judgment must be used to preserve clarity, precision, logic, and consistency within any particular document.

Bailar and Mosteller[2] and Higham[3] provide broad discussions of numeric presentations in text. The general international standard for expressing quantities and units is Standard 80000 of the International Organization for Standardization.[4]

12.1.1 Arabic and Roman Numerals

Numerals are the symbols that are used alone or in combination to represent numbers. There are 10 arabic numerals: 0, 1, 2, 3, 4, 5, 6, 7, 8, and 9. In most scientific contexts, these numerals are combined according to the decimal system, which expresses numbers in base 10. One of these 10 digits is used in each place, and each place value is a power of 10 (e.g., 0.45, 13, 352). In some contexts, another base may be used, such as the hexadecimal system (base 16) in some computer applications.

Roman numerals are not used for presenting data but do appear in some situations, such as pagination of certain parts of books or journals or numbering within certain classification systems. Roman numerals consist of combinations of 7 letters, each of which has a specific value (Table 12.1). A given letter may appear up to 3 times within a roman numeral, and the corresponding values are summed. For values that would entail 4 or more of the same letter, the roman numeral is formed by combining multiple letters. When a letter of smaller value follows one of equal or higher value, the values are summed. When a letter of smaller value precedes one of higher value, the smaller value is subtracted from the larger.

XXI = 21 XXIV = 24 XXVI = 26 XLIV = 44 iv *not* iiii

Pages in the front matter of books and other publications are often numbered with lowercase roman numerals: chapters or sections are usually indicated with uppercase roman numerals.

12.1.2 Numerals or Words—Modern Scientific Number Style

The conventions presented here revise what has often been called the "scientific number style". That style generally used words for single-digit whole numbers and numerals for larger numbers, a distinction many people found arbitrary. The revised or modern scientific number style treats numbers more consistently by extending the use of numerals to most single-digit whole numbers that were previously expressed as words. This style allows all quantities to be expressed in a similar manner, and because numerals are more distinctive than words, it increases the profile of quantities in running text. Emphasizing quantity with numerals is also made easier by using words for numbers

Table 12.1 Values of roman numerals

Roman numeral[a]	Arabic equivalent	Roman numeral[a]	Arabic equivalent
I, i	1	C, c	100
V, v	5	D, d	500
X, x	10	M, m	1,000
L, l	50		

[a] A bar over a letter multiplies its value by 1,000; for example, $\overline{\text{V}}$ = 5,000. Upper- and lowercase letters should not be combined in a single expression.

that are only secondarily quantitative, such as in idioms (e.g., the sixty-four-dollar question).

The numbers zero and one present additional challenges. For these numbers, applying consistent logic (numerals for quantities and words otherwise) often increases the difficulty of determining correct usage and creates an inconsistent appearance, primarily because "one" has a variety of functions and readers might not quickly grasp the logic. For example, "one" can be used in ways in which quantity is irrelevant: as a personal pronoun or synonym for "you" (e.g., "one must never forget that") or as an indefinite pronoun ("this one is preferred"). "Zero" and "one" are also used in ways that are more like figures of speech than precise quantifications (e.g., "in one or both of the . . .", "in any one year", "a zero-tolerance policy"). In addition, the numeral "1" can be easily confused with the letters "l" and "I", particularly in running text, and the value "0" can be confused with the letter "O" or "o" used to designate a variable. Therefore, simplicity and consistent appearance have been given priority for these 2 numbers (see Section 12.1.2.1, items 3 and 4).

12.1.2.1 Cardinal Numbers

Use numerals rather than words to express whole and decimal numbers in scientific text, titles, headings, tables, and figure captions. This practice increases their visibility and distinctiveness and emphasizes their enumerative function.

3 hypotheses	7 samples	52 trees
328 amino acids	4 times	0.5 mm

Also, use numerals to designate mathematical relationships, such as ratios and multiplication factors.

5:1   at 100× magnification   3-fold *not* threefold

Use words to represent numbers in 4 categories of exceptions:

1) If a number begins a sentence, title, or heading, then spell out the number or, if possible, reword so that the number appears elsewhere in the sentence or join the sentence to the previous sentence.

   Twenty milligrams is the desired amount, but 15 mg is enough.
   The desired amount is 20 mg, but 15 mg is enough.
   The drug is administered in a single dose; 20 mg is the desired amount, but 15 mg is enough.

2) When 2 numbers are adjacent, spell out one of the numbers and leave the other as a numeral; in general, retain as a numeral any number that occurs with a unit of measurement. A better option may be to reword the sentence to separate the numbers.

   The sample was divided into eight 50-g aliquots *not* The sample was divided into 8 fifty-gram aliquots.
   The sample was divided into 8 aliquots of 50 g each.

3) For most general uses, spell out zero and one.

   zero-based budgeting
   values approaching zero
   the zeros were . . .
   one of the subspecies
   at one time
   on the one hand

one doctor
in one such instance
Using a leave-one-out procedure . . .
. . . were bimodally distributed, one peak at 10 to 13 cm, the other at 24 to 28 cm.
In supporting scientific ethics, one is obliged to . . .
Of the possible avenues of research, this one is the most promising.
Here was one alternative that we should have examined.

However, express the whole numbers zero and one as numerals when they are directly connected to a unit of measure or when they specify assigned or calculated values.

1 year	1 J	when β is less than 1	when *z* = 0
1 mm	0 A1-digit numbers	with *q* fixed at 1	a mean of 0

Similarly, express zero and one as numerals when they are part of a series or are closely or intermittently linked with other numbers.

Series: 0, 1, 5, and 9 were . . .
Closely linked: 1 of 4 subspecies
2 applications instead of 1 were . . . between 0 and 2
Intermittently linked: Of these, 3 samples were . . . , and 1 sample was . . . The last 5 were . . .

4) When a number is used idiomatically or within a figure of speech, spell it out. However, figures of speech may be inappropriate because they may not be readily understood by readers whose first language is not English. Recasting the phrase is generally the better option.

Idiomatic Prose	**Rewording**
This fact tells us a thing or two about	This fact tells us several things about
Of two minds	Undecided
A thousand and one possibilities	Innumerable possibilities
The one and only reason that	The only reason *or* The specific reason
Third-world countries	Developing countries

A number may be used in such a way that the exact numeric quantity is secondary to the overall meaning. In scientific writing, reword to avoid the number altogether. Otherwise, use either the word or the numeral.

Idiomatic Prose	**Rewording**
Among the four of us	Among our group *but* Among our group of 4
We three	We
The two of them	Both of them
One or two of these	Some of these
An additional week or two of growth	An additional 1 to 2 weeks of growth

The decision between idiomatic and enumerative uses may not always be clear. For example, in the phrase "We deleted those five data points", "five" could be considered more descriptive (indicating *which* data points) than enumerative and as such may be better expressed as a word, although the numeral form could be used, if preferred. In contrast, in the phrase "We deleted 5 data points", the "5" is clearly enumerative (it indicates *how many* data points, but not which ones in particular) and therefore should be expressed as a numeral. The word "the" or "those" immediately preceding a number generally indicates ambiguous situations such as these.

12.1.2.2 Ordinal Numbers

Ordinal numbers generally convey rank order, not quantity. Rather than being expressly enumerative (answering the question "How many?"), ordinals often describe "which", "what", or "in what sequence". Because this function of ordinals is more prose-oriented than quantitative, distinctiveness within the text is less important for ordinal numbers, and undisrupted reading flow and comprehension take precedence. Potential confusion between the numeral "1" and the letters "l" and "I" is also a consideration.

1) Spell out single-digit ordinals (corresponding to the numbers 1 to 9) used as adjectives or adverbs.

   the ninth time a third wave of immigrants
   were first discovered the first ducklings emerged

2) The numeric form of 2-digit ordinals (corresponding to the numbers 10 and higher) is less likely to reduce comprehension, and the practice of using the numeric form for such ordinals is well established. Therefore, express these larger ordinals as numerals.

   for a 10th time the 98th test run the 19th century

3) Express single-digit ordinals in numeric form if they appear in a series or are intermittently linked with larger ordinals.

   The 5th, 8th, and 10th hypotheses were tested *not*. . . fifth, eighth, and tenth *or* 10th
   [Note: rewording to "We tested hypotheses 5, 8, and 10" might be better.]
   We developed 12 hypotheses . . . tested the 1st . . . The 11th was . . .
   not We developed 12 hypotheses . . . tested the first . . . The eleventh was . . .

4) To provide visual cues to comprehension, express single-digit ordinals in the numeric form if they are used repeatedly.

   Of those 6, we first examined the 4th subject, who . . . Next, we looked at the 5th subject . . . We finally returned to review the 1st, 2nd, and 3rd subjects . . .

   Although the general policy for ordinals dictates that words be used here, the numeric form provides more distinction for the references to the individual subjects. "Subject 1", "subject 2", and so on, would accomplish the same thing. The numeric ordinals also enhance contrast with the adverbial use of "first" in this example. Whichever style is chosen in this situation—numeric ordinals or the spelled-out form—use it consistently.

12.1.2.3 Fractions

In general, spell out fractions in running text. Hyphenate all 2-word fractions, whether used as adjectives or nouns.

one-half *or* half of the subjects nearly three-quarters of the population
a third of the study plots a two-thirds majority

For fractional quantities greater than 1, use mixed fractions if the precise value is not intended. Set a built-up fraction (e.g., $^{1}/_{3}$; or a special character, ⅓) close to the whole number, but insert a space if the fraction is not built up (e.g., 2 5/8).

was followed for $3^{1}/_{3}$ years about 1¼ km away approximately 2 5/8 yards

When the precise value must be conveyed, the decimal or percent form is preferred.

3.5 L 27% of the . . . a study area measuring 1.25 × 3.0 km

12.1.3 Format of Numbers

12.1.3.1 Breaks within a Number

Separate every three digits with a comma except in numbers after a decimal. Use a period (not a comma) as a decimal point.

3,968 145,000 243,568,981 3.14 0.32147 47,938.275

For certain numbers, a comma between digits is inappropriate—for example, US ZIP codes, address numbers, and telephone numbers in some countries (see Section 5.3.3.2).

12.1.3.2 Zeros before Decimals and Decimals after Integers

For numbers less than 1.0, always use an initial zero to the left of the decimal point. The initial zero removes any ambiguity about possibly erroneous omission of a digit before the decimal point; it also improves the readability of numbers in tables. Also, see Section 12.5 for this requirement in statistical reporting. Never follow an integer with a decimal point. A period does, of course, follow an integer ending a sentence.

0.497 *not* .497   *P* = 0.04 *not P* = .04   74 dogs *not* 74. dogs

12.1.3.3 Significant Digits

Numerals in text, tables, and figures should usually reflect the precision with which the number was derived. In estimating numbers of ducks within flocks, a biologist might count in units of 10 individuals, so estimates of the average flock size should reflect that level of precision. Likewise, instruments used to derive numeric measurements measure accurately to some finite level, and that level should be reflected in the numbers provided (e.g., a duck's wingspread might be accurately measured in centimeters and should be expressed in the same units [76 cm], not tenths of a centimeter [75.8 cm]; body mass determined by means of a scale that is accurate to 0.1 g should be expressed in tenths of grams [5.4 g], not hundredths [5.43 g]).

Express calculated values based on these raw measurements (e.g., means and standard deviations) to not more than one significant digit beyond the accuracy of the original measurement. In the case of duck wingspread measured to the nearest centimeter, the mean might be mathematically calculated as 75.3333 cm but should be rounded to 75.3 cm or even 75 cm for reporting purposes. For the average size of duck flocks measured in tens of individuals, the standard deviation of mean flock size would be reported in whole numbers (e.g., 34, not 33.5).

In reporting a quantity, the number of significant digits must be commensurate with scientific importance, as well as the precision of the measurement. It is unnecessary and distracting to report significant digits that have little scientific meaning or import. For example, percentages expressed to 1 decimal place may be acceptable if sample sizes are large (at least 100); otherwise, round them to whole numbers.

12.1.3.4 Rounding Numbers

Round a calculated quantity (e.g., a conversion to SI units) to the appropriate number of significant digits (see Section 12.1.3.3 for a discussion of appropriate level of precision). The following rules and examples illustrate the rounding of a number in which 3 significant digits are retained.

1) If the digit to the right of the third digit is less than 5, leave the third digit unchanged.

   4.282 becomes 4.28

2) If the digit to the right of the third digit is 5 or greater, increase the third digit by 1.

   4.286 becomes 4.29   4.275 becomes 4.28

3) Rounding is not the same as truncating, in which unwanted digits are simply deleted without changing the remaining number. Report whether values have been rounded or truncated.

   4.286 rounded becomes 4.3 but truncated, it becomes 4.2

When communicating numbers, round them to 2 significant digits unless the science requires more precision. Round numbers only for reporting purposes, not during analysis. For an alternative system, consult "Guide to the Rounding of Numbers", Annex B in ISO standard 31-0.[5]

12.1.4 Units and Symbols with Numbers and Ranges

12.1.4.1 Numbers Combined with Units of Measure or Symbols

Symbols may have an alphabetic basis (e.g., mm, kV, g, qt, ft) or a nonalphabetic basis (e.g., %, $, ≥, ±). Generally, neither type of symbol should be used without an accompanying numeral. Rules differ for separating numerals from alphabetic and nonalphabetic symbols.

1) Separate a number from a following alphabetic symbol with a single space.

   the overall length of 130 mm
   closure occurred 106 s after

2) Generally, close up a number and a nonalphabetic symbol, whether the symbol precedes or follows the number (except if the symbol is being used as a mathematical operator). In a few cases, nonalphabetic symbols are separated from the number by a space (as in degrees Celsius; see Section 12.2.1.1, item 1).

   =3 mm $98 12° [for angles]
   44% ≤5 *but* $P < 0.001$ 12 °C [for temperature; see Section 12.2.1.1, item 1]

3) Geographic coordinates are closed up over the entire coordinate. Designating latitude and longitude is optional because the compass direction identifies such (see also Section 14.2).

   lat 38°45′N 38°45′N
   long 77°12′26″E 77°12′26″E

4) If the number and associated symbol or unit start a sentence, write out the complete term in words.

   Five milliliters of supernatant was extracted. *not* Five mL of supernatant was extracted.
   We extracted 5 mL of . . . *not* We extracted five milliliters of . . . *or*. . . 5 milliliters of . . .
   Fifty-seven percent of the samples were contaminated. *not* Fifty-seven % of the samples were contaminated.

12.1.4.2 Numeric Ranges, Dimensions, Series, and Placement of Units

1) When expressing a range of numbers in text, always use the word "to" or "through" to connect the numbers, not a hyphen or a dash. In tables, when space is limited, 2 numbers in a range can be separated with a hyphen if they are not also separated by words, mathematic operators, or symbols; this avoids confusion with the minus symbol. (One exception is the page range in a reference, where the context makes such confusion unlikely.)

   yielded −0.3 to +1.2 differences *not* −0.3–+1.2 differences
   7 June to 15 June *not* 7 June–15 June
   (7 to12 aliquots) *not* (7–12 aliquots)
   from 240 to 350 participants *not* from 240–350 participants

2) The hyphen or dash represents only the word "to", so when the word "between" precedes a range, use the word "and" (not "to" or an en dash) between the numbers.

   between 1 and 12 June *not* between 1–12 June

3) When the range includes numbers of several digits, do not omit leading digits from the second number in the range. (The exception is indicating page ranges in references, where omitting duplicate leading digits is an option.)

   1938 to 1954 *not* 1938 to 54 1466 to 1472 km *not* 1466–72 km

   For page ranges in reference citations, use a hyphen or a dash instead of the word "to" (see Section 29.3.6.9).

4) A range of numbers and the accompanying unit can be expressed with a single unit symbol after the second number of the range, except when the symbol must be closed up to the number (e.g., the percent symbol). Alternatively, the unit symbol may be presented with both numbers of the range. Whichever style is selected, use it consistently.

   23 to 47 kV *or* 23 kV to 47 kV
   50 to 250 W/m^2 *or* 50 W/m^2 to 250 W/m^2
   2 to 7 °C *or* 2 °C to 7 °C [see also Section 12.2.1.1, item 1]
   10% to 15% *not* 10 to 15%

5) If a range begins a sentence, spell out the first number and write the second as a numeral, or recast the sentence. Generally, any accompanying units will appear only after the second number and may be abbreviated. If the unit must also appear after the first number (as for the percent symbol), then write out both units or recast the sentence.

   Twenty-three to 25 km . . . *not* Twenty-three to twenty-five kilometers . . .
   [Alternative: The test range was 23 to 25 km]
   Twelve to 15 g was added, *not* Twelve to fifteen grams was added.
   Seven percent to 11 percent of all samples . . . *not* Seven percent to 11% of all samples
   [Alternative: Of all the samples, 7% to 11% . . .]

6) To prevent misunderstanding, avoid using the word "by" before a range; this construction may convey an amount of change from an original value, rather than a range of values.

   Growth increased 0.1 to 0.3 g/d [a range] *not* Growth increased by 0.1 to 0.3 g/d. [to avoid implication that growth increased by 0.1 g/d to a new level of 0.3 g/d]

   Use "from" to indicate an initial and a final value.

   Growth increased from 0.1 g/d to 0.3 g/d

7) To prevent misinterpretation, be careful when expressing 2 numbers preceded by words such as "increase", "decrease", or "change". In those situations a range may be intended but the reader might misinterpret the first value as an initial value and the second as the new value.

   increased from 10 g/d to 18 g/d [Was the increase 10 to 18 g or was the increase 8 g?]

   Unless context makes the meaning unambiguous, qualification may be needed.

   increased by a range of 10 to 18 g/d *versus* increased from an initial value of 10 g/d to a final value of 18 g/d

   When changes are from one range to a new range, placing an en dash or hyphen within each range may be easier for readers to interpret quickly.

   increased from 10–23 mm to 12–27 mm *not* increased from 10 to 23 mm to 12 to 27 mm

8) For dimensions, use a multiplication symbol (not a small letter "x") or the word "by" to separate the measurements (also see Section 12.2.1.2, item 6).

   10 × 55 × 5 mm 10 mm × 55 mm × 5 mm 10 by 55 by 5 mm

9) For a series of numbers, present the unit after the last numeral only, except when the unit symbol must be set close to the number.

 12, 17, 43, and 66 kV
 diameters of 5 and 9 mm
 categories of <3, 3–7, and >7 g [However, use symbols to substitute for words only in parenthetical statements: "the total was less than 7 g" *not* "the total was < 7 g"; *but* "(total < 7 g)".]
 $15, $22, or $31
 38%, 55%, and 29%

12.1.4.3 Rates

Express simple rates (1 or 2 units) with a slash and express complex rates (more than 2 units) with negative exponents. Using negative exponents for simple rates is not incorrect (see Sections 12.2.1.1, item 4, and 12.2.1.2, item 11), but it involves unnecessary keystrokes. Some units are defined in the form of rates (e.g., A/m, a measure of magnetic field strength); such units should not be written with negative exponents.

 256/h [see Section 12.2.1.2, item 4]
 $0.15\ g{\cdot}m^{-1}{\cdot}d^{-1}$
 5 A/m *not* $5\ A{\cdot}m^{-1}$ [because A/m is a single unit; see above]

12.1.5 Scientific Notation

In scientific writing, express very large or very small numbers in powers of 10 (scientific notation).

 2.6×10^4 *not* 26,000 4.23×10^8 *not* 423,000,000 7.41×10^{-6} *not* 0.000,007,41

When expressing a range of values in scientific notation, write out both limits of the range in full or enclose the range within parentheses.

 2.6×10^4 to 9.7×10^4 *or* $(2.6$ to $9.7) \times 10^4$ *not* 2.6 to 9.7×10^4

If the limits of the range have different exponents, both limits must be shown in full.

 3.7×10^2 to 5.9×10^6

Present a mean of 7.4×10^3 with a standard deviation of 0.4×10^3 as follows:

 7.4×10^3 (SD, 0.4×10^3) *or* 7.4 (SD, 0.4) $\times 10^3$ *not* 7.4 (SD, 0.4×10^3)

For large numbers that are not expressed with high precision (i.e., those with a number of trailing zeros, such as 3,000,000), the text form may be a combination of numerals and words, particularly if the numbers do not represent experimental quantities. However, do not use such expressions with scientific units of measure.

 3 million people $13.9 million 1.5×10^6 km *not* 1.5 million km

The combination of numerals and words can cause confusion because words for numbers greater than 1 million can have differing meanings. For example, the term "billion" means 1,000 million (1,000,000,000) in the United States but may mean 1 million million (1,000,000,000,000) in the United Kingdom.[6] In documents that may have an international readership, write the number out completely or use scientific notation if appropriate.

12.1.6 Monetary Units

For monetary sums, use the appropriate currency symbol and either a combination of numerals and words or the full numeric form. To avoid potential confusion, use numerals alone for amounts above millions (see Section 12.1.5).

$245 million $245,000,000

To designate thousands, millions, or larger amounts, do not mix units of currency with the prefixes of the SI system or with scientific notation.

$58,000 *not* $58K *or* $5.8 × 10^4

Maintain this distinction in tables, illustrations, and other places where abbreviated forms are appropriate (see Chapter 30).

$, millions $, 000,000   millions of $ *not* $M *or* $, M *or* M$

Close up nonalphabetic currency symbols to the numeral; for alphabetic-only currency abbreviations, separate the abbreviation from the numeral with a space. Table 12.2 provides symbols of many national currency units.[7]

$749   £749   F 749   DM 749   Col$749   €749

Many nations use the dollar as their unit of currency. Generally, introduce the dollar symbol being used by including the country qualifier or prefix on the first usage; if 2 or more dollar currencies are being used, include the prefix with each currency usage (see Table 12.2).

A$749 [Australia]   Can$749 [Canada]
NZ$749 [New Zealand]   US$331 [United States]

The pound sterling symbol (£) may be similarly prefixed to designate the pound currencies of various countries.

12.2 UNITS AND SYSTEMS OF MEASURE

12.2.1 Système International (SI)

In scientific writing, metric measure is the accepted form of expression for physical and chemical quantities. The currently recommended system is the Système International d'Unités (the International System of Units, abbreviated SI), as developed by the International Committee for Weights and Measures (the Comité International des Poids et Mesures or CIPM). Additional information on SI units and conversions is available through the Bureau International des Poids et Mesures[9] and other publications.[10–12]

The SI provides a coherent system of measure constructed from 7 base units, each of which is precisely defined (Table 12.3). From these base units, several additional units have been mathematically derived, some of which have been given special names and their own symbols (e.g., pascal, joule, hertz; Table 12.4). Other SI-derived units have not been assigned names of their own and are known only by algebraic combinations of the base units (e.g., square meter or m^2 for area, cubic meter per second or m^3/s for flow, amperes per meter or A/m for magnetic field strength) or of the named derived units (e.g., pascal second or Pa·s; joule per mole Kelvin, or J/(mol·K); newton meter, or N·m).

Table 12.2 Symbols for selected currencies[a]

Country	Monetary unit[b]	Symbol: Common	Symbol: Replaced by[c]
Argentina	Argentine peso	$	
Australia[d]	Australian dollar	A$, $A	
Bahamas	Bahamian dollar	B$	
Belgium	Belgian franc	BF	euro, €
Bermuda	Bermuda dollar	Bd$	
Brazil	Brazilian real	R$	
Canada	Canadian dollar	Can$	
Chile	Chilean peso	Ch$	
China	yuan renminbi	¥	
Colombia	Colombian peso	Col$	
Costa Rica	Costa Rican colón	₡	
Denmark	Danish krone	DKr	
Dominican Republic	Dominican peso	RD$	
Ecuador	sucre	S/	
Egypt	Egyptian pound	£E	
Finland	markka	mk	euro, €
France	French franc	F	euro, €
Germany	deutsche mark	DM	euro, €
Greece	drachma	*Dp*	euro, €
Guyana	Guyana dollar	G$	
Hungary	forint	Ft	
India	rupee	Rs	
Ireland	Ireland pound	IR£	euro, €
Italy	Italian lira	£	euro, €
Japan	yen	¥	
Mexico	Mexican peso	Mex$	
New Zealand[d]	New Zealand dollar	NZ$, $NZ	
Norway	Norwegian krone	NKr	
Peru	sol	S/	
Poland	zloty	Zl	
Russia	ruble	R	
Spain	pesata	Pts	euro, €
Sweden	Swedish krona	Sk	
Switzerland	Swiss franc	CHF	
Taiwan	Taiwanese dollar	NT$	
Turkey	Turkish lira	TL, T£	
United Kingdom[e]	pound sterling	£ stg.	
United States of America[f]	United States dollar	$, US$	

[a] More extensive tables are available in the ISO currency standard.[7]
[b] Capitalization in this column reflects correct use of uppercase and lowercase letters.
[c] In total, 18 countries have converted to the euro as of January 1, 2014. For more information on the euro, see http://www.euro.ecb.int.
[d] The first form is recommended for consistency in style among the many symbols for different national currencies that are based on a dollar unit. The form $A is recommended by the Australian Government Publishing Service.[8]
[e] The second symbol listed is uncommon.
[f] The first form is recommended for US publications having mainly a domestic readership; the second form is recommended for US documents that refer to several dollar currencies, whether they have a domestic or an international readership.

Table 12.3 SI base units and their symbols

Quantity[a]	Unit[b]	Symbol[b]	Definition
Length	meter	m	Length of the path traveled by light in a vacuum during a time interval of 1/299,792,458 of a second
Mass[c]	kilogram	kg	Unit of mass equal to the mass of the international prototype of the kilogram
Time	second	s	Duration of 9,192,631,770 periods of the radiation corresponding to the transition between the 2 hyperfine levels of the ground state of the cesium 133 atom
Electric current	ampere	A	Constant current that, if maintained in 2 straight parallel conductors of infinite length, of negligible circular cross-section, and placed 1 m apart in a vacuum, would produce between these conductors a force equal to 2×10^{-7} N/m
Thermodynamic temperature	kelvin	K	Unit of thermodynamic temperature defined as 1/273.16 of the thermodynamic temperature of the triple point of water
Amount of substance	mole	mol	Amount of substance of a system that contains as many elementary entities as there are atoms in 0.012 kg of carbon 12 (when used, the elementary entities must be specified: atoms, molecules, ions, electrons, other particles, or specified groups of such particles)
Luminous intensity	candela	cd	Luminous intensity, in a given direction, of a source that emits monochromatic radiation of frequency 540×10^{12} Hz and that has a radiant intensity in that direction of 1/683 W/sr

[a] SI units are often used to derive other units. Some examples (name and symbol in parentheses) are area (square meter, m^2), volume (cubic meter, m^3), density or mass density (kilograms per cubic meter, kg/m^3), and current density (amperes per square meter, A/m^2). Some other units derived from SI units have been given special names of their own (see Table 12.4).

[b] Capitalization in these columns reflects correct use of uppercase and lowercase letters.

[c] The word "weight" is often used for "mass" in everyday language, but in science "weight" is a force, for which the SI unit is the newton (see Table 12.5).

Some additional units of measure, although not themselves part of the SI, have been recommended by CIPM for use with SI units, and still other units are currently accepted because they are widely used and important in the sciences (Table 12.5).

Standard prefixes are used with the SI units to designate quantities much larger or smaller than a given unit (Table 12.6). Prefixes are combined with the symbol for the base SI unit or the derived SI unit. An exception, the base unit kilogram, is already prefixed; in this case, the prefixes are attached to the unit stem "gram" rather than to "kilogram". With few exceptions (liter, curie, and roentgen), prefixes are not used with the non-SI units in Table 12.5.

Several units of measurement traditionally used in medicine are still preferred to their newer SI equivalent units. For example, blood pressure, traditionally measured in millimeters of mercury (mm Hg), would be measured in kilopascals (k Pa). Normal blood pressure, traditionally expressed as 120/80 mm Hg, would be expressed as 16/10.6 k Pa. Concentrations traditionally expressed as ng/mL or mg/mL would both be expressed as mmol/L in SI. Likewise, calories would be expressed in Joules and pH as nmol/L of

Table 12.4 Derived SI units that have been given special names and symbols

Derived quantity	Name[a]	Symbol	Derivation via SI base units	Derivation via SI-derived units
Activity (of a radionuclide)	becquerel	Bq	s^{-1}	
Electric charge, quantity of electricity	coulomb	C	s·A	
Celsius temperature	degree Celsius[b]	°C	K	
Capacitance	farad	F	$m^{-2}·kg^{-1}·s^4·A^2$	C/V
Absorbed dose, specific energy (imparted), kerma	gray	Gy	$m^2·s^{-2}$	J/kg
Inductance	henry	H	$m^{-2}·kg·s^{-2}·A^{-2}$	Wb/A
Frequency	hertz	Hz	s^{-1}	
Energy, work, quantity of heat	joule	J	$m^{-2}·kg·s^{-2}$	N·m
Catalytic activity	katal	kat	s^{-1}·mol	
Luminous flux	lumen	lm	$m^2·m^{-2}$·cd = cd	cd·sr
Illuminance	lux	lx	$m^2·m^{-4}$·cd = m^{-2}·cd	lm/m^2
Force	newton	N	$m·kg·s^{-2}$	
Pressure, stress	pascal	Pa	$m^{-1}·kg·s^{-2}$	N/m^2
Electric resistance	ohm	Ω	$m^2·kg·s^{-3}·A^{-2}$	V/A
Plane angle	radian[c]	rad	$m·m^{-1}$ = 1	
Electric conductance	siemens	S	$m^{-2}·kg^{-1}·s^3·A^2$	A/V
Dose equivalent[d]	sievert	Sv	$m^2·s^{-2}$	J/kg
Electric potential difference, electromotive force	volt	V	$m^2·kg·s^{-3}·A^{-1}$	W/A
Power, radiant flux	watt	W	$m^2·kg·s^{-3}$	J/s
Magnetic flux	weber	Wb	$m^2·kg·s^{-2}·A^{-1}$	V·s

[a] Capitalization in this column reflects correct use of uppercase and lowercase letters.
[b] This unit may be used in combination with SI prefixes—for example, millidegree Celsius or m°C.
[c] The radian and steradian may be used advantageously in expressions for derived units to distinguish between quantities of a different nature but the same dimension. In practice, the symbols rad and sr are used where appropriate, but the derived unit "1" is generally omitted. In photometry, the unit name steradian and the unit symbol sr are usually retained in expressions for derived units.
[d] Other quantities expressed in sieverts are ambient dose equivalent, directional dose equivalent, personal dose equivalent, and organ equivalent dose.

hydrogen. Concern about health care providers not understanding the SI units and making errors in health care delivery has led to the continued use of some traditional units.

12.2.1.1 SI Rules

The CIPM has adopted rules for writing unit symbols with numerals and for combining SI units with SI prefixes.

1) The symbols for SI units are printed in lowercase roman type except for the units with a name derived from a proper name, in which case the first letter of the symbol is capitalized (e.g., watt is W, tesla is T, hertz is Hz).

14 s	45 °C	17.6 W	0.076 Pa
4 L	386 K	37 lx	0.59 cd

    The symbol for degrees Celsius, °C (not simply C), is separated from the number on its left by one space, whereas the degree symbol for a plane angle (e.g., a 45° angle)

Table 12.5 Units accepted for use with SI units[a]

Name[b]	Symbol[b]	Value in SI units
Units accepted for use with SI		
minute (time)	min	1 min = 60 s
hour	h	1 h = 60 min = 3,600 s
day	d	1 d = 24 h = 86,400 s
degree (angle)	°	1° = (π/180) rad
minute (angle)	′	1′ = (1/60) ° = (π /10,800) rad
second (angle)	″	1″ = (1/60) ′ = (π /648,000) rad
liter	L	1 L = 1 dm^3 = $10^{-3}m^3$
metric ton or tonne	t	1 t = 10^3 kg
neper	Np	1 Np = 1
bel[c]	B	1 B = (1/2) ln 10 Np
electron volt	eV	1 eV = 1.60218 × 10^{-19} J, approximately[d]
unified atomic mass unit	u	1 u = 1.66054 × 10^{-27} kg, approximately[d]
astronomical unit	ua	1 ua = 1.49598 × 10^{11} m, approximately[d]
Units accepted for use with SI but not encouraged[e]		
nautical mile		1 nautical mile = 1,852 m
knot		1 nautical mile per hour = (1,852/3,600) m/s
are	a	1 a = 1 dam^2= 10^2 m^2
hectare	ha	1 ha = 1 hm^2 = 10^4 m^2
bar	bar	1 bar = 0.1 MPa = 100 kPa = 1,000 hPa = 10^5 Pa
ångström	Å	1 Å = 0.1 nm = 10^{-10} m
barn	b	1 b = 100 fm^2 = 10^{-28} m^2
curie[f]	Ci	1 Ci = 3.7 × 10^{10} Bq
roentgen[f]	R	1 R = 2.58 × 10^{-4} C/kg
rad[f]	rad	1 rad = 1 cGy = 10^{-2} Gy
rem[f]	rem	1 rem = 1 cSv = 10^{-2} Sv

[a] The units in this table are accepted by the SI to accommodate the needs of specialized scientific, legal, and commercial interests (except curie, roentgen, rad, and rem; see footnote f).
[b] Capitalization in these columns reflects correct use of uppercase and lowercase letters.
[c] More commonly used with the SI prefix deci- (as decibel).
[d] This value can only be obtained by experiment and is therefore not known exactly.
[e] The SI recommends that the relationship of these units to SI units be defined in every document in which they appear. Although nautical mile and knot are accepted by Comité International des Poids et Mesures, use of both these units is discouraged by the International Association for the Physical Sciences of the Ocean (see Section 25.13).
[f] The curie, roentgen, rad, and rem are widely used in the United States and are accepted by the National Institute of Standards and Technology for use with the SI (also see Table 12.7).

and for longitude and latitude (e.g., 45°30′N) is not separated from the numeral by a space.

2) SI unit symbols are identical for singular and plural representations, and they are not followed by periods, unless the unit appears at the end of a sentence.

 10 g *not* 10 gs
 6 mm *not* 6 mms
 We added 4 mL. *or*. . . 4 mL was added. *not*. . . 4 mL. was added.

3) Indicate a compound unit that is the product of 2 or more SI units with a centered dot (raised period or half-high dot) without spaces around it.[13] A single space between the symbols is no longer acceptable; the centered dot reduces possible confusion between

Table 12.6 SI prefixes[a]

Factor[b]	Name[c]	Symbol	Factor[b]	Name[c]	Symbol
10^{-1}	deci	d	10^{1}	deka	da
10^{-2}	centi	c	10^{2}	hecto	h
10^{-3}	milli	m	10^{3}	kilo	k
10^{-6}	micro[d]	μ	10^{6}	mega	M
10^{-9}	nano	n	10^{9}	giga	G
10^{-12}	pico	p	10^{12}	tera	T
10^{-15}	femto	f	10^{15}	peta	P
10^{-18}	atto	a	10^{18}	exa	E
10^{-21}	zepto	z	10^{21}	zetta	Z
10^{-24}	yocto	y	10^{24}	yotta	Y

[a] Scientific notation may be used instead of prefixes (e.g., 4 MW can be written 4×10^{6} W).
[b] That is, 10^{-1} is 1/10, 10^{-2} is 1/100, etc., and 10^{1} is ×10, 10^{2} is ×100, etc.
[c] Capitalization in these columns reflects correct use of uppercase and lowercase letters.
[d] In medical practice, the symbol μ is increasingly replaced with "mc" because in pharmaceutical prescriptions, μ has been misinterpreted as M (meaning mega), posing life-threatening consequences. This misinterpretation has been exacerbated by the fact that μ is the only SI prefix that is a Greek letter.

compound units and SI units with prefixes (e.g., m N means meter newton, whereas mN means millinewton). Do not use a times symbol (×) in place of the centered dot.

V·s *not* V s *or* V×s

4) Use a slash, a horizontal line, or a negative exponent to express a derived unit formed from 2 other units by division.

m/s $\mathrm{m{\cdot}s^{-1}}$

5) Set prefixes in roman (upright) type with no space between the prefix and the unit symbol.

ng   kV   mJ   MN

6) Use only one prefix per unit symbol, and never use a prefix on its own.

8 ng *not* 8 mμg 1×10^{6}/s *not* 1 M/s

12.2.1.2 Other Recommendations for Writing SI Units

In addition to the CIPM rules (Section 12.2.1.1), the following rules (many of them as suggested by the National Institute of Standards and Technology[14]) are recommended.

1) Spell out the names of units when they are used in text without an accompanying numeric value. However, to save space in tables, illustrations, and parenthetic text notations, unit symbols may be used without accompanying numeric values.

Lengths, in millimeters, were . . . *or* Lengths (mm) were . . . *not* Lengths (in mm) were . . .

2) When a numeric value with an SI unit starts a sentence, spell out both the number and the SI unit; the alternative is to reword the sentence (see Sections 12.1.2.1, item 1, and 12.1.4.1, item 4). Numeric values with SI units anywhere else in a sentence should be in arabic numerals (not spelled out) and separated from the accompanying SI symbol (not the SI name) by a space.
3) Present both a prefix and the unit it modifies as either symbols or words, depending on usage.

kV *or* kilovolt *not* kvolt *or* kiloV

4) Avoid placing other words between 2 or more SI symbols or names, unless the information is an essential part of the unit, in which case, use the SI words rather than their symbols. Also, place the other words carefully for the correct association with the intended SI unit.

 larval density averaged 30.2 g/m^2 [*not* 30.2 g of larvae/m^2 *or* 30.2 grams of larvae per square meter] *but* counts averaged 30.2 larvae/m^2 ["larvae" is not between 2 SI units]
 larval density of 30.2 grams per lineal transect kilometer ["lineal transect" is essential to "kilometer"] *not* larval density of 30.2 g/lineal transect km *or* larval density of 30.2 grams/lineal transect kilometer *or* 30.2 grams of larvae/lineal transect kilometer *or* 30.2 g of larvae/lineal transect kilometer
 at 30.2 user days per square kilometer *or* user days, at 30.2/km^2 *not* at 30.2 user days/km^2
 containing MIB at 34 ng/kg of sample *not* samples containing 34 ng MIB/kg
 samples containing MIB (34 ng/kg) *not* samples containing 34 ng/kg MIB

5) Do not mix unit symbols and unit names. Do not apply mathematical operations to unit names.

 10 m^3/s
 Ten cubic meters per second . . . [start of a sentence]
 [*not* 10 cubic meters/s *or* 10 m^3/second *or* 10 cubic meters/second *or* 10 m^3 per s]

6) Units for 2 or more related quantities must be unambiguous. If necessary, include the unit with every numeral; if possible, avoid mixing units. Even when unnecessary, it is acceptable to present the unit with every numeral, provided that this style is consistently applied.

 from 10 s to 75 s *or* from 10 to 75 s *not* 10 s to 1.25 min
 10 m × 30 m *or* 10 × 30 m *not* 1 dm × 30 m
 1 cm by 133 m *not* 0.01 by 133 m [see item 7 below]

7) When a prefix is required for a derived unit, attach the appropriate prefix to the term in the numerator (e.g., 4,000,000 N/m^2 becomes 4 MN/m^2, not 4 N/mm^2). Use SI prefixes to express numerical values less than 0.1 and greater than 1,000. Alternatively, express very large and very small numbers with scientific notation (e.g., 4×10^6 N/m^2), in which case 10^x or 10^{-x} substitutes for the SI prefix. This alternative is especially useful for non-SI units, to which prefixes cannot be applied.

 70.5 km *not* 70,500 m

8) Use SI units instead of parts per thousand (ppt), parts per million (ppm), or parts per billion (ppb) (e.g., 14 mL/L rather than 14 ppt). The following conversions may be helpful.

Concentration
Parts per thousand
Parts per million
Parts per billion

Weight : Weight
g/kg *or* mg/g
mg/kg *or* µg/g
µg/kg *or* ng/g

Weight : Volume
g/L *or* mg/mL
mg/L *or* µg/mL
µg/L *or* ng/mL

Volume : Volume
mL/L *or* µL/mL
µL/L *or* nL/mL
nL/L *or* pL/mL

9) When spelling out the names of compound units, separate the words by a space or a hyphen, not a centered dot[13] (also see Section 12.2.1.1, item 3).

   volt second *or* volt-second *not* volt·second

10) For a compound unit that includes a quotient, spell out the word "per" when the unit name is written in text without a number; however, the symbol may be allowed within parentheses, in which case use the slash rather than the word "per".

    was measured in kilojoules per hour *not* was measured in kilojoules/hour
    was measured (kJ/h) just before *not* was measured in kJ/h just before

11) For a compound unit that includes a quotient, use a slash or a negative exponent rather than the word "per" or the letter "p" (for "per").

    kJ/h *or* kJ·h^{-1} *not* kph [for kilojoules per hour]

Multiple slashes in a mathematical expression are ambiguous (e.g., 8/2/4 might be (8/2)/4 = 1 or 8/(2/4) = 16). Therefore, when an expression has more than one unit in the denominator, only one slash may be used. In this situation, use parentheses and negative exponents in combination with the slash to correctly express the unit.

0.3 kg/(mg·h) or 0.3 kg·mg^{-1}·h^{-1} *not* 0.3 kg/mg/h

12.2.2 Deprecated Units

Several non-SI units (Table 12.7), some based on the CGS system (the centimeter-gram-second system, as opposed to the subsequent meter-kilogram-second system that itself was replaced by SI units), are now outdated, and CSE recommends not using them. If they are used, their relation to the corresponding SI units should be defined.

12.2.3 Other Measurement Systems

Other systems of measurement are still in use, despite support for the SI in almost all countries. For example, parts of the avoirdupois, apothecaries, and troy weight systems are still in common use, particularly in the United States (see Table 12.8). Although these other units of measure may be appropriate for certain audiences, CSE recommends using only SI units for all scientific writing.

For further discussion of the metric system and other common systems of measure, see Jerrard and McNeill.[11] For an extensive set of unit conversion factors, see Horvath.[15] Alternatively, most dictionaries have tables of weights and measures that provide conversions, and conversion programs are now available on the Internet.[9]

12.3 MATHEMATICAL SYMBOLS AND EXPRESSIONS

The standard for mathematical symbols in the natural sciences and technology is the ISO standard on quantities and units.[16] The previous publication standard[17] and publications from the American Mathematical Society[18,19] may also be helpful.

New and idiosyncratic notations, which are often needed to advance the field of mathematics, should be carefully defined.

Table 12.7 Deprecated non-SI units[a]

Name	Symbol	Value in SI units
CGS-based units		
erg	erg	1 erg = 10^{-7} J
dyne	dyn	1 dyn = 10^{-5} N
poise	P	1 P = 1 dyn·s/cm^2 = 0.1 Pa·s
stokes	St	1 St = 1 cm^2/s = 10^{-4} m^2/s
gauss[b]	G	1 G ≙ 10^{-4} T
oersted[b]	Oe	1 Oe ≙ (1,000/4π) A/m
maxwell[b]	Mx	1 Mx ≙ 10^{-8} Wb
stilb	sb	1 sb = 1 cd/cm^2 = 10^4 cd/m^2
phot	ph	1 ph = 10^4 lx
gal[c]	Gal	1 Gal = 1 cm/s^2 = 10^{-2} m/s^2
Other non-SI units		
curie[d]	Ci	1 Ci = 3.7 × 10^{10} Bq
roentgen[d]	R	1 R = 2.58 × 10^{-4} C/kg
rad[d,e,f]	rad	1 rad = 1 cGy = 10^{-2} Gy
rem[d,f]	rem	1 rem = 1 cSv = 10^{-2} Sv
X unit[g]		1 X unit ≈ 1.002 × 10^{-4} nm
gamma[f]	γ	1 γ = 1 nT = 10^{-9} T
jansky	Jy	1 Jy = ×10^{-26} W·10^{-2}·Hz^{-1}
fermi		1 fermi = 1 fm = 10^{-15} m
metric carat[h]		1 metric carat = 200 mg = 2 × 10^{-4} kg
torr	Torr	1 Torr = (101,325/760) Pa
standard atmosphere	atm	1 atm = 101,325 Pa
calorie[i]	cal	. . .
micron[j]	µ	1 µ = 1 µm = 10^{-6} m

[a] These deprecated units are listed here because they are often found in older publications and to emphasize that they are not appropriate for current scientific writing; some are centimeter-gram-second (CGS) units.
[b] This unit (which is in 3 dimensions) cannot strictly be compared with the corresponding SI unit (which is in 4 dimensions when only mechanical and electric quantities are considered), so the unit is linked to the SI unit by the mathematical symbol for "corresponds to" (≙).
[c] The gal, used in geodesyand geophysics, expresses acceleration due to gravity.
[d] The curie (a special unit used in nuclear physics to express the activity of radionuclides), the roentgen (which expresses exposure to x or γ radiation), the rad (which expresses an absorbed dose of ionizing radiation), and the rem (used in radioprotection to express dose equivalent) are accepted by the National Institute of Standards and Technology for use with SI because they are widely used in the United States; however, these units are not accepted by the Comité International des Poids et Mesures.
[e] When confusion with the symbol for radian is possible, use the alternative symbol for rad, rd.
[f] This non-SI unit is exactly equivalent to an SI unit with an appropriate submultiple prefix.
[g] The X unit was used to express the wavelengths of x rays.
[h] The metric carat was adopted for commercial dealings in diamonds, pearls, and precious stones.
[i] Several "calories" have been in use: 1) a calorie labeled "at 15 °C": 1 cal_{15} = 4.1855 J; 2) a calorie labeled "IT" (International Table): 1 cal_{IT} = 4.1868 J; and 3) a calorie labeled "thermochemical": 1 cal_{th} = 4.184 J.
[j] The micron, originally adopted in 1879, was abolished in 1968. In its place use the appropriate SI unit and symbol: micrometer, µm.

12.3.1 Font Style Conventions

Use arabic numerals for mathematical expressions. Present scalar variables and constants represented by a single letter in italics in both displayed equations and text (e.g., *A*, *M*, *x*, *y*). Such variables and constants may incorporate subscripts or superscripts, which should be set without a separating space. Use roman (not italic) type for abbrevia-

Table 12.8 Selected other weights and measures used in the United States and the United Kingdom and their metric (SI) equivalents

Unit[a]	Metric equivalent	Other equivalents
acre	0.4047 ha	4,840 square yards
board foot (bd ft)	2.36 dm^3	1 ft × 1 ft × 1 in
bushel		
US	35.2 dm^3	32 quarts, 4 pecks
imperial	36.4 dm^3	4 imperial quarts
calorie (cal)[b]	4.18 J	
carat, metric	0.200 g	3.086 grains
cord	3.625 m^3	128.01 ft^3
drachm (apothecaries; UK)		
dry[c]	3.888 g	60 grains
fluid	3.552 mL	
dram (apothecaries; US)		
dry[c]	3.888 g	60 grains
liquid	3.697 mL	0.23 in^3
fathom	1.829 m	6 ft
		2 yards
foot (ft)	30.48 cm	12 in
furlong	201.2 m	40 rods
		660 ft
gallon (gal)		
US, liquid	3.785 L	4 qt
UK	4.546 L	4 imperial quarts
grain (gr)[d]	0.065 g	
inch (in)[e]	2.54 cm	1,000 mils
mile (mi)		
statute	1.609 km	5,280 ft
		1,760 yards
nautical[f]	1.852 km	1.151 statute miles
		6,076 ft
mil	0.0524 mm	0.001 in
ounce (oz)		
apothecaries	31.10 g	480 grains
avoirdupois	28.35 g	437.5 grains
troy	31.10 g	480 grains
US fluid	29.57 mL	1.805 in^3
UK fluid	28.41 mL	
peck, dry		
US	8.810 dm^3	8 qt
UK	9.092 dm^3	
pint (pt)		
US dry	550.6 cm^3	33.6 in^3
US liquid	0.4732 L	16 fluid ounces
UK	568.3 cm^3	
pound		
apothecaries (lb)	373.2 g	12 oz
		5,760 grains

Table 12.8 Selected other weights and measures used in the United States and the United Kingdom and their metric (SI) equivalents (*continued*)

Unit[a]	Metric equivalent	Other equivalents
avoirdupois (lb)	453.6 g	16 oz
		7,000 grains
troy (lb)	373.2 g	12 oz
		5,760 grains
quart (qt)		
US liquid	0.9464 L	2 pints
		57.75 in^3
US dry	1,101 cm^3	2 pints
		67.20 in^3
UK dry	1,137 cm^3	69.35 in^3
rod	5.029 m	16.5 ft
ton (ton; avoirdupois)		
long	1,016 kg	2,042.5 lb
short	907.2 kg	2,000 lb
yard	0.9144 m	3 ft
		36 in

[a] Capitalization in these columns reflects correct use of uppercase and lowercase letters.
[b] The value of a calorie in SI units beyond the second decimal place depends on the definition of "calorie" being used (see Table 12.7).
[c] The avoirdupois dram in both the United States and the United Kingdom is equal to 1.772 g.
[d] The grain is the smallest or base unit in the avoirdupois, apothecaries, and troy weight systems of measure; in all 3 systems, it is 0.0648 g.
[e] The unit "inch" is preferably spelled out. The abbreviation (in) without a period is acceptable but only if it is accompanied by an arabic numeral.
[f] Approved for use with SI (see Table 12.5) and included here only for completeness and for conversions.

tions or symbols consisting of several letters. Similarly, abbreviations for mathematical functions (such as "d" for "derivative"), should be set in roman, not italic, type; see examples in Table 12.9.

The symbols for many mathematical quantities and operations take distinguishing accessory marks (embellishments) above or below the character. Many of these marks are difficult, expensive, and sometimes impossible to set in type, so alternatives have been developed for some of the most common marks. For example, vectors, traditionally represented by a right arrow above the variable, $\vec{v}$, are represented in print as a boldface italic letter ($\boldsymbol{v}$) instead. Similarly, the boundaries of sigma-class symbols are often represented by indices traditionally placed directly above and below the symbol, the normal practice in displayed mathematics. However, when these symbols must be used in running text, the indices can be placed adjacent to the symbol as subscripts and superscripts, which allows a better fit in text lines.

$$\sum_{i=1}^{n} \text{ becomes } \textstyle\sum_{i=1}^{n}$$

$$\int\limits_{\pi}^{\infty} \text{ becomes } \int_{\pi}^{\infty}$$

Although occasionally 2, sometimes more, levels of embellishments are used by an author (e.g., $\overline{\widetilde{\mathrm{P}}}$), authors and editors should try to replace such notation with alternatives.

Table 12.9 Mathematical functions

Symbol[a]	Meaning	Remarks
exp x *or* e^x	Exponential of x	
$\log_a x$	Logarithm to the base a of x	When the subscript is omitted, the base 10 is assumed.
ln x *or* $\log_e x$	Natural (Napierian) logarithm of x	
sin x, cos x, tan x, cot x, sec x, csc x	Trigonometric functions of x	
sinh x, tanh x, etc.	Hyperbolic functions of x	
arcsin x or $\sin^{-1}x$, arctan x, arcsec x, etc.	Inverse trigonometric functions of x	
arsinh x or $\sinh^{-1}x$, artanh x, arsech x, etc.[b]	Inverse hyperbolic functions of x	
$\lim_{x \to a} y$	The limit of y as x approaches a	$\overline{\lim}$, least upper limit $\underline{\lim}$, greatest lower limit

[a] For any of these functions, parentheses may be used for greater clarity—for example, exp(x), ln($x + y$), cos(x_1). Capitalization in this column reflects correct use of uppercase and lowercase letters.
[b] For the inverse hyperbolic functions, the International Union of Pure and Applied Physics[4] recommends dropping the letter "c" of "arc"; the symbols thus become "arsinh", "arsech", and "arcosh", etc.

More examples of alternative embellishments can be found in Swanson[19] or American Mathematical Society.[18]

Some mathematical presentations use roman letters in a script-type face (e.g., $\mathcal{R}$) or non-English characters. Greek letters are commonly used, but occasionally Fraktur (German black letter) or Hebraic letters are needed. For further discussions of such uses see Burton[20] and Swanson.[19]

12.3.1.1 Mathematical Operators and Spacing

When used with numbers or variable symbols, set off common mathematical operators (plus, minus, times, and division) and all the equality and inequality symbols (e.g., <, >, =, ≥) (see Table 12.10) from variables and numerals by a space or a thin space. Do not use these symbols between 2 words in running text. When these symbols are modifying a number rather than serving as operators, close them up to the numeral that follows or write them out. Two or more mathematical operators should not appear side by side. (However, see Section 12.3.1.3 for uses of operators with vectors. See Section 12.3.2.2 for additional spacing information.)

biomass of <1,500 kg
the +2.5 difference
$x > y < z$ at á = 0.05 *but* when á is >0.05
at 50× magnification
at 50× lens objective *but* at 50 × gravity
at >2 °C *or* at greater than 2 °C
at greater than −2 °C *not* at >−2 °C
the target zone equals the optimum plus the . . . *not* the target zone = the optimum + the . . .

Close up the following operators: the centered dot or product dot indicating multiplication (·), parentheses to denote multiplication, exponents designating power, the slash to denote division, and the factorial symbol (!). Avoid using the asterisk to denote multiplication except when used appropriately as a computer programming symbol.

Represent the minus symbol by the special minus symbol available in most word-

Table 12.10 Common operators in arithmetic, algebra, and number theory[a]

Symbol	Meaning	Remarks
$+$	Plus	
$-$	Minus	
$\times$ [or] $\cdot$	Times	$x \times y$ [or] $x \cdot y$ [also shown by juxtaposition of the quantities, xy]
$/$ [or] $\div$	Divided by	x/y [or] $\frac{x}{y}$ [or] xy^{-1} [also shown as] $x \div y$
$=$	Equals; is equal to	
$\neq$	Does not equal; is not equal to	
$\equiv$	Is identical with; identically equal to	
$\cong$	Is approximately equal to; is congruent with	Physics and plane geometry use $\approx$
$\simeq$	Asymptotically equal to	Physics
$\sim$	Is similar to	Plane geometry
$\backsim$	Is equivalent to	Matrix calculus
$>$	Is greater than	Double symbols, $>>$, mean "much greater than" in physics
$<$	Is less than	Double symbols, $<<$, mean "much less than" in physics
$\geq$	Is greater than or equal to	
$\leq$	Is less than or equal to	
$\propto$	Is proportional to	
!	Factorial	Example: 4! represents $1 \times 2 \times 3 \times 4$

[a] Some mathematical symbols are frequently called "signs", but this manual does not use the word "sign" for operators (see Section 4.2).

processing software. If such a symbol is not available, use a hyphen. A slash through an equality, identity, or congruency symbol produces the negative of that symbol (e.g., equals, $=$; does not equal, $\neq$).

Wavy lines ($\approx$, meaning "is approximately equal to") are used to indicate approximate rather than exact relationships. Do not use the symbol "$\sim$" in prose to mean "about" or "approximately"; use it only for specific mathematical applications in plane geometry (is similar to) and matrix calculus (is equivalent to). The addition of such lines to symbols sometimes has a specific meaning (e.g., in physics, $\simeq$ means "asymptotically equal to"). The preferred indicator of approximation in text usage is the appropriate word, spelled out.

> The temperature of the system was approximately 45 °C.
> *not* The temperature of the system was ~45 °C.

Do not use "circa" or its abbreviation "ca." for "about" or "approximately" in mathematical expressions.

12.3.1.2 Calculus

The symbols used in calculus are shown in Table 12.11, and some of the common notations for set theory are presented in Table 12.12. More complete lists and definitions of mathematical symbols can be found in Zwillinger[21] and James.[22] The mathematical symbols used in physics may be specialized (see the recommendations of the International

Table 12.11 Calculus symbols

Symbol	Meaning	Remarks
Sigma-class (often used with the upper and lower limits displayed)		
Σ	Summation of terms	
Π	Product of terms	
$\int$	Integration of terms	
$\oint$	Curvilinear integration	
Other classes		
Δ	Delta; a finite increment of a function	As in Δx
D or d	Derivative symbol; an infinite increment of a function[a]	As in $\mathrm{d}x$
∂	Partial derivative symbol; a variation in a function	
lim	Limit	
$\mathrm{d}y/\mathrm{d}x$ [or] $\mathrm{D}_x y$	Derivative of y with respect to x	Where $y = f(x)$
$\partial u/\partial x$ [or] $\mathrm{D}_x u$	Partial derivative of u with respect to x	Where $u = f(\mathrm{x,y})$
$\partial^2 u/\partial \mathrm{xy}$ [or] $\mathrm{D}_y(\mathrm{D}_x u)$	The second partial derivative of u, the first with respect to x and the second with respect to y	
∇	Del [or] nabla	Del is used as an operator on vector functions or, with superscript 2, as the Laplacian operator
Grad f [or] ∇f	Gradient of f	
div A [or] $\nabla \cdot A$	Divergence of A	
$\nabla \times v$	Curl of v	
∇^2 [or] Δ	The Laplacian operator	

[a] The derivative symbol should be in roman type in accordance with the recommendation of the International Union of Pure and Applied Physics.[23] The italic form, occasionally used in the United States, is discouraged because the derivative symbol represents a function, not a variable or quantity.

Table 12.12 Notations used in set theory

Symbol	Meaning	Remarks
$\in$	Is an element of	$x \in M$; x is an element of set M
$\notin$	Is not an element of	$y \notin M$; y is not an element of set M
$\ni$	Contains as an element	$M \ni z$; set M contains z as an element
$\supset$	Contains as a proper subclass	$M \supset N$; set M contains set N as a proper subclass
$\subset$	Is contained as a proper subclass within	$M \subset N$; set M is contained as a proper subclass within set N
$\supseteq$	Contains as a subclass	$C \supseteq E$; set C contains set E as a subclass
$\subseteq$	Is contained as a subclass within	$C \subseteq E$; set C is contained within set E as a subclass
$\cup$	Union or sum of	$A \cup B$; the union of set A and set B
$\cap$	Intersection of	$A \cap B$; the intersection of set A and set B
$\wedge$ or $\varnothing$	The empty (or null) set	A set containing no members

Union of Pure and Applied Physics[23]). The recommendations of the Association of American Publishers[24] include not only detailed directions for markup of manuscripts in Standard Generalized Markup Language (SGML) but also extensive tables of mathematical symbols, entity references for markup, and verbal descriptions of the symbols.

12.3.1.3 Vectors, Scalars, and Tensors

The symbols for vectors are usually lowercase letters set in bold upright (not italic) type; the components of a vector are scalars, which are lowercase letters set in italics (not bold) and of the same typeface as vectors. Tensors are sometimes represented by bold characters, but this presentation can be misread as the notation for vectors; expressing tensor symbols in bold italics of a sans serif typeface is clearer.[23]

Vector: $\mathbf{f}$
Scalar components of vector: f_x, f_y
Tensor: $\boldsymbol{S}$

This representation of tensors is not universal in the mathematics and physics literature; exceptions and other symbols are common.

In the multiplication of vectors and tensors, some of the traditional mathematical symbols, such as centered dots and multiplication symbols, have special meanings.

$\mathbf{a} \cdot \mathbf{b}$	scalar product of 2 vectors	[multiplication dot is used]
$\mathbf{a} \times \mathbf{b}$	vector product of 2 vectors	[multiplication cross is used]
$\mathbf{ab}$	dyadic product of 2 vectors	[closed up, no multiplication symbol]
$\boldsymbol{S} \cdot \mathbf{b}$	product of a tensor and a vector	
$\boldsymbol{S} : \boldsymbol{U}$	scalar product of 2 tensors	
$\boldsymbol{P} \otimes \boldsymbol{U}$	tensor product of 2 tensors	[also sometimes $P \otimes U$, i.e., italics and not bold]

12.3.1.4 Matrices and Determinants

Matrices and determinants are arrays of elements in columns and rows. In text, an overall symbol for the matrix may be used (e.g., matrix **A**), or general symbols representing the array may be shown between double vertical bars or, alternatively, within parentheses or braces (Table 12.13). Double vertical bars, brackets, and sometimes large parentheses are used to enclose displayed matrices.

In text **In display**

$$\mathbf{A} = \left\| a_i b_i c_i \right\| = \begin{bmatrix} a_1 & b_1 & c_1 \\ a_2 & b_2 & c_2 \\ a_3 & b_3 & c_3 \end{bmatrix} = \begin{Vmatrix} a_1 & b_1 & c_1 \\ a_2 & b_2 & c_2 \\ a_3 & b_3 & c_3 \end{Vmatrix}$$

A determinant is represented by a similar array of elements set between single vertical bars. Because vertical bars are also used to indicate the absolute value of a real number or the modulus of a complex number, the text notation for a determinant should be of the form "det **A**" rather than |**A**|.

$$\det \mathbf{B} = \det\left(x_{ij}\right) = \begin{vmatrix} x_{1,1} & x_{1,2} & x_{1,3} & x_{1,4} \\ x_{2,1} & x_{2,2} & \cdots & \\ x_{3,1} & x_{3,2} & \cdots & \\ x_{4,1} & x_{4,2} & x_{4,3} & x_{4,4} \end{vmatrix}$$

Table 12.13 Matrix notations

Symbol	Meaning
A [or] (a_{ij}) [or] $\|\|a_{ij}\|\|$ [in physics, *A*]	Symbol for a matrix
$\mathbf{A}^{-1}$	Inverse of matrix **A**
A′ [or] $\mathbf{A}^{T}$ [in physics, $A^{\sim}$]	Transpose of matrix **A**
$\bar{\mathbf{A}}$ [in physics, A*]	Complex conjugate of matrix **A**
$\mathbf{A}^{H}$ [in physics, A'; do not use A^{+}]	The Hermitian conjugate of matrix **A**
det **A** [or] |**A**|	Determinant of matrix **A**
tr **A** [in physics, Tr *A*]	Trace of matrix **A**

12.3.2 Aggregating and Spacing of Symbols

12.3.2.1 Enclosures or Fences

Enclosures such as parentheses and brackets, called "fences" in mathematics, are used to aggregate groups of symbols. In general, the order of use of the common fences in mathematics—parentheses within brackets within braces—is the reverse of their use in nonmathematical prose—braces within brackets within parentheses.

mathematics: { [( )] }
prose or nonmathematics: ([{}])

If more levels of fences are needed, use larger parentheses, brackets, and braces in the order shown.

Other types of fences include angled brackets and single and double vertical bars, which have special meanings and should not be used to extend the basic sets shown above. Avoid using angled brackets for aggregation, as in $\langle a + b \rangle$. Such usage could be ambiguous because angled brackets may also be used 1) as an alternative to the horizontal bar above a symbol to mean an average value; 2) to represent an ordered set of objects, as in $\langle x, y, z \rangle$; and 3) as the "less than" or "greater than" operators. Fences are usually used in matching pairs, but in physics and higher mathematics, unlike members may be used as pairs (e.g., $|z\rangle$ for the state *z* of a system), and occasionally only the left-hand fence is used.

12.3.2.2 Spacing of Mathematical Symbols

Spaces are not appropriate in a variety of mathematical expressions. In the following situations, the terms are closed up (no spaces):

1) between quantities multiplied together when the multiplication operator is not shown

   $2b$ ac $6yz\beta$

2) between fences and the variables on either side of them

   $(a-1)y$

   $(4p-4bc)(1-a)$

   $a|x|$

3) between terms and their subscripts or superscripts or between subscripts or superscripts and the following terms

$\cos^3 y$

$(a-1)y^3 z$

$c^{x-2}d$

4) between the symbols for plus, minus, or plus/minus and the numbers or variables to which they apply, when used to designate positive or negative values of those numbers or variables

$-2x$

the values $+13$, -7, or ± 2

5) between superscripts and subscripts and the variables to which they apply; if there is both a subscript and a superscript, set the superscript directly above the subscript, unless this is typographically impossible, in which case either the subscript or the superscript may be set close and the other spaced from the variable. (In higher mathematics [e.g., tensors], the alternatives below are not allowed, and no spaces may appear between the variable and either index; such notations should not be altered without querying the author.)

Usual

m_n^{2r}

Alternatives

$m_n{}^{2r}$

$m^{2r}{}_n$

Place a space before and after the operator symbols listed in Table 12.10 when used to represent mathematical operations.

$x = -4y - 1$ *not* $x{=}{-}4y{-}1$

$0 < y < zw$ *not* $0{<}y{<}zw$

$(x + p)a \geq y^3 z(1-3r)$

Also use a space on either side of symbols for trigonometric functions, logarithms, and exponential and limit functions. However, no space is used if the quantities preceding or following these symbols are enclosed by fences; the function carries a superscript or subscript; or the function itself is part of a superscript, subscript, or limit of a sigma-class symbol.

$b \sin x$

$(ac)\sin^3 2y$

$\log x$

$\exp(a+2b)$

$y^{\sin x}$

12.3.3 -Fold, Factor, and Times

Expressing increases or decreases from an original value (i.e., the base value) that use the terms "-fold", "factor", or "times" can easily be misinterpreted. That is, sentence structure and word choice often produce doubt about whether the author is referring to the amount of the increase or decrease (i.e., the difference between the new and the old) or is referring to the new value in terms of the multiplier.

In the following examples of an increase, the author is trying to describe a base volume of 10 mL and a final volume of 30 mL, but the construction can be wrongly interpreted to mean that the amount of the increase was 10 mL × 3, or 30 mL, which could lead the reader to believe the new volume was 40 mL instead of 30 mL. Such ambiguity can generally be overcome by careful word choice, one of the easiest ways being to avoid using the word "increase" altogether and instead let the mathematics do the explaining. Another option is to include the actual values.

Weak

the volume increase was 3-fold [was the final volume 3 times as high as the initial volume, or was the size of the increase 3 times the initial value?]

Unambiguous

the final volume was 3 times the initial volume
the final volume was 300% of the initial volume
the final volume was 3 times the initial volume of 10 mL
the final volume represented a 3-fold increase, from 10 to 30 mL

Because decreases cannot exceed 100% of the original value (except in discussions involving positive and negative numbers), "-fold", "factor", and "times" should be replaced by percentages. However, "-fold" (which has a logarithmic logic, whereby 2-fold = 50%, 4-fold = 25%, etc.) may be appropriate in some situations because any given "-fold" value is the same for both positive and negative changes. For example, if a base value is decreased by 50%, a 100% increase is required to get back to the original value; however, when this sequence is expressed in terms of "-fold", the same value applies in either direction (i.e., 2-fold). As with the word "increase" for positive changes, use of "decrease" for negative changes is best avoided, but when used, include the actual values.

Unambiguous

the final volume was 66% of [or 0.66 times] the initial volume
the final volume decreased 2-fold [*or* 50%], from 60 to 30 mL

12.3.4 Ratios, Proportions, and Percentages

Ratios, proportions, and percentages convert data to a common scale that simplifies data comparisons.[25] When presenting ratios, proportions, or percentages, be sure both the numerator and denominator are readily apparent. If they are not, the percentage cannot be fully interpreted: 50% can mean 2 of 4 or 5,000 of 10,000. Knowing the actual values used to calculate the percentage is necessary for full and accurate interpretation.

A ratio presents a comparison as 2 numbers separated by a colon. Ratios can be expressed in a variety of ways, but the relevant data should be included if the ratio is reduced to its simplest form. Close up the 2 numbers around the colon. Ratios should not be expressed as a single number, where the "1" is understood.

the ratio of positive to negative results was 35:105
the ratio of positive to negative results was 1:3 (35:105)
not the ratio of positive to negative results was 3 (35:105) [in this case the 1 is understood]

A proportion is calculated by division, and generally the quantity in the numerator is a subset of the quantity in the denominator.

the proportion of subjects with a negative result was 0.75 (105/140)
0.25 (35/140) of the subjects had a positive result

A percentage is a proportion multiplied by 100. It is often used to avoid the necessity of presenting the leading zero and decimal point and to shorten very small proportions (e.g., 0.014 becomes 1.4%). As above, when presenting proportions or percentages, the numerator and denominator must be readily apparent to avoid misleading readers.[25]

> 75% of the 140 subjects tested had a negative result
> 75% (105/140) of the subjects tested had a negative result
> Of the 140 subjects tested, 25% had a positive result

Report proportions and percentages to the degree of precision appropriate for the measurement. For denominators less than about 20, reporting the actual numeric data may be more useful or less misleading. Generally, when proportions or percentages derived from a mix of large (>100) and smaller (<100) denominators occur in close proximity, report all to the decimal level appropriate for the smallest denominator. Indicate clearly whether a change is a difference between 2 percentages or, as shown below, a change expressed as a percentage of the initial value; in the latter case, include the numerator and denominator to avoid misinterpretation. When reporting change as a percent, use the formula [(Final value − Initial value) /Initial value] and multiply the result by 100. The following examples indicate how to express the relationship between the values of 20% and 80%.

If the initial value is 20% and the final value is 80%, the percent change is 300%: (80 − 20)/20 = 3; × 100 = a 300% increase.

This change can be expressed as:

> a 300% increase [The difference of 60 is 300% of 20]
> an increase of 60 percentage points [preferred]
> a final value 4 times as great as [*not* greater than] the initial value
> the final value increased to 80% from 20%

If the initial value is 80% and the final value is 20%, the percent change is 75%: (20 − 80)/80 = −0.75; × 100 = a 75% decrease.

This change can be expressed as:

> a 75% reduction
> a decrease of 60 percentage points [preferred]
> an initial value that was 4 times as large as [*not* larger than] the final value
> a final value one-fourth of the initial value
> the final value decreased to 20% from 80%

12.4 MATHEMATICS IN TEXT AND DISPLAY

12.4.1 Guidelines for Text

When mathematical expressions are set within lines of text, make every effort to limit the vertical dimensions of fractions to maintain the spacing and appearance of the lines. This means limiting the fractions, complex exponents, and large symbols used in the expressions. Present fractions and other quantities with slashes or exponents rather than vertical stacking; convert graphic symbols such as that for square root to exponential notations. Keep in mind, however, that the author may prefer a graphic symbol.

$$a/b \text{ or } ab^{-1} \text{ not } \frac{a}{b}$$

$$(b-d)^{1/2} \text{ not } \sqrt{b-d}$$

Only one slash may be used in any given expression.

$$\frac{a/b}{c} \text{ or } (a/b)/c \text{ not } a/b/c$$

Exponents containing more than one level become unwieldy within text lines. The exponential form can be converted to the "exp" form for use in text; even then, large exponents (as in the second example that follows) may be better shown in display rather than text format.

In Display	In Text
e^{x^2-1}	$\exp(x^2-1)$
$e^{\frac{a-b}{c+d}}$ $\exp\left(\frac{a-b}{c+d}\right)$	$\exp[(a-b)/(c+d)]$

When an equation set within the text carries over from one line to the next, the choice of where to break the equation requires knowledge of mathematics. Breaks after operators are preferred as a hint to the reader that more of the expression is still to come. Such breaks are more important when the text lines are left-justified than when they are fully justified. The most basic guidelines are given here; for a more complete treatment, see Swanson.[19] In text, avoid breaking an equation within a pair of fences whenever possible. Break equations in the following order of preference:

1) before or after an equals symbol

   All of the variables in the following equation should be set in italic: $x - 2 =$
   $3by^3z - 4m$.

2) before or after a plus or a minus symbol, but not if the symbol occurs within a set of fences

   For our goal of an approximation to the solution, $b_st_r = -K[(u/y) + (v/r)]^n +$
   $ba[(u_m/y) + (v_m/r)]$.

3) before or after a multiplication symbol, or between sets of fences, in which case a multiplication symbol should be displayed between the sets of fences

   The authors represented this complex equation as $R_e = (d^n u^{2-n} \rho / 8^{n-1} K) \times$
   $(4n/3n + 1)^n$.

4) before a sigma-class symbol

12.4.2 Guidelines for Display

In general, break displayed equations according to the same principles as for text equations (see Section 12.4.1), except that a single term can include mathematical operations involving several or more lines. In breaking expressions, the main differences are that vertical alignment is a consideration in displayed equations but not running text, and breaks occur before, not after, operators in displayed equations.

1) For a sequence of equations in which the left side is unchanged, align the equal symbol (=) in each line.

$$\begin{aligned} 2u_o v_o &= u_0^2 + v_0^2 - (u_o - v_o)^2 \\ &= k - (u_o - v_o)^2 \end{aligned}$$

2) For continued expressions in which the left side is long, vertically align the equals symbol with the first operator in the first line.

$$\begin{aligned} &\left[\left(a_1 + ia_2 + \left(a_{11}s_1 + a_{21}s_2\right)\right]/\left[\left(b_1 + ib_2 + \left(b_{11}s_1 + b_{21}s_2\right)\right] \\ &\quad = f(x)g(y) + \ldots \end{aligned}$$

3) For continued expressions in which the right side is long, align the continuing operator with the first term to the right of the equals symbol.

$$\begin{aligned} f(x) = {} & 2k(a^2 + 5b_1)(3c - b^2c) \\ & + 4ac\{a_1b_1 + [(4 - b^2)^2(ab + 4ck = b_2c)]\} \end{aligned}$$

4) If you must split within fences, vertically align the continuing operator with the first symbol within the enclosed group.

$$\begin{aligned} f(x)g(y) = \sin ab[&R(2k\cos b) - 2R_0(2k\cos b) \\ &+ R_1(b\sin ab) + \cos b] \end{aligned}$$

12.4.3 Punctuation and Spacing

When mathematics is written in sentences or highlighted in display, the appropriate use of punctuation such as commas and periods is just as essential to comprehension as it is in nonmathematical text. Syntax, not presentation format, should determine punctuation after mathematics. Punctuation spacing should emphasize the mathematics, as follows:

1) Use a space before punctuation that follows mathematics within a line of text.

. . . was represented by $y + (y - t) = z$.

2) Use a space before punctuation that follows mathematics in display.

. . . was represented by
$y + (y - t) = z$.

. . . was represented by
$y + (y - t) = z$,
where y is the . . .

In text and display, indicate an ellipsis in a mathematical expression by 3 dots or periods. For operations and relations, the dots or periods are centered, as in the first example. For lists, the dots or periods are set on the baseline, and each item in the list is set off by a comma. Avoid using the ellipsis character available in most word-processing programs.

$x_1 + x_2 + \cdots + x_n$ $\qquad$ $x_1, x_2, \ldots, x_n$

12.5 STATISTICS

The international standard for statistical definitions and symbols is ISO 3534-1.[26] Several aspects of statistical usage not covered in that standard merit comment here. Common symbols are given in Table 12.14; additional tables of statistical symbols can be found in

Table 12.14 Symbols used in statistics[a]

Population symbols	Sample symbols	Explanation
	F	Variance ratio (F test)
H_0		Null hypothesis
H_1		Alternative hypothesis
N		Number of individuals or subjects (population or lot size)
	n	Number of individuals or subjects (sample)
	P	Probability of obtaining a more extreme test value for a hypothesis test
	R	Coefficient of multiple correlation, range of a sample
ρ	r	Coefficient of correlation[b]
	r^2	Coefficient of determination[c]
	R^2	Coefficient of multiple determination[c]
σ	s	Standard deviation[d]
σ^2	s^2	Variance
	$s_{\bar{x}}$	Standard error of the mean[e]
	CV	Coefficient of variation
	t	Statistic derived in Student t test
$\bar{X}$	$\bar{x}$	Arithmetic mean
	α	Probability of a type I error; significance level
	β	Probability of a type II error
	$1 - \beta$	Statistical power
	x^2	Statistic derived in chi-square test
	v	Number of degrees of freedom[f]
θ	$\hat{\theta}$	Parameter (unknown constant) and estimator (function of a data set)

[a] For additional symbols, see ISO standard 3534-1.[26]
[b] There are several measures of correlation: Pearson product–moment correlation coefficient, r, for 2 normally distributed, continuous variables; Spearman rank correlation coefficient, rho (ρ), for 2 continuous variables of any distribution; Kendall rank correlation coefficient, tau (τ), for 2 ordinal variables or 1 ordinal and 1 continuous variable; point biserial correlation coefficient for a continuous variable and a categorical variable with 2 levels ("recovery status": recovered or not); and intraclass and interclass correlation coefficients to assess agreement within and between observers.
[c] The coefficient of determination, r^2, indicates the amount of variation in the response variable that is explained by the explanatory variable in a simple linear regression model. The coefficient of multiple determination, R^2, indicates the amount of variation in the response variable that is explained by the explanatory variables in a multiple linear regression model.
[d] The abbreviation "SD" is often used and is acceptable.
[e] The abbreviation "SE" is often used and is acceptable; the abbreviation "SEM" is not recommended for "standard error of the mean".
[f] The abbreviation "df" is often used and is acceptable.

Bailar and Mosteller[2] and the ISO standard.[26] Lang and Altman[27] provide a set of statistical reporting guidelines suitable for inclusion in a publisher's or journal's instructions for authors. Lang and Secic[25] have reviewed and summarized common errors and pitfalls in statistical reporting.

12.5.1 Descriptive Statistics

Descriptive statistics summarize data sets to simplify their reporting and sometimes their analysis. Categorical data are summarized with the category name and the number, proportion, or percentage of observations in each category. Continuous data (measurements on a scale of equal intervals) form distributions, which are most often described with two numbers, a "measure of central tendency" indicating the peak of the distribution (most commonly, the mean, median, and mode) and a "measure of dispersion" indicating the variability of the distribution (most commonly, the range, interpercentile [usually interquartile] range, and standard deviations).

Measurements that are approximately normally distributed are appropriately described with means and standard deviations and sometimes ranges. Measurements that are markedly nonnormally distributed are usually best described with medians and interpercentile ranges and sometimes ranges. Avoid describing skewed data (measurements forming asymmetrical distributions) with means and SDs.

All measurements have variability. Variability is often reported in the text; the units and measure of dispersion used should be unambiguous (see Section 12.1.4.2). In particular, indicate the measure of variability as SD, range, or interpercentile range.

> mean (SD) = 44% (3%) *or* mean of 44% (SD, 3%)
>
> *not*
>
> SD = 44 F 3%
>
> *and not*
>
> 44% (±3%) [the ± sign is redundant because the SD extends on either side of the mean by definition]
>
> mean of 104 mm (SD, 11 mm) [*not* mean of 104 F 11 mm *or* mean of 44 mm (SD F11)]
> Median (25th percentile to 75th percentile) survival was 7 years (4.5 to 9.5 years)

The standard error of the mean (SE or $s_{\bar{x}}$) is not a descriptive statistic for indicating the variability of measurements but rather an inferential statistic for indicating the precision of an estimate. Nevertheless, the tradition in some fields of science is to use it as a descriptive statistic. In that situation, label it appropriately to prevent it from being confused with the SD.

12.5.2 Estimates and Confidence Intervals

An estimate is a probable value for a population that is projected from a measured value of a sample from that population. The results of many studies are (or can be) estimates and should therefore be accompanied by a measure of precision, usually a confidence interval (CI). The most common CI is probably the 95% CI, although other confidence coefficients can be used. (The standard error mentioned above is actually a measure of precision; it is about a 68% CI.)

Confidence intervals can be fit around any outcome: differences; measures of correlation; percentages; effort-to-yield measures (e.g., the number needed to treat); risk, odds, and hazard ratios; regression coefficients, and so on. Because a CI is expressed in the same units as the estimate it accompanies, it keeps attention on the estimate itself and away from any associated *P* value, which has no scientific interpretation aside from being a measure of chance as an explanation. Thus, confidence intervals are often preferred to *P* values, although *P* values may still be reported.

> The increase in mean survival was 7 months (95% CI, 5.3 to 8.7 months)
> Median concentration increased by 112 mg/dL, from 43 to 155 mg/dL (95% CI, 64 to 160 mg/dL; $P = 0.09$)
> The odds ratio for structural failure with untreated material was 16.5 (95% CI, 12.5 to 20.5)

12.5.3 Hypothesis Testing

The ultimate product of hypothesis testing is the probability, or *P*, value. The *P* value should not be confused with the α level that defines the threshold of statistical significance (the most common value being α = 0.05). *P* values should be reported for all analyses (i.e., give the value and not $P > 0.05$ or "NS" for not significant) and should be expressed as an equality, not an inequality. The smallest *P* value that need be reported for most applications is $P < 0.001$. *P* values for genetic associations, however, may be orders of magnitude smaller[1] (see Section 12.1.3.3).

> $P = 0.002$ *not* $P < 0.05$
> $P = 0.09$ *not* P = NS
> widths did not differ greatly ($P = 0.13$) *not* ($P \approx 0.13$)

In scientific writing, "significance" is reserved for its statistical meaning; use "marked", "important", "substantial", or a similar word for nonstatistical descriptions. Sentences containing significant *P* values do not need to the describe the result as "significant".

> The association was important *not* The association was significant
> The difference was 72 mL ($P = 0.03$) *not* The 72-mL difference was significant ($P = 0.03$)

In addition, editors and authors should understand the importance of reporting the following often overlooked or ignored information.

1) The purpose of the analysis
2) The name of the statistical test or procedure
3) Whether the test was one- or two-tailed
4) Whether the data conformed to the assumptions of the test(s) used to analyze them, especially whether skewed data were successfully transformed to a more normal distribution or whether they were analyzed with nonparametric tests
5) The result of the comparison (e.g., differences between groups, the absolute risk reduction, the slope of a regression line)
6) 95% confidence intervals for each major result
7) The practical relevance and implications of the result (see Section 12.5.3)
8) The statistical software package used in the analysis

Statistical results are often reported parenthetically in the text. The individual elements of a test are normally separated by commas. (The example below includes the test statistic and the degrees of freedom [df], which are less often reported aspects of some statistical tests.)

> . . . the output was greater in the treatment group ($x^2 = 18.2$, df = 2, $P < 0.001$) . . .

Place a zero to the left of the decimal when reporting *P* values and correlation coefficients (also see Section 12.1.3.2).

> The means differed by 17.8 g ($P = 0.23$). *not* by 17.8 g ($P = .23$)

CITED REFERENCES

1. Altman GA, Bland JM. Presentation of statistical data. BMJ. 1996;312(7030):572.
2. Bailar JC, Mosteller F. Medical uses of statistics. 2nd ed. Boston (MA): NEJM Books; 1992. p. 385–389.

3. Higham NJ. Handbook of writing for the mathematical sciences. 2nd ed. Philadelphia (PA): Society for Industrial and Applied Mathematics; 1998.

4. International Organization for Standardization and International Electrotechnical Commission. ISO/IEC 80000 2009 ISO/IEC international standards 80000, parts 1 to 14. Geneva (Switzerland): International Organization for Standardization; 2009.

5. International Organization for Standardization. Quantities and units. Part 0: General principles. 3rd ed. Geneva (Switzerland): ISO; 1992. (ISO 31-0(E)). Amendment 1, 1998; Amendment 2, 2005.

6. Billion bites the dust [editorial]. Nature. 1992;358(6381):2.

7. International Organization for Standardization. Codes for the representation of currencies and funds. Geneva (Switzerland): ISO; 2001. (ISO 4217).

8. Australian Government Publishing Service. Style manual for authors, editors and printers. 6th ed. Canberra (Australia): AGPS Press; 2002.

9. Bureau International des Poids et Mesures. The international system of units (SI). 8th ed. [Paris (France)]: Organisation Intergouvernementale de la Convention du Mètre; 2006. English translation available in PDF format at http://www.bipm.org/utils/common/pdf/si_brochure_8_en.pdf.

10. ASTM Committee E-43 on Metric Practice. Standard practice for use of the International System of Units (SI): the modernised metric system. Philadelphia (PA): American Society for Testing and Materials; 1993. (ASTM E380-89a).

11. Jerrard HG, McNeill DB. A dictionary of scientific units: including dimensionless numbers and scales. Springer; 2013.

12. Taylor BN, editor. The international system of units (SI). 2001 ed. Gaithersburg (MD): National Institute of Standards and Technology; 2001 Jul. 74 p. (NIST special publication; SP 330 2008 ED).

13. American National Standards Institute. Metric practice. New York (NY): ANSI; 1992. (ANSI/IEEE standard 268-1992).

14. National Institute of Standards and Technology (US), Physics Laboratory. The NIST reference on constants, units, and uncertainty. Gaithersburg (MD): NIST; 1998 Feb [updated 2003 Dec]. Definitions of the SI base units; [accessed 2013 Jul 28]; [about 2 p.]. http://physics.nist.gov/cuu/Units/current.html.

15. Horvath AL. Conversion tables of units in science and engineering. New York (NY): Elsevier; 1986.

16. International Organization for Standardization. Quantities and units. Part 2: Mathematical signs and symbols to be used in the natural sciences and technology. Geneva (Switzerland): ISO; 1992. (ISO 80000-2:2009(E)).

17. International Organization for Standardization. Quantities and units. 3rd ed. Geneva (Switzerland): ISO; 1993. (ISO standards handbook).

18. American Mathematical Society. A manual for authors of mathematical papers. Providence (RI): AMS; 1990.

19. Swanson E. Mathematics into type: updated edition. Providence (RI): American Mathematical Society; 1999.

20. Burton BW. Dealing with non-English alphabets in mathematics. Tech Commun. 1992;39(2): 219–225.

21. Zwillinger D. CRC standard mathematical tables and formulae. 32nd ed. Boca Raton (FL): CRC Press; 2011.

22. James R. Mathematics dictionary. 5th ed. New York (NY): Van Nostrand Reinhold; 1992.

23. International Union of Pure and Applied Physics, Commission for Symbols, Units, and Nomenclature, Atomic Masses and Constants. Symbols, units, nomenclature and fundamental constants in physics: 1987 revision. Physica. 1987;146A(1–2):1–68. Prepared by ER Cohen, P Giacomo.

24. Association of American Publishers. Markup of mathematical formulas. Ver. 2.0. Revised ed. Dublin (OH): AAP; 1989.

25. Lang TA, Secic M. How to report statistics in medicine: annotated guidelines for authors, editors, and reviewers. 2nd ed. Philadelphia (PA): American College of Physicians; 2006.

26. International Organization for Standardization. Vocabulary and symbols. Part 1: General statistical terms and terms used in probability. Geneva (Switzerland): ISO; 2006. (ISO 3534-1).

27. Lang T, Altman D. Basic statistical reporting for articles published in clinical medical journals: the SAMPL Guidelines. In: Smart P, Maisonneuve H, Polderman A, editors. Science Editors' Handbook, European Association of Science Editors, 2013. Listed as a Key Reporting Guideline on the EQUATOR website, http://www.equator-network.org.

ADDITIONAL REFERENCES

AIP Publication Board. AIP style manual. 4th ed. New York (NY): American Institute of Physics; c1990–1997 [accessed 2012 Jan 11]. http://www.aip.org/pubservs/style/4thed/toc.html.

American National Standards Institute. IEEE/ASTM standard for use of the International System of Units (SI): the modern metric system. New York (NY): ANSI; 2002. (SI10-2002).

Baron DN. Units, symbols, and abbreviations: a guide for biological and medical editors. 4th ed. London (UK): Royal Society of Medicine (GB); 1988.

Cook JL. Conversion factors. New York (NY): Oxford University Press; 1991.

Darton M, Clark J. The Macmillan dictionary of measurement. New York (NY): Macmillan; 1994.

Hansen WR, editor. Suggestions to authors of the reports of the United States Geological Survey. 7th ed. Reston (VA): Geological Survey (US); 1991 [accessed 2012 Jan 11]. 289 p. http://www.nwrc.usgs.gov/lib/lib_sta.htm.

McGraw-Hill dictionary of mathematics. 2nd ed. New York (NY): McGraw-Hill; c2003. 307 p.

Monteith JL. Consistency and convenience in the choice of units for agricultural science. Exp Agric. 1984;20:105–107.

Oxford Dictionaries. http://oxforddictionaries.com/definition/english/number.

Young DS. Implementation of SI units for clinical laboratory data. Ann Intern Med. 1987;106(1):114–129.

Young DS, Huth EJ, editors. SI units for clinical measurement. Philadelphia (PA): American College of Physicians; 1998.

13 Time, Dates, and Age Measurements

13.1 STANDARDS

The International Organization for Standardization (ISO) has a comprehensive standard for time and date representations,[1] as well as a standard for quantities and units of space and time.[2] *The Chicago Manual of Style*[3] provides detailed guidance on time and dates in published documents.

This chapter provides general guidance on dates and time, as well as specialized conventions that are applied in some disciplines. (See Section 25.2 for a discussion of geologic time and Section 26.6 and its subsections for a discussion of astronomical time.)

13.2 TIME

13.2.1 Units and Symbols

In general, spell out units of time in running text when they appear without a numeric value. Use standard symbols (usually the International System of Units [SI]) for units of time accompanied by numeric values (see Section 12.2.1 and Tables 12.3 and 12.5). The Bureau International des Poids et Mesures[4] has designated 3 additional units as acceptable for use in conjunction with SI units, even though they are not officially part of the system: day (d), hour (h), and minute (min). Although SI prefixes may be used with "second" (e.g., "millisecond" or "ms"), an integral unit in the SI system, these prefixes are never used with day, hour, or minute. Other, larger units of time (specifically week, month, and year) are not mentioned in the SI system. As such, the SI offers no suggestions about the symbols that would be acceptable for these units. However, where abbreviated forms are needed, it is appropriate to use "wk", "mo", and "y" for "week", "month", and "year", respectively. It is best to check with publisher guidelines concerning when abbreviations should be used (e.g., never in text or in text as well as in tables and figures/figure legends).

The year can be measured accurately to millionths of a second.
The samples were heated for 30 min and kept refrigerated for 6 h before analysis.

If the numeric value involves only one unit of time, insert a space between the number and symbol. If the value calls for more than one unit, set the numbers and symbols close, with symbols either on the line or superscripted. Again, check with the publisher as to the appropriate convention.

27 h
22h3min
2h15min4s
The current position of the north galactic pole is near right ascension 12^h52^m, declination 27°8′, and the galactic center is located at about right ascension 17^h45^m, declination −28°56′.

For fractions of hours, minutes, and seconds, use decimal fractions.

3.21 s	15.8 min = 15min48s *not* 15 min and 8 s
10min14.6s	12.3 h = 12h18min *not* 12 h and 30 min
3h5min37.5s	

For values of time, do not use the prime and double prime symbols for minute (′) and second (″) that are used for geographic coordinates (latitude and longitude).

13.2.2 Clock Time

Two systems are used to designate the hour of the day. In the 12-h system, the hours of each day are divided into two 12-h portions: 1 through 12 noon and 1 through 12 midnight. In the 24-h system, the hours are numbered consecutively, 1 through 24 (with 24 representing midnight).

In the 12-h system, the abbreviations AM (ante meridiem [i.e., before noon]) and PM (post meridiem [i.e., after noon]) are used to distinguish between the halves of the day. Minutes are separated from hours by a colon, and a leading zero is added if necessary so that the minutes are presented as a 2-digit number.

12:01 in the morning = 1 min after midnight *or* 12:01 AM
12:01 in the afternoon = 1 min after noon *or* 12:01 PM

The abbreviations AM and PM may be presented in capital letters, small capitals, or lowercase, but capital letters are recommended to avoid the extra coding required for small capitals and to make the abbreviations stand out from the running text. Separate the time from the abbreviation with a space but no intervening punctuation.

With the 12-h system, express time using the AM and PM notation or the informal (and ambiguous) "o'clock", but do not use both.

10 PM *or* 10 o'clock *not* 10 o'clock PM

If you choose the informal "o'clock", it is useful to add "in the morning", "in the afternoon", or "in the evening" to aid reader understanding.

6 o'clock in the morning = 6 AM

The 24-h system, commonly known as "military time" in the United States (US), obviates ambiguities as to which part of the day is meant. Time is expressed as a 4-digit number (including zeros as appropriate) without punctuation; no period or colon between hours and minutes should be added. The day begins at 0000 (midnight) and ends at 2359 (1 min before the next day's midnight); thus, 2400 of one day is equivalent to 0000 of the next day. The last 2 digits represent minutes, not decimal fractions of an hour; therefore, do not use the abbreviation "h" after the 4-digit number. However, if the context of a time designation in the 24-h format would produce ambiguity, add "hours"; in this context, spell out the word "hours".

0602 = 6:02 AM 1802 = 6:02 PM

0028 = 28 minutes past midnight 1228 = 28 minutes past noon

We stopped counting at 1530 hours. [Without "hours", it is unclear whether 1530 refers to a counted number or the time.]

The ISO standard[1] recommends a format representing hours, minutes, and seconds separated by colons (hh:mm:ss), with decimalization for fractions of a second. Use a period (full stop) to represent the decimal point (see Section 12.1.3.1). The format can be truncated for less precise representations.

23:59:59.5 = 0.5 s before midnight 23:59 = 1 min before midnight

13.2.3 Time Zones

International time zones were legally established in the 19th century. The standard time system, fixed in 1883, partitioned natural continuous time (which gains 1 min every 22.4 km [14 miles] traveled from east to west) into 24 international time zones in increments of 15° of longitude. (One degree of longitude covers about 111 km [69 miles] at the equator.) Greenwich, United Kingdom (UK), is at 0°, the prime meridian. Mean solar time (formerly known as Greenwich mean time) is determined by this meridian and is equivalent to coordinated universal time (also known as universal time coordinated [UTC]), which is the time scale available from civilian broadcast time signals (see also Section 26.6.2.2). In general, each time zone differs from UTC by a whole number of hours (or, in some instances, half hours).

In the contiguous United States, 4 meridians are designated for standard time: 75° (eastern), 90° (central), 105° (mountain), and 120° (Pacific) west of Greenwich, UK. Alaska, except for the Aleutian Islands, is in the Alaska time zone (135° west of Greenwich). Hawaii and the Aleutian Islands are in the Hawaii–Aleutian time zone (150° west of Greenwich). Samoa is in the Samoa time zone (165° west of Greenwich, UK). In addition to the 4 meridians for the contiguous United States, Canada contains the 135° meridian (Yukon), the 60° meridian (Atlantic), and the Newfoundland zone within the Atlantic zone, which encircles the island of Newfoundland and has a time 0.5 h later than Atlantic time. These time zones deviate from the meridians because of political boundaries. When the time in a zone is advanced in the spring by 1 h to lengthen the period during the evening that is in daylight, replace "standard" with "daylight" in the term for the time zone.

Do not capitalize the names of time zones when written out in full, except for a proper noun (e.g., Pacific standard time) or at the beginning of a sentence. Capitalize the abbreviations for time zones, without periods, when they immediately follow the time, and separate the time from the abbreviation with a space but no intervening punctuation.

eastern standard time   Pacific daylight time
11 AM CST   0500 AST

When it is noon EST (eastern standard time), it is 11:00 AM CST (central standard time), which is equivalent to noon CDT (central daylight time). When it is 0100 in British Columbia (Pacific standard time), it is 0500 in New Brunswick (Atlantic standard time) and 0530 in Newfoundland and Labrador. See http://www.time.gov/images/worldzones .gif for a map of the world time zones.

13.3 DATES

13.3.1 Days and Months

Capitalize and spell out the names of days of the week and months, which may be abbreviated to their first 3 letters if they appear in tables, graphs, end references, and other locations where short forms are needed.

For a complete date, use the full name or the 3-letter abbreviation for the month, rather than the numeric equivalent, to avoid any ambiguity related to differing conventions. (See Section 13.3.3 for recommendations regarding the order of elements in a date).

1999 Jan 5 *or* 1999 5 January
not 5/1/99 [European style] *or* 1/5/99 [US style]

The julian date, used in astronomy, is the interval in mean solar days since 4713 BC January 1 at Greenwich mean noon (for more detail, see Section 26.6.1.1). Do not use the term "julian day" to designate the days of the calendar year by number.[4] Instead, use an expression such as "day of the year".[5,6]

Day of the year June 24, 2012 = Julian day 2455562.5000

13.3.2 Years

Express years in numerals. In informal writing, a year is sometimes abbreviated to 2 digits, preceded by an apostrophe (e.g., class of '91), but avoid this form in scientific publications.

This species was first reported in 1986.
the 1980s and 1990s *not* the 1980s and '90s *or* the '80s and '90s

13.3.3 Sequence of Date Elements

In scientific publications, write dates in the sequence of year, month, and day (a form common in astronomical and aeronautical literature and in references) or day, month, and year (a form common in Europe, Canada, and US military organizations). Do not truncate the year to 2 digits.

2002-04-23 *or* 2002 April 23 *or* 2002 Apr 23
23-04-2002 *or* 23 April 2002 *or* 23 Apr 2002

The ISO[1] recommends the numeric format of year-month-day (YYYY-MM-DD) and variations on it (including week designations and designations of a specified time on a specified day).

1985-04-12 *for* 12 April 1985
1985-W15-5 *for* 5th day of 15th week of 1985
1985-04-12T15:30:00 *for* 3:30 PM on 12 April 1985

Avoid the sequence month, day, and year in the all-numeric form. If this form must be used, spell out the month and set the year off with commas.

The total solar eclipse of July 11, 1991, passed directly over the world's largest telescopes on Mauna Kea volcano, Hawaii.

If only the month and year are used, there is no intervening punctuation.

An eclipse passed over the South Atlantic Ocean in June 1992.

See Section 29.3.6.7 for information about dates in end references.

13.4 SPANS OF TIME

13.4.1 Divisions of Historic Time

Use numerals for year numbers, preceded or followed by era designations (see Table 13.1), if necessary. In general, use capital letters for the era designations to avoid the extra coding required for small capitals and for consistency with the usage recommended for abbreviations in general (see Chapter 11); however, some typographers prefer small capitals if available in the typeface. The choice of era designation should agree with the author's culture or with the convention used by the publisher to which the manuscript is submitted.

The era designations of AD, AC, and AH precede the year; all others follow the year. Note that there is no year zero.

AD 1492 2050 BC

See Section 13.5.1 for information on chronometric dating and Section 25.2 for information on geologic time units.

13.4.2 Names of Periods

The so-called 3-age system, published in 1836 by Christian Jürgensen Thomsen[7] and used primarily in archaeology, divides prehistory into 3 successive periods, defined

Table 13.1 Eras: names and abbreviations

Name of era	Abbreviation
After Christ	AC
Anno Domini (in the year of our Lord)	AD
Anno hegirae (in the year of the [Muhammad's] Hegira AD 622) or anno Hebraico (in the Hebrew year)	AH
Anno mundi (in the year of the world)	AM
Anno salutis (in the year of salvation)	AS
Ab urbe condita (from the founding of the city [Rome] in 753 BC)	AUC
Before Christ	BC
Before the common (Christian) era (= BC)	BCE
Before present (with "present" designated as 1950)	BP
Of the common (Christian) era (= AD)	CE

Table 13.2 The 3-age chronologic system for describing prehistory[a]

Age	Artifacts and cultural characteristics
Stone Age	
Paleolithic or Old Stone Age	Chipped stone tools only
Upper Paleolithic	
Middle Paleolithic (Mousterian)	
Lower Paleolithic	
Mesolithic or Middle Stone Age	Chipped stone tools, microliths, beginnings of pottery and ground stone in some areas
Neolithic or New Stone Age	Ground and polished stone tools, pottery, agriculture
Chalcolithic, Eneolithic, Copper Age[b]	Period between the Neolithic and the Bronze Age when unalloyed copper was used rather than true tin bronze
Bronze Age	Emergence of an urban way of life
Iron Age	Tools and weapons of iron

[a] Some cultures remained in "prehistory" in historical periods of the main civilizations—for example, the aboriginal culture of Australia.
[b] Terms used interchangeably.

by the main material in the period for making tools: stone, bronze, or iron (see Table 13.2). These periods are defined by cultural developments, and the actual dates differ among regions. Because these developments are evinced by artifacts, "Upper", as in "Upper Paleolithic", refers to strata closer to the surface, which are newer; hence, "upper" is equivalent to "late" and "lower" to "early" (see also Section 25.2). The 3-age system is not useful in Africa, where bronze was not used south of the Sahara, or the Americas, where bronze was never important and iron was not used until introduced by Europeans.

Capitalize the names of cultural periods, eras, and epochs (spans of time associated with dates). In addition, capitalize the words "age", "early", "middle", and "late" in this context, but do not capitalize the word "period". These words are also used to describe geologic time, in which capitalization is used to distinguish formal from informal names (see Section 25.2).

Late Bronze Age   Chalcolithic Age   Early Woodland period   Viking period

13.4.2.1 Ice Ages

The generic terms in names of ice ages are set in lowercase type, except for the Glacial Epoch (i.e., the Pleistocene Epoch).

Illinoian glaciation   Riss glaciation

The term "neoglacial" refers to the small-scale glacial advance that occurred during the "Holocene" (see Section 13.4.2.2). The most recent advance was the Little Ice Age, AD 1550 to 1850. "Neoglaciation" is an informal term used to designate glacial expansions later than the Holocene climatic optimum.

13.4.2.2 Recent and Present Time

The term "Holocene" refers to the epoch of the last 10,000 years, which replaced "Recent". Now the term "recent" is always set in lowercase type and connotes any recent time of unspecified duration. For any geochronologic date that has been correctly expressed (e.g., 1990 BP or 1990 Ma BP), do not add a term such as "ago" or "before the present". Abbreviations for the number of years without reference to the present differ widely and do not necessarily follow SI guidelines (see Section 25.2).

13.5 AGE MEASUREMENT

13.5.1 Chronometric Dating

Absolute (chronometric) dating is the measurement of age with reference to a specific time scale that dates from or to a fixed reference point (see Table 13.1). In the predominantly Christian world, the convention is to use the birth of Christ, set in the year AD 1 (= 1 CE), as that point. Years are counted back before Christ (BC) and forward after Christ (AD for *Anno Domini* [Latin for "in the year of our Lord"], AC for after Christ, or CE for of the common [Christian] era). There is no year zero.

An international system, without reference to a particular calendar system, counts years before the present (BP). AD 1950 is usually considered to represent the "present", so a date of 400 BP is about 450 years ago (AD 1550).

13.5.2 Radiometric Dating

Radiometric dating refers to all methods of age measurement that rely on the nuclear decay of naturally occurring radioactive isotopes, whether they are short-life radioactive elements (e.g., carbon-14 [^{14}C]) or long-life radioactive elements and their decay products (e.g., the ratio of potassium [^{40}K] to argon [^{40}Ar]; the uranium [U] series, involving the decay of ^{238}U and ^{235}U to thorium [^{230}Th] and protactinium [^{231}Pa]).

Radiocarbon is still the most frequently used tool for dating organic samples as old as perhaps 50,000 years, whereas potassium–argon dating (determining the ratio of potassium to argon) is common for determining the age of rock samples at least 100,000 years old. For strata or artifact assemblages between those ages, dating is far less reliable.

In early practice, archaeomagnetic dates were reported in the same form as ^{14}C dates, as a mean value plus or minus a certain number of years (AD 1015 ± 35). It is now considered more accurate to report dates as ranges (e.g., AD 980 to 1050). When such dates are cited, follow the original exactly. Include with a plus-or-minus designation a symbol indicating what the numeric value following the symbol represents: "*s*" for "standard deviation", "2*s*" for "2 standard deviations", or "S_x" for "standard error of the mean" (see Section 12.5.1).

AD 1015 ± 35(*s*)

The conversion of radiocarbon ages to calendar dates must take into account temporal variations in the natural radiocarbon content of atmospheric carbon dioxide. Consult the website of the journal *Radiocarbon*[8] (http://www.radiocarbon.org/Info/index.html) for up-to-date calibrations.

Uncalibrated dates (i.e., dates not checked against those derived from another system) are not linked to any system of reckoning in calendar years. If the date has not been calibrated with another dating method (e.g., dendrochronology), the era is sometimes presented in lowercase letters (e.g., ad, bc, bp); alternatively, the calibrated date may be presented as "Cal BC" or "Cal AD".

uncalibrated:	ad 1500	3000 bc	400 bp
calibrated:	AD 1400	3700 BC	Cal AD 1550

CITED REFERENCES

1. International Organization for Standardization. Data elements and interchange formats—information interchange—representation of dates and times. Geneva (Switzerland): International Organization for Standardization; 2004. (ISO 8601).

2. International Organization for Standardization. Quantities and units. Part 3: Space and time. Geneva (Switzerland): ISO; 2006. (ISO 80000-3).

3. The Chicago manual of style: the essential guide for writers, editors, and publishers. 16th ed. Chicago (IL): University of Chicago Press; 2010. Also available at http://www.chicagomanualofstyle.org.

4. Bureau International des Poids et Mesures. Le Système internationale des unités (SI). 8th ed. Paris (France): Organisation Intergouvernementale de la Convention du Metre; 2006. English translation available in PDF format at http://www.bipm.org/utils/common/pdf/si_brochure_8_en.pdf.

5. Wilimovsky NJ. Misuses of the term "Julian Day" [letter]. Trans Am Fish Soc. 1990;119(2):162.

6. Kendall RL. Misuses of the term "Julian Day" [letter]. Trans Am Fish Soc. 1990;119(2):162.

7. Spjeldnaes N. Christian Jürgensen Thomsen. In: Dictionary of scientific biography. Vol. 13. New York (NY): Charles Scribner Sons; 1976. p. 357–358.

8. Radiocarbon. Tucson (AZ): Arizona Board of Regents. Vol. 1, 1959– . Available at http://www.radiocarbon.org. Up-to-date information on radiocarbon dating and calibration is provided at http://www.radiocarbon.org/Info/index.html.

14 Geographic Designations

14.1 GEOGRAPHIC NAMES

The geographic location of a study area and descriptions of geographic features are important components of the reports of many scientific studies. The spelling of all geographic place names should be verified by the author or by the manuscript editor. There are printed sources for verifying spelling and accuracy of naming, such as *Merriam-Webster's Geographical Dictionary*[1] and *The Columbia Gazetteer of the World*,[2] but in some countries and regions of the world names change so frequently that it may be best to refer to continually updated geographic name databases on the Internet. These resources and others are described below.

14.1.1 Geographic Names in Text

For guidance on capitalization of geographic names, see Section 9.7.3.

14.1.1.1 States and Similar Divisions

In general, include the unabbreviated name of the state, territory, possession, or province with the name of a city or town when the names are used in running text. Where abbreviations of US state or Canadian province names are needed for reasons of brevity, as in tables and end references, use the 2-letter postal abbreviations; see Table 14.1 and Section 29.3.6.5.

Depending on the anticipated readership, a journal or publisher may choose to omit the name of the state or other regional subunit for well-known cities, but do not omit this information if there is a reasonable potential for confusion (e.g., London, Ontario,

Table 14.1 US and Canadian government units and their postal symbols

Unit	Symbol	Unit	Symbol
	United States		
Alabama	AL	Montana	MT
Alaska	AK	Nebraska	NE
American Samoa	AS	Nevada	NV
Arizona	AZ	New Hampshire	NH
Arkansas	AR	New Jersey	NJ
California	CA	New Mexico	NM
Colorado	CO	New York	NY
Connecticut	CT	North Carolina	NC
Delaware	DE	North Dakota	ND
District of Columbia	DC	Northern Mariana Islands	MP
Federated States of Micronesia	FM	Ohio	OH
Florida	FL	Oklahoma	OK
Georgia	GA	Oregon	OR
Guam	GU	Palau	PW
Hawaii	HI	Pennsylvania	PA
Idaho	ID	Puerto Rico	PR
Illinois	IL	Rhode Island	RI
Indiana	IN	South Carolina	SC
Iowa	IA	South Dakota	SD
Kansas	KS	Tennessee	TN
Kentucky	KY	Texas	TX
Louisiana	LA	Utah	UT
Maine	ME	Vermont	VT
Marshall Islands	MH	Virgin Islands	VI
Maryland	MD	Virginia	VA
Massachusetts	MA	Washington	WA
Michigan	MI	West Virginia	WV
Minnesota	MN	Wisconsin	WI
Mississippi	MS	Wyoming	WY
Missouri	MO		
	Canada		
Alberta	AB	Nunavut	NU
British Columbia	BC	Ontario	ON
Manitoba	MB	Prince Edward Island	PE
New Brunswick	NB	Quebec	QC
Newfoundland and Labrador	NL	Saskatchewan	SK
Northwest Territories	NT	Yukon Territory	YT
Nova Scotia	NS		

Canada; London, United Kingdom). Use commas to separate the name of a city from the name of a state or other regional subunit. Do not abbreviate the names of regional subunits when they appear on their own (i.e., without the name of a city or town).

> Evidence of a new fault line was found 8 km south of Turkington, Missouri.
> The Children's Hospital in Denver, Colorado, conducts extensive research in this area.
> The energy industry in Alberta has developed new methods to search for natural gas deposits.

14.1.1.2 Regions and Geographic Features

Capitalize any generic geographic term, such as "river", that is part of a place name (as verified in an appropriate resource, such as an atlas or gazetteer), but do not capitalize the generic term if it appears on its own or if it is plural and follows 2 or more proper names.

Mississippi River [but "the river" in a subsequent text reference]
Great Dismal Swamp [but "the swamp" in a subsequent text reference]
Mississippi and Missouri rivers

Some geographic names contain foreign words that are the equivalents of generic terms (e.g., "rio" means "river"; "mauna" and "yama" mean "mountain"; "sierra" means "mountains" or "mountain range"); therefore, do not add the generic English term.

Fujiyama *or* Mount Fuji	*not*	Fujiyama Mountain *or* Mount Fujiyama
Mauna Loa	*not*	Mauna Loa Mountain *or* Mount Mauna Loa
Sierra Madre	*not*	Sierra Madre Mountains
Rio Grande	*not*	Rio Grande River
the Sahara	*not*	Sahara Desert

Many regions and localities have informal names that are recognized as representing a specified area and may or may not be recognized as proper nouns. Such terms should be used with care in scientific publications. If such a term is used to describe a collecting area, readers unfamiliar with the boundaries of the informal regions could be deprived of important information; furthermore, the definition of the area may not be absolute. Therefore, whenever possible, specify locations by geographic coordinates (see Section 14.2) or, more rarely, by map-indicated political boundaries (depending on the context) to eliminate any possibility of confusion. If such regional names are used, capitalize them.

the Great Plains [plains region in Canada and the United States east of the Rockies]
the National Capital [USA]
the National Capital Region [Canada]
the Maritimes [Nova Scotia, New Brunswick, and Prince Edward Island, Canada]

Many regional names include descriptive terms (e.g., "east", "west", "lower", "upper", and "central"), which should be capitalized only if they are consistently used to denote a defined region that is recognized as a proper name. *The Chicago Manual of Style*[3] lists a variety of such terms and their variations; print and online atlases, gazetteers, and dictionaries may also be used to determine whether such terms are considered proper names.

Middle East	eastern Europe
western California	Southern California
Lower Canada [historical]	lower St Lawrence River
Central America	central Manitoba
the Upper Peninsula [Michigan]	

The Arctic is the region of the earth north of the Arctic Circle. The word "arctic" is used adjectivally in several ways: when it refers to the geographic region, capitalize the adjective (Arctic communities); when it refers to very low temperatures, use lowercase (arctic gale). Established names of Arctic flora and fauna are usually lowercased (arctic char).

14.1.1.3 Countries

For countries with names that include the article "the", do not capitalize the article, except for The Bahamas and The Gambia.[4]

the Netherlands	the Philippines

Spell out United Kingdom and United States (or United States of America) when used as nouns, but "UK" and "US" may be used as adjectives.

Owing to internal and external strife and country reorganization, the names of some countries change over time. It is wise to check with online geographic databases or, perhaps better, with the official website of the country in question to be sure that the name used is current.

14.1.1.4 Multinational Regions

Multinational regions typically have some common feature, such as geography, language, or culture, but these regions are not always strictly defined, and some may overlap in composition. For example, North America includes all the land from, and including, Panama north to the Arctic, encompassing the West Indies (Caribbean islands) and Greenland. Central America, which is part of North America, includes Guatemala, Belize, El Salvador, Honduras, Nicaragua, Costa Rica, and Panama but not the West Indies.

The Middle East has no precise definition but is generally thought to include Turkey, Cyprus, Syria, Lebanon, Israel, Jordan, Egypt, Iraq, Iran, and the countries of the Arabian Peninsula. At its maximum, it would extend from Morocco to Bangladesh and would include North Africa, the Horn of Africa, Greece, and Bulgaria. "Near East" is a dated term for the same area that may appear in historical references.

Consult a geographic dictionary or database to determine the composition of multinational regions. The names of most recognized regions are capitalized. If in doubt, consult a geographic dictionary or database, or other resource.

Middle East   North America   West Bank   Micronesia   Far East

14.1.2 Authorities for Names

14.1.2.1 United States

The United States Board on Geographic Names formulates the principles, policies, and procedures determining the use (especially by the US government) of domestic (and foreign) geographic names and the names of undersea and Antarctic features. The Geographic Names Information System (GNIS),[4] developed by the United States Geological Survey in cooperation with the Board on Geographic Names, contains about 2 million records for names of physical and cultural geographic features, both historical and current. Official names, as specified by state geographic names boards, are identified as such, but the database also includes records of named features that do not have official status. The GNIS is available as a searchable online database (http://geonames.usgs.gov/). For geographic names in the United States, use this database as the authoritative source for spelling and capitalization unless there is a specific reason for using a different form.

14.1.2.2 Canada

In Canada, the official body that accepts or rejects geographic names is the government of the province or territory in which the feature lies, except for federally administered lands, such as Indian reserves, national parks, and military reserves. The Geographical Names Board of Canada, a national coordinating body, acts as a central registry for all approved names in Canada. Its secretariat enters all official, and some unofficial, names into the Canadian Geographical Names Data Base,[5] which covers all types of geographic features, including undersea features. It is available as a searchable online database (http://www.nrcan.gc.ca/earth-sciences/search/index_e.php). The Commission de toponymie Québec maintains an online database for names of places in the province of Quebec (http://www.toponymie.gouv.qc.ca/ct/accueil.aspx).[6] For geographic names in Canada, use these databases as the authoritative sources for spelling and capitalization unless there is a specific reason for using a different form; for example, publishing offices within the US government will follow the preferences of the Foreign Names Committee of the Board on Geographic Names; see Section 14.1.2.3.

14.1.2.3 Other Parts of the World

Many other countries also have official bodies with responsibility for standardizing geographic names, both domestic and foreign (e.g., the Permanent Committee on Geographical Names in the United Kingdom, searchable at http://www.pcgn.org.uk/); many have searchable online databases.

In the United States, the Foreign Names Committee of the Board on Geographic Names is responsible for standardizing the use of place names outside the borders of the United States and its possessions. For some foreign place names, the board has approved anglicized forms.

Jordan River *for* ahr al Urdunn
Rome *for* Roma
Danube River *for* Donau (Austria, Germany) *or* Duna (Hungary)
Vatican City *for* Città del Vaticano

The National Geospace Intelligence Agency offers a searchable online database (http://geonames.nga.mil/ggmagaz/) that serves as the official repository of foreign places.

If an author has a strong preference regarding the name of a country (such as Republic of China or Taiwan), disregarding that preference is perhaps insensitive when the only reason for doing so is that the US Board on Geographic Names prefers a different spelling or name. For instance, non-Malawian authors may spell the country name "Malawi", but in the official language of that country (Chichewa) the spelling is "Malaŵi" (meaning flames); without the circumflex over the "w", the word is meaningless and hence not accepted nationally. In the past, editorial offices might have rejected an author's preference because creating a character not typically used in English was difficult with standard equipment; however, modern word-processing programs and publication software can now produce a wide range of characters not used in English, and thus this should not present an impediment.

14.1.3 Punctuation

Few North American place names include punctuation (see Section 14.1.2 for resources).

Richardsons Creek *not* Richardson's Creek
Baileys Crossroads *not* Bailey's Crossroads
Martha's Vineyard

14.1.4 Numbers

For most place names that include a number, spell out the number and do not hyphenate the name.

Two Rivers   Three Mile Island   Fourteen Mile Point

For cities named with dates, do not spell out the number, but alphabetize such names as though they were spelled out in English.

20 de Junio, Argentina   25 de Diciembre, Peru

14.1.5 Multiword Names

Capitalize all elements of multiword geographic names except prepositions, conjunctions, and articles.

Fond du Lac, Wisconsin   Truth or Consequences, New Mexico
Point of Rocks, Arizona

For multiword place names in the province of Quebec, Canada, hyphenate all components of the name.[6]

Saint-Augustin-de-Demaures   Sainte-Anne-de-la-Pérade

14.1.6 Prefixes

Spell out and capitalize most prefixes to geographic names, including generic geographic terms that precede the proper name.

Point Lobos   Port Arthur   San Diego

For English-language place names, "Saint" and "Mount" may be abbreviated, respectively, to "St" and "Mt" if they do not occur together.

St Louis   Mt Washington *but* Mount St Helens

For place names in the province of Quebec, do not abbreviate "Saint", "Sainte", or "Mont".[6]

Saint-Lawrence   Sainte-Foy   Saint-Henri

14.2 GEOGRAPHIC COORDINATES

14.2.1 Latitude and Longitude

Latitude is the distance north or south of the equator, designated by parallels that have traditionally been measured in degrees, minutes, and seconds, beginning with 0° at the equator and progressing to 90° north or south of the equator. Longitude is the distance east or west of the prime meridian Greenwich, designated by meridians and traditionally measured in degrees, minutes, and seconds from 0° to 180°. Spell out these terms when they appear alone in text (without specific numeric designations), but abbreviate them (to "lat" and "long") when they appear as part of a coordinate.

Several conventions exist for presenting a geographic coordinate in terms of latitude and longitude. Some of the variants have been developed recently to facilitate the storage of data in electronic databases. For example, the coordinate may be stripped of nonnumeric characters. However, other aspects of the presentation are standard, such as presenting latitude first.

In text where space and special characters are not limiting factors, present a geographic coordinate with latitude first, followed by a comma and then longitude. For both latitude and longitude, the abbreviation precedes its coordinate, and the directional designation (N, S, E, W) follows. Present the numbers and their symbols without spaces and use a leading zero for degrees, minutes, and seconds less than 10. Although the directional indicators N and S imply latitude, and E and W imply longitude, use of the abbreviations "lat" and "long" may aid in understanding. Use the prime symbol (not a single quotation mark or apostrophe) for minutes and the double prime symbol (not a double quotation mark) for seconds.

lat 43°15′09″N, long 116°40′18″E
or 43°15′09″N, 116°40′18″E

lat 04°59′17″S, long 01°02′03″W
or 04°59′17″S, 01°02′03″W

According to another convention, latitudes north of the equator and longitudes east of the prime meridian are considered positive (although the plus sign is not used), and latitudes south of the equator and longitudes west of the prime meridian are considered negative. If this convention is used, omit both abbreviations "lat" and "long" and the directional indicator.

43°15′09″, 116°40′18″   -04°59′17″, -01°02′03″

If breaking a coordinate at the end of a line is unavoidable, do so after the symbol for a coordinate unit and use a hyphen to indicate the break.

long 116°-   lat 45°29′-
40′18″E   14″S

Latitude and longitude may be reported in "decimal degrees" instead of degrees, minutes, and seconds. Decimal degrees are easier to use in databases and spreadsheets. In this system, the degree value appears to the left of the decimal point, and the minutes and seconds are converted to a decimal value that appears to the right of the decimal point. To compute the decimal value, divide the minutes by 60, divide the seconds by 3600, and sum the 2 resulting values. The degree symbol is omitted, as is the plus sign for northern latitudes and eastern longitudes; the minus sign is included for southern latitudes and western longitudes. Some software applications have built-in functions for changing latitude and longitude from one system to the other, and websites are available to perform the same function; an example can be found at http://transition.fcc.gov/mb/audio/bickel/DDDMMSS-decimal.html.[7]

lat 38°45′N *becomes* lat 38.75   long 38°52′30″ *becomes* long 38.875

Latitude–longitude coordinates for many US geographic entities are available in the Geographic Names Information System (http://geonames.usgs.gov/).[4]

14.2.2 Other Coordinate Systems

Some other coordinate systems (e.g., plane rectangular coordinates, polar coordinates) are not based on the latitude–longitude system. Nonetheless, any coordinate system must be based on a known reference point, stated at the outset if necessary. For an overview of various coordinate systems, see Maling.[8]

14.2.2.1 Plane Coordinate Systems

Plane coordinate systems (also called plane rectangular coordinate systems or plane cartesian coordinate systems) consist of a pair of fixed axes in 2 directions from which linear measurements are made. When grid lines drawn from points along the axes intersect at right angles, the coordinates identifying the intersection are referred to as "rectangular coordinates". Rectangular coordinates are commonly written as a pair of numbers separated by a comma, within parentheses: (x, y). Grid units in rectangular coordinates are arbitrary, but ground measurements are usually in meters. Coordinates follow the mathematical sign convention whereby coordinates on the x-axis are positive to the right of the y-axis and negative to the left of the y-axis. Similarly, coordinates on the y-axis are positive above the x-axis (toward the top of the page) and negative below the x-axis.

Usually the first number in a coordinate pair represents a point along the x or east–west axis, commonly called the "easting". The second number is a point along the y or north–south axis, usually called the "northing". The directions represented by the coordinates in a plane coordinate system are opposite those described in geographic coordinates (where the first designation in the coordinate pair is the latitude, representing a north–south direction, and the second designation is longitude, representing an east–west direction). However, in studies of geodesy, surveying, and map projections, this convention may not be observed. The capital letters E (for easting) and N (for northing) are recommended if the coordinates do not follow the usual order of x (easting) followed by y (northing).

The Universal Transverse Mercator (UTM) coordinate system is a commonly used plane coordinate system in which the world is divided into 60 north–south zones (6° of longitude each). In the plane coordinate systems for US states, the origin is shifted so that most of the state or locality lies to the right of the y-axis and above the x-axis. Most coordinates in these systems are positive.

14.2.2.2 Polar Coordinate Systems

Polar coordinates also define position by using a pair of coordinates (r, θ), but the first coordinate in the pair is a linear measurement (r) and the second is an angular measurement (θ). The 2 axes in a planar coordinate system are replaced in a polar coordinate system by a single axis, the polar axis or initial line, which passes through the origin or pole. The position of a point is defined as the distance of the radius from the pole to the point and the angle that that radius makes with the polar axis. Angles in polar coordinates usually are expressed in degrees or centesimal (grad or grade) units, although radians are required in some theoretical map projections. In mathematics, angles measured in a counterclockwise direction from the polar axis are positive, but surveyors, cartographers, and navigators usually refer to angles measured clockwise from the axis as positive.

A small area or locale may be designated on the basis of the US Public Lands Survey inaugurated in 1785. This system has as its reference points a number of principal north–south meridians and east–west baselines established by the Public Lands Survey. The intersection of a meridian and a baseline is designated an "initial point". Townships (6 miles square) are numbered north and south of the baseline with a township number and east and west of the meridian with a range number. Townships are divided into 36 square sections (1 mile square) numbered from 1 to 36, beginning with the northeasternmost section. Sections are divided into quarter sections. Topographic maps of the US Geological Survey carry symbolization relevant to identifying tracts covered by the system of public lands subdivisions. Further details on the style for identifying such tracts are provided in Section 25.9.

Some maps of the British Ordnance Survey carry grid-square alphanumeric identifiers based on the British National Grid that enable a location in Great Britain to be specified to the nearest 100 m; for example, the grid reference for Treglossik, a village in Cornwall, is SW 787 236. SW is the designation for the grid square covering a large fraction of Cornwall; the first 2 digits of each 3-digit number (78 and 23, respectively) represent small squares within SW, and the third digits (7 and 6, respectively) represent easting and northing estimates from the grid markings on the Ordnance Survey map.

14.3 ADDRESSES

A scientific journal article usually designates one author to receive correspondence on behalf of all coauthors and provides this person's mailing address, fax number, and email address (on either the first or the last page of the article). Ensure that the mailing address includes the following information:

1) the department and institution name
2) the street address
3) the city or town
4) the state, province, or other regional unit, if applicable (for addresses in the United States and Canada, use the 2-letter postal abbreviation for the state or province, as listed in Table 14.1)
5) the postal code (in the United States the ZIP code, for Zone Improvement Plan), an alphanumeric designation assigned to specific addresses by the postal agency within each country, whose placement varies by country and which sometimes precedes or follows the city name, rather than following the regional unit
6) the country, without abbreviation (except "USA" for United States of America and "UK" for United Kingdom); if in doubt about the proper form of a country name, use the spelling specified by the post office of origin of the publication

Many national postal agencies—such as the US Postal Service (http://pe.usps.com/businessmail101/addressing/deliveryaddress.htm), Canada Post (http://www.canadapost.ca/tools/pg/manual/PGaddress-e.asp), and the Royal Mail (in the United Kingdom [http://www.royalmail.com/personal/help-and-support/How-do-I-address-my-mail-correctly])—provide detailed guidance on addressing mail for domestic and international delivery. Other websites, such as Bitboot Systems International Mailing Address Formats and Other International Mailing Information (http://bitboost.com

/ref/international-address-formats.html), provide information on addressing mail to multiple countries.

An email address should be included as part of the author correspondence information, because it is the most likely way in which the journal will correspond with authors. Because most such addresses include internal periods, choose a form of presentation in which no punctuation is needed after the email address (e.g., email address in parentheses, on its own line, or at the end of a run-on paragraph).

CITED REFERENCES

1. Merriam-Webster's geographical dictionary. 3rd ed. Springfield (MA): Merriam-Webster; 2007.

2. Cohen SB, editor. The Columbia gazetteer of the world. 2nd ed. New York (NY): Columbia University Press; 2008. Also available at http://cup.columbia.edu/static/gazonline.

3. The Chicago manual of style: the essential guide for writers, editors, and publishers. 16th ed. Chicago (IL): University of Chicago Press; 2010. Also available at http://www.chicagomanualofstyle.org.

4. Geographic Names Information System (GNIS). Reston (VA): US Geological Survey; [date unknown]. http://geonames.usgs.gov/.

5. Geographical names of Canada. Ottawa (ON): Natural Resources Canada; [date unknown]. http://www.nrcan.gc.ca/earth-sciences/search/index_e.php.

6. Banque de noms de lieux du Québec. Québec (QC): Gouvernement du Québec, Commission de toponymie Québec; 2011. http://www.toponymie.gouv.qc.ca/ct/accueil.aspx.

7. Degrees, Minutes, Seconds and Decimal Degrees Latitude/Longitude Conversions. Washington (DC): Federal Communications Commission; 2011. http://transition.fcc.gov/mb/audio/bickel/DDDMMSS-decimal.html.

8. Maling DH. Coordinate systems and map projections. 2nd ed. Oxford (UK): Pergamon Press Ltd.; 1992.

ADDITIONAL REFERENCES

Hansen WR, editor. Suggestions to authors of the reports of the United States Geological Survey. 7th ed. Reston (VA): United States Geological Survey; 1991 [site modified 2012 May 30; accessed 2012 June 20]. http://www.nwrc.usgs.gov/lib/lib_sta.htm.

National atlas of the United States. Washington (DC): US Department of the Interior; [updated 2012 June 1; accessed 2012 June 22]. http://nationalatlas.gov/.

National Geographic atlas of the world. 9th ed. Washington (DC): National Geographic Society; 2010.

United States Government Printing Office style manual: main page. 30th ed. Washington (DC): The Office; 2008 [accessed 2013 Jul 28]. http://www.gpo.gov/fdsys/pkg/GPO-STYLEMANUAL-2008/pdf/GPO-STYLEMANUAL-2008.pdf.

3

Special Scientific Conventions

15 The Electromagnetic Spectrum

15.1 UNITS OF MEASURE

The electromagnetic spectrum is a continuum of radiated electromagnetic energy characterized by the basic relation $c = \lambda v$, where c is the speed of light (approximately 3×10^8 m/s), λ is the wavelength of the radiation in meters (m), and v is the frequency of the radiation in hertz (Hz). The practical bounds of the electromagnetic continuum covered in this chapter are from extremely low frequency radio waves (about 10 Hz, or wavelengths of about 3×10^7 m) to the gamma rays produced by primary cosmic rays (about 3×10^{22} Hz [30 ZHz], wavelength about 10^{-14} m). Given that the spectrum is a continuum, none of the boundaries between segments of the spectrum is exact, and each segment overlaps with neighboring segments (see Table 15.1).

Electromagnetic waves and quantized photons (or quantized particles associated with light propagation) are equivalent concepts because of wave–particle duality. The energy of a photon is given by $E = hv$, where h is Planck's constant. As the value of the light frequency (in hertz) increases, the energy of the radiation quanta (i.e., discrete units) increases. Convenient dimension sets vary with discipline, but a common one relates wavelength in micrometers (µm) to photon energy in electron volts (eV) by the equation $E = 1.234/\lambda$. (The electron volt is a non-SI unit of energy; 1 eV = 0.160 aJ.)

Radio waves are usually characterized by frequency (in hertz) or wavelength (in meters or millimeters). Visible, infrared, and ultraviolet light is often described by its wavelength in nanometers (nm) or micrometers (µm). Infrared spectroscopists, however, often express wavelength in wave numbers ($\bar{v}$), the number of wavelength units in 1 cm, where $\bar{v} = 1/\lambda$ and is given in units of reciprocal centimeters (cm^{-1}). Spectroscopists sometimes express X-ray wavelength in units of ångströms (Å), where 10 Å = 1 nm. Astronomers and other physicists usually characterize X-rays and gamma rays by their energy, in thousands or millions of electron volts (keV, MeV). Use the units that are standard for a particular discipline.

Table 15.1 The electromagnetic spectrum

Region	Frequency range		Wavelength		Other common units
	Conventional units	Hertz	Conventional units	Meters	
Radiowaves					
Marine communications					
Extremely low frequency (ELF)	3 Hz to 30 Hz	3×10^0 to 3×10^1	100 Mm to 10 Mm	10^8 to 10^7	
Super low frequency (SLF)	30 Hz to 300 Hz	3×10^1 to 3×10^2	10 Mm to 1 Mm	10^7 to 10^6	
Ultra low frequency (ULF)	300 Hz to 3 kHz	3×10^2 to 3×10^3	1 Mm to 100 km	10^6 to 10^5	
Very low frequency (VLF)	3 kHz to 30 kHz	3×10^3 to 3×10^4	100 km to 10 km	10^5 to 10^4	
Broadcasting, communications					
Low frequency (LF)	30 kHz to 300 kHz	3×10^4 to 3×10^5	10 km to 1 km	10^4 to 10^3	
Amplitude modulation (AM) radio broadcasts					
Medium frequency (MF)	300 kHz to 3 MHz	3×10^5 to 3×10^6	1 km to 100 m	10^3 to 10^2	
Shortwave radio					
High frequency (HF)	3 MHz to 30 MHz	3×10^6 to 3×10^7	100 m to 10 m	10^2 to 10^1	
Frequency modulation (FM radio and TV broadcasts)					
Very high frequency (VHF)	30 MHz to 300 MHz	3×10^7 to 3×10^8	10 m to 1 m	10^1 to 10^0	

TV broadcasts					
Ultra high frequency (UHF)	300 MHz to 3 GHz	3×10^8 to 3×10^9	1 m to 100 mm	10^0 to 10^{-1}	
Microwaves					
Communication, radar					
Super high frequency (SHF)	3 GHz to 30 GHz	3×10^9 to 3×10^{10}	100 mm to 10 mm	10^{-1} to 10^{-2}	
Extremely high frequency (EHF)	30 GHz to 300 GHz	3×10^{10} to 3×10^{11}	10 mm to 1 mm	10^{-2} to 10^{-3}	
Micrometer waves (submillimeter or THz waves)					
Instrumentation applications	300 GHz to 3 THz	3×10^{11} to 0.003×10^{11}	1 mm to 0.1 mm	10^{-3} to 0.010^{-3}	
Optical wavelengths					
Infrared	300 GHz to 400 THz	3×10^{11} to 4×10^{14}	1 mm to 760 nm	10^{-3} to 7.6×10^{-7}	10,000 to 10 cm^{-1} (wavelength units)
Visible light	400 THz to 750 THz	4×10^{14} to 7.5×10^{14}	760 nm to 400 nm	7.6×10^{-7} to 4×10^{-7}	
Ultraviolet	750 THz to 3 PHz	7.5×10^{14} to 3×10^{15}	400 nm to 100 nm	4×10^{-7} to 10^{-7}	
X-ray	30 PHz to 6 EHz	3×10^{16} to 6×10^{18}	10 nm to 50 pm	10^{-8} to 5×10^{-11}	1 to 100 keV
					0.1 to 100 Å
Gamma ray	6 EHz to 600 EHz	6×10^{18} to 6×10^{20}	50 pm to 500 fm	5×10^{-11} to 5×10^{-3}	100 keV to 10 MeV

15.2 LONGER WAVELENGTHS

The long wavelengths at the low-energy end of the spectrum (wavelengths longer than 1×10^{-2} m) include wavelengths that are used for communications, electronics, and instrumentation, such as radio and television transmission bands and microwave and submillimeter bands. The segments of this end of the spectrum and their corresponding use, abbreviations, and approximate frequencies and wavelengths are given in Table 15.1 (from extremely low to extremely high frequency). The submillimeter band from 0.3 to 3 THz merges into the longest wavelengths of the thermal infrared region of the electromagnetic spectrum. Allocations for using specific bands in the radio-frequency spectrum vary by country.[1–3]

15.3 OPTICAL WAVELENGTHS

Electromagnetic radiation with frequencies between 3×10^{11} Hz (300 GHz, $\lambda = 1$ mm) and 3×10^{15} Hz (3 PHz, $\lambda = 100$ nm) is termed optical radiation. This range includes the infrared (IR), visible, and ultraviolet (UV) regions of the spectrum. Optical radiation is more often referred to by wavelength than by frequency.

IR radiation has wavelengths between 1 mm and 760 nm (frequency 3×10^{11} Hz and 4×10^{14} Hz). The IR spectrum may be divided into 3 segments, labeled C, B, and A, with the following approximate boundaries (use of a hyphen in the segment designations facilitates recognition of the standard abbreviation for "infrared").

IR-C	1 mm (1×10^{6} nm) to 3,000 nm
IR-B	3,000 to 1,400 nm
IR-A	1,400 to 760 nm

Conventional IR spectroscopy of vibrational modes in molecules is performed in the midrange of 4,000 to 200 wave numbers (cm^{-1}), which corresponds roughly to the range of 2.5 to 50 μm. The term "near infrared" refers to wavelengths shorter than this range; "far infrared" denotes longer wavelengths. Common techniques used for IR spectroscopy include absorption spectroscopy, Raman spectroscopy, and Fourier transform IR spectroscopy (FTIR).

Visible light occupies the segment of the electromagnetic spectrum with wavelengths of approximately 760 to 400 nm. The human eye can perceive wavelengths between approximately 700 and 400 nm, the wavelengths of the familiar spectrum of colors from red to violet and of photosynthetically active radiation (i.e., usable by photosynthetic organisms). The colors of the visible spectrum overlap, so boundaries for color ranges often vary slightly among scientific disciplines. The approximate ranges of wavelengths are as follows:

Color	Wavelength
red	700–630 nm
orange	630–590 nm
yellow	590–530 nm
green	530–480 nm
blue	480–440 nm
violet	440–400 nm

UV radiation occupies the segment of the electromagnetic spectrum between visible light and the X-ray region, with boundaries of approximately 400 nm on the visible light side and approximately 100 nm on the X-ray side. The UV region has traditionally been subdivided into the following 4 ranges:

near UV	400–300 nm
middle UV	300–200 nm
far UV	200–100 nm
extreme UV	<100 nm

The short-wavelength limit of extreme UV is variously cited as between 40 and 10 nm. Because wavelengths below 200 nm are strongly absorbed in air, studies of this region must be conducted in another gas or in a vacuum, resulting in the alternative name "vacuum ultraviolet". The region from 185 to 120 nm is also called the Schuman region, after its first investigator.

For biology,[4] the UV spectrum has conventionally been subdivided differently into the following 3 regions:

near UV	400–315 nm
actinic UV	315–200 nm
vacuum UV	<200 nm

Other biologically oriented subdivisions have been designated by the International Commission on Illumination[5] on the basis of the interactions of the wavelengths with biological materials. These subdivisions have been labeled A, B, and C, with the original wavelength boundaries listed below.

UV-A	400–315 nm
UV-B	315–280 nm
UV-C	280–100 nm

In practice, the boundary between UV-A and UV-B is often regarded as 320 nm (e.g., for dermatologic applications).

Although the symbols for the UV spectrum segments are often written as 3 unspaced capital letters, the use of a hyphen (as shown above) facilitates recognition of the standard abbreviation for "ultraviolet". To prevent possible misreading as subdivisions of the UV region, avoid using the symbols UVR and UVL to designate "ultraviolet radiation" and "ultraviolet light"; instead, use "UV radiation" and "UV light".

15.4 SHORTER WAVELENGTHS

Beyond the UV portion of the spectrum are X-rays, gamma rays, and cosmic rays. Hyphenate the term "X-ray" and capitalize the "X"[6]; do not hyphenate "gamma ray" (or "γ ray") except when it is used as an adjective.

X-rays occupy the spectrum segment from about 3×10^{16} Hz (30 PHz, $\lambda = 10$ nm) to about 6×10^{18} Hz (6 EHz, $\lambda = 50$ pm). X-rays of lower frequencies (3×10^{16} Hz to about 3×10^{18} Hz) are known as "soft" X-rays; those of shorter wavelengths are "hard" X-rays. X-rays are distinguished from the forms of electromagnetic radiation described in Sections 15.2 and 15.3 by having enough energy to cause ionization on passing through matter (ionizing radiation).

X-rays are produced by several techniques. Some of these techniques yield X-rays of a characteristic single energy; others produce a continuum of energies. One of the most common techniques is bombardment of a metal target with a high-energy electron beam in a vacuum enclosure. Bombardment of metal atoms ionizes the inner atomic orbitals; subsequent transitions of electrons from outer to inner atomic orbitals release X-rays that have a characteristic energy equal to the energy difference between the orbital levels. When the transition is to the *K* shell of an atom, the series of spectral lines produced by these X-rays is called the *K* series (or *K* lines); other series are named after the appropriate orbital shell, also using italic capital letters[6]: *L*, *M*, *N*, *O*, and *P*. The *K* series of virtually all elements consists of 4 major lines, which are traditionally named γ (or β_2), ß (consisting of β_1 and β_3), α_1, and α_2. In the traditional (Siegbahn) designation for a particular line, present the capital letter for the series, followed (without a space) by the line name: for example, $K\alpha_2$ or $L\beta_3$. The International Union of Pure and Applied Chemistry[7] has recommended a different system, which uses the starting and ending shell letters, presented as roman capital letters instead of italic capitals. In this system, $K\alpha_2$ becomes $\mathrm{K\text{-}L_2}$, and $L\beta_3$ becomes $\mathrm{L_1\text{-}M_3}$. This system is preferred to avoid using italic capital *K*, which is also used in statistics.

Gamma rays occupy the next segment of the electromagnetic spectrum. Soft gamma rays share an ambiguous boundary with hard X-rays, starting near 6×10^{18} Hz (E = 25 keV or λ = 50 pm) and extending to perhaps 6×10^{19} Hz (E = 250 keV, λ = 5 pm); hard gamma rays extend to frequencies of approximately 6×10^{20} Hz (600 EHz, E = 2.5 MeV, λ = 500 fm), which is roughly the lower limit of energy for the gamma rays produced from atmospheric collisions of primary cosmic rays.

CITED REFERENCES

1. Industry Canada, Spectrum and Radio Policy, Telecommunications Policy Branch. Canadian table of frequency allocations 9 kHz to 275 GHz. 2009 ed. Ottawa (ON): Industry Canada; 2009 [accessed 2012 May 15]. http://www.ic.gc.ca/spectrum.

2. Department of Commerce (US), National Telecommunications and Information Administration, Office of Spectrum Management. United States frequency allocations: the radio spectrum. Washington (DC): The Department; 2003 [accessed 2012 May 15]. http://www.ntia.doc.gov/osmhome/allochrt.pdf.

3. Rostami A, Rasooli H, Baghban H. Terahertz technology: fundamentals and applications (lecture notes in electrical engineering). Heidelberg (Germany): Springer; 2011.

4. McKinlay AF. Ultraviolet radiation: potential hazards. In: McAinsh TF, editor. Physics in medicine & biology encyclopedia: medical physics, bioengineering, and biophysics. Vol. 2. Oxford (UK): Pergamon; 1986. p. 859–862.

5. International Commission on Illumination; Commission Internationale de L'Eclairage. ILV: International lighting vocabulary. Vienna (Austria): Commission Internationale de L'Eclairage; 2011. (Publ. no. CIE S 017/E:2011).

6. International Union of Pure and Applied Physics, Commission for Symbols, Units, and Nomenclature, Atomic Masses and Constants. Symbols, units, nomenclature and fundamental constants in physics: 1987 revision (reprinted 2010). Physica. 1987;146A(1–2):1–68. Prepared by ER Cohen, P Giacomo. Also available at http://metrology.wordpress.com/measurement-process-index/iupap-red-book/, with additional 2007 references.

7. International Union of Pure and Applied Chemistry, Analytical Chemistry Division, Commission on Spectrochemical and Other Optical Procedures for Analysis. Nomenclature, symbols, units, and their usage in spectrochemical analysis — VIII. Nomenclature system for X-ray spectroscopy (recommendations 1991). Pure Appl Chem. 1991;63(5):735–746. Prepared for publication by R Jenkins, R Manne, R Robin, C Senemaud. Also available at http://pac.iupac.org/publications/pac/pdf/1991/pdf/6305x0735.pdf.

ADDITIONAL REFERENCES

National Institute of Standards and Technology. NIST Reference on Constants, Units, and Uncertainty. http://physics.nist.gov/cuu/Constants/.

A guide to SI units in radiation protection. Health Phys. 2004;87(2 Suppl):S18.

Haynes WM, editor. CRC handbook of chemistry and physics. 93rd ed. Boca Raton (FL): CRC Press; 2012.

International Commission on Non-Ionizing Radiation Protection. General approach to protection against non-ionizing radiation. Health Phys. 2002;82(4):540–548. Also available at http://www.icnirp.de/documents/philosophy.pdf.

International Commission on Radiation Units and Measurements. Fundamental quantities and units for ionizing radiation (revised). Washington (DC): The Commission; 2011. (ICRU report; 85a, revised).

International Commission on Radiation Units and Measurements. Quantities, units and terms in radioecology. Washington (DC): The Commission; 2001. (ICRU report; 65).

International Commission on Radiation Units and Measurements. Quantities and units in radiation protection dosimetry. Washington (DC): The Commission; 1993. (ICRU report; 51).

Kathren RL, Petersen GR. Units and terminology of radiation measurement: a primer for the epidemiologist. Am J Epidemiol. 1989;130(6):1076–1087.

Matthes R, Bernhardt JH, McKinlay AF, editors. Guidelines on limiting exposure to non-ionizing radiation. Oberschleissheim (Germany): International Commission on Non-Ionizing Radiation Protection; 1999.

Meinhold CB. Quantities and units in radiation protection. Radiat Prot Dosimetry. 1995;60(4):343–346.

National Council on Radiation Protection and Measurements (US). SI units in radiation protection and measurements. Bethesda (MD): The Council; 1985. (NRCP report; 82).

National Institute of Standards and Technology. The International System of Units (SI). Gaithersburg, MD: National Institute of Standards and Technology; Special Publication 330; 2008a.

Sinclair WK. The present system of quantities and units for radiation protection. Health Phys. 1996; 70(6):781–786.

16 Subatomic Particles, Chemical Elements, and Related Notations

16.1 SOURCES OF RECOMMENDATIONS

The authorities for nomenclature, symbolization, and other notations for subatomic particles and chemical elements are the Particle Data Group,[1] the International Union of Pure and Applied Chemistry (IUPAC),[2] and the International Union of Pure and Applied Physics (IUPAP).[3]

16.2 ELEMENTARY AND COMPOSITE PARTICLES

The smallest units of matter, the elementary particles, can be categorized as gauge bosons, leptons, quarks, and hadrons (see Table 16.1). Each lepton and quark has a corresponding antilepton or antiquark, which has the same mass as the lepton or quark but opposite quantum numbers, such as the electric charge or the lepton number. Quarks and antiquarks may combine to create protons and neutrons. Mesons and baryons, collectively known as hadrons, are also composed of quarks and antiquarks, but because they cannot be taken apart into their constituents, they are also considered elementary particles. Elementary and composite particles are represented by Greek or Roman letters, some with superscript or subscript modifiers.

Table 16.1 presents typical symbols for elementary particles. For additional examples of particle symbols, consult the "The Review of Particle Physics",[1] which is published biennially, with updates posted online between printed editions (http://pdg.lbl.gov/). "The Review of Particle Physics" presents symbols for particles in "italic (or slanted) characters" except for nonalphabetic superscripts and subscripts. This convention is intended to allow particle symbols to be distinguished readily in text and is followed by most US physics journals. However, there is no accepted standard for the use of roman or italic characters for particle symbols, and some European journals use roman type (as in the IUPAP recommendations for symbols and units[3]). The CSE recommends "The

Table 16.1 Elementary particles: examples of symbols

Particle	Symbols[a]	Charge
Gauge bosons	γ, Z, g	0
	W	+1, -1
Leptons		
Electron neutrino	ν_e	0
Electron	e	-1
Muon neutrino	ν_μ	0
Muon	μ	-1
Tau neutrino	ν_τ	0
Tau, tauon	τ	-1
Quarks	See Table 16.2	
Hadrons (particles having quarks and antiquarks as constituents)		
Mesons (containing 1 quark and 1 antiquark)		
Containing only *u* and *d* particles	π^0, π^+, π^-	0, +1, -1
Containing 1 *s* particle[b]	$K^+, K^0, K^-, \bar{K}^0$	+1, 0, -1, 0
Containing *s* and $\bar{s}$	Φ	0
Containing 1 *c* particle[b]	D^+, D^0, D_s^+	+1, 0, +1
Containing *c* and $\bar{c}$	Ψ	0
Containing 1 *b* particle[b]	$B^+, B^0, B^-, \bar{B}^0, B_s, B_c$	+1, 0, -1, 0, 0, +1
Containing *b* and $\bar{b}$	Y	0
Baryons (containing 3 quarks)	p, n, Λ_c^+	+1, 0, +1
	$\Lambda, \Sigma^+, \Sigma^0, \Sigma^-$	0, +1, 0, -1
	$\Xi^0, \Xi^-, \Xi_c^0, \Xi_c^+$	0, -1, 0, +1
	Ω^-	-1

[a] The superscript plus and minus symbols and the superscript zero character indicate charge. The subscripts *s* and *c* indicate strange and charmed quarks. Distinct antiparticles are indicated by a bar or overline.
[b] In combination with 1 other particle.

Review of Particle Physics" use of italics; however, authors must be aware of and follow the preferred style of the journal to which they are submitting.

Quarks are known by their names (e.g., up, down, strange) or their symbols (e.g., *u*, *d*, *s*) (see Table 16.2).

Supersymmetric particles are hypothetical particles that appear in some extensions of the Standard Model, which is aimed at unifying particle physics and gravity. Use a superimposed tilde to indicate a supersymmetric particle—for example, ẽ.

16.3 NUCLEAR REACTIONS

The symbols for the particles that are projectiles or products in nuclear reactions are set as Greek and Roman letters[3] (see Table 16.3). The general designation of a heavy ion is HI.

Indicate the electric charge of such a particle by adding a superscript plus or minus symbol or the zero character as in π^0, π^+, π^-, e^-.

The symbols *p* and *e* without an indication of charge refer to the positive proton and

Table 16.2 Symbols, names, and charges of quarks[a]

Symbol	Name	Charge
d	Down	−⅓
u	Up	+⅔
s	Strange	−⅓
c	Charmed	+⅔
b	Bottom; *also* beauty	−⅓
t	Top; *also* truth	+⅔

[a] Antiquark symbols are the same as the quark symbols, with an overbar; the charge of an antiquark has the opposite sign to that of the corresponding quark.

Table 16.3 Projectiles and products in natural and artificial nuclear reactions

Name	Nuclear symbol(s)	Chemical symbol
Photon, gamma	γ	
Neutrino	$\nu, \nu_e, \nu_\mu, \nu_\tau$	
Electron	e, β	
Positron	e^+	
Tau, tauon	τ	
Pion	π	
Nucleon[a]	N	
Neutron	n	
Muon	μ^+	Mu^+
Muonium	μ^+e^-	$Mu^\cdot$
Muonide	$\mu^+(e^-)_2$	Mu^-
Proton	p^+	$^1H+$
Protium	p^+e^-	$^1H^\cdot$
Protide	$p^+(e^-)_2$	$^1H^-$
Deuteron	d^+	$^2H^+$ (D^+)
Deuterium	d^+e^-	$^2H^\cdot$ $(D^\cdot)$
Deuteride	$d^+(e^-)_2$	$^2H^-$ (D^-)
Triton	t^+	$^3H^+$ (T^+)
Tritium	t^+e^-	$^3H^\cdot$ $(T^\cdot)$
Tritide	$t^+(e^-)_2$	$^3H^-$ (T^-)
Hydron		H^+
Hydrogen		$H^\cdot$
Hydride		H^-
Helion	h	$^3He^{2+}$
Alpha particle	α	$^4He^{2+}$

[a] Often used either with mass within parentheses after the symbol for the resonances—for example, $N(1440)$, or with other particle symbols to indicate the products of a reaction—for example, $\pi\pi N$.

negative electron, respectively; however, the charge is often used for the latter (e^-). A bar above the symbol for a particle indicates the corresponding antiparticle. Both $\bar{e}$ and e^+ are used for the positron, the positive electron, but the latter is preferred. For the antiproton, the symbol is $\bar{p}$.

An electron (e^-) and a positron (e^+) occurring in a bound state are collectively known as positronium (Ps). A hydrogen-like atom in which the proton is replaced by a positive muon (i.e., a μ^+ with an e^- bound to it) is known as muonium and is given the notation $Mu^{\cdot}$ ("Mu" with superscript dot).

The notation for a nuclear reaction incorporates the following sequence of symbols, set closed, with no spaces before, after, or within the parentheses: the initial nuclide (^{14}N in the example below), followed in parentheses by the incoming particle or photon (α in the example), a comma, the outgoing particle(s) or photon(s) (p in the example), and the final nuclide (^{17}O in the example).

$^{14}N(\alpha,p)^{17}O$

16.4 CHEMICAL ELEMENTS

The symbols for most of the chemical elements are derived from their names (English, Latin, Greek, or systematic) and consist of 1, 2, or 3 letters (e.g., H from hydrogen, K from kalium [for potassium], Na from natrium [for sodium], Ca from calcium, Uuu from unununium [the numeric or systematic name for element 111]). Print symbols for the elements in roman type (except in text or headings that are in italic), with an initial capital letter. Do not capitalize the spelled-out names of elements except at the beginning of a sentence or for other appropriate reasons, such as capitalization in book titles.

Three-letter symbols representing interim systematic names were used for elements of atomic number 103 through 111 until IUPAC adopted names and corresponding symbols for these elements; these 3-letter symbols are still in use for elements with atomic numbers above 111. For the approved names and symbols of the chemical elements through atomic number 111, as well as systematic names and alternative names that may appear in the literature, see Table 16.4.

In 1979 the following roots were selected for the formation of interim systematic names for elements with atomic number above 100. At that time, approved names already existed for elements 101 through 103, and the systematic names for these 3 elements have never been used.

1	un
2	bi
3	tri
4	quad
5	pent
6	hex
7	sept
8	oct
9	enn
0	nil

A systematic name is formed by joining the roots corresponding to the digits of the atomic number and adding the ending "ium". The symbol is formed by joining the first letters of each of the 3 roots and capitalizing the first letter. For example, the systematic name and symbol for element 106 are formed as follows: un [1] + nil [0] + hex [6] = unnilhexium (Unh). Systematic names for elements 104 through 109 were used until 1997, for element 110 until 2003, and for element 111 until 2004, when IUPAC adopted names for these elements (see Table 16.4).

Table 16.4 Atomic numbers, names, and symbols of the chemical elements

Atomic number	Name[a]	Symbol	Atomic number	Name[a]	Symbol
1	hydrogen	H	46	palladium	Pd
2	helium	He	47	silver (argentum)	Ag
3	lithium	Li	48	cadmium	Cd
4	beryllium	Be	49	indium	In
5	boron	B	50	tin (stannum)	Sn
6	carbon	C	51	antimony (stibium)	Sb
7	nitrogen	N	52	tellurium	Te
8	oxygen	O	53	iodine	I
9	fluorine	F	54	xenon	Xe
10	neon	Ne	55	cesium[b], caesium[c]	Cs
11	sodium (natrium)	Na	56	barium	Ba
12	magnesium	Mg	57	lanthanum	La
13	aluminum[b], aluminium[c]	Al	58	cerium	Ce
			59	praseodymium	Pr
14	silicon	Si	60	neodymium	Nd
15	phosphorus	P	61	promethium	Pm
16	sulfur[b], sulphur[c]	S	62	samarium	Sm
17	chlorine	Cl	63	europium	Eu
18	argon	Ar	64	gadolinium	Gd
19	potassium (kalium)	K	65	terbium	Tb
20	calcium	Ca	66	dysprosium	Dy
21	scandium	Sc	67	holmium	Ho
22	titanium	Ti	68	erbium	Er
23	vanadium	V	69	thulium	Tm
24	chromium	Cr	70	ytterbium	Yb
25	manganese	Mn	71	lutetium	Lu
26	iron (ferrum)	Fe	72	hafnium	Hf
27	cobalt	Co	73	tantalum	Ta
28	nickel	Ni	74	tungsten (wolfram)	W
29	copper (cuprium)	Cu	75	rhenium	Re
30	zinc	Zn	76	osmium	Os
31	gallium	Ga	77	iridium	Ir
32	germanium	Ge	78	platinum	Pt
33	arsenic	As	79	gold (aurum)	Au
34	selenium	Se	80	mercury (hydrargyrum)	Hg
35	bromine	Br			
36	krypton	Kr	81	thallium	Tl
37	rubidium	Rb	82	lead (plumbum)	Pb
38	strontium	Sr	83	bismuth	Bi
39	yttrium	Y	84	polonium	Po
40	zirconium	Zr	85	astatine	At
41	niobium	Nb	86	radon	Rn
42	molybdenum	Mo	87	francium	Fr
43	technetium	Tc	88	radium	Ra
44	ruthenium	Ru	89	actinium	Ac
45	rhodium	Rh	90	thorium	Th

Table 16.4 Atomic numbers, names, and symbols of the chemical elements (*continued*)

Atomic number	Name[a]	Symbol	Atomic number	Name[a]	Symbol
91	protactinium	Pa	105	dubnium[d,g]	Db
92	uranium	U		unnilpentium[f]	Unp
93	neptunium	Np	106	seaborgium[d]	Sg
94	plutonium	Pu		unnilhexium[f]	Unh
95	americium	Am	107	bohrium[d]	Bh
96	curium	Cm		unnilseptium[f]	Uns
97	berkelium	Bk	108	hassium[d]	Hs
98	californium	Cf		unniloctium[f]	Uno
99	einsteinium	Es	109	meitnerium[d]	Mt
100	fermium	Fm		unnilennium[f]	Une
101	mendelevium	Md	110	darmstadtium[h]	Ds
102	nobelium	No		ununnilium[f]	Uun
103	lawrencium	Lr	111	roentgenium[i]	Rg
104	rutherfordium[d,e]	Rf		unununium[f]	Uuu
	unnilquadium[f]	Unq			

[a] Names in parentheses are Latin or other names on which the symbol is based.
[b] American spelling.
[c] British spelling.
[d] Name approved by International Union of Pure and Applied Chemistry in 1997.
[e] Also known, before approval of the official name in 1997, as kurchatovium (Ku).
[f] Interim systematic name (see Section 16.4).
[g] Also known, before approval of the official name in 1997, as hahnium (Ha) and nielsbohrium (Ns).
[h] Name approved by International Union of Pure and Applied Chemistry in 2003.
[i] Name approved by International Union of Pure and Applied Chemistry in 2004.

Detailed information on each element, including historical data and tabulations of key isotopes, can be found in books by Emsley (1998)[4] and Krebs (2006)[5] and in the Royal Society of Chemistry's *Visual Elements Periodic Table*,[6] as well as numerous web resources (e.g., http://periodic.lanl.gov).

The symbols for the elements can be modified with additional symbols to indicate atomic number, mass number, charge or oxidation number, numbers of atoms per molecule, and other information. In the example below, E represents the element symbol, and the positions of the modifying symbols indicate their meaning.

$${}^{b}_{a}\mathrm{E}^{c}_{d}$$

a = atomic number
b = mass number (nucleon number, baryon number)
c = charge number (or oxidation number or other information)
d (in a molecular formula) = the number of atoms of this element in the molecule

When no left superscript appears, the symbol is read as including all isotopes in natural abundance. Ionic charge is denoted by a right superscript consisting of a number and the appropriate sign, but not multiple plus or minus symbols; omit the number when it is 1.

Na^{+} Al^{3+} *not* Al^{+++} S^{2-} *not* S^{--}

Oxidation numbers are indicated by positive or negative roman numerals or by zero, in superscript.

Mn^{VII} O^{-II} Pt^{0}

A nuclide is an atom of specified atomic number (proton number) and mass number (nucleon number). Isotopic nuclides (isotopes) are nuclides with the same atomic number but different mass numbers. For example, the 3 nuclides of carbon all have atomic number 6 ($_6C$), but they have different mass numbers: ^{12}C, ^{13}C, ^{14}C. Isobaric nuclides or isobars are nuclides that have the same mass number but different atomic numbers. For example, ^{14}C and ^{14}N are isobars with mass number 14, although their atomic numbers are 6 and 7, respectively.

16.5 CHEMICAL FORMULAS

Chemical formulas represent entities composed of more than one atom (e.g., molecules, complex ions, and groups of atoms).

N_2 $CaSO_4$ $PtCl_4{}^{2-}$ $Fe_{0.91}S$ CH_3OH

The consecutive sequencing of the subscript and superscript designations in the third example above ($PtCl_4{}^{2-}$) is preferable to stacking (i.e., with the superscript $^{2-}$ directly over the subscript $_4$), primarily because it conveys more accurate information. To illustrate the difference in meaning conveyed by different sequences of subscript and superscript, consider the representation $I_3{}^-$, which indicates the entity I_3 bearing a single negative charge. In contrast, $I^-{}_3$ represents a molecule consisting of 3 I^- entities. That distinction would be lost in a stacked representation.

Parentheses, square brackets, and braces may be needed to identify complex groups and to prevent ambiguity. The specific enclosure is often mandated by the type of compound, complex, or group being represented. Those recommendations have been detailed by IUPAC (see also Section 17.2.3).

See Section 17.2 and its subsections for more detailed information about chemical formulas.

16.6 SYMBOLS FOR ATOMIC STATES

The electronic state of an atom is labeled by the value of the angular momentum quantum number *L* for the state. The *L* values 0, 1, 2, 3, 4, 5, 6, 7 . . . are represented by capital letters S, P, D, F, G, H, I, K . . . respectively. The corresponding lowercase letters (see below) are used for the orbital angular momentum of single electrons. For a many-electron atom, the electron spin multiplicity ($2S + 1$) may be indicated as a superscript to the left of the letter and the value of the total angular momentum *J* as a subscript to the right of the letter. If either *L* or *S* is zero, only one value of *J* is possible, and the subscript is then usually omitted.

$^2P_{1/2}$ $^3P_0$ ^4S

Indicate the electronic configuration of an atom by giving the population of each orbital as in the examples below. The designations s and p (and d, f, and other designations as required) are set in roman type.

boron	$(1s)^2(2s)^2(2p)^1$	[$1s^22s^22p^1$ is acceptable]
carbon	$(1s)^2(2s)^2(2p)^2$	[$1s^22s^22p^2$ is acceptable]
nitrogen	$(1s)^2(2s)^2(2p)^3$	[$1s^22s^22p^3$ is acceptable]

16.7 SYMBOLS FOR MOLECULAR STATES

The electronic states of molecules are labeled with the symmetry species of the wave function in the molecular point group. Greek or Roman nonitalic capital letters are used for these labels.

The electronic configurations of orbitals are labeled with lowercase letters in a manner analogous to that for atoms.

ground state of OH $(1\sigma)^2(2\sigma)^2(3\sigma)^2(1\pi)^3$

CITED REFERENCES

1. Beringer J, Arguin JF, Barnett RM, Copic K, Dahl O, Groom DE, Lin CJ, Lys J, Murayama H, Wohl CG, et al. (Particle Data Group). The review of particle physics. Phys Rev D. 2012;86:010001. Also available at http://pdg.lbl.gov.

2. Cohen ER, Cvitaš T, Frey JG, Holström B, Kuchitsu K, Marquardt R, Mills I, Pavese F, Quack M, Stohner J, et al. Quantities, units and symbols in physical chemistry. 3rd ed. 2nd Printing. Cambridge (UK): International Union of Pure and Applied Chemistry & Royal Society of Chemistry Publishing; 2008. Also available at http://media.iupac.org/publications/books/gbook/IUPAC-GB3-2ndPrinting-Online-22apr2011.pdf. Also known as IUPAC Green Book.

3. International Union of Pure and Applied Physics, Commission for Symbols, Units, Nomenclature, Atomic Masses, and Fundamental Constants. Symbols, units, nomenclature and fundamental constants in physics: 1987 revision. Physica. 1987;146A(1–2):1–68. Prepared by ER Cohen, P Giacomo. [Reprinted 2010].

4. Emsley J. The elements. 3rd ed. Oxford (UK): Oxford University Press; 1998.

5. Krebs RE. The history and use of our Earth's chemical elements: a reference guide. 2nd ed. Westport (CT): Greenwood; 2006.

6. Royal Society of Chemistry. Visual elements periodic table. Cambridge (UK): Royal Society of Chemistry Publishing; 2012. Interactive table available at http://www.rsc.org/periodic-table.

ADDITIONAL REFERENCES

Haynes WM, editor. CRC handbook of chemistry and physics. 93rd ed. Boca Raton (FL): CRC Press; 2012.

Leigh GJ, editor. Principles of chemical nomenclature: a guide to IUPAC recommendations. Cambridge (UK): Royal Society of Chemistry Publishing; 2011.

International Union of Pure and Applied Chemistry. IUPAC periodic table of elements. International Union of Pure and Applied Chemistry; 2012. http://www.iupac.org/fileadmin/user_upload/news/IUPAC_Periodic_Table-1Jun12.pdf. Updates of the table are available at http://iupac.org/reports/periodic_table/.

17 Chemical Formulas and Names

17.1 RESOURCES

Most of the nomenclature rules, recommendations, and publication styles for inorganic, organic, and biochemical compounds are established by Commissions of the International Union of Pure and Applied Chemistry (IUPAC) and the International Union of Biochemistry and Molecular Biology (IUBMB), formerly known as the International Union of Biochemistry (IUB). Rules and recommendations can be found in *Nomenclature of Inorganic Chemistry*,[1] *Nomenclature of Organic Chemistry*,[2] and *A Guide to IUPAC Nomenclature of Organic Compounds*,[3] as well as in more recent issues of *Pure and Applied Chemistry* and online at an IUPAC nomenclature website[4] (http://www.chem.qmul.ac.uk/iupac/). Recommendations for polymer nomenclature and terminology have been collected as a *Compendium of Macromolecular Nomenclature*.[5] Many documents of particular importance to biochemical nomenclature have been collected in the publication *Biochemical Nomenclature and Related Documents, a Compendium*[6] and at an IUBMB nomenclature website (http://www.chem.qmul.ac.uk/iubmb/). Two widely used reference handbooks, *CRC Handbook of Chemistry and Physics*[7] and *Lange's Handbook of Chemistry*,[8] as well as *The ACS Style Guide*,[9] have sections on chemical nomenclature and symbolism.

The Chemical Abstracts Service compiles a database of information on naturally occurring and synthetic substances, the *CAS Registry*,[10] in which each substance has a unique identification number. Registry numbers can be used to link names (trivial, systematic, semisystematic, and trade names), formulas, diagrams, and other identifying data, regardless of the nomenclature system being used. Authors are encouraged to include the Registry numbers as part of clear, concise, unambiguous descriptions of the chemicals under study.

17.2 FORMULAS

17.2.1 Empirical Formulas

Empirical formulas are the simplest possible formulas for expressing chemical composition and are therefore often used in indexes, or when knowledge of the exact molecular composition is not necessary, or when the identity and ratio of atoms in the compound is known, but not the precise molecular composition. In the Hill system they are formed by citing the sequence of element symbols together with appropriate subscripts in alphabetic order unless organic groups are present or other ordering criteria apply. In the presence of organic groups, cite C and H first.

ClHg OSi $\mathrm{C_{10}H_{10}CaClFeO_4S}$

17.2.2 Molecular Formulas

Molecular formulas for compounds consisting of discrete molecules denote their atomic composition. The order of citation is based on relative electronegativities, with the more electropositive constituents being cited first. If the compound contains more than one electropositive or electronegative constituent, the sequence within each class is the alphabetic order of their symbols. The numbers of atoms or groups, designation of oxidation states, and indication of ionic charge follow the recommendations in Chapter 16.

17.2.3 Structural Formulas

Structural formulas (stereoformulas and line formulas) provide information about the way the atoms of a molecule are connected. Stereoformulas indicate the 3-dimensional distribution of atoms in space, whereas line formulas provide certain structural information that may be enhanced by means of structural or stereochemical descriptors.

No firm rules exist for representing single and double bonds in text. For line formulas, single bonds need not be represented; if bonds must be represented, use an en dash for a single bond and the equals symbol for a double bond. A triple bond is available as a special symbol in many word-processing programs.

$\mathrm{CH_3CH_2CH_2CH_3}$
$\mathrm{CH_3{-}CH{=}CH{-}CH_2{-}C{\equiv}CH}$

Use parentheses to indicate groups that branch above (or below) the line. In inorganic formulas, always use square brackets to enclose coordination entities. The nesting order for enclosing marks for coordination formulas is [(..)], [{(..)}], [({(..)})], [{({(..)})}], etc.[11] In inorganic names the nesting order is slightly different—that is, ({[(..)]}).

$\mathrm{CH_3CH_2{-}P(S)(OH){-}OCH}$
$\mathrm{Pb^{II}{}_2Pb^{IV}O_4}$
$\mathrm{[Pt(-CH_2-CH_2NH_2)_2]Cl_2}$
$\mathrm{[P^{V}{}_2Mo_{18}O_{62}]^{6-}}$
$\mathrm{[PCl_4]^{+}{}_4[Fe(CN)_6]^{4-}}$
$\mathrm{[Co(NH_3)_6]_2(SO_4)_3}$
$\mathrm{[\{Fe(CO)_3\}_3(CO)_2]^{2-}}$

In organic formulas, use square brackets to enclose multiples of chain segments and parentheses around atoms and groups not in the chain. In organic names the nesting order for enclosing marks is {[({[(..)]})]}.[3]

$CH_3[CH_2CH(CH_3)]_4CH_3$

Displayed structural formulas provide more detailed information about the connection of atoms. Lines of normal weight denote bonds approximately in the plane of the ring or paper. For a bond extending above the plane of the paper, IUPAC recommends using a thick line (▬) or a solid wedge (◀); for consistency in this manual, a thick line is used. For a bond extending below the plane of the paper, IUPAC recommends a hashed thick line (||||||||||).[11]

Many authors use solid and hashed wedges (◀, ||||||||||) to denote bonds above and below the plane of the ring or paper, respectively, with the points of the wedges being atoms in—or pointing toward atoms in—the plane. However, different readers interpret the perspective artistically, making the results potentially ambiguous. Therefore, IUPAC has discouraged the use of hashed wedges,[11] although it allows the use of solid wedges. If wedges are used, authors should define the meaning of each type of wedge, including the position of the point.

Unknown stereochemistry can be indicated by a wavy line. If stereochemistry is implied, indicate double bonds with accurate angles (approximately 120°) if possible. Otherwise, use linear representation. See the IUPAC rules[11] for additional guidelines for graphic representation of 3-dimensional structures.

$(C_6H_5)_3P$ Cl Pt $(C_6H_5)_3P$ Cl

H_3C H C=C H CH_3

NO_2 H_3N NO_2 Co H_3N NO_2 NH_3

H_2 C Cl H_2C C H H C CH_2 C Cl H_2

Proper orientation of fused ring systems is essential for proper numbering. The rules are complex and are only summarized here. Orient fused ring systems with as many *ortho*-fused rings with vertical common bonds as possible in the same horizontal row. When additional rings are fused to the horizontal row, give preference to orientations that do not involve overlapping bonds, to those with the maximum number of rings in the upper right quadrant, to those with the minimum number of rings in the lower left quadrant, and to those that allow low locants to fusion positions. A 6-membered ring may be modified on one side to create a 5-membered ring, or stretched in the middle to insert additional pairs of horizontal bonds, as needed. See the IUPAC rules[12] for detailed examples and guidelines.

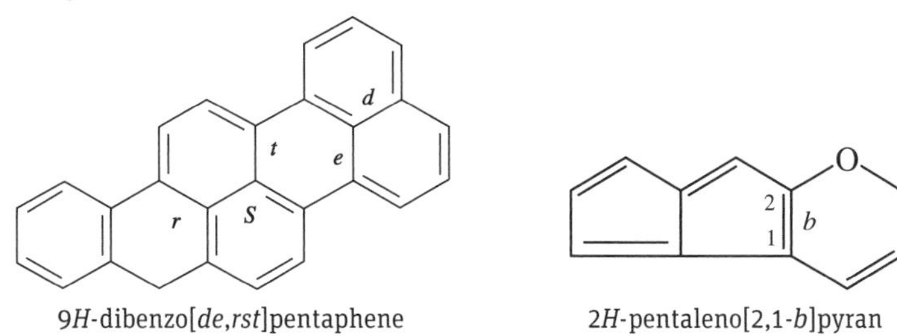

9*H*-dibenzo[*de*,*rst*]pentaphene

2*H*-pentaleno[2,1-*b*]pyran

When a family of chemical structures is being discussed, R may be used to designate part of the molecule. In some texts, it may represent the moiety that is the same in all members of the family, thus allowing the discussion to focus on the functional group that changes from one member of the family to another. For example, the structural formula below represents the common moiety of retinoids. Here, R can be used to represent the organic structure without a characteristic group; thus retanoic acid would be R–COOH. Compounds related to retanoic acid can be represented by R plus the appropriate functional group, examples of which are listed.

R–CH_2OH, retinol
R–CHO, retinal
R–COOH, retinoic acid
R–CH=NOH, retinal oxime

where R =

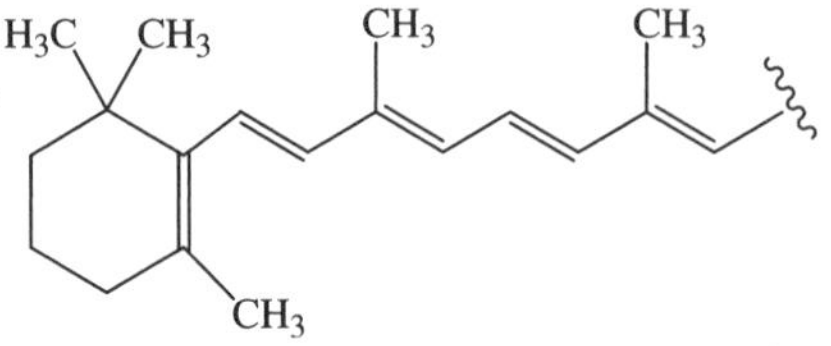

17.3 GENERAL CONVENTIONS OF NOMENCLATURE

17.3.1 Case and Type Style

Do not capitalize a chemical name in text unless it begins a sentence or section heading or appears in a title in which the preferred style is all initial capital letters. In such titles, capitalize the first letter of each word, but not of any descriptor, positional prefix, or locant other than elemental symbols. Capitalize trade names wherever they appear unless they were trademarked with initial lowercase letters; use such names as adjectives with the appropriate nouns. Set trivial, systematic, and semisystematic names in roman type except for those affixes and locants that the nomenclature rules specify should be in italic or Greek letters.[9]

17.3.2 Prefixes

Numerical or multiplicative prefixes derived from Greek or Latin number names, when used to describe a multiplicity of identical features in the name of a chemical compound (see Table 17.1), are closed up to the rest of the name and are not italicized. Structural prefixes that are integral parts of the name are also closed up to the rest of the name and are not italicized. Nonintegral structural descriptors are italicized and set off by hyphens.

dibenzo[*c*,*e*]oxepine	cyclohexane	*p*-aminobenzoic acid
polyester	isopentane	8-*iso*-prostaglandin E_1

17.3.3 Enclosing Marks

Chemical nomenclature uses 3 kinds of enclosing marks: braces, { }; square brackets, [ ]; and parentheses, ( ). These enclosing marks are used slightly differently in formulas and in coordination and organic names. How they are used in each area is described at the appropriate point in this chapter (see Section 17.2.3 for use of enclosing marks in structural formulas).

Table 17.1 Simple numerical terms

Number	Prefix	Number	Prefix	Number	Prefix	Number	Prefix
1	mono-	10[a]	deca- *or* deci-	100	hecta-	1,000	kilia-
		11	undeca-	101	henhecta-	1,001	henkilia-
		12	dodeca-	102	dohecta-	1,002	dokilia-
2	di- *or* bi-	20[b]	icosa- *or* eicosa-	200	dicta-	2,000	dilia-
		21	henicosa-	201	hendicta-	2,001	hendilia-
		22	docosa-	202	dodicta-	2,002	dodilia-
3	tri- *or* ter-	30	triaconta-	300	tricta-	3,000	trilia-
4	tetra- *or* quarter-	40	tetraconta-	400	tetracta-	4,000	tetralia-
5	penta- *or* quinque-	50	pentaconta-	500	pentacta-	5,000	pentalia-
6	hexa- *or* sexi-	60	hexaconta-	600	hexacta-	6,000	hexalia-
7	hepta- *or* septi-	70	heptaconta-	700	heptacta-	7,000	heptalia-
8	octa- *or* octi-	80	octaconta-	800	octacta-	8,000	octalia-
9	nona- *or* novi-	90	nonaconta-	900	nonacta-	9,000	nonalia-

[a] When alone, the numerical term for 1 is "mono". In association with other numerical terms, the term for 1 is "hen", except in the case of 11, where the term is "undeca".

[b] When alone, the numerical term for 2 is "di". In association with other numerical terms, the term for 2 is "do", except in the case of "dicta"and "dilia".

17.3.4 Locants

Use arabic numbers, Greek letters, and italicized Latin letters and words to describe particular structural features or positions in chemical structures. Set them off from the rest of the name by hyphens or square brackets, and/or from each other by commas or periods, with no space around the comma or period.

N-methylethanamine	*O*-phosphono-*enol*-pyruvate	4,7-epoxyfuro[2,3-*d*]oxepin
2*H*-pyran-3(4*H*)-thione	benz[*a*]anthracene	spiro[4.5]dec-6-ene
myo-inositol	5α-androstane	

Capitalize symbols for atoms being added or substituted and set them in italic type—for example, *O*-methylhydroxylamine to indicate substitution on the oxygen atom of hydroxylamine or 6*H*-1,2,5-thiadiazine to indicate the presence of an additional hydrogen atom at the site specified by the locant. Never capitalize structural descriptors such as "*o*" or "*as*" at the beginning of a sentence; capitalize instead the nonitalicized part of the name—for example, *o*-Methylphenol or *as*-Triazine.

17.3.5 Levels of Complexity in Nomenclature

Historically, different chemical nomenclature conventions have been developed for different levels of molecular complexity.

17.3.5.1 Types of Names

1) Trivial names are the common names of chemicals, including traditional names (e.g., acetone, benzene, urea), trade names (e.g., aspirin, Clorox, Lucite), and those given for convenience to compounds of uncertain structure, especially those of biological origin (e.g., solanine from members of the plant family Solanaceae, penicillin from members

of the genus *Penicillium*). They carry very little (if any) structural information, are usually derived from the source of the material, and—after their structures have been determined—should be used only when they will adequately and unambiguously describe a well-established compound.[9]

2) Systematic names are those based on a set of rules; for organic compounds, they are based on appropriate names of parent hydrides, stereodescriptors, prefixes, and suffixes,[2] e.g., phosphane, (2*E*)-but-2-ene, cyclobutane. Depending on the context and the nomenclature convention used, a compound may have more than one appropriate systematic name.
3) Semisystematic names are usually simplified alternatives to the systematic names, e.g., methane, cubane, benzoic acid, phenol.

Any paper describing a complex molecule—especially a new complex molecule—should include a full systematic name at least once at the beginning and may otherwise use the semisystematic name, a trivial name, or an abbreviation.

17.3.5.2 Approaches to Nomenclature

Systematic conventions include binary nomenclature, coordination nomenclature, and substitutive nomenclature, which are described in the following sections, as well as various other operations, such as subtractive, multiplicative, and conjunctive operations.

17.4 BINARY NOMENCLATURE

The binary approach to nomenclature bases names on the ionic nature of the constituents, with electropositive components being listed first, followed by electronegative components. Separate all terms for ions and neutral segments by spaces. Although the term "binary" literally refers to 2 components, this nomenclature system is readily extended to multiple components. Represent compounds in text by either the chemical formula or the binary name, but do not use hybrids.

NaCl	sodium chloride *not* Na chloride
MgKF	magnesium potassium fluoride *not* Mg potassium fluoride
Zn(OH)I	zinc hydroxide iodide
$Pt(NH_3)_2Cl_2$	platinum diammine dichloride
N_2O	dinitrogen oxide
OF_2	oxygen difluoride

17.5 COORDINATION NOMENCLATURE

In coordination nomenclature, compounds are regarded as having central atoms to which other ions or neutral atoms or groups are attached. It is an additive system (associated ligand names added to the name of a central atom) in which the central atom is cited last in the name segment, preceded by the ligands (ions, atoms, or groups) in alphabetic sequence. Use italic letter locants and the Greek letter κ to specify precisely the atoms attached to the metal.[1,13]

$Pt(NH3)_2Cl_2$	diamminedichloroplatinum(II)
$[Fe(CN)_5(NO)]^{2-}$	pentacyanonitrosylferrate(2−)

$K_4[Fe(CN)_6]$	tetrapotassium hexacyanoferrate(4−)
$[Co(NH_3)_5Cl]Cl_2$	pentamminechlorocobalt(2+) dichloride

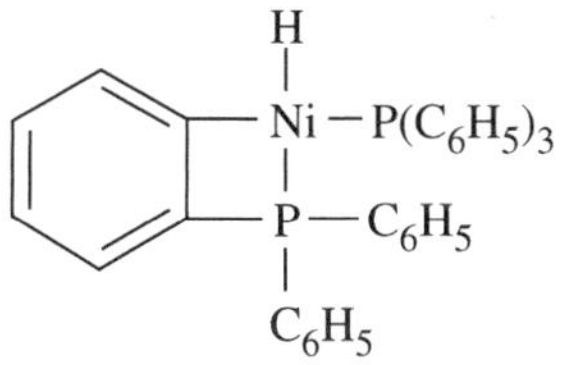

2-(diphenylphosphanyl-κ*P*)phenyl-κ*C*]hydrido(triphenylphosphane)nickel

17.6 SUBSTITUTIVE NOMENCLATURE

The substitutive approach to naming chemical compounds is based on the concept of parent hydrides, which are compounds with a skeletal structure consisting of carbon atoms, atoms of certain metallic or nonmetallic elements (called "heteroatoms"), and attached hydrogen atoms, the number of which is consistent with the skeletal structure and the bonding requirements of each skeletal atom. Heteroatoms in organic compounds and inorganic parent hydrides have standard bonding numbers; other bonding numbers are indicated by the λ-convention.[3] Hydrogen atoms of parent hydrides or other parent compounds are substituted (replaced) by other atoms or groups called "substituents".

For organic compounds, the underlying principle is to dissect the structure into characteristic groups and fragments that can be represented by the parent hydride structures discussed above. One of the parent hydride structures is chosen as the main part of the name and other parts are cited as substituents to it. A substituent prefix is formed from the name of a parent hydride by adding the suffixes in Table 17.2.

Characteristic atoms and groups attached to a parent hydride that are not themselves derived from parent hydrides are cited either as prefixes or suffixes to the name of the parent hydride. Table 17.3 gives the names of some compulsory prefixes—that is, characteristic groups that are cited only as prefixes. Table 17.4 gives the names of suffixes and prefixes for some common characteristic groups.

When the mononuclear parent hydrides (Table 17.5) are modified, the names of the resulting compounds reflect the substitution. For example, 3 chlorine atoms may be substituted for the 3 hydrogen atoms of phosphane (PH_3) to create trichlorophosphane (PCl_3). Likewise, in λ^5-phosphane (PH_5), 3 of the 5 hydrogen atoms may be replaced by

Table 17.2 Suffixes for derivation of prefix names from parent hydrides[a]

Monovalent	Divalent	Trivalent	Tetravalent
-yl	-diyl	-triyl	-tetrayl
	-ylidene	-ylidyne	-ylylidyne
		-ylylidene	-diylidene
			-diylylidene

[a] Suffixes beyond the tetravalent are modeled on these forms.

Table 17.3 Characteristic groups cited only as prefixes

Characteristic group	Prefix	Characteristic group	Prefix
–Br	bromo-	–IO	iodosyl-
–Cl	chloro-	$–IO_2$	iodyl-
–ClO	chlorosyl-	$=N_2$	diazo-
$–ClO_2$	chloryl-	$–N_3$	azido-
$–ClO_3$	perchloryl-	–NO	nitroso-
–F	fluoro-	$–NO_2$	nitro-
–I	iodo-		

Table 17.4 Characteristic group suffixes and prefixes

Formula	Suffix	Prefix
–COOH	-carboxylic acid	carboxy-
–(C)OOH	-oic acid	. . .
$–COO^-$	-carboxylate	carboxylato-
$–(C)OO^-$	-oate	. . .
$–SO_2OH$	-sulfonic acid	sulfo-
$–SO_2O^-$	-sulfonate	sulfonato-
–COX[a]	-carbonyl halide	halocarbonyl- carbonohalidoyl-
–(C)OX[a]	-oyl halide	. . .
$–CONH_2$	-carboxamide	carbamoyl- aminocarbonyl-
$–(C)ONH_2$	-amide	. . .
$–C(=NH)NH_2$	-carboximidamide -carboxamidine	carbamimidoyl- amidino-
$–(C)(=NH)NH_2$	-amidine	. . .
–CN	-carbonitrile	cyano-
–(C)N	-nitrile	. . .
–CHO	-carbaldehyde -carboxaldehyde	formyl- . . .
–(C)HO	-al	. . .
›(C)=O	-one	oxo-
–OH	-ol	hydroxy-
$–O^-$	-olate	oxido-
–SH	-thiol	sulfanyl- mercapto-
$–S^-$	-thiolate	sulfido- sulfanidyl-
$–NH_2$	-amine	amino-
=NH	-imine	imino-

[a] X = F, Cl, Br, I, N_3, CN, NC, NCO, NCS, NCSe, NCTe.

Table 17.5 Mononuclear parent hydrides

Formula	Name	Formula	Name
BH_3	borane	SH_4	λ⁴-sulfane
CH_4	methane	SH_6	λ⁶-sulfane
SiH_4	silane	SeH_2	selane[a]
GeH_4	germane	SeH_4	λ⁴-selane
SnH_4	stannane	SeH_6	λ⁶-selane
PbH_4	plumbane	TeH_2	tellane[a]
NH_3	azane[a]	TeH_4	λ⁴-tellane
PH_3	phosphane (phosphine[b])	TeH_6	λ⁶-tellane
PH_5	λ⁵-phosphane (phosphorane[b])	FH_3	λ³-fluorane
AsH_3	arsane (arsine[b])	ClH_3	λ³-chlorane
AsH_5	λ⁵-arsane (arsorane[b])	ClH_5	λ⁵-chlorane
SbH_3	stibane (stibine[b])	BrH_3	λ³-bromane
SbH_5	λ⁵-stibane (stiborane[b])	BrH_5	λ⁵-bromane
BiH_3	bismuthane (bismuthine[b])	IH_3	λ³-iodane
BiH_5	λ⁵-bismuthane	IH_5	λ⁵-iodane
OH_2	oxidane[a]	IH_7	λ⁷-iodane
SH_2	sulfane[a]		

[a] Not normally used as a parent hydride.
[b] Less preferred names still in use.

chlorine atoms to yield trichloro-λ⁵-phosphane (H_2PCl_3) or trichlorophosphorane (a name that is less preferred, but still in use).

17.6.1 Mononuclear Parent Hydrides

Names of mononuclear parent hydrides are given in Table 17.5.

17.6.2 Acyclic Parent Hydrides

Names of unsubstituted acyclic (straight-chain) parent hydrides have a stem, a characteristic element term (except for carbon chains), and a characteristic ending, -ane, designating fully saturated chains, and -ene and/or -yne, designating unsaturated chains. The -ene ending indicates the presence of a double bond, the -yne ending the presence of a triple bond; numerical prefixes describe the number of each, and locants indicate the position of the unsaturation in the chain when known. Examples are shown in Table 17.6.

17.6.3 Cyclic Parent Compounds

Cyclic structures in which the ring contains only carbon and hydrogen atoms are called "hydrocarbons". Cyclic structures in which the ring contains atoms other than carbon are called "heterocycles". Atoms other than carbon in such a structure are called "heteroatoms". Structures with only one ring containing carbon and noncarbon atoms are called "heteromonocycles" or "heteromonocyclic rings".[12] Names and locants of known organic ring systems can be found by name or by ring analysis in the *Chemical Abstracts Service Ring Systems Handbook*[14] and its supplements.

Table 17.6 Representative acyclic parent hydrides

Name	Stem	Element term	Ending	Number of chain atoms; Degree of saturation
ethane	eth[a]		ane	2 carbon atoms; fully saturated
propane	prop[a]		ane	3 carbon atoms; fully saturated
butane	but[a]		ane	4 carbon atoms; fully saturated
pentane	pent		ane	5 carbon atoms; fully saturated
disilane	di	sil	ane	2 silicon atoms; fully saturated
diazene	di	az	ene	2 nitrogen atoms; 1 double bond
tetrasulfane	tetra	sulf	ane	4 sulfur atoms; fully saturated
octa-1,3-diene	octa[b]		diene	8 carbon atoms; 2 double bonds at positions 1 and 3
nona-2,4,7-triyne	nona[b]		triyne	9 carbon atoms; 3 triple bonds at positions 2, 4, and 7
deca-1,3-dien-7-yne	deca[b,c]		dienyne	10 carbon atoms; 2 double bonds at positions 1 and 3; 1 triple bond at position 7
tetraaza-1,2-diene	tetra	aza	diene	4 nitrogen atoms; 2 double bonds at positions 1 and 2

[a] Trivial (nonnumeric) stems are retained for compounds containing 1–4 carbon atoms.
[b] An "a" is inserted between the stem and the characteristic ending (for euphony) when the stem ends in a consonant and the ending begins with a consonant.
[c] An "e" in -endings denoting the state of saturation is elided when followed by a vowel or the letter "y" (e.g., decadienyne = dien(e) + yne, with the "e" enclosed in parentheses elided).

17.6.3.1 Monocyclic Rings

All saturated monocyclic rings may be named by attaching the prefix cyclo- to the name of the corresponding acyclic hydrocarbon. Unsaturation is described by the endings -ene and -yne.[3]

cyclo + butane = cyclobutane
cyclo + octane + 1,3,5-trien-7-yne = cycloocta-1,3,5-trien-7-yne
cyclo + hexasilane = cyclohexasilane

Heteromonocyclic rings are named using 3 different methods:

1) by the Hantzsch-Widman system,[3] combining terms for each heteroatom in the ring in the order given in Table 17.7, followed by an ending that indicates the ring size and whether the ring is saturated or unsaturated, as given in Table 17.8.

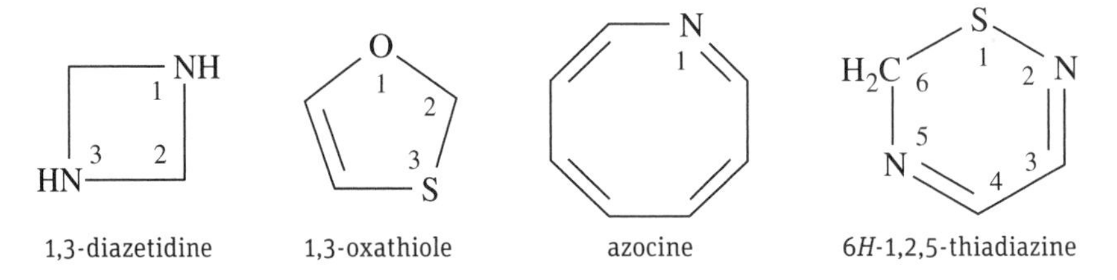

1,3-diazetidine 1,3-oxathiole azocine 6*H*-1,2,5-thiadiazine

2) by skeletal replacement ("a") nomenclature, in which replacement of the carbon atoms of a monocyclic hydrocarbon is denoted by the "a" prefixes given in Table 17.7. The system may also be used for silicon heterocycles (e.g., silabenzene) and is mandatory for monocyclic rings with more than 10 ring members (e.g., 1,4-dioxacyclododecane).

Table 17.7 Prefixes for Hantzsch–Widman system of skeletal replacement ("a")

Element	Bonding number (valence)	Prefix	Element	Bonding number (valence)	Prefix
fluorine	1	fluora-	arsenic	3	arsa-
chlorine	1	chlora-	antimony	3	stiba-
bromine	1	broma-	bismuth	3	bisma-
iodine	1	ioda-	silicon	4	sila-
oxygen	2	oxa-	germanium	4	germa-
sulfur	2	thia-	tin	4	stanna-
selenium	2	selena-	lead	4	plumba-
tellurium	2	tellura-	boron	3	bora-
nitrogen	3	aza-	mercury	2	mercura-
phosphorus	3	phospha-			

Table 17.8 Stems for Hantzsch–Widman system

Ring size	Unsaturated	Saturated	Ring size	Unsaturated	Saturated
3	irene (irine)	irane (iridine)	6C[c]	inine	inane
4	ete	etane (etidine)	7	epine	epane
5	ole	olane (olidine)	8	ocine	ocane
6A[a]	ine	ane	9	onine	onane
6B[b]	ine	inane	10	ecine	ecane

[a] 6A elements: O, S, Se, Te, Bi, Hg.
[b] 6B elements: N, Si, Ge, Sn, Pb.
[c] 6C elements: B, F, Cl, Br, I, P, As, Sb.

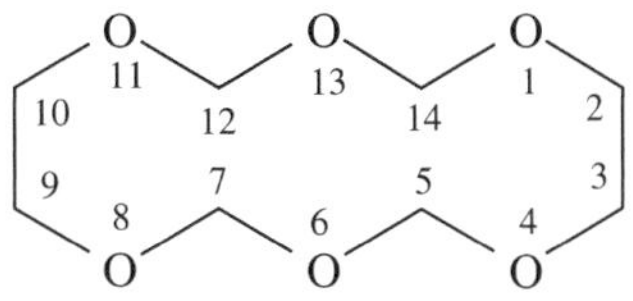

1,4,6,8,11,13-hexaoxacyclotetradecane

3) by a trivial system known as "cyclo(ab)$_x$", a method where "a" is a heteroatom that occurs later in the list in Table 17.7 and "b" is a heteroatom that occurs earlier in Table 17.7. A numerical term defines the number of the "a" heteroatoms.

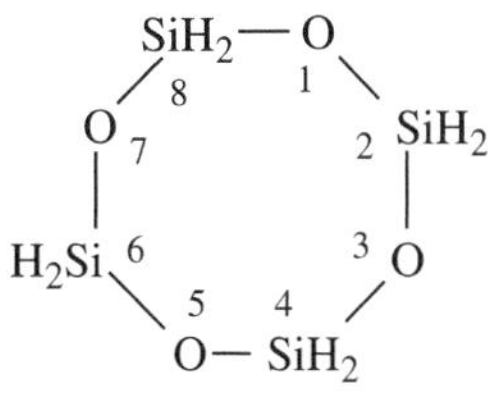

cyclotetrasiloxane

17.6.3.2 Bicyclic and Polycyclic Ring Systems

There are several nomenclature systems for bicyclic and polycyclic ring systems.

1) *Fused Ring Systems.*[12] Select one component ring or ring system as the parent system and letter its bonds in italics as *a*, *b*, *c*, etc., corresponding to the bonds between carbon atoms 1 and 2, 2 and 3, 3 and 4, etc. If there are more than 26 sides, number the additional sides with subscript numbers a_1, b_1, c_1, etc., then a_2, b_2, c_2, etc. Indicate bonds common to a pair of components by the appropriate numerical locants for the attached components and letter locants for the parent component; separate them by a hyphen, all within square brackets. Run together letter locants if there is more than one, and list them in the direction in which they were assigned. Separate numerical locants by commas and follow the same direction as the letters. If there is no ambiguity, locants may be omitted in simple structures. For more information, see the IUPAC rules.[12]

   benz[*a*]anthracene
   1*H*-benzo[*a*]cyclopent[*f*]anthracene
   9*H*-dibenzo[*de*,*rst*]pentaphene [see Section 17.2.3 for structure]
   2*H*-pentaleno[2,1-*b*]pyran [see Section 17.2.3 for structure]

2) *Bridged Ring Systems.* There are 2 types of bridged ring systems: bridged fused ring systems[12] and polycyclic ring systems that cannot be treated as bridged fused ring systems, the so-called von Baeyer ring systems,[15] each having its own style and format considerations.

   Bridged fused ring systems are named by attaching names for bridges to fused ring system names formed as described in Section 17.6.3.2(1). Locants of the fused ring system locate the bridge.

   1,4-methanonaphthalene   4,7-epoxyfuro[2,3-*d*]oxepin

   Polycyclic bridged ring systems that do not contain a fused ring are named using the extended von Baeyer system,[16] beginning with prefixes that indicate the number of rings (e.g., bicyclo, tricyclo). Numbers indicating the bridge lengths are in decreasing order, are separated by periods (rather than commas), and are enclosed in square brackets. Locants of the attachment points for secondary bridges in polycyclic systems are indicated by pairs of superscript numbers, the lowest one of each set cited first, separated by commas. Unsaturation is indicated by endings like -ene and -diene, and heteroatoms are denoted by skeletal replacement prefixes (see Table 17.7).

   bicyclo[3.2.1]octane   1,3-diazatetracyclo[2.2.0.$0^{2,6}$.$0^{3,5}$]hexane
   tricyclo[3.3.2.$0^{2,4}$]dec-6-ene

3) *Spiro Ring Systems.*[16] A spiro ring system has 2 or more rings, at least 2 of which have only a single atom in common, the spiro atom. For nomenclature purposes there are 2 distinct types of spiro compounds.

   A spiro compound composed only of saturated monocyclic rings has the same name as the corresponding acyclic hydrocarbon, prefixed by terms like spiro- or dispiro-, according to the number of spiro atoms present, and followed by a series of numbers separated by periods and enclosed in brackets giving the numbers of atoms between the spiro atoms. Cite these numbers in ascending order within each ring, starting with one of the terminal rings and proceeding to the other terminal ring and then back to the first ring. Number the system in the same order that the descriptors are cited in such a way that the spiro atoms receive the lowest locants. In trispiro and higher systems, each time a spiro atom is reached for a second time, cite its locant as a superscript to the number giving the linking atoms. Such treatment is necessary in higher polyspiro systems to distinguish between linear systems and branched systems. Unsaturation is indicated by endings like -ene and -diene, and hetero atoms are denoted by skeletal replacement prefixes (see Table 17.7). For additional information, see the IUPAC rules.[16]

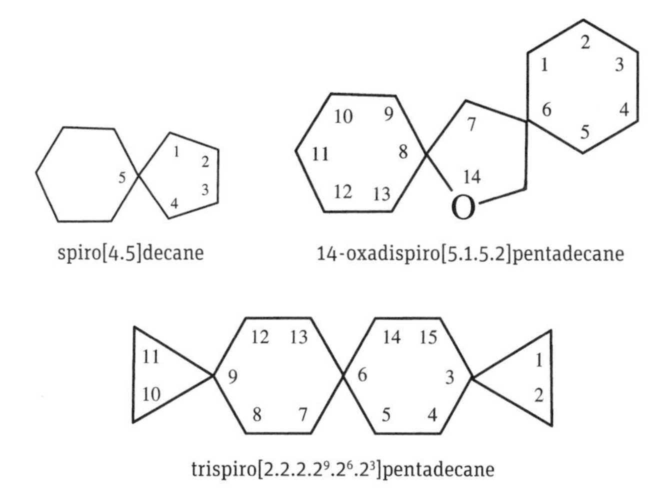

spiro[4.5]decane

14-oxadispiro[5.1.5.2]pentadecane

trispiro[2.2.2.2^9.2^6.2^3]pentadecane

Name spiro compounds in which at least one component is a polycyclic ring system by enclosing the names of the components within brackets cited in order beginning with the component that is earlier alphabetically and adding prefixes like spiro- or dispiro-, according to the number of spiro atoms. Serially prime locants of the second, third, and higher ring or ring system. Denote the positions of the spiro atoms by the locants of each ring, separate them by commas, and insert them between the names of the appropriate rings or ring systems.

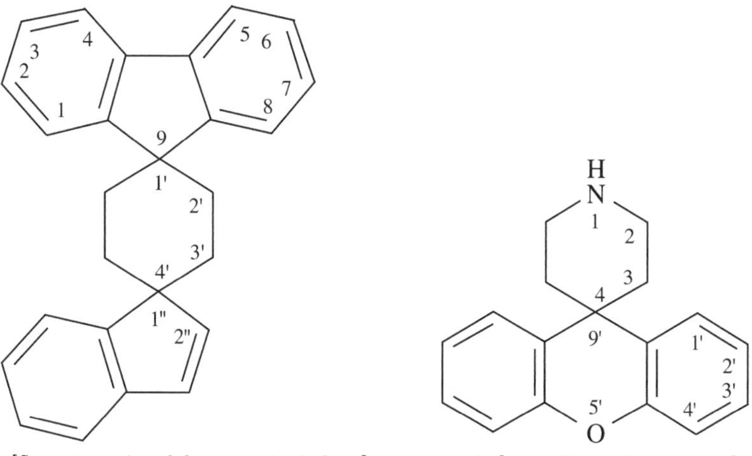

dispiro[fluorene-9,1′-cyclohexane-4′,1-indene]

spiro[piperidine-4,9′-xanthene]

4) *Ring Assemblies.*[3] Two or more rings or ring systems directly joined to each other by only single or by only double bonds are classified as ring assemblies when the number of direct ring connections is one less than the number of rings or ring systems involved. Assemblies of identical rings or ring systems are named by prefixing the name of a substituent prefix (see below) or the name of the ring or ring system by numerical prefixes such as bi-, ter-, or quarter- (see Table 17.1). Locants denoting the attachment points on each component are cited in front of the numerical prefixes, and sets of locants are separated by colons. Traditionally,[3] locants for the first component are unprimed and locants for succeeding components are primed serially. However, to avoid the need for long strings of primes, the rings or ring systems may be numbered as if they were chains, and

the positions on each component may be denoted by superscript numbers corresponding to the locant numbers of the ring system.

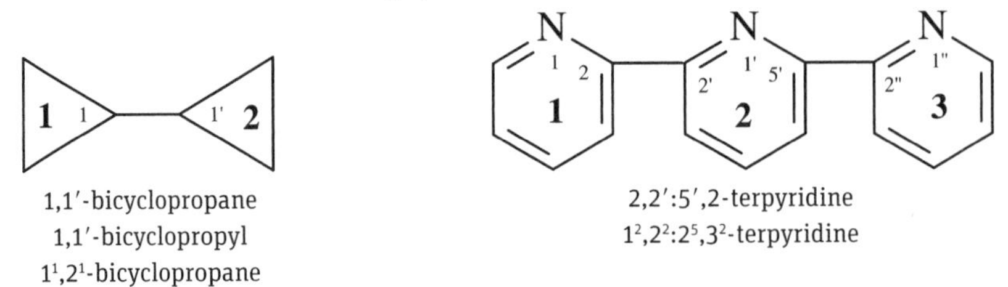

1,1′-bicyclopropane
1,1′-bicyclopropyl
$1^1,2^1$-bicyclopropane

2,2′:5′,2-terpyridine
$1^2,2^2:2^5,3^2$-terpyridine

Nonidentical ring assemblies are named by substitutive nomenclature principles (see Section 17.6). One ring or ring system is chosen as the parent, and the other rings or ring systems are cited as substituents.

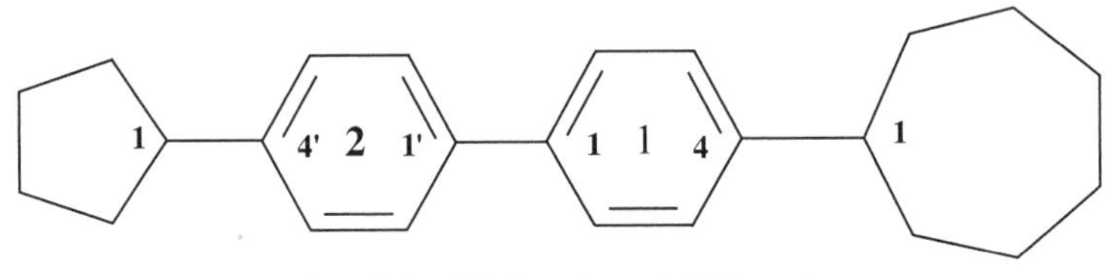

4-cycloheptyl-4′-cyclopentylbiphenyl

17.6.4 Fullerenes

Fullerenes are closed-cage, fully unsaturated carbon macrocycles having more than 20 carbon atoms, each carbon being bound to 3 other carbon atoms (therefore devoid of any hydrogen atoms), and informally called "buckyballs". Nomenclature for fullerenes and their derivatives is based on organic nomenclature principles.

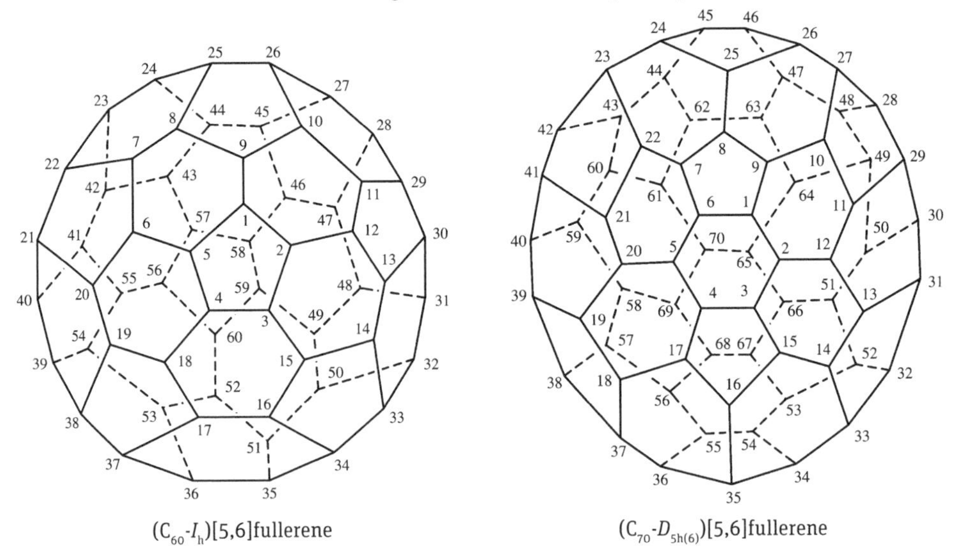

(C_{60}-I_h)[5,6]fullerene

(C_{70}-$D_{5h(6)}$)[5,6]fullerene

17.6.5 Phane Systems

Phane nomenclature may be used for the large number and variety of large rings and/or ring systems directly connected or linked together by atoms or chains of atoms. In this system, rings and ring systems are "collapsed" to a single "atom", called a "superatom", represented by large solid dots in the phane skeletal structure. The resulting monocyclic or polycyclic ring or ring system is named and numbered as a saturated monocyclic hydrocarbon, a saturated bicyclic or polycyclic hydrocarbon, or a saturated spiro hydrocarbon (see Section 17.6.3.2(3)), except that the ending of the name is -phane rather than -ane. In the phane name, prefixes derived from the names of the ring or ring system define the nature of the superatoms.

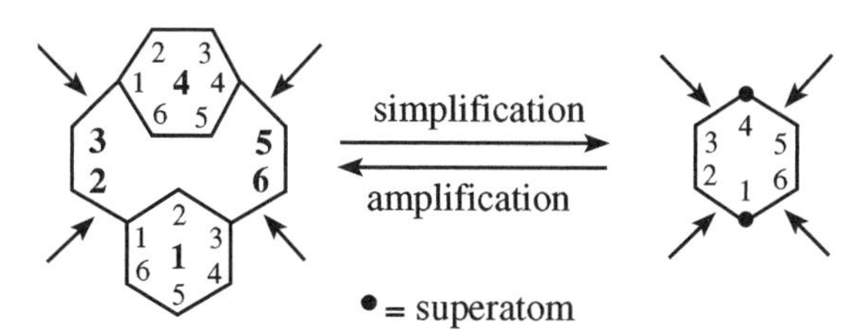

Simplified skeletal name: cycloheptaphane
Phane parent hydride name: 1(1,3),4(1,4)-dibenzenacycloheptane

17.6.6 Structural Affixes

The following structural affixes are used for organic compounds. Except for epi-, iso-, poly-, and pseudo-, which are set in roman type and closed up to the name, set the affixes in italics with a hyphen separating them from the rest of the name.

aci-	the acid form, as in *aci*-acetoacetate, $CH_3C(OH)=CH-C(O)O-R$ and *aci*-nitro, $HON(O)=$
as-	asymmetric, as in *as*-trichlorobenzene and *as*-triazine
endo-	prefix indicating a bridging group
epi-	prefix indicating a bridging group
gem-	geminal, 2 groups attached to the same atom, as in *gem*-diol or *gem*-dimethyl-
iso-	denoting a single, simple branch at the end of a straight chain, as in isopentane and isobutyl
m-	*meta*, a 1,3-positional relationship in benzene
n-	abbreviation for normal, as in *n*-butane
o-	*ortho*, a 1,2-positional relationship in benzene
p-	*para*, a 1,4-positional relationship in benzene
peri-	a 1,8-positional relationship in naphthalene
poly-	many, as in polysulfide
pseudo-	indicating resemblance or relation, as in pseudohalogen
s-, *sym-*	symmetric, as in *s*-trichlorobenzene and *s*-triazine
sec-	secondary, indicating a single, simple branch next to the termination of a hydrocarbyl group, as in *sec*-butyl
t-, *tert-*	tertiary, indicating a double branch next to the termination of a hydrocarbyl group, as in *t*-Bu or *tert*-butyl
v-, *vic-*	vicinal, 3 adjacent groups, as in *vic*-triazole

17.7 FUNCTIONAL REPLACEMENT NOMENCLATURE

Functional replacement nomenclature uses either prefixes or infixes to indicate the replacement of oxygen atoms or hydroxy groups by other atoms or groups in parent compounds (see Table 17.9). Set symbols for atoms being added or substituted in italics.

Table 17.9 Functional replacement infixes and prefixes

Infix	Prefix	Replaced atom(s)	Replacing atom(s)
-amido-	amido-	–OH	$-NH_2$
-azido-	azido-	–OH	$-N_3$
-bromido-	bromo-	–OH	–Br
-chlorido-	chloro-	–OH	–Cl
-cyanatido-	cyanato-	–OH	–OCN
-cyanido-	cyano-	–OH	–CN
-dithioperoxo-	dithioperoxy-	–OH	–SS–
-fluorido-	fluoro-	–OH	–F
-iodido-	iodo-	–OH	–I
-isocyanatido-	isocyanato-	–OH	–NCO
-isocyanido-	isocyano-	–OH	–NC
-nitrido-	nitrido-	=O *and* –OH	=N– {*or* –N< / *or* =N–}
-thiocyanatido-	thiocyanato-	–OH	–SCN
-isothiocyanatido-	isothiocyanato-	–OH	–NCS
-imido-	imido-	=O *or* –O–	=NH *or* –NH–
-hydrazido-	hydrazido-	–OH	$-NHNH_2$
-peroxo-	peroxy-	–OH	–OO–
-seleno-	seleno-	=O *or* –O–	=Se *or* –Se–
-telluro-	telluro-	=O *or* –O–	=Te *or* –Te–
-thio-	thio-	=O *or* –O–	=S *or* –S–
-thioperoxo-	thioperoxy-	–O–	–OS *or* –SO–

Structure	Name
C_6H_5–C(O)SH	thiobenzoic *S*-acid *or* benzenecarbothioic *S*-acid
$CH_3CH_2CH_2CHCH_2C(=NH)OH$	hexanimidic acid
$CH_3P(O)(NH_2)(SeH)$	*P*-methylphosphonamidoselenoic *Se*-acid

17.8 OTHER NOMENCLATURE SYSTEMS

Although substitutive nomenclature is the basis for forming systematic names for organic compounds, other systems are also used, sometimes in combination with substitutive nomenclature. One of the most common is functional class nomenclature, in which compounds are named on the basis of a class of compounds, such as alcohols, amines, ethers, or ketones. Specific compounds are identified by using the class name and the appropriate form of the name of the parent hydride. In the following examples, note the spaces in the names. Each of the names contains a class name representing a functional group: –OH, –O–, or –CO–; no hydrogen atoms are replaced. However, the organic part of the structure is usually expressed by means of a substituent group name, which used to be called a "radical" and led to the older term "radicofunctional" for this type of nomenclature.

$CH_3{-}OH$ methyl alcohol
$CH_3CH_2{-}O{-}CH_2CH_3$ diethyl ether
$CH_3CH_2{-}CO{-}CH_3$ ethyl methyl ketone

17.9 RADICALS AND IONS

Suffixes and substituent prefix endings for describing radical and ionic centers were formalized by IUPAC in recommendations dealing with classical valence structures, but not including delocalization or concepts such as paired and unpaired electronic configurations.[17] Indicate the radical by a superscript dot and the appropriate charge sign, if needed. In complex formulas, use square brackets or parentheses. Suffixes and substituent prefix endings for radicals and ions in substitutive nomenclature and some simple examples are listed in Table 17.10. For more complex examples, see the comprehensive publication.[17]

$H^{\cdot}$ $^{\cdot}CH_3$ $(SO_2)^{\cdot-}$ $[CH_2{=}CH{-}CH_3]^{\cdot+}$

Zwitterions (i.e., structures with internally compensating charges) are named by combining appropriate operational suffixes at the end of the name of a neutral parent hydride in the order -ium, -ylium, -ide, and -uide.

$(CH_3)_3\overset{+}{\underset{2}{N}}{-}\overset{-}{\underset{1}{N}}H$ 2,2,2-trimethylhydrazin-2-ium-1-ide

17.10 ISOTOPICALLY MODIFIED COMPOUNDS

An isotopically modified compound has an isotopic ratio of nuclides for at least one element that deviates measurably from that occurring in nature. It is either an isotopically substituted or an isotopically labeled compound.[3]

17.10.1 Isotopically Substituted Compounds

An isotopically substituted compound is one in which essentially all the molecules of the compound have only the indicated nuclide at each of the designated positions. Write the formulas for these compounds with the appropriate nuclide symbol. Write the names by including the nuclide symbol in parentheses, with locants if necessary, before the name of the compound or the group being substituted.

$^{14}CH_4$ (^{14}C)methane
$CH_3CH^2H{-}OH$ (1-$^2H_1$)ethanol

17.10.2 Isotopically Labeled Compounds

An isotopically labeled compound may be either specifically labeled or selectively labeled. In formulas and names for both, use square brackets around the appropriate nuclide symbol(s) and multiplying subscript(s), if any.

A specifically labeled compound is one in which a unique isotopically substituted compound is formally added to the analogous isotopically unmodified compound.

A selectively labeled compound is one in which a mixture of isotopically substituted

Table 17.10 Suffixes and substituent prefix endings for radicals and ions derived from parent hydrides

Class	Operation	Suffix	Examples	Substituent prefix ending or prefix	Examples	Replacement nomenclature suffix	Examples
Radicals	Loss of H•	-yl, etc. (see Table 17.2)	methyl CH_3•	ylo-	ylomethyl $-CH_2$•	. . .	. . .
Cations	Addition of H^+	-onium	ammonium	-oniumyl	ammoniumyl	-onia	azonia $>N^+<$
			NH_4^+		$-NH_3^+$		thionia $>S^+-$
			sulfonium		sulfoniumyl		
			SH_3^+		$-SH_2^+$		
		-ium	methanium	-iumyl	methaniumyl		
			CH_5^+		$-CH_4^+$		
			azanium		azaniumyl		
			NH_4^+		$-NH_3^+$		
	Loss of H^-	-ylium	methylium	-yliumyl	methyliumyl	-ylia	λ^5-phosphanylia
			CH_3^+		$-CH_2^+$		$>P^+<$
			λ^5-phosphanylium		λ^5-phophanyliumy		
			PH_4^+		$-PH_3^+$		
Anions	Loss of H^+	-ide	methanide	-idyl	methanidyl	-ida	λ^5-phosphanida
			CH_3^-		$-CH_2^-$		$>P^-<$
			λ^5-phosphanide		λ^5-phosphanidyl		
			PH_4^-		$-PH_3^-$		
		-ate	R-carboxylate	-ato	carboxylato	. . .	. . .
			R-COO^-		$-COO^-$		
		-olate	methanolate	-ido	oxido	. . .	. . .
			CH_3-O^-		$-O^-$		
	Addition of H^-	-uide	boranuide	-uidyl	boranuidyl	-uida	boranuida
			BH_4^-		$-BH_3^-$		$-B^--$ (with vertical bonds above and below)

compounds is formally added to the isotopically unmodified compound in such a way that the position(s) but not necessarily the number of each labeling nuclide is defined.

	Formula	Name
Specifically labeled:	$[^{14}C]H_4$	[14C]methane
	$CH_2[^2H_2]$	$[^2H_2]$methane
	$CH_3CH[^2H]$–OH	$[1\text{-}^2H_1]$ethanol
	CH_3CH_2–$[^{18}O]H$	$[^{18}O]$ethanol
	CH_3CH_2–$O[^2H]$	ethan$[^2H]$ol
Selectively labeled:	$[^2H]CH_4$	$[^2H]$methane
	$[1\text{-}^{14}C,^{18}O]CH_3CH_2$–OH	$[1\text{-}^{14}C,^{18}O]$ethanol

A nonselectively labeled compound is one in which both the position(s) and the number of the labeling nuclide(s) are undefined.

$[^{14}C,^2H]CH_3CH_2$–OH $[^{14}C,^2H]$ethanol

In general labeling, all positions of the designated element are labeled, but not necessarily in the same isotopic ratio. The nonitalic symbol G may be used in place of locants to indicate general labeling.

$[G\text{-}^{14}C]$pentanoic acid D-$[G\text{-}^{14}C]$glucose

In uniform labeling, all positions or all specified positions of the designated element are labeled in the same isotopic ratio. The nonitalic symbol U may be used in place of locants to indicate uniform labeling.

$[U\text{-}^{14}C]$pentanoic acid D-[U-14C]glucose D-$[U\text{-}1,3,5\text{-}^{14}C]$ glucose

17.10.3 Isotopically Deficient Compounds

An isotopically deficient compound is one in which one or more of the nuclides are present in less than the natural ratio. Indicate the deficiency with the italic term *def* in square brackets, immediately preceding, without a hyphen, the appropriate nuclide symbol.

[*def*]$^{13}CHCl_3$ [*def*^{13}C]chloroform

17.11 STEREOCHEMICAL NOMENCLATURE

Many special, italicized symbols and terms are used in chemical nomenclature to describe the stereochemical configuration of a variety of compounds. Some specialized areas have developed their own local systems.

17.11.1 Organic Compounds

Descriptors frequently used in organic stereochemical nomenclature are listed in Table 17.11. For details on the use of these descriptors, consult the glossary "Basic Terminology of Stereochemistry".[11]

17.11.1.1 Double Bonds

Configurations at double bonds are described by the italic letters *Z* and *E*, enclosed in nonitalic parentheses and placed immediately in front of the part of the name to which they refer. "(*Z*)" (for *zusammen*, together) describes the configuration where the preferred atoms or groups attached to each atom of the double bond lie on the same side of

Table 17.11 Descriptors for organic stereochemical nomenclature

Descriptor	Definition/use	Examples
(+)-	rotation of the plane of polarization to the right	(+)-glucose
(−)-	rotation of the plane of polarization to the left	(−)-fructose
(±)-	racemic form	(±)-glucose (±)-4-(2-aminopropyl)phenol
ambo-	molecule with 2 (or more) chiral elements that are present as a mixture of the 2 racemic diastereoisomers in unspecified proportions; set in italics	L-alanyl-*ambo*-leucine [the dipeptide formed from L-alanine and DL-leucine]
anti-	similar to *trans* (i.e., on the opposite side of a defined structural feature); set in italics	*anti*-benzaldoxime *anti*-7-bromobicyclo[2.2.1]hept-2-ene
cis-	atoms or groups on the same side of a reference plane; set in italics	*cis*-cyclohexane-1,4-diol *cis*-but-2-ene
D-	absolute configuration to the right at the α-carbon in an amino acid compared to serine or at the highest-numbered center of chirality in a carbohydrate compared to glyceraldehyde when displayed in a Fischer projection; set as a small capital letter; compare with L-	D-serine D-glucose
DL-	mixture of equimolar amounts of D- and L-enantiomers; set as small capital letters	DL-leucine DL-glucose
d, dextro-	**obsolete term** for the rotation of the plane of polarization to the right; the preferred prefix is "(+)-"	[no longer used]
endo-	inner position as opposed to an outer position; used mainly in bi- and polycyclic ring systems where an *endo* substituent points toward the interior part of the structure; set in italics; compare with *exo-*	*endo*-2-methylbicyclo[2.2.1]heptane
ent-	abbreviation for enantio or enantiomers, indicating that the steric configurations are the opposite of that described or implied by the descriptors in the name; set in italics	*ent*-symplocosigenol
epi-	abbreviation for epimer, indicating diastereoisomers that have the opposite configuration at only 1 of 2 or more tetrahedral stereogenic centers present in the respective molecular entities; set in italics [This prefix has other meanings in other contexts. See Appendix K of the *Naming and Indexing of Chemical Substances for Chemical Abstracts*.[18]]	16-*epi*-vobasan 2-*epi*-α-tocopherol
exo-	outer position as opposed to an inner position; used mainly in bi- and polycyclic ring systems where an exo substituent points away from the interior part of the structure; set in italics; compare with *endo-*	*exo*-2-dimethylbicyclo[2.2.2]octane
L-	absolute configuration to the left at the α-carbon in an amino acid compared to serine or at the highest-numbered center of chirality in a carbohydrate compared to glyceraldehyde when displayed in a Fischer projection; set as a small capital letter; compare with D-	L-alanine L-arabinose
l, levo-	**obsolete term** for the rotation of the plane of polarization to the left; the preferred prefix is "(−)-"	[no longer used]

Table 17.11 Descriptors for organic stereochemical nomenclature (*continued*)

Descriptor	Definition/use	Examples
meso-	a term describing achiral member(s) of a set of diastereoisomers that also includes one or more chiral members	meso-tartaric acid
rac-	abbreviation for racemic, a mixture of equal amounts of enantiomers; set in italics	*rac*-leucine
syn-	similar to *cis*, i.e., on the same side of a defined structural feature; set in italics	*syn*-benzaldoxime *syn*-7-bromobicyclo[2.2.1]hept-2-ene
trans-	atoms or groups on opposite sides of a reference plane; set in italics	*trans*-but-2-ene *trans*-cinnamic acid

a reference plane that is perpendicular to the plane defined by the atoms of the double bond and its adjacent atoms; "(*E*)" (for *entgegen*, opposite) describes the configuration where the preferred atoms or groups attached to each atom of the double bond lie on opposite sides of the reference plane. When there is no ambiguity, the prefixes *cis*- and *trans*- (Table 17.11) may be substituted for *Z* and *E*, respectively. This symbolism was adopted by IUPAC in its recommendations for stereochemical nomenclature.[2,3]

H_3C, H, C=C, H, CH_3

(*E*)-but-2-ene
trans-but-2-ene

H_3C, Cl, C=C, H, CH_3

(*Z*)-2-chlorobut-2-ene

17.11.1.2 Chiral Compounds

The chiral center (atom) is described as *R* or *S* (in italics) (or *P* or *M* in helical molecules), depending on whether the order of the ranked groups a > b > c traces a clockwise or counterclockwise path, respectively, when viewing a tetrahedral model from the side opposite the least preferred atom or group—that is, d.[2,3]

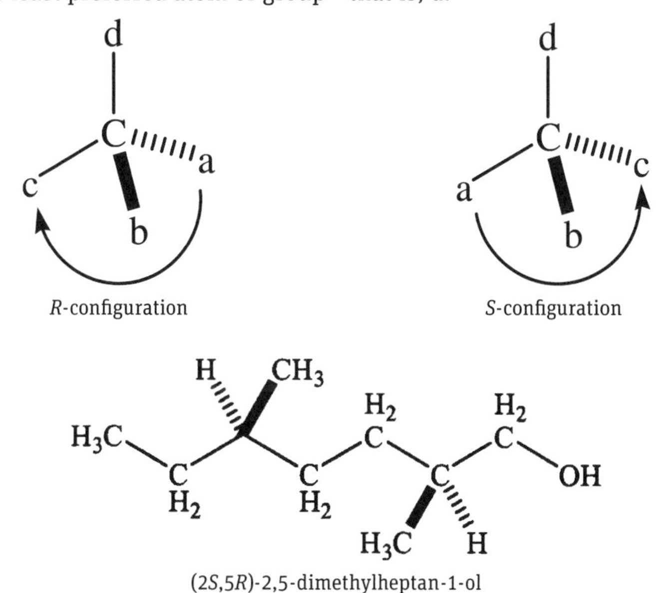

R-configuration

S-configuration

(2*S*,5*R*)-2,5-dimethylheptan-1-ol

The *R*/*S* symbols describe absolute configuration. If the absolute configuration is not known, the relative configuration can be given by the symbols *R** (spoken *R*-star) and *S** assigned in the same manner as above for *R*/*S*, but in such a way that the first cited chiral center is *R**.

17.11.2 Inorganic Compounds

Inorganic compounds exhibit many more geometries than organic compounds; therefore, descriptions of stereochemical configurations for inorganic compounds are more complex. Not only the geometry of the polyhedra but also the order of the attached ligating atoms must be described. In addition, organic compounds having their own stereochemical configurations can be ligands in inorganic coordination compounds.

17.11.2.1 Geometrical Descriptors for Inorganic Polyhedra

The geometrical descriptors for the polyhedra of inorganic coordination compounds are listed in Table 17.12.[1]

17.11.2.2 Configuration Index

The configuration index portion of the inorganic stereochemical descriptor is a single digit or a pair of digits that identifies the positions of the ligating atoms on the polyhedron. Consult the IUPAC recommendations[1] for the detailed procedure for deriving the digits.

Table 17.12 Inorganic polyhedra

Coordination polyhedron	Coordination number	Polyhedral symbol
Linear	2	*L*-2
Angular	2	*A*-2
Trigonal plane	3	*TP*-3
Trigonal pyramid	4	*TPY*-4
Tetrahedron	4	*T*-4
Square plane	4	*SP*-4
Square pyramid	5	*SPY*-5
Trigonal bipyramid	5	*TBPY*-5
Octahedron	6	*OC*-6
Trigonal prism	6	*TPR*-6
Pentagonal bipyramid	7	*PBPY*-7
Monocapped octahedron	7	*OCF*-7
Square face monocapped trigonal prism	7	*TPRS*-7
Cube	8	*CU*-8
Square antiprism	8	*SAPR*-8
Dodecahedron	8	*DD*-8
Hexagonal bipyramid	8	*HBPY*-8
Trans bicapped octahedron	8	*OCT*-8
Triangular face bicapped trigonal prism	8	*TPRT*-8
Square face bicapped trigonal prism	8	*TPRS*-8
Square face tricapped trigonal prism	9	*TPRS*-9
Heptagonal bipyramid	9	*HBPY*-9

17.11.2.3 Chirality Symbol

The final component of the inorganic coordination stereochemical descriptor is the chirality symbol. It is determined by orienting the polyhedron in a way specific to each polyhedron and noting the direction, going from a lower-priority ligating atom to a higher-priority one along the edge of the polyhedron. If the direction is clockwise, the chirality symbol is an italic *C*; if it is anticlockwise, the chirality is *A*.

(*OC*-6-32-*C*)-diaminedibromo(ethane-1,2-diamine-*N*,*N*′)cobalt(1+)

17.11.2.4 Descriptors for Inorganic Structural and Stereochemical Nomenclature

Other descriptors used in systematic inorganic stereochemical and geometric nomenclature include the following:

antiprismo-	8 atoms forming a rectangular antiprism
cis-	2 groups occupying adjacent positions in a coordination sphere
dodecahedro-	8 atoms forming a triangular dodecahedron
fac-	3 groups occupying the corners of the same face of an octahedron
hexahedro-	8 atoms forming a hexahedron, e.g., a cube
hexaprismo-	12 atoms forming a hexagonal prism
icosahedro-	12 atoms forming a triangular icosahedron
mer-	meridonal, 3 groups occupying the vertices of an octahedron in such relationship that one is *cis* to the 2 others, which are themselves *trans*
octahedro-	6 atoms forming an octahedron
pentaprismo-	10 atoms forming a pentagonal prism
quadro-	4 atoms forming a quadrangle, e.g., a square
tetrahedro-	4 atoms forming a tetrahedron
trans-	2 groups directly across a central atom from each other, i.e., in the polar positions of a sphere
triangulo-	3 atoms forming a triangle
triprismo-	6 atoms forming a triangular prism

17.12 POLYMER NOMENCLATURE

Polymers are characterized by multiple repetitions of atoms or groups of atoms linked to each other in amounts sufficient to provide a set of properties that would not differ significantly with the addition or deletion of one or a few of the repeating units.[19]

17.12.1 Source-Based Nomenclature

Polymers have long been named by attaching the prefix "poly" to the name of the real or assumed monomers from which they are derived.[19] Enclose the monomer name in parentheses when it consists of 2 or more words or is prefixed by a descriptor.

polystyrene
poly(methyl acrylate)
poly(L-lysine)

Source-based nomenclature for polymers is misleading in that the chemical structure of the monomer is different from the chemical structure of the monomeric unit in the polymer (e.g., a monomer CH_2=CHX versus the polymer unit –CH2–CHX–).

Copolymers are polymers derived from more than one species of monomer. A series of italicized infixes are used to provide information about the structure of the copolymer, as given below.[5]

-alt-	alternating, e.g., poly(styrene-*alt*-maleic anhydride)
-block-	block, e.g., polystyrene-*block*-polybutadiene
-co-	unspecified, e.g., poly(styrene-*co*-methyl methacrylate)
-graft-	graft, e.g., polybutadiene-*graft*-polystyrene
-per-	periodic, e.g., poly[formaldehyde-*per*-(ethylene oxide)-*per*-(ethylene oxide)]
-ran-	random, e.g., poly[ethylene-*ran*-(vinyl acetate)]
-stat-	statistical, e.g., poly(styrene-*stat*-butadiene)

Other infixes to describe polymer structures have been published by IUPAC.[19]

17.12.2 Structure-Based Nomenclature

For regular organic polymers (i.e., polymers that have only one species of repeating group), names are formulated by using the pattern "poly(constitutional repeating unit)", with the constitutional repeating unit (CRU) being the name of a bivalent organic group or a combination of bivalent groups[5] and placed within enclosures.

poly(methylene) [structure-based name]	polyethylene [source-based name]
poly(propylene) [structure-based name]	polypropene [source-based name]
poly(1-acetoxyethylene) [structure-based name]	poly(vinyl acetate) [source-based name]

Names for regular single-strand and quasi-single-strand inorganic and coordination polymers are formed in a similar way to names for organic polymers.[5] The italic prefix catena- is used before "poly" to emphasize the linear nature of the polymer.

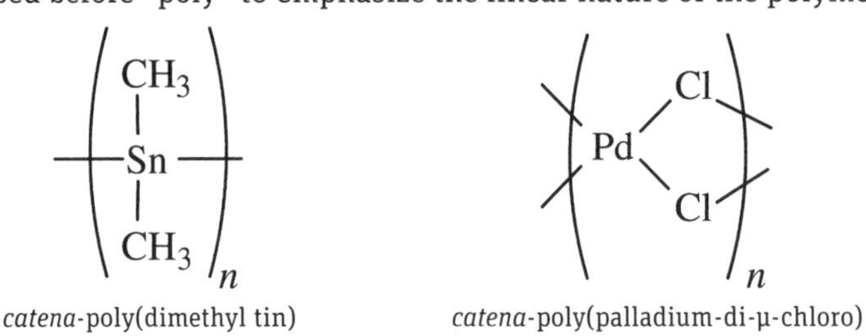

catena-poly(dimethyl tin) *catena*-poly(palladium-di-μ-chloro)

17.13 COMPOUNDS OF BIOCHEMICAL IMPORTANCE

Style, format, and nomenclature conventions for many compounds of biochemical importance are based on recommendations developed by IUPAC for natural products.[20] The major compound classes, including alkaloids, steroids, and terpenoids, along with some relatively minor classes, such as carotenoids, bilins, flavans, corrinoids, porphyrins, penicillins, and prostaglandins, fit cleanly into these general rules (which are outlined in Section 17.13.1). For other compound classes, specific guidelines and recommendations are available, as described in Sections 17.13.2 through 17.13.19.

17.13.1 Natural Products

The IUPAC rules for natural products[20] provide guidance for naming and numbering. Unless otherwise specified in the rules, methods of name construction and the principles of organic nomenclature, such as addition of suffixes and prefixes, are applied to fundamental parent structures. For each class, a parent structure is selected that reflects the basic skeletal structure (including nonterminal heteroatoms, heterogroups, and stereochemistry) that is common to most compounds in the class and is given a name derived to the extent possible from a trivial name of a member of the class and numbered to the extent possible according to the numbering established for the class.

The following special operations denoted by nonitalic prefixes are applied to all fundamental parent structures:

abeo- migration of one end of a single bond from its original position in a fundamental parent structure to another position; the prefix is preceded by a descriptor *x*(*y*→*z*), where "*x*" is the locant of the unchanged end of the bond, "*y*" is the locant of the original position of the moving bond, and "*z*" is the locant of the new position of the moving bond, e.g., 3α*H*-5(4→3)-abeopodocarpane

cyclo- creation of an additional ring by means of a direct link between any 2 nonadjacent atoms of a fundamental structure; preceded by locants of the positions joined, e.g., 16β*H*-1,16-cyclocorynan

de- removal of an atom or group from a parent structure, e.g., deoxy (removal of an oxygen atom) and dehydro (removal of a hydrogen atom), sometimes with subsequent replacement by hydrogen atoms, e.g., demethylmorphine

des- removal of a terminal ring from a fundamental parent structure with addition of an appropriate number of hydrogen atoms; followed by an italic letter designating the ring that is removed, e.g., des-*A*-androstane

homo- addition of a methylene ($-CH_2-$) group between 2 skeletal atoms of a fundamental parent structure; preceded by the locant of the added atom, e.g., 10a-homotaxane, 5(6)a-homoergoline

nor- removal of an unsubstituted skeletal atom, saturated or unsaturated, from a ring or from an acyclic portion of a fundamental parent structure; preceded by the locant of the atom removed, e.g., 1,20-dinorprostane

seco- cleavage of a ring bond (saturated or unsaturated) with the addition of an appropriate number of hydrogen atoms at each new terminus thus created; preceded by the locants at the ends of the removed bond, e.g., 2,3-secoyohimban

Combinations of these prefixes provide the means for even more drastic changes of the structure of the fundamental parent structure—for example, 13(17)a-homo-12,18-dinor-5α-pregnane.

Heteroatoms may be introduced by skeletal replacement nomenclature (e.g., 1-thiaergoline), and rings or ring systems can be added to the fundamental structures by adaptation of fusion nomenclature (e.g., benzo[2,3]-5α-androstane), bridging (e.g., 3α,8-epidioxy-5α,8α-androstane), or functionalization (e.g., aspidospermidine-3α,4α-diyl carbonate).

Some compound classes (e.g., electron-transfer proteins, folates) have no unique style and format issues, even though they may be very important biologically. For compounds of interest not listed in the sections below, see the IUPAC Nomenclature website[4] for guidance.

17.13.2 Amino Acids, Peptides, and Polypeptides

17.13.2.1 Amino Acids

Amino acids are compounds containing both an amino group and a carboxylic acid group. Details for naming amino acids have been published.[6]

An α-amino acid is a carboxylic acid substituted by an amino group at the carbon atom immediately adjacent to the carboxylic acid group—that is, the α-carbon. Most naturally occurring amino acids are α-amino acids. The simplest amino acid is glycine, $H_2N–CH_2–COOH$ (2-aminoacetic acid). All other α-amino acids—for example, alanine, $CH_3CH(NH_2)COOH$ (2-aminopropanoic acid or 2-aminopropionic acid), contain at least one chiral center (the α-carbon atom is asymmetric) and thus have 2 optically active forms or enantiomers, termed D and L. All naturally occurring α-amino acids (with the exception of glycine, which has no chiral center) in "higher" organisms are L-enantiomers; D-amino acids are often found in microorganisms. See Table 17.13 for a list of those frequently found in proteins.

Semisystematic names of substituted α-amino acids are formed according to the general principles of organic nomenclature, including stereochemistry, by attaching the name of the substituent group, with the appropriate numerical locant, to the trivial name of the amino acid.

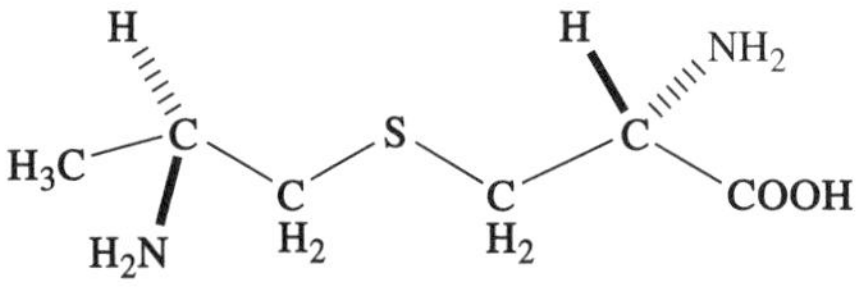

trans-4-hydroxy-L-proline

Trivial names for unmodified amino acids are preferred in straight text. Standard 3-letter symbols, consisting of a capital letter followed by 2 lowercase letters, are frequently used in tables, diagrams, peptide sequences, and terms indicating residue numbers (e.g., Tyr-110 for the tyrosine residue 110 in a protein sequence). They represent the L- configuration of the chiral amino acids unless otherwise indicated by the presence of D- or DL- before the symbol. When needed for emphasis, the L- may be included. The corresponding 1-letter symbols should be restricted to tables and sequences and should never be used in text.

17.13.2.2 Peptides and Polypeptides

Peptides are compounds containing 2 or more amino acids joined through a peptide linkage (–CO–NH–) (a peptide bond) between the carboxylic acid group of one amino acid and the amino group of another. Peptides with fewer than about 10 to 20 amino acid residues are termed oligopeptides; those with more amino acid residues are polypeptides.[6] A systematic name of any peptide with one or more peptide linkages is formed by combining the acyl forms of the names of all the constituent amino acids (Table 17.13), in sequence from left (the amino, or N-terminal) to right (the carboxy, or C-terminal), until the C-terminal amino acid is reached; the trivial name of the last amino acid is retained. Details for naming peptides have been established.[6]

L-valyl-L-alanylglycyl-D-phenylalanyl-L-tryptophyl-D-proline

[Note the omission of the hyphen between L-alanyl and glycyl, the latter being the only amino acid without a chiral carbon atom. The example displayed is a hexapeptide (a peptide containing 6 amino acid residues).]

17.13.2.3 Sequence Conventions for 3-Letter Symbols

When the sequences of polypeptides or short proteins are being represented, it is customary in text and tabular material to use the 3-letter symbols for the amino acids. The hyphens between the configurational prefix and the amino acid symbol may be omitted for brevity, as in the second example below.

L-Val-L-Ala-Gly-D-Phe-L-Trp-D-Pro
LVal-LAla-Gly-DPhe-LTrp-DPro

Other conventions for writing peptides are as follows, shown by example:

Ala-Leu	a hyphen between amino acid abbreviations represents a peptide linkage
-Ala	a hyphen preceding an amino acid abbreviation represents loss of H from the NH_2 of the amino group
Ala-	a hyphen following an amino acid abbreviation represents loss of OH from the COOH of the amino acid (the acyl form of the amino acid)
-Ala-	a hyphen preceding and following an amino acid abbreviation represents loss of H from the NH_2 and loss of OH from the COOH
| Lys *or* Lys |	a vertical bar represents a linkage through the side chain functional group (here, the amino group at C-6 with lysine; with cysteine, the SH group would be involved; with serine, the OH group; with glutamic and aspartic acids, the second carboxylic acid group)
|5 Lys	represents substitution at C-5 of lysine

17.13.2.4 Sequence Conventions for 1-Letter Symbols

The most frequent uses of 1-letter symbols are in the display of polypeptide or protein sequences and in the alignment of homologous sequences, where the spacing of symbols and punctuation is critical. Use an equal-space font in which each letter and each punctuation mark occupy the same amount of horizontal space. When reporting a single sequence, insert a space after every 10 symbols.

```
GDVEKGKKIF IMKCSQCHTV EKGGKHKTGP N
```

When presenting a comparison between sequences, insert hyphens instead of spaces to maintain alignment.

human	`GDVEKGKKIFIMKCSQCHTV------EKGGKHKTGPN`
Neurospora crassa	`GFSAGDSKKGAKLFKTRCAQCHTL------EEGGGNKIGPA`
Euglena gracilis	`GDAERGKKLFESRAAQCHSA------QKG-VNSTGPS`
Paracoccusi denitrificans	`NEGDAAKGEKEFN-KCKACHMIQAPDGTDI-KGGKTGPN`

17.13.2.5 Side Chains

The term "side chain" refers to C-3 and higher-numbered carbon atoms and their substituents. Substitutions involving side chains may also be presented by placing the symbol of the substituent within parentheses immediately after the symbol of the amino acid being substituted.

Cys(Et)	*S*-ethylcysteine
Cys-OEt	cysteine ethyl ester
Ser(Ac)	O^3-acetylserine
Ac-Glu(OEt)-OMe	O^5-ethyl methyl *N*-acetylglutamate *or* O^5-ethyl O^1-methyl *N*-acetylglutamate

Table 17.13 Names and symbols of the common α-amino acids[a]

Trivial name	Symbols[b]		Systematic name	Acyl radical	
	3-letter	1-letter		Structure	Name
alanine	Ala	A	2-aminopropanoic acid	$CH_3CH(NH_2)CO-$	alanyl
arginine	Arg	R	2-amino-5-guanidinopentanoic acid	$H_2NC(=NH)NH(CH_2)_3CH(NH_2)CO-$	arginyl
			2-amino-5-(carbamimidoylamino)pentanoic acid		
asparagine	Asn	N	2-amino-3-carbamoylpropanoic acid	$H_2NCOCH_2CH(NH_2)CO-$	asparaginyl
			2,4-diamino-4-oxobutanoic acid		
aspartic acid	Asp	D	2-aminobutanedioic acid	$-COCH_2CH(NH_2)CO-$	aspartoyl
				$HOOCCH_2CH(NH_2)CO-$	α-aspartyl
				$HOOCCH(NH_2)CH_2CO-$	β-aspartyl
cysteine	Cys	C	2-amino-3-sulfanylpropanoic acid	$HSCH_2CH(NH_2)CO-$	cysteinyl[c]
glutamic acid	Glu	E	2-aminopentanedioic acid	$-COCH_2CH_2CH(NH_2)CO-$	glutamoyl
				$HOOCCH_2CH_2CH(NH_2)CO-$	α-glutamyl
				$HOOCCH(NH_2)CH_2CH_2CO-$	γ-glutamyl
glutamine	Gln	Q	2-amino-4-carbamoylbutanoic acid	$H_2NCOCH_2CH_2CH(NH_2)CO-$	glutaminyl
glycine	Gly	G	2,5-diamino-5-oxopentanoic acid	H_2NCH_2CO-	glycyl

			2-aminoacetic acid		
histidine	His	H	2-amino-3-(1*H*-imidazol-4-yl)propanoic acid	$(N_2C_3H_2)CH_2CH(NH_2)CO-$	histidyl
isoleucine	Ile	I	2-amino-3-methylpentanoic acid	$CH_3CH_2CH(CH_3)CH(NH_2)CO-$	isoleucyl
leucine	Leu	L	2-amino-4-methylpentanoic acid	$(CH_3)_2CHCH_2CH(NH_2)CO-$	leucyl
lysine	Lys	K	2,6-diaminohexanoic acid	$H_2N(CH_2)_4CH(NH_2)CO-$	lysyl
methionine	Met	M	2-amino-4-(methylsulfanyl)butanoic acid	$CH_3SCH_2CH_2CH(NH_2)CO-$	methionyl
phenylalanine	Phe	F	2-amino-3-phenylpropanoic acid	$C_6H_5-CH_2CH(NH_2)CO-$	phenylalanyl
proline	Pro	P	pyrrolidine-2-carboxylic acid	$NHCH_2CH_2CH_2CHCO-$	prolyl
serine	Ser	S	2-amino-3-hydroxypropanoic acid	$HOCH_2CH(NH_2)CO-$	seryl
threonine	Thr	T	2-amino-3-hydroxybutanoic acid	$CH_3CH(OH)CH(NH_2)CO-$	threonyl
tryptophan	Trp	W	2-amino-3-(1*H*-indol-3-yl)propanoic acid	$(C_8H_6N)CH_2CH(NH_2)CO-$	tryptophyl
tyrosine	Tyr	Y	2-amino-3-(4-hydroxyphenyl)propanoic acid	$4\text{-}HOC_6H_4CH_2CH(NH_2)CO-$	tyrosyl
valine	Val	V	2-amino-3-methylbutanoic acid	$(CH_3)_2CHCH(NH_2)CO-$	valyl
unspecified	Xaa	X			

[a] The trivial name refers to the L, D, or DL amino acid. For those that are chiral, only the L form is used for protein biosynthesis.
[b] Use of 1-letter symbols should be restricted to the comparison of long sequences.
[c] An exception to the rule for naming of acyl radicals in order to differentiate it from the cysteic acid radical.

Lack of substitution at the amino and carboxy groups may be emphasized by applying the principles stated above.

H-Ala Ala-OH

17.13.2.6 Modification of Named Peptides

Amino acids in named peptides are numbered starting from the terminal amino acid with the free amino group. When one or more of these amino acids is replaced, the name or symbol of the replacement amino acid preceded by the number of the amino acid being replaced is placed within square brackets immediately preceding the name of the peptide. Commas separate multiple amino acids. In the abbreviated form, the residue numbers of the amino acids being replaced are superscript.

[8-citrulline]vasopressin *or* [Cit^{8}]vasopressin
[5-isoleucine,7-alanine]angiotensin II *or* [Ile^{5},Ala^{7}]angiotensin II

Indicate replacement of a peptide bond by another group by placing the Greek uppercase letter psi, Ψ, between the 2 residue numbers as superscripts, followed by a comma and the symbol for the group, all within square brackets before the name of the peptide. In the abbreviated form, however, use parentheses around the symbol for the group and hyphens to represent the bonds attaching the group to the rest of the peptide.

[$^{3}\Psi^{4}$, CH_2-S]oxytocin *or* . . . -Ψ (CH_2-S)- . . .

Extensions of the peptide chain are named, for the amino terminal, by preceding the name of the peptide with the acyl form of the extending amino acid (Table 17.13) and, for the carboxy terminal, by adding "-yl" to the name of the peptide and attaching that "yl" form to the name of the extending amino acid.

valylvasopressin *or* Val-vasopressin
(angiotensin II)ylglycine *or* angiotensin II-Gly

Insertion and deletion of amino acid residues are indicated by the terms "endo" (in italics) and "des" (not in italics), respectively.

endo-4a-tyrosine-angiotensin II *or* *endo*-Tyr^{4a}-angiotensin II
des-7-proline-oxytocin *or* des-Pro^{7}-oxytocin

17.13.3 Antibiotics and Drugs

See Drugs and Pharmacokinetics, Chapter 20.

17.13.4 Carbohydrates

The principles for naming carbohydrates have been elaborated by IUPAC-IUBMB.[21] Monosaccharide classes are listed in Table 17.14.

17.13.4.1 Nomenclature

Systematic names have 4 components:

1) a configurational symbol D or L in small-capital roman letters, linked by a hyphen to the name of the monosaccharide

Table 17.14 Monosaccharide classes[a]

General formula	X	Y	Class	Characteristic ending
	CHO	CH_2–OH	aldose	ose
X	CHO	CHO	dialdose	dialdose
|	CHO	COOH	uronic acid	uronic acid
$[CH–OH]^n$	CH_2–OH	CH_2–OH	alditol	itol
|	COOH	CH_2–OH	aldonic acid	onic acid
Y	COOH	COOH	aldaric acid	aric acid
X				
|				
$[CH–OH]^n$	CH_2–OH	CH_2–OH	ketose	ulose
|				
CO	CHO	CH_2–OH	aldoketose	osulose
|				
$[CH–OH]^m$	COOH	CH_2–OH	ketoaldonic acid	ulosonic acid
|				
Y				

[a] Adapted with permission from Fox and Powell. *Nomenclature of Organic Compounds: Principles and Practice*, 2nd ed., Table 31.3. Copyright 2001 American Chemical Society.

2) a configurational prefix, representing the relative positions of the hydroxy groups, in lowercase italic letters
3) the stem name, indicating the number of carbon atoms
4) the characteristic ending, indicating the type of monosaccharide (e.g., aldose, ketose)

D-*gluco*-hexose   D-*ribo*-pentose   L-*glycero*-D-*manno*-heptose

17.13.4.2 Abbreviations

Because the systematic names of substituted carbohydrates are quite long, abbreviations are used extensively. The following examples represent extended and condensed forms.

D-Glc*p*, DGlc*p*

[Glc is used for glucose; italic *p* represents a pyranose ring; thus, these are symbols for D-glucopyranose]

L-Rib*f*, LRib*f*

[Rib is used for ribose; italic *f* represents a furanose ring; thus, these are symbols for L-ribofuranose]

β-D-Xyl*p*-(1→4)-α-D-Gal*p*-(1→6)-α-D-Man*p*-(1→2)-β-D-Fru*f*

DXyl*p*(β1–4)DGal*p*(αl–6)DMan*p*(αl–2β) DFru*f*

[Both of these condensed forms represent the same compound, *O*-D- xylopyranosyl-(1→4)-α-D-galactopyranosyl-(1→6)-α-D-mannopyranosyl β-D- fructofuranoside. Note that en dashes replace arrows between anomeric descriptors in the second example.]

17.13.4.3 Stereoisomers

Racemates are indicated by the prefix DL- in small capital letters. Optically inactive structures that have a plane of symmetry are indicated by the italicized prefix *meso*-. Optical rotation, or lack thereof, under specified conditions may be indicated by adding (+)-, (−)-, or (±)- before the configurational prefix. Do not use the obsolete terms *d*-, *l*-, and *dl*-.

DL-glucose   (+)-D-glucose   (±)-glucose   *meso*-galactitol

17.13.5 Carotenoids

Carotenoids, acyclic tetraterpenes, a class of hydrocarbons called carotenes and their oxygenated derivatives (xanthophylls), consist of 2 groups of 4 isoprenoid units, each joined at their terminal carbon atoms such that the methyl groups of one 4-unit group are in opposite configurational relationships to the methyl groups of the second 4-unit group.[6] Each group of isoprenoid units is numbered from 1 to 20 and 1′ to 20′ as shown below.

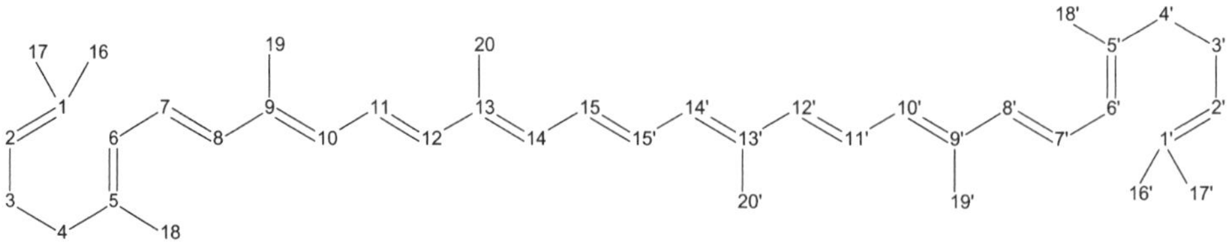

Specific names are constructed by adding 2 Greek-letter prefixes to the stem name "carotene". The appropriate Greek letters depend on the type of C-9 end groups (positions 1–6, 16–18, 1′–6′, and 16′–18′). The prefixes are in Greek-letter alphabetical order, separated by commas, and connected to the stem names with hyphens. See Table 17.15 for the C-9 end groups designations. Substituents on the carbon chains are named according to the general rules of organic nomenclature, including those applied to natural products.

1-ethoxy-1,2,7′,8′-tetrahydro-ω,ω-carotene   2,2′-dinor-β,β-carotene
3-hydroxy-3′-oxo-β,ε-caroten-16-oic acid   2,3-seco-ε,ε-carotene

Carotenoid derivatives shortened by removal of fragments from one or both ends of the molecule are noted with the nonitalic prefixes "apo" or "diapo", preceded by locants to indicate the point of cleavage.

9-apo-β-caroten-9-one   6′-apo-β-carotene   2′-apo-β,ψ-caroten-2′-al

Carotenoid derivatives in which all single and double bonds of a conjugated polyene system are shifted by one position are denoted by a pair of locants followed by the italic descriptor "retro" (see "apo" and "diapo").

4′,11-*retro*-β,ψ-carotene   3-hydroxy-6′,7-*retro*-β,ε-caroten-3′-one

17.13.6 Cyclitols

Cyclitols are cycloalkanes that have one hydroxy group on each of 3 or more ring atoms; they usually consist of 5- and 6-membered rings. Cyclitols and related compounds have stereochemical features characteristic of their class, so special methods of designating that stereochemistry are used. In other respects, their nomenclature should follow the general rules of organic nomenclature.[6,22]

Cyclitols with one hydroxy group on each atom of a cyclohexane ring (i.e., 1,2,3,4,5,6-hexahydroxycyclohexane) are called "inositols". Individual inositol stereoisomers are distinguished by italicized prefixes or a series of numbers enclosed in parentheses representing the positions of the hydroxy groups. A slash separates the numbers indicating hydroxy groups above and below the pseudo plane of the cyclohexane ring; numbers

Table 17.15 Carotenoid C9 end groups[a]

End group type	Formula	Structure	Greek letter prefix
acyclic	C_9H_{15}	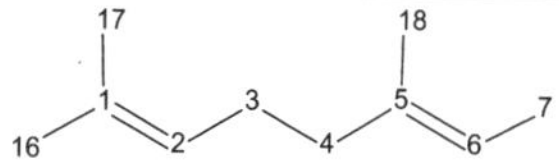	Ψ (psi)
cyclohexene	C_9H_{15}	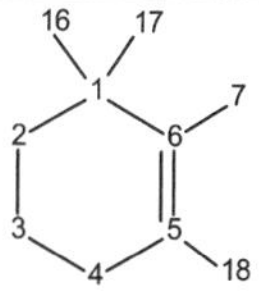	β (beta)
cyclohexene	C_9H_{15}	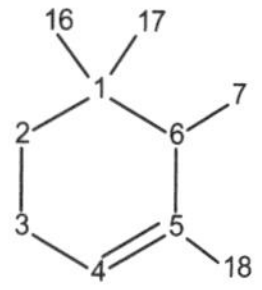	ε (epsilon)
cyclohexane	C_9H_{15}	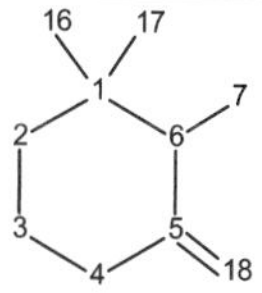	γ (gamma)
aryl	C_9H_{11}	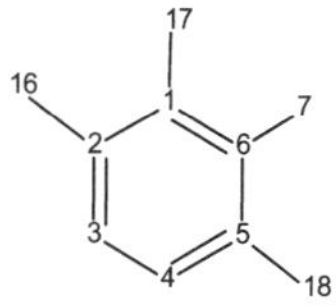	φ (phi)
aryl	C_9H_{11}	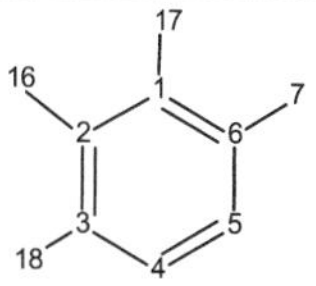	χ (chi)
cyclopentane	C_9H_{17}	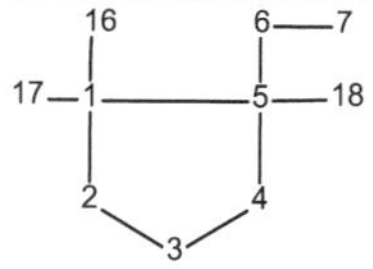	κ (kappa)

[a] Adapted from International Union of Pure and Applied Chemistry; International Union of Biochemistry. Pure Appl Chem. 1975;41(3):405–431, Rule 3.2.

preceding the slash indicate the hydroxy groups that are above, those following the slash are below.

cis- (1,2,3,4,5,6/0)	*neo*- (1,2,3/4,5,6)	*chiro*- (1,2,4/3,5,6)
epi- (1,2,3,4,5/6)	*myo*- (1,2,3,5/4,6)	*scyllo*- (1,3,5/2,4,6)
allo- (1,2,3,4/5,6)	*muco*- (1,2,4,5/3,6)	

Several unusual features occur in cyclitol nomenclature, some of which are common to carbohydrate nomenclature. Among these features, note the following in particular:

1) Because the principal functional group is most commonly hydroxy, variations are often emphasized.

   (2,3,4/1(COOH),5)-2,3,4,5-tetrahydroxycyclopentane-1-carboxylic acid

2) Because most carbon atoms have hydroxy substituents having hydrogen atoms that are replaceable, substitution of the hydrogen atoms directly attached to carbon atoms should be emphasized.

   2-*C*-methyl-*myo*-inositol

3) Configurational D and L symbols (small capitals) are often preceded by a locant. Absence of a prefix D-, L-, or DL- indicates a *meso* configuration; therefore, the prefixes should not be omitted unless the *meso* form is intended.

   1L-1-*O*-methyl-6-mercapto-6-deoxy-*chiro*-inositol

The recommended rules use italic configurational prefixes, but older, traditional names in the field deviate from these recommendations—for example, "D-chiro-inositol" instead of "(+)-chiroinositol". Also, on occasion, the locants in the recommended rules may differ from those in older traditional names, and as such, errors may be introduced by attempts to edit the latter to the former.

17.13.7 Enzymes

Enzymes are proteins with catalytic activity. Multienzymes are proteins with more than one catalytic function contributed by distinct parts of a polypeptide chain ("domains") or by distinct subunits.

Enzyme names have been formulated according to the recommendations of the Nomenclature Committee of IUBMB.[23] Most enzyme names end in -ase; however, the long-established names of peptidases, such as chymotrypsin, thrombin, coagulation factor Xa, and subtilisin, are permitted as exceptions.

Restriction endonucleases have a standardized trivial nomenclature of their own in addition to that given them by the IUBMB. Italicization of the names was abandoned in 2003.[23] Each name consists of a nonitalic 3-letter abbreviation for the source organism followed by nonitalic designations that may be derived from the name of a strain or the number of enzymes from the source organism (see also Section 21.2.8).

Bce1229 TaqI XmaII

17.13.7.1 Classification

Enzymes are classified on the basis of the reaction they catalyze into 6 main divisions, each of which has subclasses. These classifications serve as the basis for assigning 4 code numbers (separated by periods as shown below) to each enzyme:

1) a number representing one of the 6 main divisions (classes) to which the enzyme belongs; the numbers and divisions are:

   Class 1, oxidoreductases, which catalyze oxidoreduction reactions. The substrate is the hydrogen donor. The name often ends with "dehydrogenase" or "reductase". Use "oxidase" only when O_2 is the acceptor. "Oxidoreductase" is normally used only in systematic names.

   EC 1.1.1.1, alcohol dehydrogenase

   Class 2, transferases, which transfer a group from one compound to another. The name often ends with "transferase".

   EC 2.4.1.162, aldose β-D-fructosyltransferase
   EC 2.1.1.1, nicotinamide *N*-methyltransferase

   Class 3, hydrolases, which cleave C–O, C–N, C–C, and some other bonds. Names of hydrolases are often formed by appending "-ase" to the substrate name.

EC 3.2.1.15, polygalacturonase

Class 4, lyases, which cleave C–O, C–N, C–C, and some other bonds by elimination, leaving double bonds or rings, or by adding groups to double bonds. Names often end in "decarboxylase", "aldolase", or "dehydratase".

EC 4.1.2.13, fructose-bisphosphate aldolase *not* fructose–biphosphate aldolase

Class 5, isomerases, which catalyze geometric or structural changes within individual molecules. Names vary with the type of change.

EC 5.3.1.9, glucose-6-phosphate isomerase

Class 6, ligases, which catalyze the coupling of 2 molecules and the hydrolysis of a pyrophosphate bond in ATP or similar triphosphate. Most ligases have names in the form, X–Y ligase.

EC 6.3.1.1, aspartate–ammonia ligase

2) a number representing the subclass
3) a number representing the sub-subclass
4) a serial number within each sub-subclass

The website for Enzyme Nomenclature[23] gives a complete list of the subclasses and sub-subclasses. Below are 2 examples of common and systematic enzyme names.

adenosinetriphosphatase, EC 3.6.1.3 [common name]
ATP phosphohydrolase [systematic name]

[EC is the abbreviation for "Enzyme Commission". Adenosinetriphosphatase is a hydrolase (class 3) that catalyzes the hydrolytic cleavage of acid anhydrides (subclass 6 of hydrolases) in phosphorus-containing hydrides (sub-subclass 1). It was the third enzyme thus categorized.]

asparagine synthase (glutamine-hydrolyzing), EC 6.3.5.4 [common name]
L-asparatate:L-glutamine amido–ligase (AMP-forming) [systematic name]

[Asparagine synthase (glutamine-hydrolyzing) is a ligase (class 6) that catalyzes the formation of carbon–nitrogen bonds (subclass 3) involving an amido–N donor (sub-subclass 5). It was the fourth enzyme thus categorized.]

17.13.7.2 Domain and Association Symbols

Catalytic domains (part of a polypeptide chain with a catalytic function) are indicated by capital letters from early in the alphabet (A, B, C . . .); substrate-carrier domains are indicated by capital letters from late in the alphabet (R, S, T . . .).[6] Regulatory domains are given lowercase letters from early in the alphabet (a, b, c . . .). Domains in the same polypeptide chain are placed within the same pair of parentheses. Thus, "(ABC)" represents a multienzyme polypeptide, and "(A)(BC)" represents a multienzyme complex. Braces, { }, may be used to indicate stable association. For example, tryptophan synthase from *Escherichia coli*, "(A)2(B)2", may be portrayed as "{(A)2(B)2}", which indicates that the association is stable, or as "{(A)(B)}2", which indicates that each A chain binds one B chain tightly, but that the 2 {(A)(B)} units are more loosely associated.

17.13.8 Lignans and Neolignans

Lignans and neolignans are large groups of natural products characterized by the coupling of 2 C_6C_3 units, each consisting of a propylbenzene structure. If the 2 C_6C_3 units are connected at the β position of each propyl group, the compound is lignane (the parent structure for all lignans). If the 2 C_6C_3 units are joined at any other position, the compound is neolignane (the parent structure of all neolignans), the locants of the bond linking the 2 units are given in front of the name, and the second locant is primed.

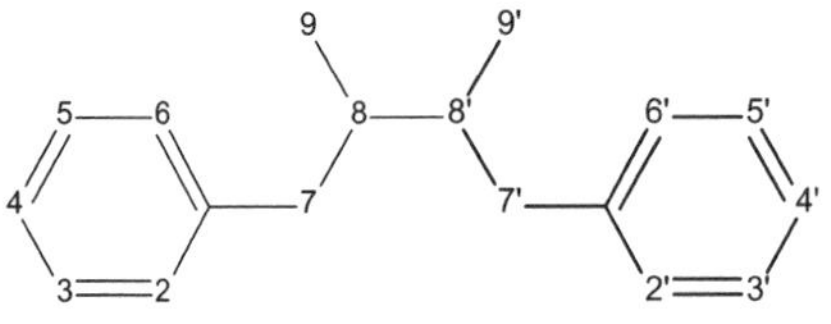

lignane

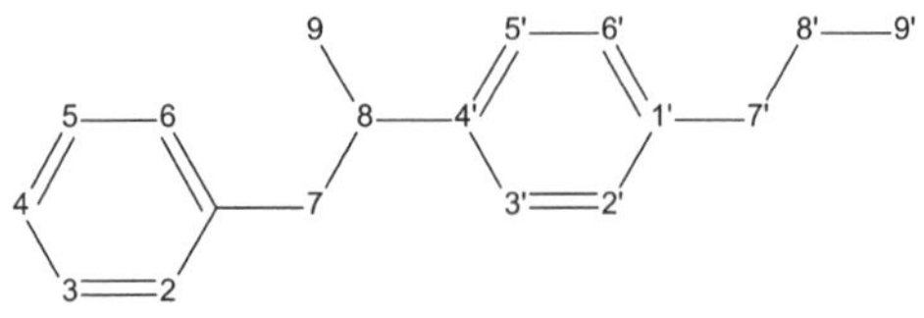

8,4′-neolignane

Nomenclature for lignans and neolignans follows the general principles for naming natural products[20] (see also Section 17.13.1). Substituents are described by substitutive nomenclature principles. Several different numbering systems are in use, so the writer should indicate which is being used in trivial, systematic, and semisystematic names and their modifications.

9-norlignane
7,9′:7′,9-diepoxylignane
5′H-2V-oxa-8,3′-neolignane
4′,7-epoxy-8,3′-neoligna-7,8′-diene
3,3′,4,4′-tetramethoxy-7,9′-epoxylignan-7′-yl acetate

Higher analogs of the lignans and neolignans composed of 3 or more C_6C_3 units are called sesquineolignans, dineolignans, etc., by analogy with the terminology of terpenes. One of the terminal units has unprimed numbers as locants, and locants for the other units are primed serially.

17.13.9 Lipids

17.13.9.1 Fatty Acids, Neutral Fats, Long-Chain Alcohols, and Long-Chain Bases

Fatty acids are the aliphatic monocarboxylic acids obtained by hydrolysis of naturally occurring fats and oils. Neutral fats are esters of glycerol and fatty acids and are appropriately termed “mono-acylglycerol”, “diacylglycerol”, and “triacylglycerol”; use of the terms “glyceride”, “diglyceride”, and “triglyceride” is discouraged. “Long-chain alcohols” refers to alcohols whose carbon chain length is greater than 10. “Long-chain base” refers to any base containing a long-chain aliphatic radical. Sphinganine (2*S*,3*R*)-2-aminooctane-1,3-diol) and its homologs, stereoisomers, and derivatives are called “sphingoids” or “sphingoid bases”.[6]

The prefix *sn* (for “stereospecifically numbered”) indicates that the carbon atom appearing at the top of a Fischer projection showing a vertical carbon chain with the C-2 hydroxy group to the left is designated as C-1. Use the lowercase italic symbol *sn* immediately before the glycerol term and separated from it by a hyphen.

The following examples are representative of these natural product classes:

4D-hydroxysphinganine	*also*	(2*S*,3*S*,4*R*)-2-aminooctadecane-1,3,5-triol
		phytosphingosine
sphingosine	*also*	(4E)-sphingenine
		trans-4-sphingenine
		(2*S*,3*R*,4*E*)-2-aminooctadec-4-ene-1,3-diol
tristearoylglycerol	*also*	*tri-O*-stearoyl glycerol
		glycerol tristearate
glycerol 2-phosphate	*also*	2-phosphoglycerol
sn-glycerol 1-phosphate	*also*	L-(glycerol 1-phosphate)
		D-(glycerol 3-phosphate)
sn-glycerol 3-phosphate	*also*	D-(glycerol 1-phosphate)
		L-(glycerol 3-phosphate)

Do not abbreviate the term "fatty acid(s)". In tables and/or text in which several fatty acids are described, each fatty acid can be designated by the numbers of carbon atoms and double bonds, separated by a colon. When 2 or more fatty acids have the same notation, they can be identified unambiguously by adding the positions of the double bonds in parentheses. Radicals of fatty acids can be identified by adding "acyl" instead.[6]

16:0 *for* palmitic acid
18:1 *for* oleic acid
18:3(9,12,15) *for* α-linolenic acid
18:3(6,9,12) *for* γ-linolenic acid
16:0(acyl) *for* the acyl radical of palmitic acid

17.13.9.2 Phospholipids

A phospholipid is any lipid containing phosphoric acid as a monoester or diester. The prefix phospho- may be used as an infix to designate phosphodiester bridges, as in "glycerophosphocholine", or as an alternative to *O*-phosphono- or *N*-phosphono-.

Phosphatidic acids are derivatives of glycerol phosphate in which both of the hydroxy groups of glycerol are esterified with fatty acids. The position of the phosphate group may be emphasized by stereospecific numbering.[6]

lecithin [trivial name]
3-*sn*-phosphatidylcholine [semisystematic name]
1,2-diacyl-*sn*-glycero-3-phosphocholine [systematic name]

17.13.9.3 Glycolipids

Glycolipids are compounds in which one or more monosaccharide residues are linked through a glycosyl group to a lipid group. The structures are often complex and the use of abbreviations is prevalent (but "acyl" should not be abbreviated). D- may be omitted from abbreviated formulas unless needed to avoid ambiguity. Hyphens may replace left-to-right arrows between anomeric descriptors. See the IUPAC documents on lipids[24] and glycolipids[25] for details.

1,2-diacyl-3-β-D-galactosyl-*sn*-glycerol *or* 1,2-diacyl-*sn*-glycerol 3-β-D-galactoside mucotriaosylceramide *or* $McOse_3Cer$ *or* Gal(β1→4)Gal(β1→4)Glc(1→4)Cer

17.13.10 Nucleic Acids and Related Compounds

17.13.10.1 Nucleic Acids

Nucleic acids are compounds consisting of nucleotide units, purine or pyrimidine bases attached to a ribosyl or deoxyribosyl group joined by phosphoric acid residues in ester linkages with the hydroxy groups of the monosaccharides. The 5′-phosphates (mono, di, tri) of the common nucleosides may be represented by commonly used abbreviations of the form (for adenosine) AMP, ADP, ATP. For single nucleoside and nucleic acid representations, a 3-letter formulation is preferred (e.g., ATP becomes Ado-5′*PPP*), whereas for chains single-letter symbols are used, with connecting hyphens or the lower case letter "p" representing phosphate linkages.

Designate the purine and pyrimidine bases and their nucleosides by 3-letter symbols, each having an initial capital letter followed by 2 lowercase letters (Table 17.16). Use these symbols whenever single bases or short sequences of bases are being discussed. Restrict the corresponding 1-letter symbols to tables and sequences, and never use them in text.[6]

Table 17.16 Purines, pyrimidines, and nucleosides: recommended symbols[a]

Purine or pyridimine	3-letter symbol	Nucleoside	3-letter symbol	1-letter symbol
adenine	Ade	adenosine	Ado	A
		guanosine	Guo	G
guanine	Gua	inosine	Ino	I
		thioinosine	Sno	
xanthine	Xan	xanthosine	Xao	X
hypoxanthine	Hyp	. . .	. . .	. . .
thymine	Thy	ribosylthymine	Thd	T
cytosine	Cyt	cytidine	Cyd	C
		uridine	Urd	U
uracil	Ura	thiouridine	Srd	S
		pseudouridine	Ψrd	Ψ[b]
orotate	Oro	orotidine	Ord	O
unknown purine	Pur	unknown purine nucleoside	Puo	R
unknown pyrimidine	Pyr	unknown pyrimidine nucleoside	Pyd	Y
unknown base	. . .	unknown nucleoside	Nuc	N

[a] Adapted from International Union of Pure and Applied Chemistry; International Union of Biochemistry.[27]
[b] Q may replace Ψ for computer work.[27]

17.13.10.2 General Principles for Describing Chains

1) *Chain Direction.* The conventional representation of a polynucleotide chain is from the 5′ terminus to the 3′ terminus. A nucleotide, the repeating unit of a polynucleotide chain, is composed of 3 parts: the D-ribose or 2-deoxy-D-ribose monosaccharide ring, the phosphate group, and the purine or pyrimidine base, with the monosaccharide ring and the phosphate forming the backbone of the chain. The unit is defined by the sequence of atoms from the phosphorus atom at the 5′-end to the oxygen atom at the 3′-end of the monosaccharide. The following are equivalent representations for a 6-nucleotide unit.

   GAATTC *or* G-A-A-T-T-C *or* pGpApApTpTpC

2) *Numbering.* Designate specific nucleotide units by a letter or number in parentheses, starting with the first nucleotide residue in the sequence.

   A(1)   pU(5)   C(10)p   pG(*i*)p

   Atoms in each of the constituents follow the standard numbering practice.

   C2, C2(5)   N3, N3(10) O^5 P(*i*+1)   H1   H2′   O5′H

17.13.10.3 Functional Forms of DNA and RNA

When used as general terms or in reference to preparations of specific molecules, different types of DNA and RNA and the methods that use them may be abbreviated as follows[26,27]:

cDNA	complementary DNA
cRNA	complementary RNA
dsDNA	double-stranded DNA
dsRNA	double-stranded RNA

mRNA	messenger RNA
miRNA	micro-RNA
mtDNA	mitochondrial DNA
nRNA	nuclear RNA
RNAi	RNA interference
rRNA	ribosomal RNA
siRNA	small interfering RNA
snRNA	small nuclear RNA
snoRNA	small nucleolar RNA
ssDNA	single-stranded DNA
ssRNA	single-stranded RNA
tRNA	transfer RNA

17.13.10.4 Transfer RNAs

Transfer RNAs that accept specific amino acids are named by attaching the name of the amino acid or its 3-letter symbol to the stem "tRNA" in the following format[6,27]:

alanine tRNA *or* $tRNA^{Ala}$ (for the nonacylated form)
alanyl-tRNA *or* Ala-tRNA *or* Ala-$tRNA^{Ala}$ (for the aminoacylated form)

When 2 or more tRNAs accept the same amino acid, differentiate the isoacceptors by subscript numbers. Sources may be included in parentheses after the symbol. Write the codon triplet as a short sequence of symbols for the nucleic acids with no punctuation (e.g., the codons for histidine are CAU and CAC).

$tRNA_1^{Ala}$ $tRNA_2^{Ala}$ alanyl-$tRNA_2^{Ala}$ (*E. coli*)

17.13.10.5 Conventions for 1-Letter Symbols

DNA and RNA sequences are usually presented from left to right indicating the order from the 5′ to the 3′ end of the molecule. If another order is used, indicate the correct orientation. In the alignment of homologous sequences, where the spacing of symbols and punctuation is critical, use an equal-space font in which each letter and each punctuation mark occupies the same amount of horizontal space. When reporting a single sequence, insert a space after every 10 symbols.

```
GAATTCCAUT CCTAGTAAGG GTCTCATCCT AGT
```

When presenting a comparison between sequences, insert hyphens instead of spaces to maintain the alignment.

Bacillus subtilis	`GAATTC-AUT-CCTAGTAAGGGTCTCA-----CTAG`
Staphylococcus aureus	`GAATTCCAUT-CCTAGTAAGGGTCTGA---TCCTAGT`
Tetrahymena caudatus	`GAAATCCAUTCCCTAGTAACGGTCTCATCCTAGT`

Indicate noncovalent associations between base pairs by a single raised dot (e.g., an A·T base pair) and ratios between bases in the format (A+T)/(G+C), not by AT/GC or A+T/G+C.

17.13.11 Peptide Hormones

Many peptide hormones have well-established trivial names, some of which are so long that the hormones are known mainly by abbreviations. To preclude the proliferation of additional abbreviations and to decrease the use of many of the abbreviations employed, IUPAC–IUB[6] proposed alternative names based on the following 3 principles:

1) New names for hormones of the adenohypophysis should end with "tropin".

   corticotropin *for* adrenocorticotropic hormone, ACTH

2) Hypothalamic releasing factors should end with "liberin".

   corticoliberin *for* corticotropin-releasing factor, CRF

3) Hypothalamic release-inhibiting factors should end with "statin".

   somatostatin [somatotropin release-inhibiting factor]
   [Note that the abbreviations ACTH and CRF are not recommended for biochemical literature; they are given here only for identification purposes.]

The peptide sequences of these hormones vary significantly between species; therefore, the names of the hormones are essentially generic terms that do not denote distinct chemical entities. Authors should indicate the source of each hormone when appropriate in each publication.

17.13.12 Phosphorus-Containing Compounds

The nomenclature of phosphorus-containing compounds in inorganic and organic chemistry is well defined. However, strict application of the rules to biochemically important compounds usually results in complicated names that are inconvenient for most biochemists and biologists. The following principles are recommended for use in biochemical, biological, and medicinal fields.[6]

1) Phosphoric esters, RO–PO(OH)2, are named as O-substituted phosphoric acids or as substituted alcohols.

   glycerol 1-phosphate *or* glycero-1-phosphate [the second, condensed, form is acceptable]
   choline *O*-(dihydrogen phosphate) ["phosphate" may be used instead of "dihydrogen phosphate"]
   O-phosphonocholine ["phosphono" may be contracted to "phospho"; "phosphoryl" is incorrect for this group]

2) The prefixes "bis", "tris", and so on should be used to indicate 2 or more independent phosphoric acid residues; the prefixes "di" and "tri" and similar prefixes representing higher numbers are used to denote phosphate chains.

   fructose 1,6-bisphosphate *not* fructose 1,6-diphosphate   adenosine diphosphate

3) Phosphodiesters, -PO(OH)-, are named by using the infix "phospho".

   glycerophosphocholine

4) Nucleoside triphosphate analogs (i.e., compounds in which methylene groups, imido groups, or sulfur atoms replace an oxygen atom bridging 2 phosphorus atoms) may be named by indicating, within brackets (implying replacement), the locants of the phosphorus atoms being bridged, followed by the name of the replacement group.

   adenosine 5′-[α,β-methylene]triphosphate
   or
   adenosine 5′-α,β-μ-methylene-triphosphate

17.13.13 Prenols

Prenols are alcohols containing one or more isoprene units. The term "prenol" (from iso***pren***oid alcoh***ol***) is recommended[6] to describe the structure shown below. The carbon adjacent to the hydroxy group is numbered 1. The carbon of the methyl group attached to C-3 is numbered "31" (note this unusual representation; it occurs in many natural products containing rings and chains with fixed numbering systems). The repeating C_5H_8 unit (enclosed within parentheses) is called an "isoprene unit" or an "isoprene residue", and compounds containing this unit are called "isoprenoids".

$$\mathrm{H{-}(\underset{4}{C}H_2{-}\underset{3}{C}(CH_3){=}\underset{2}{C}H{-}\underset{1}{C}H_2)_{\mathit{n}}{-}OH}$$

If the value of *n* is known, the compounds and their derivatives are named accordingly.

hexaprenol [*n* = 6]   heptaprenyl diphosphate [*n* = 7, ester of diphosphoric acid]

17.13.14 Prostaglandins

Prostaglandins are icosanoids, naturally occurring compounds derived from the parent C_{20} acid, prostanoic acid. Systematically, prostaglandins are named on the basis of the parent hydride names prostane and thromboxane[21]; the structure-modifying prefixes given in Section 17.13.1 may be applicable. Trivial names for prostaglandins are divided into A, B, C, D, E, F, G, H, and I families. The families are distinguished by the presence and position of unsaturation and oxo and hydro groups giving rise to forms such as prostaglandin A_1 (PGA_1) and prostaglandin A_2 (PGA_2). The F family is subdivided further into α and β groups to distinguish between 2 stereochemical forms; a typical designation is "prostaglandin $F_{2\alpha}$". The subscript numbers with family-letter designations designate particular double-bond configurations.

Subscript Number	Double-Bond Configuration
1	13*E*
2	5*Z* and 13*E*
3	5*Z*, 13*E*, and 17*Z*

Because of the complexity of prostaglandin compounds and the cumbersome quality of their systematic names derived by the rules for naming organic compounds, the use of trivial names has been proposed for common prostaglandin compounds as a basis for further substitution and modification.[28]

Trivial Name	Semisystematic Prostane Names (see Section 17.3.1)
ent-prostanoic acid	(8α,12β)-prostan-1-oic acid
4,5,6-trinor-3,7-*inter*-*m*- phenylene-3-oxaprostaglandin A_1	(13*E*,15*S*)-15-hydroxy-9-oxo-3-oxa-4,5,6-trinor-3,7-*inter*-[1,3]phenyleneprosta-10,13-dien-1-oic acid [The infix *inter* has no official recognition for semisystematicprostane names]
	2-[3-({(1S,5R)-5-[(3S)-3-hydroxyoct-1-en-1-yl] cyclopent-3-en-1-yl}methyl)phenoxy]acetic acid (substitutive name; see Section 17.6)
(19*R*)-19- hydroxyprostaglandin B_1	(13*E*,15*S*,19*R*)-15,19-dihydroxy-9-oxoprosta-8(12),13-dien-1-oic acid
(18*Z*)-18,19-didehydroprostaglandin E_1	(13*E*,15*S*,18*Z*)-11α,15-dihydroxy-9-oxoprosta 13,18-dien-1-oic acid
11,15-anhydro-11-*epi*-prostaglandin E_1	(13*E*,15*S*)-11β,15-epoxy-9-oxoprost-13-en-1-oic acid
2,3,4,5-tetranorprostaglandin E_1	(13*E*,15*S*)-11α,15-dihydroxy-9-oxo-4,5,6,7-tetranorprost-13-en-1-oic acid
8-*iso*-prostaglandin E_1	(13*E*,15*S*)-11α,15-dihydroxy-9-oxo-8α-prost-13-en-1-oic acid

[Note the italicization of *ent*, *epi*, *inter*, and *iso* in the trivial names: *ent* = inversion of all stereochemical features; *epi* = inversion of the normal configuration of a particular substituent at the numbered position; *inter* = replacement of the carbons at the numbered positions by the specified infix; *iso* = inversion of the normal chirality at the numbered center.]

17.13.15 Proteins

Proteins are polypeptides of specific sequences of more than about 50 amino acids; there is disagreement among authors on where this term should start.[6] In contrast to the standardized nomenclature for enzymes (see Section 17.13.7), there is no broadly applicable system for naming nonenzyme proteins. Current standardization efforts generally involve specific families of proteins in which a consensus has formed on the membership of individual proteins. Some trends in the naming of newly discovered proteins are apparent in the biochemical literature.

17.13.15.1 Protein Designations Based on Open Reading Frames

Gene sequencing has led to identification of potential protein-coding regions referred to as open reading frames (ORFs). Existence of the translated protein may or may not be experimentally shown. An identified protein may be designated as "protein ORFnumber" with this designation including the number of amino acids in the protein or their sequential numbering (e.g., "protein ORF216" would indicate a protein of 216 amino acids in length). This system of designation refers to both the DNA region containing the open reading frame and the protein resulting from the translation coded by the DNA region.

17.13.15.2 Protein Designations Based on Molecular Weight

One of the earliest experimental findings for a protein, its molecular weight, may be used as the basis for the name of the protein. For example, p21 and p75 are proteins of molecular weights of 21 and 75 kDa, respectively; posttranslationally modified forms of these proteins may be referred to as "gp21" and "pp75", indicating a glycoprotein of 21 kDa and a phosphoprotein of 75 kDa.

Designations by molecular weight may be further modified by including the gene symbol as a superscript. For example, gp160^{env} designates the 160-kDa glycoprotein encoded by the *env* gene in the human immunodeficiency virus (HIV). This form in turn may be used to identify proteins that are processed to smaller molecular-weight forms encoded by a single gene, such as the processing of HIV's gp160^{env} to gp120^{env} and gp41^{env}.

Molecular-weight designations may also be used in combination with the class of the protein, especially if the protein has multiple isoforms or multiple subunits. Give the class as an abbreviation in superscript following the molecular-weight designation (e.g., p75^{NGFR} is a 75-kDa isoform of a nerve growth-factor receptor).

17.13.15.3 Protein Designations Based on Gene Names

Many protein names are derived from the symbols for the corresponding gene names. Variations of this system can be found throughout the literature. When gene symbols are in lowercase italic letters, designate a protein by capitalizing the initial letter, or all letters, of the relevant gene symbol and set the resulting designation in roman letters. When nonmutant gene symbols are fully capitalized, as for yeast, a protein is sometimes indicated by the gene symbol in roman characters (all capitals or only the initial letter) followed by a "p" suffix.

Gene Symbol	Encoded Protein
ras	Ras
myc	Myc
NPL3	NPL3p or Npl3p

[Note the importance of distinguishing between these closely related symbol pairs by using italic characters for gene symbols and roman characters for protein designations.]

Names for proteins representing mutants characterized by the replacement of a single amino acid are also sometimes derived from the corresponding gene symbol. For example, an amino acid replacement with valine at position 12 of the Ras protein may be indicated by Ras^{Val12}.

17.13.15.4 Protein Sequences

Present protein sequences in tables and other abbreviated formats using the 3-letter or 1-letter symbols for the constituent amino acids; see Sections 17.13.2.3 and 17.13.2.4 for conventions.

17.13.16 Retinoids

Retinoids are compounds consisting of 4 isoprenoid units[6] joined to form a cyclohexene ring attached to an acyclic chain with 4 double bonds (see structure below) terminated by a characteristic group (Table 17.17).

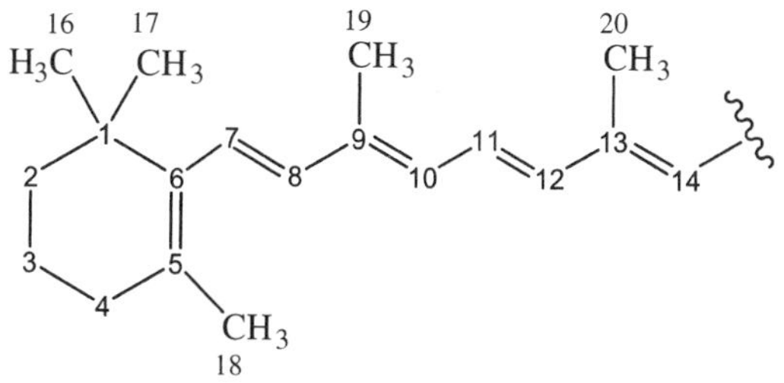

"Vitamin A" is a generic descriptor for retinoids having the biological activity of retinol, a retinoid with a terminal hydroxy functional group. Recommended names are based on 3 defined parent compounds: retinol, retinal, and retinoic acid.

Substituted derivatives are named using prefixes according to the general rules of organic nomenclature. The stereoparent name implies that the polyene chain has the *trans* configuration around all double bonds unless the contrary is indicated. Use the stereochemical prefixes *Z* and *E* for all double bonds whenever *cis* and *trans* might be ambiguous.

neovitamin A *or* 13-*cis*-retinol *or* (7*E*,9*E*,11*E*,13*Z*)-retinol

17.13.17 Steroids

Steroids are compounds having the skeletal structure of cyclopenta[*a*]phenanthrene, with or without methyl groups at positions C-10 and C-13, and compounds derived therefrom by bond scissions, ring expansions or contractions, or bond rearrangements. The names of steroids and their derivatives are based on the standard principles of the nomenclature of organic chemistry.[6]

Table 17.17 Characteristic groups of retinoids

Characteristic group	Systematic name	Other names
$-CH_2OH$	retinol	vitamin A; vitamin A alcohol vitamin A_1; vitamin A_1 alcohol axerophthol; axerol
$-CHO$	retinal	vitamin A aldehyde vitamin A_1 aldehyde retinene; retinene$_1$
$-COOH$	retinoic acid	vitamin A acid vitamin A_1 acid tretinoin
$-CH_3$	deoxyretinol [*not* retinene *or* retinane]	axerophthene
$-CH_2-O-CO-CH_3$	retinyl acetate	
$-CH_2-NH_2$	retinylamine	
$-CH_2=NOH$	retinal oxime	
$-CO-O-CH_2CH_3$	ethyl retinoate	

The basic structure with its numbering is shown below. Indicate atoms or groups projecting below the plane of the paper in the structure by broken lines (IIIIIIIII or -----) and in the name by "α"; indicate those above the paper in the structure by solid wedges or thick lines (◀ or ▬) and in the name by "β".[6]

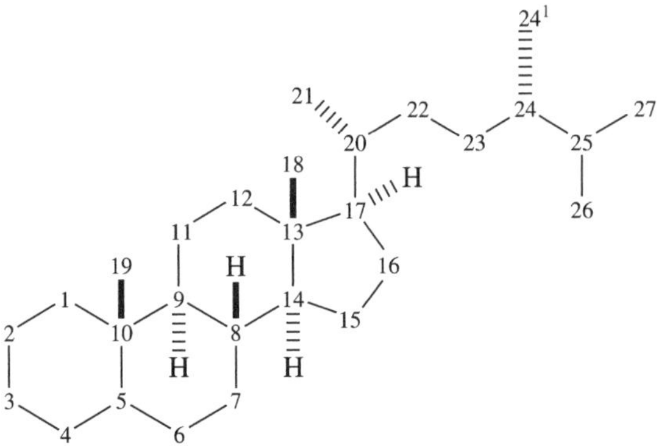

The absolute stereochemistry of some chiral centers of steroids is defined by the name of the parent; that of other chiral centers is indicated by the use of α, β, *R*, and *S*, with ξ (xi) being used when the configuration is not known. Examples are given in Table 17.18.

17.13.18 Terpenes

Terpenes are hydrocarbons of biological origin with carbon skeletons formally derived from isoprene [$CH_2=C(CH_3)CH=CH_2$]. They are subdivided into classes on the basis of the number of isoprene units: hemiterpenes, C_5; monoterpenes, C_{10}; sesquiterpenes, C_{15}; diterpenes, C_{20}; sesterterpenes, C_{30}; tetraterpenes (carotenes), C_{40}; and polyterpenes, C_{5n}. Derivatives, especially oxygenated derivatives, are called terpenoids. Terpenes may be structurally modified significantly by the prefixes described in Section 17.13.1, giving rise to a wide variety of organic structures.[20]

Table 17.18 Representative steroids

CH_3 at C-19?	CH_3 at C-18?	Side chain at C-17	Implied configurations	Name
No	No	None	8β,9α,10β,13β,14α	gonane
No	Yes	None	8β,9α,10β,13β,14α	estrane
Yes	Yes	None	8β,9α,10β,13β,14α	androstane
Yes	Yes	$-CH_2CH_3$	8β,9α,10β,13β,14α	pregnane
Yes	Yes	H, Me, 20, 22, 23, 24, 17	8β,9α,10β,13β,14α,20*R*	cholane
Yes	Yes	H, Me, 20, 22, 23, 24, 25, 26, 27, 17	8β,9α,10β,13β,14α,20*R*	cholestane
Yes	Yes	H, Me, 20, 22, 23, H, Me, 24, 25, 26, 27, 17	8β,9α,10β,13β,14α,20*R*,24*S*	ergostane
Yes	Yes	H, Me, 20, 22, 23, H, CH_2–CH_3, 24, 25, 26, 27, 17	8β,9α,10β,13β,14α,20*R*,24*S*	poriferastane (IUPAC) stigmastane (CAS)

17.13.19 Vitamins

"Vitamin" is a generic term for many unrelated organic compounds with specific biological activities essential in minute quantities for normal metabolism and growth of organisms. Its use in phrases like "vitamin A activity" and "vitamin B-6 deficiency" is appropriate. Trivial names for individual vitamins are widely used, are often ambiguous, and should, therefore, be used only if there is no chance of misinterpretation. Names with number designations are often written as subscript rather than on the line with a hyphen (e.g., as vitamin B_6 instead of B-6); either form is acceptable as long as the author is consistent.

Systematic names of vitamins and descriptions of their stereochemistry are determined by the rules of the individual classes of compounds to which the vitamins belong.[6]

Vitamin	Class of Compound
A	Retinoids
B-6	Pyridoxines
B-12	Corrinoids, tetrapyrroles
D	Steroids
E	Tocopherols
K	Prenols

CITED REFERENCES

1. International Union of Pure and Applied Chemistry, Division of Chemical Nomenclature and Structure Representation in collaboration with Inorganic Chemistry Division. Nomenclature of inorganic chemistry (the Red Book), recommendations 2005. Cambridge (UK): Royal Society of Chemistry Publishing; 2005. Also available at http://old.iupac.org/publications/books/rbook/Red_Book_2005.pdf.

2. International Union of Pure and Applied Chemistry, Commission on the Nomenclature of Organic Chemistry. Nomenclature of organic chemistry (the Blue Book). 1979 ed. Rigaudy J and Klesney SP, editors. Oxford (UK): Pergamon Press; 1979. Sections A, B, C, D, E, F, and H. Correction. Pure Appl Chem. 1999;71(7):1327–1330.

3. International Union of Pure and Applied Chemistry, Organic Chemistry Division, Commission on Nomenclature of Organic Chemistry. A guide to IUPAC nomenclature of organic compounds: recommendations 1993. Oxford (UK): Blackwell Scientific Publications; 1993.

4. International Union of Pure and Applied Chemistry. Recommendations on organic & biochemical nomenclature, symbols & terminology etc. London (UK): Queen Mary University of London, Department of Chemistry; [date unknown]. http://www.chem.qmul.ac.uk/iupac.

5. International Union of Pure and Applied Chemistry, Macromolecular Division, Commission on Macromolecular Nomenclature. Compendium of macromolecular nomenclature. Oxford (UK): Blackwell Scientific Publications; 1991.

6. International Union of Biochemistry and Molecular Biology. Biochemical nomenclature and related documents: a compendium. 2nd ed. London (UK): Portland Press; 1992.

7. Haynes WM, editor. CRC handbook of chemistry and physics. 93rd ed. Boca Raton (FL): CRC Press; 2012.

8. Speight JA, editor. Lange's handbook of chemistry. 16th ed. New York (NY): McGraw-Hill; 2004.

9. Coghill A, Garson L, editors. The ACS style guide: a manual for authors and editors. 3rd ed. Washington (DC): American Chemical Society; 2006.

10. CAS REGISTRY. Columbus (OH): Chemical Abstracts Service. http://www.cas.org/content/chemical-substances.

11. International Union of Pure and Applied Chemistry, Organic Chemistry Division, Commission on Nomenclature of Organic Chemistry; International Union of Pure and Applied Chemistry, Organic Chemistry Division, Commission on Physical Organic Chemistry. Basic terminology of stereochemistry. (IUPAC recommendations 1996). Pure Appl Chem. 1996;68(12):2193–2222. Also available at http://www.chem.qmul.ac.uk/iupac/stereo/ (prepared by GP Moss).

12. International Union of Pure and Applied Chemistry, Organic Chemistry Division, Commission on Nomenclature of Organic Chemistry. Nomenclature of fused and bridged fused ring systems. (IUPAC recommendations 1998). Pure Appl Chem. 1998;70(1):143–216. Also available at http://www.chem.qmul.ac.uk/iupac/fusedring/ (prepared by GP Moss) and http://www.iupac.org/publications/pac/1998/pdf/7001x0143.pdf.

13. Leigh GJ, editor. Principles of chemical nomenclature: a guide to IUPAC recommendations. Cambridge (UK): Royal Society of Chemistry Publishing; 2011.

14. Chemical Abstracts Service. Ring systems handbook. Columbus (OH): American Chemical Society; 2003.

15. International Union of Pure and Applied Chemistry, Organic Chemistry Division, Commission on Nomenclature of Organic Chemistry. Extension and revision of the von Baeyer system for naming polycyclic compounds (including bicyclic compounds). (IUPAC recommendations 1999). Pure Appl Chem. 1999;71(3):513–529. Also available at http://www.chem.qmul.ac.uk/iupac/vonBaeyer/ and http://pac.iupac.org/publications/pac/pdf/1999/pdf/7103x0513.pdf.

16. International Union of Pure and Applied Chemistry, Organic Chemistry Division, Commission on Nomenclature of Organic Chemistry. Extension and revision of the nomenclature for spiro compounds. (IUPAC recommendations 1999). Pure Appl Chem. 1999;71(3):531–558. Also available at http://www

.chem.qmul.ac.uk/iupac/spiro (prepared by GP Moss) and http://pac.iupac.org/publications/pac/pdf/1999/pdf/7103x0531.pdf.

17. International Union of Pure and Applied Chemistry, Organic Chemistry Division, Commission on Nomenclature of Organic Chemistry. Revised nomenclature for radicals, ions, radical ions and related species. (IUPAC recommendations 1993). Pure Appl Chem. 1993;65(6):1357–1455. Also available at http://www.chem.qmul.ac.uk/iupac/ions.

18. Chemical Abstracts Service. Naming and indexing of chemical substances for chemical abstracts 2007. Columbus (OH): American Chemical Society; 2008 [accessed 2013 July 19]. http://www.cas.org/File%20Library/Training/STN/User%20Docs/indexguideapp.pdf.

19. International Union of Pure and Applied Chemistry, Macromolecular Division, Commission on Macromolecular Nomenclature. Source-based nomenclature for non-linear macromolecules and macromolecular assemblies. (IUPAC recommendations 1997). Pure Appl Chem. 1997;69(12):2511–2521. Also available at http://pac.iupac.org/publications/pac/pdf/1997/pdf/6912x2511.pdf.

20. International Union of Pure and Applied Chemistry, Organic Chemistry Division, Commission on Nomenclature of Organic Chemistry. Revised section F: natural products and related compounds. (IUPAC recommendations 1999). Pure Appl Chem. 1999;71(4):587–643. Also available at http://www.chem.qmul.ac.uk/iupac/sectionF (prepared by GP Moss). Corrections and modifications (2004). Pure Appl Chem. 2004;76(6):1283–1292. Prepared for publication by HA Favre, PM Giles Jr, KH Hellwich, AD McNaught, GP Moss, WH Powell. Also available at http://pac.iupac.org/publications/pac/pdf/2004/pdf/7606x1283.pdf.

21. International Union of Pure and Applied Chemistry and International Union of Biochemistry and Molecular Biology, Joint Commission on Biochemical Nomenclature. Nomenclature of carbohydrates. (Recommendations 1996). Pure Appl Chem. 1996;68(10):1919–2008. Also available at http://www.chem.qmul.ac.uk/iupac/2carb/ (prepared by GP Moss).

22. International Union of Pure and Applied Chemistry and International Union of Biochemistry and Molecular Biology. Nomenclature of cyclitols (1973 recommendations). Pure Appl Chem. 1974;37(1–2):283–297. Also available at http://www.iupac.org/publications/pac/37/1/0283/ and http://www.chem.qmul.ac.uk/iupac/cyclitol/.

23. International Union of Biochemistry and Molecular Biology, Nomenclature Committee. Enzyme nomenclature: recommendations of the Nomenclature Committee of the International Union of Biochemistry and Molecular Biology on the nomenclature and classification of enzyme-catalysed reactions. London (UK): Queen Mary University of London, Department of Chemistry; [updated 2013 Jul 24; accessed 2013 Jul 28]. http://www.chem.qmul.ac.uk/iubmb/enzyme/.

24. IUPAC-IUB Commission on Biochemical Nomenclature. The nomenclature of lipids. Recommendations (1976). Lipids. 1977;12(6):455–468. Also available at http://www.chem.qmul.ac.uk/iupac/lipid/ (prepared by GP Moss).

25. International Union of Pure and Applied Chemistry and International Union of Biochemistry and Molecular Biology, Joint Commission on Biochemical Nomenclature. Nomenclature of glycolipids. (IUPAC recommendations 1997). Pure Appl Chem. 1997;69(12):2475–2487. Also available at http://www.chem.qmul.ac.uk/iupac/misc/glylp.html (prepared by GP Moss).

26. Chitty M. Genomics glossaries & taxonomies. Newton Upper Falls (MA): Cambridge Healthtech Institute; [updated 2013 Jul 15; accessed 2013 Jul 28]. http://www.genomicglossaries.com.

27. International Union of Pure and Applied Chemistry; International Union of Biochemistry. Abbreviations and symbols for nucleic acids, polynucleotides and their constituents. Rules approved 1974. Pure Appl Chem. 1974;40(3):277–290. Also available at http://www.chem.qmul.ac.uk/iupac/misc/naabb.html.

28. Nelson NA. Prostaglandin nomenclature. J Med Chem. 1974;17(9):911–918.

18 Chemical Kinetics and Thermodynamics

18.1 INTRODUCTION

Kinetic studies are important in interpreting scientific reactions and in extrapolating the results of scientific studies to in vivo and in vitro systems. Because of the long history of accurate notation in the physical sciences, consistency in terminology is much more evident in the field of kinetics than in the life sciences.[1–4] The need for greater systemization and consistency across fields resulted in a multicommission effort in biothermodynamics, with subsequent recommendations for presentation of data and results.[5,6]

18.2 UNITS IN BIOTHERMODYNAMICS

Quantities used specifically in thermodynamics are given in Table 18.1.

18.2.1 Mass

The term "specific" preceding the name of a quantity means "divided by mass" (for which the SI unit is the kilogram). The term "molar" preceding the name of a quantity generally means "divided by amount of substance" (for which the SI unit is the mole), but occasionally it means "divided by amount of substance concentration". "Molar" in this usage is typically represented by a lowercase subscript letter "m". For a quantity represented by a capital letter (e.g., *V* for volume), the symbol for the specific quantity is often the corresponding lowercase letter (e.g., *v* for specific volume).

specific volume ($v = V/m$), with units $m^3 \cdot kg^{-1}$
molar volume (V_m), with units $m^3 \cdot mol^{-1}$

Table 18.1 Symbols and units for some thermodynamic quantities[a]

Quantity name	Quantity symbol	SI unit	SI unit symbol
Volume	V	cubic meter	m^3
Force	F	newton	$N = m \cdot kg \cdot s^{-2}$
Density	ρ	kilograms per cubic meter	$kg \cdot m^{-3}$
Pressure	p	pascal	$Pa = N \cdot m^{-2}$
Viscosity (dynamic viscosity)	η	pascal second	Pa·s
Energy	E	joule	J = N·m
Heat	q, Q	joule	J
Work	w, W	joule	J
Internal energy	U	joule	J
Enthalpy	H	joule	J
Gibbs energy	G	joule	J
Helmholtz energy	A	joule	J
Entropy	S	joule per kelvin	J/K *or* $J \cdot K^{-1}$
Power	P	watt	W
Heat capacity			
At constant pressure	C_p	joule per kelvin	J/K *or* $J \cdot K^{-1}$
At constant volume	C_V	joule per kelvin	J/K *or* $J \cdot K^{-1}$
Osmotic pressure	Π	pascal	Pa
Chemical potential of B	μ_B	joule per mole	J/mol *or* $J \cdot mol^{-1}$
Absolute activity of B	λ_B	[dimensionless]	
Activity coefficient of B			
With reference to Raoult's law	f_B	[dimensionless]	
With reference to Henry's law			
Molality basis	$\gamma_{m,B}$	[dimensionless]	
Concentration basis	$\gamma_{c,B}$	[dimensionless]	
Mole fraction basis	$\gamma_{x,B}$	[dimensionless]	
Osmotic coefficient	φ	[dimensionless]	

[a] Symbols for physical quantities represented by single Greek or Roman letters should be printed in italic type; symbols for units should be in roman type.

Where possible, report thermodynamic quantities in molar terms. However, for research with macromolecules and research in quantum chemistry, either molecular mass or molar mass may be used. The values of molecular mass and molar mass are numerically identical, but the units differ.

1) For molecular mass, use either the unified atomic mass unit (u) or the dalton (Da), which are alternative names for the same unit: 1 u = 1 Da = $m(^{12}C)/12$, where $m(^{12}C)$ is the mass of the carbon atom. Neither the unified atomic mass unit nor the dalton is an SI unit, but they are appropriate units for the mass of an atom and have been accepted for use in conjunction with the SI.[7]
2) For molar mass, the coherent SI unit is kilograms per mole ($kg \cdot mol^{-1}$), but grams per mole ($g \cdot mol^{-1}$) is commonly used.

Do not use "dalton" in referring to molecular weight; the term "molecular weight" is considered a synonym for "relative molecular mass" and is therefore dimensionless. The following forms of expression are acceptable.

The molecular weight of subunit A is 76,000.
The molecular mass of subunit A is 76,000 Da.
The molecular mass of subunit A is 76,000 u.
The molar mass of subunit A is 76,000 $g \cdot mol^{-1}$.

18.2.2 Volume, Temperature, Energy, and Density

The SI unit for volume is the cubic meter (m^3). However, in the expressions for most units, the use of submultiples, including "deci" and "centi", is acceptable. The cubic centimeter (cm^3, not cc) and the cubic decimeter (dm^3) are also acceptable. It is acceptable to use "liter" (L) as a specialized name for the cubic decimeter, along with its submultiples the milliliter and the microliter.

In thermodynamic calculations and kinetic studies, express temperature in the SI base unit, the kelvin (K). However, experimental temperatures may be expressed in degrees Celsius (°C).

The SI unit for energy is the joule (J). The use of "calorie" is generally discouraged. The term "calorie" has been used in the field of nutrition to mean "kilocalorie", but this usage is now discouraged. Thus 1 calorie as formerly used by nutritionists is equal to 4.184 kJ.

The SI unit for density is kilograms per cubic meter ($kg \cdot m^{-3}$), but a more convenient unit is grams per cubic centimeter ($g \cdot cm^{-3}$).

18.3 NOTATIONS FOR VARIABLES, STATES, AND PROCESSES

18.3.1 Variables

Symbols for quantities in general (including both variables and fixed quantities like the speed of light or the mass of a proton) should generally be single letters of the Greek or Roman alphabet. Both capital and lowercase letters may be used. Single-letter variables should be printed in italic type[6] (see also Section 4.2.2). Where necessary, the symbol may be modified by a subscript or a superscript of specified meaning, or it may be qualified by further information in parentheses.[3] For example, a symbol for a variable may be appended within parentheses to the symbol for a thermodynamic function.

$\mu^{\circ}{}_{\mathrm{B}}(T)$ [the standard chemical potential of substance B at temperature T; superscript ° = standard state]

18.3.2 States

Particular states are indicated by superscripts to symbols for thermodynamic functions (e.g., G°, representing standard Gibbs energy). The 7 most frequently encountered states are represented by the following superscript symbols.

activated complex	‡
apparent	app
excess	E
ideal	id
infinite dilution	∞
pure substance	*
standard state	° *or* ⊖

States of aggregation are indicated by an abbreviation that is set in roman type, enclosed in parentheses, and placed after the symbol for the property[3] (e.g., V^{*}(cr), representing the volume of a substance in its pure crystalline state). The symbols below specify the states of aggregation.

amorphous solid	am
aqueous solution	aq
crystalline solid	cr
fluid	f
gas	g
liquid	l
liquid crystal	lc
solid	s
solution	sln
vitreous substance	vit

The following examples show representations using these symbols.

C_V (f)	constant volume heat capacity of a fluid
NaCl(sln)	sodium chloride in solution
NH_3(aq)	ammonia in aqueous solution

18.3.3 Processes

Use the Greek capital letter delta (Δ) before a thermodynamic symbol to indicate a change in thermodynamic quantity—that is, the final value minus the initial value. Use a subscript immediately after the Δ to indicate the type of change; for example, $\Delta_\mathrm{f} H$ is the change in enthalpy of formation. Symbols for the most frequently occurring processes are listed below.[3]

combustion	c
formation	f
melting (fusion)	fus
mixing of fluids	mix
reaction (except combustion)	r
solution (dissolution)	sol
sublimation (evaporation)	sub
transition (solid–solid)	trs
vaporization (evaporation)	vap

The following examples are representations using these symbols.

$\Delta_\mathrm{f} S°$($HgCl_2$, cr, 298.15 K)

[change in (molar) standard entropy due to the formation of crystalline mercuric chloride from its elements at a temperature of 298.15 K]

$\Delta_\mathrm{r} G°$(1,000 K)

[change in (molar) standard Gibbs energy due to a chemical reaction at standard pressure and a temperature of 1,000 K]

18.4 MICROBIAL PROCESSES

The International Union of Pure and Applied Chemistry (IUPAC) has published a detailed list of terms important in the biotechnology of microbial processes.[8] Many of these terms, definitions, and symbols are the same as those in other IUPAC documents (e.g., the IUPAC Green Book[3]) or have been brought into agreement with those recommendations. General quantities, intensive quantities, rate quantities, concentrations, and amounts

Table 18.2 Selected terms, symbols, and units related to microbial processes[a]

Quantity	Symbol	SI unit	Customary unit
Vapor pressure	p^*	Pa	atm *or* bar
Molar activation energy	E	$J \cdot mol^{-1}$	$J \cdot mol^{-1}$
Molar activation energy for specific growth rate	E_μ		$kJ \cdot mol^{-1}$
Molar chemical energy[b]	U, H, G, A, and others	$J \cdot mol^{-1}$	$kJ \cdot mol^{-1}$
Area per volume	a	m^{-1}	cm^{-1}
Yield of cell biomass per amount of ATP produced in cells	Y_{ATP}	$kg \cdot mol^{-1}$	$g \cdot mol^{-1}$
Gas hold-up (volume of gas divided by volume of liquid)	ε	L *or* l	L *or* l
Doubling time, biomass	t_d	s	min *or* h *or* d
Growth rate, colony radial (rate of extension of biomass colony on a surface)	K_r	$m \cdot s^{-1}$	$m \cdot h^{-1}$
Specific mass metabolic rate	q	s^{-1}	h^{-1}
Mutation rate	w	s^{-1}	h^{-1}

[a] Based on International Union of Pure and Applied Chemistry, Applied Chemistry Division, Commission on Biotechnology.[8]
[b] For types, see entries under "Energy" in Table 18.1.

are described and defined, and their recommended symbols and units (both SI units and non-SI units in common use [e.g., atmosphere, bar, hour]) are provided (see Table 18.2 for some representative examples).

18.5 ION MOVEMENT

Research in physiology involves aspects of solute transport studied by convection, diffusion, and permeation across membranes (ion movement and transport). Use the standardized symbols and terms recommended by Bassingthwaighte et al.[9]; see Table 18.3 for some of the important symbols.

18.6 SOLUTIONS

A solution is a liquid or solid phase containing more than one component. The term "solvent" is generally used for the substance in largest concentration, and the other components are called "solutes". However, in biological solutions, "solvent" is often used to refer to a fixed mixture of components (e.g., water plus buffer components), with all other substances being the solutes.

Use the following notations for the composition of solutions, recommended symbols, and appropriate units.

Composition	Symbol	Unit
amount of substance B	n_B	mol
concentration of solute substance B	c_B	$mol \cdot dm^{-3}$ *or* $mol \cdot L^{-1}$
mass concentration of substance B	P_B *or* ρ_B	$g \cdot dm^{-3}$
molality of solute substance B	m_B *or* b_B	$mol \cdot kg^{-1}$
mole fraction of substance B	x_B	[dimensionless]
mass fraction of substance B	w_B	[dimensionless]
volume fraction of substance B	φ_B	[dimensionless]

Table 18.3 Selected symbols for ion mass transport and exchange[a]

Symbol	Definition or description	Unit (if applicable)
Principal symbol[b]		
a	Relative activity, molar	
A	Area of indicator concentration–time curve excluding recirculation	$mol \cdot s \cdot L^{-1}$
C	Concentration	$mol \cdot L^{-1}$
CV	Coefficient of variation	[dimensionless]
D	Diffusion coefficient	$cm^2 \cdot s^{-1}$
E	Electrical potential	V
F	Flow	$cm^3 \cdot s^{-1}$
F_B	Blood flow to an organ	$cm^3 \cdot g^{-1} \cdot min^{-1}$
$h(t)$	Transport function	s^{-1}
Hct	Hematocrit, the fraction of the blood volume composed of erythrocytes	[dimensionless]
J	Flux	$mol \cdot s^{-1} \cdot cm^{-2}$
P	Permeability coefficient for a solute traversing a membrane	$cm \cdot s^{-1}$
RD	Relative dispersion	[dimensionless]
Subscript symbol		
A	Arterial	
B	Blood	
C *or* cap	Capillary, or the region of blood–tissue exchange	
i, j	Indices in series, in summations, or in elements of arrays	
in, i	Into, inside, inflow	
ISF, I	Interstitial fluid space (extravascular, extracellular fluid)	
m	Membrane	
out, o	Out of, outside, outflow	
p	Plasma	
RBC	Red blood cell	
S	Solute	
T	Total	
V	Venous	

[a] Based on Bassingthwaighte et al.[9]
[b] Single-letter symbols for physical quantities should be printed in italic type; multiletter symbols should be in roman type.

Concentration (meaning amount-of-substance concentration) is sometimes called "molarity", and a solution of 1 $mol \cdot dm^{-3}$ may be referred to as a 1 molar solution or a 1 M solution.[3] The symbol m_B can refer to both mass of B and molality of B; to avoid ambiguity, the second of the 2 symbols shown above for molality is recommended.[3]

18.7 ENZYME KINETICS

Practices in the field of enzyme kinetics follow closely those in chemical kinetics.[1,4] However, the rigorous detail associated with all variables in chemical kinetics is generally not necessary with enzyme-catalyzed reactions, which most frequently occur in a liquid phase at constant pressure.

Table 18.4 Selected terms, symbols, and units recommended for enzyme kinetics[a]

Term	Symbol	Customary unit[b]
Concentration of substrate A	[A]	$mol \cdot dm^{-3}$
Concentration of inhibitor I	[I]	$mol \cdot dm^{-3}$
Concentration of product Y	[Y]	$mol \cdot dm^{-3}$
Catalytic constant	k_0	s^{-1}
Rate constant of any order *n*	k	$(mol \cdot dm^{-3})^{1-n} \cdot s^{-1}$
Forward and reverse rate constants for *i*th step	k_i, k_{-i}	as k
Inhibition constant	K_i	$mol \cdot dm^{-3}$
Michaelis constant	K_m	$mol \cdot dm^{-3}$
Michaelis constant for substrate A	K_{mA}	$mol \cdot dm^{-3}$
Time	t	s
Rate (or velocity) of reaction	v	$mol \cdot dm^{-3} \cdot s^{-1}$
Initial rate of reaction	v_0	$mol \cdot dm^{-3} \cdot s^{-1}$
Rate of conversion	ξ	kat = $mol \cdot s^{-1}$

[a] Based on International Union of Biochemistry, Nomenclature Committee.[1]
[b] In all cases, "dm^3" may be replaced with "L" or "l" (for "liter").

18.7.1 Kinetic Equations

Definitions of consumption and formation rates, reaction rates, and elementary and composite reactions have been published by the International Union of Biochemistry (now the International Union of Biochemistry and Molecular Biology).[1] Of special interest are the kinetic equations describing Michaelis–Menten and non-Michaelis–Menten relationships.

1) Michaelis–Menten: Express the relationship between the rate of an enzyme-catalyzed reaction and the substrate concentration in the following form.

   $v = V[A]/(K_{mA} + [A])$

2) Non-Michaelis–Menten: The following form is an example.

   $v = V'[A]/(K'_{mA} + [A] + [A]^2/K_{iA})$

See Table 18.4 for a selected list of symbols for kinetic reactions.

18.7.2 Enzyme Activity

The catalytic activity of an enzyme is measured as the increase in the rate of conversion (i.e., the rate of reaction expressed as an extensive quantity, measured as the increase in amount of substance per unit of time) of a specified chemical reaction that the enzyme produces in a specific assay system.[1] It may be expressed using the SI-derived unit "katal" (symbol "kat"), which is equivalent to moles per second ($mol \cdot s^{-1}$). One katal is the catalytic activity that will raise the rate of reaction by 1 mole per second in a specified assay system.[2] The katal was approved by the 1999 Conférence Général des Poids et Mesures as a special name for the unit "moles per second" within the SI.[7]

CITED REFERENCES

1. International Union of Biochemistry, Nomenclature Committee. Symbolism and terminology in enzyme kinetics: recommendations 1981. Eur J Biochem. 1982;128(2–3):281–291. Prepared by A Cornish-

Bowden, HBF Dixon, KJ Laidler, IH Segel, J Ricard, SF Velick, EC Webb. Correction in Eur J Biochem. 1993;213(1):1. Also available at http://www.chem.qmul.ac.uk/iubmb/kinetics.

2. International Union of Biochemistry, Nomenclature Committee. Units of enzyme activity: recommendations 1978. Eur J Biochem. 1979;97(1):319–320.

3. International Union of Pure and Applied Chemistry, Physical and Biophysical Chemistry Division. Quantities, units and symbols in physical chemistry. 3rd ed. Cambridge (UK): RSC Publishing; 2007. Prepared for publication by ER Cohen, T Cvitas, JG Frey, B Holmström, K Kuchitsu, R Marquardt, I Mills, F Pavese, M Quack, J Stohner, HL Strauss, M Takami, and AJ Thor. Also available at http://www.iupac.org/fileadmin/user_upload/publications/e-resources/ONLINE-IUPAC-GB3-2ndPrinting-Online-Sept2012.pdf. Also known as IUPAC Green Book.

4. International Union of Pure and Applied Chemistry, Physical Chemistry Division, Subcommittee on Chemical Kinetics. Symbolism and terminology in chemical kinetics. Pure Appl Chem. 1981;53(3):753–771. Prepared for publication by KJ Laidler.

5. Interunion Commission on Biothermodynamics. Recommendations for measurement and presentation of biochemical equilibrium data. J Biol Chem. 1976;251(22):6879–6885. Also available at http://www.jbc.org/cgi/reprint/251/22/6879.

6. International Union of Pure and Applied Chemistry, International Union of Pure and Applied Biophysics, and International Union of Biochemistry Interunion Commission on Biothermodynamics. Recommendations for the presentation of thermodynamic and related data in biology (1985). Eur J Biochem. 1985;153(3):429–434. Prepared for publication by I Wadso.

7. Bureau International des Poids et Mesures. Le Système internationale des unités (SI). 8th ed. Paris (France): Organisation Intergouvernementale de la Convention du Mètre; 2006. English translation available in PDF format at http://www1.bipm.org/utils/common/pdf/si_brochure_8_en.pdf.

8. International Union of Pure and Applied Chemistry, Applied Chemistry Division, Commission on Biotechnology. Selection of terms, symbols and units related to microbial processes (IUPAC recommendations 1992). Pure Appl Chem. 1992;64(7):1047–1053. Prepared for publication by VK Eroshin. Also available at http://www.iupac.org/publications/pac/1992/pdf/6407x1047.pdf.

9. Bassingthwaighte JB, Chinard FP, Crone C, Goresky CA, Lassen NA, Reneman RS, Zierler KL. Terminology for mass transport and exchange. Am J Physiol. 1986;250(4 Pt 2):H539–H545.

ADDITIONAL REFERENCES

International Union of Pure and Applied Chemistry, Physical Chemistry Division. Abbreviated list of quantities, units and symbols in physical chemistry. Research Triangle Park (NC): The Union; 2000 [modified 2000 Jun 12; accessed 2011 Nov 17]. Prepared for publication by KH Homann. http://old.iupac.org/reports/1993/homann/index.html.

Laidler KJ. A glossary of terms used in chemical kinetics, including reaction dynamics. Pure Appl. Chem. 1996;68:149–192.

Speight JG, editor. Lange's handbook of chemistry. 16th ed. New York (NY): McGraw-Hill; 2004.

Ulicky L, Kemp T, editors. Comprehensive dictionary of physical chemistry. Englewood Cliffs (NJ): Prentice Hall; 1993.

19 Analytical Chemistry

19.1 FUNDAMENTALS AND SOURCES

Analytical chemistry (chemical analysis) can be defined as the science of obtaining, processing, and communicating information about the composition of matter. The information may be qualitative (What is there?) or quantitative (How much is there?). Accordingly, much of analytical chemistry is built around methods, which may involve a variety of techniques.

In general, the form and style of presenting information in analytical chemistry are the same as those for any of the physical sciences. American Society for Testing and Materials (ASTM) International offers relevant guidance in its style manual, *Form and Style for ASTM Standards* (especially Part G).[1] The American Chemical Society style guide[2] provides detailed information on the presentation of chemical information, including the results of analytical studies.

When established methods have been used, it is sufficient to present the experimental details in summary form:

> . . . was determined by the Lineweaver–Burk methodology, as modified by Smith.

If new techniques have been used, however, or important modifications of basic methods have been introduced, then the experimental procedures should be presented in sufficient detail to allow duplication by other experienced researchers:

> . . . titanium was determined by precipitation of the hydrous oxide followed by filtration and ignition to TiO_2.

The International Union of Pure and Applied Chemistry (IUPAC) and other organizations have developed terminology covering the broad range of techniques used by analytical chemists.

Quantities, Units and Symbols in Physical Chemistry,[3] known as the IUPAC Green Book, is a collection of all terms recommended by IUPAC in the field of physical chemistry. In addition to defining a wide range of physical quantities and units, the IUPAC Green

Book lists fundamental physical constants, provides conversion factors for units, and defines abbreviations and acronyms used in physical chemistry.

IUPAC has also published collections of all its terminology recommendations in the fields of analytical chemistry (the IUPAC Orange Book[4]) and clinical laboratory sciences (the IUPAC Silver Book[5]). A complete compilation of all IUPAC terminology recommendations has additionally been published as the IUPAC Gold Book.[6,7]

19.2 ANALYTICAL TECHNIQUES AND NOMENCLATURE

19.2.1 Types of Techniques

IUPAC recognizes 13 types of analytical techniques.[4] Some are qualitative, some are quantitative, and some may be both. The relevant terminology for each, with units (if applicable), is available in the IUPAC Orange Book.[4]

1) Thermoanalytical and enthalpimetric methods encompass thermal analysis (in which a physical property of a substance is measured as a function of temperature) and enthalpimetric analysis (in which the enthalpy change of a chemical reaction is measured).
2) In titrimetric analysis, the concentration of one substance is determined by adding known quantities of another substance with which it reacts; the result depends on the existence of a measurable end point (often visual). Examples include acid–base and oxidation–reduction titration.
3) Automatic analysis reduces human intervention in analytical methods. It may be achieved by mechanization, instrumentation, or automation (each incorporating some degree of nonhuman decision making).
4) Electrochemical analysis is based on electrochemical processes or phenomena.
5) Analytical separation methods encompass chromatography, ion exchange, liquid–liquid distribution (solvent extraction), precipitation, electrophoresis, and centrifugation. Chromatography refers to the separation of components in a mobile phase from a stationary component. Chromatographic methods may be classified according to the shape of the chromatographic bed (column, planar), the physical state of the mobile and stationary phases (gas–liquid, liquid–liquid, gas–solid, liquid–solid), the mechanism of separation (adsorption, partition, ion exchange, exclusion, affinity), or special methods (including gradient elution, isothermal chromatography, reversed-phase chromatography).
6) Spectrochemical analysis involves techniques in which the components of the sample emit or absorb radiation (including arc emission spectrometry, X-ray spectrometry, flame emission spectrometry).
7) Other optical methods encompass photochemistry and light scattering.
8) Mass spectroscopy involves the separation of beams of ions according to the quotient mass/charge.
9) Diffraction methods (X-ray, electron, and neutron diffraction) are used to investigate crystalline structures (see Section 25.6).
10) Magnetic methods take advantage of the magnetic properties of substances to investigate their molecular structure.
11) Kinetic methods use rates of reaction to determine concentration of substances.
12) In radioanalytical methods, the substance under investigation may itself be radioactive, or alternatively, a radioactive substance or ionizing radiation may be used as an investigative tool.
13) Surface analysis is conducted primarily by means of spectroscopic methods, including

several types of photoelectron emission spectroscopy, electron impact spectroscopies, Auger electron spectroscopies, and field emission electron spectroscopies. For each type of spectroscopy, the IUPAC Orange Book[4] lists the incident and detected particles.

14) Electron microscopy (transmission, scanning, and others) uses the interaction between an electron beam and a prepared sample to produce a high-resolution image.
15) Lab-on-a-chip is a fast and portable microelectromechanical tool that miniaturizes tests typically performed in a laboratory.

19.2.2 Nomenclature of Surface Analytical Techniques

The method of analysis is often abbreviated, but multiple abbreviations for the same term have arisen, which creates confusion. IUPAC[4] recommends a systematic method of generating the abbreviation for a given technique. The abbreviation should consist of one or more letters representing descriptive adjectives, a letter for the type of probe or particle, and a letter for the type of technique. The following is a selection of terms and their abbreviations:

Descriptive Adjective	Probe or Particle	Technique
A (absorption)	P (photon, potential)	S (spectroscopy, scattering)
E (emission)	E (electron)	R (resonance)
R (reflection)	N (neutron)	D (diffraction)
T (transmission)	A (atom)	M (microscopy)
F (far)	I (ion)	I (ionization)
M (mid)	F (field)	
N (near)	R (radiation)	
H (high)		
L (low)		
S (scanning, surface)		
AR (angle-resolved)		
X, XR (X-ray)		
U, UV (ultraviolet)		

Examples:

LEED *for* low-energy electron diffraction
STEM *for* scanning transmission electron microscopy

Some well-established abbreviations for techniques used in surface analysis do not follow this pattern; in such cases, use the established abbreviation:

NMR *for* nuclear magnetic resonance *not* MNR *for* magnetic nuclear resonance

See Table 17.5 (Index of techniques and abbreviations) in the IUPAC Orange Book[4] for recommended abbreviations.

CITED REFERENCES

1. ASTM International. Form and style for ASTM standards. West Conshohocken (PA): ASTM International; 2010 [accessed 2011 Nov 16]. http://www.astm.org/COMMIT/Blue_Book.pdf.

2. Coghill AM, Garson LR, editors. The ACS style guide: effective communication of scientific information. 3rd ed. Washington (DC): American Chemical Society; 2006.

3. International Union of Pure and Applied Chemistry, Physical Chemistry Division. Quantities, units and symbols in physical chemistry. 3rd ed. Cambridge (UK): RSC Publishing; 2007. Prepared for publication by ER Cohen, T Cvitaš, JG Frey, B Holmström, K Kuchitsu, R Marquardt, I Mills, F Pavese, M Quack,

J Stohner, HL Strauss, M Takami, AJ Thor. Also available at http://media.iupac.org/publications/books/gbook/IUPAC-GB3-2ndPrinting-Online-22apr2011.pdf. Also known as IUPAC Green Book.

4. International Union of Pure and Applied Chemistry, Analytical Chemistry Division. Compendium of analytical nomenclature, definitive rules 1997. 3rd ed. Oxford (UK): Blackwell Scientific Publications; 1998. Prepared for publication by J Inczedy, T Lengyel, AM Ure. Also available at http://www.iupac.org/publications/analytical_compendium/. Also known as IUPAC Orange Book.

5. International Union of Pure and Applied Chemistry, Clinical Chemistry Division, Commission on Quantities and Units in Clinical Chemistry; International Federation of Clinical Chemistry, Scientific Division, Committee on Quantities and Units. Compendium of terminology and nomenclature of properties in clinical laboratory sciences (recommendations 1995). Oxford (UK): Blackwell Scientific Publications; 1995. Prepared for publication by JC Rigg, SS Brown, R Dybkaer, H Olesen. Also known as IUPAC Silver Book.

6. International Union of Pure and Applied Chemistry. Compendium of chemical terminology. 2nd ed. Oxford (UK): Blackwell Scientific Publications; 1997. Prepared for publication by AD McNaught, A Wilkinson. Also known as IUPAC Gold Book.

7. International Union of Pure and Applied Chemistry. Compendium of chemical terminology. Research Triangle Park (NC): The Union; c2005–2011 [accessed 2011 Nov 14]. http://goldbook.iupac.org/. Updated online version of Compendium of Chemical Terminology (reference 6), also known as IUPAC Gold Book.

ADDITIONAL REFERENCES

ASTM International. ASTM dictionary of engineering, science and technology. 10th ed. West Conshohocken (PA): ASTM International; 2005.

International Organization for Standardization. International vocabulary of metrology—basic and general concepts associated with terms (VIM). 3rd ed. Geneva (Switzerland): ISO; 2007.

20 Drugs and Pharmacokinetics

20.1 DRUG NOMENCLATURE

A detailed summary of the various kinds of drug designations and their bases appears in the preface to the *USP Dictionary of USAN and International Drug Names*,[1] issued annually by the United States Pharmacopeial Convention (see Section 20.1.1.1).

20.1.1 Nonproprietary (Generic) Names

Most drugs have complex molecular structures. Hence, the systematic chemical names of most of them are also complex (see Section 20.1.3). In literature not concerned specifically with the chemical characteristics of drugs and in clinical medicine, shorter, more convenient names are used. This practice has given rise to national and international mechanisms for coining names that imply no commercial ownership but make accurate comprehension of the pharmacological and medical literature universally possible.

20.1.1.1 United States Adopted Names and Other National Names

A nonproprietary (generic) name for a drug in the United States is generally proposed to the United States Adopted Names Council by the company that developed the therapeutic agent. Proposed names are expected to conform to the criteria set forth in "Guiding Principles for Coining U.S. Adopted Names for Drugs", published as an appendix to the *USP Dictionary of USAN and International Drug Names*.[1] This document gives rules for coining names and provides lists of approved stem terms and terms that represent contractions for radicals (chemical names) and adducts. It also provides guidelines for naming interferons, interleukins, somatotropins, colony-stimulating factors, erythropoietins, and monoclonal antibodies.

Most nonproprietary names consist of a single word, but those representing chelates, complexes, esters, and salts are 2-word terms:

rolitetracycline	magnesium salicylate
rolitetracycline nitrate	fluocinolone acetonide

Radiopharmaceuticals are multiword terms (see Section 20.1.5).

Adopted names are published annually in the *USP Dictionary*[1] issued by the US Pharmacopeial Convention under the auspices of its Expert Committee on Nomenclature and Labeling. The *USP Dictionary* also includes information relevant to other designations for drugs; see Table 20.1 for a list, with examples, of the types of information provided. In its entry terms the *USP Dictionary* capitalizes the initial letters of nonproprietary names; however, other publications do not capitalize these names, except as specified by the rules for capitalizing initial words in sentences and nouns in titles (see Sections 9.2.1 and 9.3.1). New US Adopted Names are published in the annual issues of the *USP Dictionary* and in the "New Names" section of *Pharmacopeial Forum*,[2] a periodical published by the US Pharmacopeial Convention; until November 1994 they were also published in the journal *Clinical Pharmacology & Therapeutics*.[3] Another valuable source of chemical information on drugs is *The Merck Index*.[4]

Table 20.1 Information in entries of the *USP Dictionary of USAN and International Drug Names*[a]

Type of information	Example, as presented in *USP Dictionary*[b,c]
United States Adopted Name (USAN)	mirincamycin hydrochloride
Year of publication of the USAN	*[1963]*
Pronunciation guide	(sul fa dyé a zeen)
Official compendium	USP
Molecular formula and weight	$C_{25}H_{31}NaO_8$. 482.51
Alternative names in other systems of nonproprietary names	[paracetamol is INN and BAN]
Chemical name(s)	(1) 1*H*-imidazole-1-ethanol, 2-methyl-5-nitro-; (2) 2-methyl-5-nitroimidazole-1-ethanol
Chemical Abstracts Service (CAS) registry number	*CAS-7660-71-1*
Indication that the USAN is also official in other systems	INN; BAN; DCF; MI
Pharmacologic and/or therapeutic category	*Vasodilator; relaxant (smooth muscle)*
Brand name(s), manufacturer(s)	Lidanar (Sandoz)
Code designation	*NIH 7607*
Graphic formula	[Graphic molecular structure or amino acid sequence when known]

[a] Based on *USP Dictionary of USAN and International Drug Names*.[1]

[b] The examples given are from different entries, not a single entry. The categories of information are presented in the sequence and format in which they appear in entries. The *USP Dictionary of USAN and International Drug Names*[1] capitalizes chemical and nonproprietary drug names that in fact should not be capitalized except in titles and other settings that require capitalization of common nouns; see Section 20.1.1.1. Such names are presented in this table without an initial capital rather than in the capitalized forms used in the source.

[c] USP = *United States Pharmacopeia*, INN = International Nonproprietary Name, BAN = British Approved Name, DCF = Dénomination Commune Française, MI = The Merck Index, NIH = National Institutes of Health.

Many other countries have similar means for establishing nonproprietary names, such as those of the British Pharmacopoeia Commission. These names are identified in a number of authoritative sources, including the *USP Dictionary*.[1] The primary Canadian source for both nonproprietary and brand names is the *Compendium of Pharmaceuticals and Specialties*.[5] For the names established in European countries, comprehensive sources include *Martindale: The Complete Drug Reference*[6] and *Index Nominum*.[7]

20.1.1.2 International Nonproprietary Names

A committee of the World Health Organization (WHO) serves a function akin to that of the United States Adopted Names Council and establishes the International Nonproprietary Name (INN) of each drug. Newly accepted names are published from time to time in the journal *WHO Drug Information*,[8] and they are identified in various authoritative sources, such as the *USP Dictionary of USAN and International Drug Names*.[1] In countries without their own national authority for nonproprietary names, the INNs should be used as properly established names. Cumulative lists of INNs are published at about 5-year intervals; Cumulative List 10 was published in 2002 as a multilingual CD-ROM.[9] Information about proposed and recommended INNs is also available through Mednet, the Medicines Information Service of the WHO (http://mednet.who.int).

20.1.1.3 Pharmacy Equivalent Names

For convenient and brief designations of dosage forms containing 2 or more therapeutic substances, the US Pharmacopeial Convention allows the use of pharmacy equivalent names (also known as PENs) derived from stems in the separate drug names, combined with the prefix "co-", which indicates a combination dosage form.[1] Specific short prefix and suffix units are suggested for the formation of the stem terms (those following "co-"):

co-trimoxazole [from "trimethoprim" and "sulfamethoxazole" with the prefix "co-"]
co-codAPAP [representing codeine phosphate and acetaminophen]

An additional intent in recommending these names is to prevent the proliferation of trivial names and ad hoc abbreviations for these dosage forms. These pharmacy equivalent names are not included as official names in *The United States Pharmacopeia and the National Formulary*,[10] although they are represented by entries in the *USP Dictionary of USAN and International Drug Names*.[1]

20.1.1.4 Abbreviations of Chemical Names and Multidrug Regimens

Some drugs become known by abbreviations of their chemical names:

AZT or ZDV *for* zidovudine   5-FU *for* fluorouracil

These names are frequently not standardized, and their use can result in confusion and error. Avoid such abbreviations and instead use the US Adopted Name, if established.

Some drug treatments use 2 or more single drugs in a complex regimen. These have frequently been identified by abbreviations based on, usually, initial letters of the individual drug names (some of which may have been superseded by a later name):

CMF *for* **c**yclophosphamide, **m**ethotrexate, **f**luorouracil
MOPP *for* **m**echlorethamine, **O**ncovin (vincristine sulfate), **p**rednisone, **p**rocarbazine hydrochloride

If editorial policy allows abbreviations without definition in the title of an article, include the full drug names at least in the abstract and give the explanation at the first use of the abbreviations in the text or in a table of abbreviations on the first page of the article.

20.1.1.5 Endocrinologic and Metabolic Drugs

Some US Adopted Names represent hormones or vitamins that in contexts other than diagnosis or treatment (such as endocrinology and nutrition) are referred to by other names. When referring to these drugs as treatments, use the US Adopted Name:

USAN	Other Name
calcitriol	1,25-dihydroxycholecalciferol
leucovorin calcium	calcium folinate

20.1.2 Proprietary Names (Trade Names, Trademarks)

Proprietary names are those established by manufacturers and vendors of drugs to represent their own products. Because such names are proper nouns, they must be capitalized (see Sections 9.7.1 and 9.7.7). Authors other than manufacturers and inventors are not required to use a superscript symbol indicating trademarking or registration of a proprietary name. Proprietary names can be verified in sources such as the *USP Dictionary of USAN and International Drug Names*,[1] *Compendium of Pharmaceuticals and Specialties*,[5] the *American Drug Index*,[11] and the *European Drug Index*[12] (see also Section 20.1.1.1).

20.1.3 Chemical Names

Most drugs are complex chemical compounds. The nonproprietary names are not intended to indicate the details of chemical structure; these details are properly indicated only by systematic chemical names developed by the principles set forth in various documents of the International Union of Pure and Applied Chemistry (IUPAC); see Chapter 17 of this manual. The *USP Dictionary of USAN and International Drug Names*[1] provides 2 chemical names for each nonproprietary name: the inverted Chemical Abstracts Service index name, which lends itself better to electronic storage and retrieval, and an uninverted IUPAC chemical name, which is preferred by the WHO:

inverted form: hydrazinecarboximidamide, 2-[2-(2,6-dichlorophenoxy) ethyl]-, sulfate (2:1)
uninverted form: 2-[2-(2,6-dichlorophenoxy)ethyl]hydrazinecarboximidamide sulfate (2:1)

The index (inverted) form can readily be converted to the uninverted form.

20.1.4 Chemical Formulas

Chemical molecular formulas for drugs indicate the relative proportions of elements making up compounds and give little or no indication of chemical structure:

nonproprietary name: clemastine
chemical formula: $C_{21}H_{26}ClNO$

Structural formulas of drugs are illustrated in the 2 major US reference sources on drugs, *The Merck Index*[4] and the *USP Dictionary of USAN and International Drug Names*.[1] Structural formulas (see Section 17.2.3) represent the stereochemistry of the drug molecule better than chemical molecular formulas.

20.1.5 Radiopharmaceuticals

The nonproprietary names of pharmaceuticals carrying radioisotopes do not follow the conventions applied elsewhere (see Section 17.10). Place the symbol for the radioactive isotope after the carrier name and then the number for its atomic weight on the line:

albumin, chromated Cr 51 serum
iodopyracet I 125
sodium pertechnetate technetium Tc 99m

20.1.6 Chemical Abstracts Service Registry Numbers

The rapid handling of data, as in databases, may be facilitated by numbering systems for identifying complex chemical entities, including drugs. The most prominent of such systems is that of Chemical Abstracts Service (CAS; http://www.cas.org). CAS registry numbers are randomly assigned to substances, and each number is unique for the substance it identifies. In the online CAS Registry File, available through STN International (http://www.stnweb.org/ or http://stnweb.cas.org/), all the names for a single substance are brought together and may be displayed under the unique CAS registry number with an indication of the related bibliographic references to the chemical literature. Some entries in the *USP Dictionary of USAN and International Drug Names*[1] carry more than one CAS registry number. For example, an anhydrous compound will have one number and its hydrate another:

theophylline monohydrate CAS 5967-84-0   theophylline [anhydrous] CAS 58-55-9

Other sources of CAS registry numbers are the MEDLINE database and *Martindale: The Complete Drug Reference.*[6]

20.1.7 Code Designations

During developmental stages in the study and testing of a potential new drug, the organization carrying out the work often assigns an alphanumeric designation to the compound. The alphabetic component is usually derived from the organization's name.

NIH 7607 [the code designation for etonitazene assigned by the National Institutes of Health]

The Merck Index[4] and an appendix to the *USP Dictionary of USAN and International Drug Names*[1] list a number of these code designations.

20.2 USE OF NONPROPRIETARY (GENERIC) OR TRADE NAMES

Decisions on when to use nonproprietary names or trade names depend in part on the audience for the publication and the context in which drug names are used. Because of pervasive drug advertising in medical journals, many physicians are more accustomed to trade names, especially those for relatively new drugs still available only from the originating manufacturer, than to nonproprietary names. However, there are 3 important reasons for generally preferring nonproprietary names: nonproprietary names are more likely to represent the chemical characteristic of the drug; trade names often differ greatly from one country to another, whereas nonproprietary names are more likely to be identical in different countries; and most indexing services index drug-related publications by nonproprietary names and not by trade names.

When using nonproprietary names, identify the specific drug used in the reported research by trade name and manufacturer, at least in the methods section of the paper:

> title: Sucralfate Compared with Placebo in the Treatment of Peptic Ulcer
> methods section: The treatment group received sucralfate (Carafate, Aventis Pharmaceuticals).

If the reported research involved comparing 2 drugs with the same nonproprietary name, use the trade name of each in the paper's title, abstract, and methods section along with the nonproprietary name.

20.3 DRUG AND OTHER RECEPTORS

Symbols for receptors of drugs and other humoral mediators have not been systematically standardized. Those in wide use represent a general pattern of alphanumeric symbols: Roman or Greek characters and additional alphanumeric or numeric designators, which are usually subscript:

alpha-adrenoreceptors	á$_{1A}$	á$_{1B}$	á$_{1C}$	á$_{1D}$	á$_{2A}$	á$_{2B}$	á$_{2C}$
adenosine receptors	A_1	A_{2b}					
bradykinin receptors	B_1	B_2					
cholecystokinin and gastrin receptors	CCK_A	CCK_B					
dopamine receptors	D_1	D_2	D_4	D_5			
purine nucleotide receptors	P_{2x}	P_{2u}	P_{2t}				

Additional symbols, along with previous symbolizations and other related information, can be found in the nomenclature supplement of *Trends in Pharmacological Science*[13]; also see Abbracchio et al.[14] Some of these symbols might be confused with other symbols (such as those for some vitamins), but they are unlikely to appear in potentially confusing contexts.

20.4 UNITS FOR DRUG CONCENTRATIONS IN BIOLOGICAL FLUIDS

Drug concentrations in biological fluids are generally measured and reported in gravimetric units (such as milligrams per liter [mg/L] or nanograms per milliliter [ng/mL]). The Scientific Committee of the Association of Clinical Biochemists (United Kingdom)[15] has recommended SI units (see Section 12.2.1), the liter (L) being the reference volume and millimole (mmol) or nanomole (nmol) the unit for amount of drug or metabolite measured; the resulting units would be mmol/L and nmol/L. The use of molar rather than gravimetric units has strong scientific rationales and is used universally in some countries, such as Canada, but this change has not been widely adopted in the United States, largely because dosages there are expressed in gravimetric units.

20.5 PHARMACOKINETICS

Symbols for variables in pharmacokinetic studies are various combinations of capital and lowercase Roman and Greek letters,[16,17] as exemplified in Table 20.2. Use subscript for qualifiers specifying sites of measurement, organs, and routes of administration and elimination; use superscript for other modifiers for general conditions, such as "SS" for "steady state" (but note that many publications use subscript for all such modifiers).

Italicize single-letter symbols for variables in accordance with international style for symbols of quantities and for chemical kinetics (see Section 4.2.2). Use roman type for multi-letter symbols to distinguish them from multiplied single-letter variables.

rate of administration = $\mathrm{CL} \cdot C^{\mathrm{SS}}$, where CL = clearance, C^{SS} = steady-state drug concentration
CL = dose/AUC, where CL = clearance, AUC = area under the curve

Table 20.2 **Examples of common pharmacokinetic symbols and qualifiers**[a]

Symbol	Term symbolized
Measured or calculated variables (roman type for multilettered symbols; italic type for single-lettered symbols[b]**)**	
A^{SS}	Amount of drug in the body at steady state
AUC^{t}	Area under plasma concentration–time curve from time zero to time *t*
AUC^{∞}	Area under plasma concentration–time curve from time zero to infinity
C	Drug concentration in plasma at any time *t*
CL	Total body clearance of drug from plasma
$\mathrm{CL}_{\mathrm{CR}}$	Creatinine clearance
C^{SS}	Steady-state concentration of drug in plasma during infusion at a constant rate
E	Organ extraction ratio (specify organ by subscript; see elsewhere in this table)
f, F	Fraction of the dose systemically available; bioavailability
f_{a}	Fraction of administered dose absorbed
f_{u}	Fraction of drug unbound in plasma
k	First-order rate constant
K_{m}	Michaelis–Menten constant
Q_{R}	Renal blood flow; use qualifier for plasma flow
R_{0}	Constant infusion rate (zero-order)
t_{max}	Time to reach peak or maximum concentration after administration of drug
$t_{1/2}$	Elimination half-life
$t_{1/2\mathrm{a}}$	Absorption half-life
τ	Dosage interval
V_{c}	Pharmacokinetic volume of central or plasma compartment
V_{d}	Apparent volume of distribution
V^{SS}	Apparent volume of distribution at steady state
V_{max}	Maximum rate of metabolism by an enzyme-mediated reaction
Modifiers representing sites of measurement (roman type, subscript)	
b	Blood
p	Plasma
s	Serum
t	Tissue
u	Unbound species
ur	Urine
Modifiers representing organs or elimination routes (roman type, subscript)	
e	Excreted into urine
H	Hepatic

Table 20.2 Examples of common pharmacokinetic symbols and qualifiers[a] (*continued*)

Symbol	Term symbolized
Modifiers representing organs or elimination routes (roman type, subscript)	
m	Metabolized
NR	Nonrenal
R	Renal
Modifiers representing routes of administration (roman type, subscript)	
im	Intramuscular
ip	Intraperitoneal
iv	Intravenous
o	Oral
po	Oral
pr	Rectal
sc	Subcutaneous
sl	Sublingual
top	Topical

[a] Based on American College of Clinical Pharmacology, Committee for Pharmacokinetic Nomenclature[16] and *Clinical Pharmacokinetics* preferred symbols.[17] The first of these sources also presents applicable units, previously used symbols, and recommended pharmacokinetic equations.

[b] The international convention of italicizing single-letter symbols for variables has been applied in this table.

CITED REFERENCES

1. USP dictionary of USAN and international drug names. Rockville (MD): United States Pharmacopeial Convention; 2014. Annual. Also available as an online database by subscription at http://store.usp.org/.

2. Pharmacopeial Forum. Rockville (MD): United States Pharmacopeial Convention. Vol. 1, Jan–Feb 1975–present.

3. Clinical Pharmacology & Therapeutics. New York (NY): Elsevier. Vol. 1, Jan–Feb 1960–Nov 1994.

4. O'Neil MJ, Heckelman PE, editors. The Merck index: an encyclopedia of chemicals, drugs, and biologicals. 15th ed. Cambridge (UK): Royal Society of Chemistry; 2013.

5. Repchinsky C, editor. Compendium of pharmaceuticals and specialties. Ottawa (ON): Canadian Pharmacists Association; 2014. Annual.

6. Sweetman SC, editor. Martindale: the complete drug reference. 37th ed. London (UK): Pharmaceutical Press; 2011.

7. Swiss Pharmaceutical Society, editor. Index nominum: international drug directory. 19th ed. Stuttgart (Germany): Medpharm Scientific Publications; 2008. Accompanied by: 1 CD-ROM.

8. WHO Drug Information. Geneva (Switzerland): World Health Organization. Vol. 1, 1987–present.

9. International Nonproprietary Names (INN) for pharmaceutical substances CD-ROM. Lists 1–101 of Proposed INN and lists 1–62 of Recommended INN. Cumulative List No. 13 [CD-ROM]. Geneva (Switzerland): World Health Organization; 2009.

10. The United States Pharmacopeia 35th and the National Formulary 30th. Rockville (MD): United States Pharmacopeial Convention; 2012. Also available as a database by subscription at http://www.usp.org/usp-nf.

11. Billups NF, Billups SM, editors. American drug index. 58th ed. St Louis (MO): Lippincott Williams & Wilkins; 2014. Annual.

12. Muller NF, Dessing RP, editors. European drug index. 4th ed. Stuttgart (Germany): Deutscher Apotheker Verlag; 1997.

13. Alexander S, Mathie A, Peters J, editors. TiPS nomenclature supplement. 12th ed. London (En-

gland): Elsevier Science Ltd.; 2001. Endothelin receptors; p. 42–43. Published as a supplement to Trends in Pharmacological Sciences.

14. Abbracchio MP, Cattabeni F, Fredholm BB, Williams M. Purinoceptor nomenclature: a status report. Drug Dev Res. 1993;28(3):207–213.

15. Ratcliffe JG, Worth HGJ. Recommended units for reporting drug concentrations in biological fluids. Lancet. 1986;1(8474):202–203.

16. American College of Clinical Pharmacology, Committee for Pharmacokinetic Nomenclature. Manual of symbols, equations & definitions in pharmacokinetics. J Clin Pharmacol. 1982;22(7):1S–23S.

17. Clinical Pharmacokinetics preferred symbols. Clin Pharmacokinet. 2001;40(1):73–75.

21 Genes, Chromosomes, and Related Molecules

21.1 GENERAL RULES AND PRINCIPLES OF GENETIC NOMENCLATURE

For some organisms, detailed systems for the symbolic representation of mutant and wild-type genes and of normal and abnormal chromosomes are available. For other organisms, little or no work has been done to describe either the genes or the chromosomes. The conventions presented in this chapter represent all taxa for which published, generally accepted rules and guidelines were available at the time of printing. They do not cover all conventions for all organisms that have been studied with genetic or molecular methods, but they do cover the most-studied organisms. Many of these sets of conventions are similar; they differ only in minor details of symbolization and notation.

21.1.1 Standardization Efforts

The first formal system of genetic nomenclature and symbolization appears to have been the rules on symbols for genes of the laboratory mouse.[1] A later set of recommendations[2] for wider application influenced the systems developed for many organisms.

Although there is no one system for the genetic nomenclature of different organisms, considerable standardization exists, thanks to the effort and cooperation of the nomenclature committees involved, beginning with the International Committee on Gene Symbols and Nomenclature in 1957.[2] All of the recommendations proposed then are still in use today for most organisms covered in this manual, although some have been modified (especially in the use of hyphens, superscripts, and subscripts), and many have been expanded by Demerec et al.,[3] the American Society for Microbiology,[4] the International Standing Committee on Human Cytogenetic Nomenclature (for the latest report, see ISCN),[5] the HUGO Gene Nomenclature Committee (for the latest report, see http://www.genenames.org),[6] the International Committee on Standardized Genetic Nomenclature for Mice (for the latest report, see http://www.informatics.jax.org/),[7,8] the Commission on Plant Gene Nomenclature (for the latest report, see Price and Reardon[9]), and other less well-known committees. The general recommendations are summarized in Table 21.1.

The Gene Ontology Consortium (GO Consortium) is designed to supplement gene nomenclature and the different sets of rules for different organisms. The goal is to produce dynamic controlled vocabularies for molecular functions, biological processes, and cellular components that can be applied to all organisms, even as knowledge of gene and protein roles in cells accumulates and changes.[10]

Table 21.1 General rules and principles: recommendations for symbolization[a]

Feature	Convention
Genes	
Gene name	Use languages of higher internationality. Traditionally, gene names have described, with a concise word or phrase, the main diagnostic feature of the mutant phenotype, the name of the protein encoded, the metabolic requirement, or the sensitivity or resistance to a drug or other agent. Few gene names are based on the wild-type phenotype. Some of the newer guidelines advocate naming genes on the basis of their gene family and the similarity of the sequences to each other (see Section 21.14.1).
Gene symbol	Derive the symbol from the full, original name by shortening the name, using the initials of a multiword term, or otherwise devising a symbol that is recognizable and, preferably, pronounceable. Although many of the older gene symbols are only 1 or 2 letters, and some are as many as 5 letters, most guidelines now call for 3-letter symbols in italics. Most avoid the use of Greek letters, roman numerals, superscripts, and subscripts; they make minimal use of commas, colons, and semicolons. If they do call for superscripts or subscripts, they make provisions for substitutions. Almost all guidelines call for the symbol to be in italics; unapproved symbols for oat are the lone exception among those outlined in this manual.
Dominant trait	Whenever unambiguous, name and symbol begin with a capital letter.
Recessive trait	Whenever unambiguous, name and symbol begin with a lowercase letter.
Allele series	Gene symbol with superscripts to represent the different members
Standard, wild-type alleles	Gene symbol with superscript nonitalic plus sign, or a plus sign with the gene symbol as a superscript. In formulas the plus sign may be used alone.
Different loci with similar phenotypes	
Nonalleles (mimics, polymeric genes, etc.)	Gene symbol followed by an additional letter or arabic numeral, either on the same line after a hyphen or as a subscript
Alleles of independent mutational origin	Gene symbol with a subscript
Enhancers, inhibitors, lethals, and suppressors	*En*, *I*, *L*, or *Su* for dominant traits (*en*, *i*, *l*, or *su* for recessive traits), followed by a hyphen and the symbol of the allele affected
Sterility or incompatibility	*S* for dominant trait (*s* for recessive trait), followed by a hyphen and the symbol of the allele affected
Genic formulas	Write as fractions with maternal alleles first or above. Each fraction corresponds to a single linkage group. Order the groups in numerical sequence separated by semicolons. Enclose unlocated genes in parentheses at the end of the formula. In euploids and aneuploids, repeat the gene symbol as many times as there are homologous loci.
Extrachromosomal factors	Enclose symbols in square brackets and list them at the beginning of a genic formula.
Chromosomes	
Autosomes	Designate by arabic numerals in ascending order from longest to shortest.
Sex chromosomes	Animals in which the male is heterogametic have sex chromosomes designated X and Y; therefore, females are XX, males are XY. Those in which the female is heterogametic have sex chromosomes designated Z and W; therefore, females are ZW, males are ZZ.

Table 21.1 General rules and principles: recommendations for symbolization[a] *(continued)*

Feature	Convention
Chromosomes	
Designations	Use arabic numerals to designate individual chromosomes, roman numerals for linkage groups. Set the designations for the chromosomes, their bands, and other markers in roman type, not in italics.
Karyotypes, ideograms	Arrange the autosomes vertically in horizontal rows on the page, in decreasing order by length, with the short arm on top, and with the sex chromosomes at the end of the series.
Chromosomal aberrations	Use the following abbreviations for chromosomal aberrations: Df or Def, deficiency Del, deletion Dp or Dup, duplication In or Inv, inversion T or Tran, translocation Tp, transposition Rules vary on whether these abbreviations should be in italics and which of a pair of choices should be used.
Chromosome count	Indicate the zygotic number of chromosomes by 2*n*, the gametic number by *n*, and the basic number by *x*.

[a] Based on the report of the International Committee on Genetic Symbols and Nomenclature[2]; Demerec et al.[3]; and the patterns prevailing throughout the guidelines outlined in this chapter.

21.1.2 Guidelines for Compliance with Existing Standards

The use of standard gene names and symbols for each major taxon facilitates communication among researchers. While the scientific community works toward a standard nomenclature for all taxa, CSE recommends compliance with the existing guidelines set by the individual nomenclature committees for their respective organisms. When comparing genes and alleles from different organisms, use the appropriate symbols for each, including attention to case and type style. Avoid beginning titles or sentences with gene symbols or names that begin with a lowercase letter.

> The *TP53* gene in humans and *Trp53* in mice both code for protein p53.
> "Chlorophyll-Deficiency Chlorina Is Coded by av in Oat and *chl2* in Rice" *not*
> "av in Oat and *chl2* in Rice Both Code for Chlorophyll-Deficiency Chlorina" [Some names for oat genes should not be in italics; those for rice genes should be.]
> Insertion mutations in the *b* gene . . . *not b* Gene insertion mutations . . .

In the title of an article and in tables, use only gene symbols and names that have been approved by the appropriate committee. In the abstract and introduction, enclose well-known synonyms for those same genes in parentheses after the approved symbol or name; to avoid ambiguity, separate 2 synonyms with "and", and 3 or more with commas, not slashes. Use the approved name throughout the rest of the text. Submit information to the relevant database before publication and include citations to the database in the text.

> *RPS14B* (also known as *CRY2*)   *CDKN1A* (also known as *CIP1*, *WAF2*, and *p21*)

Researchers who believe they have discovered a new gene or allele should search the relevant gene lists and consult with the appropriate nomenclature committee or gene registry before publishing a new gene name and symbol. The discovery may be a new member of an existing gene family or one that already has an established symbol, or the proposed symbol may already be in use for a different gene family. Most committees require statistically valid segregation or sequence data to support the existence of a "new" gene before a symbol will be assigned. When the same symbol is assigned to different genes, or when 2 or more symbols are assigned to the same gene, priority in publication is usually the primary criterion for establishing the preferred symbol in each case. Treat symbols not given priority as synonyms. Some publishers require newly determined sequences to be made available in public databases on or before the date of publication of the article describing the sequences.

21.1.3 Institute for Laboratory Animal Research

The Institute for Laboratory Animal Research (ILAR) maintains an international registry of laboratories and assigns unique codes to them. Each code identifies an investigator, laboratory, or institution. The codes range from 1 to 4 letters (2 or 3 letters preferred), but only the first letter is capitalized. They were designed for use in designations for substrains of mice, rats, and rabbits but are now used in the symbols for transgenes, DNA loci, targeted and induced mutations, and chromosome anomalies as well. They are referred to in this manual as "laboratory codes". Information on the registry and its assignment of codes is available from the ILAR website[11] or by email (ilar@nas.edu).

21.1.4 Terminology

In the rules outlined in this chapter, the terminology of the published rules has been retained. Comparable rules for certain features may, therefore, be worded differently.

21.2 GENETIC UNITS, MEASURES, AND TOOLS

21.2.1 Base Pairs

The length of a DNA sequence (i.e., the distance between 2 loci on a contig map) can be designated by the number of base pairs encompassed. Use the symbol bp (for base pairs) for relatively short sequences (fewer than 1,000 bases); use kb (for kilobases) or Mb (for megabases) for longer sequences or to indicate the total number of base pairs.

> The total length of DNA sequenced from the clone is 255 bp.
> The sequence-tagged site sY80 is 1.075 kb long.
> The total amount of mapped sequence is now more than 10.3 Mb.

21.2.2 Centimorgans

The genetic distance between 2 loci on a chromosome can be expressed in centimorgans (cM). The centimorgan numerically equals the statistically corrected recombination frequency expressed as a percentage; thus, distances of more than 1 M cannot exist. For example, if the recombination frequency between 2 given markers is 15%, they are separated by 15 cM. The genetic distance in human of 1 cM approximates, on average, the physical distance of 1 million base pairs (1 Mb), but this correlation varies throughout the genome, especially at telomeres, where recombination is more frequent. The M in cM is capitalized, the symbol having been named for Thomas Hunt Morgan, who studied the function of the chromosome in heredity.

21.2.3 Sequence-Tagged Sites

A sequence-tagged site (STS) is a short DNA segment that can be located by polymerase chain reaction (PCR) techniques and that identifies a unique location on a chromosome.

sY163 [sequence 163 on the Y chromosome]

Such DNA sequences of known map locations serve as markers for genetic and physical mapping of genes.[12] Sequence-tagged sites include microsatellites, sequence-characterized amplified regions (SCARs), cleaved amplified polymorphic sequences (CAPS), and inter-simple sequence repeats (ISSRs). The National Center for Biotechnology Information maintains databases (dbSTS, part of GenBank, http://www.ncbi.nlm.nih.gov/dbSTS/; UniSTS, http://www.ncbi.nlm.nih.gov/sites/entrez?db=unists) of STSs that cover many species. The symbolization of STSs is not standardized within or across species, but many STSs are given D segment symbols (see the discussion of anonymous DNA segments in Section 21.2.6).

21.2.4 Contigs

Overlapping, contiguous cloned DNA sequences that together constitute a continuous DNA segment of a genome are known as a contig, which may be either clone based or generated from whole genome shotgun (wgs) sequencing projects. If clone based, the components of the contig are generally represented by the symbols of the clones carrying the overlapping nucleotide sequences, but the symbolization of clones is not standardized. In some cases, the clones are represented by the sequence database accession numbers of the components of the contig. The contig itself may also be given a name and accession number.

If the contig is based on a wgs sequencing project, the designation is usually a 12-character database accession number composed of a 4-letter designation for the project, a 2-digit version number, and a 6-digit designation for each individual contig[13] (e.g., CAAA01119629). Related contigs can be aligned to form a scaffold, and related scaffolds to represent a chromosome.

21.2.5 Loci and Markers

A locus is the location on a chromosome (or plasmid or other genetic molecule) of a gene, an anonymous DNA segment, a fragile site, a breakpoint, an insertion, or another distinguishable sequence by which the location can be specified. For genes, the terms "gene" and "locus" are often used interchangeably. For sections of genetic code that do not specify gene products but serve regulatory roles, are attachment sites, or encompass a cluster of genes, use the term "locus" instead.

When possible, specify the type of the locus—for example, gene locus or restriction fragment length polymorphisms (RFLP) locus. In a genome database, a locus entry is likely to include some location information (genetic or cytogenetic), the official name and symbol, the mapping method(s) used, and the homologous genes of a reference organism. Genes are represented by gene symbols. The locus tag (also called the "ORF name" [for "open reading frame"] or the "systematic name") is often used instead of the gene name, especially in descriptions of the whole genome or when the specific locus is known but the gene product has not yet been determined. The style of the locus tag varies with the organism or organelle.

YKR123C [where Y = yeast, K = chromosome 11 converted to a letter, R = right arm, 123 = the open reading frame, and C = the Crick strand]
ycf42 [temporary designation for the hypothetical chloroplast ORF that was later renamed *bas1*]
C2orf1 [temporary designation for chromosome 2, open reading frame 1, in human]

A marker can be a gene locus associated with a particular, and usually readily identifiable, phenotype, a subdivision of a gene, or an anonymous DNA segment.

21.2.6 Anonymous DNA Segments

An anonymous DNA segment is one with no known functional identity. These segments are represented by unique alphanumeric D-number symbols assigned by the genome-specific database (human, Human Genome Database[14]; mouse, Mouse Genome Informatics[15]; rat, Rat Genome Database[16]) or by the laboratory performing the sequencing. A human DNA segment symbol consists of the letter D to designate an anonymous DNA segment, the number (1 to 22) or letter (X or Y) of the chromosome on which the segment resides (or 0 if unknown), a 1-letter code to designate the type of sequence (S, Z, or F), and a unique sequential number (e.g., D7S645, the 645th single-copy sequence assigned to chromosome 7). The letter S denotes a unique DNA segment, followed by a sequential number unique for the chromosome on which the sequence resides. The letter Z denotes a repetitive DNA segment found at a single chromosome site, followed by a sequential number unique for the chromosome on which the sequence resides (e.g., DYZ3, the third repetitive sequence assigned to Y). The letter F denotes a small, undefined family of homologous sequences found on multiple chromosomes, followed by a sequential number unique in the genome for the sequence family, then by the letter S and a sequential number, unique in the family, designating the specific member of the family (e.g., D14F23S1, the first member of the 23rd sequence family assigned in the genome; this member of the family is on chromosome 14; D14F23S2, the second member of the 23rd family, also on chromosome 14; and D2F23S3, the third member of the 23rd family; this member is on chromosome 2). The letter E is added to the end of a sequence if it is known to be expressed (e.g., D9S220E, the 220th single-copy sequence assigned to chromosome 9; the sequence is expressed). Note that the fact that it is expressed means that the sequence is likely to be a gene, but it cannot be given a gene symbol because its function is unknown.[17]

Mouse and rat DNA segment symbols differ slightly from those of human and consist of the letter D, the number or letter of the chromosome, the laboratory code (instead of the type of sequence), and a unique sequential number (e.g., D1Mit1 is the D symbol for the first sequence for mouse chromosome 1 from the Massachusetts Institute of Technology [MIT]). All D segments are also loci.

21.2.7 Probes

A probe is a DNA, RNA, protein, or other cellular constituent used to visualize a genetic target object or structure. Probe names are expected to be distinct from gene symbols and names of loci and are assigned by the originating laboratories. No standardized nomenclature for probes exists, although many probes may be identified by a D segment symbol, as described in Section 21.2.6. Superscripts and subscripts must not be used. Lowercase-letter prefixes can indicate vectors: p for plasmid, c for cosmid, l for lambda phage, and y for yeast.

21.2.8 Single-Nucleotide Polymorphisms

A single-nucleotide polymorphism (SNP, pronounced "snip") is a small change (often a single nucleotide) that can occur within the genome. SNPs occur frequently throughout the genome of a species and thus serve as useful biological markers, having an identical physical location that can be tracked and used to construct genome maps showing the relative position of known genes or markers. The increasing number of SNP association studies and the improvements in genome annotation necessitate a standardized nomenclature for unequivocal and correct SNP identification. The National Center for Biotechnology Information facilitates the identification and classification of SNPs through its public domain SNP database, dbSNP (http://www.ncbi.nlm.nih.gov/projects/SNP/). Records of SNPs in the dbSNP are linked to other NCBI sources, including GenBank records, the Entrez gene database, dbSTS, and PubMed. As a researcher submits a new SNP, it is assigned an "ss" number (submitted SNP). When the submitted SNP(s) is (are) mapped to the correct location on the genome, NCBI assigns the SNP (or cluster of SNPs) an "rs" number (reference SNP cluster) or RefSNP. Many journals now recommend the use of ss or rs numbers for SNPs in published articles.

21.2.9 Banding Patterns

Chromosomes can be stained by a variety of methods, each producing a different type of banding pattern. "A band is defined as the part of a chromosome that is clearly distinguishable from its adjacent segments by appearing darker or lighter with one or more banding techniques" (page 7 in ISCN[5]). The results of staining, which can be photographed and compared, vary considerably among organisms. Most karyotypes and ideograms are based primarily on Giemsa banding (usually referred to as G-banding), with supporting data from the other types (C, constitutive; Q, quinacrine; R, reverse; T, telomeric; and NORs, nucleolus organizing regions).

21.2.10 Restriction Endonucleases

Although restriction endonucleases are not genetic material themselves, they play a key role in genetic research and have a standardized nomenclature. A 3-letter italic abbreviation represents the source organism: an initial uppercase letter that is the first letter of the genus name is followed by the first 2 letters of the specific epithet. Nonitalicized strain designations consisting only of arabic numerals are spaced; other designations are closed up.

Bce 1229 [an enzyme from *Bacillus cereus*, strain IAM 1229]
*Xma*II [the second enzyme from *Xanthomonas malvacearum*]

See Roberts[18] for more details and REBASE[19] for a catalog of restriction enzymes.

21.3 TRANSCRIPTION-RELATED PROTEINS

Initiation factors needed to start protein synthesis are designated with the nonitalic symbol IF followed by a hyphen and a numeric identifier. The symbols for bacterial factors (e.g., those for *Escherichia coli*) have no prefix; those for eukaryotic factors have the lowercase, nonitalic prefix e. A hyphenated number indicates the group assignment. Symbols for new initiation factors in a group add the next uppercase letter available among the group symbols. Subunits of initiation factors are represented by adding Greek letters: α, β, γ, and so on.

IF-2 eIF-2 eIF-4B eIF-3β

Elongation factors have similar representations; the symbols for prokaryotic factors differ in minor ways.

EF-Ts EF-Tu eEF-1α eEF-1βγ

For further details on both groups of factors, consult the recommendations drafted by Safer.[20]

21.4 INSERTIONS AND TRANSPOSONS

Simple bacterial insertion elements are represented by the nonitalic symbol IS (for "insertion sequences") followed by an italic arabic numeral. Bacterial transposons are represented by the nonitalic symbol Tn followed by specific alphanumeric designators in italics.[4,21]

The symbols for transposons in eukaryotes vary with the organism, but most are not italicized. Transposons in filamentous fungi are given gene names that begin with an uppercase letter and are displayed in italics.[22]

IS*2*, IS*4* [bacterial insertion elements]
Tn*1*, Tn*5 lac* [bacterial transposons]
Tcr1, Tcr2 [*Chlamydomonas reinhardtii* transposons]
Ty1, Ty2 [yeast transposons]
Hidaway, *Ans1*, *MGR586* [filamentous fungal transposons]

21.5 TRANSSPECIES GENE FAMILIES

Efforts to coordinate gene family nomenclature are being led by the International Nomenclature Workshop. The goals are to develop guidelines for the rules, to consider strategies for dealing with the data, and to redefine the point at which the nomenclature committees become involved. In contrast to the traditional mutant phenotype–based method of naming genes, many of the gene nomenclature committees have recommended that gene names and symbols be based on function whenever possible. The concept of gene families that transcend taxonomic and functional boundaries represents yet another

Table 21.2 Cytochrome P450 gene superfamily, families, and subfamilies: symbols[a]

Feature	Convention	Examples
Gene root symbol for the superfamily	Italicized symbol; in all uppercase for human and some other species, initial capital only for others	*CYP* for cytochrome P450 [for human] *Cyp* for cytochrome P450 [for the mouse and *Drosophila*]
Cytochrome P450 family	Root symbol with an italic arabic numeral for the family	*CYP2* *Cyp7*
Subfamily	Family symbol with an italic capital letter for the subfamily (lowercase letter for mouse)	*CYP11B* *Cyp7a*
Individual gene	Subfamily symbol with an arabic numeral for the gene	*CYP2B2* *Cyp7a1*, *Cyp7b1*
Pseudogene	A closing italic capital letter *P* (*ps* for mouse and *Drosophila*)	*CYP2B7P*
Gene product	The gene symbol in uppercase nonitalic characters	CYP1A2 [from *CYP1A2*] CYP24 [from *Cyp24*]
Trivial names	Genes known by their trivial names before they are known to encode cytochrome P450 enzymes may continue under their original names with each *CYP* name being an official synonym. If the original names continue being used in publications, include the official names in the title, the summary, or a footnote. Be consistent within a publication.	*olf1* = *CYP2G1* [in rat] *IIC16* = *CYP2C16* [in rabbit]

[a] Based on Nelson et al.[24]

approach, one in which the names and symbols are based on sequence similarities and the presence of various motifs or domains, which may imply possible function but are not conclusive.

An example of such a system of nomenclature and symbolization is that proposed for the cytochrome P450 enzymes in eukaryotes and prokaryotes by Nebert et al.[23] and updated by Nelson et al.[24] Table 21.2 summarizes their system. Committees for other gene families are listed on the websites for Mouse Genome Informatics (MGI)[15] and the HUGO Gene Nomenclature Committee[6]; see Section 21.14.1 for more details of the system in plants.

21.6 ONCOGENES

Symbols for oncogenes were originally 3-letter italic symbols based on the retrovirus in which each oncogene was first identified, and other oncogenic proteins have been named by the same convention. Nonitalic single lowercase-letter prefixes specify the immediate origin of the gene; uppercase-letter prefixes indicate the cellular localization of the protein in eukaryotic cells.

v-*fes* [viral gene, from *Feline sarcoma virus*]
c-*myc* [cellular, from myelocytomatosis]
N-*erb A* [nucleus localization, from avian erythroblastosis]
S-*sis* [secreted, from *Simian sarcoma virus*]

Nomenclature for human and mouse cellular oncogene sequences now follows the standard nomenclature for genes of those species.[25] When referring specifically to the human locus, use the name of the homologous retroviral oncogene, but without the "v-" or "c-" prefixes; for example, use "*JUN*" instead of "v-*jun* sarcoma virus 17 oncogene homolog (avian)". Name mouse loci in a similar fashion, but capitalize only the initial letter; for example, use "*Hras1*", not "*c-Hras1*", for Harvey rat sarcoma-1 oncogene.

The names and symbols of oncogenes should be regarded as provisional until the true functions of the genes become known and the genes are renamed. For example, *Erbb* has become *Egfr*, epidermal growth factor receptor; and *Sis* has become *Pdgfb*, platelet-derived growth factor, beta polypeptide.

21.7 PLASMIDS

The notations for plasmid loci, genes, and alleles follow, in general, the rules set forth by Demerec et al.[3] for bacteria; see Section 21.10. Detailed recommendations were published by Novick et al.[26] along with definitions of relevant terms, but symbols may vary in completeness and spacing. Naturally occurring plasmids often have descriptive names (e.g., the circular plasmid in *Saccharomyces cerevisiae*, 2 μm). In general, each newly described plasmid and newly isolated genotypic modification of a known plasmid is given a unique alphanumeric designation with the form pXY1234 in which p stands for plasmid, XY are initials for the laboratory or the reporting scientist, and 1234 represents the laboratory's numeric designation. Indicate a deletion by an uppercase Greek delta (Δ) followed by unique serial numbers and a list of deleted genes. Indicate insertions, transpositions, and translocations by an uppercase Greek omega (Ω) followed by unique serial numbers and a list of translocated genes.

The phenotypic notation for a plasmid gene consists of an uppercase letter and a lowercase letter (or 2 letters if necessary) and should reflect the phenotypic trait for which the gene is responsible. The genotype is given in the form recommended by Demerec et al.[3] and the American Society for Microbiology[4] for bacteria (see Section 21.10). The complete genotypic identification includes as a prefix the name of the plasmid on which the gene was found.

21.8 VIRUSES

21.8.1 Bacteriophages (Phages)

Rules for the symbolization of bacteriophage genes vary with the phage. In general, the symbols for genes are 1, 2, or 3 italic letters. Specify the phage itself with a prefix for the phage (e.g., λ for phage lambda, Mu for phage Mu, P1 for phage P1), spaced or unspaced from the gene symbol proper. Because phages do not have metabolism outside bacterial cells, genotype and phenotype are not differentiated. Symbols may be combined, spaced or unspaced, to represent mutations. Superscripts indicate hybrid genomes.

λ*che*22
Mu dII345 [phage Mu]
P1 *vir* [phage P1]
λcI857*int*2*red*114*susA*11 [mutations in genes cI, *int*, and *red* and a suppressible (*sus*) mutation in gene *A*]
λ *att*^{434} *imm*^{21} [hybrid of phage λ carrying the attachment (*att*) region of phage 434 and the immunity (*imm*) region of phage 21]

See the Instructions to Authors in the *Journal of Bacteriology*[27] for more details and resources for individual phages.

21.8.2 Human Retroviruses

Various italic 1-, 2-, or 3-letter and alphanumeric symbols have represented genes of the human immunodeficiency viruses (HIV-1 and HIV-2) and the human T-cell leukemia viruses (HTLV-1 and HTLV-2), formerly designated HTLV-III, IV, I, and II, respectively.

p40x *tel* *tat-3* *sor* *X*

Gallo et al.[28] have proposed consistent use of 3-letter italic symbols, with letters reflecting gene functions.

tax$_1$ [transactivator]
nef [negative factor]
rev [regulator of expression of virion proteins]

21.9 ARCHAEA

No rules for genetic nomenclature have been published for archaea as a group separate from bacteria. Until or unless they are, the gene names for archaea are the same as those for bacteria (Donald P Nierlich, personal communication, University of California, Los Angeles, 2002 Mar 29).

21.10 BACTERIA

A unified system for genetic nomenclature for bacterial genes and phenotypes was proposed by Demerec et al.[3] in 1966 and adopted by committees for other groups as well. The editors for the American Society for Microbiology (ASM) journals have updated and expanded the original proposal. Table 21.3[9,30] summarizes the major conventions. For more details and examples, see the Instructions for Authors in the *Journal of Bacteriology*[27] or Berlyn et al.[31]

21.11 YEASTS

21.11.1 *Saccharomyces cerevisiae*

1) Genes. Gene symbols for *Saccharomyces cerevisiae* are based on the proposals of Demerec et al.[3] as updated by Sherman[32] and the *Saccharomyces* Genome Database (SGD).[33] Table 21.4[34] summarizes the major conventions. Search the SGD Gene Name Registry[33] and refer to the SGD Gene Naming Guidelines before naming newly discovered genes and designating symbols for them. Gene names and symbols can be reserved while manuscripts are in press.

Table 21.3 Bacteria: symbols for genes and phenotypes[a]

Feature	Convention	Examples
Gene or locus symbol	3-letter, lowercase italic symbol with added capital italic letter. For genes, the terms "gene" and "locus" are used interchangeably; for regions that do not specify a gene product but serve regulatory roles, are attachment sites, or encompass a cluster of genes, the term "locus" is used.	*araA araB* [genes or loci controlling arabinose utilization] *oriC* locus [origin of replication]
Mutation (allele)	Locus symbol with added italic serial isolation number (allele number).	*araB1*
	If the exact locus is unknown, the capital letter is replaced with a hyphen.	*ara-2*
	When written as part of a genotype, the 3-letter gene symbol and capital locus letter (with or without a superscript minus) imply an unnumbered mutation. Such notation is not correct, although it is frequently encountered. Authors should assign an allele number or use phenotypic notation [below].	*araB*
Wild-type gene or allele	Gene or locus symbol with superscript plus sign. Do not use a superscript minus sign to designate a mutation.	ara^+ $araB^+$
Promoter site	Locus symbol with added italic *p*; subscript numerals (nonitalic) may be added to distinguish multiple promoters	*lacZp* $glnAp_1$ $glnAp_2$
Terminator site	Locus symbol with added italic *t*	*lacAt*
Operator site	Locus symbol with added italic *o*	*lacZo*
Attenuator site	Locus symbol with added italic *a*	*trpAa*
Operon	Gene symbol followed by more than one locus letter	*trpEDCBA* or *trp* operon
Phenotype	Generally, 3 or fewer nonitalic characters with the first capitalized. With very few exceptions, a superscript character is required to indicate the nature of the phenotype.	Lac^+ Ara^- Ts
Wild-type or positive phenotype	Phenotype symbol with added superscript plus sign; locus letters are not used	Lac^+
Mutant or negative phenotype	Phenotype symbol with added superscript minus sign	Lac^-
Drug-resistance trait carried on host chromosome	3-letter nonitalic symbol for drug with added superscript s (sensitive) or r (resistant). Superscript is mandatory.	Amp^r

Table 21.3 Bacteria: symbols for genes and phenotypes[a] (*continued*)

Feature	Convention	Examples
Drug-resistance trait carried on plasmid	2-letter nonitalic symbol for drug with added superscript s (sensitive) or r (resistant); the superscript s is mandatory; r is optional. A 3-letter symbol is permissible if necessary for clarity.	Ap^s or Ap or Ap^r Asa [arsenate resistance] Asi [arsenite resistance]
Phenotypic property or gene characteristic	Phenotypic symbol in parentheses added to the end of the gene symbol for the trait described	*araA230*(Am) *hisD21*(Ts) *rpsL20*(Str^r)
Plasmids and episomes	Generally, 3 or 4 nonitalic letters, the first a lowercase p, followed by numbers; enclosed in parentheses when written as part of a strain name or bacterial genotype. Names of older, naturally occurring plasmids may have various names.	pUC19 pBR322 *E. coli* (pXY1234) Col E1 F^+ F128
Deletions	Uppercase Greek delta before the symbol for the gene or region being deleted. Parentheses around the region are not in italics; the number after them is in italics.	Δ*trpA432* Δ(*aroP-aceE*)*419*
Inversions	Uppercase nonitalic symbol IN before the symbols for the genes being inverted. Parentheses around the region are not in italics; the number after them is in italics.	IN(*rrnD-rrnE*)*1*
Insertions	Symbol for the locus affected, followed by a double colon and the symbol for the gene or other element being inserted.	*galT236*::Tn5
	For complex insertions into plasmids, use uppercase Greek omega followed by a description in parentheses.	pSC101 Ω(0kb::K-12*hisB*)*4* [*E. coli his* gene into pSC101 at 0 kb]
Fused genes (fusions)	Uppercase Greek phi before the symbols of the fused genes. Parentheses around the region are not in italics; the number after them is in italics. Indicate a truncated gene in a fusion by an apostrophe. Fused genes are sometimes indicated by a double colon [see Insertions and Promoter fusions].	Φ(*ara-lac*)*95* Φ(*ara'-lac*)*96*
Promoter fusions	Symbol for the promoter site (including the lowercase *p*), a double colon, and the symbol for the unrelated structural gene fused to the promoter, all in italics[b]	*lacZp*::*trp*
Transposons	2-letter nonitalic symbol Tn followed by italic numbers or letters	Tn*5*, Tn*5-1* Tn*A*, Tn*phoA* zef-123::Tn*5*

Table 21.3 Bacteria: symbols for genes and phenotypes[a] (*continued*)

Feature	Convention	Examples
Foreign genes	Indicate the source or origin of a gene by a subscript mnemonic of the scientific name or name and strain designation. Use the scientific name if ambiguity exists.	$lacZYA_{Eco}$ [the *lac* operon of *E. coli*] $lacZ_{EcoO157}$

[a] Based on Demerec et al.[3]; Low[30]; Novick et al.[26]; Chumley et al.[29]; Campbell et al.[21]; the Instructions to Authors, *Journal of Bacteriology*[27] (available from http://www.journals.asm.org); Mary Berlyn, Yale University, personal communications, 2001 Jan 11, 2001 Jan 25, and 2003 Sep 3; Donald P Nierlich, University of California, Los Angeles, personal communications, 2002 Mar 27, 2002 March 29, and 2003 Oct 20.

[b] The fusion of a promoter to an unrelated gene is rarely written in a conventional style. The American Society for Microbiology discourages the form widely used (e.g., *Plac-trp*, *Ptac-araBAD-ftsQ*) because it can easily be mistaken for a plasmid or protein name. In either case, explain the mnemonic used for the promoter (American Society of Microbiology[4]; Mary Berlyn, Yale University, personal communication, 2003 Sep 3; Donald P Nierlich, University of California, Los Angeles, personal communication, 2003 Oct 20).

Table 21.4 *Saccharomyces cerevisiae*: symbols for genes and phenotypes[a]

Feature	Convention	Examples
Gene name	3 italic letters	*ARG*, *arg*
Gene symbol	Gene name followed by an arabic numeral. For genes, the terms "gene" and "locus" are often used interchangeably; for open reading frames, regions that do not specify a gene product but serve regulatory roles, are attachment sites, or encompass a cluster of genes, the term "locus" is used.	*ARG2*, *arg2* *ADE12* *YJL191W*
Allele designation	Gene symbol followed by a hyphen and an italic arabic numeral. (Gene numbers are those of original assignments, but allele numbers may be those of a specific laboratory.)	*arg2-14*
Gene cluster[b]	Gene symbol followed by an italic uppercase letter	*his4A*, *his4B*
Dominant allele	Gene symbol in uppercase italic letters	*SUP6*
Recessive allele	Gene symbol in lowercase italic letters	*arg2*
Wild-type gene	Gene symbol followed by a superscript plus sign. When there is no confusion, wild-type genes may be designated simply as +.	$ARG2^+$ $sup6^+$
Gene conferring resistance or susceptibility	Gene symbol with nonitalic superscript R or S inserted after the 3-letter gene symbol	$CUP^{R}1$ $CAN^{S}1$
Mating type	Wild-type mating types are designated *MAT*a and *MAT*α in genotype descriptions. The phenotypes are simply a and α.	. . .

Table 21.4 ***Saccharomyces cerevisiae*: symbols for genes and phenotypes[a] (*continued*)**

Feature	Convention	Examples
Homothallic genes	Wild-type homothallic genes at the *HMR* and *HML* loci are *HMR*a, *HMR*α, *HML*a, and *HML*α. Mutations at these loci are denoted *hmr*a-*1*, *hml*α-*1*, and so forth.	. . .
Informational suppressors	Symbol *SUP* or *sup* followed by a locus designation. Frameshifts take the symbols *SUF* and *suf*.	*SUP4*, *sup35* *SUF1*, *suf11*
Deletions	Insert an uppercase Greek delta after the hyphen in the symbol for the allele being deleted.	*arg2-Δ1*
Insertions	Symbol for the gene or locus affected, followed by a double colon and the symbol for the gene or other element being inserted	*ARG2::LEU2* *arg2-10::LEU2* *yjl031c::KanMX4*
Transposons (also called "Ty elements")	Nonitalic symbol Ty and a number	Ty1, Ty5 YERCTy1-1
Non-Mendelian and mitochondrial genes	Enclose the symbols in square brackets when necessary to distinguish them from chromosomal genes. Avoid Greek letters for newly discovered genes, but retain the symbols ρ^+, ρ^-, Ψ^+, and Ψ^- or their transliterations *rho*$^+$, *rho*$^-$, *psi*$^+$, and *psi*$^-$.	[ρ^+] or [*rho*$^+$] [*COX1*], [*cox1*]
Phenotypes	Same characters as gene symbol in roman type, with initial capital letter, followed by superscript plus or minus sign to indicate the independence of, or requirement for, a substance, respectively	Arg$^+$ [does not require arginine] Arg$^-$ [requires arginine]

[a] Based on Sherman[32] (http://dbb.urmc.rochester.edu/labs/sherman_f/yeast/index.html); *Saccharomyces* Genome Database[33] (http://www.yeastgenome.org/gene_guidelines.shtml); Cherry.[34]
[b] Also for complementation groups within a gene and for domains within a gene having different properties.

The ORF name (also called the "systematic name") is often used instead of the gene name, especially in descriptions of the whole genome or when the specific locus is known but the gene product has not yet been determined. The designation consists of 3 nonitalic uppercase letters (Y for yeast, the chromosome number converted to letters A through P, and the arm designation, L or R), a 3-digit number (for the open reading frame, counting from the centromere), and C or W (for the Crick or Watson strand).[33] The Crick strand is the sense or reverse complementary strand; the Watson strand is the antisense, coding, or template strand.

YJL191W [where Y = yeast, J = chromosome 10, L = left arm, 191 = the open reading frame, and W = the Watson strand]

2) Chromosomes. The chromosomes of *S. cerevisiae* are designated by roman numerals I to XVI; the arms are labeled R (right) and L (left).

Table 21.5 *Schizosaccharomyces pombe*: symbols for genes[a]

Feature	Convention	Examples
Gene symbol	3 lowercase italic letters	*arg*
Gene locus	Gene symbol followed by an arabic numeral	*arg1, arg2*
Allele	Locus symbol followed by a hyphen and an arabic numeral or by a combination of letters and numerals; all characters set in italics. No distinction is shown between dominant and recessive alleles.	*arg1-230* *ade6-M210*
Complementation group	Locus symbol followed by an italic uppercase letter	*trp1A, trp1B*
Wild-type gene	Locus symbol followed by a superscript plus sign. When there is no confusion, wild-type genes may be designated simply as +.	$arg1^{+}$ $cdc19^{+}$
Gene conferring resistance or susceptibility	Allele symbol with italic superscript *r* or *s* at the end	$can1\text{-}1^{r}$
Gene conferring temperature sensitivity	Allele symbol with italic superscript *ts* (heat-sensitive) or *cs* (cold-sensitive) at the end	$cdc2\text{-}5^{ts}$
Phenotype	Same characters as the gene symbols, but not italicized and not capitalized. Superscript plus and minus signs indicate the wild type and the mutant.	arg^{+} arg^{-}
Mating type	Homothallic (wild-type) strains are designated h^{90}; heterothallic strains are h^{+} and h^{-}	
Mitochondrial and other non-Mendelian genes	The same rules apply for nuclear and non-Mendelian genes. Enclose the symbols in square brackets when necessary to distinguish them from chromosomal genes.	*ana1* or *[ana1]*
Plasmid designations	Nonitalic letters and numerals, beginning with lowercase p	pFL20
Transposons (also called "retrotransposons")	Nonitalic symbol Tf and a number	Tf1-107 Tf2
Open reading frames	When only an open reading frame with unassigned biological function is known, it may be named *orf* until a function is assigned	*orf42*
Ribosomal RNA	Symbols for ribosomal protein genes begin with *rpl* (ribosomal protein, large subunit) or *rps* (ribosomal protein, small subunit)	*rps6*
Gene products	Gene or allele designation in nonitalic characters followed by a lowercase p (for protein)	Cdc19p

[a] Based on Kohli[35]; Fantes and Kohli.[36]

21.11.2 *Schizosaccharomyces pombe*

1) Genes. Symbols for *Schizosaccharomyces pombe* genes are based on the proposals of Demerec et al.[3] Kohli's proposed adaptation explains the deviations in practice for bacteria and *Saccharomyces cerevisiae*.[35] Table 21.5 summarizes the major conventions. For more details and examples, see Fantes and Kohli.[36]
2) Chromosomes. Chromosomes of *S. pombe* are designated by roman numerals I, II, and III; the arms are labeled R (right) and L (left).

21.12 FILAMENTOUS FUNGI

21.12.1 Ascomycete Mating Type Genes

Nomenclature for mating type genes of filamentous ascomycetes, based on the type of protein encoded, has been proposed by Turgeon and Yoder.[37] Mating type alleles are designated *MAT1-1* (for the idiomorph encoding a protein with an α box motif) and *MAT1-2* (for the idiomorph encoding a protein with a high-mobility group motif). Each gene is indicated by the symbol followed by a hyphen and a number (e.g., *MAT1-2-1*). The number assigned to any particular gene within an idiomorph should correspond to that of its homolog in other fungi that have characterized *MAT* genes. Conventions for *Neurospora* and *Podospora* are specifically excluded from the proposal. For details and examples, see Turgeon and Yoder.[37]

21.12.2 *Aspergillus nidulans*

1) Genes. Table 21.6[38,39] summarizes the major conventions for *Aspergillus nidulans*. For more details and examples, see Clutterbuck.[40] The linkage map[41] and other genome data[42] are available online.
2) Chromosomes/linkage groups. The 8 chromosomes/mitotic linkage groups are numbered I to VIII in the order of their discovery.[43] Indicate chromosomal aberrations by a nonitalic symbol for the aberration (T for translocation, Inv for inversion), a serial number, and the relevant chromosome(s) in parentheses in numerical order separated by a semicolon. Indicate a mutation associated with an aberration by attaching its symbol with a hyphen.

   T1(III;VIII)   Inv2(I)-*areB405*

21.12.3 *Neurospora crassa*

1) Genes. The conventions for *Neurospora* genes[44] are similar to those for *Drosophila*, but they predate and differ significantly from those of many other organisms. Table 21.7 summarizes the major conventions. Consult the Fungal Genetics Stock Center[45] before assigning new names, symbols, locus numbers, or allele-number prefixes to avoid duplication.

   Oak Ridge strains OR23-1A and ORS-6a have been adopted as the standard wild-type strains by most laboratories.[44]
2) Chromosomes/linkage groups. Linkage groups (LGs) are designated by roman numerals I to VII,[44] chromosomes (which are more difficult to identify) by arabic numerals 1 to 7, and left and right arms by L and R. Aberrations are indicated by the italic symbols *Df*, *Dp*, *In*, *T*, and *Tp* for deficiency, duplication, inversion, translocation, and transposition on the same chromosome, respectively. Each description includes the symbol for the aberration, the linkage group(s) with arm designation, and an identification number. The entire symbol is italicized with no intervening spaces; semicolons are added as needed to separate linkage groups; arrows indicate the direction of action. If a rearrangement breakpoint is inseparable from the mutant phenotype of an associated gene, the gene symbol follows the rearrangement symbol and is separated from it by a space with no comma.

   T(IIIR;VR)P1226 [a translocation of the right arm of LGIII and the right arm of LGV]
   T(IL→IIR)39311 [a translocation of the left arm of LGI to the right arm of LGII]
   Dp(IL→IIR)39311 [a duplication of the left arm of LGI in the right arm of LGII]
   T(IR;IIR)4637 al-1 [a translocation between the right arms of LGI and LGII at the *al-1* locus]

Table 21.6 *Aspergillus nidulans*: symbols for genes and phenotypes[a]

Feature	Convention	Examples
Gene symbol	Italic 3-letter symbol for new mutants. Older symbols remain unchanged and may be longer or shorter.[b]	*pro* *panto*
Nonallelic loci with the same primary symbol	Gene symbol with added capital letter	*proA* *proB*
Allele	Locus symbol with added serial number	*proA1*, *proA2*, *proB3*
Undetermined allelic relationships of a mutant	Allele symbol in which a hyphen replaces the capital letter	*pro-99*
Wild-type allele	Locus symbol with superscript plus sign	$proA^{+}$
Dominant mutant	Described in written text, not as part of the symbol[c]	. . .
Multimarked strains	Allele symbols separated by spaces; linkage groups in the order I to VIII and separated by semicolons; diploids shown as fractions, preferably but not always one linkage group per fraction	*proA1 biA1* $\frac{proAI\ BiA1}{+\quad +}$; $\frac{+}{pyroA4}$ or *proA1/+ biA1/+*; *pyroA4/+*
Specific property of mutants	Optional, temporary superscript letter, not to be regarded as a permanent or essential part of the symbol	$amdR6^{r}$
Suppressor	Symbol *su* with locus letters and mutant number followed by the symbol of the mutant being suppressed. Alternatively, standard gene symbols may be used.	*suA1proA1* *snxA*
Phenotype	The unabbreviated word or phrase from which the gene symbol is derived is preferred. The nonitalic version of the gene symbol with an initial capital letter is an alternative.	requires proline Pro
Proteins	Nonitalic version of gene symbol with initial capital letter[d]	ProA
Mitochondrial gene symbol	Use same gene symbol as for a nuclear gene; enclosed in square brackets when written as part of a genotype.	*camA* *proA1 biA1 [camA]*
Transposon name	No specific system for *Aspergillus*[d]; standard fungal nomenclature calls for a symbol in italic letters and numbers with an initial capital[22]	*Afut1* *F2PO8*

[a] Based on Clutterbuck.[38]

[b] Where appropriate, use the symbols proposed by Demerec et al.[3] or Sermonti.[39] Pronounceable symbols are preferred. Older, unchanged symbols may be 1 to 5 italic letters; for example, *panto* for pantothenic acid requirement.

[c] If the author wishes to draw attention to the dominance of a mutant, it can be done either by a superscript or by an initial italic capital letter. It should be understood, however, that *AcrA1* and *acrA1* refer to the same mutation.

[d] Recommendation from A John Clutterbuck, University of Glasgow, personal communication, 2003 Aug 4.

Table 21.7 ***Neurospora crassa*****: symbols for genes and phenotypes**[a]

Feature	Convention	Examples
Gene name	Name of mutant phenotype in italics	*arginine*
Gene symbol	1 to 4 italic letters or arabic numerals, with 3-letter symbols preferred; all lowercase unless the name is based on a dominant allele	*inv* *Asm* *tom22*
Different loci with similar phenotypes	Gene symbol followed by a hyphen and an arabic numeral. When the name applies to only one locus, the number 1 is optional.	*arg-1, arg-2, arg-3* *hsp70-1, hsp70-2* *inv* or *inv-1*
Suppressors	Symbol *su* followed by the symbol of the suppressed gene in parentheses and the locus number (if known); su^+ is the wild type	*su(met-7)* *su(met-7)-2*
Enhancers	Symbol *en* followed by the symbol of the enhanced gene in parentheses and the locus number	*en(am)-1*
Mating types	Either *mat A* or *mat a*; may be abbreviated *A* or *a* if the context is clear; usually printed at the end of a genotype	*leu-3 cr-1 al-2 A* *cot-1 a*
Transposons	Symbol in italic letters with initial capital	*Tad* *Pogo*
Recessive mutant alleles	All lowercase italic letters for name and symbol	*albino, al*
Dominant mutant allele	Italic letters with initial capital for name and symbol	*Banana, Ban*
Codominant alleles	All lowercase italic letters for name and symbol	*heterokaryon incompatibility, het*
Anonymous genes	Symbol *anon* followed by a unique isolation designation in parentheses	*anon(NP6C9)*
Ectopic genes	Gene symbol followed by *(EC)*	$am^+(EC)$
Gene fusions	Double colon between the symbols of the genes being fused	*am::Tad* $mtr::Asm\text{-}1^+(EC)$
Genotypes	Gene symbols in linear order for each linkage group separated by spaces; linkage groups separated by semicolons and spaces. Diploids are shown as fractions, each fraction representing one linkage group.	*cr-1 al-2; am inl; nic-3* $\frac{al\text{-}2\ \ arg\text{-}6}{al\text{-}2^+\ \ +}\ ;\ \frac{am\ \ inl}{am\ \ inl}$
Wild-type allele	Locus symbol with superscript plus sign	Bml^+
Mutant allele	Locus symbol without a superscript	*Bml*
Multiple alleles at one locus	Locus symbol with italic superscript serial numbers; use italic letters if indicating resistance or sensitivity to a toxic agent. Use square brackets when superscripts are not available.	frq^1, frq^2 $cyh\text{-}1^R$, $cyh\text{-}1^S$ *frq[1], frq[2], frq[3]* *cyh-1[R], cyh-1[S]*
Deletion of part of a gene	Treat as an allele at the gene locus	
Deletion of an entire gene	Precede the gene symbol with an uppercase Greek delta	*Δam*

Table 21.7 *Neurospora crassa*: **symbols for genes and phenotypes**[a] ***(continued)***

Feature	Convention	Examples
Pseudogenes	Gene symbol followed by superscript *ps*. Use square brackets when superscripts are not available.	*Fsr63*^{ps} *Fsr63[ps]*
Isolation numbers (allele numbers)	Gene or locus symbol followed by a nonitalic laboratory code in parentheses; used only when necessary to distinguish between alleles	*pyr-3*(KS43)
Heterokaryon	Separate symbols for component nuclei with a plus sign; enclose the whole genotype in parentheses.	*(col-2 A + ad-3B cyh1 A)*
Phenotype	Gene symbol without italics, with an initial capital letter, and with superscript plus, minus, or other allele designation	Al^- Arg^+ Cpl^S
Gene product	Symbols for gene products are the same characters as the gene symbol, but in all capital letters and without italics. When the name is spelled out, capital letters are not necessary.	INV, invertase
Centromeres, telomeres	Symbol *Cen-* or *Tel-* followed by a roman numeral for the linkage group; all characters in italics	*Cen-III* *Tel-VIR*
Mitochondrial genes	Lowercase letters and arabic numerals in italics, without hyphen or space	*cox* *atp6*
Mutant mitochondrial genomes	Italicized names or symbols enclosed in nonitalicized square brackets	[*mi-2*] [*poky*] [*stp*]

[a] Based on Perkins.[44] (Also available from http://www.fgsc.net/fgn46/perkins.htm.)

21.12.4 *Phycomyces* spp.

1) Genes. The genetic nomenclature of *Phycomyces*[46] follows the recommendations of Demerec et al.[3] Genes are named with 3 lowercase italic letters that refer to the gene's general function, followed without interruption by an uppercase letter to specify the gene. Mutant alleles are designated with the name of the gene and a number; when the specific gene is unknown, the uppercase letter is replaced by a hyphen. The genotype is a list of alleles that differ from the wild type. Mating types are indicated by plus and minus signs within parentheses. Heterokaryons are indicated by using the names or genotypes of the components separated by an asterisk.

   car [genes related to carotene]
   carB [gene for phytoene dehydrogenase]
   carB10 (−) [one of the alleles of *carB*, mating type minus]
   C115*NRRL1555 [heterokaryon of strains C115 and NRRL1555]

2) Chromosomes/linkage groups. *Phycomyces* has at least 11 linkage groups designated by roman numerals I to XI; the total number of linkage groups and chromosomes is unknown.[47] NRRL1555 is the standard wild-type strain for genetic studies.

21.12.5 Plant-Pathogenic Fungi

1) Genes. To standardize genetic nomenclature and symbolization for plant-pathogenic fungi, the Genetics Committee of the American Phytopathological Society prepared recommendations[48] based mainly on the conventions for yeast current at that time (i.e., Sherman[32]). Gene loci are designated by unique 3-letter italic symbols based on the name of the mutant phenotype they represent. The symbol for the locus itself begins with an uppercase letter and is followed by 2 lowercase letters. A dominant allele is indicated by 3 uppercase letters, a recessive allele by 3 lowercase letters. Wild-type alleles can be identified by a plus sign following the locus number, mutant alleles by a minus sign. Mutations at different loci yielding similar phenotypes can be identified with identical letter symbols followed by a number unique to the locus. Resistance and sensitivity associated with a particular allele are indicated by adding italic *R* or *S*, respectively, to the gene symbol.

   Met [for methionine auxotrophy]
   MET [dominant allele]
   met [recessive allele]
   Met1, *Met2*, *Met3* [different loci with similar phenotypes]
   Met+ [wild type, methionine-independent growth]
   Cyh1S [locus 1 confers sensitivity to cyclohexamide]

   Mating type alleles are designated *MAT1-1* and *MAT1-2*. All letters in the symbols for both alleles are uppercase because both alleles are needed for activity.
   Symbols for cytoplasmically inherited genes follow the same rules as for nuclear genes. They are enclosed in brackets when it is necessary to distinguish them from nuclear genes, as in the genotype of a strain.
   Similar conventions are applied for phenotypes, but the symbols are in roman type.

   Met+ [methionine-independent growth]   Met− [methionine auxotrophy]

2) Chromosomes/linkage groups. Linkage groups are indicated by roman numerals in the order in which they are recognized or, preferably, by chromosome size (largest to smallest) as determined by gel electrophoresis. Linked genes in a genotype are separated by spaces, unlinked genes by semicolons. See Yoder et al.[48] and Turgeon and Yoder[37] for more details.

21.13 PROTISTS

21.13.1 *Chlamydomonas reinhardtii*

1) Genes. Table 21.8[49,50] summarizes the major conventions for *Chlamydomonas reinhardtii*.
2) Chromosomes. The 17 chromosomes of *Chlamydomonas* are designated by roman numerals I to XIX (XII/XIII and XVI/XVII have been consolidated). Detailed gene maps have been published.[51]

21.13.2 *Dictyostelium discoideum*

1) Genes. Gene names and symbols for *Dictyostelium discoideum* follow the rules proposed by Demerec et al.[3] for bacterial genes (summarized in Table 21.3). Kay et al.[52] summarize the rules as adapted for *Dictyostelium*. The basic symbol is 3 lowercase italic letters, followed by uppercase letters to indicate different loci for the same phenotype, and then by arabic numerals to indicate different alleles at the same locus. Until the exact locus is known, the uppercase letter is replaced by a hyphen, but the isolation number is

retained. Wild-type and mutant alleles can be distinguished by superscript plus and minus signs. There is no established system for distinguishing between allele types.

acpA, *acpB* *aggA2* *aga*$^+$, *aga*$^-$

2) Chromosomes. *Dictyostelium discoideum* has 6 chromosomes, designated by arabic numerals 1 to 6.

21.13.3 *Leishmania* spp. and *Trypanosoma* spp.

Gene nomenclature for *Leishmania* spp. and *Trypanosoma* spp., based on the rules for *Saccharomyces cerevisiae*, was proposed by Clayton et al.[53] in 1998. Table 21.9 summarizes the major conventions. Genes are assumed to be wild type unless a mutation has caused a significant loss of function.

Table 21.8 *Chlamydomonas reinhardtii*: symbols for genes and phenotypes[a]

Feature	Convention	Examples
Gene name	Name of mutant phenotype, not in italics	thiostrepton resistant
Nuclear gene symbol	3 uppercase italic letters based on gene name; no hyphens, superscripts, or subscripts	*TSB*
Different loci with similar phenotypes	Gene symbol with added arabic numerals	*TSB1* *TUB2*
Wild-type allele	Same as the locus symbol; no plus sign	*TSB1*
Mutant allele	Locus symbol in lowercase italics followed by a hyphen and an arabic numeral	*maa1-1*
Allele made by insertion of transforming DNA	Symbol for the initial allele followed by a double colon and the symbol for the introduced DNA	*abc1-1::NIT1*
Allele made by insertion of a transposon	Symbol for the initial allele followed by a double colon and the symbol for the transposon	*abc1-1::Tcr1*
Gene product	Gene symbol without italics	TSB1
Phenotype	Gene symbol without italics, only the initial letter capitalized	Arg
Chloroplast gene	3 lowercase italic letters followed by an arabic numeral or an uppercase letter	*psaA* *rps7*
Mitochondrial gene	3 lowercase italic letters followed by an arabic numeral	*atp1*
Mating type locus	Mating types are designated MT on maps and *mt+* or *mt−* in genotypes. (This is an exception to the uppercase rule for gene names.)	MT *mt+* *mt−*
Genotypes		
Linked genes	Locus symbols separated by spaces	*ac17 nit2 tua1*
Unlinked genes	Locus symbols separated by semicolons and spaces	*arg7; ac17; tar1*
Diploids	Symbols for linked genes followed by a slash and the symbols for the homologous genes; unlinked sets separated by semicolons	*ac17 NIT2/AC17 nit2* *tar1/ TAR1; ac17/AC17*
Transposons	If the name can be pronounced as a word, set it in italics. If the name begins with the symbol Tcr, set it in roman when it stands alone, in italics when it is part of an allele name.	*Gulliver* *Pioneer* *TOC* Tcr1, Tcr2, Tcr3 *abc1-1::Tcr1*

[a] Based on Dutcher and Harris[50]; Chlamy DB[49] (http://www.chlamy.org/).

Table 21.9 ***Leishmania*** **spp. and** ***Trypanosoma*** **spp.: symbols for genes and phenotypes**[a]

Feature	Convention	Examples
Wild-type gene name	3 to 6 italic, uppercase letters; no hyphens	*ABC* *DHFRTS*
Mutant gene	Wild-type gene name in lowercase, even when the mutant gene is dominant	*abc* *dhfrts-3*
Members of a gene family	Gene name with added numbers or letters	*ABC1*, *ABC2* or *ABCA*, *ABCB*
Different alleles of the same gene	Gene name followed by hyphenated numbers	*ABC-1*, *ABC-2*, *ABC-3*
Deletions	Gene name preceded by uppercase Greek delta	*Δdhfrts*
Replacements	Uppercase Greek delta followed by the name of the gene being replaced, a double colon,and the gene being inserted	*Δdhfrts::NEO*
Fused genes	Gene names in 5′→3′ order; double colon between the names	*DHFRTS::GFP*
Insertional inactivation	Caret (^) followed by the name of the gene being inactivated, a double colon, and the name of gene being inserted	*^dhfrts::NEO*
Genes from different taxa	Gene name with a prefix consisting of a code for the taxon. Use only when needed to compare the genes of different taxa.	*LmjDHFRTS* (for *L. major*), *LmxDHFRTS* (for *L. mexicana*)
Diploids, triploids	Gene names for homologous genes separated by a slash	*dhfrts-3/DHFRTS Δdhfrts/ Δdhfrts/DHFRTS*
Extrachromosomal elements	Names and symbols in square brackets, all in italics. A high copy number may be added as a subscript.	*[pX HYG GFP]*$_{85}$
Wild-type protein	Gene name without italics. A species prefix, if used, should be in italics.	ABC DHFRTS *Lmj*DHFRTS
Mutant protein	Gene name without italics, with initial capital letter	Dhfrts-3

[a] Based on Clayton et al.[53]

21.13.4 *Paramecium* spp.

1) Genes. Table 21.10 summarizes the major conventions for *Paramecium* spp. The strain(s) from which genes are derived should be identified for each gene by an alphanumeric designation within square brackets after the locus name 1.

   nd2-1[d4-2] [mutant ND2 locus from inbred strain d4-2]

   The macronucleus contains about 800 copies of each gene. If all copies of a particular gene in the macronucleus are the same, the gene name is listed only once in the genotype. If there are 2 or more alleles, they are separated by slashes. Such use of the slash does not indicate the number of copies of each allele in the macronucleus and is not meant to imply that the macronucleus is diploid.

2) Chromosomes. Each of the 2 micronuclei (i.e., the germline nuclei) of *Paramecium* has approximately 50 pairs of chromosomes and is, therefore, diploid. The single macronucleus (i.e., the somatic nucleus) is derived from the micronuclei during conjugation and is not diploid.

Table 21.10 *Paramecium* spp. and *Tetrahymena* spp.: symbols for genes and phenotypes[a]

Feature	Convention	Examples
Micronuclear Genes		
Gene symbol	3 italic letters; all uppercase for the wild type	*BTU*
Locus name	Gene symbol followed by an arabic numeral	*BTU2*
Induced mutants	Allele name in all lowercase with a hyphenated number	*chx1-1*
Insertions	Mutant allele name followed by a double colon and the name of the inserted element	*btu2-1::neo2*
Deletions	Mutant allele name followed by uppercase Greek delta and the range of amino acids deleted	*btu2-3Δ1-40*
Mutant alleles with known modifications	Mutant allele name followed by appropriate designations	*btu2-2A251K* [allele of *BTU2* in which alanine at position 251 is replaced by lysine]
Genotypes	Symbols for alleles separated by a slash; linked genes may be grouped together, separated by commas on the same side of a slash; symbols for unlinked genes separated by semicolons and spaces	*BTU2/BTU2*; *CHX1, EST1/CHX1, EST1*
Variations for *Paramecium* spp.		
Locus name	Gene symbol followed by 3-digit arabic numerals, the first digit indicating the gene family; set in italics	*TUB101*, *TUB102* [β-tubulin loci in the same family] *TUB201* [β-tubulin locus in a different family]
Strain of origin	Allele names followed by the designation of the strain of origin in square brackets	*CHX[d4-2]*
Variations for *Tetrahymena* spp.		
Strain of origin	Assumed to be strain B unless otherwise indicated; allele names from other strains followed by the designation of the strain of origin in square brackets	*CHX* [no strain designation needed if strain B was used] *CHX1[C3]*
Randomly applied polymorphic DNAs (RAPDs)	Symbols in italics, including a numeric code for the laboratory of origin, 2 initials of the person who discovered the polymorphism, and a sequential number [see Allen et al.[54] for additional details and examples]	*1JB11*, *1JB12*

Table 21.10 *Paramecium* **spp. and** *Tetrahymena* **spp.: symbols for genes and phenotypes**[a] **(*continued*)**

Feature	Convention	Examples
Macronuclear Genes and Phenotypes		
Placement within genotype	Designations in italics enclosed within a single pair of nonitalic parentheses after the micronuclear genotype	*BTU2/BTU2*; *CHX1*, *EST1/CHX1*, *EST1* (*CHX1/chx1-1*)
Gene names	Symbols conform with those of micronuclear genes. In *Paramecium* an O or E is used to indicate macronuclear mating type alleles.	*CHX* *chx1-1[C3]* *MPR1* *tam6-1[d4-2]*
Genotypes	If more than one allele is present for a gene, symbols for alleles separated by a slash; linked genes may be grouped together, separated by commas on the same side of a slash; symbols for unlinked genes separated by semicolons and spaces. If all copies of the gene are the same, only one allele need be indicated.	(*CHX1/chx1-1*) (*CHX1*, *PJB1/chx1-1*, *PJB1*; *MPR1*)
Phenotypes	2- or 3-letter, nonitalic designations separated by commas; drug phenotypes followed by hyphenated s or r for sensitive or resistant; listed after the genotype	(exo$^-$) (exo$^+$) (cy-s, pm-r)
Mating types	In *Tetrahymena*, mating types in uppercase roman numerals; listed after the genotype and/or phenotype. In *Paramecium*, mating types are either o (odd) or e (even); listed in roman (not italic) type after the genotype and/or phenotype.	(*CHX1*, *PJB1/chx1-1*, *PJB1*; *MPR1/MPR1*; cy-r, mp-s, IV) (*tam6-1[d4-2]*; *O*; exo$^-$, o)

[a] Based on Allen et al.[54] and notes from Marsha I Altschuler, Williams College, personal communications, 2003 Aug 26 and Oct 10.

21.13.5 *Tetrahymena* spp.

1) Genes. Table 21.10 summarizes the major conventions for *Tetrahymena* spp. Unless otherwise indicated, genes are assumed to be from inbred strain B, the reference strain for *Tetrahymena thermophila*. If any other strain is used, identify the strain from which genes are derived for each gene by an alphanumeric designation within square brackets after the locus name.[54]

   CHX1[C3] [wild-type *CHX1* locus from inbred strain C3]

   The macronucleus contains about 50 copies of each gene. If all copies of a particular gene in the macronucleus are the same, list the gene name only once in the genotype. If there are 2 or more alleles, separate them by slashes. Such use of the slash does not indicate the number of copies of each allele in the macronucleus and is not meant to imply that the macronucleus is diploid.

2) Chromosomes. The single micronucleus (i.e., the germ-line nucleus) of *Tetrahymena* has 5 pairs of chromosomes and is, therefore, diploid. The single macronucleus (i.e., the somatic nucleus) is derived from the micronuclei during conjugation[55] and is not diploid.

21.14 PLANTS

The following guidelines are based on published rules for genes and chromosomes. Most are adaptations of the rules proposed in 1957 by the International Committee on Genetic Symbols and Nomenclature[2]; others are based on gene families. Tables 21.11[55–60] and 21.12[57,61–63] summarize and index the following sections.

21.14.1 Gene Families

The Commission on Plant Gene Nomenclature (CPGN) developed a common nomenclature for sequenced plant genes on the basic principle that all plant genes that encode similar products and have similar coding sequences are members of the same gene family and will be assigned the same gene family number and gene family name.[9] The CPGN recommendations were used to build the Mendel database (now UKCropNet),[64] which authors should consult before publishing new gene names.

Gene family names of nuclear, endomorph, and viral genes are in the form *XyzN*. Sets of genes that encode products with similar functions but whose sequences contain distinct motifs are numbered in the form *Xyz1*, *Xyz2*, *Xyz3*, and so on. Organelle gene names begin with lowercase letters: *xyzA*, *xyzB*, *xyzC*, and so on. The *Xyz* or *xyz* portion of the gene family name is usually a mnemonic that reflects the function of the gene product, is no longer than 8 characters, and is printed in italics. Consult the Mendel website[64] for details on each group or family.

The plant species of origin is identified by a 5-letter abbreviation (issued by Swiss-Prot) that is used as a prefix to—but not part of—the gene name.

ARAth;Chia2 [*Arabidopsis thaliana* chitinase, PR-3, class II]

Members of multigene families within a species are identified by individual numbers that are assigned by the relevant working group.

ARAth;Chia2;2 [the second member of the family encoding *A. thaliana* chitinase, PR-3, class II]

Designations for alleles follow the conventions of the relevant plant species. Mutant alleles are usually represented in lowercase and are separated by a hyphen from the gene name.

ZEAma;Adh1;1 [wild-type gene in maize for alcohol dehydrogenase]
ZEAma;Adh1;1-Fm335 [*Fm335* allele of the wild-type gene for maize alcohol dehydrogenase]

21.14.2 Cereals

21.14.2.1 Barley

1) Genes. Rules for names of barley genes were outlined by Franckowiak et al.[65] Gene names should be as descriptive as possible of the phenotype. Names and symbols of dominant genes begin with uppercase letters, those of recessive genes with lowercase letters. A barley gene symbol consists of 3 letters that designate the character, a number

Table 21.11 Plants: gene nomenclature

Plant	Section or table[a]	Swiss–Prot ID code[b]	Model	Reference(s)
Alfalfa, *Medicago sativa*	Section 21.14.5.1	MEDSA	CPGN[c]	Price and Reardon[9]
Arabidopsis, *Arabidopsis thaliana*	Section 21.14.3, Table 21.17	ARATH	. . .	Meinke and Koornneef[57]
Barley, *Hordeum vulgare*	Section 21.14.2.1	HORVU	. . .	Franckowiak et al.[65]
Cotton, *Gossypium* spp.	Section 21.14.7.1	[varies with species][b]	ICGSN[c]	Kohel[92]
Crucifers, *Brassica* spp.	Section 21.14.3	. . .	*Arabidopsis*	Crucifer Genetics Cooperative[78]
Cucumber, *Cucumis sativus*	Section 21.14.4, Table 21.18	CUCSA	cucurbits	Cucurbit Genetics Cooperative[80]
Cucurbits: *Citrullus*, *Cucumis*, and *Cucurbita* spp.	Section 21.14.4, Table 21.18	[varies with species][b]	ICGSN[c] and tomato	Cucurbit Genetics Cooperative[80]
Lettuce, *Lactuca sativa*	Section 21.14.7.2	LACSA	ICGSN[c]	Robinson et al.[104]
Maize, *Zea mays*	Section 21.14.2.2, Table 21.13	MAIZE	CPGN[c]	A standard for maize genetics nomenclature[67]
Melon, *Cucumis melo*	Section 21.14.4, Table 21.18	CUCME	cucurbits	Cucurbit Genetics Cooperative[80]
Oat, *Avena* spp.	Table 21.14	AVESA	ICGSN[c]	Simons et al.[59]; Marshall and Shaner[69]
Pea, *Pisum* spp.	Section 21.14.5.2, Table 21.19	[varies with species][b]	. . .	Ellis and Ambrose[85]
Pepper, *Capsicum* spp.	Section 21.14.6.1	[varies with species][b]	. . .	CENL[89]
Pumpkin, *Cucurbita* spp.	Section 21.14.4, Table 21.18	[varies with species][b]	cucurbits	Cucurbit Genetics Cooperative[80]
Rice, *Oryza sativa*	Section 21.14.2.4, Table 21.15	ORYSA	. . .	Kinoshita[70]
Rye, *Secale cereale*	Section 21.14.2.5	SECCE	wheat	Sybenga[74]
Soybean, *Glycine* spp.	Section 21.14.5.3, Table 21.20	[varies with species][b]	. . .	Soybean Genetics Committee[87]
Squash, *Cucurbita* spp.	Section 21.14.4, Table 21.18	[varies with species][b]	cucurbits	Cucurbit Genetics Cooperative[80]
Tomato, *Lycopersicon* spp.	Section 21.14.6.2, Table 21.21	[varies with species][b]	. . .	Tomato Genetics Cooperative[90]; Livingstone et al.[56]
Watermelon, *Citrullus lanatus*	Section 21.14.4, Table 21.18	CITLA	cucurbits	Cucurbit Genetics Cooperative[80]
Wheat, *Triticum aestivum*	Table 21.16	WHEAT	. . .	McIntosh et al.[75]; Raupp et al.[58]; GrainGenes[55]

[a] Entries refer to the section or table for the model organism when there are no separate rules for the plant named in the first column.

[b] When the common name of the organism in English is 5 characters or fewer, the Swiss–Prot identification code is usually the common name (e.g., MAIZE). When the common name is longer, the prefix is usually the first 3 letters of the genus name and the first 2 letters of the epithet (e.g., ARATH for *Arabidopsis thaliana*). See the ExPASy Proteomics Server (http://www.dsimb.inserm.fr/~fuchs/M2BI/AnalSeq/Annexes/Serveurs_db/ExPASy%20Proteomics%20Server.htm) for details and the list of organisms.

[c] Commission on Plant Gene Nomenclature (CPGN); International Committee on Genetic Symbols and Nomenclature[2] (ICGSN).

Table 21.12 Plants: cytogenetic nomenclature

Plant	Section	Chromosome (chr) or linkage group (LG)[a]	Reference(s)
Alfalfa, *Medicago sativa*	Section 21.14.5.1	*n* = 8; autotetraploid with 4 nearly identical chromosomes	Bauchan and Hossain[83]
Arabidopsis, *Arabidopsis thaliana*	Section 21.14.3	*n* = 5	Fransz et al.[79]; Meinke et al.[61]
Barley, *Hordeum* spp.	Section 21.14.2.1	1H to 7H; *n* = 7	Linde-Laursen[66]
Cotton, *Gossypium* spp.	Section 21.14.7.1	A1 to A13 and D1 to D13; *n* = 13; tetraploid	Kohel[92]
Cucumber, *Cucumis sativus*	Section 21.14.4	*n* = 7	Cucurbit Genetics Cooperative[80]
Cucurbits: see cucumber, melon, pumpkin, squash, and watermelon in this list for basic numbers	Section 21.14.4	[varies with species]	Cucurbit Genetics Cooperative[80]
Lettuce, *Lactuca sativa*	. . .	*n* = 9	Robinson et al.[104]
Maize, *Zea mays*	Section 21.14.2.2	*n* = 10	A standard for maize genetics nomenclature[67]
Melon, *Cucumis melo*	Section 21.14.4	*n* = 12	Cucurbit Genetics Cooperative[80]
Oat, *Avena* spp.	. . .	*n* = 7; genomes A, B, C, and D	Thomas[63]
Onion, *Allium cepa*	Section 21.14.7.3	1C to 8C; *n* = 8	Kalkman[106]; de Vries[105]
Pea, *Pisum* spp.	Section 21.14.5.2	chr 1 to chr 7; LGI to LGVII; *n* = 7[b]	Ellis and Ambrose[85]
Pepper, *Capsicum* spp.	. . .	*n* = 12	Livingstone et al.[56]
Potato, *Solanum tuberosum*	. . .	*n* = 12	Livingstone et al.[56]
Pumpkin, *Cucurbita* spp.	Section 21.14.4	*n* = 20	Cucurbit Genetics Cooperative[80]
Rice, *Oryza sativa*	Section 21.14.2.4	*n* = 12	Kinoshita[70]
Rye, *Secale cereale*	Section 21.14.2.5	1R to 7R; *n* = 7	Sybenga[74]
Soybean, *Glycine* spp.	Section 21.14.5.3	*n* = 20	Soybean Genetics Committee[87]
Squash, *Cucurbita* spp.	Section 21.14.4	*n* = 20	Cucurbit Genetics Cooperative[80]
Tomato, *Lycopersicon* spp.	Section 21.14.6.2	*n* = 12	Tomato Genetics Cooperative[90]; Livingstone et al.[56]
Watermelon, *Citrullus lanatus*	Section 21.14.4	*n* = 11	Cucurbit Genetics Cooperative[80]
Wheat, *Triticum* spp.	Section 21.14.2.6	1A to 7D; *n* = 7; genomes A, B, and D	McIntosh et al.[75]

[a] Chromosomes and linkage groups are designated by sequential arabic numerals unless specified otherwise.

[b] The numbers of the chromosomes do not correspond to those of the linkage groups.

that represents a particular locus, and 1 or more lowercase letters that represent a particular allele or mutational event at that locus. A period separates the locus from the allele symbol. The entire symbol is written on the line in italics. Wild-type alleles are not designated with a plus sign as they are in many other species.

Mdh1 [a malate dehydrogenase locus]   *Mdh1.a*, *Mdh1.b* [alleles of the *Mdh1* locus]

Inhibitors, suppressors, and enhancers are designated by *I*, *Su*, and *En*, respectively, if they are dominant (by *i*, *su*, and *en* if they are recessive), followed by a hyphen and the symbol of the allele affected.

Genic formulas are written as fractions with the maternal alleles given first or above. Each fraction corresponds to a single linkage group, and the groups are listed in numerical order separated by semicolons. In euploids and aneuploids, the gene symbols are repeated as many times as there are homologous loci. Symbols for extrachromosomal factors are placed within square brackets before the genic formula.

2) Chromosomes. Barley chromosomes are designated 1H to 7H in accord with those of other Triticeae species, the short and long arms by S and L without a space after the chromosome number.[66]

2HS   6HL

Indicate chromosomal aberrations with the nonitalic symbols Df, Dp, In, T, and Tp for deficiency, duplication, inversion, translocation, and transposition, respectively.

The genome of barley cultivar Betzes is the reference genome in the Triticeae to which definitions of translocations and short arm/long arm reversals are standardized in all species.[66]

21.14.2.2 Maize (Corn)

1) Genes. The definitions and standards for gene nomenclature and symbolization are published periodically in the *Maize Genetics Cooperation Newsletter* (*MNL*).[67] Table 21.13[68] summarizes the major conventions. A catalog of symbols for named genes is published annually in the *MNL*. No specific wild-type strain has been designated for maize; therefore, many naturally occurring variants are not properly termed "mutants".

Although the Nomenclature Subcommittee strongly recommends that all newly identified genes be given a 3-letter symbol, they also recommend that names and symbols that have been used in the past be retained, even though many have only 1 or 2 letters. Name newly detected genes previously known in other species with reference to the list of gene families compiled by the Commission on Plant Gene Nomenclature (see also Section 21.14.1). Give a newly identified locus a unique name and designate it as 1, even though there is no 2 yet.

2) Chromosomes/linkage groups. The basic complement of A maize chromosomes are designated by arabic numerals 1 to 10 in decreasing order by length, the short and long arms by S and L. Linkage groups are represented beginning with position zero at the distal end of the short arm, which is termed the "left" or "top" or "north" end on the linkage map. Unless otherwise specified, translocations and other aberrations are assumed to affect the A chromosomes. Designate reciprocal translocations by T, followed by the numbers of the rearranged chromosomes separated by a hyphen, followed by a letter or an isolate number, the latter within parentheses.

4S   9L   T1-2a   T1-2b   T1-9(4995)

The supernumerary, B chromosomes are highly variable, unnumbered, and devoid of structural genes. Designate translocations of A chromosomes with B chromosomes by TB followed by a hyphen, the number of the A chromosome and the arm translocated, and a letter or an isolate number. Designate inversions, deletions, or duplications by Inv,

Table 21.13 Maize (corn): symbols for genes and phenotypes[a]

Feature	Convention	Examples
Gene name	Lowercase italic characters	*defective kernel*
Gene symbol	3-letter italic symbols based on gene name are recommended, but older, shorter names are retained.	*dek*
Different loci with similar phenotypes	Gene symbol with added arabic numeral[b]	*dek12*
Dominant allele	Italicized locus symbol with initial capital letter	*Dek12* *Cg1*
Codominant alleles	Italicized locus symbol with initial capital letter followed by a hyphen and an arabic allele number	*Pgm2-5*, *Pgm2-7*
Recessive allele	Italicized locus symbol in all lowercase letters	*dek12* *cgl*
Reference allele	Locus symbol followed by *-Ref* or *-R*	*bz1-Ref* or *bz1-R*
Newly identified allele	Allele symbol followed by a hyphen and a laboratory number[c]	*sh2-6801*
Provisional symbol for new mutation at given phenotype	Symbol for known gene followed by an asterisk and the laboratory number[c]	*bt*-8711*
Gene product and phenotype	Symbol in all capital letters; name of product neither capitalized nor italicized	ADH1, alcohol dehydrogenase
Nonmutant alleles	Name of the inbred strain may be included as part of the allele designation.	*Bz1-W22*
Mutations as result of deletion	Symbol(s) for gene(s) deleted in parentheses preceded by def, indicating a deficiency; symbols for genes being deleted separated by double periods	*def(wx1)-C34* *def(an1..bz2)-6923* *def(bz1..sh1)-X2*
Mutation from transposable element insertion	Double colons or single apostrophe	*wx-m1::Ds1* *Bz1'-7801* *dnap2094::Ac*
Genotypes		
Homozygous unlinked genes	Semicolons separate gene symbols.	*a1*; *a2*; *c1*; *c2*; *r*
Homozygous linked genes	Spaces separate gene symbols.	*C1 sh1 bz1 Wx1*
Heterozygous unlinked genes	Slashes separate homologous genes; semicolons separate sets of unlinked gene symbols.	*Sh2/sh2*; *Bt2/bt2*
Heterozygous linked genes	Slashes separate sets of linked gene symbols.	*C1 Bz1/cl1 bz1*
Restriction fragment length polymorphisms (RFLPs) and random amplified polymorphic DNAs (RAPDs)	Lowercase 3- or 4-letter laboratory code followed by a laboratory number; cDNA or subclones of a gene in parentheses after the RFLP locus designation	*umc000(a1)*
Transposons	Names in italics; names of defective transposons begin with lowercase italic *d*	*Ac* *Ds* *dSpm(1)*

[a] Based on *A Standard for Maize Genetics Nomenclature*.[67]

[b] No hyphen separates the gene name from the numerical suffix; the hyphen is reserved for symbols for mutant alleles to separate the allele designation from a suffix specifying the particular allele.

[c] See MaizeGDB[68] for a list of laboratory codes in current use. The laboratory number could be a number indicating the date of identification; in the example shown, 6801 would represent January 1968.

Del, or Dup, respectively, followed by the chromosome number and a letter or an isolate number. For additional details, see *A Standard for Maize Genetics Nomenclature*.[67]

TB-1La TB-5Sc Inv1c Inv2(8865)

21.14.2.3 Oat

Tables 21.12 and 21.14[60] summarize the major conventions for oat. Proper display of oat gene symbols depends on approval by the Committee on Nomenclature and Cataloging of Oat Genes. Present the approved symbols in italics; present those not yet approved in roman type.[69]

21.14.2.4 Rice

1) Genes. Rules for symbolizing rice genes were described by Kinoshita[70] and updated by Kinoshita and Rothschild.[71] Table 21.15[72] summarizes the major conventions. The standard strain may be any strain selected by the researcher as long as it is named and its genetic formula is made explicit.
2) Chromosomes/linkage groups. Nomenclature for rice chromosomes and linkage groups has been described by Kinoshita and Rothschild.[70,71,73] Chromosomes of rice are numbered 1 (longest) through 12 (shortest) according to the length at the pachytene stage. The short arm of each chromosome is designated as the left arm, and the zero position is the distal end of the short arm.

   Represent aberrations and other structural changes by the italic symbols *Dp*, *In*, *T*, and *Tp* (for duplication, inversion, translocation, and transposition, respectively) and the nonitalic chromosome number. Distinguish similar aberrations involving the same chromosome with lowercase nonitalic letters following the chromosome number. Monosomics and trisomics are designated by the number of the affected chromosome.

   In(2)a *T*(1–2)b Mono-1 Triplo-2

21.14.2.5 Rye

1) Genes. The recommendations for symbolizing genes of rye are based on those for wheat as reported by Sybenga.[74] Symbols are usually 2 letters and correspond to the wheat symbols when there is no possibility for misunderstanding. When a proposed rye symbol might be confused with a wheat symbol, a different one can be used. Gene symbols, which are derived from the original gene names, are in italic letters. When unambiguous, the name and symbol of a dominant allele begin with an uppercase letter and those of a recessive allele with a lowercase letter. All letters and numerals in the symbol are on the same line.

   Two or more genes producing similar phenotypes are designated by the same basic symbol. Nonallelic loci (such as mimics and polymeric genes) are designated by arabic numerals added to the locus symbol or by a hyphen and the genome symbol.

   Inhibitors, suppressors, and enhancers are designated by *I*, *Su*, or *En*, respectively, if they are dominant (by *i*, *su*, or *en* if they are recessive), followed by a space and the symbol of the allele affected.
2) Chromosomes. Rules for rye chromosome nomenclature were published by Sybenga.[74] Chromosomes are designated 1R to 7R, their short and long arms by S and L. When the *Secale* species must be designated to indicate the derivation of a chromosome arm, a 3-letter superscript is added to the chromosome symbol (e.g., $2R^{cer}$ L represents the long arm of chromosome 2R of *Secale cereale*). The standard rye chromosome set are those of cultivar Imperial added to those of the wheat cultivar Chinese Spring.

Table 21.14 Oat: symbols for genes and chromosomes[a]

Feature	Convention	Examples
Gene symbol	Approved symbols in italic letters; unapproved symbols in nonitalic letters; initial capital letter. Except for genes for reactions to living organisms, the symbol stands for the dominant allele where only 2 alleles, 1 dominant, are known.	*Sg-1* Dw-7a
Dominant allele	Same as gene symbol	Cda-4
Recessive allele	Gene symbol with a lowercase letter	cda-1
Alleles	Gene symbol followed by an arabic numeral; the first locus is number 1, the second 2, and so forth; members of allelic series distinguished by lowercase letters after the locus number	*Tc-1* *Tc-1a* *Tc-1b* *Tc-2*
Reactions to living organisms	Allele symbol with uppercase initial letter; alleles ending in lowercase letter a indicate resistance, those ending in b indicate susceptibility	Tg-1a
Genotype		
Homozygous unlinked genes	Semicolons separate gene symbols; linkage groups in numerical order	*Cda-6*; *Cda-5*; *Cda-4* [representing unlinked genes on chromosomes 2, 15, and 21, respectively]
Homozygous linked genes	Spaces separate gene symbols	*cda-1 ma-1 L-1*
Heterozygous unlinked genes	Slashes separate homologous gene symbols, with the maternal gene written first; semicolons separate sets of unlinked gene symbols; linkage groups in numerical order	. . .
Heterozygous linked genes	Slashes separate sets of linked gene symbols, with the maternal genes written first	*A-1/Ba-1*
Unlocated genes	Enclosed in parentheses at the end of genic formulas	. . .
Euploids, aneuploids	Gene symbols repeated in genic formula as many times as there are homologous loci	. . .
Extrachromosomal factors	Enclosed in square brackets preceding the genic formula	. . .
Aberrations	Use abbreviations: Df (deficiency), Dp (duplication), In (inversion), T (translocation), and Tp (transposition)	. . .

[a] Based on Simons et al.[59]; Marshall and Shaner.[69]

21.14.2.6 Wheat

1) Genes. Table 21.16 summarizes the major conventions for wheat.
2) Chromosomes/linkage groups. Linkage groups and the corresponding chromosomes of wheat are designated by arabic numerals 1 to 7 followed by an uppercase Roman letter representing genome A, B, or D.[75] Short and long chromosome arms are designated by S and L. The designation of a particular chromosomal band requires 5 symbol elements[76]: chromosome number, genome designation, arm symbol, region number, and the band number within the region.

   1BL21 [chromosome 1, genome B, long arm, region 2, band 1]

Table 21.15 Rice: symbols for genes and phenotypes[a]

Feature	Convention	Examples
Gene name	Use an international language, preferably English, to describe the trait	Long kernel glutinous endosperm Chromogen
Gene symbol	Abbreviation of gene name in 2 or 3 italic characters; commonly used symbols should be retained even if they do not fit the rule	*Lk* *wx* *C*
Dominant alleles	Name and symbol begin with capital letter	Activator, *A* Acid phosphatase-4, *Acp4*
Recessive alleles	Name and symbol in all lowercase	alkali degeneration, *alk* glutinous endosperm, *wx*
Wild-type allele	Gene symbol with a superscript plus sign	al^{+}
Multiple alleles	Gene symbol with letter or number superscripts	$d18^{+}$, $d18^{k}$, $d18^{h}$ $Pgi1^{1}$, $Pgi1^{2}$
Nonalleles with similar phenotypes	Gene symbols differentiated by number or letter suffixes, which may be written on the line or as superscripts	*d1*, *d2*, *d3* D^{1}, d^{2}, d^{3}
Enhancers, inhibitors, modifiers, and suppressors	*En*, *I*, *M*, and *Su* (if dominant) or *en*, *i*, *m*, and *su* (if recessive), followed by a hyphen and the symbol of the gene affected	*En-Se1(t)* *M-Piz* *I-Pl-1* $Su\text{-}g^{1}$
Cytoplasmic genes	Names and symbols follow the same rules as nuclear genes	
Uncertain allelic relationships	Gene symbol with lowercase italic t in parentheses (for "tentative") added until the allelic relationship is determined	*d50(t)*
Quantitative trait loci (QTLs)	Lowercase q followed by a 2- to 5-letter name in all uppercase to describe the trait being measured, followed by a hyphen and the number of the chromosome to which the trait has been mapped, followed by another hyphen and a number indicating the individual QTL	qDTH-2-3 [3rd QTL for the days-to-heading trait reported on chromosome 2]

[a] Based on Kinoshita[70]; Kinoshita and Rothschild[71]; McCouch et al.[72]; Kinoshita.[73]

The cultivar Chinese Spring has the standard chromosome arrangement. Chromosomal aberrations are indicated by the nonitalic abbreviations Df, Dp, In, T, and Tp for deficiency, duplication, inversion, translocation, and transposition, respectively. For a gene not in the standard chromosome position, insert the new chromosome designation within parentheses following the gene designation.

Hp (Tp 6D) [the introgressed *hairy neck* gene on chromosome 6D instead of the standard chromosome 4A]

21.14.3 Crucifers, Including *Arabidopsis thaliana*

1) Genes, *Arabidopsis*. Table 21.17[58] summarizes the major conventions for *Arabidopsis thaliana*. When a locus has long been known by the mutant phenotype rather than by

Table 21.16 Wheat and related species: symbols for genes and phenotypes[a]

Feature	Convention	Examples
Gene symbol	Abbreviation of gene name in 2, 3, or 4 italic letters, all on the line; no superscripts or subscripts	*Pc* [purple culm] *co* [corroded] *Adh* [alcohol dehydrogenase]
Dominant alleles	Name and symbol begin with capital letter.	*A*, *Ce*
Recessive alleles	Name and symbol in all lowercase (except for reactions to pathogens)	*a*, *mlre*
Gene pairs with similar effects	Use same gene symbol.	. . .
Mimics, polymeric genes	Sequential arabic numerals added to the gene symbol; different alleles of independent origin indicated by lowercase letters following the locus designation	*Sr9* *Sr9a*
Orthologous sets of genes	Gene symbol followed by a hyphen, followed by the accepted genome symbol and the set number represented by an arabic numeral; different alleles of independent origin indicated by lowercase letter following the locus designation	*Adh-A1*, *Adh-B1* *Adh-A1a*
Temporary symbols	Gene symbol followed by abbreviation for the line and an arabic numeral referring to the gene	*SrFr1*, *SrFr2* [2 genes for reaction to *Puccinia graminis* in cultivar Federation]
Enhancers, inhibitors, and suppressors	*En*, *I*, and *Su* (if dominant) or *en*, *i*, and *su* (if recessive), followed by a space and the symbol of the allele affected	. . .
Genetic formulas	Written as fractions, with maternal alleles given first or on top; each fraction corresponds to a single linkage group	. . .
Extrachromosomal factors	Symbols enclosed in parentheses before the genetic formulas	. . .
Locus symbol	Consists of italicized 2-, 3-, or 4-letter abbreviation of the trivial name of the macromolecule affected; initial capital letter	*Adh*
Phenotypes	Same as symbol for the locus except in all capital letters, not italicized	ADH-1
Pathogen-reaction genes	All gene symbols begin with capital letter, even though they may behave as recessive alleles.	*Sr17*

[a] Based on McIntosh et al.[75] (http://wheat.pw.usda.gov/ggpages/wgc/2003/catalogue.pdf).

gene function, retain and use the original gene symbol even after the gene is cloned. When the mutant gene symbol is vague, misleading, or not widely known, petition the curator of gene symbols to change the symbol to one that does reflect the gene function. Further information is available from The *Arabidopsis* Information Resource, known as TAIR.[77]

2) Genes, other crucifers. The Crucifer Genetics Cooperative has adopted the guidelines used by the *Arabidopsis* research community for the designation of genotypes and phenotypes of *Brassica* spp.[78] Symbols for phenotypic traits may be followed by a number

Table 21.17 ***Arabidopsis thaliana*** **and other crucifers: symbols for genes and phenotypes**[a]

Feature	Convention	Examples
Gene name	Wild type in uppercase italics; mutant allele in lowercase italics	*ALPHABETICA* *alphabetica*
Gene symbols	Wild type in 3 uppercase italic letters; mutant alleles in lowercase italics; well-known symbols may have only 2 letters	*ABC*, *abc*
Different loci with the same symbol	Gene symbol followed by an arabic numeral	*abc1*, *abc2*
Different alleles of the same locus	Gene symbol followed by a hyphen and an arabic numeral.	*abc4-1*, *abc4-2*
	If only one allele is known, the hyphen and number are not needed.	*abc3* = *abc3-1*
	Add an italic D to the allele number to indicate that the allele shows dominance relative to the wild type.	*abc5-2D* [allele 2 is dominant to the wild type]
Phenotype	Gene symbol in roman type; initial capital letter; superscript plus sign for wild type, minus sign for mutant	Abc^{+} [wild type] Abc^{-} [mutant]
Protein product of gene	Gene symbol in all uppercase letters, no italics	ABC

[a] Based on Meinke and Koornneef[57]; Crucifer Genetics Cooperative.[78]

from 0 to 9 in parentheses to indicate the level of expression, with 0 indicating trait absent and 9 indicating very high expression.

Ant(0) [no expression of anthocyanin pigment]
Ant(8) [high expression of the anthocyanin pigment]

3) Chromosomes/linkage groups. At the pachytene stage, *Arabidopsis* has 5 bivalent chromosomes that conform to known linkage groups and are designated by arabic numerals 1 to 5.[79] The ideogram is not in descending order by length as they are in most other organisms.

21.14.4 Cucurbits

1) Genes. The rules for cucurbit gene names and symbols developed by the Cucurbit Genetics Cooperative (CGC) Gene List Committee are published in each CGC Report.[80] Table 21.18 summarizes the major conventions. The diversity of cultivars precludes the selection of one of them to represent the norm, so each researcher must distinguish between the normal and mutant alleles of the genes being named and choose appropriate gene and allele symbols.[81]
2) Chromosomes. The chromosomes of cucurbits are small and not easily differentiated, so their morphology is not well characterized; see Table 21.12 for haploid numbers of each group.

Table 21.18 Cucurbits[a]: symbols for genes and phenotypes[b]

Feature	Convention	Examples
Gene name	Short description in italics of the characteristic feature of the mutant	*bitter* *cream*
Gene symbol	Minimum number of italic letters, the first letter being the same as that of the name of the gene. The symbol applies to allelic genes in compatible species.	*Bi* *cr*
Dominant alleles	Gene symbol with initial capital letter	*Bi*
Recessive alleles	Gene symbol with all lowercase letters	*cr*
Wild-type normal allele	Gene symbol with initial capital letter; alternatively, plus sign may be used, or the symbol for the mutant gene followed by a superscript plus.	cr^{+}
Multiple alleles	Gene symbol followed by a distinguishing superscript roman letter or arabic numeral	yg^{w}
Mimics, polymeric genes	May have distinctive names and symbols or be assigned the same gene symbol followed by a hyphen and an arabic numeral or roman letter. The suffix *-1* is used, or may be understood and not used, for the original gene in the series.	*ms-1* *ms-2*
Indistinguishable alleles	Identical phenotypes should have identical symbols. If assigning distinctive symbols to recurrences of the same mutation, assign the same symbol with distinguishing numbers or letters in parentheses as superscripts.	$ms\text{-}2^{(PS)}$
Intensifiers, inhibitors, and suppressors	Gene symbol followed by a hyphen and the symbol of the allele affected. Alternatively, assign a distinctive name unaccompanied by the symbol of the gene affected.	*In-F* *Hi* *bi* [*bitterfree* suppresses *Bt* gene for bitterness]

[a] Cucurbits include cucumber (*Cucumis sativus*), melon (*Cucumis melo*), pumpkin (*Cucurbita* spp.), squash (*Cucurbita* spp.), watermelon (*Citrullus lanatus*), and other genera of the Cucurbitaceae.
[b] Based on Cucurbit Genetics Cooperative Gene List Committee.[80]

21.14.5 Legumes

21.14.5.1 Alfalfa

1) Genes. In the absence of a coordinated effort to standardize gene names and symbols for alfalfa, authors are encouraged to use the guidelines developed by the Commission on Plant Gene Nomenclature.[9]
2) Chromosomes. Cultivated alfalfa, *Medicago sativa*, has purple flowers and is tetraploid ($2n = 4x = 32$). There are diploid progenitors ($2n = 2x = 16$) to alfalfa, *M. sativa* ssp. *caerulea* (with purple flowers) and *M. sativa* ssp. *falcata* (with yellow flowers).[82] The standard karyotype of cultivated alfalfa shows 32 tiny chromosomes that are similar in length and morphology. C-banding studies have identified all of the chromosomes on the basis of their unique patterns. The chromosomes are paired in 8 sets, which are numbered 1 to 8, the first 7 in descending order by length, the eighth being the satellite chromosome.[83,84]

21.14.5.2 Pea

1) Genes. Table 21.19 summarizes the major conventions for pea. The rules presented here and by Ellis and Ambrose[85] have not been formalized or approved by the *Pisum*

Table 21.19 Pea: symbols for genes and phenotypes[a]

Feature	Convention	Examples
Gene name	Italic letters, initial uppercase letter for dominant traits, all lowercase for recessive traits	*Argentum* *ageotropum*
Gene symbol	Usually 3 italic letters derived from gene name; established symbols may have 1 to 4 letters	*Np* *Gty* *bot* *coch*
Dominant allele	Gene symbol with initial uppercase letter	*Np* *Gty*
Recessive allele	Gene symbol with all lowercase letters	*bot* *coch*
Multiple alleles	Gene symbol with superscript lowercase letters or arabic numerals. Alternatively, follow the gene symbol with a hyphen and the allele-specific designation.	Cry^{s} D^{co} *rug3-a* *lv-1*
Phenotype	Gene name or gene symbol, with or without italics	*unifoliata* or unifoliate
Different loci with similar phenotypes	Gene symbol followed by an arabic number	*rug3*, *rug4*
Isozyme loci	Italic abbreviation of the isozyme name with initial uppercase letter, followed by a hyphen and code for the cellular compartment when appropriate	*Mdh* *Aat-p* [*-p* indicates the plastid]
Genes of known function but no known phenotype	Gene symbol with initial uppercase letter followed by an arabic numeral when more than one locus is involved	*Rbcs1*, *Rbcs2*
Genes identified by homology with those in other organisms	Prefix *Ps* (for *Pisum sativum*) followed by the symbol for the homolog	*PsENOD40*
Pseudogenes	Uppercase Greek letter psi followed by the gene symbol	Ψ*LegD*

[a] Based on Ellis and Ambrose.[85]

Genetics Association (PGA). The PGA maintains the searchable PGene Pisum Gene List database.[86]

2) Chromosomes/linkage groups. Pea chromosomes are numbered 1 to 7; their short and long arms are designated S and L. The linkage groups (which do not correspond by number to the chromosomes) are designated by roman numerals I to VII. The standard karyotype is that of line JI4. Symbols for chromosome aberrations consist of a prefix for the type of aberration, the numbers of the chromosomes affected in parentheses, and a unique lowercase suffix.

   T(3L-7S)a [the first reported translocation between the long arm of chromosome 3 and the short arm of chromosome 7]

21.14.5.3 Soybean

1) Genes. Rules for gene symbols for soybean were updated by the Soybean Genetics Committee (SGC) in 1996 and, until 2010, were published in the *Soybean Genetics Newslet-*

ter.[87] The *Soybean Genetics Newsletter* ceased publication in 2010; online resources for soybean genetics researchers are now available at Soybase.org. Table 21.20 summarizes the major conventions. The standard strain may be any strain selected by the researcher as long as the strain being used and its genetic formula are made explicit. The distinction between traits that are to be symbolized with identical, similar, or unrelated base letters is necessarily not clear-cut. The decision for intermediate cases is at the discretion of the author but should agree with previous practice for the trait.

Table 21.20 Soybean: symbols for genes, alleles, and cytoplasmic factors[a]

Feature	Convention	Examples
Gene symbol	1 to 3 italic letters to which superscripts and/or subscripts may be appended	*P* [pubescence type] *L* [leaf shape]
Allele group base symbol	Each group of alleles has the same base letter(s) as a symbol.	. . .
Gene pairs with similar effects	Sets of duplicate, complementary, and polymeric genes have the same base letter symbol differentiated by numerical suffixes assigned in order of publication. Suffixes may be printed as subscripts or on the same line as the base. This is the only use of numerals.	Y_1, Y_2 . . . or *Y1*, *Y2* . . . *Rps1*, *Rps2*
Alleles	Of the first pair published, the dominant gene has an initial capital letter, the recessive has all lowercase. Subsequent alleles are designated by 1 or 2 lowercase superscript letters. This is the only use of superscripts. The entire symbol may be written on the same line if a hyphen is inserted.	*R*, *r* *R*, r^m, *r* *R*, *r-m*, *r* *Ap-a*, *Ap-b*, *Ap-c* *Rps1-b*, *Rps1-k*
Dominant alleles	Initial capital letter for dominant or partially dominant allele	*Ab* dominant over *ab* *Br1*, *br2*
Codominant alleles	Capitalized symbol assigned at author's discretion	. . .
Tentative alleles	Italicized gene symbol followed by a nonitalic identifying designation in parentheses	*ms1* (Tonica) *ms1* (Ames 2)
Any allele	An underscore in place of a gene symbol represents any allele at the indicated locus.	*A_* = *AA* or *Aa*
Unknown alleles	A question mark in place of a symbol, followed in parentheses by the name of the line in which the gene was identified	*Rps?* (Harosoy) [an allele at an unknown locus] *Ap-?* (T160) [an unknown allele at the *Ap* locus]
Standard strain	Plus sign in place of assigned gene symbols of a designated homozygous strain	. . .
Isozyme variants	3-letter gene symbol for the enzyme; include the EC name and number of the specific enzyme activity investigated in the article	*Adh* [alcohol dehydrogenase; EC 1.1.1.1]

Table 21.20 Soybean: symbols for genes, alleles, and cytoplasmic factors[a] (*continued*)

Feature	Convention	Examples
Genotypes	Precede the linked genes with the linkage group number and list the gene symbols in the order they occur on the chromosome. Separate linkage groups with a semicolon and space.	*5E3Dt1*; *15Pgm1ms2*
Probe-detected loci	Nonitalic locus designations consist of a prefix that indicates the origin of the probe, a hyphen, a string of letters or integers that identify the probe used, a code for the restriction endonuclease used, and (if needed) a hyphen followed by an integer to distinguish between loci detected by the same probe.	IaSU-B317I-1, IaSU-B317T-2 [2 loci detected at Iowa State University, the first digested with *Eco*RI (I), the second with *Taq*I (T)]
Random amplified polymorphic DNA (RAPD) loci	Nonitalic locus designations consist of an uppercase letter identifying the primer, a code for the primer name, and the fragment size in base pairs as a subscript.	$\mathrm{OA14}_{800}$ [an 800-bp fragment amplified with Operon Technologies primer14 from kit A]
Simple sequence repeat (SSR) loci	Nonitalic locus designations consist of a prefix that indicates the origin of the probe, a hyphen, and a string of letters that identify the core nucleotide repeat of the SSR followed by an identifying number. The code after the hyphen should not exceed 8 characters.	IaSU-at275 BARC-gata3412
Cytoplasmic factors	Symbol *cyt-* followed by 1 or more letters for the trait consistent with letters for nuclear gene traits	*cyt-G* [maternal green cotyledons] *cyt-Y* [maternal yellow cotyledons]

[a] Based on Soybean Genetics Committee.[87]

2) Chromosomes/linkage groups. Rules for soybean cytogenetics were also updated by the SGC in 1996.[87] Classic linkage groups and the corresponding chromosomes are represented by arabic numerals 1 (longest) through 20 (shortest), molecular linkage groups by uppercase Roman letters.[88] Symbols for chromosomal aberrations include an abbreviation for the aberration followed by the chromosome number or numbers (and a letter for each additional aberration on a chromosome). The nonitalic abbreviations for deficiency, inversion, translocation, and primary trisomics are Def, Inv, Tran, and Tri, respectively. Cytoplasmic factors are represented by the hyphenated prefix *cyt-* followed by one or more italic letters.

Tran 1-2a [the first case of reciprocal translocations between chromosomes 1 and 2]
Tran 1-2b [the second case of such translocations]
cyt-G [cytoplasmic factor for maternal green cotyledons]

21.14.6 Solanaceous Plants

21.14.6.1 Pepper

Rules for the gene nomenclature of *Capsicum* were published by the Capsicum and Eggplant Newsletter Committee in 1994.[89] Gene names concisely describe a characteristic feature of the mutant type in English or Latin. Symbols are a maximum of 3 italicized letters, the first being the same as the first letter of the name. If the mutant is dominant, the first letter of the name and symbol is capitalized; if the mutant is recessive, the name and symbol are in all lowercase. Mimics (i.e., mutants with similar phenotypes that are controlled by different genes) may either have distinctive names and symbols or be assigned the same symbol followed by a unique arabic numeral or Roman letter printed on the same line. The original gene in the series is numbered *1*. Multiple alleles have the same symbol, followed by a superscript letter or arabic numeral.

Bzt *ms-1* *ms-2* et^{a} et^{f}

Modifying genes may have a symbol for an appropriate name (such as intensifier, suppressor, or inhibitor), followed by a hyphen and the symbol of the allele affected, or the gene may be given a distinctive name unaccompanied by the symbol of the gene being modified.

Mo-A [modifier of *A*]
t [interacts with *B* to produce high levels of β-carotene]
B [interacts with *t* to produce high levels of β-carotene]

21.14.6.2 Tomato

1) Genes. The Tomato Genetics Cooperative (TGC) outlined the rules for tomato gene nomenclature[90] summarized in Table 21.21. The cultivar Marglobe is the standard, normal type. Also in 1992,[91] the TGC published a proposal to adopt the *Arabidopsis* rules for gene nomenclature (see Section 21.14.4 and Table 21.17) even though there are major differences between the guidelines. Although much discussion ensued, there has been no agreement to change the existing TGC rules summarized here.
2) Chromosomes. Rules for tomato cytogenetic nomenclature are intended for all members of the genus *Lycopersicon*.[91] Chromosomes are designated by arabic numerals in descending order by length, 1 (longest) through 12 (shortest), short and long arms by S and L. The short arm of each chromosome is designated the left arm, and the zero locus position is the distal (left) end of the short arm.

   Aberrations are designated by the nonitalic symbols Df, In, and T for deficiency, inversion, and translocation, respectively. To distinguish between aberrations involving the same chromosomes, lowercase letters are used following the chromosome numbers.

   T (1–2)a, T (1–2)b, T (1–2)c

   Aneuploids are designated according to the missing or extra chromosome. Those with whole-arm interchanges are symbolized by the component arms, a connective dot representing the centromere.

   triplo-1 [primary trisomic of chromosome 1]
   haplo-12 [monosome for chromosome 12]
   1S·12L [interchange between the short arm of chromosome 1 and the long arm of chromosome 12]

Table 21.21 Tomato: symbols for genes and phenotypes[a]

Feature	Convention	Examples
Mutant name	Appropriate descriptive name in italics, preferably in Latin or English, referring to the main diagnostic feature of the phenotype	*curly mottled*
Mutant gene symbol	Abbreviation of mutant name in one or more italic letters or numbers, the initial letter being the same as that of the name[b]	*cm*
Standard, normal allele	Mutant gene symbol with superscript plus sign. When the context is clear, the plus sign alone is sufficient.	sp^+ l^+ +
Dominant alleles	Name and symbol begin with capital letter	*Curl*, *Cu*
Recessive alleles	Name and symbol in all lowercase	*aurea*, *au*
Additional alleles	Designate a dominant allele of a normally recessive gene with a superscript uppercase *D*; designate other alleles with appropriate letter or number superscripts, numbers being preferred. For the first member of a series, the number *1* is understood, but not used.	*sp* [normal allele at *sp* locus] sp^D [dominant allele appearing later] d, d^x, d^{cr}, d^+
Indistinguishable alleles of independent origin (supposed recurrences)	Avoid unique symbols. Instead, use the existing mutant gene symbol with a series symbol as superscript. The series symbol, which consists of an arabic numeral and a laboratory code assigned by the Gene List Committee of the Tomato Genetics Cooperative, should be in parentheses.	$ag^{(1K)}$ and $ag^{(2K)}$ [first and second recurrences of *ag* found by Knowles]
Mimics (mutants with different loci and indistinguishable phenotypes)	Designate mimics by different names and symbols or in a numbered series with the same basic symbol followed by hyphenated numbers. For the first member of a series, the number *1* is understood, but not used.	*uniform*, *u* *uniform gray-green*, *ug ms*, *ms-2*, *ms-3*
Enhancers, modifiers, and suppressors	Symbol of the gene affected in parentheses, preceded by *Enh*, *Mo* (if dominant) or *mo* (if recessive), or *Sup*, respectively. Add a numerical suffix for subsequent nonallelic modifiers.	*Enh(l)* *Mo(l)*, *Mo(l)2* *Sup(l)2*

[a] Based on Tomato Genetics Cooperative.[90]

[b] Because of their long-term, widespread use, symbols *c*, *r*, *s*, and *y* are retained for the genes *potato leaf*, *yellow flesh*, *compound inflorescence*, and *unpigmented fruit epidermis*, respectively, even though they were derived from the normal standard (not mutant) lines and are not abbreviations of the gene names.

21.14.7 Other Plants

21.14.7.1 Cotton

1) Genes. Recommendations for gene names and symbols for cotton are based on those by Percy and Kohel.[92] The nomenclature was based on classical genetics that dealt with morphological mutants and chromosome aberrations. Names of mutant genes describe the main diagnostic feature of the mutant in a concise word or short phrase. Names and symbols of dominant mutants begin with an uppercase letter, those of recessive mutants with a lowercase letter. The first letter of the symbol is the same as that of the name. Newly discovered alleles at a given locus are assigned the original locus symbol followed by a superscript lowercase letter. New mutants with phenotypes similar to previously described mutants are designated with the symbol of the original mutant followed by a numeric subscript. The original mutant is considered number 1, and subsequent mutants are numbered serially. Gene symbols are italic letters.

   as_1, as_2 [asynapsis] B_1 [Blight resistance] *ml* [mosaic leaf] Sm_3 [Smooth leaf]

2) Chromosomes/linkage groups. Chromosome nomenclature of cotton was outlined by Kohel.[93] Chromosomes are designated by arabic numerals 1 to 26, and linkage groups by roman numerals. The genomes are designated by uppercase letters A and D. Aberrations are designated by the italic symbols *Del*, *Df*, *Dp*, *In*, *T*, and *Tp* for deletion, deficiency, duplication, inversion, translocation, and transposition, respectively, followed by the symbols for the involved chromosomes. Different inversions and translocations in the same chromosomes are distinguished with lowercase letters.

   In1a, *In1b* [inversions *a* and *b* on chromosome 1]
   T1–2a, *T1–2b* [translocations *a* and *b* on chromosome 2]

   Working designations for primary monosomics and monotelodisomics are M# and Te# until A–D chromosome homoeology is established (Russell J Kohel, USDA-ARS, personal communication, 2012 Aug 23). Primary monosomics and trisomics are designated by the italic terms *mono* and *triplo* followed by the number of the chromosome.

   mono-6 *triplo-6*

   The standard for *Gossypium hirsutum* shall be a type similar to cultivar Texas Marker 1; the standard for *G. barbadense* shall be similar to cultivar Pima S-4 for the pima cottons.[93] The research community has established a doubled haploid from the cultivar 3-79 as the standard for cultivated extra-long staple *G. barbadense* (Russell J Kohel, USDA-ARS, personal communication, 2012 Aug 23). The latest qualitative genetic summary was by Percy and Kohel.[93]

   Cotton genetics changed with the advent of DNA markers. The first DNA marker map utilized restriction fragment length polymorphisms (RFLP) in an interspecific hybrid population.[94] The application of DNA markers revealed a low level of polymorphism within *G. hirsutum*, so that mapping populations have been constructed primarily from hybrid populations between *G. hirsutum* and *G. barbadense*. As new types of DNA markers were developed, they were applied to cotton; however, the low level of polymorphism slowed the development of genetic maps and restricted their development to interspecific populations.[95–101] Finally, with the use of DNA markers and cytological markers, the chromosomes and linkage groups were aligned and identified with the original nomenclature.[102]

   The construction of cotton genetic recombination maps with different genetic populations and with different genetic markers prevents the development of a consensus recombinant map. Yu et al.[103] applied a mathematical algorithm to construct a genome-wide comprehensive reference map based on the relative location of markers and not recombination values.

The cotton genetics community needs to establish a nomenclature system for DNA markers that resolves current marker multiple identities and establishes a system by which new markers will be identified and integrated with existing markers.

21.14.7.2 Lettuce

The gene nomenclature of lettuce and its closely related species is adapted from the International Committee on Genetic Symbols and Nomenclature[2] to meet the needs of lettuce researchers.[104] Although names of newly discovered genes should conform to the basic guidelines, gene names and symbols already in widespread use should not be changed. Genes with capitalized German names will remain in use despite the apparent contradiction between the allele name and the allele symbol. Genes named for the normal or wild-type phenotype rather than the mutant phenotype are also retained. Where there is diversity in *Lactuca sativa* and the primitive form cannot be readily identified, the predominant form in *L. serriola* will represent the wild-type (+) allele.

21.14.7.3 Onion

The nomenclature and notation for the chromosomes of *Allium cepa* represent the standard for *Allium* species that cross-fertilize with *A. cepa* to yield F_1 hybrids.[105,106] In an ideogram, the longest chromosome is represented on the left, the shortest is on the right, the centromeres are at the same level, and the short arms are uppermost. The chromosomes are numbered, beginning at the left of the ideogram, with arabic numerals and an unspaced uppercase letter representing the specific epithet. Thus, the chromosomes for *A. cepa* are numbered 1C, 2C, and so on to 8C.

21.15 ANIMALS

The following guidelines are based on published rules for genes and chromosomes. Most conform to the human and/or mouse rules, but there are several significant exceptions. Some organisms have elaborate, detailed sets of rules; for others, there are only minor differences from the model. Tables 21.22[107–113] and 21.23[114–126] summarize and index the following sections.

21.15.1 *Caenorhabditis elegans*

1) Genes. The original recommendations for genetic nomenclature of *Caenorhabditis elegans* by Horvitz et al.[127] have been expanded by the curators of the *Caenorhabditis elegans* WWW Server.[128] Table 21.24 summarizes the major conventions. The wild-type, standard strain is Bristol N2. Consult with the *Caenorhabditis* Genetics Center (CGC)[129] when proposing new gene or allele names or when registering a new laboratory code.
2) Chromosomes. *Caenorhabditis elegans* has 5 pairs of autosomes (I, II, III, IV, and V) and 1 pair of sex chromosomes (X). Hermaphrodites are XX; males (which are rare) are XO. Aberrations are indicated by the italic symbols *Df*, *Dp*, *In*, and *T* for deficiency, duplication, inversion, and translocation, respectively. The names consist of a laboratory mutation prefix in lowercase letters, the abbreviation for the aberration, an arabic number, and optionally the number(s) of the affected linkage group(s).[128]

    mnDp2 *sDf3(I)* *mnDp1(X;V)*

Table 21.22 Animals: gene nomenclature

Animal	Section or table[a]	Prefix named in rules	Swiss–Prot species ID code[b]	Model	Reference(s)
Caenorhabditis elegans	Section 21.15.1, Table 21.24	*Ce-*	CAEEL	. . .	*Caenorhabditis elegans* website[128]
Cat, *Felis catus*	Table 21.30	*FCA*	FELCA	mouse	Committee on Standardized Genetic Nomenclature for Cats[108]; Committee on Standardized Genetic Nomenclature for Mice[109]
Cattle, *Bos* spp.	Section 21.15.5.1, Table 21.29	*BBO*	BOVIN	ruminants	Broad et al.[145]
Chicken, *Gallus domesticus*—see also Poultry and other avian species	Table 21.28	. . .	CHICK	. . .	. . .
Deer mouse, *Peromyscus* spp.	Section 21.15.6.1, Table 21.30	. . .	[varies with species][b]	mouse	Lyon[158]
Dog, *Canis familiaris*	Section 21.15.5, Table 21.29	*CFA*	CANFA	domestic animals	Ruvinsky and Sampson[113]
Domestic animals: ruminants, cattle, sheep, goats, deer, camelids, pigs, horses	Section 21.15.5, Table 21.29	. . .	[varies with species][b]	ruminants	Broad et al.[145]
Drosophila, *Drosophila melanogaster*	Section 21.15.2, Table 21.25	*Dmel*	DROME	Neurospora	FlyBase[130]
Fish (not including zebra danio)	Section 21.15.3.1, Tables 21.26 and 21.32	. . .	[varies with species][b]	human	Shaklee et al.[133]
Goat, *Capra hircus*	Table 21.29	*CHI*	CAPHI	ruminants	Broad et al.[145]
Horse, *Equus caballus*	Section 21.15.5.2, Table 21.29	*ECA*	HORSE	ruminants	Broad et al.[145]; Dolling[111]
Human, *Homo sapiens*	Section 21.15.7, Table 21.32	*HSA*	HUMAN	human	Wain et al.[17]
Mouse, *Mus* spp.	Section 21.15.6.2, Table 21.30	*MMU*	MOUSE	mouse	Mouse Genome Informatics[15]
Pig, *Sus scrofa*	Table 21.29	*SSC*	PIG	ruminants	Broad et al.[145]
Poultry and other avian species	Section 21.15.4, Table 21.28	. . .	[varies with species][b]	human	Crittenden et al.[142]; Burt[107]
Rabbit, *Oryctolagus cuniculus*	Table 21.30	*OCU*	RABIT	mouse	Fox[112]
Rat, *Rattus norvegicus*	Table 21.30	*RNO*	RAT	mouse	Levan et al.[161]
Sheep, *Ovis aries*	Table 21.29	*OOV*	SHEEP	ruminants	Dolling[110]
Zebra danio, *Danio rerio* (also known as *Brachydanio rerio*)	Section 21.15.3.2, Table 21.27	. . .	BRARE	. . .	ZFIN[137]

[a] Entries refer to the section and/or table for the model organism when there are no separate rules for the animal named in the first column.

[b] When the common name of the organism in English is 5 characters or fewer, the Swiss-Prot identification code is usually the common name (e.g., HUMAN, PIG); when the common name is longer, the prefix is usually the first 3 letters of the genus name and the first 2 letters of the epithet (e.g., DROME for *Drosophila melanogaster*). Exceptions are made when a short common name can be applied to more than one species (e.g., goat) or when the name can be shortened without ambiguity (e.g., BOVIN for bovine). See the ExPASy Proteomics Server (http://www.dsimb.inserm.fr/~fuchs/M2BI/AnalSeq/Annexes/Serveurs_db/ExPASy%20Proteomics%20Server.htm) for details and the list of organisms.

Table 21.23 Animals: cytogenetic nomenclature

Animal	Section or table[a]	Chromosomes or linkage groups (LG)	Model	Reference(s)
Caenorhabditis elegans	Section 21.15.1	I, II, III, IV, V, X	. . .	*Caenorhabditis elegans* website[128]
Cat, *Felis catus*	. . .	A1, A2, A3, B1, B2, B3, B4, C1, C2, D1, D2, D3, D4, E1, E2, E3, F1, F2, X, Y	human	Cho et al.[120]; ISCN 1985[125]; ISCN 1995[126]
Cattle, *Bos* spp.	Section 21.15.5.1	1 to 29, X, Y	domestic animals	ISCNDA 1989[148]; Popescu et al.[149]
Chicken, *Gallus domesticus*—see Poultry and other avian species	. . .	. . .	. . .	. . .
Deer mouse, *Peromyscus* spp.	Section 21.15.6.1	1 to 23, X, Y	human	Greenbaum et al.[160]; *Pero*Base[159]
Dog, *Canis familiaris*	. . .	1 to 38, X, Y	. . .	Switonski et al.[117]; Reimann et al.[116]; Breen et al.[119]
Domestic animals (see also specific animal)	Section 21.15.5	. . .	human	ISCNDA 1989[148]
Drosophila, *Drosophila melanogaster*	Section 21.15.2	1 to 4, X, Y	. . .	Lindsley and Zimm[132]; FlyBase[130]
Fox, *Vulpes fulvus*	. . .	1 to 18, X, Y [including 2 pairs of microsomes]	. . .	Makinen[114]
Goat, *Capra hircus*	. . .	1 to 29, X, Y	ruminants	Popescu et al.[149]
Hamster, *Cricetulus griseus*	. . .	1 to 10, X, Y	. . .	Ray and Mohandas[115]
Horse, *Equus caballus*	Section 21.15.5.2	1 to 31, X, Y	human	Richer et al.[156]; Bowling et al.[157]; ISCN 2009[5]
Human, *Homo sapiens*	Section 21.15.7, Tables 21.33 and 21.34	1 to 22, X, Y	human	ISCN 2009[5]
Mouse, *Mus* spp.	Section 21.15.6.2, Table 21.31	1 to 19, X, Y	mouse	Mouse Genome Informatics[15]
Pig, *Sus scrofa*	. . .	1 to 18, X, Y	human	Gustavsson[123]; Young[118]
Poultry and other avian species	Section 21.15.4	1 to 38, Z, W [those smaller than chromosome 8 are considered microchromosomes]	human	Shows et al.[141]; Crittenden et al.[142]
Rabbit, *Oryctolagus cuniculus*	. . .	1 to 21, X, Y	human	Committee for Standardized Karyotype of *Oryctolagus cuniculus*[122]

Table 21.23 Animals: cytogenetic nomenclature (*continued*)

Animal	Section or table[a]	Chromosomes or linkage groups (LG)	Model	Reference(s)
Rat, *Rattus norvegicus*	. . .	1 to 20, X, Y	human	Committee for a Standardized Karyotype of *Rattus norvegicus*[121]; Levan et al.[161]
River buffalo, *Bubalus bubalis*	. . .	1 to 24, X, Y	domestic animals and human	ISCN 1985[125]; ISCNDA 1989[148]; Iannuzzi[124]
Sheep, *Ovis aries*	. . .	1 to 26, X, Y	human	ISCN 2009[5]; Ansari et al.[150]
Zebra danio, *Danio rerio* (also known as *Brachydanio rerio*)	Section 21.15.3.2	LG01 to LG25		ZFIN[137]

[a] Entries refer to the section and/or table for the model organism when there are no separate rules for the animal named in the first column.

Table 21.24 *Caenorhabditis elegans*: symbols for genes and phenotypes[a]

Feature	Convention	Examples
Gene names	3 lowercase italic letters derived from the description of the mutant phenotype, followed by a hyphen and an italic arabic numeral; may include an italic roman numeral indicating the linkage group of the gene, separated by a space from the gene symbol	*dpy-18* *mlc-2* *dpy-18 III* *lon-2 X* *wrn-1* [corresponds to *WRN1* in human]
Homologous genes in related species	Gene name preceded by 2 italic letters referring to the species and a hyphen	*Cb-tra-1* [homolog of the *C. elegans* gene *tra-1* in *C. briggsae*] *Ce-snt-1* [specifies the *C. elegans* gene]
Homologous set of genes related to a single gene from another species	Gene name followed by decimal numbers	*sir-2.1*, *sir-2.2*, *sir-2.3*, *sir-2.4* [all correspond to *SIR2* in *Saccharomyces cerevisiae*]
Pseudogenes	Gene name followed by *ps*	*msp-48ps*
Different loci conferring similar phenotypes	Gene name prefix followed by serial, hyphenated italic arabic numerals	*mlc-1*, *mlc-2*, *mlc-3*
Alleles and mutations	Mutation symbols consist of an italic laboratory code and an italic arabic numeral for the specific mutation; may be followed by a lowercase nonitalic symbol for the description of the mutation. When they are used together, the gene name is followed by the mutation symbol in parentheses.	*e1348*rl [*e* for MRC Lab in England, rl for recessive lethal] *dpy-5(e61)*
Wild-type allele	Gene name followed by a plus sign or by a plus sign in parentheses	*sma-2+* or *sma-2(+)*

Table 21.24 ***Caenorhabditis elegans*: symbols for genes and phenotypes[a] (*continued*)**

Feature	Convention	Examples
Chromosomal duplication	Symbol for the chromosomal aberration followed optionally by the name of one or more of the duplicated genes in square brackets	*mnDp1(X;V)[unc-3(+)]*
Extrachromosomal array	Italicized laboratory code, the letters *Ex*, and a number; may be followed by description of the transgene in square brackets	*eEx3* *stEx5* *stEx5[sup-7(st5) unc-22(+)]*
Integrated transgenes	Italicized laboratory code, the letters *Is*, and a number	*eIs2*
Gene fusions	Gene name followed by double colon and the name of the reporter gene	*pes-1::lacZ* *mab-9::GFP*
Genotypes		
Homozygotes	Gene symbols in order by chromosome number; chromosomes separated by a semicolon and space	*dpy-5(e61) I; bli-2(e768) II*
Heterozygotes	Same as for homozygotes, but with slashes between homologous chromosomes; use plus sign for wild-type alleles as needed	*dpy-5/unc-13* *dpy-5/+ I; +/bli-2 II*
Transposons	Not italicized except when included in a genotype; insertions indicated by adding *::Tc* to the relevant mutation symbol	Tc1, Tc2, . . . *unc-54(r293::Tc1)*
Phenotypes	Description in words or nonitalic 3-letter gene symbol with initial letter capitalized	dumpy animal or Dpy uncoordinated or Unc
Wild-type phenotype	In comparisons of the wild type to the mutant, use prefix non- before the mutant phenotype	non-Dpy [wild type is not dumpy]
Protein product	Gene symbol in nonitalic capital letters	UNC-13 [protein encoded by the *unc-13* gene]
Mitochondria	Gene name using the standard nomenclature; follows the nuclear genes in a genotype, with *M* as the abbreviation for the linkage group	*dpy-5/+ I; +/bli-2 II; cyt-1 M*

[a] Based on the *Caenorhabditis elegans* website[128] (http://elegans.som.vcu.edu/Genome/Nomencl2001w.htm).

21.15.2 *Drosophila melanogaster*

1) Genes. The conventions for *Drosophila melanogaster* genes are similar to those for *Neurospora crassa* (see Section 21.12.3), but differ significantly in some ways from those of many other organisms. Gene names must be italicized, concise, descriptive, unique, and inoffensive. Symbols derived from them should contain no spaces, non-Roman characters, or subscripts. Superscripts are reserved—and required—for alleles, including the plus sign for the wild type.[130,131] Table 21.25 summarizes the major conventions. When necessary to distinguish a *D. melanogaster* gene from one of another organism that would otherwise have the same symbol, use *Dmel* as a prefix; other species have similar 4-letter symbols. The Swiss-Prot species identification code for *D. melanogaster*, DROME, is not used in FlyBase.
2) Chromosomes. The symbols representing chromosomal aberrations in *Drosophila* spp. and the rules for their use were summarized by Lindsley and Zimm[132] and are updated

Table 21.25 ***Drosophila melanogaster*: genes and phenotypes**[a]

Feature	Convention	Examples
Gene name	Italicized description of the gene's function or mutant phenotype	*white, Deformed* "Flies were scored for *white* mutations."
Gene symbol	Italic letters derived from the gene name	*w* [symbol for *white*]
Different loci with similar function	Gene symbol followed by italic arabic numerals and capital letters. Hyphens are not used except to separate numbers or letters that would lose their descriptive content otherwise.	*Act5C, Act42A*
Gene named for an allele recessive to wild type	Name and symbol begin with lowercase letter.	*awd* [for *abnormal wing disc*]
Gene named for an allele dominant to wild type	Name and symbol begin with uppercase letter.	*R* [for *Roughened*]
Gene named for a protein	Symbol begins with uppercase letter.	*Adh* [for alcohol dehydrogenase]
Gene named for a tRNA	Symbol begins with *tRNA:*	*tRNA:S7:23Ea*
Gene named for small-nuclear RNA	Symbol begins with *snRNA:*	*snRNA:U6:96Aa*
Mitochondrial gene	Symbol begins with *mt:*	*mt:ND4, mt:tRNA:L:UUR*
Transposons	Symbol consists of the designations for the *ends{genes=construct-symbol}*, all in italics	$P\{w^{+mC}ovo^{D1\text{-}18}=ovoD1\text{-}18\}$ $H\{Lw2\}dpp^{151H}$
Alleles	Gene symbol followed by superscript characters; square brackets may be used instead of superscripting.	b^{+} sc^{1}, sc^{2} or *sc[1], sc[2]*
Recessive allele of a gene named for a dominant mutation	Gene symbol with superscript or bracketed *r* for recessive	Hn^{r2} or *Hn[r2]*
Dominant allele of a gene named for a recessive mutation	Gene symbol with superscript or bracketed *D* for dominant	ci^{D} or *ci[D]*
Genotypes	Separate genes on the same chromosome with spaces; those on different chromosomes with semicolons and spaces. Separate homologs with slashes.	$y^{1}\ w^{1}\ f^{1}\ B^{1}$ *bw; e; ey* *y w f/B*
Genes identified during a genomic sequencing project	Nonitalic symbol consists of BG: or EG:, the clone name, a decimal, and a serial number[b]	BG:DS07851.5 EG:152A3.3
Genes identified during annotation of the genomic sequencing projects	Nonitalic symbol consists of CG and a unique number	CG10809
Phenotypes	Nonitalic words for the description	"Flies with white eyes were selected."
Protein products	Gene symbol in nonitalic capital letters	HH [protein encoded by *hh* (*hedgehog*) gene]

[a] Based on FlyBase[130] (http://flybase.org/static_pages/docs/nomenclature/nomenclature3.html).

[b] BG stands for the Berkeley Drosophila Genome Project, EG for the European Drosophila Genome Project; the clone name is the name or number of the clone on which the gene was first identified; the integer after the decimal is a serial number, with no implication of gene order on the clone.

online as needed in FlyBase,[131] which includes a catalog of genes, chromosomal aberrations and their names, the inducing agents, alternative symbols or names or both, and specific instructions for genetic nomenclature for *D. melanogaster* and related species.

The autosomes of *D. melanogaster* are represented by arabic numerals 1 to 4, the sex chromosomes by X and Y. The symbols *X* and *1* for the X chromosome are synonymous, with *1* being preferred. Aberrations are indicated by the italic prefixes *Df*, *Dp*, *In*, *R*, *T*, and *Tp* for deficiency, duplication, inversion, ring chromosomes, translocation, and transposition on the same chromosome, respectively, followed by the chromosome designation within parentheses and a specific designation that identifies the rearrangement. Superscripts are allowed only in names for synthetic inversions and when the aberration is named with an allele symbol; no spaces or subscripts are allowed.

Dp(1;1) *In(2R)C72* *T(1;2;3)OR14*

21.15.3 Fish

21.15.3.1 Fish Proteins

A formal system published by Shaklee et al.[133] established the nomenclature and symbolization for genes coding for enzymes and other proteins of fish. The system is similar to that for human genes and draws on the enzyme nomenclature of the International Union of Biochemistry and Molecular Biology.[134] Table 21.26 summarizes the major conventions. This system is quite different from the one adopted for zebra danio (see Section 21.15.3.2 and Table 21.27).

21.15.3.2 Zebra Danio (Zebrafish)

Researchers working with the species as a laboratory animal prefer the name "zebrafish" to "zebra danio". However, the term "zebrafish" is ambiguous because it has been applied to *Melambaphes zebra* (now known as *Girella zebra*) as well as *Danio rerio* (now also known as *Brachydanio rerio*). This manual uses "zebra danio", the name approved by the American Fisheries Society and others, for *Danio rerio*.

1) Genes. Rules for zebra danio, *Danio rerio*, gene nomenclature were developed in 1992, were published by Mullins in 1995,[135] and are updated as needed in ZFIN: the Zebrafish Information Network.[136–139] Table 21.27 summarizes the major conventions. Members of gene families follow existing naming conventions. Symbols should not begin with *Z* or *Zf*. Register (with ZFIN) new locus and allele names, their symbols, and codes for the laboratories researching the genes.

   When different laboratories describe the same new gene or mutant allele of a previously known gene and give it a different name, the first name in print is given priority. When a mutation is found in a previously cloned zebra danio gene that has been published, the mutant takes the name of the cloned gene. If both the cloned gene and the mutation are already known by different names and at some point are found to be the same gene, the name of the mutation usually takes priority. The unique name of the mutation may be more relevant to the function of the gene than the cloned gene's name, and this unique mutant name may be easier to remember or use than the gene name. Exceptions include cases where orthologous relationships with genes of other species are well established and these orthologs have a commonly used name.
2) Chromosomes. The 25 original linkage groups, designated LG01 to LG25, are now designated chromosomes (Chr)01 to 25. Note the use of 2 digits so that computers can order

Table 21.26 Fish: symbols for genes and phenotypes[a] (except as noted below)

Feature	Convention	Examples
Gene symbol	Uppercase italic letters[b]; preferably ends with *; all characters on the baseline; no superscript or subscript characters	*IDHP** *MPI**
Multiple loci	Hyphen and italic arabic numeral (for unknown relationships) or italic uppercase letter (for established orthologies)	*IDHP-1** *LDH-A**
Isoloci	Gene symbol with a comma between the 2 locus numbers	*GPI-B1,2**
Regulatory locus suffix	Gene symbol with a lowercase italic *r*	*LDH-Ar**
Subcellular prefix	Gene symbol with a lowercase-letter prefix[c]	*sMDH-B**
Allele	Gene symbol with an italic arabic numeral, lowercase letter, or relative electrophoretic mobility with a preceding asterisk. (The preferred convention is sequentially assigned number codes to designate alleles.)	*MDH-B*1* *ADA-1*a* *EST-2*75*
Enzyme	IUBMB-specified name and number[b]	L-lactate dehydrogenase, 1.1.1.27
Enzyme symbol	Uppercase roman letters; with numbers, Greek-letter stereochemical isomer symbols, or hyphens as necessary	LDH
	Lowercase prefix for subcellular localization[c]	mIDHP

[a] Based on Shaklee et al.[133] See Section 21.15.3.2 and Table 21.27 for discussion and symbols for zebra danio, *Danio rerio*, which has its own gene nomenclature.
[b] Same alphanumeric symbols as the abbreviations for the coded proteins, which are derived from the International Union of Biochemistry and Molecular Biology (IUBMB), Nomenclature Committee[134] names (see http://www.chem.qmul.ac.uk/iubmb/enzyme/). Italicization and asterisk-marking distinguish the symbols from those for enzymes and proteins. The asterisk follows the locus symbol and precedes an allele designation.
[c] The prefix indicates the subcellular location of the enzyme coded for by the gene: the letter l for lysosomal, m for mitochondrial, p for peroxisomal, and s for cytosolic (supernatantor soluble). Ignore the prefixes when alphabetizing a list.

them correctly. The upper and long arms are indicated by U and L without a space after the chromosome number.[137] Chromosome rearrangements are indicated by the prefixes *Df*, *Dp*, *In*, *Is*, *T*, and *Tg* for deficiency, duplication, inversion, insertion, translocation, and transgene, respectively. Chromosomal differences have not been observed between males and females. Detailed instructions for describing chromosomes and their aberrations are available at ZFIN.[137]

21.15.4 Poultry, Domestic Fowl, and Other Avian Species

1) Genes. Rules for poultry gene nomenclature were developed by the Poultry Science Association and published by Somes.[140] A later review prompted the Poultry Gene Nomenclature Committee of the International Society for Animal Genetics to adopt a system based on human locus and allele nomenclature[141] but adapted to fit the needs of the poultry research community. Table 21.28 summarizes the major conventions. See Crittenden et al.[142] for additional details and examples. In addition, the Chicken Gene Nomenclature Consortium (CGNC) is an international group of researchers interested in providing standardized gene nomenclature for chicken genes.[143]

Although allele symbols should be brief and not attempt to summarize all known information about a gene, they may convey information about morphological characteristics,

Table 21.27 Zebra danio: symbols for genes and phenotypes[a]

Feature	Convention	Examples
Gene name	Descriptive name in lowercase italics. Genes identified as orthologs of mammalian genes take the mammalian name and symbol.	*engrailed1a, engrailed2b*
Gene symbol	Short alphanumeric string derived from the gene name in lowercase italics; when the gene has a mammalian ortholog, add it in parentheses after the zebra danio symbol	*eng1a, eng2b* *syu (shh)*
Duplicate genes	Use the approved name of the mammalian ortholog followed by a and b	*hoxa13a, hoxa13b*
Mutants and genes identified by mutation	Gene nomenclature for mutants is the same as that for genes identified by mutation	
Mutant/gene name	1- or 2-word term that describes the mutant phenotype, in lowercase italics; mutations in different genes with similar phenotypes get dissimilar names, not a number appended to the first-named gene	*Cyclops fused eyes* [not *cyclops2*]
Mutant/gene symbol	Unique 3-letter abbreviation derived from the mutant/gene name	*cyc*
Wild-type allele	Symbol for mutant with superscript plus sign	*cyc*$^{+}$ *brs*$^{+}$
Mutant allele (recessive allele)	Symbol for mutant with superscript laboratory code and allele designation	*cyc*^{b16} *cyc*^{m101}
Dominant alleles	Symbol for mutant or allele with a *d* in the first position of the superscript	*lof*^{dt2}
Transgenes	Symbol in the format *TG(xxx)*^{allele}, where *TG* is the symbol for transgene, *(xxx)* describes the salient features of the transgene, and *allele* is the symbol for the allele	*TG(abc)*^{b000} *TG(Lg03,ERE:nic1)*^{b512} *TG(NBT:MAPT-GFP)*^{zc1}
Genotypes		
Homozygotes	List gene symbols in order of linkage group; separate genes in the same linkage group with spaces and those in different groups with semicolons	*rse*^{b140} *leo*^{t1}; *gol*^{b16}
Heterozygotes	Same as for homozygotes, but separate loci on homologous chromosomes with slashes	*rse*^{b140}/*rse*$^{+}$; *gol*^{b16}/*gol*^{m592}
Unmapped loci	List gene symbols in alphabetical order within braces after the genotypes of the mapped loci on other chromosomes	*rse*^{b140}; *{edit*^{z253}; *esr*^{tj236}*}*
Poorly resolved loci on the same chromosome	List gene symbols in alphabetical order within braces	*{abc* ^{b000} *def*^{m000}*}* *rse*^{b140}*{abc*^{b000}*def*^{m000}*}* *leo*^{t1}
Protein products	Nonitalic gene symbol, initial letter capitalized	Shh, Gsc, Exorh, Dlx4
Phenotypes	Mutant/gene name or symbol in italics	*cyclops, cyc*

[a] Based on ZFIN[137] (https://wiki.zfin.org/display/general/ZFIN+Zebrafish+Nomenclature+Guidelines).

Table 21.28 Poultry, domestic fowl, and other avian species: symbols for genes and phenotypes[a]

Feature	Convention	Examples
Gene symbol (locus designation)	No more than 5 capital italic letters or a combination of letters and arabic numerals derived from the gene name; no superscripts, subscripts, Greek letters, or roman numerals allowed in gene or allele symbols	*PA* [pre-albumin] *MM7* [micromelia VII] *GPDA* [α-glycerol phosphate dehydrogenase-liver] *HBA* [α-hemoglobin] *HBB* [β-hemoglobin]
Different loci with similar phenotypes	Same symbol with added arabic numerals	*PA2*, *PA3*
Mutation site	Mutation indicated by an italic letter or number for the characteristic	*HBB*6V*
Alleles	Capital italic letter or arabic numerals added to the locus designation; allele characters are separated from the locus characters by an asterisk. After initial full identification, alleles may be designated without the locus characters.	*OV*A*, *OV*B* *EAA*1*, *EAA*7* **1*, **7*
Heterozygotes		
at a single locus	Allele symbols are separated by a horizontal line or, if on the same line, by a slash.	$\frac{EAA*1}{EAA*2}$ or *EAA*1/EAA*2* or *EAA*1/*2*
at multiple loci	Linked alleles are separated by spaces over the same horizontal line.	$\frac{EAJ*1\ SE*N}{EAJ*2\ SE*S}$ or *EAJ*1 SE*N/EAJ*2 SE*S*
	Unlinked alleles are separated by semicolons.	$\frac{EAA*1}{EAA*2}$; $\frac{EAB*1}{EAB*1}$; $\frac{EAP*1}{EAP*2}$ or *EAA*1/EAA*2*; *EAB*1/EAB*1*; *EAP*1/EAP*2* or *EAA*1/*2*; *EAB*1/*1*; *EAP*1/ *2*
for sex-linked traits	Genotypes distinguish between males (which are heterozygous) and females (which are hemizygous).	male: $\frac{GHR*A}{GHR*B}$ or *GHR*A/GHR*B* female: $\frac{GHR*A}{W}$ or *GHR*A/W*

[a] Based on Crittenden et al.[142]

subcellular locations, control properties, or amino acid substitutions. Indicate lack of activity of an allele with an italic uppercase letter *O*. Present complex data on dominance, recessiveness, and wild type in tables. Use an italic *N* to indicate the normal allele.

The laboratory that first conducts the genetic segregation analysis or assigns a gene to a specific chromosomal location also names the loci and alleles. Follow the naming conventions presented in Crittenden et al.[142] and include the population in which the allele was found. When poultry genes are homologous to human genes, or there is strong evidence for homology, use the same full gene name as the human gene listed by the HUGO Gene Nomenclature Committee (see Section 21.15.7).

2) Chromosomes. Karyotypes for domestic fowl (*Gallus domesticus*) and other avian species follow the system developed by Ladjali-Mohammedi et al.[144] Chickens have 38 pairs of autosomes (numbered 1 to 38) and a pair of sex chromosomes (Z and W). The female is the heterogametic sex. The 8 largest autosomes have been differentiated by morphology and banding patterns, but the others are too small to be distinguished microscopically and await definition by molecular methods. In karyotypes, the autosomes are arranged in descending order by length with the short arm on top; the Z and W chromosomes are shown at the end. G-band landmarks remain the standard; R-, Q-, and C-bands are useful for some studies, but they are too variable for mapping.

21.15.5 Domestic Mammals

1) Genes. Genetic loci in ruminants and other domestic mammals are represented by conventions developed by the Committee on Genetic Nomenclature of Sheep and Goats (COGNOSAG).[145] The rules are recommended for all ruminants (including cattle, sheep, goats, deer, and camelids), other farmed mammals (such as pigs and horses), and dogs. The guidelines are generally consistent with those of human and mouse. Table 21.29[113] summarizes the major conventions. In addition, Online Mendelian Inheritance in Animals[146] (OMIA; http://omia.angis.org.au/), maintained at the University of Sydney, is a database of inherited disorders and traits in many animal species (but excluding human and mouse). The database contains textual information, references, and links to PubMed and Gene records at the National Center for Biotechnology Information.

   Choose new locus and allele names and symbols that are as brief as possible (but not a single letter), accurately convey the character affected or the function by which the locus is recognized, and reflect interspecies homology, being careful not to duplicate or confuse existing nomenclature. For keratins and keratin-associated proteins, use the names and symbols proposed by Powell and Rogers.[147]

   Ensure that names of new alleles at a recognized locus conform to the established nomenclature for that locus. In loci detected by biochemical, serological, or nucleotide methods, the allele name and symbol may be identical; do not use either the plus or the minus sign for alleles detected by those methods. The plus sign identifies the standard wild type for alleles having visible effects. Designate null alleles by the number zero.

2) Chromosomes. Karyotypes of domestic animals were standardized for cattle, goats, and sheep with both G- and R-bands.[148] More specific recommendations have been published since then for cattle[149] and sheep.[150] Autosomes are numbered beginning with chromosome 1; the sex chromosomes are designated X and Y (see Table 21.23).

21.15.5.1 Cattle

1) Genes. Guidelines for genetic loci in cattle are essentially those developed by COGNOSAG for all ruminants.[145] Table 21.29 summarizes the major conventions for domestic animals, including cattle. When necessary to distinguish a bovine locus from one of another organism that would otherwise have the same symbol, use *BBO* as a prefix; the Swiss-Prot species identification code is BOVIN.

   Nomenclature for DNA segments was proposed by Dolling.[151] The symbols for unmapped DNA segments with no known homologs or official names or symbols will (whenever possible) be the same as those first reported except that all letters will be in uppercase. The symbols for mapped DNA segments with no known homologs or official names or symbols will (whenever possible) consist of the same symbol assigned to the unmapped component followed by the D number assigned by BovMap[152] or CGD: the

Table 21.29 Domestic mammals (ruminants and other farmed mammals): symbols for genes and phenotypes[a]

Feature	Convention	Examples
Locus name	Name in italic letters or combination of letters and arabic numerals; initial letters of all words capitalized. Greek letters are spelled out and moved to the end of the name.	*Agouti* *Ear Length* *Horns* *Haemoglobin Beta*
Locus symbol	Symbol in italic letters or combination of roman and arabic numerals; initial letter capitalized and same as the initial letter of the locus name; no superscripts, subscripts, Greek letters, or roman numerals	*A [Agouti]* *B [Haemoglobin Beta]*
Allele name	Allele name in italics, initial lowercase letter preferred, including for dominant alleles	*hornless* [recessive] *polled* [dominant]
Allele symbol	Symbol in italic letters and/or arabic numerals; abbreviates the allele name and begins with the same letter; all lowercase unless the allele is dominant or codominant. No Greek letters or roman numerals allowed.	Ho^{hl} or *Ho**hl* Ho^{P} or *Ho**P* $ALOP^{alop1}$ or *ALOP**alop1*
	The allele symbol is always written with the locus symbol, either as a superscript or following an asterisk on the same line; no spaces between locus and allele symbols.	
	For loci other than coat color and other visible traits, the symbol is in all uppercase letters or a combination of uppercase roman letters and arabic numerals.	
Wild-type allele symbol	Plus sign may be used alone for alleles having visible effects.	E^{+} or +
	Null alleles are designated by the number zero.	E^{0}
Genotype		
Homologs	Symbols separated by slash	Ho^{P}/*HoP*, *Ho**hl*/*Ho**hl* $ALOP1^{alop1}/ALOP1^{alop1}$
Unlinked loci	Symbols separated by semicolons	*INHBB*; *INFW*
Linked loci	Symbols separated by spaces and listed in alphabetical order when gene order and/or phase not known	
X-linked loci	Symbol for the X-linked locus followed by /X	*FecX**1/X*
Y-linked loci	Symbol for the Y-linked locus followed by /Y	
Phenotype symbols	Same characters as locus and allele symbols; no italics or underlining; space instead of an asterisk inserted between locus and allele designations if all characters are on the line. Alternatively, enclose the symbols in square brackets, leaving no space between locus and allele designations if the allele symbol is superscript.	Ho P or [Ho^{P}] ALOP1 Alop1 or [$\mathrm{ALOP1}^{\mathrm{Alop1}}$]

[a] Based on Broad et al.[145] The Committee on Genetic Nomenclature of Sheep and Goats (COGNOSAG) recommends that its rules be adopted not only for the cattle, sheep, goats, deer, and camelids, but also for other farmed animals such as pigs and horses. Ruvinsky and Sampson[113] recommend that the same rules apply to dogs.

Cattle Genome Database.[153] When unnamed, newly mapped loci are identified, compare them with the database maintained by the HUGO Gene Nomenclature Committee.[6]

RM095 [unmapped] *RM095(D1S13)* [the same marker after it was mapped]

Loci are divided into 4 categories by coat color, visible trait other than coat color, blood and milk polymorphisms, and mapped loci and other genetic systems. Locus symbols for blood and milk polymorphisms are written in all uppercase letters, but recessive alleles are in all lowercase.[149]

2) Bovine leukocyte antigens. The nomenclature for bovine leukocyte antigens (BoLA) was compiled by a standing committee of the International Society for Animal Genetics and is based on the human leukocyte antigen (HLA) system.[154] Names of alleles are based on amino acid sequences and consist of 4 or 5 digits: the first and second digits indicate the major type, the third and fourth indicate the subtype, and the fifth (if used) indicates any unexpressed variations. Class I alleles of a single major type differ by no more than 4 amino acids in the first and second domains; Class II alleles differ by no more than 4 amino acids in the first domain. If a name is given based on a partial sequence, the first full-length sequence that includes the original partial sequence will assume the allele name. Minor sequence errors will be corrected when identified.[155]
3) Chromosomes. Karyotypes of cattle, sheep, and goats were standardized by the Second International Conference on Standardization of Domestic Animal Karyotypes.[146] The rules for cattle were revised in 1995 as the "Texas nomenclature".[149] All ideograms were retained from International System for Cytogenetic Nomenclature of Domestic Animals (ISCNDA) 1989,[148] but the chromosomes were renumbered in a different order. The nomenclature for chromosomal abnormalities and markers follows that for human chromosomes.[146,147] Autosomes are numbered 1 to 29; the sex chromosomes are X and Y.

21.15.5.2 Horse and Other Equines

1) Genes. Tables 21.22 and 21.29 summarize the major conventions for horse and other equines.
2) Chromosomes. Cytogenetic nomenclature for the domestic horse and other equines is based on the Paris Standard[156] as endorsed by the International Committee for the Standardization of the Domestic Horse Karyotype in 1996.[157] Karyotypes arrange the 13 pairs of metacentric/submetacentric autosomes in one group and the 18 pairs of acrocentrics in another. Within each group, the autosomes are arranged in descending order by length. The sex chromosomes are placed in the middle after the smallest bi-armed autosomes. The scheme for numbering regions and bands corresponds to ISCN 2009.[5]

21.15.6 Rodents

21.15.6.1 Deer Mouse

1) Genes. The gene nomenclature conventions published by Lyon[158] for the mouse are generally applied to the deer mouse, *Peromyscus* spp., with a few exceptions; specific rules have not been published. A list of genes is available on *Pero*Base.[159]
2) Chromosomes. The chromosomes of the deer mouse are described and symbolized to the extent possible by the same conventions used for *Homo sapiens*. However, differences between the karyotypes of the single species *H. sapiens* and the many species of *Peromyscus* prompt various departures from the rules for humans and adoption of some of the rules for mice.[160]

   Autosomes are numbered Chr 1 to Chr 23 in descending order by size; the sex chromosomes are X and Y. The ideogram of *P. boylii* is the reference for establishing G-band nomenclature. Bands are grouped into regions, which are designated alphabetically;

within regions, bands are numbered sequentially distal to the centromere. Decimal points denote subdivisions within bands. For details, see Greenbaum et al.[160]

Chr 20 band 3C1 band 2C1.4

21.15.6.2 Mouse and Rat

The International Committee on Standardized Genetic Nomenclature for Mice and the Rat Genome and Nomenclature Committee unified the rules and guidelines in 2003 for gene, allele, and mutation nomenclature in mouse and rat; the rules were updated in 2011.[7]

1) Genes, mouse. Rules and guidelines for mouse gene nomenclature were approved most recently by the International Committee on Standardized Genetic Nomenclature for Mice in 2011.[7] Mouse Genome Informatics (MGI) provides information on genes, mapping, multispecies homology, probes, clones, sequences, strain characteristics, and nomenclature rules, as well as links to other databases. The complete guidelines, instructions for registering a new locus symbol, and advice on nomenclature are available from the MGI.[15]

   Symbols for unique genes and loci are composed of italic letters and arabic numerals with no punctuation except hyphens in certain circumstances. Hyphens may be used in gene symbols only when needed for clarity to separate 2 numbers that would be in adjacent positions, related sequence and pseudogene designations from gene symbols, characters for loci in a complex from the complex symbol, and components of mutant allele symbols. Table 21.30 summarizes the major conventions.

   Within the symbols, laboratory codes have initial uppercase letters. Genes encoded by the opposite (antisense) strand have their own symbols. Alternative transcripts from the same gene are not given different symbols. When describing cross-hybridizing DNA segments, the species code is uppercase. Homologous genes in different vertebrate species (orthologs) should be given the same gene nomenclature if possible. To distinguish between mRNA, genomic DNA, and cDNA, write the relevant prefix in parentheses; e.g., (mRNA) *Rbp1*. When necessary to distinguish a mouse gene from one of another organism that would otherwise have the same symbol, use *MMU* as a prefix; the Swiss-Prot species identification code is MOUSE.

2) Genes, rat. Guidelines for rat gene nomenclature follow those of mouse.[161,162] Anonymous DNA segment symbols are assigned by RGD.[162]

3) Chromosomes, mouse. Rules and guidelines for mouse cytogenetic nomenclature published by the International Committee on Standardized Genetic Nomenclature for Mice[8] are summarized here and in Table 21.31. The 19 pairs of autosomes are designated by nonitalic, arabic numerals; the sex chromosomes are X and Y. The short and long arms are designated p and q, respectively. Chromosome bands are numbered from the centromere toward the telomeres and are designated by the chromosome number, p or q arm, the region number, and the band-specific letter. (Reference to the q arm may be omitted when the meaning is clear.) Further subdivisions are designated by numbers following a decimal point. Symbols for chromosome anomalies are not italicized.

   Distinguish each successive anomaly in a series by a symbol consisting of a number followed by the laboratory code of the person or laboratory that discovered the anomaly, i.e., the same code used for the designation of inbred substrains or sublines. Indicate the chromosome(s) involved in the anomaly by adding the appropriate arabic numeral(s) in parentheses between the initial letter and the series symbol; see examples in Table 21.31.

4) Chromosomes, rat. Rat chromosome nomenclature[16] follows that for humans (see Section 21.15.7). The 20 pairs of autosomes are designated by nonitalic, arabic numerals; the sex chromosomes are X and Y.[16]

Table 21.30 Mouse: symbols for genes and phenotypes[a]

Feature	Convention	Examples
Gene name	Unique, brief, nonitalic description of the character by which the gene is recognized or of the protein encoded	dwarf hemoglobin β-chain
Gene symbol	2-, 3-, or 4-character abbreviation[b] of the gene name in italic letters; arabic numerals if needed. The symbol must begin with the same letter as the gene name. Hyphens may be used only to separate numbers that would be in adjacent positions or components of mutant alleles.	*dw* [dwarf] *Hbb* [hemoglobin β-chain] *G6pd* [glucose-6-phosphate dehydrogenase] *Hba-ps3* [pseudogene of α-globin located away from the *Hba* complex]
Gene families	Same gene symbol for all members plus an identifying numeral	*H1*, *H2* *Es1*, *Es2* *Adam1*, *Adam2*
Allele	Gene symbol with superscript[c] italic letters; no more than 10 characters	*Hbb*^{d}, *Mi*^{wh} *Tyr*^{c} [albino] *Tyr*$^{c\text{-}ch}$ [chinchilla allele of *Tyr*^{c}]
Recessive allele	Gene symbol with lowercase initial letter for the mutant gene	*a* [nonagouti] *gus*^{mps} [recessive mutation in the *Gus* gene]
Viral expression/immune response	Gene symbol with superscript italic *a* for presence of virus or immune response; superscript italic *b* for absence of the trait	*Mtv1*^{a} *Mtv1*^{b}
Resistance/susceptibility	Gene symbol with superscript italic *r* for resistance to infectious organisms or other agents; superscript italic *s* for susceptibility to the agent	*Pla2g2a*$^{Mom1\text{-}r}$ *Pla2g2a*$^{Mom1\text{-}s}$
Cloned gene	When a mutant gene is cloned, the symbol for the structural gene becomes the gene symbol and the former mutation symbol is hyphenated as an allele symbol.	*Mbp*$^{shi\text{-}mld}$ [*shi* is the structural gene for myelin basic protein (Mbp)]
Targeted (knockout) mutation	Gene/allele symbol plus italic superscript *tm*, the number of the knockout, and the laboratory code	*Cftr*^{tm1Unc} [first targeted mutation in this gene made at the University of North Carolina]
Transgene	The transgene symbol consists of: *Tg(YYY)###Zzz*, where *Tg* is the mode, *(YYY)* is the insert designation, ### is the laboratory assigned number, and *Zzz* is the laboratory code. After the full symbol has been used in a paper, it may be abbreviated by omitting the insert designation.	*Tg(GPDHI)1Bir* [insertion of the GPDH gene, the first transgenic line produced by Birkenmeier; may be abbreviated later as *Tg1Bir*]
Wild type	Plus sign (not italic) with superscript italic gene symbol. A plus sign may be used alone when the context leaves no doubt which locus is represented.	$+^{pe}$ $+^{Myo5a\text{-}d}$ $+^{Tyr\text{-}c}$
Reversion to wild type	Symbolization reversed from the wild-type symbol	*pe*$^{+}$ *Myo5a*$^{d\text{-}+2J}$ *Tyr*$^{c+}$

Table 21.30 Mouse: symbols for genes and phenotypes[a] (*continued*)

Feature	Convention	Examples
Mutant genes with definite wild type	Gene symbol for the first discovered mutant allele is used for both the gene and the allele	*Ca* [caracul, the first mutant] Ca^{J} [a re-mutation at the caracul locus]
Mitochondrial genes	Prefix *mt-* before the gene symbol	*mt-Cytb*
Pseudogenes	Gene symbol followed by *ps* and an appropriate serial number	*Hba-ps3*, *Hba-ps4* [pseudogenes of α-globin located away from the *Hba* complex] *Pgk1-ps2* [second pseudogene to the functional Pgk1 locus]
Related loci	Gene symbol followed by *rs* and an appropriate serial number	*Ela1-rs1* [first related locus identified by the elastase-1 gene probe]
Expressed sequence tags (ESTs)	Symbolized by GenBank sequence identification number	AA066038
Sequence tagged sites (STSs)	No symbols are assigned other than those given by the laboratory that produced or used them.	
Loci recognized by anonymous DNA probes	D-number symbols composed of D for DNA, the chromosome designation, the laboratory code, and a unique serial number; set in italics	*D1Pas5* [fifth D-locus developed and mapped on Chr 1 at the Pasteur Institute]
Phenotype (antigen loci, enzyme loci)	Same elements as for gene symbol but with capital, nonitalic letters, and superscripts lowered to the type line	GPI1A GPI1B

[a] Based on International Committee on Standardized Genetic Nomenclature for Mice[7] (http://www.informatics.jax.org/mgihome/nomen/gene.shtml).

[b] In general, the initial letter is uppercase and the following letters are lowercase. However, for genes recognized only by a recessive mutation, the symbol begins with a lowercase letter. Numerals may be used if they occur in the name or abbreviation on which the symbol is based; roman numerals and Greek letters are not allowed. The complete gene or allele symbol, including superscripts, must not exceed 10 characters.

[c] When superscript symbols are not available (as in computer printouts), the superscript letter(s) may be replaced by an asterisk or angle brackets and the letter(s)—for example, *Hbb*d* or *Hbb<d>* rather than Hbb^{d}.

21.15.7 Human

1) Genes. Guidelines for human gene nomenclature have been developed and published by the HUGO Gene Nomenclature Committee.[25,163] Table 21.32[164–166] summarizes the major conventions. Consult the committee before publication of any new human gene symbol to ensure that the symbol is unique and suitable and that it can be reserved for the gene in question. When necessary to distinguish a human gene from one of another organism that would otherwise have the same symbol, use the nonitalic designation HUMAN in parentheses before the gene symbol; for example (HUMAN) *ABCA1*. Designations and orthography for messenger RNA (mRNA) and complementary DNA (cDNA) symbols generally follow those for the corresponding genes.
2) Human leukocyte antigen (HLA). The human major histocompatibility complex has its own committee, the World Health Organization Nomenclature Committee for Factors of the HLA System,[167] and its own database.[168,169] A gene name begins with HLA-, followed by a locus symbol and a specific allele designation.

   *HLA-DRB1*13* *HLA-DRB1*1301* *HLA-DRB1*13012*

Table 21.31 Mouse chromosome terminology: conventions and examples[a]

Symbol	Feature	Convention	Examples
. . .	bands	Chromosome bands on the short (i.e., p) and long (i.e., q) arms are numbered from the centromere toward the telomeres and are designated by the chromosome number, p or q arm, the region number, and the band-specific letter. Further subdivisions are numbered following a decimal point.	17B 17B1 17B1.1
		Reference to the q arm may be omitted when the meaning is clear.	15qE1 *or* 15E1
		When the positions of the chromosomal breakpoints relative to G-bands are known, add the band numbers after the appropriate chromosome numbers.	T(2H1;8A4)26H [reciprocal translocation, breaks in band H1 of Chr 2 and band A4 of Chr 8]
. . .	series symbol	Designate successive anomalies discovered at the same laboratory by adding the series number and laboratory code to the end of the description of the anomaly.	Dp(1)6H [sixth duplication found by Harwell in Chr 1] In(2)5Rk [fifth inversion found by Roderick in Chr 2]
Cen	centromere	Use Cen when referring to the centromere itself. In Robertsonian translocations, indicate the centromere with a period.	Rb(9.19)163H
Chr	chromosome	Begin the word "chromosome" with a capital letter when referring to a specific chromosome; abbreviate it to Chr after the first use.	Chromosome 2, Chr 2
Del	deletion	Use Del to define interstitial losses that are often, but not always, visible. Do not use it for single-gene deletions; treat them as alleles instead.	Del(7E1)c8R1 [deletion of band 7E1 manifesting as a mutation to albino, Tyr^{c}]
Dp	duplication	Follow Dp with the number of the chromosome involved in parentheses.	Dp(1)6H [duplication, Chr 1, Harwell 6]
Hc	pericentric heterochromatin	Follow Hc with the number of the chromosome involved.	Hc14 [pericentric heterochromatin on Chr 14]
In	inversion	Follow In with the chromosome number in parentheses; separate bands with semicolons.	In(2)5Rk [fifth inversion found by Roderick in Chr 2] In(5C2;15E1)Rb3Bnr 1Ct [first inversion found by Cattanach in Rb3Bnr and involving bands 5C2 and 15E1]
. . .	inversion, pericentric	Use the symbols pq and/or the appropriate band numbers.	In(Ypq) [pericentric inversion involving the Y chromosome] In(8pqA2) [pericentric inversion with breaks in the short arm and in band A2 of the long arm]
Is	insertion	List the number of the chromosome donating the inserted portion first.	Is(7;1)40H [an insertion of part of Chr 7 into Chr 1] Is(In7F1-7C;XF1)1Ct [an inverted insertion of segment 7F1-7C of Chr 7 into X at band F1]

Table 21.31 Mouse chromosome terminology: conventions and examples[a] (*continued*)

Symbol	Feature	Convention	Examples
Ms	monosomy	Follow Ms with the number of the chromosome involved.	Ms12 [an animal with only1 copy of Chr 12]
	multiple anomalies	When an animal has 2 or more anomalies that are potentially separable by recombination, list them all.	Rb(16.17)Bnr T(1;17)190Ca /++ [an animal with a Robertsonian translocation and a reciprocal translocation involving Chr 17] Rb(5.15)3Bnr +/+ In(5)9Rk [an animal with a Robertsonian translocation and an inversion involving Chr 5 in repulsion]
		When one anomaly is contained within another or is inseparable from it, combine the symbols.	T(In1;5)44H [translocation between Chrs 1 and 5, Chr 1 segment inverted]
Ns	nullisomy	Follow Ns with the number of the chromosome involved.	Ns2 [an animal with no copies of Chr 2]
p	short arm	Breaks in the short arm must be designated with a p	T(Yp;5)## [translocation involving a break in the short arm of Y and the long arm of Chr 5]
	polymorphic loci	Designate polymorphic loci within the ribosomal DNA region with gene symbol, Rnr, and the chromosome number.	Rnr12 [ribosomal DNA region on Chr 12]
q	long arm	Rearrangements are assumed to be in the long arm unless stated otherwise.	T(Yp;5)## [translocation involving a break in the short arm of Y and the long arm of Chr 5]
Rb	Robertsonian translocation	Indicate the centromere with a period.	Rb(9.19)163H
T	translocation	Separate the chromosomes involved with a semicolon.	T(4;X)37H T(2H1;8A4)26H [reciprocal translocation, breaks in band H1 of Chr 2 and band A4 of Chr 8]
Tel	telomere	Use Tel when referring to the telomere itself. Follow the symbol with the number of the chromosome involved. Use roman type in karyotype descriptions, italics when treating the telomere as a locus.	Tel14 Tel14p1 [first telomere sequence mapped at the centromeric end of Chr 14] Tel19q2 [second telomere sequence mapped to the distal end of Chr 19]
Tet	tetrasomy	Follow Tet with the number of the chromosome involved.	Tet14 [an animal with 4 copies of Chr 14]
Tp	transposition	Follow Tp with the number of the chromosome involved in parentheses.	Tp(Y)1Ct [transposition, Chr Y, Cattanach 1]
Ts	trisomy	Follow Ts with the number of the chromosome involved.	Ts16 [an animal with 3 copies of Chr 16]

[a] Based on International Committee on Standardized Genetic Nomenclature for Mice[8] (http://www.informatics.jax.org/mgihome/nomen/anomalies.shtml).

Table 21.32 Human genes: symbols for genes and phenotypes[a]

Feature	Convention	Examples
Gene name	Brief description of the character or function of the gene in American spelling; begin with lowercase letter unless named for an eponymous disease, syndrome, or phenotype; nonitalic	alcohol dehydrogenase aconitase 1, soluble Allan-Herndon-Dudley syndrome
Gene symbol	Uppercase italic letters[b] or a combination of uppercase letters and arabic numerals based on gene name; first letter of name is also the first letter of the symbol; all elements on the line.	*ADH* *ACO1* *AHDS*
	Greek letters, roman numerals, superscripts, subscripts, and punctuation are not acceptable.	β GAL-1 [becomes] *GLB1*
Allele	Same as for genes; allele characters (4 or fewer)separated from gene symbol by an asterisk[c]	*ADA*1 HBB*6V*
Gene cluster	The character @ at the end of the symbol represents a gene cluster.	*IFN@* [interferon type 1 cluster]
Oncogenes	Symbols correspond to the homologous retroviral oncogene, but without the v- or c- prefixes. When referring specifically to the human locus, omit the prefix and capitalize the symbol.	*MYC, HRAS*
Putative genes	Genes predicted from expressed sequence tags (EST) clusters or genomic sequence alone are regarded as putative and are designated[d] by the chromosome of origin and an arbitrary number in the format C#orf#	*C2orf1* [chromosome 2 open reading frame 1]
DNA segment	Arbitrary DNA fragments and loci can be designated[e] with a capital letter D followed by chromosome number, uppercase letter indicating sequence type, and arbitrary numbers.[See Section 21.2.6 for details.] When known to be an expressed sequence, the suffix E can be added.	*D9S14* *D1F10S1* *DXS1004E*
Pseudogenes	Pseudogenes have been identified with the gene symbol of the structural gene followed by a P for pseudogene, and if necessary, a number. In future, pseudogenes will be assigned the next number in the relevant symbol series, followed by P or PS if requested.	*PGK1P1* *RPL32P* *OR5B12P*
Transposable elements	Transposable elements have the stem symbol TE followed by a character string derived from the name of the element; set in italics.	*TEMAR1* [transposable element mariner-type 1]

[a] Based on Wain et al.[17]

[b] Italic letters need not be used in gene catalogs. Where an italic typeface is not available, a gene symbol can be indicated by following it with an asterisk, and an allele symbol by preceding it with an asterisk. Base the names and symbols for genes encoding enzymes on the names recommended by the International Union of Biochemistry and Molecular Biology.[134]

[c] Allele terminology is now the responsibility of the Human Genome Variation Society[166] (http://www.hgvs.org/). The system has been described in Antonarakis[164] and den Dunnen and Antonarakis.[165]

[d] Obtain *C#orf* numbers from the HUGO Gene Nomenclature Committee, which keeps records of all assigned numbers: nome@galton.ucl.ac.uk.

[e] Obtain D numbers from the GDB: Genome Database: help@gdb.org.

3) Chromosomes. Human chromosome nomenclature is based on the recommendations of the International Standing Committee on Human Cytogenetic Nomenclature (ISCN). The most recent report,[5] which summarizes the current nomenclature, incorporates and supersedes all previous ISCN recommendations.

3a) *Chromosome number and morphology*. The autosomes are numbered from 1 to 22 in decreasing order of length. The sex chromosomes are referred to as X and Y. When the chromosomes are stained by methods that do not produce bands, they can be arranged into 7 readily distinguishable groups (A through G) based on descending order by size and the position of the centromere.

After chemical treatment and staining, each chromosome displays a continuous series of light and dark bands. A band is defined as that part of a chromosome that is clearly distinguishable from its adjacent segments by appearing darker or lighter with one or more banding techniques. Thus, by definition, there are no interbands. The bands are allocated to various regions along the chromosome arms; the regions are delimited by specific landmarks (consistent and distinct morphological features, including the ends of the chromosome arms, the centromere, and certain bands). A region is defined as an area of a chromosome lying between 2 adjacent landmarks.

Regions and bands are numbered consecutively from the centromere outward along each chromosome arm. The symbols p and q are used to designate, respectively, the short and long arms of each chromosome. The centromere (cen) itself is designated 10; the region facing the short arm is p10, the region facing the long arm is q10. The 2 regions adjacent to the centromere are labeled as 1 in each arm; the next, more distal regions as 2; and so on.

In designating a particular band, 4 items are required: the chromosome number, the arm symbol, the region number, and the band number within that region. These items are given in order without spacing or punctuation.

1p31 [chromosome 1, short arm, region 3, band 1]

Whenever an existing band is subdivided, a decimal point is placed after the original band designation and is followed by the number assigned to each sub-band. The sub-bands are numbered sequentially from the centromere outward. For example, if the original band 1p31 is subdivided, the sub-bands are labeled 1p31.1, 1p31.2, and so on. If a sub-band is subdivided, additional digits, but no further punctuation, are used; for example, sub-band 1p31.1 might be further subdivided into 1p31.11, 1p31.12, and so on.

3b) *Karyotype designation*. In the description of a karyotype, the first item to be recorded is the total number of chromosomes, including the sex chromosomes, followed by a comma, followed by the sex chromosome constitution. The autosomes are specified only when an abnormality is present.

46,XX [normal female] 46,XY [normal male]

In the description of chromosome abnormalities, sex chromosome aberrations are presented first, followed by abnormalities of the autosomes listed in numerical order irrespective of aberration type. Each abnormality is separated from the next by a comma. The only exception to this rule is the convention to designate constitutional numerical sex chromosome abnormalities by listing all sex chromosomes after the chromosome number—for example, 48,XXXY.

Letter designations are used to specify rearranged (i.e., structurally altered) chromosomes. All symbols and abbreviated forms used to designate chromosome abnormalities in humans are listed in Table 21.33, and examples of karyotypic descriptions are presented in Table 21.34. In single chromosome rearrangements, the chromosome involved in the change is specified within parentheses immediately following the symbol identifying the type of rearrangement; for example, inv(2), del(4), r(18). If 2 or more chromosomes have been altered, a semicolon is used to separate their designations. If one of the rearranged chromosomes is a sex chromosome, then it is listed first; otherwise, the chromosome having the lowest number is always specified first; for example, t(X;3) or t(2;5). An exception to this rule involves certain 3-break rearrangements in which part of one chromosome is inserted at a point of breakage in another chromosome. In this event, the recipient chromosome is specified first, regardless of whether it is a sex chromosome or an autosome with a number higher or lower than that of the donor chromosome; for example, ins(5;2).

For balanced translocations involving 3 separate chromosomes, with one breakpoint in each chromosome, the rule is still followed that the sex chromosome or autosome with the lowest number is specified first. The chromosome listed next is the one that receives a segment from the first chromosome, and the chromosome specified last is the one that donates a segment to the first listed chromosome. The same rule is followed in 4-break and more complex balanced translocations. To distinguish homologous chromosomes, one of the numerals may be underlined (single underlining).

Breakpoints are specified within parentheses immediately following the designation of the type of rearrangement and the chromosome(s) involved. The breakpoints are identified by band designations and are listed in the same order as the chromosomes involved, again separated by semicolons—for example, t(2;5)(q22;p14). No semicolon is used between breakpoints in single chromosome rearrangements—for example, inv(2)(p21q31).

A derivative chromosome (der) is a structurally rearranged chromosome generated either by a rearrangement involving 2 or more chromosomes or by multiple aberrations within a single chromosome. The term always refers to chromosomes with intact centromeres. Derivative chromosomes are specified in parentheses, followed by all aberrations involved in the generation of the derivative chromosome. The term "Philadelphia chromosome" is retained for historical reasons to describe the derivative chromosome 22 generated by the translocation t(9;22)(q34;q11). The abbreviation Ph (formerly Ph1) may be used in text but not in the description of the karyotype, where der(22)t(9;22)(q34;q11) is recommended.

A plus or minus sign is placed before a chromosome or an abnormality designation to indicate additional or missing normal or abnormal chromosomes; for example, +21, −7, +der(2). When normal chromosomes are replaced by structurally altered chromosomes, the normal ones should not be recorded as missing. The plus or minus signs placed after a chromosome arm symbol (p or q) may be used in text to indicate an increase or decrease in the length of a chromosome arm (e.g., 4p+, 5q−) but should not be used in the description of karyotypes. The multiplication sign (×) can be used to describe multiple copies of a rearranged chromosome but should not be used to denote multiple copies of normal chromosomes.

Uncertainty in chromosome or band designation may be indicated by a question mark or an approximate sign (~). The term "or" is used to indicate alternative interpretations of an aberration.

The same rules for designating chromosome aberrations are followed in the description of constitutional and acquired chromosome aberrations. When an acquired chromosome abnormality is found in an individual with a constitutional chromosome anomaly, the latter is indicated by the lowercase nonitalic letter c after the abnormality designation. The karyotype designations of different clones are separated by a slash. The absolute number of cells in each clone is presented within square brackets, placed after the karyotype description.

Table 21.33 Human chromosomes: symbols and abbreviations[a]

Symbol	Description
AI	first meiotic anaphase
AII	second meiotic anaphase
ace	acentric fragment
add	additional material of unknown origin
approximate sign (~)	denotes intervals and boundaries of a chromosome segment
arrow (→ or ->)	from – to; separates the unaltered karyotype from the altered karyotype, the direction of the arrow indicating the direction of change
b	break
brackets, angle (‹ ›)	surround the ploidy level
brackets, square ([ ])	surround number of cells
c	constitutional anomaly
cen	centromere
chi	chimera
chr	chromosome
cht	chromatid
colon, single (:)	break
colon, double (::)	break and reunion
comma (,)	separates chromosome numbers, sex chromosomes, and chromosome abnormalities
cp	composite karyotype
cx	complex chromatid interchanges
decimal point (.)	denotes sub-bands
del	deletion
de novo	designates a chromosome abnormality that has not been inherited
der	derivative chromosome
dia	diakinesis
dic	dicentric
dip	diplotene
dir	direct
dis	distal
dit	dictyotene
dmin	double minute
dup	duplication
e	exchange
end	endoreduplication
equal sign (=)	number of chiasmata
fem	female
fis	fission, at the centromere
fra	fragile site
g	gap
h	heterochromatin, constitutive
hsr	homogeneously staining region
i	isochromosome
idem	denotes the stemline karyotype in subclones
ider	isoderivative chromosome
idic	isodicentric chromosome
inc	incomplete karyotype
ins	insertion

Table 21.33 Human chromosomes: symbols and abbreviations[a] (*continued*)

Symbol	Description
inv	inversion or inverted
lep	leptotene
MI	first meiotic metaphase
MII	second meiotic metaphase
mal	male
mar	marker chromosome
mat	maternal origin
med	medial
min	minute acentric fragment
minus sign (–)	loss
ml	mainline
mn	modal number
mos	mosaic
multiplication sign (×)	multiple copies of rearranged chromosomes
oom	oogonial metaphase
or	alternative interpretation
p	short arm of chromosome
PI	first meiotic prophase
pac	pachytene
parentheses ( )	surround structurally altered chromosomes and breakpoints
pat	paternal origin
pcc	premature chromosome condensation
pcd	premature centromere division
Ph	Philadelphia chromosome
plus sign (+)	gain
prx	proximal
psu	pseudo-
pvz	pulverization
q	long arm of chromosome
qdp	quadruplication
qr	quadriradial
question mark (?)	questionable identification of a chromosome or chromosome structure
r	ring chromosome
rcp	reciprocal
rea	rearrangement
rec	recombinant chromosome
rob	Robertsonian translocation
roman numerals I–IV	indicate univalent, bivalent, trivalent, and quadrivalent structures
s	satellite
sce	sister chromatid exchange
sct	secondary constriction
sdl	sideline
semicolon (;)	separates altered chromosomes and breakpoints in structural rearrangements involving more than one chromosome
sl	stemline
slant line (/)	separates clones

Table 21.33 Human chromosomes: symbols and abbreviations[a] (*continued*)

Symbol	Description
spm	spermatogonial metaphase
stk	satellite stalk
t	translocation
tan	tandem
tas	telomeric association
tel	telomere
ter	terminal (end of chromosome)
tr	triradial
trc	tricentric chromosome
trp	triplication
underlining (single)	used to distinguish homologous chromosomes
upd	uniparental disomy
v	variant or variable region
xma	chiasma(ta)
zyg	zygotene

[a] Adapted from ISCN 2009.[5]

3c) *In situ hybridization.* Results obtained by in situ hybridization (ISH) are described using the symbol ish. If a standard cytogenetic observation has been made, it may be given followed by a decimal point, followed by the abbreviation ish and the ISH results. If a standard cytogenetic observation has not been made, the ISH observations only are given; for details, see ISCN 2009.[5]

Observations on structurally abnormal chromosomes are expressed by the symbol ish, followed by the symbol for the structural abnormality (whether seen by standard techniques and ISH or only by ISH), followed in separate parentheses by the chromosome(s), the breakpoint(s), and the locus or loci for which probes were used, designated according to GDB (Genome Database),[14] and ordered from pter to qter. If no GDB locus is available, the probe name can be used. The locus designations (in uppercase letters but not in italics) are separated by commas, and the status of each locus is given immediately after the locus designation—for example, 46,XY.ish del(22)(q11.2q11.2)(D22S75–).

Observations on normal chromosomes are expressed by the symbol ish followed by the chromosome, region, band, or sub-band designation of the locus or loci tested (not in parentheses), followed in parentheses by the locus (loci) tested, a multiplication sign (×), and the number of signals seen—for example, 46,XY.ish 22q11.2(D22S75×2).

ISCN 2009[5] further clarifies, and provides examples of, nomenclature based on newer ISH technologies: interphase or nuclear ISH (nuc ish), extended chromatin or fiber ISH (fib ish), reverse ISH (rev ish), comparative genomic hybridization (cgh), chromosome painting, and microarray techniques, including multiple ligation-dependent probe amplification (MLPA).

21.15.8 Transgenic Animals

The International Committee on Standardized Genetic Nomenclature for Mice has approved the symbolic representation of transgenic animal lines. The guidelines are posted on the Mouse Genome Informatics (MGI) website.[7]

Use the symbol *Tg* to designate genetically engineered transgenic events that result

Table 21.34 Human chromosomes: examples of karyotypic descriptions[a]

Karyotype	Explanation
45,X	Karyotype with one X chromosome (Turner syndrome)
45,XX,–7	Female karyotype with monosomy 7
47,XXY	Karyotype with 2 X chromosomes and 1 Y chromosome (Klinefelter syndrome)
47,XY,+21	Male karyotype with trisomy 21
47,XY,+mar	One additional marker chromosome
46,XX,inv(3)(q21q26)	Inversion of one chromosome 3 with breakpoints in bands 3q21 and 3q26
46,XX,del(5)(q13)	Terminal deletion with a break in band 5q13
46,XX,del(5)(q13q33)	Interstitial deletion with breakage and reunion of bands 5q13 and 5q33
46,XX,del(5)(q?)	Deletion of the long arm of chromosome 5, but it is unclear whether it is a terminal or an interstitial deletion, and the breakpoint(s) are unknown
46,XX,dup(1)(q22q25)	Direct duplication of the segment between bands 1q22 and 1q25
46,XY,add(19)(p13)	Additional material of unknown origin attached to band 19p13
46,XX,ins(5;2)(p14;q22q32)	Direct insertion. The long-arm segment between bands 2q22 and 2q32 has been inserted into the short arm of chromosome 5 at band 5p14. The original orientation of the inserted segment has been maintained in its new position (i.e., 2q22 remains more proximal to the centromere than 2q32). Note that the recipient chromosome is specified first.
46,XY,t(2;5)(q21;q31)	Reciprocal translocation between chromosomes 2 and 5. Breakage and reunion have occurred at bands 2q21 and 5q31. The segments distal to these bands have been exchanged.
46,XX,t(2;7;5)(p21;q22;q23)	The segment on chromosome 2 distal to 2p21 has been translocated onto chromosome 7 at band 7q22, the segment on chromosome 7 distal to 7q22 has been translocated onto chromosome 5 at 5q23, and the segment of chromosome 5 distal to 5q23 has been translocated onto chromosome 2 at 2p21.
46,XX,der(1)t(1;3)(p22;q13)	A derivative chromosome 1 has resulted from a translocation of the chromosome 3 segment distal to3q13 to the short arm of chromosome 1 at band 1p22. The der(1) replaces a normal chromosome 1 and there is no need to indicate the missing chromosome. There are 2 normal chromosomes 3. The karyotype is unbalanced with loss of the segment 1p22-pter and gain of 3q13-qter.
46,XY,del(8)(q21) or i(8)(p10)	A deletion of the long arm of chromosome 8 with a breakpoint in 8q21 or an isochromosome for the short arm of chromosome 8. Note that there should be a space before and after the symbol "or".

[a] Table courtesy of Felix Mitelman.

from random insertion of DNA into the genome. The full symbol for the transgenic line consists of 4 parts in italic characters in the form *Tg(YYY)###Zzz*, where *Tg* indicates a transgenic insertion by microinjection, *(YYY)* indicates the inserted sequence in the official nomenclature for that gene, ### is the laboratory's transgene line designation or a serial number, and *Zzz* is the laboratory code assigned by the Institute for Laboratory Animal Research (ILAR).[11] The strain background is independent of the nomenclature, but should precede it whenever possible.

C57BL/6J-*Tg(CD8)1Jwg* [The human CD8 sequence inserted into C57BL/6 mice from Jackson Laboratory (J). The first mouse was reported by author Jon W Gordon (Jwg).]

Whether the transgene inserted "benignly" into the genome or disrupted a locus is irrelevant to the nomenclature but should be reported elsewhere. However, if a disrupted locus is identified, the transgene becomes an allele of that locus.

> *Tg(Crya1)2Ove* [The mouse α1 crystallin gene was inserted by microinjection and reported by Paul Overbeek, his second transgenic line. In a later publication, he reported the insertion is allelic with the downless (*dl*) mutant mouse; the symbol for the allele became $dl^{Tg(Crya1)2Ove}$. In yet another publication, the *dl* locus and the locus disrupted by the transgene insertion were identified as the gene encoding the ectodysplasin-A receptor (*Edar*). The symbol changed again to $Edar^{dl\text{-}Tg(Crya1)2Ove}$.]

When a construct is composed of roughly equal parts of 2 genes (i.e., a fusion gene), the symbols of the genes are separated by a slash.

> *Tg(TCFE2A/HLF)1Mlc* [The human transcription factor E2Aα and the hepatic leukemia factor were expressed as a chimeric cDNA and reported by ML Cleary.]

When the construct is a reporter (e.g., GFP, LacZ) or a recombinase (e.g., Cre), the promoter is an essential construct element. Thus, the promoter is included in the parentheses, separated from the inserted element by a hyphen.

> *Tg(Zp3-Cre)3Mrt* [The *Zp3* promoter plus *Cre*; the third transgenic line from the laboratory of G Martin, Mrt.]

Transgene symbols can be abbreviated by omitting the insert; for example, the full symbol *TgN(GPDHIm)1Bir* would be abbreviated *TgN1Bir*. The full symbol should be used the first time the transgene is mentioned in a publication and the abbreviation used thereafter.

CITED REFERENCES

1. Dunn LC, Grünberg H, Snell GD. Report of the Committee on Mouse Genetics Nomenclature. J Hered. 1940;31:505–506.

2. International Committee on Genetic Symbols and Nomenclature. Report of the International Committee on Genetic Symbols and Nomenclature. Union Int Sci Biol, Ser B. 1957;(30):1–6.

3. Demerec M, Adelberg EA, Clark AJ, Hartman PE. A proposal for a uniform nomenclature in bacterial genetics. Genetics. 1966;54(1):61–76.

4. American Society for Microbiology. ASM style manual for journals and books. Washington (DC): American Society for Microbiology; 1991. This edition no longer available.

5. ISCN 2009. An international system for human cytogenetic nomenclature. Shaffer LG, Slovak ML, Campbell LJ, editors. Basel (Switzerland): Karger; 2009.

6. HUGO Gene Nomenclature Committee. Cambridge (UK): HUGO Nomenclature Committee; updated 2005 May 31 [accessed 2011 Nov 19]. http://www.genenames.org/.

7. International Committee on Standardized Genetic Nomenclature for Mice. Rules and guidelines for gene, allele, and mutation nomenclature. In: Mouse genome informatics. Bar Harbor (ME): Jackson Laboratory; 2005 [revised 2011 Sep; accessed 2011 Nov 19]. http://www.informatics.jax.org/mgihome/nomen/gene.shtml.

8. International Committee on Standardized Genetic Nomenclature for Mice. Rules for nomenclature of chromosome aberrations. In: Mouse genome informatics. Bar Harbor (ME): Jackson Laboratory; 2005 [revised Jan 2005; accessed 2011 Nov 19]. http://www.informatics.jax.org/mgihome/nomen/anomalies.shtml.

9. Price CA, Reardon EM. Mendel, a database of nomenclature for sequenced plant genes. Nucleic Acids Res. 2001;29(1):118–119.

10. Gene Ontology Consortium. [Place unknown]: Gene Ontology Consortium; c1999–2005 [accessed 2011 Nov 19]. http://www.geneontology.org.

11. Institute for Laboratory Animal Research. Washington (DC): National Academy of Sciences; c2005 [accessed 2011 Nov 19]. http://dels.nas.edu/ilar/.

12. Olson M, Hood L, Cantor C, Botstein D. A common language for physical mapping of the human genome. Science. 1989 Sep 29;245(4925):1434–1435.

13. National Center for Biotechnology Information. Bethesda (MD): National Center for Biotechnology Information; 2004 [accessed 2011 Nov 19]. Whole genome shotgun submissions. http://www.ncbi.nlm.nih.gov/Genbank/wgs.html.

14. GDB: The Human Genome Database. Washington (DC): US Department of Energy; 2005 [accessed 2011 Nov 19]. http://www.ornl.gov/sci/techresources/Human_Genome/home.shtml.

15. Mouse Genome Informatics (MGI). Bar Harbor (ME): Jackson Laboratory; 2011 [accessed 2011 Nov 19]. http://www.informatics.jax.org/.

16. Rat Genome and Nomenclature Committee. Milwaukee (WI): Medical College of Wisconsin; [accessed 2011 Nov 19]. Rat idiogram. http://rgd.mcw.edu/nomen/nomen.shtml.

17. Wain HM, Bruford EA, Lovering RC, Lush MJ, Wright MW, Povey S. Guidelines for human gene nomenclature. Genomics 2002;79(4):464–470. Updated version. [accessed 2012 Aug 26]. http://www.genenames.org/guidelines.html.

18. Roberts RJ. Restriction enzymes and their isoschizomers. Nucleic Acids Res. 1989;17(suppl): 347–387.

19. REBASE: The Restriction Enzyme Database. Beverly (MA): New England BioLabs; c2003 [accessed 2011 Nov 19]. http://rebase.neb.com.

20. Safer B. Nomenclature of initiation, elongation and termination factors for translation in eukaryotes: recommendations 1988. Eur J Biochem. 1989;186(1–2):1–3.

21. Campbell A, Berg DE, Botstein D, Lederberg EM, Novick RP, Starlinger P, Szybalski W. Nomenclature of transposable elements in prokaryotes. Gene. 1979;5(3):197–206.

22. Kempken F, Kück U. Transposons in filamentous fungi—facts and perspectives. BioEssays. 1998;20(8):652–659.

23. Nebert DW, Adesnik M, Coon MJ, Estabrook RW, Gonzalez FJ, Guengerich FP, Gunsalus IC, Johnson EF, Kemper B, Levin W, et al. The P450 gene superfamily: recommended nomenclature. DNA. 1987;6(1):1–11.

24. Nelson DR, Koymans L, Kamataki T, Stegeman JJ, Feyereisen R, Waxman DJ, Waterman MR, Gotoh O, Coon MJ, Estabrook RW, et al. P450 superfamily: update on new sequences, gene mapping, accession numbers and nomenclature. Pharmacogenetics. 1996;6(1):1–42.

25. Yoder OC, Valent B, Chumley F. Genetic nomenclature and practice for plant pathogenic fungi. Phytopathology. 1986;76(4):383–385.

26. Novick RP, Clowes RC, Cohen SN, Curtiss R III, Datta N, Falkow S. Uniform nomenclature for bacterial plasmids: a proposal. Bacteriol Rev. 1976;40(1):168–189.

27. Instructions for Authors. Journal of Bacteriology. Washington (DC): American Society for Microbiology. http://jb.asm.org/site/misc/ifora.xhtml.

28. Gallo R, Wong-Stahl F, Montagnier L, Haseltine WA, Yoshida M. HIV/HTLV gene nomenclature [letter]. Nature. 1988;333(6173):504.

29. Chumley FG, Menzel R, Roth JR. Hfr formation directed by Tn10. Genetics. 1979;91(4):639–655.

30. Low KB. *Escherichia coli* K-12 F-prime factors, old and new. Bacteriol Rev. 1972;36(4):587–607.

31. Berlyn M, Rudd K, Chater K. Bacteria. In: Wood R, editor. Genetic nomenclature guide. West Sussex (UK): Elsevier; 1998. p. S.1–S.4. Supplement to Trends in Genetics.

32. Sherman F. An introduction to the genetics and molecular biology of the yeast *Saccharomyces cerevisiae*. In: University of Rochester Medical Center. Rochester (NY): University of Rochester; 1998 [updated 2000 Sep 19; accessed 2011 Nov 19]. http://dbb.urmc.rochester.edu/labs/sherman_f/yeast/index.html.

33. SGD: *Saccharomyces* Genome Database. Stanford (CA): Stanford University; c1997–2011 [accessed 2011 Nov 19]. SGD gene naming guidelines. http://www.yeastgenome.org/gene_guidelines.shtml.

34. Cherry JM. *Saccharomyces cerevisiae*. In: Wood R, editor. Genetic nomenclature guide. West Sussex (UK): Elsevier; 1998. p. S.10–S.11. Supplement to Trends in Genetics.

35. Kohli J. Genetic nomenclature and gene list of the fission yeast *Schizosaccharomyces pombe*. Curr Genet. 1987;11(8):575–589.

36. Fantes P, Kohli J. *Schizosaccharomyces pombe*. In: Wood R, editor. Genetic nomenclature guide. West Sussex (UK): Elsevier; 1998. p. S.7–S.9. Supplement to Trends in Genetics.

37. Turgeon BG, Yoder OC. Proposed nomenclature for mating type genes of filamentous ascomycetes. Fung Genet Biol. 2000;31(1):1–5.

38. Clutterbuck AJ. Gene symbols in *Aspergillus nidulans*. Genet Res. 1973;21(3):291–296.

39. Sermonti G. Genetics of antibiotic-producing microorganisms. London (UK): Wiley-Interscience; 1969.

40. Clutterbuck AJ. *Aspergillus nidulans*. In: Wood R, editor. Genetic nomenclature guide. West Sussex (UK): Elsevier; 1998. p. S.12–S.13. Supplement to Trends in Genetics.

41. Clutterbuck J, compiler. The *Aspergillus nidulans* linkage map. Glasgow (Scotland): University of Glasgow; [updated Nov 2008; accessed 2011 Nov 19]. http://www.fgsc.net/Aspergillus/gene_list/index.htm.

42. *Aspergillus nidulans* Database. Cambridge (MA): Massachusetts Institute of Technology, Broad Institute, Center for Genome Research; 2003 [accessed 2011 Nov 19]. http://www.broadinstitute.org/annotation/genome/aspergillus_group/MultiHome.html.

43. Clutterbuck AJ. The validity of the *Aspergillus nidulans* linkage map. Fungal Genet Biol. 1997;21(3):267–277.

44. Perkins DD. *Neurospora* genetic nomenclature. In: Fungal Genetics Newsletter 46:31–41 [accessed 2011 Nov 19]. http://www.fgsc.net/fgn46/perkins.htm.

45. Fungal Genetics Stock Center. Kansas City (KS): University of Missouri–Kansas City; c2005 [updated 2011 Sep 21; accessed 2011 Nov 19]. http://www.fgsc.net.

46. Cerda-Olmedo E. Carotene mutants of *Phycomyces*. Methods Enzymol. 1985;105:220–243.

47. Eslava AP, Alvarez MI. Genetics of *Phycomyces*. In: Bos CJ, editor. Fungal genetics: principles and practice. New York (NY): Marcel Dekker; 1996. p. 385–406.

48. Yoder OC, Valent B, Chumley F. Genetic nomenclature and practice for plant pathogenic fungi. Phytopathology. 1986;76(4):383–385.

49. Chlamy Database. Durham (NC): Duke University; 2005 Aug 18 [accessed 2011 Nov 19]. http://www.chlamy.org.

50. Dutcher S, Harris E. *Chlamydomonas reinhardtii*. In: Wood R, editor. Genetic nomenclature guide. West Sussex (UK): Elsevier; 1998. p. S.18–S.19. Supplement to Trends in Genetics.

51. Lefebvre P. Molecular and genetic maps of the nuclear genome. 2002. In: Chlamy Center. Durham (NC): Duke University; 2004 [accessed 2011 Nov 19]. http://www.chlamy.org/nuclear_maps.html.

52. Kay R, Loomis B, Devreotes P. *Dictyostelium discoideum*. In: Wood R, editor. 1998. Genetic nomenclature guide. West Sussex (UK): Elsevier; 1998. p. S.5–S.6. Supplement to Trends in Genetics.

53. Clayton C, Adams M, Almeida R, Baltz T, Barrett M, Bastien P, Belli S, Beverley S, Biteau N, Blackwell J, et al. Genetic nomenclature for *Trypanosoma* and *Leishmania*. Mol Biochem Parasitol. 1998;97(1–2):221–224.

54. Allen SL, Altschuler MI, Bruns PJ, Cohen J, Doerder FP, Gaertig J, Gorovsky M, Orias E, Turkewitz A. Proposed genetic nomenclature rules for *Tetrahymena thermophila*, *Paramecium primaurelia* and *Paramecium tetraurelia*. Genetics. 1998;149(5):459–462.

55. GrainGenes: A Database for Triticeae and Avena. Albany (CA): US Department of Agriculture; 2003 [accessed 2011 Nov 19]. http://wheat.pw.usda.gov/GG2/index.shtml.

56. Livingstone DL, Lackney VK, Blauth JR, van Wijk R, Jahn MK. Genome mapping in Capsicum and the evolution of genome structure in the Solanaceae. Genetics. 1999;152(3):1183–1202.

57. Meinke D, Koornneef M. Community standards for *Arabidopsis* genetics. Plant J. 1997;12(2): 247–253.

58. Raupp WJ, Friebe B, Gill BS. Suggested guidelines for the nomenclature and abbreviation of the genetic stocks of wheat, *Triticum aestivum* L. em Thell., and its relatives. Wheat Inf Serv. 1995;81:51–55. Also available at http://wheat.pw.usda.gov/ggpages/nomenclature.html.

59. Simons MD, Martens JW, McKenzie RIH, Nishiyama I, Sadanaga K, Sebesta J, Thomas H. Oats: a standardized system of nomenclature for genes and chromosomes and catalog of genes governing characters. Agriculture Handbook Number 509. Washington (DC): US Department of Agriculture; 1978.

60. UniProt Knowledgebase. Geneva (Switzerland): Swiss Institute of Bioinformatics; 2005 [released 2011 Nov 16; accessed 2011 Nov 19]. UniProtKB/Swiss-Prot protein knowledgebase documentation, controlled vocabulary of species. http://www.uniprot.org/docs/speclist.

61. Meinke DW, Cherry JM, Dean C, Rounsley SD, Koornneef M. *Arabidopsis thaliana*: a model plant for genome analysis. Science. 1998;282(5389):662–682. Includes a foldout genome map.

62. Robinson RW, McCreight JD, Ryder EJ. The genes of lettuce and closely related species. Plant Breeding Rev. 1983;1:267–293.

63. Thomas H. Cytogenetics of *Avena*. In: Marshal HG, Sorrells ME, editors. Oat science and technology. Madison (WI): American Society of Agronomy; 1992. p. 473–507. Agronomy; no. 33.

64. Mendel-GFDb. Aberystwyth, Norwich, Loughborough, and Dundee (UK): UK CropNet; 2003 [accessed 2011 Nov 19]. http://ukcrop.net/perl/ace/search/Mendel-GFDb.

65. Franckowiak JD, Lundqvist U, Konishi T. New and revised names for barley genes. Barley Genet Newsl. 1997 May [accessed 2011 Nov 19];26:4–8. http://wheat.pw.usda.gov/ggpages/bgn/26/text261a .html#4.

66. Linde-Laursen I. Recommendations for the designation of the barley chromosomes and their arms. Barley Genet Newsl. 1997 May [accessed 2011 Nov 19];26:1–3. http://wheat.pw.usda.gov/ggpages/bgn /26/text261a.html.

67. A standard for maize genetics nomenclature. Maize Genet Coop Newsl. 1995;69:182–184. Updates made in 1996, 2000, 2002, and 2006. http://www.maizegdb.org/maize_nomenclature.php.

68. MaizeGDB: Maize Genetics and Genomics Database. Ames (IA): US Department of Agriculture, Agricultural Research Service; 2005 [updated 2011 Oct 31; accessed 2011 Nov 19]. http://www.maizegdb.org.

69. Marshall HG, Shaner GE. Genetics and inheritance in oat. In: Marshal HG, Sorrells ME, editors. Oat science and technology. Madison (WI): American Society of Agronomy; 1992. p. 509–571. Agronomy; no. 33.

70. Kinoshita T. Report of the Committee on Gene Symbolization, Nomenclature and Linkage Groups. Rice Genet Newsl. 1986;3:4–5.

71. Kinoshita T, Rothschild G. Report of the Coordinating Committee of Rice Genetics Cooperative. Rice Genet Newsl. 1995;12:5.

72. McCouch SR, Cho YG, Yano M, Paul E, Blinstrub M, Morishima H, Kinoshita T. Report on QTL nomenclature. Rice Genet Newsl. 1997;14:11.

73. Kinoshita T. Linkage mapping using mutant genes in rice. Rice Genet Newsl. 1998;15:13–74.

74. Sybenga J. Rye chromosome nomenclature and homoeology relationships: workshop report. Z Pflanzenzüchtg. 1983;90(4):297–304.

75. McIntosh RA, Yamazaki Y, Devos KM, Dubcovsky J, Rogers WJ, Appels R. Catalogue of gene symbols for wheat: 2003 supplement. [Washington (DC)]: Department of Agriculture (US); 2003 [accessed 2011 Nov 19]. Part I, Introduction: 1. Recommended rules for gene symbolization in wheat. http://wheat .pw.usda.gov/ggpages/wgc/2003/Catalogue.pdf.

76. Gill B. Chromosome banding methods, standard chromosome band nomenclature, and applications in cytogenetic analysis. In: Heyne E. Wheat and wheat improvement. Madison (WI): American Society of Agronomy, Crop Science Society of America, Soil Science Society of America; 1987. p. 243–254.

77. TAIR: The *Arabidopsis* Information Resource. Stanford (CA): Stanford University; 2005 [accessed 2011 Nov 19]. http://www.arabidopsis.org/.

78. Crucifer Genetics Cooperative. CrGC Information Catalogue. Madison (WI): University of Wisconsin–Madison; 2003. p. 9–10.

79. Fransz P, Armstrong S, Alonso-Blanco C, Fischer TC, Torres-Ruiz RA, Jones G. Cytogenetics of the model system *Arabidopsis thaliana*. Plant J. 1998;13(6):867–876.

80. Cucurbit Genetics Cooperative Gene List Committee. Gene nomenclature for the Cucurbitaceae. Cucurbit Genet Coop Rep. 2000;23:139.

81. Hutton MG, Robinson RW. Gene list for *Cucurbita* spp. Cucurbit Genet Coop Rep. 1992;15:102–109.

82. Quiros CF, Bauchan GR. The genus *Medicago* and the origin of the *Medicago* sativa complex. In: Alfalfa and alfalfa improvement. Monograph Series 29. Madison (WI): ASA-CSSA-SSSA; 1988. p. 93–124.

83. Bauchan GR, Hossain MA. Advances in alfalfa cytogenetics. In: One hundred years of alfalfa genetics, the alfalfa genome. Proceedings of the Alfalfa Genome Conference; 1999 Aug 1–3; Madison (WI), Stillwater (OK): North American Alfalfa Improvement Conference; 1999. http://www.naaic.org/TAG/TAGpapers/Bauchan/advcytog.html.

84. Bauchan GR, Hossain MA. Karyotypic analysis of C-banded chromosomes of diploid alfalfa: *Medicago sativa* ssp. *caerulea* and ssp. *falcata* and their hybrid. J Hered. 1997;88(6):533–537.

85. Ellis TN, Ambrose M. Pea. In: Wood R, editor. Genetic nomenclature guide. West Sussex (UK): Elsevier; 1998. p. S.22–S.23. Supplement to Trends in Genetics.

86. http://data.jic.bbsrc.ac.uk/cgi-bin/pgene/default.asp. Accessed 2011 Nov 26. Maintained by MJ Ambrose, Biotechnology and Biological Sciences Research Council, UK.

87. Soybean Genetics Committee. Rules for genetic symbols. Soybean Genet Newsl. 1996;23:19–23.

88. Cregan PB, Jarvik T, Bush AL, Shoemaker RC, Lark KG, Kahler AL, Kaya N, VanToai TT, Lohnes DG, Chung J, Specht JE. An integrated genetic linkage map of the soybean genome. Crop Sci. 1999;39(5):1464–1490.

89. CENL Committee for *Capsicum* Gene Nomenclature. Rules for gene nomenclature of Capsicum. Capsicum Eggplant Newsl. 1994;13:11–14.

90. Tomato Genetics Cooperative. Rules for nomenclature in tomato genetics. TGC Rep. 1992;42:6–7.

91. *Arabidopsis* rules proposed for tomato [editorial]. TGC Rep. 1992;42:5.

92. Kohel RJ. Genetic nomenclature in cotton. J Hered. 1973;64(5):291–295.

93. Percy RG, Kohel RJ. Qualitative genetics. In: Smith CW, Cothren JT, editors. Cotton: origin, history, technology, and production. New York: John Wiley & Sons; 1999.

94. Reinisch MJ, Dong J, Brubaker CL, Stelly DM, Wendel JF, Paterson AH. A detailed RFLP map of cotton, *Gossypium hirsutum* × *Gossypium barbadense*: chromosome organization and evolution in a disomic polyploidy genome. Genetics. 1994;138:829–847.

95. Frelichowski JE Jr, Palmer MB, Main D, Tomkins JP, Cantrell RG, Stelly DM, Yu J, Kohel RJ, Ulloa M. Cotton genome mapping with new microsatellites from Acala 'Maxxa' BAC-ends. Mol Genet Genomics. 2006;275:479–491.

96. Guo W, Cai C, Wang C, Han Z, Song X, Wang K, Niu X, Lu K, Shi B, Zhang T. A microsatellite-based, gene-rich linkage map reveals genome structure, function and evolution in *Gossypium*. Genetics. 2007;176:527–541.

97. Lacape JM, Jacobs J, Arioli T, Derijcker R, Forestier-Chiron N, Llewellyn D, Jean J, Thomas E, Viot C. A new interspecific, *Gossypium hirsutum* × *G. barbadense*, RIL population: towards a unified consensus linkage map of tetraploid cotton. Theor Appl Genet. 2009;119:281–292.

98. Lacape JM, Nguyen TB, Thibivilliers S, Bojinov B, Courtois B, Cantrell RG, Burr B, Hau B. 2003. A combined RFLP-SSR-AFLP map of tetraploid cotton based on a *Gossypium hirsutum* × *Gossypium barbadense* backcross population. Genome. 46:612–626.

99. Nguyen TB, Giband M, Brottier P, Risterucci AM, Lacape JM. 2004. Wide coverage of the tetraploid cotton genome using newly developed microsatellite markers. Theor Appl Genet. 109:167–175.

100. Yu JZ, Kohel RJ, Fang DD, Cho J, Van Dyenze A, Ulloa M, Hoffman SM, Pepper AE, Stelly DM, Jenkins JN, et al. A high-density simple sequence repeat and single nucleotide polymorphism genetic map of the tetraploid cotton genome. Genes Genomes Genet. 2012;2(1):43–58.

101. Van Deynze A, Stoffel K, Lee M, Wilkins TA, Kozik A, Cantrell R, Yu J, Kohel R, Stelly D. Sampling nucleotide diversity in cotton. BMC Plant Biol. 2009;9:125.

102. Wang K, Song X, Han Z, Guo W, Yu JZ, Sun J, Pan J, Kohel RJ, Zhang T. Complete assignment of the chromosomes of *Gossypium hirsutum* L. by translocation and fluorescence in situ hybridization mapping. Theor Appl Genet. 2006;113:73–80.

103. Yu J, Kohel RJ, Smith CW. The construction of a tetraploid cotton genome wide comprehensive reference map. Genomics. 2010;95(4):230–240.

104. Robinson RW, McCreight JD, Ryder EJ. The genes of lettuce and closely related species. Plant Breeding Rev. 1983;1:267–293.

105. de Vries JN. Onion chromosome nomenclature and homoeology relationships: workshop report. Euphytica. 1990;49(1):1–3.

106. Kalkman ER. Analysis of the C-banded karyotype of *Allium cepa* L.: standard system of nomenclature and polymorphism. Genetica. 1984;65(2):141–148.

107. Chick BD. 1998. In: Wood R, editor. Genetic nomenclature guide. West Sussex (UK): Elsevier; 1998. p. S.34–S.36. Supplement to Trends in Genetics.

108. Committee on Standardized Genetic Nomenclature for Cats. Standardized genetic nomenclature for the domestic cat. J Hered. 1968;59(1):39–40.

109. Committee on Standardized Genetic Nomenclature for Mice. A revision of the standardized genetic nomenclature for mice. J Hered. 1963;54:159–162.

110. Dolling CHS. Standardized genetic nomenclature for sheep. In: Piper L, Ruvinsky A, editors. The genetics of sheep. Wallingford (UK): CABI; 1997. p. 593–601.

111. Dolling CHS. Standardized genetic nomenclature for the horse. In: Bowling AT, Ruvinsky A, editors. The genetics of the horse. Wallingford (UK): CABI; 2000. p. 499–506.

112. Fox RR. Taxonomy and genetics. In: Manning PJ, Ringler DH, Newcomer CE, editors. The biology of the laboratory rabbit. New York (NY): Academic Press; 1974. p. 1–22.

113. Ruvinsky A, Sampson J. Standardized genetic nomenclature for the dog. In: The genetics of the dog. Wallingford (UK): CABI; 2001. p. 537–540.

114. Makinen A. The standard karyotype of the silver fox (*Vulpes fulvus* Desm.). Committee for the Standard Karyotype of *Vulpes fulvus* Desm. Hereditas. 1985;103(2):171–176.

115. Ray M, Mohandas T. Proposed banding nomenclature for the Chinese hamster chromosomes (*Cricetulus griseus*). Cytogenet Cell Genet. 1976;16(1–5):83–91.

116. Reimann N, Bartnitzke S, Nolte I, Bullerdick J. Working with canine chromosomes: current recommendations for karyotype description. J Hered. 1999;90(1):31–34.

117. Switonski M, Reimann N, Bosma AA, Long S, Bartnitzke S, Pienkowska A, Moreno-Milan MM, Fischer P. Report on the progress of standardization of the G-banded canine (*Canis familiaris*) karyotype. Committee for the Standardized Karyotype of the Dog (*Canis familiaris*). Chromosome Res. 1996;4(4):306–309.

118. Young LD. Standard nomenclature and pig genetic glossary. In: Rothschild MF, Ruvinsky A, editors. The genetics of the pig. Wallingford (UK): CABI; 1998. p. 541–550.

119. Breen M, Switonski M, Binns MM. Cytogenetics and physical chromosome maps. In: The genetics of the dog. Wallingford (UK): CABI; 2001. p. 299–328.

120. Cho KW, Youn HY, Watari T, Tsujimoto H, Hasegawa A, Satoh H. A proposed nomenclature of the domestic cat karyotype. Cytogenet Cell Genet. 1997;79(1–2):71–78.

121. Committee for a Standardized Karyotype of *Rattus norvegicus*. Standard karyotype of the Norway rat, *Rattus norvegicus*. Cytogenet Cell Genet. 1973;12(3):199–205.

122. Committee for Standardized Karyotype of *Oryctolagus cuniculus*. Standard karyotype of the laboratory rabbit, *Oryctolagus cuniculus*. Cytogenet Cell Genet. 1981;31(4):240–248.

123. Gustavsson I. Standard karyotype of the domestic pig. Committee for the Standardized Karyotype of the Domestic Pig. Hereditas. 1988;109(2):151–157.

124. Iannuzzi I. Standard karyotype of the river buffalo (*Bubalus bubalis* L., 2n=50). Report of the

committee for the standardization of banded karyotypes of the river buffalo. Cytogenet Cell Genet. 1994;67(2):102–113.

125. ISCN 1985. An International System for Human Cytogenetic Nomenclature. Harnden DG, Klinger HP, editors. Birth Defects: Original Article Series, vol 21(1). New York (NY): March of Dimes Birth Defects Foundation; 1985.

126. Mitelman F, editor. ISCN 1995: An International System for Human Cytogenetic Nomenclature. Recommendations of the International Standing Committee on Human Cytogenetic Nomenclature, Memphis, Tenn., October 1994. Basel (Switzerland): Karger AG; 1995. Published in collaboration with Cytogenetics and Cell Genetics.

127. Horvitz HR, Brenner S, Hodgkin J, Herman RK. A uniform genetic nomenclature for the nematode *Caenorhabditis elegans*. Mol Gen Genet. 1979;175(2):129–133.

128. *Caenorhabditis elegans* nomenclature. Minneapolis (MN): Caenorhabditis Genetics Center, University of Minnesota; [accessed 2011 Nov 19]. Genetic nomenclature for *Caenorhabditis elegans*. http://elegans.som.vcu.edu/Genome/Nomencl2001w.htm.

129. Caenorhabditis Genetics Center. Minneapolis (MN): *Caenorhabditis* Genetics Center; 2004 [accessed 2011 Nov 19]. http://www.cbs.umn.edu/CGC/.

130. FlyBase: A database of *Drosophila* genes and genomes. Bloomington (IN): Indiana University; c1993–2011 [updated 2011 Nov 18; accessed 2011 Nov 19]. Genetic nomenclature for *Drosophila melanogaster*. http://flybase.org/.

131. FlyBase Consortium. *Drosophila melanogaster*. In: Wood R, editor. Genetic nomenclature guide. West Sussex (UK): Elsevier; 1998. p. S.28–S.31. Supplement to Trends in Genetics.

132. Lindsley DL, Zimm GG. The genome of *Drosophila melanogaster*. San Diego (CA): Academic Press; 1992.

133. Shaklee JB, Allendorf FW, Morizot DC, Whitt GS. Gene nomenclature for protein-coding loci in fish. Trans Am Fish Soc. 1990;119(1):2–15 and published erratum in Trans Am Fish Soc. 1990;119(4):unnumbered page after 790.

134. International Union of Biochemistry and Molecular Biology, Nomenclature Committee. London (UK): IUBMB; 2005 [accessed 2011 Nov 19]. Enzyme nomenclature. http://www.chem.qmul.ac.uk/iubmb/enzyme/.

135. Mullins M. Genetic methods: conventions for naming zebrafish genes. In: Westerfield M, editor. The zebrafish book: a guide for the laboratory use of zebrafish *Danio* (*Brachydanio*) *rerio*. 4th ed. Eugene (OR): University of Oregon Press; 2000 [accessed 2011 Nov 19]. http://zfin.org/zf_info/zfbook/chapt7/7.1.html.

136. Westerfield M. The zebrafish book, a guide for the laboratory use of zebrafish (*Danio rerio*). 5th ed. Eugene (OR): University of Oregon Press; 2007.

137. ZFIN: the Zebrafish Information Network. Eugene (OR): University of Oregon; c1994–2005 [updated 2011 May 9; accessed 2011 Nov 19]. Zebrafish nomenclature guidelines. https://wiki.zfin.org/display/general/ZFIN+Zebrafish+Nomenclature+Guidelines.

138. Bradford Y, Conlin T, Dunn N, Fashena D, Frazer K, Howe DG, Knight J, Mani P, Martin R, Moxon SA, et al. ZFIN: enhancements and updates to the zebrafish model organism database. Nucleic Acids Res. 2011;39(suppl 1):D822–D829.

139. Sprague J, Bayraktaroglu L, Clements D, Conlin T, Fashena D, Frazer K, Haendel M, Howe DG, Mani P, Ramachandran S, et al. The Zebrafish Information Network: the zebrafish model organism database. Nucleic Acids Res. 2006;34:D581–585.

140. Somes RG Jr. Alphabetical list of the genes of domestic fowl. J Hered. 1980;71(3):168–174.

141. Shows TB, McAlpine PJ, Boucheix C, Collins FS, Conneally PM, Frezal J, Gershowitz H, Goodfellow PN, Hall JG, Issitt P, et al. Guidelines for human gene nomenclature: an international system for human gene nomenclature (ISGN, 1987). Cytogenet Cell Genet. 1987;46(1–4):11–28.

142. Crittenden LB, Bitgood JJ, Burt DW, Ponce de Leon FA, Tixier-Boichard M. Nomenclature for nam-

ing loci, alleles, linkage groups, and chromosomes to be used in poultry genome publications and databases. Genet Sel Evol. 1996;28(3):289–297.

143. Burt DW, Carrë W, Fell M, Law AS, Antin PB, Maglott DR, Weber JA, Schmidt CJ, Burgess SC, McCarthy FM. The Chicken Gene Nomenclature Committee report. BMC Genomics. 2009;10(suppl 2):S5.

144. Ladjali-Mohammedi K, Bitgood JJ, Tixier-Boichard M, Ponce de Leon FA. International system for standardized avian karyotypes (ISSAK): standardized banded karyotypes of the domestic fowl (*Gallus domesticus*). Cytogenet Cell Genet. 1999;86(3–4):271–276.

145. Broad TE, Dolling CHS, Lauvergne JJ, Miller P. Revised COGNOSAG guidelines for gene nomenclature in ruminants 1998. Genet Sel Evol. 1999;31(3):263–268.

146. Online Mendelian Inheritance in Animals, OMIA. Faculty of Veterinary Science, University of Sydney; [accessed 2012 Aug 26]. http://omia.angis.org.au/.

147. Powell BC, Rogers GE. Differentiation in hard keratin tissues: hair and related structures. In: Leigh IM, Lane EB, Watt FM. The keratinocyte handbook. Cambridge (UK): Cambridge University Press; 1994. p. 401–436.

148. ISCNDA 1989. International system for cytogenetic nomenclature of domestic animals 1989. Di Berardino D, Hayes H, Fries R, Long S, editors. Cytogenet Cell Genet. 1990;53(2–3):65–79.

149. Popescu CP, Long S, Riggs P, Womack J, Schmutz S, Fries R, Gallagher DS. Standardization of cattle karyotype nomenclature: report of the committee for the standardization of the cattle karyotype. Cytogenet Cell Genet. 1996;74(4):259–261.

150. Ansari HA, Bosma AA, Broad TE, Bunch TD, Long SE, Maher DW, Pearce PD, Popescu CP. Standard G-, Q-, and R-banded ideograms of the domestic sheep (*Ovis aries*): homology with cattle (*Bos taurus*). Report of the Committee for the Standardization of the Sheep Karyotype. Cytogenet Cell Genet. 1999;87(1–2):134–142.

151. Dolling CHS. Standardized genetic nomenclature for cattle. In: Fries R, Ruvinsky A, editors. The genetics of cattle. Wallingford (UK): CABI; 1999. p. 657–666.

152. Bovmap Database. Jouy-en-Josas (France): Institut National de Recherche Agronomique; [accessed 2011 Nov 19]. http://dga.jouy.inra.fr/cgi-bin/lgbc/main.pl?BASE=.

153. CGD: Cattle Genome Database. Brisbane (Australia): Queensland Biosciences Precinct; 2003 [revised 2002 Dec 10; accessed 2011 Nov 19]. http://www.cgd.csiro.au/.

154. Bodmer JG, Marsh SGE, Albert ED, Bodmer WF, Bontrop RE, Charron D, Dupont B, Erlich HA, Mach B, Mayr WR, et al. Nomenclature for factors of the HLA system. Tissue Antigens. 1995;46(1):1–18.

155. BoLA Nomenclature. Midlothian (Scotland): Roslin Institute. ISAG BoLA Nomenclature Committee; c2002 [accessed 2013 Nov 19]. Sequence-based nomenclature of BoLA alleles. http://www.ebi.ac.uk/ipd/mhc/bola/nomenclature.html.

156. Richer CL, Power MM, Klunder LR, McFeely RA, Kent MG. Standard of the domestic horse (*Equus caballus*). Committee for standardized karyotype of Equus caballus. The Second International Conference for Standardization of Domestic Animal Karyotypes, INRA, Jouy-en Josas, France, 22nd–26th May 1989. Hereditas. 1990;112(3):289–291.

157. Bowling AT, Breen M, Chowdhary BP, Hirota K, Lear T, Millon LV, Ponce de Leon FA, Raudsepp T, Stranzinger G. International System for Cytogenetic Nomenclature of the Domestic Horse. Report of the Third International Committee for the Standardization of the Domestic Horse Karyotype, Davis, CA, USA, 1996. Chromosome Res. 1997;5(7):433–443. Also called: ISCNH (1997).

158. Lyon MF. Rules and guidelines for gene nomenclature. In: Lyon MF, Searle AG, editors. Genetic variants and strains of the laboratory mouse. 2nd ed. Oxford (UK): Oxford University Press; 1989. p. 1–11.

159. *Pero*Base. Columbia (SC): University of South Carolina; c2001 [accessed 2011 Nov 19]. *Peromyscus* gene list. http://wotan.cse.sc.edu/perobase/genetics/genelist.htm.

160. Greenbaum IF, Gunn SJ, Smith SA, McAllister BF, Hale DW, Baker RJ, Engstrom MD, Hamilton MJ, Modi WS, Robbins LW, et al. Cytogenetic nomenclature of deer mice, *Peromyscus* (Rodentia): revision and review of the standardized karyotype. Cytogenet Cell Genet. 1994;66(3):181–195.

161. Levan G, Hedrich HJ, Remmers EF, Serikawa T, Yoshida MC. Standardized rat genetic nomenclature. Mamm Genome. 1995;6(7):447–448.

162. RGD: Rat Genome Database. Milwaukee (WI): Medical College of Wisconsin, Bioinformatics Program; [accessed 2011 Nov 19]. http://rgd.mcw.edu/.

163. Seal RL, Gordon SM, Lush MJ, Wright MW, Bruford EA. genenames.org: the HGNC resources in 2011. Nucleic Acids Res. 2011;39(Database issue):D519-9.

164. Antonarakis SE; Nomenclature Working Group. Recommendations for a nomenclature system for human gene mutations. Hum Mutat. 1998;11(1):1–3.

165. den Dunnen JT, Antonarakis SE. Mutation nomenclature extensions and suggestions to describe complex mutations: a discussion. Hum Mutat. 2000;15(1):7–12. Also available at http://onlinelibrary.wiley.com/doi/10.1002/(SICI)1098-1004(200001)15:1%3C7::AID-HUMU4%3E3.0.CO;2-N/pdf.

166. Human Genome Variation Society. Carlton South (Australia): Human Genome Variation Society; c2012 [updated 2005 Apr 18; accessed 2013 Jul 26]. http://www.hgvs.org/.

167. Marsh SGE, Albert ED, Bodmer WF, Bontrop RE, Dupont B, Erlich HA, Geraghty DE, Hansen JA, Mach B, Mayr WR, et al. Nomenclature for factors of the HLA system. Tissue Antigens 2002;60:407–464. Also available at http://www.anthonynolan.com/HIG/lists/nomenc.html.

168. IMGT/HLA Sequence Database. Cambridge (UK): European Bioinformatics Institute; 2005 [updated 2005 Jun 1; accessed 2011 Nov 19]. http://www.ebi.ac.uk/imgt/hla/.

169. Robinson J, Waller MJ, Parham P, de Groot N, Bontrop R, Kennedy LJ, Stoehr P, Marsh SGE. IMGT/HLA and IMGT/MHC: sequence databases for the study of the major histocompatibility complex. Nucleic Acids Res. 2003;31(1):311–314. Also available at http://www.ebi.ac.uk/imgt/hla/citations.html.

ADDITIONAL REFERENCES

Breen M, Switonski M, Binns MM. Cytogenetics and physical chromosome maps. In: Ruvinsky A, Sampson J, editors. The genetics of the dog. Wallingford (UK): CABI Publishing; c2001. p. 299–328.

King RC, Stansfield WD. A dictionary of genetics. 5th ed. New York (NY): Oxford University Press; 1997.

O'Brien SJ, editor. Genetic maps: locus maps of complex genomes. 6th ed. Plainview (NY): Cold Spring Harbor Laboratory Press; 1993.

Rieger R, Michaelis A, Green MM. Glossary of genetics: classical and molecular. 5th ed. New York (NY): Springer-Verlag; 1991.

Singh RJ. Plant cytogenetics. Boca Raton (FL): CRC Press; 1993.

Singleton P, Sainsbury D. Dictionary of microbiology and molecular biology. 3rd ed. New York (NY): Wiley; 2001.

Thompson MW, McInnes RR, Willard HF. Thompson & Thompson genetics in medicine. 5th ed. Philadelphia (PA): WB Saunders; 1991.

Wood R, editor. Genetic nomenclature guide. West Sussex (UK): Elsevier; 1998. Supplement to Trends in Genetics.

22 Taxonomy and Nomenclature

22.1 SPECIES CONCEPT

Living organisms are classified into hierarchical systems. The species is generally considered the basic unit in these systems. The concept of species changes as more is learned about the evolution of life on earth.[1] Perhaps the earliest concept of the species was what is now termed the typological concept, whereby species were recognized solely by morphological criteria. Individual variation was regarded as the imperfect manifestation of an underlying ideal type. Following the publication in 1859 of *Origin of Species* by Charles Darwin, scientists began to regard species not as entities that had existed since the beginning of time but rather as entities with a phylogenetic history. Since then, a tremendous amount of research and writing has been devoted to the subject of what constitutes a species.

Despite this long history of taxonomic research directed toward defining the term "species", there is still no consensus.[2] A species definition strives to meet 3 important criteria: universality, applicability, and theoretical significance.[2] The dichotomy between applicability and theoretical significance has not been satisfactorily resolved. Thus, several concepts are in current use, some stressing morphological and biochemical criteria and others stressing evolutionary relatedness.

A concept widely accepted in zoology is the biological species. The principal tenet of this concept is that species are a reflection of genetic relatedness. Members of a species represent a reproductive community, such that all members of a species are capable or potentially capable of mating to produce viable and fertile offspring. A biological species also represents an interconnecting gene pool and an ecological unit. All of its members interact with the environment and with other species as a unit. Mayr offers this definition: "Species are groups of interbreeding populations that are reproductively isolated from other such groups."[3]

The application of the biological species concept presents problems for organisms that reproduce exclusively or mainly by asexual means (as is the case for many microorganisms) and for immobile organisms, specifically plants. Some morphologically distinct species of plants in different parts of the world, when brought together, readily form viable hybrids. Even among sympatric species of some groups (e.g., oak), formation of interspecific hybrids is common.

The concept of species is of academic interest not only to biologists who specialize in classification but also to anyone who conducts research with organisms. Most scientific articles in all fields of biology refer to species as subjects of investigation. Interpretation and utilization of this research depend on correct identification of species by authors and on meaningful classification of organisms into species. Even in the absence of consensus as to what constitutes a species, systems of classification and nomenclature based on the species unit have been developed. These systems are described in the sections that follow.

22.2 TAXONOMY AND NOMENCLATURE ACROSS KINGDOMS

22.2.1 Overview of Taxonomy

Taxonomy is the study of classification. It includes "rules, theories, principles, and procedures".[4] Data on variation in organisms are used to arrange those organisms in a system of classification.[4–6] Some workers regard systematics and taxonomy as essentially synonymous, whereas others consider systematics a broader term that includes taxonomy, identification, and nomenclature.[4] Nomenclature is the naming of organisms and the establishment, interpretation, and application of regulations governing the system of naming. Identification is the assignment of an unknown specimen to a previously recognized taxonomic group, often at the species level.

Systems of nomenclature have been developed for all groups of organisms on the earth. These systems have some features in common, which are discussed here. Features that differ among groups of organisms are covered later in this chapter.

The 7 basic systematic ranks or taxa (singular, "taxon") are, in descending order, kingdom, phylum (plural, "phyla") or division, class, order, family, genus, and species. Every species is a member of a genus, every genus is a member of a family, every family is a member of an order, and so forth. Botanists generally use the term "division" rather than "phylum", the term preferred by zoologists and bacteriologists. Any of these taxa may be prefixed with "sub" or "super" to provide additional taxonomic categories. Other supplementary taxa (such as tribe—a taxon below subfamily) are sometimes used to further extend the structure of the hierarchy.

In most cases, write the names of taxa at the level of genus and below (including subgenus, species, subspecies, variety, and forma) in italics. Write the names of taxa for family and above in roman type. See Sections 22.3.4 on bacteria and 22.3.5 on viruses for important exceptions to this general rule. Use an initial capital letter for the names of all taxa down through genus; any of these may stand alone. When the term for the taxon precedes the name in text, write the taxon term in all lowercase letters, for example, "the phylum Protozoa", "the family Terebratulidae". Treat names of families and higher-order taxa as plural.

> When the Papilionaceae are regarded as a family . . . [Papilionaceae is a family name]

Genus names are treated as nouns in the nominative singular.

> *Poa* is widespread in temperate North America.
> *Poa* species are widespread in temperate North America.

Species are treated as singular.

> *Castanea dentata* was dominant in much of the eastern hardwood forest of the United States until chestnut blight virtually eliminated it.

22.2.2 Kingdoms

Early works on taxonomic classifications recognized 2 kingdoms: Plants and Animals. More recently, 5 kingdoms have been recognized.[7] In this system, living organisms are placed into 2 superkingdoms (Prokarya and Eukarya) and 5 kingdoms (Bacteria in Prokarya; Protoctista, Animalia, Plantae, and Fungi in Eukarya). The Bacteria are divided into 2 subkingdoms: Archaea and Eubacteria.

With the discovery and characterization of the Archaea, a new group of bacterium-like organisms, yet another system has been proposed. In this system, 3 domains (a category in popular use and equivalent to superkingdom, but not formally recognized by international codes of nomenclature) have been proposed: 2 for Prokaryotes (Archaea and Eubacteria) and 1 for the Eukaryotes (Eukaryotes). The Archaea consist of organisms that are morphologically indistinguishable from bacteria but genetically quite distinct. The Eubacteria are the true bacteria, and Eukaryotes consist of protists, fungi, plants, and animals. More detail about the 3-domain system of classification can be found at the Tree of Life Web Project website (http://tolweb.org/tree/phylogeny.html).[8] Both of these classification systems are in current use.

22.2.3 Latin (Scientific) Names (Species and Lower)

Note: Much of the material here does not apply to viruses. See Section 22.3.5 for details about virus names.

22.2.3.1 Binomial System of Species Names

In formal taxonomy, a binomial system is used for the Latinized name of a species.[9] The species name consists of 2 Latin words, the genus name and the specific epithet (in botany and bacteriology) or the specific name (in zoology). In plant and bacterial taxonomy, the word "epithet" has a particular biological meaning, which is unrelated to the more frequent meanings of a substitute for a proper name (such as "The Great Emancipator" for Abraham Lincoln) or a derogatory tag (such as "a racial epithet"); "specific" is derived from "species". In zoological nomenclature, the species group encompasses all nominal taxa at the ranks of species and subspecies. The specific epithet or name of species and the names of infraspecific taxa are written in all lowercase letters and cannot stand alone.

> *Quercus alba* [*Quercus* is the genus, *alba* is the specific epithet, and *Quercus alba* is the species]
> *Equus caballus* [*Equus* is the genus, *caballus* is the specific name, and *Equus caballus* is the species name]

Genus names and specific epithets and names may not actually be Latin words, but they are treated as if they were and they have Latin endings. The first letter of the genus name is capitalized and the entire name is italicized, whereas the specific epithet or name (as well as the names of any infraspecific taxa) are in all lowercase letters and italicized, even if the name is derived from a proper noun (to honor an individual or based on a geographic name). Apply these conventions wherever a species name is used, including running text, titles, tables, indexes, and dictionary entry terms.

> *Homo sapiens*, *Drosophila melanogaster*, *Shigella boydii* [named after Sir John Boyd]
> The Metabolism of *Drosophila melanogaster* [a title]
> *Musca domestica* 25–37 [an index entry]

A species name may appear in a title that, for reasons of document design, must be italicized. An example is a book title italicized in running text to distinguish it from the rest of the text. In such a title, write the species name in roman type to maintain the distinction between the title and the species name. Unfortunately, in many instances, this convention may confuse the reader as to the full extent of the title.

> His major monograph was *The Metabolism of* Drosophila melanogaster: *A New Treatise.*

The rules of nomenclature for plants, animals, bacteria, and viruses are set forth in various nomenclatural codes. These codes are discussed in the sections that follow for each group of organisms.

22.2.3.2 Use of Binomials in Scientific Writing

When referring to an organism in scientific writing, use its Latin name. If the organism is widely known by a vernacular name, this may be used if, at the first reference to the organism, the vernacular name is presented in clear association with the Latin name. When the Latin name is first mentioned in the text (but not the title), it is also customary to append the name of the author or authors who validly published this name. Details of presenting author names differ among plants, bacteria, and animals and are covered in the appropriate sections of this chapter (Sections 22.3.1, 22.3.2, 22.3.3, and 22.3.4). Some scientific journals no longer require mention of the authors of Latin names unless the subject of the article concerns taxonomy or nomenclature.

22.2.3.3 Monomials

The names of genera and higher ranks may stand alone as monomials—for example, "strains or species of *Rhizobium*", "the genus *Erwinia*", "a member of the Poaceae". The same applies to certain subordinate categories of genus: subgenus, section, and series. However, specific epithets or names and the names of any infraspecific taxa are not meaningful on their own, and hence use of monomials is not permitted for categories at the rank of species and below. In referring to one or more species of a genus for which the specific epithets or names are unknown, the genus name may be followed by the term "species", "sp." (singular), or "spp." (plural); for example, *Phaseolus* sp. (to refer to a single unidentified species of *Phaseolus*) or *Phaseolus* spp. (to refer to 2 or more species of *Phaseolus*). Do not italicize the abbreviations "sp." and "spp." See also Section 22.2.3.5.

22.2.3.4 Abbreviation of Genus and Species Names

When a species name is first used in a document, spell out in full both the genus name and the specific epithet or specific name; subsequently, the genus name may be abbreviated, as outlined below. Do not use the abbreviated form of the genus name if the specific epithet or name is not given.

> We included several species of *Festuca* in our study.
> *not* We included several species of *F.* in our study.

For a species name that has been presented in full at first mention in a particular text, the genus name may be abbreviated, to the first letter and a period, at second and subsequent mentions; for example, *Salmo clarki* becomes *S. clarki*. If 2 or more species in the same genus are mentioned, the genus name may be abbreviated at first mention of the second species, provided there is no possibility of confusion with any other genus already mentioned; for example, in the same paper, *Salmo salar* could be presented as *S. salar* at first mention, provided no other genera starting with "S" have been mentioned.

Abbreviated genus names could create confusion if 2 or more genera begin with the same letter. For example, *Mesocricetus auratus* and *Meriones unquiculatus* would be presented as *M. auratus* and *M. unquiculatus*. Recommendation 25A of the *International Code of Zoological Nomenclature* ("Zoological Code") allows multiletter abbreviations of animal genus names in this situation, so *Mesocricetus auratus* and *Meriones unquiculatus* would become *M. auratus* and *Mer. unquiculatus*, with the first-used name taking the single-letter abbreviation.[10] Advisory Note A of the *International Code of Nomenclature of Bacteria* ("Bacteriological Code") states that "later use of the name of the species previously cited usually has the genus abbreviated, commonly to the first letter of the generic name."[11] However, genera of *Halomonadaceae* and phototrophic bacteria have 3-letter abbreviations that are officially sanctioned by the relevant taxonomic subcommittees of the International Committee on Systematics of Prokaryotes (ICSP) (see http://www.bacterio.cict.fr/abbreviations.html). Although the *International Code of Nomenclature for algae, fungi, and plants* ("Melbourne Code") makes no recommendation on this matter, the use of only single-letter abbreviations is implied by examples for plant names.[12] CSE discourages the use of multiletter abbreviations for genus names to avoid variation from one document to another in the abbreviated name of a given genus. In most cases, the specific name will serve to uniquely identify each species, even with single-letter abbreviations of genus names beginning with the same letter (provided each name is spelled out in full at first mention). Otherwise, it is never wrong to write out the genus name in full at each mention, and this approach is preferable to creating a unique abbreviation for each genus. The space saved by using multiletter abbreviations may not justify the potential for confusion.

Some journals allow a genus name to be abbreviated when it starts a sentence; others do not. Where no confusion results from abbreviating a genus name at the beginning of a sentence, there seems no reason to avoid this practice, but authors should defer to the journal's style.

22.2.3.5 Proper Use of sp. and spp.

When able to identify a genus but unable or unwilling to identify to the species level, use "sp." (meaning the identity is believed to be one species) or "spp." (the identities probably include 2 or more species) to stand in for the actual specific epithet. Unlike the specific epithet, however, those abbreviations for "species" are set roman rather than in italics (e.g., *Angelica* sp. or *Angelica* spp.).

Do not use sp. or spp. when simply referring to the genus as a whole; instead, use the genus name alone (e.g., *Ceriodaphnia* occurs in . . .). The reason for making such a distinction between unidentified species and a genus as a whole is to avoid possible confusion. Although context would generally preclude such misunderstanding, there are situations where the intended meaning could be obfuscated. For example, the phrase "*Ceriodaphnia* sp. [or 'spp.'] populations peak" means that this (or these) particular unidentified species peaks. However, if the author intends to refer to the genus as a whole (i.e., "*Ceriodaphnia* populations peak") and instead writes "*Ceriodaphnia* spp. populations peak", then the meaning is compromised.

22.2.3.6 Plurals of Genus Names

Names of genera do not have plural forms. If reference is made to a group of species in one genus, write the genus name followed by the abbreviation for "species" (spp.). See Section 22.2.3.5 for more information about the use of this abbreviation.

Poa spp. not *Poas*
Iris spp. not *Irises* [note that *Iris* here is a genus name, not a common name, as indicated by the italic and capitalized first letter]
Erwinia spp. not *Erwinias*

Similarly, the binomial name is always singular.

Escherichia coli was . . . [but *E. coli* strains were . . .]

Editors of some journals discourage the use of Latin binomials as adjectives, as in the example above ("*E. coli* strains were . . ."). However, the adjectival use of Latin binomials seems to be gaining acceptance.

22.2.3.7 Diacritical Marks

The nomenclatural codes for plants, animals, and bacteria (discussed in more detail in the later sections of this chapter) all prohibit the use of diacritical marks in Latin binomials (with one exception in botany, below). If the word that forms the basis for a Latin name contains such marks, the name is modified to eliminate them:

ä, ö, and ü become ae, oe, and ue, respectively
é, è, and ê become e
ø, æ, and å become oe, ae, and aa, respectively
ñ becomes n (Melbourne Code and Zoological Code)

The Melbourne Code[12] allows use of a diaeresis, indicating that a vowel is to be pronounced separately; for example, Cephaëlis or Isoëtes.

The Virological Code[13,14] does not address the issue of diacritical marks.

22.2.3.8 Validity of Scientific Names

For the name of a binomial or any other taxonomic rank to be legitimate, it must be published according to rules set forth in the various governing codes (Melbourne Code, Zoological Code, Bacteriological Code, Virological Code). The legitimate name is the correct name to use when referring to a taxon. In general, the legitimate name is the name that was first validly published. Because names applied to fossils are often based on fragmentary material, the same organism may have more than one name when such names have been applied to different parts, life-history stages, or preservational states. It is conventional in many journals that when a new taxon is being described, the author's name appears after the first mention of the new name, followed by a comma and an abbreviated designation indicating the rank of the new name: sp. nov. (species novum) for a new species, gen. nov. (genus nova) for a new genus, and fam. nov. (familia nova) for a new family. In botany, the term nothogen. (nothogenus) or nothosp. (nothospecies) may be used to designate hybrids.

> Gilbertellaceae Benny, fam. nov.
> *Dictyocoprotus* Krug & Khan, gen. nov.
> *Piper allardii* Yuncker, sp. nov.
> *Strotheria gypsophila* B.L. Turner, gen. et sp. nov.
> *Dryostichum singulare* W.H. Wagner, nothogen. et nothosp. nov.

22.2.3.9 Nomenclatural Types

Nomenclatural codes for plants, animals, and bacteria all use the concept of nomenclatural type. The concept of type applies to the taxonomic categories of family and below. When a new species of a plant or animal is named, a physical, preserved specimen is designated as the nomenclatural type and the name remains permanently attached to this specimen. The type specimen for a bacterium is a living pure culture descended from a strain designated as the nomenclatural type. A description, illustration, or preserved specimen may also serve as the type only if the type strain is no longer extant. The nomenclatural type for a genus is the type species, and the nomenclatural type for a family is the type genus. There are various kinds of type specimens—for example, holotype, neotype, lectotype. Distinctions among these can be found in the various nomenclatural codes.

22.2.3.10 Grammar of Latin (Scientific) Names

Latin binomials should conform to the rules of Latin grammar, but a name that is validly published cannot be rejected if it fails to conform to these rules. This is an issue primarily for articles in which a new name is proposed. If a Latin diagnosis (description) is required for valid publication of a new name (e.g., for a new plant name), it must also conform to the rules of Latin grammar. Scientific Latin is more formalized in style and vocabulary than classical Latin. As in classical Latin, all adjectives used in nomenclature must agree in gender, number, and case with the nouns they modify. Latin and Greek words adopted as genus names generally retain their classical gender unless an exception is specified within the various nomenclatural codes. Some of the challenges in determining gender are explained by Manara.[15]

1) Latin names ending in "us" are masculine, unless they are trees recognized by ancient Romans, which are feminine; thus, *Helianthus annuus* L. is masculine, but *Pinus resinosa* Aiton. is feminine with a masculine ending (*Pinus*).
2) Genus names ending in "on" are masculine (e.g., *Rhododendron canadense* [L.] Torr.), and genus names ending in "ma" are neuter (e.g., *Alisma subcordatum* Raf.).
3) Genus names ending in "a" are feminine (e.g., *Cicuta maculata* L.).
4) Genus names ending in "is" are either feminine or masculine but are treated as feminine (e.g., *Amerorchis rotundifolia* Banks ex Pursh.).
5) Some other feminine endings are "ago", as in *Plantago cordata* Lam.; "es", as in *Prenanthes aspera* Michx.; "ix", as in *Larix laricina* (DuRoi) K. Koch; "odes", as in *Erythrodes querceticola* (Lindl.) Ames; and "oides", as in *Typhoides arundinacea* Moench.
6) Genus names ending in "um" are neuter (e.g., *Viburnum dentatum* L.).
7) Genus names ending in "e" are neuter (e.g., *Daphne mezereum* L.).

Latin grammar and gender are covered in detail by Gledhill,[16] Manara,[15] Nybakken,[17] and Stearn.[18] Gotch[19] provides meanings for many Latin names applied to animals. Genders of bacterial genera, as well as the meanings of many words used in the names of prokaryotes, are covered in detail on Jean Euzéby's website (http://www.bacterio.cict.fr/gender.html).

22.2.4 Vernacular (Common) Names

Unlike Latin names, vernacular names are in local languages, and the discussion here applies only to English names. Many organisms have vernacular names. Some of these names are widely known—for example, chimpanzee, dandelion, white pine, cobra; others are probably known only to people with special interest in a group—for example, plain chachalaca (*Ortalis vetula*, a bird), Utah bladderpod (*Lesquerella utahensis*, a plant). In scientific writing, once the scientific name and its vernacular equivalent have been presented, the vernacular name may be used, especially if it is widely recognized.

Vernacular names of animals and plants are generally written in lowercase, except for proper name components, which are capitalized. This style is widely followed for nearly all plants, invertebrates (including insects and crustaceans), fish, and mammals and is reflected in a variety of vernacular name lists developed by professional societies (e.g., the American Fisheries Society lists for fishes, mollusks, decapod crustaceans, and Cnidaria and Ctenophora) and in common dictionaries (e.g., *Webster's New World Dictionary of American English*).

hoary marmot	Alaska marmot
common juniper	Vancouver groundcone
golden retriever	Chesapeake Bay retriever
confused flour beetle	Japanese beetle

That style, however, is not universal. Vernacular names of birds, reptiles, and amphibians are uppercased (e.g., Brown Shrike, Diamondback Terrapin) by a few professional societies: the Society of Amphibians and Reptiles, the Herpetological League, and the American Ornithological Union (but not the Audubon Society). They advocate this style to enable readers to readily distinguish between an adjective modifying a vernacular name and the vernacular name itself. For example, does "a white heron" mean a heron that is white ("white" is an adjective) or is it the full vernacular name? If uppercased (White Heron), the potential ambiguity is usually avoided, but not always; for example, does "an Arctic Tern" refer to a tern living in the Arctic or is it the full vernacular name? Because neither style prevents possible confusion, authors and editors must be alert to the problem and use context and wording that is unambiguous.

Unfortunately, no umbrella organization on biological nomenclature has addressed this uppercase/lowercase style inconsistency. Left up to individual societies, editorial offices, and authors, this style inconsistency has created dilemmas for authors and editors, especially when one style converges with the other in the same publication or the same article—or even the same sentence, table, or figure.

Until a cross-disciplinary authority (e.g., International Union of Biological Sciences) makes a style recommendation for all the biological sciences, following the predominant style choice of journals, dictionaries, books, and periodicals will promote consistency and lessen confusion. That is, apart from those journals that specifically require uppercasing of vernacular names, use lowercase for vernacular names of all animals and plants, but use uppercase for any proper name components (see examples at the start of this section).

On first reference, avoid connecting the name of a race, strain, or stock to the vernacular name of a species; this will preclude possible misinterpretation of the vernacular name.

> populations of striped bass (*Morone saxatilis*) from the Chesapeake
> *not* populations of Chesapeake striped bass (*Morone saxatilis*)

After that introduction, use of Chesapeake striped bass is acceptable.

22.2.4.1 Problems with Vernacular Names

The use of vernacular names presents several problems for precise scientific communication: the same species may have more than one vernacular name, the same vernacular name may be applied to quite different species, vernacular names are specific to a language and often to a region, many species do not have vernacular names, and vernacular names may be misleading as to their relations. When a vernacular name is used in scientific writing, give the Latin (scientific) name at least once in the abstract and once in the text of the manuscript, and associate it with the vernacular name. See Sections 22.3.1.7, 22.3.3.7, 22.3.4.3, and 22.3.5.3 for more information about vernacular names for specific groups.

22.2.4.2 Equivalence of Genus or Family Names and Vernacular Names

Many genus names are also used as vernacular names. A genus name used in this way is neither italicized nor capitalized: gorilla, *Gorilla*; octopus, *Octopus*; python, *Python*; rhododendron, *Rhododendron*; aster, *Aster*; fuchsia, *Fuchsia*.

Families often have vernacular names reflecting the vernacular names of only some species within the family. For example, the family Centrarchidae includes the black bass, sunfish, and crappies. Therefore, a vernacular name for such a family becomes either cumbersome (mention of all 3) or potentially ambiguous if just one species name is used as the family's vernacular name (e.g., if "sunfish" were used, then presumably use of "sunfish" alone would mean just the species of sunfish but use of "the sunfish family" would include all 3).

A widespread practice therefore evolved that converts the scientific name of an animal family to a quasi-vernacular name that is simple to use and unambiguous: The "idae" ending for the family scientific name is shortened to "id" or "ids" (depending on usage) and the leading capital letter is lowercased. Hence, "Centrarchidae" becomes "centrarchid" or "centrarchids". For plant families, the "aceae" ending may likewise be replaced by "id" or "ids"—for example, bromeliad (for Bromeliaceae) or orchid (for Orchidaceae). In scientific writing, if these shortened forms of family names are used, give the Latin name at least once.

Care must be taken that quasi-vernacular family names refer to valid scientific families. For example, the snailfish genera *Liparis*, *Lipariscus*, *Paraliparis*, and a few others have been grouped into the family Liparidae, from which the quasi-vernacular name liparid is derived. Other taxonomists have placed these genera in the Cyclopteridae, along with genera of lumpsuckers. If an author follows this latter classification, then use of "liparid" would be inappropriate. "Liparid" should not be used to refer to the genus *Liparis* (see below). The quasi-vernacular name can be singular or plural.

this centrarchid is evolutionarily . . . *or* this species of Centrarchidae is evolutionarily . . .
centrarchids are evolutionarily . . .

Because endings of scientific names for genera vary, no such system has evolved or become recognized for eukaryotes. For genera, use either the scientific name of the genus or its official or generally accepted vernacular name. Avoid applying the "id" or "ids" endings used for families to genera. Also avoid applying a plural ending to a scientific genus name; the scientific name of the genus is always singular. The following examples apply to the genus *Lepomis* or sunfish within the Centrarchidae.

these sunfish are found in . . . *or* these species of *Lepomis* are found in . . . *or Lepomis* is found in . . .
not these lepomids are found in . . .

Vernacular names may be included with genus-level identifications.

black bass (*Micropterus*) populations *or Micropterus* (black bass) populations

However, bear in mind that vernacular names within a genus often include 2 or more vernacular names; for example, *Oncorhynchus* includes some "salmon" and some "trout". In this case, use a format that conveys the vernacular and scientific names unambiguously.

The genus *Oncorhynchus* (some salmon/trout) is . . .
Pacific salmon (genus: *Oncorhynchus*) stocks [note that this and the example below exclude trout]
Pacific salmon stocks, which are of the genus *Oncorhynchus*, . . .
not Pacific salmon (*Oncorhynchus*) stocks [because this might lead someone to conclude that the genus *Oncorhynchus* is limited to just Pacific salmon]

Remember that if the species is unresolved, then so is the vernacular name (unless information other than taxonomic characters [e.g., geographic distinction] provides the vernacular name). Therefore, when referring to unidentified species of a genus encompassing several or more vernacular names, exclude the vernacular name or include all the possibilities.

Oncorhynchus sp. [or spp.] larvae
larval salmon or trout (*Oncorhynchus* sp.) *not* larval trout (*Oncorhynchus* sp.)
larval salmon and/or trout (*Oncorhynchus* spp.) *not* larval salmon (*Oncorhynchus* spp.)

22.3 TAXONOMY AND NOMENCLATURE WITHIN KINGDOMS

The codification of nomenclature for plants predates that for other groups of organisms, and therefore plants, including algae, fungi, and lichens, are discussed first in this manual, followed by animals, bacteria, and viruses. Viruses are covered last because the nomenclature for these entities differs in several respects from that for organisms.

22.3.1 Plants

22.3.1.1 Scientific Names of Plants

For the name of a species or any other taxonomic rank of plants to be legitimate, it must be validly published. Requirements for valid publication are set forth in the *International Code of Nomenclature for algae, fungi, and plants* (ICN), formerly the *International Code of Botanical Nomenclature* (the "Botanical Code"). The Code was most recently revised and ratified in 2011 (the Melbourne Code) and was published in 2012.[12] The Code is published in English, with translations to several other languages available. If any differences are perceived among the language versions, the English version is considered official. The ultimate authority for the ICN rests with the International Botanical Congress, which is held every 6 years. The latest edition of the Code supersedes all previous editions. The rules and recommendations in it are intended to bring uniformity and clarity to future nomenclature. They apply to the names of all organisms treated as plants (including fungi, algae, and lichen-forming fungi but not bacteria), whether fossil or nonfossil. Rules of nomenclature for fossil plants are generally the same as for nonfossil plants with some modification of the rule for priority (Article 11.7), because a name may be applied to fragmentary material.

Family, genus, or species names found to be incorrect according to the Melbourne Code can be conserved or retained when the substitution of correct names would cause confusion and considerable inconvenience, especially for scientists in different disciplines. Conserving a name, however, requires specific action by the International Botanical Congress. Such names are categorized as nomina conservanda (nom. cons.) and are listed in an appendix to the Melbourne Code.

> *Maclura* Nutt., nom. cons. [the genus for osage-orange]

22.3.1.2 Authors of Plant Names

The person who publishes a scientific name is the author of that name. Citation of the author name is optional according to the Melbourne Code, and the author is not part of the scientific name. When an author name does appear, it must be that of the person who validly published the name under the provisions of the Melbourne Code.

> *Proserpinaca palustris* L.
> *Myriophyllum tenellum* Bigelow

If an author name is originally written in a non-Roman alphabet, it should be romanized.

22.3.1.2.1 TRANSFER OF A SPECIES TO ANOTHER GENUS

As more is learned about taxonomic relationships, it may become necessary to move a species originally placed in one genus to another genus. For plants and other organisms covered by the Melbourne Code, when such a change is made, the name of the author who first validly described the species is enclosed in parentheses, followed by the name of the author who moved the species to another genus. The same applies to infraspecific taxa.

Greene validly described *Tetraneuris herbacea*, so the name was originally given as *Tetraneuris herbacea* Greene. Cronquist moved this species to the genus *Hymenoxys*, so the name became *Hymenoxys herbacea* (Greene) Cronquist.

Variety *rydbergii* St.-Yves of *Festuca ovina* L. was reclassified by Cronquist in the genus *Festuca*, and so became *Festuca brachyphylla* Schult. var. *rydbergii* (St.-Yves) Cronquist.

Kartesz published a list of synonyms for names of plants found in the United States, Canada, and Greenland.[20]

22.3.1.2.2 ABBREVIATION OF AUTHOR NAMES

Botanical nomenclature begins with *Species Plantarum* published by Linnaeus in 1753.[21] For names published by Linnaeus and still considered valid, the author is designated simply as "L." The names of other authors, especially those who were responsible for many scientific names, are often abbreviated. Names of more recent authors are usually written out. The accepted abbreviations for author names can be found in *Authors of Plant Names*[22] and *Taxonomic Literature*.[23,24] Abbreviations and corresponding full names are also often listed in an appendix in plant identification manuals. Abbreviations can also be found online in the International Plant Names Index (IPNI; http://www.ipni.org/ipni/plantnamesearchpage.do).

22.3.1.2.3 MULTIPLE AUTHORS

When a name has been published by 2 authors, cite the names of both authors and link them by the Latin word "et" or by an ampersand.

Panicum ravenelii Scribn. et Merr. *or Panicum ravenelii* Scribn. & Merr.

In older literature, "in" between 2 names means that the first-named person described the plant in a work authored or published by the second person.

Solanum sarrachoides Sendtn. in Mart.

Beginning with the 1995 edition of the Botanical Code (the "Tokyo Code"), "in" is taken to be part of the bibliographic citation and not part of the name. Therefore, do not use "in" unless it is followed by the complete bibliographic citation.

The Latin word "ex" between 2 names means that the first author proposed the name but did not validly publish it, and the second author was the person who validly published it.

Cypripedium candidum Muhl. ex Willd.

Omit author citations for scientific names from the title of an article unless the article refers to the use of different authors for the same taxon.

22.3.1.2.4 AUTHORS OF INFRASPECIFIC TAXA

Use authors for infraspecific ranks such as subspecies (subsp. or ssp.), varieties (var.), and forms (f.). The form is the lowest rank in formal taxonomy that is governed by the Melbourne Code.

Medicago sativa L. subsp. *falcata* (L.) Arcangeli
Clematis flammula L. var. *maritima* (L.) DC
Salix candida Flüggé f. *denudata* (Andersson ex DC) Rouleau

If an infraspecific taxon is created within a species, the nomenclatural type (a physical specimen—a dried plant preserved in a herbarium—to which the name is permanently attached) automatically becomes an autonym, a name in which the infraspecific epithet repeats the epithet. For an autonym, cite authors only after the species name, not after the autonym epithet.

Rorippa palustris (L.) Besser var. *palustris*

22.3.1.3 Infrageneric and Infraspecific Taxa

An epithet referring to an infrageneric or infraspecific rank must be preceded by a word, often abbreviated, indicating the taxonomic rank. The word or abbreviation indicating rank is not italicized and is written without an initial capital letter (e.g., ser.).

Costus subg. Metacostus ["subg." is the abbreviation for "subgenus"; *Metacostus* is the subgenus epithet]
Ricinocarpos sect. *Anomodiscus* ["sect." is the abbreviation for "section"; *Anomodiscus* is the section epithet]
Desmodium ser. *Stipulata* Schub. ["ser." is the abbreviation for "series"; *Stipulata* is the series name]
Rorippa palustris (L.) Besser var. *fernaldiana* (Butters & Abbe) Stuckey ["var." is the abbreviation for "variety"; *fernaldiana* is the varietal name]

The name of any entity below the rank of species can be shortened to a trinomial encompassing the genus name, the specific epithet, and the name at the lowest rank. Thus, the earlier example can be shortened to this form:

R. palustris var. *fernaldiana* (Butters & Abbe) Stuckey

In this situation, abbreviate only the genus name, not the specific epithet, even though the name of a lower rank follows. Moreover, only the author for the ultimate combination need be cited; the author for the specific epithet may be omitted (see Section 22.3.1.2 for a discussion of authors of plant names). The *International Code of Nomenclature for algae, fungi, and plants* (formerly the *International Code of Botanical Nomenclature*), unlike the *International Code of Zoological Nomenclature*, does not allow "undesignated trinomials". Thus, do not write *Rorippa palustris* var. *hispida* as *Rorippa palustris hispida*.

22.3.1.4 Names of Families and Higher Taxa

The names of plant families end in "aceae", which is appended to the type genus name—for example, Cyperaceae (where the type genus is *Cyperus*). The Melbourne Code permits alternative names for 8 families: Compositae or Asteraceae, Cruciferae or Brassicaceae, Gramineae or Poaceae, Guttiferae or Clusiaceae, Labiatae or Lamiaceae, Leguminosae or Fabaceae, Palmae or Arecaceae, and Umbelliferae or Apiaceae. Use of either name is acceptable; however, be consistent within a given work in the choice of these alternative family names.

Order names end in "ales", class names end in "opsida", and subclass names end in "idea".

Names at the rank of family and above are plural in form and therefore require plural verbs and pronouns.

Polypodiopsida include the royal ferns.
The Rosales are estimated to comprise 6,600 species.
The Liliaceae are very diverse and have been separated into numerous smaller families by some botanists.

22.3.1.5 Names of Hybrids (Nothotaxa)

Rules for naming hybrids (nothotaxa) are presented in Appendix I to the Melbourne Code and apply to naturally occurring hybrids as well as to agronomic and horticultural ones. Hybrids are designated in 2 ways: by a formula or by a name.

In a formula, the names of the known or putative parents are separated by a multiplication symbol (×) with a space before and after them; the symbol is not italicized. Because a formula consists of existing names, it requires none of the formalities for new names. List the names in alphabetical order. Alternatively, the seed (female) parent can be listed first if this convention is explained.

Digitalis lutea L. × *D. purpurea* L. [for the hybrid between *D. lutea* and *D. purpurea*]
Fatsia × *Hedera* [for the hybrid between a species of *Fatsia* and a species of *Hedera*]

When a new name is applied to a hybrid, it is published under essentially the same rules as any other name of the same rank; in this case, the multiplication symbol is placed just before the hybrid name without a following space.

Mentha ×*piperita*
×*Stiporyzopsis* [for *Stipa* × *Oryzopsis*]
×*Mahoberberis* sp. [for a hybrid between species of the genus *Mahonia* and species of the genus *Berberis*]

If the multiplication symbol (×) is not available, the lowercase letter "x" may be used instead, but in this situation a space is needed between "x" and the following name to avoid ambiguity.

A naturally occurring nothospecies is *Quercus* x *asheana*, a natural hybrid between *Quercus cinerea* and *Q. laevis*.

The multiplication symbol is not used for infraspecific ranks, but the abbreviations denoting their ranks carry the prefix "notho"— for example, *Polypodium vulgare* L. nothosubsp. *mantoniae* Schidlay in Futák.

22.3.1.6 Cultivated Plants

The *International Code of Nomenclature for Cultivated Plants* (ICNCP), which supplements the Melbourne Code, was published in 1953 and has been revised several times, most recently in 2009.[25] The International Commission for the Nomenclature of Cultivated Plants of the International Union of Biological Sciences has responsibility for new editions of this code, and only the Commission can modify the code. The aim of the ICNCP is to promote uniformity, accuracy, and stability in the naming of agricultural, horticultural, and silvicultural cultivated varieties (cultivars).

Plants brought from the wild into cultivation retain the names that were applied to the same taxa growing in nature. The ICNCP deals with additional, independent designations for plants used in agriculture, forestry, and horticulture. However, nothing precludes the use of names published in accordance with the requirements of the Melbourne Code for cultivated plants (Article 28 of the Melbourne Code).

To ensure that a cultivar has only one correct name, the ICNCP requires that each cultivar be named by means of priority acts, which consist of publication and registration of the name in documents that are dated and distributed to the public. Appendices I, II, and IV of the ICNCP contain a list of international cultivar registration authorities, a directory of statutory plant registration authorities, and a list of places maintaining nomenclatural standards. Because the ICNCP has no legal status, the commercial interests of plant breeders are protected by national and international law. The Council of the International Union for the Protection of New Varieties of Plants (http://www.upov.int/) is an international organization that coordinates laws regarding protection of plant cultivars among member countries.

A cultivar is an assemblage of plants that has been selected for a particular attribute or combination of attributes and that is clearly distinct, uniform, and stable in these characteristics and that when propagated by appropriate means retains these characteristics (Article 21 of the ICNCP eighth edition).[25] Trademarks are not considered names and are not covered by the ICNCP.

The term "cultivar" encompasses clones derived vegetatively from a single parent; lines of selfed or inbred individuals; and collections of individuals that are resynthesized only by crossbreeding—for example, F_1 hybrids. Cultivar names can include a forestry provenance or a particular growth-habit form that can be retained by appropriate methods of propagation.

Indicate cultivar names, sometimes referred to as "fancy" names, by placing them within single quotation marks. Write cultivar names with initial capital letters and roman type.

Rubus arcticus L. var. *grandiflorus* Ledeb. [a botanical variety]
Rubus flagellaris Willd. 'American Dewberry' [a cultivated variety or cultivar]
Triticum aestivum L. 'Era'
Hordeum vulgare L. 'Proctor'
Juniperus communis L. var. *depressa* Pursh. 'Plumosa'

Authors have sometimes used the abbreviation "cv." (for "cultivar"), placing it after the Latin name and before the cultivar name. The ICNCP specifically prohibits this practice and states that "such use is to be corrected" (Article 14.1).

Triticum aestivum L. 'Era' *not Triticum aestivum* L. cv. Era

Single quotation marks are not necessary when cultivar names are written alone (without the species name), unless their absence would be confusing.

Era is a widely grown cultivar of wheat.
Hedera helix L. 'Chicago' is a popular ivy because of its fast growing habit and good keeping qualities. It resembles *H. helix* 'Pittsburgh' except that plants of 'Chicago' grow more slowly than those of 'Pittsburgh'.

A Group is a formal category for assembling cultivars, individual plants, or combinations thereof on the basis of defined similarity (Article 3.1 of ICNCP).

Allium cepa L. Shallot Group
Rosa L. Polyantha Group

Each element of the group name, as well as the word "Group", begins with a capital letter. If the Group name is followed by a cultivar name, enclose the Group name in parentheses or square brackets:

Hydrangea macrophylla Ser. (Hortensia Group) 'Ami Pasquier'
or
H. macrophylla [Hortensia Group] 'Ami Pasquier'

A chimera (plural, "chimeras") is an individual plant or organ consisting of tissues of different genetic constitution resulting from a graft union (graft chimera). If a name is coined from elements of the names of the species that were grafted, this name is preceded by a plus sign; for example, +*Crataegomespilus* 'Dardarii' is the name of a graft chimera between *Crataegus monogyna* and *Mespilus germanica* (example 4, page 38, ICNCP). Alternatively, a graft chimera can be designated by writing the accepted names of its component taxa with a plus sign between them—for example, *Crataegus* + *Mespilus*, with a space on either side of the plus sign. Because it is not a true hybrid, it is not represented by a multiplication sign. Instead, a plus sign appears before the genus name or the specific epithet.

+*Laburnocytisus* 'Adamii' for the graft chimera of *Cytisus purpureus* and *Laburnum anagyroides* (example 3, page 38, ICNCP)

When component taxa belong to the same genus, the graft chimera can be designated by the genus name and a cultivar name—for example, *Camellia* 'Daisy Eagleson' (example 5, page 38, ICNCP).

22.3.1.7 Vernacular Names for Plants

There is no authoritative source of guidance on writing vernacular names for plants, and there is inconsistency in how vernacular names are presented. CSE recommends that authors and editors consult an authoritative dictionary or plant identification manual for guidance on use of hyphens and joining of names in vernacular names for plants, bearing in mind that similar names may be treated differently in a single source and that different sources may disagree regarding presentation of the same name. Some of these problems and a recommended general approach to vernacular names are outlined below.

In terms of general form, vernacular names may consist of a group name corresponding to a genus or family. Simple group names consist of a single word.

ash	fern	lily	mustard	tulip
aster	grass	mallow	orchid	willow
clover				

A modifier that corresponds to a species may precede a simple group name (e.g., soapwort gentian, Indian grass, three-birds orchid). Some vernacular names do not include a group name—for example, adder's-mouth (for the orchid *Pogonia ophioglossoides*).

Common names are not capitalized unless derived from proper nouns.

English ivy	Jerusalem artichoke	Blue Ridge gayfeather
Dutchman's-pipe	Good King Henry	

A genus name used as a vernacular name is neither capitalized nor italicized and forms a plural as in English.

aster for the genus *Aster*	camellia, *Camellia*	iris, *Iris*	crocus, *Crocus*
asters	camellias	irises	crocuses [*not* croci]

In some cases, botanists have adopted a practice similar to that used by entomologists (see Section 22.3.3.7) by inserting a hyphen before a group name that is misapplied (e.g.,

Douglas-fir for *Pseudotsuga menziesii*, which is not a true fir) and not using a hyphen when the name is properly applied (e.g., white fir for *Abies concolor*, a true fir). Writers, however, do not follow this principle consistently. For example, the vernacular name for *Xerophyllum* (a member of the lily family; not a true grass) is given as beargrass in one manual,[26] as bear-grass in another,[27] and as bear grass in a compilation of plant names.[28] Even when the vernacular name reflects the true taxonomic position of a plant (e.g., use of "grass" for members of the Poaceae), there is inconsistency as to whether to present the elements of the name as separate words, to join them with a hyphen, or to join them into a single word. For example, the following vernacular names for various true grasses are given in a manual of plants for the northeastern United States.[29]

moor-matgrass [*Nardus strictus*]
Indian ricegrass [*Oryzopsis hymenoides*]
needle-and-thread grass [*Stipa comata*]
porcupine-grass [*Stipa spartea*]
bluegrass [*Poa* spp.]
meadow-fescue [*Festuca pratensis*]
western fescue [*Festuca occidentalis*]

Were botanists to follow the guidelines of the entomologists, the word "grass" or "fescue" would be a separate word in all of these names because all are true grasses.

Recourse to authoritative dictionaries will not necessarily eliminate inconsistency. For example, the vernacular name for *Podophyllum peltatum* is given as mayapple in *Merriam-Webster's Unabridged Dictionary* and as May apple in *Webster's New World College Dictionary*.

Despite these challenges, authors and editors should strive for consistent style when vernacular names are used in a document. The following guidelines, derived mainly from Rickett[30] (who developed a short set of rules nearly 50 years ago) and from material presented in the introduction of *Scientific and Common Names of 7,000 Vascular Plants in the United States*[31] and in Kartesz and Thieret,[32] are offered as recommendations to achieve greater consistency.

1) Join words of 1 or 2 syllables that refer to plants in general or to some part of a plant to the preceding word unless the result would be long and unwieldy. Examples of such words are plant, tree, vine, wort, nut, berry, flower, and weed. Examples are bladderwort, mugwort, knapweed, and duckweed.
2) If the second part of a vernacular name is correctly applied (according to taxonomic groups), do not join it to the preceding word with a hyphen (this corresponds to the practice of entomologists); for example, orange lily, tumbling mustard. Note that by long practice, elements of some such names are joined: bluegrass. These will generally be indicated in authoritative dictionaries.
3) If a name is misapplied according to taxonomic principles, hyphenate it: rue-anemone, poison-oak, Douglas-fir. However, do not hyphenate names that begin with "false" because this prefix already indicates misapplication. Rickett considered the modifier "Indian" to be similar to "false" in indicating misapplication of the group name (e.g., Indian bean); therefore, do not hyphenate this word.
4) If the second part of the name has nothing to do with plants or plant parts, hyphenate the name: blazing-star, adder's-mouth.
5) For names composed of more than 2 words, hyphenate throughout: Jack-in-the-pulpit, Star-of-Bethlehem, Joe-Pye-weed, Queen-Anne's-lace.
6) Do not use the word "common" in vernacular names.
7) Follow the presentation in an authoritative dictionary for capitalization.
8) Follow Kartesz and Thieret[32] for vernacular names of plants that refer to animals.

Vernacular group names correspond roughly to genera or higher taxa. They need not have a modifier. Group names may consist of single words describing a particular family, genus, subgenus, tribe, or section.

More than half a century ago, the American Joint Committee on Horticultural Plant Nomenclature attempted to bring standardization to vernacular names applied to plants involved in horticultural trade, forestry, and agriculture.[33] This group recommended joining words into a single word wherever possible. Thus, *Pseudotsuga taxifolia* (now *P. menziesii*) was written as Douglasfir, *Rhus toxicodendron* was written as common poisonivy, and *Ipomoea hederacea* was written as ivyleaf morningglory. More recent works (e.g., Gleason and Cronquist[29]) use ivy-leaved morning-glory, presumably to adhere to the standard English practice of hyphenating compound adjectives and to avoid the non-English double g in "morningglory". *Composite List of Weeds*, published by a subcommittee of the Weed Science Society of America,[34] makes recommendations similar to those of the American Joint Committee on Horticultural Plant Nomenclature,[33] doing so "because of the constraints of the computer sorting program used in the preparation of the list". CSE recommends not joining words to create vernacular names, unless such compounds are already well established in the language—for example, bluegrass.

Distinguish the vernacular name of the crop from the harvest product.

oat [crop], oats [the grain]	grape, grapes	beet, beets
bean, beans	apple, apples	banana, bananas
pea, peas	olive, olives	

22.3.2 Fungi, Algae, Lichens, and Related Organisms

22.3.2.1 Fungi

The Melbourne Code[12] applies to fungi as well as to plants in the usual sense. Fungi were included in the Plant kingdom in earlier taxonomic works based on a 2-kingdom classification (see Section 22.2). In modern classifications, fungi have their own kingdom (Fungi); they are nongreen eukaryotic organisms with cell walls that differ chemically from those of green plants, and they have different modes of nutrition. There are 3 divisions within the kingdom Fungi: Zygomycota (zygomycetes), Basidiomycota (basidiomycetes), and Ascomycota (ascomycetes). Some divisions in the kingdom Protista (including Myxomycota, Oomycota, Plasmodiophora, and Chytridiomycota) were formerly considered to be fungi, and the Melbourne Code covers nomenclature for these.

As with plants, the binomial method is used for the Latin (scientific) name of a fungus. The genus name and specific epithet are written in Latin and are italicized; the author of the name is written in roman type. For example, in the name *Amanita phalloides* Fr. (death cap), *Amanita* is the genus name and *phalloides* the specific epithet. After first mention of the complete genus name, it can subsequently be abbreviated to a single capital letter followed by a period, unless there is potential for ambiguity as to the genus.

> *Amanita phalloides* Fr. is a poisonous mushroom, but *A. caesarea* (Fr.) Schw. is generally regarded as nonpoisonous.

Names of fungal genera do not have plural forms. If reference is made to a group of species in a genus, write the genus name followed by the abbreviation for the plural word "species". (See Section 22.2.3.5 for a discussion of the proper use of "sp." and "spp.")

Pythium spp. *not* pythia
Fusarium spp. *not* fusaria
Helminthosporium spp. *not* helminthosporia

Do not apply adjectival endings to names of fungal genera.

Fusarium head blight [*not* Fusarial head blight *or* fusarial head blight]

22.3.2.1.1 ANAMORPH, TELEOMORPH, AND HOLOMORPH NAMES

For non–lichen-forming fungi with a pleiomorphic life cycle belonging to *Ascomycota* or *Basidiomycota*, Article 59 of the Melbourne Code[12] has removed the provision that allows the use of separate names for the anamorph (asexual morph) and the teleomorph (sexual morph) at the genus and species levels. Before January 2013, the correct name of the holomorph (the organism in all of its life stages) had been that of its teleomorph. Now, all legitimate names, regardless of the life history stage, will be treated equally to establish priority as the valid name of the organism on the basis of the date of publication. Often the anamorph stage was the first discovered and validly named; therefore, that name has priority. However, the Melbourne Code[12] also accepted a new set of rules that will allow lists of widely used names to be protected en masse or names of uncertain application to be rejected en masse (http://www.seaweed.ie/nomenclature/pdf/taxon_60_major_changes_xviii_ibcm.pdf). The online database Index Fungorum (http://www.indexfungorum.org/names/names.asp), the *Dictionary of the Fungi*,[41] and the searchable database (http://speciesfungorum.org/Names/fundic.asp) based on the classification in the *Dictionary* are good sources to search for names and publication dates.

For example, *Fusarium graminearum* Schwabe (1839) (anamorph) and *Gibberella zeae* (Schwein.) Petch (1936) (teleomorph) are names for the same fungus, and the valid name for the holomorph, the organism in all its stages, was the teleomorph name. Now, under the "one fungus, one name" rule, the earlier published name, *Fusarium graminearum* Schwabe, is the valid name.

22.3.2.1.2 SANCTIONED NAMES

Although the starting point for names of fungi is the same as that for plants (the 1753 date of publication of *Species Plantarum*), fungal taxa names adopted by Persoon in his works on Uredinales, Ustilaginales, and Gasteromycetes, published in December 1801, or by Fries for all other fungi except Myxomycetes, in his works of January 1821, are "sanctioned" as privileged names with special priority and typification status.[35,36] When Fries or Persoon took up a name applied prior to these works, a colon, with a space before and after, is placed between the originating and the sanctioning author.

Boletus piperatus Bull : Fr. [the name *Boletus piperatus* predates Fries's publication, but Fries included the name in his publication, so it is valid]

In fungal nomenclature, this use of the colon has virtually replaced use of the Latin word "ex"; the use of "ex" is now limited to referring to a person who first validly publishes a name ascribed to another person.

Ramichloridium Stahel ex De Hoog [Stahel coined the genus name but did not provide a Latin diagnosis; De Hoog validly published the name with Latin diagnosis]

22.3.2.1.3 INFRASPECIFIC NAMES FOR FUNGI

Fungi have several infraspecific ranks, including subspecies, variety, and form (abbreviated as "f."). The form is the lowest rank in formal taxonomy that is governed by the Melbourne Code. Recognition of these infraspecific taxa is usually based on morphological characteristics.

An infraspecific rank used with plant pathogenic fungi is the special form (forma specialis; "f. sp.", singular; "ff. sp.", plural), which is characterized by physiological criteria (host adaptation), not morphological characteristics. The Botanical Code does not govern nomenclature for this rank, and no Latin diagnosis or author citation is required.

Blumeria graminis f. sp. *hordei*
[*hordei* is the forma specialis epithet; this designates a form of the species *B. graminis* that parasitizes barley]

Another category used for fungi pathogenic to plants is variously referred to as race, physiologic race, pathogenic race, pathotype, or biotype. Scientists working with particular groups establish the rules and methods for race nomenclature, and there is considerable variation among species. Races are usually identified by the reactions of several host cultivars, each carrying a different gene or genes for resistance to the pathogen species. Each unique pattern of reaction of these "differential" cultivars defines a race. An isolate of the fungus is placed in a race according to which differentials are susceptible and which are resistant.

In some instances, races are named according to the host resistance genes (designated by numbers, e.g., *R1*, *R2*) that they "overcome". For example, race 1,2 (or R_1R_2) of *Phytophthora infestans* can successfully infect potato cultivars that have either the *R1* or *R2* gene for resistance but not cultivars that carry other *R* genes. Races for many other pathogens are simply numbered consecutively as they are discovered, starting with the number 1. Wheat stem rust workers developed a special system to deal with the large number of host resistance genes used to differentiate races.[37] They use letter codes to designate patterns of reaction on subsets of differentials. Thus, a race of *Puccinia graminis* f. sp. *tritici* might be designated TNM, which indicates a pattern of reaction on each of 3 sets of 4 differential cultivars. A similar system exists for leaf rust of wheat.[38]

22.3.2.1.4 FOSSIL FUNGI

The Melbourne Code also deals with fossil fungi. Fossil fungi may be given names formed by adding a suffix (usually "ites") to currently used names.

Clasterosporites Pia [fossil, genus name based on *Clasterosporium* Schwein, a living fungus]
Graphiolites Fritel [fossil, genus name based on *Graphiola* Poit., a living fungus]
Pleuricellaesporites v.d. Hammen *Fungites* Hallier [both are fossil fungi, but their names are not derived from current names]

Fossil fungi may be named with new form-generic names the same way as for living fungi and, in fact, the same way as any fossil plants are named.

Grilletia Renault & Betrand *Plectosclerotia* Stach & Pickh

22.3.2.1.5 YEASTS

A yeast is a unicellular fungus that reproduces by budding or fission. In some genera, pseudomycelia or mycelia may be present. Yeasts that produce spores have teleomorphs in the divisions Ascomycotina or Basidiomycotina; those without spores are placed in the artificial class Hyphomycetes. The nomenclature and typification of yeasts follow the guidelines for other fungi in the Melbourne Code.[12]

Saccharomyces cerevisiae Meyen ex E. Hansen var. *ellipsoideus* (Hansen) Dekker is a common yeast.

The term "yeast" refers to a growth form and has no taxonomic standing.[39] The term is usually applied to fungi in the order Saccharomycetales and includes such genera as *Candida*, *Cryptococcus*, *Saccharomyces*, *Schizosaccharomyces*, and *Torulopsis*. The term is sometimes applied to fungal genera such as *Aspergillus* and *Neurospora*, probably because these genera have a yeast-like phase with budding cells at some stage in the life cycle. Baker's or brewer's yeast is *Saccharomyces cerevisiae* Meyen ex E. Hansen. "Black yeast" is jargon referring to yeast-like states of *Aureobasidium*, *Cladosporium*, *Moniliella*, and other nonyeast fungi. Industrial yeasts, such as those used in distilling, brewing, and winemaking, are often not identified by genus or species.

If the word "yeast" is used, identify the organism by its Latin name at least once in the abstract and once in text. Although the word "yeast" is a noun, it may be used as an adjective if only one kind of yeast is described in the paper.

yeast cells yeast DNA yeast culture yeast colony

22.3.2.1.6 MOLDS

The term "mold" usually refers to any fungal growth but may be used in nontechnical writing for fungi that grow in damp areas of dwellings and other structures. The term is also applied to a group of organisms known as slime molds. Historically, mycologists considered slime molds to be fungi[39] and applied nomenclature consistent with the Melbourne Code.[12] However, Olive[40] considered their status as fungi to be uncertain, and he classified them in the kingdom Protista (which includes protozoa and unicellular algae) and gave names consistent with usage in the *International Code of Zoological Nomenclature*.[10] Margulis and Schwartz[7] classified these organisms in the kingdom Protoctista. Slime molds are now placed in the division Myxomycota, which contains 3 orders: Myxomycetes (true slime molds), Dictyostelimycetes (cellular slime molds), and Protosteliomycetes (protostelid slime molds).[41]

22.3.2.2 Algae

Algae (singular, "alga") is a nontaxonomic word for a group of taxa that have been classified traditionally as plants but are now regarded as being of diverse origin. The blue-green algae (Cyanobacteria) are a separate division in the prokaryotes.[42] The remaining algae are dispersed among 7 divisions, plus a tentative division (Chlorarachniophytes) that consists of amoeba-like organisms containing chloroplasts.[43] An alga is any plant-like organism that carries on photosynthesis and differs structurally from ordinary land plants, such as mosses, ferns, and seed plants. The rules for nomenclatural style and format given for plants also apply to the algae, and the Melbourne Code governs names of algae. *Freshwater Algae of North America: Ecology and Classification*[43] provides descriptions; information on classification, biology, and ecology; keys to genera; and guidance to sources of information for species identification for algae found in fresh waters of North America.

22.3.2.3 Lichens

Lichens represent a biological, not a systematic, group. A lichen is a stable, self-supporting association (consortium) of a fungus (mycobiont) and an alga or cyanobacterium (photobiont). For nomenclatural purposes, the names given to lichens apply to their fungal components, and lichens are classified mainly in the Lecanorales of the Ascomycota. The algae in lichens are usually green algae or occasionally blue-green algae (cyanobacteria) and have separate names. *Lichens of North America*[44] contains information about lichen biology, keys for identification, and descriptions of species.

Cladonia cristatella Tuck. is a lichen in the Lecanorales of the fungi, and its algal component is *Trebouxia* sp., a green alga.

Peltula polysora (Tuck.) Wet. is a lichen, and its algal component is *Anacystis montana* (Lightf.) Dr. et Daily, a cyanobacterium.

22.3.3 Animals

22.3.3.1 Scientific Names for Animals

The rules and recommendations for giving a scientific name to a taxonomic group of animals are found in the *International Code of Zoological Nomenclature* (the "Zoological Code")[10] and its amendments. The criteria for valid publication of scientific names of animals are set out in Articles 7, 8, and 9 of the Zoological Code. The essential features of zoological nomenclature and taxonomy are explained by Mayr and Ashlock.[45] Current developments are reported in *Systematic Biology*.[46] Applications to the International Commission on Zoological Nomenclature for approval of scientific names (and comments thereon) are published in the *Bulletin of Zoological Nomenclature*,[47] which also includes official decisions of the Commission regarding names and works. The Commission produces official lists of approved names and official indexes of rejected names and works (publications).[48] The Zoological Code covers fossil animals as well as extant animals, and several articles of the Zoological Code deal specifically with fossils.

The Zoological Code is independent of codes for groups of organisms other than animals. Thus, a valid name can be published for an animal that is the same as a valid name for a plant. For example, *Sida* is the valid name of a genus of waterflea (a crustacean) and a genus of plants. The Zoological Code recommends that an author wishing to propose a new genus name first check to see if the name has been used for plants or bacteria and, if so, refrain from using it for an animal.

22.3.3.2 Authors of Animal Names

The starting point for zoological nomenclature is 1 January 1758, which is taken to be the date of publication of Linnaeus's *Systema Naturae* and Clerck's *Aranei Svecici*.[10] The author of a genus or species is not part of the scientific name, and its inclusion in the taxonomic name is optional. However, Appendix E of the Zoological Code recommends that each use of a name of a genus or taxon of lower rank in a document include the name of the author of the taxon and the year of naming—for example, *Aphis gossypii* Clover, 1877. In a taxonomic work, a full reference to the publication in which the name was published should be included in the document's list of references. The author's name and the year for a given taxon need appear only once in an article, preferably at first mention of the taxon, but should not appear in the title. The author's name follows the taxonomic name and is separated from the taxonomic name by a space. A comma and space separate the author's name from the date of publication. The author name and date are not italicized.

Taenia diminuta Rudolphi, 1819 (Rudolphi 1819) [presentation at first mention in a document with in-text references in the name–year format]
Taenia diminuta Rudolphi, 1819[10] [presentation at first mention in a document with in-text references in the citation–name or citation–sequence format, where Rudophi's description of the species is reference 10]

22.3.3.3 Transfer of a Species to Another Genus

When a species is transferred from its originally designated genus to another on the basis of new information or a new interpretation of its characters, the name of the original author is placed within parentheses. Unlike the practice in botany (see Section 22.3.1.2) and bacteria (see Section 22.3.4.2), the author of the new combination is not added for animal names (Article 51.3 in the Zoological Code).

Taenia dunubyta Rudolphi was reclassified as *Hymenolepis diminuta* (Rudolphi).

For adjectival specific names, the ending of the specific name changes if the new genus name for the organism differs in gender from that of the previous genus name: *Taeniothrips albus* (masculine) was reclassified as *Frankliniella alba* (feminine).

22.3.3.4 Abbreviation of Author Names

The Zoological Code recommends that surnames of authors of scientific names be spelled out in full (Appendix B). Although Linnaeus is commonly abbreviated as "L." or "Linn.", this practice is not consistent with the recommendation of the International Commission on Zoological Nomenclature. When 3 or more authors publish a scientific name, the surname of the first author may be cited followed by "et al." (Recommendation 51C; Appendix B).

The abbreviation "Anon." is used in place of an author's name for a species or taxon that was published anonymously (Recommendation 51D). If a species or taxon was published anonymously, but the author's name is known or inferred from other reliable information, the author's surname is given in square brackets following the species name or taxon (Recommendation 51D). For this reason, parentheses around the name of the author of a species or taxon should not be converted to square brackets, as is the common practice when a parenthetical statement is nested within another parenthetical statement.

22.3.3.5 Diacritical Signs and Other Punctuation Marks

The Zoological Code requires omission of all diacritical signs, apostrophes, diaereses, and hyphens (see Section 22.2.3.7), except when hyphens are used to set off individual letters in a compound species-group name, as in *Polygonia c-album*, so named because a white mark on the wing of the butterfly is similar to the letter c (Article 32.5.2.4.3 of the Zoological Code). Article 32.5.2 states that a name published with any diacritical sign, ligature, apostrophe, or hyphen is to be corrected. For names published before

1985 that contain an umlaut, an "e" is to be inserted after the vowel with the umlaut and the umlaut omitted. For example (Article 32.5.2.1), *nuñezi* is corrected to *nunezi*, and *mjøbergi* to *mjobergi*, but *mülleri* (published before 1985) is corrected to *muelleri*.

22.3.3.6 Infraspecific Taxa

The Zoological Code regulates the names of subspecies. "Variety", "form", "aberration", and "morph" are infrasubspecific names, and these are not regulated by the Code. However, an infrasubspecific name may be regarded as a subspecific name, depending on when it was first published (before 1961) and the ambiguity or clarity of the author regarding its designation as an infrasubspecific name (Articles 45.6.3 and 45.6.4). Undesignated trinomials may be used in zoology, contrary to the practice in botany (see Section 22.3.1.3 in this manual) and bacteriology. Thus, a subspecies can be designated with a trinomen (Article 5.2).

Panthera tigris altaica [Siberian tiger] [*altaica* and *sumatrae* are subspecies names]
Panthera tigris sumatrae [Sumatran tiger]

A subspecies that contains the species type has the same species-group name, author, and date as the species. This is known as a nominotypical subspecies (Article 47.1).

Tamias amoenus amoenus [the Klamath chipmunk, 1 of 14 subspecies of *T. amoenus*; because this subspecies includes the species type, the subspecies name is the same as the specific name]

22.3.3.7 Vernacular Names for Animals

Lists of approved vernacular names for species in a number of phyla have been published: amphibians and reptiles,[49] cnidaria and ctenophora,[50] mollusks,[51] decapod crustaceans,[52,53] fishes,[54,55] insects,[56,57] and birds.[58,59] These references link each scientific name with an approved common name. Various compilations of scientific names of mammals include common names, which are not necessarily "approved" by an authoritative body.[60–63]

The Committee on Names of Fishes of the American Fisheries Society[55] states that common names should not be capitalized except for those elements that are proper names (principle 5, page 12). Common names of other animals often follow the same guidelines (see, for example, orthography of vernacular names in publications such as the *Journal of Ecology*, the *Journal of Wildlife Science*, and *Ecology*). However, the American Ornithological Union recommends capitalization of the first letter of all words of bird names—for example, White-collared Manikan (*Manacus candei*) and Brown Shrike (*Lanius cristatus*); see also Section 22.2.4.1 for a discussion of this issue.

The Entomological Society of America publishes approved lists of common names for insects, which can be found on the society's website (http://www.entsoc.org/common-names).[56] Only proper nouns are capitalized. For 2-part insect common names, the second part is presented as a separate word when it is a systematically correct name but is combined with a preceding modifier when it is not.

bed bug [a true bug] house fly [a true fly] butterfly [not a true fly]

In common names for insect larvae, the suffix "worm" is always combined with the modifier because larvae are not true worms (annelids).

silkworm *not* silk worm striped cutworm *not* striped cut worm

22.3.3.8 Laboratory Strains of Animals

Information on inbred animal strains can be found in the International Index of Laboratory Animals.[64] The Committee on Standardized Genetic Nomenclature for Mice (CSGNM) has established a symbolic system for designating inbred strains of mice.[65] Essential elements of this system can be found online (http://www.informatics.jax.org/; Mouse Genome Informatics).[66]

For further details on codes for inbred strains and sublines and on designating recombinant inbred, coisogenic, congenic, and segregating strains, and strains preserved by freezing, consult the CSGNM document[65] and the related chapters of the same work, "Inbred Strains of Mice" and "Subline Codes for Holders and Producers". Similar conventions are recommended for inbred strains and substrains of laboratory rats.[67]

Nomenclature for outbred laboratory animals should conform to the recommendations of Festing et al.[68]

Guidelines for nomenclature of transgenic animals can be found online (http://ilarjournal.oxfordjournals.org/content/34/4/45.full).[69]

22.3.4 Bacteria and Archaea

Bacteria (singular, "bacterium") are single-celled organisms in the kingdom Prokaryotae.[70] In the 5-kingdom system,[7] bacteria are placed in the kingdom Bacteria (formerly referred to as Monera or Prokaryotae). In the most recent 3-domain system, bacteria constitute the Eubacteria domain. A morphologically similar but genetically distinct group is the Archaea, which constitute a separate domain in the 3-domain system (see Section 22.2.2).

Bacteria include rickettsiae (singular, "rickettsia"), chlamydiae (singular, "chlamydia"), mycoplasmas (singular, "mycoplasma"), cyanobacteria (formerly known as blue-green algae), and actinomycetes (singular, "actinomycete"). Rickettsiae are mainly rod-shaped, sometimes coccoid, and often pleiomorphic gram-negative microorganisms with typical bacterial cell walls and no flagella; they are, with one exception, obligate intracellular parasites of eukaryotic hosts.[71] Mycoplasmas are prokaryotes in the phylum *Aphragmabacteria* and are the smallest free-living cells without cell walls, bounded only by a lipoprotein cell membrane; they evolved from the Eubacteria.[72] Mycoplasma-like, nonculturable microorganisms without cell walls associated with certain plant diseases are referred to as phytoplasmas.[73]

The systematics of bacteria and archaea are similar to those of plants. The names of taxa from the rank of order to subtribe, inclusive, are formed by adding a designated suffix to the name of the type genus; unlike the situation for plants and animals (see Section 22.2.1), however, the names of all taxonomic ranks are italicized.

22.3.4.1 Scientific Names for Bacteria

The nomenclature of bacteria is governed by the *International Code of Nomenclature of Bacteria* (the "Bacteriological Code").[11] The Bacteriological Code deals with taxonomic ranks of subspecies through class. Names of taxa must be both effectively and validly published to be recognized. To be effectively published, the description of the taxon must appear in a printed publication that is sold to the public or to a bacteriologic institution and that provides a permanent record. Valid publication requires that the name of the taxon be published in the *International Journal of Systematic Bacteriology* (*IJSB*) or its successor, the *International Journal of Systematic and Evolutionary Microbiology* (*IJSEM*), and that the name be accompanied by a description of the taxon in the same article or by reference to a description effectively published elsewhere. The Bacteriological Code requires that descriptions of bacterial species be written in English, not Latin.

Validly published scientific names of bacteria and archaea appear in the *List of Prokaryotic Names with Standing in Nomenclature* (http://www.bacterio.cict.fr)[74]; this list is updated after the publication of each issue of *IJSEM*. An index of nomenclatural changes for bacteria and yeast appearing in the *IJSB* from 1 January 1980 to 1 January 1992 has been published.[74] Priority of publication for bacterial names dates from 1 January 1980. Names published before then are valid only if included in the *Approved Lists of Bacterial Names*[75,76] but are invalid if they are not included in this list. It is the list that is approved, not the names; thus, there can be more than one valid name for the same entity, both of which appear on the approved list. Names often carry a superscripted "AL", meaning that the name is on the Approved Lists.

> *Methanococcus vannielii* Stadtman and Barker 1951^{AL} is the type species for the genus.

A name that was first effectively published before the publication of the *Approved Lists* but not included on the lists but that is subsequently validly published should be referred to as "nom. rev." (nomen revictum or revived name) in addition to "sp. nov.", etc. In such cases, the author of the original effective publication can be acknowledged in parentheses using "*ex*" (in italics, in contrast to the practice in zoology and botany)—for example, *Pseudomonas cannabina* (*ex* Sutic and Dowson 1959) Gardan et al. 1999 sp. nov., nom. rev.

Bergey's Manual of Systematic Bacteriology[70,77–79] is a useful reference for bacterial taxonomy and classification but has no standing in nomenclature; inclusion of a name in *Bergey's Manual* does not mean, per se, that the name is validly published. *Bergey's Manual of Determinative Bacteriology* is a useful reference for identification of unknown bacteria.[80]

Names that have not been validly published should be enclosed in single, roman quotes; this includes names that have been effectively but not validly published (e.g., '*Actinomadura spinosa*' Saitoh et al.) and names that are coined, for instance, in a deposit in a sequence database such as GenBank. In the case of organisms that the author considers to be misclassified, part or all of the name may be enclosed in square brackets, such as [*Cytophaga*] latercula; the name *Cytophaga latercula* was included in the *Approved Lists* so is validly published, but the species was recognized for some time as not belonging to the genus *Cytophaga* and was reclassified as *Stanierella latercula*.

Note that, since the meetings of the International Committee on Systematics of Prokaryotes (ICSP) in 1999, orthographic or grammatical errors in names of prokaryotes can no longer be corrected. The only exception is that adjectival epithets that are in the wrong gender must be corrected if that species is reclassified in a new genus (although the gender of the epithet remains incorrect in the original genus).

Names, not the organisms to which they refer, are validly published.

22.3.4.2 Authors of Bacterial Names

In referring to a bacterial species, present the name of the author after the specific epithet along with the date of valid publication of the name (not the year of effective publication, if different), without intervening punctuation (unlike the recommendation for authors of animal names, which specifies a comma between the name and the date). If there are 2 authors, use "and" (unlike the recommendation for authors of animal names, which specifies "et" or the ampersand).

> *Erwinia quercina* Hildebrand and Schroth 1967

If there are more than 2 authors for a taxon, "et al." can be used instead of the complete list of authors.

> *Legionella pittsburghensis* Pasculle, Feeley, Gibson, Cordes, Myerowitz, Patton, Gorman, Carmack, Ezzel, and Dowling 1980
> *or*
> *Legionella pittsburghensis* Pasculle et al. 1980

If the species has been reclassified using the same specific epithet, the original author's name appears within parentheses after the specific epithet, followed by the name of the person who reclassified it.

> *Bacterium herbicola* Löhnis 1911 was reclassified by Dye, so the name becomes *Erwinia herbicola* (Löhnis 1911) Dye 1964.

If the same species is transferred again, the author of the first validly published name of the taxon (the basonym) should be given in parentheses, followed by the name of the latest person to reclassify it.

> *Alcaligenes eutrophus* Davis 1969 has subsequently been reclassified as *Ralstonia eutropha* (Davis 1969)
> Yabuuchi et al. 1996 and again as *Wautersia eutropha* (Davis 1969) Vaneechoutte et al. 2004 [*not* (Yabuuchi et al. 1996) Vaneechoutte et al. 2004].

22.3.4.3 Vernacular Names for Bacteria

Always set vernacular names for bacteria in roman lowercase letters. Plural forms may take the Latin or the English form.

brucella, brucellae	vibrio, vibrios
bacillus, bacilli	pseudomonad, pseudomonads

Avoid using "bacillus" or "bacilli" if there might be confusion as to whether the genus *Bacillus* or a nonspecific, rod-shaped bacterium is meant. (The word *Bacilli*, in italics and capitalized, is never correct.) Organisms in the genus *Treponema* can be designated by "treponema" (singular) and "treponemas" or "treponemata" (plural). The following are some additional examples of vernacular names (singular and plural).

klebsiella, klebsiellae	mycobacterium, mycobacteria
streptomycete, streptomycetes	citrobacter, citrobacters

Additional vernacular names and plural forms can be found in some general dictionaries and in medical dictionaries.

Although international rules do not cover adjectives derived from scientific names, some guidelines have been widely followed. Adjectival forms usually end in "al", but a noun form can also serve as an adjective. A formal genus name can be used as a vernacular adjective if it is in roman lowercase letters.

streptococcal infections
streptococcus infection
corynebacterium growth
corynebacterial growth
He died of pneumocystis pneumonia.

A formal genus name should not be used as a modifier unless it is meant to refer to all species of the genus. A binomial can be used as a modifier[81] when only that species is referred to.

Pseudomonas aeruginosa bacteremia [not *Pseudomonas* bacteremia]
He studied all *Pseudomonas* species.
Pneumocystis carinii pneumonia [not *Pneumocystis* pneumonia]

Adjectival forms must not be derived from specific epithets; for example, do not use "coli" for *Escherichia coli* or "amylovora" for *Erwinia amylovora*.

22.3.4.4 Nomenclatural Types

In this context, the word "type" does not mean "typical"; rather, it refers to a specimen to which the name remains permanently attached. The type of an order, suborder, family, subfamily, tribe, or subtribe is the genus, which is the basis for the name of the higher taxon. The type of a genus is the species and that of the species and subspecies is the strain.

Xanthomonas fragariae Kennedy and King 1962, type strain NCPPB 1469

If the cells of the type strain cannot be maintained in culture or have been lost, the type strain may be represented by the original description or by illustrations.

The type strain of a species is a strain on which the original description of the species is based and is the strain designated by the author or a subsequent author as the type strain. This strain should be designated by a strain number of the author and, since 1999, the strain numbers of at least 2 public service culture collections in different countries where the culture is available. For the *IJSEM*, a superscript T is required to designate the type strain at each occurrence in the text, tables, and figures.

We obtained strain NRS 966 of *Bacillus thuringensis* (ATCC 10792^{T}) from . . .

Rule 18a of the Bacteriological Code states, "A type strain is made up of living cultures of an organism that are descended from a strain designated as the nomenclatural type. The strain should have been maintained in pure culture and should agree closely in its characters with those in the original description." Bacterial strains can be designated "in any manner, e.g., by the name of an individual, by a locality, or by a number" (Bacteriological Code, Appendix D).

If the organism is a symbiont, the type culture can be maintained as a defined co-culture with another named strain.

The type of *Syntrophobacter wolinii* Boone and Bryant 1984 is strain DB, which is maintained in the DSMZ culture collection as a co-culture with *Desulfovibrio* sp. strain G-11 as DSM 2805, or as strain DSM 2805M, a monoculture isolated from the co-culture DSM 2805.

Ideally, the strain designation should be presented with the full binomial.

Xanthomonas campestris (Pammel) Dowson pathovar *campestris* NCPPB 528 [Int J Syst Bacteriol. 1990;40(4):348–369]

If the specific epithet is missing, insert the word "strain".

Pseudomonas strain B13 or *Pseudomonas* sp. strain B13

Rule 13d of the Bacteriological Code states that a subspecies that contains the type specimen of the species must bear the same epithet as the species. Thus, whenever a subspecies is created, the type automatically becomes a subspecies. For example, publication of *Bacillus subtilis* subsp. *viscosus* Chester 1904 automatically created a new subspecies, *Bacillus subtilis* subsp. *subtilis* for the type. The author of the species name is to be cited as the author of the automatically created subspecies name:

> *Bacillus subtilis* subsp. *subtilis* (Ehrenberg 1835) Cohn 1872
> [subspecies created automatically from *Bacillus subtilis* (Ehrenberg 1835) Cohn 1872]

22.3.4.5 Organisms That Cannot Be Cultured

Names of prokaryotic organisms that cannot be isolated in culture can no longer be validly published. However, the Judicial Commission of the ICSP recommends that organisms that are well characterized but that cannot be maintained in culture so as to satisfy the requirements of the Bacteriological Code for naming of new taxa should be given "*Candidatus*" status[82,83] (see also http://www.bacterio.cict.fr/candidatus.html). The category *Candidatus* is not covered by the Bacteriological Code and such names are not validly published; they should therefore be given within quotation marks, with the word *Candidatus* in italics and the name of the taxon in roman—for example, "*Candidatus* Arsenophonus triatominarum" Hypsa and Dale 1997. Such names when proposed should not be accompanied by "sp. nov.", etc. The word *Candidatus* can be abbreviated to *Ca.* at subsequent uses.

22.3.4.6 Infraspecific Taxa

Some bacteria that cannot be differentiated taxonomically at the level of subspecies are given infraspecific designations that are not covered in the Bacteriological Code and are therefore excluded from the *Approved Lists* (see Section 22.3.4.1). These designations include pathovar or pathotype (pv.); biovar or biotype (bv.); serovar or serotype (sv.); phagovar or phagotype; chemoform, chemotype, or chemovar; cultivar; forma specialis or special form; and morphovar or morphotype. The taxon variety (var.) is not used for bacteria. The suffix "var" is preferred to "type" to avoid confusion with nomenclatural types. Infrasubspecific taxa can be designated by latinized words, vernacular names or words, numbers, letters, or formulae (see Appendix 10 C of the Bacteriological Code).

The term "pathovar" is derived from "pathovarietas". The pathovar designation, but not its abbreviation (pv.), is written in italics. Pathovars (pathotypes) are distinguished primarily but not solely by their pathogenic characteristics.

> *Corynebacterium michiganense* (Smith) Jensen 1977 pv. *tritici* causes spike blight in wheat.

Because the Bacteriological Code does not govern nomenclature of pathovars, a committee of the International Society for Plant Pathology has drawn up a set of international standards for naming pathovars of phytopathogenic bacteria as well as a list of pathovars and their representative pathotype cultures.[84–86] When first described, a new pathovar is designated as "pv. nov." (pathovarietas nova).

> *Xanthomonas campestris* (Pammel 1895) Dowson 1939 pv. *caladiae* pv. nov.

Mention of the name of a pathovar reported previously should bear the name of the author of the publication in which the pathovar epithet was proposed formally, followed by the date of publication with no intervening punctuation. Full citation of the publication should include the number of the page in the body of the text (not in the summary or abstract) on which the name was proposed.

> *Xanthomonas campestris* (Pammel 1895) Dowson 1939 pv. *cannabina* Severin 1978, 13 [indicates that the pathovar was proposed by Severin in 1978 and that the proposed name can be found on page 13 of the publication cited]

Names of serovars of *Salmonella* should preferably not be used as if they are Latin binomials. For example, *Salmonella enterica* subsp. *enterica* serovar Dublin should not be referred to as *Salmonella dublin*. Such names can be referred to as *Salmonella* Dublin in material that is not of a taxonomic nature; the form *S.* Dublin should not be permitted. Note also that the names of some serovars have been validly published as specific epithets—for example, *Salmonella typhi* and *Salmonella paratyphi*. See http://www.bacterio.cict.fr/salmonellanom.html for more detailed discussion of the problems of *Salmonella* nomenclature.

22.3.5 Viruses

Viruses are neither prokaryotes nor eukaryotes; they are nucleic acid molecules that can enter cells, replicate in them, and encode proteins that form protective shells around themselves. Because they are not cellular, do not grow, have no observable activity except replication, and function only within living cells, they are not living organisms.[87] Viruses appear to be partial or degenerate forms of living systems, but they are not placed in any of the systems of categories of living things (see, e.g., Margulis and Schwartz[7]).

22.3.5.1 Classification of Viruses

The International Committee on Taxonomy of Viruses (ICTV) has classified viruses into the following taxonomic ranks: order, family, subfamily, and genus, subgenus or group, and species or individual viruses.[13,88] Viral classification has no ranks above order. Names of orders are not written in italics; the ending for an order name is "-ales". The names for families, genera, and species are written in italics. The ending for a family name is "*viridae*", for a subfamily name is "*virinae*", and for a genus "*virus*". Not all hierarchical levels need be used—for example, a species need not be assigned to a genus, and a genus need not be assigned to a family. The Virological Code does not deal with taxa below species. Specialist groups deal with serotypes, strains, and so on.

22.3.5.2 Scientific Names for Viruses

Valid scientific names of viruses are those approved by the ICTV and published in the ninth report.[14] The ICTV maintains the authoritative classification database, ICTVdB. A list of virus words and abbreviations approved by the ICTV has been prepared by Calisher and Fauquet.[89]

The orthography of virus names differs from that used for organisms. Virus species do not have Latin names, but when used in a formal taxonomic sense, the name of a virus (as approved by the ICTV) is written in italics. The first word of the species name is capitalized.

> *Maize dwarf mosaic virus* *Avian leukosis virus* *Yellow fever virus*

Do not print tentative names for species in italics until they are formally approved by the ICTV.

Capitalize other parts of a virus name only if they are proper nouns:

Murray River encephalitis virus ["River" is part of a proper noun]
Sandfly fever Naples virus

Virus species are placed in genera, but the genus name is generally not part of the species name. The genus name may be an acronym of the species name. For example, the species *Alfalfa mosaic virus* is in the genus *Alfamovirus*, and *Tobacco mosaic virus* is in the genus *Tobamovirus*. When referring to the formal names of taxa, precede the name with the term for the taxonomic unit, for example, "the family *Caulimoviridae*", "the genus *Betavirus*", and "the species *Cryphonectria hypovirus virus 1*".

Some virologists, particularly those who work with plant viruses, tend to include the genus name with the species name, in a "binomial" system—for example, *Tobacco mosaic tobamovirus*. In this example, *Tobacco mosaic* would correspond to the specific epithet of a virus in the genus *Tobamovirus*. The genus name follows the species name.

This system of orthography is not universally accepted by virologists, and adoption varies from discipline to discipline; particularly, many researchers of medically important viruses have not embraced the italic style. If italics and ICTV-approved names have not been used or have been used inappropriately (for instance, when referring to a concrete virus entity rather than a formal taxonomic unit), the practice of the individual publication should be ascertained.

22.3.5.3 Vernacular Names for Viruses

The above rules for orthography apply only when names of viruses are used in a taxonomic sense—that is, when the name refers to a taxonomic category. When referring to concrete viral entities, such as virus particles that are centrifuged from plant sap or viewed with an electron microscope, the name is considered vernacular and is written in roman script. In vernacular usage, none of the words in a virus name is capitalized unless English grammar requires it.

RNA of barley stripe mosaic virus   Fraser Point virus virions

When genus names are used in a vernacular sense, they are written in roman script and not capitalized.

potexvirus   rhabdovirus   rhinovirus
enterovirus   ilarvirus   luteovirus

22.3.5.4 Abbreviated Virus Designations

Most viruses are designated by acronyms.[13] Acronyms can be used in a paper for publication if they are clearly associated with the full virus name at the first mention.

PAdV-B [*Porcine adenovirus B*]   LCDV-2 [*Lymphocystis disease virus 2*]
TMV [*Tobacco mosaic virus*]

The Virological Code makes a distinction between sigla and acronyms, a distinction that is not recognized by standard dictionaries, such as *Webster's New World Dictionary of American English*. In the Virological Code, sigla "are names comprising letters and/or letter combinations taken from words in a compound term." The name of the genus *Comovirus* has the sigla stem "Co" taken from "cowpea" and "mo" taken from "mosaic"; the name of the genus *Reovirus* has the sigla stem "R" from "respiratory", "e" from "enteric", and "o" from "orphan". However, the definition of acronym offered by *Webster's* includes examples that contain more than the initial letter of component words.

22.3.5.5 Strain Designations

The ICTV regulates virus names only down to the level of species. "The classification and naming of serotypes, genotypes, strains, variants and isolates of virus species is the responsibility of acknowledged international specialist groups."[13] The details of the strain designation may be placed within the virus name: "influenza virus A/WS/33" or "influenza A/WS/33 virus." More detailed designations may be needed.

> A/Equine/Prague/1/56 (H7N7) [A = the virus type; Equine = the host of origin (for animal influenza isolates); Prague = geographic origin; 1 = strain number; 56 = 1956, the year of isolation; (H7N7) = antigenic description]

A simple strain name may follow the virus name or acronym, being separated from it by a dash.

> *Barley yellow dwarf virus*—PAV
> BYDV—MAV [MAV and PAV are 2 strains of *Barley yellow dwarf virus*]

22.3.5.6 Catalogs of Virus Names and Strain Designations

The ICTV gives all validly published virus names and valid synonyms.[13,14] The American Type Culture Collection publishes its catalog of plant and animal viruses online (http://www.atcc.org).[90] A catalog of plant viruses is provided in the VIDE (Virus Identification Data Exchange) database.[91]

CITED REFERENCES

1. Claridge MS, Dawah HA, Wilson MK. Practical approaches to species concepts for living organisms. In: Claridge MS, Dawah HA, Wilson MK, editors. Species: the units of diversity. New York (NY): Chapman & Hall; 1997. p. 1–15. (Systematics Association special volume series; 54).

2. Hull DL. The ideal species concept—and why we can't get it. In: Claridge MF, Dawah HA, Wilson MR, editors. Species: the units of diversity. London (UK): Chapman & Hall; 1997. p. 357–380. (Systematics Association special volume series; 54).

3. Mayr E. Populations, species, and evolution: an abridgement of animal species and evolution. Cambridge (MA): The Belknap Press; 1970.

4. Woodland DW. Contemporary plant systematics. 3rd ed. Berrien Springs (MI): Andrews University Press; 2000.

5. Jeffrey C. Biological nomenclature. 3rd ed. New York (UK): Edward Arnold; 1989.

6. Stace CA. Plant taxonomy and biosystematics. 2nd ed. London (UK): Edward Arnold; 1989.

7. Margulis L, Schwartz KV. Five kingdoms: an illustrated guide to the phyla of life on earth. 3rd ed. New York (NY): W. H. Freeman and Company; 1998.

8. Tree of Life web project. [Tucson (AZ)]: University of Arizona College of Agriculture; c1999–2005 [accessed 2011 Nov 11]. http://tolweb.org/tree/phylogeny.html.

9. Winston JE. Describing species. New York (NY): Columbia University Press; 1999.

10. International Commission on Zoological Nomenclature. International code of zoological nomenclature. 4th ed. London (UK): The Natural History Museum, International Trust for Zoological Nomenclature; 1999 [accessed 2011 Nov 11]. http://www.nhm.ac.uk/hosted-sites/iczn/code.

11. International Committee on Systematics of Prokaryotes. International code of nomenclature of bacteria: bacteriological code (1990 revision). Washington (DC): American Society for Microbiology; 1992 [updated 2010 Aug 2; accessed 2011 Nov 11]. http://www.the-icsp.org/.

12. McNeill J, Barrie FR, Buck WR, Demoulin V, Greuter W, Hawksworth DL, Herendeen PS, Knapp S, Marhold K, Prado J, et al. International code of nomenclature for algae, fungi, and plants (Melbourne code). Konigstein (Germany): International Association of Plant Taxonomy; 2012 [accessed 2012 Sep 17]. http://www.iapt-taxon.org/nomen/main.php.

13. Regenmortel MHV van, Fauquet CM, Bishop DHL, Carstens EB, Estes MK, Lemon SM, Maniloff J, Mayo MA, McGoech DJ, Pringle CR, Wickner RB. Virus taxonomy: classification and nomenclature of viruses. 7th report of the International Committee on Taxonomy of Viruses. New York (NY): Academic Press; 2000.

14. King AMQ, Adams MJ, Carstens EB, Lefkowitz EJ. Virus taxonomy: classification and nomenclature of viruses: 9th report of the International Committee on Taxonomy of Viruses. San Diego (CA): Academic Press; 2011.

15. Manara B. Some guidelines on the use of gender in generic names and species epithets. Taxon. 1991;40(2):301–308.

16. Gledhill D. The names of plants. 4th ed. Cambridge (UK): Cambridge University Press; 2008.

17. Nybakken OE. Greek and Latin in scientific terminology. Ames (IA): Iowa State University Press; 1959.

18. Stearn WT. Botanical Latin: history, grammar, syntax, terminology and vocabulary. 4th ed. Portland (OR): Timber Press Inc.; 1995.

19. Gotch AF. Latin names explained: a guide to the scientific classification of reptiles, birds, and mammals. New York (NY): Facts on File; 1996.

20. Kartesz JT. A synonymized checklist of the vascular flora of the United States, Canada and Greenland. 2nd ed. Portland (OR): Timber Press; 1994.

21. Linnaeus C. Species plantarum: a facsimile of the first edition. London: The Ray Society. Vol. 1, 1957; Vol. 2, 1959.

22. Brummitt RK, Powell CE, editors. Authors of plant names: a list of authors of scientific names of plants with recommended standard forms of their names, including abbreviations. Kew (UK): Royal Botanic Gardens; 1996.

23. Stafleu FA, Cowan RS. Taxonomic literature. Vol. 1–7. Regnum vegetabile 94, 98, 105, 110, 112, 115, 116. Utrecht (The Netherlands): Bohn, Scheltema & Holkema; 1976–1988.

24. Stafleu FA, Mennega EA. Taxonomic literature, Supplement 1. Regnum vegetabile 125. Utrecht (The Netherlands): Bohn, Scheltema & Holkema; 1992.

25. Brickell CD, Alexander C, David JC, Hetterscheid WLA, Leslie AC, Malecot V, Jin X, editors. International code of nomenclature for cultivated plants. 8th ed. Scripta Horticulturae 10. International Society of Horticultural Science; 2009.

26. Dorn RD. Vascular plants of Wyoming. 3rd ed. Cheyenne (WY): Mountain West Publishing; 2001.

27. Peck ME. A manual of the higher plants of Oregon. 2nd ed. Portland (OR): Oregon State University Press; 1961.

28. Mabberly DJ. The plant-book: a portable dictionary of the higher plants. 2nd ed. Cambridge (UK): Cambridge University Press; 1997.

29. Gleason HA, Cronquist A. Manual of vascular plants of northeastern United States and adjacent Canada. 2nd ed. New York (NY): New York Botanical Garden; 1991.

30. Rickett HW. The English names of plants. Bull Torrey Bot Club. 1965;92(2):137–139.

31. Brako L, Rossman AY, Farr DF. Scientific and common names of 7,000 vascular plants in the United States. St. Paul (MN): APS Press; 1995.

32. Kartesz JT, Thieret JW. Common names for vascular plants: guidelines for use and application. Sida Contrib Bot. 1991;14(3):421–434.

33. American Joint Committee on Horticultural Plant Nomenclature. Standardized plant names. Harrisburg (PA): J. Horace McFarland Co.; 1942.

34. Patterson DT. Composite list of weeds. Champaign (IL): Weed Science Society of America; 1989.

35. Korf RP. Citation of authors' names and the typification of names of fungal taxa published between 1753 and 1832 under the changes in the code of nomenclature enacted in 1981. Mycologia. 1982;74(2):250–255.

36. Korf RP. Simplified author citations for fungi and some old traps and new complications. Mycologia. 1996;88(1):146–150.

37. Roelfs AP, Martens JW. An international system of nomenclature for *Puccinia graminis* f. sp. *tritici*. Phytopathology. 1988;78(5):526–533.

38. Long DL, Kolmer JA. A North American system of nomenclature for *Puccinia recondita* f. sp. *tritici*. Phytopathology. 1989;79(5):525–529.

39. Alexopoulos CJ, Mims CW, Blackwell M. Introductory mycology. 4th ed. New York (NY): John Wiley & Sons; 1996.

40. Olive LS. The mycetozoans. New York (NY): Academic Press; 1975.

41. Kirk PM, Cannon OF, Stalpers JA. Dictionary of the Fungi. 10th ed. Wallingford (UK): CAB International; 2011.

42. Sze P. A biology of the algae. 3rd ed. Boston (MA): WCB/McGraw-Hill; 1998.

43. Wehr JD, Sheath RG, editors. Freshwater algae of North America: ecology and classification. Boston (MA): Academic Press; 2003.

44. Brodo IM, Sharnoff SD, Sharnoff S. Lichens of North America. New Haven (CT): Yale University Press; 2001.

45. Mayr E, Ashlock PD. Principles of systematic zoology. 2nd ed. New York (NY): McGraw-Hill; 1991.

46. Page R, editor. Systematic Biology. A quarterly of the Society of Systematic Biologists. Philadelphia (PA): Taylor & Francis. Vol. 41, No. 1, 1992– . Continues: Systematic Zoology.

47. Bulletin of Zoological Nomenclature. London (UK): British Museum, International Trust for Zoological Nomenclature. Vol. 1, 1943– . Quarterly.

48. Melville RV, Smith JDD, editors. Official lists and indexes of names and works in zoology. London (UK): International Trust for Zoological Nomenclature; 1987.

49. Collins JT, Taggart TW. Standard common and current scientific names for North American amphibians, turtles, reptiles, and crocodilians. 6th ed. Lawrence (KS): The Center for North American Herpetology; 2009.

50. Cairns SD, Calder DR, Brinkmann-Voss A, Castro CB, Fautin DG, Pugh PR, Mills CE, Jaap WC, Arai MN, Haddock HD, Opresko DM. Common and scientific names of aquatic invertebrates from the United States and Canada: Cnidaria and Ctenophora. 2nd ed. Bethesda (MD): American Fisheries Society; 2002. (Special publication; 28).

51. Turgeon DD, Quinn JF Jr, Bogan AE, Coan EV, Hockberg FG Jr, Lyons WG, Mikkelsen PM, Neves RJ, Roper CFE, Rosenberg G, et al. Common and scientific names of aquatic invertebrates from the United States and Canada: mollusks. 2nd ed. Bethesda (MD): American Fisheries Society; 1998. (Special publication; 26).

52. McLaughlin PA, Camp DK, Eldredge LG, Felder DL, Goy JW, Hobbs HH III, Kensley BA, Lemaitre R, Martin JW. Decapoda. In: McLaughlin PA, Camp DK, Angel MV, editors. Common and scientific names of aquatic invertebrates from the United States and Canada: Crustaceans. Bethesda (MD): American Fisheries Society; 2005.

53. Williams AB, Abele LG, Felder DL, Hobbs HH Jr, Manning RB, McLaughlin PA, Farnante IP. Common and scientific names of aquatic invertebrates from the United States and Canada: decapod crustaceans. Bethesda (MD): American Fisheries Society; 1989. (Special publication; 17).

54. FINS: the Fish Information Service. Fish index. Cambridge (MA): Active Window Productions, Inc.; c1993–2009 [accessed 2012 Sep 7]. http://fins.actwin.com/species/.

55. Nelson JS, Crossman EJ, Espinosa-Perez H, Findley LT, Gilbert CR, Lea RN, Williams JD. Common and scientific names of fishes from the United States, Canada, and Mexico. 6th ed. Bethesda (MD): American Fisheries Society; 2004. (Special publication; 29).

56. Entomological Society of America, Committee on the Common Names of Insects. Common names of insects and related organisms. Lanham (MD): ESA; c1995–2005 [accessed 2011 Nov 11]. http://www.entsoc.org/common-names.

57. Iowa State entomology index of Internet resources. Ames (IA): Iowa State University, Entomology Department; c2005 [accessed 2011 Nov 11]. http://www.ent.iastate.edu/list/directory/130/vid/4.

58. American Ornithologists' Union (A.O.U.) Check-list of North American birds. 7th ed. McLean (VA): American Ornithologists' Union; [accessed 2011 Nov 11]. http://www.aou.org/checklist/north/print.php.

59. Sibley CG, Monroe BL Jr. Distribution and taxonomy of birds of the world. New Haven (CT): Yale University Press; 1990.

60. Corbet GB, Hill JE. A world list of mammalian species. 3rd ed. New York (NY): Oxford University Press; 1991.

61. Hall ER. The mammals of North America. 2nd ed. New York (NY): John Wiley & Sons; 1981.

62. Nowak RM. Walker's mammals of the world. 6th ed. Baltimore (MD): The Johns Hopkins University Press; 1999.

63. Wilson DE, Cole FR. Common names of mammals of the world. Washington (DC): Smithsonian Institution Press; 2000.

64. Festing MFW. Inbred strains of rats and their characteristics. Leicester (UK): University of Leicester, MRC Toxicology Unit; [updated 1998 Apr 9; accessed 2011 Nov 11]. http://www.informatics.jax.org/external/festing/rat/INTRO.shtml.

65. Lyon MF. Rules for nomenclature of inbred strains. In: Lyon MF, Searle AG, editors. Genetic variants and strains of the laboratory mouse. 2nd ed. Oxford (UK): Oxford University Press; 1989. p. 632–635.

66. International Committee on Standardized Genetic Nomenclature for Mice; Rat Genome and Nomenclature Committee. Guidelines for nomenclature of mouse and rat strains. Bar Harbor (ME): Jackson Laboratory; [revised 2010 Sep; accessed 2011 Nov 11]. http://www.informatics.jax.org/mgihome/nomen/strains.shtml.

67. Hedrich HJ, editor. Genetic monitoring of inbred strains of rats: a manual on colony management, basic monitoring techniques, and genetic variants of the laboratory rat. New York (NY): Gustav Fischer; 1991.

68. Festing M, Kondo K, Loosli R, Poiley SM, Spiegel A. International standardized nomenclature for outbred stocks of laboratory animals. Z. Versuchstierkd. 1972;14(4):215–224.

69. National Research Council (US), Commission on Life Sciences, Institute of Laboratory Animal Resources, Committee on Transgenic Nomenclature. Standardized nomenclature for transgenic animals. ILAR J. 1992 [accessed 2012 Sep 7];34(4). http://ilarjournal.oxfordjournals.org/content/34/4/45.full.

70. Garrity G, editor. Bergey's manual of systematic bacteriology. Vol. 2, The Proteobacteria. 2nd ed. New York (NY): Springer; 2005.

71. Weiss E, Moulder JW. Genus I. Rickettsia da Rocha-Lima 1916, 567AL. In: Krieg NR, Holt JG, editors. Bergey's manual of systematic bacteriology. Vol. 1. Baltimore (MD): Williams & Wilkins; 1984. p. 688–698.

72. Maniloff J. Evolution of wall-less prokaryotes. Annu Rev Microbiol. 1983;37:477–499.

73. Seemuller E, Schneider B, Maurer R, Ahrens U, Daire X, Kison H, Lorenz K, Firrao G, Avinent L, Sears BB, Stackebrandt E. Phylogenetic classification of phytopathogenic mollicutes by sequence analysis of 16S ribosomal DNA. Int J Syst Bacteriol. 1994;44(3):440–446.

74. Euzéby JP. List of prokaryotic names with standing in nomenclature. [place unknown]: Societe de Bacteriologie Systematique et Veterinaire; 1997 [updated 2011 Nov 4; accessed 2011 Nov 11]. http://www.bacterio.cict.fr/.

75. Moore WEC, Moore LVH. Index of the bacterial and yeast nomenclatural changes: published in the International Journal of Systematic Bacteriology since the 1980 Approved Lists of Bacterial Names (1 January 1980 to 1 January 1992). Washington (DC): American Society for Microbiology; 1992.

76. Skerman VBD, McGowan V, Sneath PHA, editors. Approved lists of bacterial names. Amended ed. Washington (DC): American Society for Microbiology; 1989.

77. Boone DR, Castenholz RW, editors. Bergey's manual of systematic bacteriology. Vol. 1, The Archaea and the deeply branching and phototropic bacteria. 2nd ed. New York (NY): Springer; 2001.

78. Vos P, Garrity G, Jones D, Krieg NR, Ludwig W, Rainey FA, Schleifer K-H, Whitman WB, editors. Bergey's manual of systematic bacteriology. Vol. 3, The Firmicutes. 2nd ed. New York (NY): Springer; 2009.

79. Williams ST, editor. Bergey's manual of systematic bacteriology. Vol. 4. 2nd ed. New York (NY): Springer; 2010.

80. Holt JG, editor-in-chief. Bergey's manual of determinative bacteriology. 9th ed. Baltimore (MD): Williams & Wilkins; 1994.

81. Huth EJ. Style notes: taxonomic names in microbiology and their adjectival derivatives. Ann Intern Med. 1989;110(6):419–420.

82. Murray RGE, Schleifer KH. Taxonomic notes: a proposal for recording the properties of putative taxa of procaryotes. Int J Syst Bacteriol. 1994;44(1):174–176.

83. Murray RGE, Stackebrandt E. Taxonomic note: implementation of the provisional status Candidatus for incompletely described procaryotes. Int J Syst Bacteriol. 1995;45(1):186–187.

84. Dye DW, Bradbury JF, Goto M, Hayward AC, Lelliott RA, Schroth MN. International standards for naming pathovars of phytopathogenic bacteria and a list of pathovar names and pathotype strains. Rev Plant Pathol. 1980;59(4):153–168.

85. Young JM, Bradbury JF, Davis RE, Dickey RS, Ercolani GL, Hayward AC, Vidaver AK. Nomenclatural revisions of plant pathogenic bacteria and list of names 1980–1988. Rev Plant Pathol. 1991;70:211–221.

86. Young JM, Saddler GS, Takikawa Y, De Boer SH, Vauterin L, Gardan L, Gvozdyak RI, Stead DE. Names of plant pathogenic bacteria 1864–1995. Rev Plant Pathol. 1996;75:721–763.

87. Hull R. Matthews' plant virology. 4th ed. New York (NY): Academic Press; 2001.

88. Sander DM. All the virology on the WWW. [place unknown]: D. Sander; 1995 May [updated 2003 Jan 8; accessed 2011 Nov 11]. http://www.virology.net/garryfavweb.html.

89. Calisher CH, Fauquet C, editors. Stedman's/ICTV virus words. Baltimore (MD): Williams & Wilkins; 1992.

90. ATCC: The Global Bioresource Center. Manassas (VA): American Type Culture Collection; c2005 [accessed 2011 Nov 11]. http://www.atcc.org.

91. Brunt A, Crabtree K, Dallwitz M, Gibbs A, Watson L, Zurcher E. Viruses of plants: descriptions and lists from the VIDE database. Cambridge (UK): CAB International; 1996.

23 Structure and Function

23.1 GENERAL PRINCIPLES

The ways to describe an organism's structural organization depend on the particular organism. Microorganisms are typically described according to size and shape. Plants consist of vegetative structures (roots, stems, and leaves) and reproductive structures (flowers, fruits, and seeds), whereas most animals are organized around an exoskeleton or an endoskeleton, with organs and tissues nurtured by a circulatory system of blood or other fluids. Functionally, microorganisms are totipotent, although they may specialize in function on maturation. In contrast, functional specialization in plants and animals (exemplified by the characteristic functions of various organs) occurs early in embryogeny, and most tissues cannot revert to an adventitious state. Exceptions are plant callus tissue (a reversion) and animal stem cells, produced in embryos, bone marrow, or teeth, which can produce new kinds of cells.

Many reference works are available describing the structure and function of viruses,[1–3] microorganisms,[4,5] plants,[6,7] and animals,[8] and specialized dictionaries and glossaries[9–13] provide guidance on terminology in particular areas. The focus in this chapter is on style and format matters pertaining to the presentation of this type of information where they differ from standard text.

23.2 PLANT PHYSIOLOGY

Most journals of plant physiology use SI units of measurement (International System of Units; see Section 12.2.1). *Units, Symbols, and Terminology for Plant Physiology*[14] represents an effort to standardize terminology and other aspects of presenting data in plant physiology.

Guidelines for measuring and reporting environmental conditions in controlled-environment studies of plants were updated in 2000.[15] These guidelines reflect the wide use of plant growth chambers and other controlled chamber facilities to provide a reproducible environment for growing plants, plant tissues, or plant cells. Some typical units of measure appear in Table 23.1. Terms and units for light measurement and photosynthesis are presented in Table 23.2.[16,17]

In agriculture, growth is measured differently for each crop and is presented by means of a decimal-based, crop-specific growth-stage code from 00 (dry seed) to 99 (harvested product).[18]

23.3 FISHES: SALMONID AGES

23.3.1 Age Notation

A special notation system is used to designate the age of anadromous fish (e.g., salmon), which begin life in freshwater streams, spend time in the ocean, and return to their

Table 23.1 Measurement of selected environmental conditions in controlled environments for growing plants, plant tissues, and plant cells[a]

Condition	Unit
Radiation	
Photosynthetic photon flux (PPF), 400–700 nm	μmol $m^{-2}s^{-1}$
Energy flux (irradiance)	W m^{-2} (nm waveband)
Spectral photon flux (bandwidths ≤ 2 nm)	μmol $m^{-2}s^{-1}mm^{-1}$
Spectral energy flux (spectral irradiance)	W m^{-2}
Temperature (air, soil, liquid)	°C
Atmospheric moisture	
Relative humidity (RH)	% RH
Dew point temperature	°C
Water vapor density	g m^{-2}
Air velocity	m s^{-1}
Watering	L, l
Nutrition	mol m^{-3} or mol kg^{-1}

[a] Based on Tibbitts et al.[15]

Table 23.2 Terms for light measurements and photosynthesis[a]

Term	Abbreviation	Unit
Apparent photosynthesis	AP	none
CO_2 exchange rate[b]	CER	µmol $cm^{-2}s^{-1}$
Photosynthetic irradiance	PI	W m^{-2}
Photosynthetic photon flux density	PPFD	µmol $m^{-2}s^{-1}$
Photosynthetically active radiation[c]	PAR	PPFD or PI

[a] Terms originally defined by Shibles.[16] Updated by the American Society of Agronomy, Crop Science Society of America, and Soil Science Society of America.[17]

[b] Use this term instead of "net CO_2 exchange".

[c] For reporting PAR, photon units (PPFD) are preferred over energy units (PI), but both are acceptable. The term "light intensity" is no longer used.

natal streams to spawn and die. The freshwater phase and the ocean or seawater phase of anadromous life history are commonly represented by the "European system" as *x.y*, where *x* is the number of winters the fish spent in freshwater and *y* the number of winters spent at sea, separated by a period and no spaces. Use arabic numerals in age designations of adult and juvenile salmon, and present the ages without units. According to this system, an adult salmon aged 1.2 has lived 1 year in freshwater and 2 years at sea; the first winter spent during embryonic development is not counted as part of either phase.

Often, only one phase of the salmon life history is described. In European notation, write the age as *x.* when only the freshwater phase is described and as *.y* when only the ocean phase is described.

> Salmon age 2. emigrated from Black Lake. [freshwater phase only]
> Sockeye salmon return to their natal stream at age .3. [seawater phase only]

Because of potential ambiguity, do not use a plus symbol after the age (e.g., 2.3+) to indicate an additional summer of growth after the last winter counted in the age determination or to indicate that a group of salmon is composed of fish at least as old as the given age.[19,20]

Systems other than the European system, as well as textual descriptions, are also used to designate salmonid ages. No standard terminology for these descriptions has been adopted, and word descriptions of age can be awkward and ambiguous. For example, fish designated as age 2. in the European system have been described as "two-freshwater fish" or "two-years-in-the-lake type" fish. A salmon designated as age .3 may also be described with terms such as "3-ocean fish" to indicate a fish that has spent 3 winters at sea. The construction in the last example, although common, is potentially confusing to those outside the discipline, who might interpret "3-ocean" to mean that the salmon has lived in 3 different oceans.

23.3.2 Age in Years

The birth date for salmonids is conceptually standardized as 1 January of the year following the brood year, regardless of when the brood actually hatched.[19] The brood year is the year in which adult salmon return to freshwater to spawn, which may not be the same as the year in which the embryos produced by those spawning adults hatch.

Ages in years are reckoned differently for Atlantic salmon (*Salmo salar*) and Pacific salmon (*Oncorhynchus* spp.). Atlantic salmon ages are described as the total number of winters a fish has lived since hatching from its egg or may be determined from the number of annuli on its scales.[21] An Atlantic salmon hatched in the spring of 2002 would be 3 years old in 2005. In contrast, Pacific salmon ages are determined from their brood year. Eggs spawned by Pacific salmon in 2002 (i.e., the 2002 brood) would be 4 years old in 2005. A fish less than 1 year of age is age 0.

23.4 BIRDS, WILDLIFE, AND DOMESTIC MAMMALS

The conventions in symbolization for the structure and function of nonhuman animals have developed mainly within zoology. Some of these conventions are formally documented, but many have become established simply through usage. Where specific conventions for the animal sciences are lacking, veterinarians and veterinary medical researchers have adopted many of the style conventions established by the medical and basic science communities. The information provided here covers topics for which specific conventions exist for nonhuman animals. Otherwise, consult the relevant parts of Section 23.5 or the *American Medical Association Manual of Style*.[22]

23.4.1 Anatomic Description

Consult the *Nomina Anatomica Veterinaria*, published by the World Association of Veterinary Anatomists,[23] for the official list of anatomic terms for most nonhuman animals. Nonhuman animal anatomy incorporates a distinct set of terms indicating orientation of a body part, which differ from those used for humans (e.g., "caudal", meaning toward the tail).

The *Handbook of Avian Anatomy*[24] presents the official list of anatomic terms for birds. This publication follows the conventions established by the World Association of Veterinary Anatomists for terms of orientation to describe the relative positions of body parts. The *Atlas of Avian Anatomy*[25] has excellent figures and definitions, but because it is now out of print, accessibility is limited.

23.4.2 Clinical Chemistry

Information on the clinical chemistry of laboratory animals has been compiled[26]; this work covers the mouse, rat, guinea pig, hamster, rabbit, dog, and nonhuman primates. This publication includes extensive lists of reference values for various analyses, arranged by substance analyzed and then by species. A more general reference for domestic animals, including dog, cat, horse, cow, and pig, is also available.[27]

23.4.3 Blood Groups

Blood typing represents one of the most significant areas where human medical conventions cannot be applied in veterinary medicine.

The red blood cell antigens and blood group systems of selected animal species have been defined by specific nomenclature.[28–30] The major histocompatibility complex antigens of selected mammals and birds have also been characterized.[31]

The nomenclature system for canine blood groups used by most veterinary immunohematologists today was adopted at international workshops held in the 1970s. The blood groups are designated by the term "dog erythrocyte antigen" (DEA) followed by a number and sometimes a decimal point and another number to indicate the specific locus and alleles (e.g., DEA 1.1, DEA 1.2, DEA 3, DEA 4, DEA 5). However, this DEA nomenclature has not been universally accepted, and efforts are continuing to rename the system with alphabetical designations to conform with the designations used for other species.[28]

The blood group phenotypes in cat are A (common), B (relatively rare except in certain breeds), and AB (very rare), the latter reflecting the presence of both phenotypes on the red cell.[28]

A total of 34 factors are used to define equine blood group characteristics distributed in 7 systems (EAA, EAC, EAD, EAK, EAP, EAQ, EAU).[29] The factors or their combinations in each system are transmitted as a specific phenogroup unit and inherited as codominant traits. Allelic frequencies vary between horse breeds. Blood group typing has been used for years to define equine parentage, although this is being replaced by DNA sequencing methods.

The literature defining the blood group systems of the cow, pig, sheep, goat, and llama were reviewed by Penedo.[32] Blood grouping in these species is based on hemolytic tests, although a saline agglutination system is also used in the pig. The International Society of Animal Genetics sponsors biannual comparison testing to exchange and standardize reagents and nomenclature worldwide.

In cattle, there are 11 genetic systems of blood groups: A, B, C, F, J, L, M, S, Z, R′, and T′. More than 70 blood group factors are currently recognized in cattle; these cosegregate in specific combinations defining separate alleles and are called phenogroups. In pig, 16 genetic systems are recognized, namely A through P alphabetically. Some of these are complex, like the B and C systems of cattle. The A system is related to cattle J, human A, and sheep R systems. Both A and O systems are found in serum and saliva and are genetically related to the sheep R system. The pig N system is found in serum and milk but not saliva. The sheep has 7 genetic systems, namely A, B, C, D, M, R, and X. A total of 22 blood group factors are recognized within these 7 systems, but official nomenclature has yet to be assigned for several of them. The B, C, and R systems are homologous to cattle B, C, and J systems. The sheep R antigens are found in serum and saliva. Six blood group systems have been identified in llama (and the related alpaca species), designated alphabetically as A through F.

Blood typing is used to verify parentage according to the principle of exclusion and is mandatory in some countries to register cattle (worldwide for artificial insemination or embryo transfer programs), pig, sheep, and goat (several countries in Europe but not the United States or Canada) and llama and alpaca (required in the United States and Canada for registration). In the future, it is likely that DNA sequence technology will replace blood typing for parentage testing. Blood typing is also used to avoid neonatal isoerythrolysis in valuable breeding stock.

23.4.4 Immunologic Systems

The cluster of differentiation (CD) nomenclature developed for human leukocyte antigens has become more widely accepted for nonhuman animals, including the mouse[33] and the rat[34] (see also Section 23.5.6.3).

23.5 HUMANS

23.5.1 Anatomic Description

The official list of human anatomic terms, *Terminologia Anatomica*,[35] was published by the Federative Committee on Anatomical Terminology and the International Federation of Associations of Anatomists.

Anatomic terms derived from Latin are regarded as common nouns; therefore, do not capitalize or italicize them. The Latin form of any term is the official term approved by the International Anatomical Nomenclature Committee, but English equivalents that either translate the Latin terms directly or provide idiomatic substitutes may be used in most scientific texts. For example, "brachial plexus" may be used instead of "plexus brachialis" and "stomach" may be used instead of "ventriculus". Some terms retain the Latin form because formal equivalents are lacking in English (e.g., "cisterna chyli").

Terms describing relative orientation of body parts in humans differ from those for nonhuman animals and birds because of the upright position of adult humans.

23.5.2 Blood

23.5.2.1 Blood Groups

Blood groups have been designated within various classification systems (e.g., the ABO, Rh, Lewis, and MNS systems). The antigens and factors are usually represented by capital and lowercase letters; some of these designations are arbitrary (the A, B, and O blood groups of the ABO system), and others stand for the name of the person who was the source of the component initially identified ("Le" for "Lewis", "Fy" for "Duffy"). In systems defined since 1945, antigens, phenotypes, and genes for blood groups are all represented by the same letter or letters for a particular group, with the gene designations set in italic. However, the specifics of how each system represents subclasses differ.

Antigen	Phenotype	Gene
K	K+	*K*
A_1	A_1	A^1
Fy^a	Fy(a+)	Fy^a

Some systems, including the Kell, Lutheran, and Duffy systems, have both letter and number designations—for example, K and K1 for the same antigen, K− and K:−1 for the same phenotype, and *K* and K^1 for the same gene. One of the most complex symbolizations is that for the Rh system ("Rh" for "rhesus monkey"). Three systems have been advocated and applied: the Rh-Hr notation, the CDE notation, and the numeric system.

Rh-Hr Notation	CDE Notation	Numeric System
Rh_0	D	1
rh″	E	3
hr″	e	5
rh^G	G	12
Hr^B	Bas	34

These examples illustrate some of the uses of capital and lowercase letters and superscripts. Further details can be found in Chapter 5, "The Rh Blood Group System", of *Mollison's Blood Transfusion in Clinical Medicine.*[36]

The Committee on Terminology for Red Cell Surface Antigens of the International Society for Blood Transfusion (ISBT) has published a system of 6-digit numeric codes for the established, serologically determined blood groups. The first 3 numbers represent the system (of which there are now 29) and the last 3 numbers represent the antigenic specificity.[37] These codes are intended primarily for computer applications. For publication purposes, phenotype designations may be represented by the system symbol (all capital letters), a colon, and the appropriate numbers for the specificities; in text, any leading zeros in the 3 digits designating specificity are dropped. For example, in the ABO system, the A antigen becomes ABO1 (not ABO001), and in the Lutheran system, Lua becomes LU1 (not LU001). Antigens that are missing are preceded by a minus symbol. Genotypes are presented as the system symbol, an asterisk, and the alleles or haplotypes separated by a forward slash, all in italics (e.g., *KEL*2,3/2,4*).

System	Traditional	ISBT
Colton	Co(a+b−)	CO:1, −2
	Co^a/Co	*CO*1/0*
Kell	K− k+ Kp(a−b+)	KEL:−1,2,−3,4
	k,Kp^b,Js^b/K^0	*KEL*2,4,7/0*

Within this system, other antigens and specificities are grouped under "collections" of specificities that have serologic, biochemical, or genetic connections. These collections are given 6-digit designations beginning with "2xx" followed by 3 digits for each member of the collection. For example, the Cost collection is number 205, and the codes for the 2 antigens in this collection are 205001 and 205002. Antigens not assigned to systems or collections are put into 1 of 2 series, the 700 series for low-incidence antigens (currently 19 antigens) and the 901 series for high-incidence antigens (currently 9 antigens).

23.5.2.2 Hemoglobins

Human hemoglobins have been conventionally designated by the symbol "Hb" followed by a space and then a letter (A, C through Q, or S); new additions to the list of hemoglobin variants are now given a word indicating the location or laboratory of discovery instead of a letter.

Hb A   Hb S   Hb Providence   Hb Hopkins-2

The normal components of human hemoglobin are Hb A, Hb A_2, and Hb F. Each molecule of normal hemoglobin comprises 2 pairs of the 4 globin chains, which are designated by the lowercase Greek letters alpha (α), beta (β), gamma (γ), and delta (δ); the chains designated epsilon (ε) and zeta (ζ) are found in embryonal hemoglobins.

The composition of the normal major fraction of human hemoglobin, Hb A, is 2 alpha and 2 beta chains, $\alpha_2\beta_2$; Hb A_2 contains 2 alpha and 2 delta chains, $\alpha_2\delta_2$. Some forms of hemoglobin contain only nonalpha chains.

Hb H: β_4 Hb Portland: $\gamma_2\zeta_2$

23.5.2.3 Platelet Antigens

A nomenclature for platelet antigens has been put forward by the Working Party on Platelet Serology of the ISBT and the International Council for Standardization in Haematology.[38] The systems are designated as HPA (for human platelet antigen) and numbered in order of date of publication (e.g., HPA-1, HPA-2). Allelic antigens are designated alphabetically in order of their frequency of occurrence in the population, from high to low (e.g., HPA-1a occurs more frequently than HPA-1b).

23.5.2.4 Coagulation Factors

Human clotting factors are represented by the term "factor" with a roman numeral (from I to XIII, excluding VI; e.g., factor III, factor XI). A lowercase "a" designates the activated form. Factor VIII and von Willebrand factor have special terminology.[39]

23.5.3 Bone Histomorphometry

A system of nomenclature, symbols, and units for bone histomorphometry has been published.[30] A set of 123 abbreviations are used in combination to produce standard representations of the variables reported in histomorphometry. Symbols are given for the primary measurements in both 2-dimensional and 3-dimensional systems.

With one exception, the abbreviations consist of 1 or 2 letters. Both capital and lowercase letters are used for 1-letter abbreviations; the 2-letter abbreviations consist of an initial capital letter followed by a lowercase letter. The sole 3-letter abbreviation is all capitals (BMU). When used in combinations, single-letter abbreviations are run together and a multiletter abbreviation is separated by a period from any adjacent abbreviation.

BS *for* bone surface
B.Pm *for* bone perimeter
Ob.S *for* osteoblast surface
Ct.Th *for* cortical thickness

23.5.4 Circulation

23.5.4.1 Hemodynamics

Symbols for hemodynamic quantities should, as far as possible, follow the international conventions for symbolizing quantities (see Sections 4.2 and 10.1.1.2): italics (sloping) for single-letter symbols representing quantities and roman for subscript or superscript symbols modifying the main symbol (unless the modifier is itself a symbol for a quantity). See Table 23.3 for examples of symbols, their modifiers, and their applications for specific measurements.

Multiletter abbreviations have been widely used to represent quantities (such as "LVEDP" for "left ventricular end-diastolic pressure"), but these should be replaced as much as possible by standard symbols (Table 23.3).

Table 23.3 Examples of hemodynamic and related variables

Variable	Symbol	SI unit
Main Symbols		
Area	A	mm^2
Quantity (volume)	Q	mL, L
Flow rate (volume/unit time)	Q (V/t)	$mL\ s^{-1}$, $L\ min^{-1}$
Flow velocity	v	$cm\ s^{-1}$
Flow acceleration	v/t	$cm\ s^{-2}$
Tissue or organ mass	M	g, kg
Perfusion	Q/M	$mL\ min^{-1}g^{-1}$, $mL\ min^{-1}\ (100\ g)^{-1}$
Pressure	P	kPa (mm Hg or cm H_2O for clinical use)
Resistance	R	$kPa\ L^{-1}$
Velocity	v	$mm\ s^{-1}$, $cm\ s^{-1}$
Volume	V	mL, L
Modifying Symbols		
Arterial	a	
Venous	ven	
Capillary	cap	
Atrium	A	
Ventricle	V	
Systemic	S	
Systolic	syst	
Diastolic	diast	
Examples of Modifying Symbols		
Pressure, arterial	P_{a}	kPa (also mm Hg)
Systolic	$P_{a,\ syst}$	
Diastolic	$P_{a,\ diast}$	
Aortic	P_{aor}	
Pressure, venous	P_{ven}	
Pressure, left ventricular	P_{LV}	kPa (also cm H_2O)

23.5.4.2 Clinical Notations

Leads (recording electrodes and connecting wires) for recording electrocardiographic tracings are designated by alphanumeric characters. The standard leads are designated with roman numerals. In the unipolar lead system, the central terminal is designated by V (voltage) with the letter R (right arm), L (left arm), or F (foot). In the augmented lead system, the central terminal is designated by aV.

lead I	lead II	lead III
VR	VL	VF
aVR	aVL	aVF

Chest leads are designated by V for the central terminal and a subscript arabic numeral for the chest electrode; R refers to the right side of the chest and roman numerals to intercostal-space locations above the standard locations. Esophageal leads have the added designation E; the number represents the distance in centimeters from the nares to the esophageal electrode.

lead V_1 lead V_2 lead V_1R lead V_3III lead VE28

Electrocardiographic waves (deflections from the tracing baseline) are designated by the capital letters P, Q, R, S, T, and U. Wave complexes are represented by unspaced letters, and a minor wave in a complex is represented by a lowercase letter. Prime signs indicate waves located after their usual position. An interval between 2 waves is represented by an en dash (or hyphen) between the wave symbols.

P wave QRS complex rS rRV P–R interval

The severity of manifestations of cardiac and vascular disease is commonly represented by roman or arabic numerals.

grade IV hypertensive retinopathy mitral stenosis grade 2/6 [2 on a scale of 6]

The New York Heart Association (NYHA) uses a 2-number system for representing functional capacity (roman numeral from I to IV) and objective assessment (capital letter from A to D).[40] Class I functional capacity represents the least impairment, and class IV the greatest impairment; objective assessment A represents no objective evidence of cardiovascular diseases, and objective assessment D represents objective evidence of severe cardiovascular disease.

23.5.5 Clinical Chemistry

SI units are used in many countries for reporting clinical chemistry findings. Their use has not been widely adopted in the United States, and US journals may require reporting in terms of older metric units or dual reporting. Young[41] compiled tables of conversion factors from the older metric units to SI units, along with recommendations on significant digits and minimum increments for clinical hematology and clinical chemistry measurements.

23.5.6 Immunologic Systems

Human histocompatibility leukocyte antigens (HLA) determine a person's tissue type and are important in tissue transplantation. The gene loci for the class I antigens, found on most nucleated cells and on platelets, are lettered A, B, C, E, F, G, H, J, K, L, N, S, X, and Z and are symbolized as follows:

HLA-A *HLA-C* *HLA-E* *HLA-H*

The D region, which contains the loci for class II antigens, is divided, a second capital letter distinguishes subregions, and a third capital letter indicates an alpha (A) or beta (B) chain. If needed, a number distinguishes among several chains.

HLA-DRA *HLA-DOB* *HLA-DQA1* *HLA-DRB4* *HLA-DPA2*

The class III region of the major histocompatibility complex carries various genes, including those for some of the complement factors.

Alleles of the HLA genes are designated by the letter(s) of the locus, followed by an asterisk and then 4 numerals (or 6 numerals, if needed for synonymous variants of the allele).

A*020101 Cw*0203 DPB1*0102 DRB1*080203

Current nomenclature for the HLA system is provided in a continuing series of reports from the World Health Organization Nomenclature Committee for Factors of the HLA System. The 2004 report[42] contains complete lists of all officially recognized HLA class I and class II genes and alleles. Monthly updates are published in *Tissue Antigens*,[43] *Human Immunology*,[44] and the *International Journal of Immunogenetics*[45] (formerly the *European Journal of Immunogenetics*[46]). Resources related to HLA nomenclature are accessible through the National Marrow Donor Program.[47]

The major histocompatibility complexes have been described for other species (e.g., RT1 in the rat,[48] DLA in the dog,[49] *H-2* genes in the mouse,[50] and RhLA in the rhesus monkey). A uniform system of symbols for major histocompatibility complexes across species (except for the human and mouse HLA and H-2 designations) has been proposed.[51] Ellis[52] reviewed the major histocompatibility complex in domestic animals.

23.5.6.1 Immunoglobulins

The 5 classes of human immunoglobulins are represented by the symbol "Ig" followed by an unspaced capital letter. Subclasses are indicated by an arabic numeral after the class letter.

IgA IgD IgE IgG IgM IgG1 IgA2

Immunoglobulin molecules are composed of heavy and light polypeptide chains. The heavy chains of each immunoglobulin class are different and are symbolized by the Greek letter that corresponds to the roman capital letter representing the class (A, α; D, δ; E, ε; G, γ; and M, μ); the heavy chains of IgG1 and IgA2 would be γ1 and α2. The light chains of all classes are either kappa (κ) or lambda (λ) chains.

The basic structure of an immunoglobulin molecule consists of 4 chains: 2 heavy chains (specific to the class) and 2 light chains (either kappa or lambda chains, never both in the same molecule). Thus, an IgG molecule contains 2 γ chains and either 2 κ or 2 λ chains. IgA may exist as a dimer—for example, $(\alpha_2\lambda_2)_2$; IgM may exist as a pentamer—for example, $(\mu_2\lambda_2)_5$. Because only one light chain type is found in each molecule, another shorthand representation of the immunoglobulins is IgG-κ, IgM-λ, and so on.

The immunoglobulin molecule can be fragmented into portions that are designated with a capital F and unspaced lowercase letters.

Fab F(ab′)2 Fd Fd′ Fv Fc

Each light and heavy chain contains variable (V), joining (J), and constant (C) regions designated by capital letters, with subscripts showing the type of chain (H or L). Subgroups are designated by arabic numerals attached to the symbol. Roman numerals attached to the symbol are used for the 4 subgroups of the heavy chain variable region, the 4 subgroups of the kappa chain variable region, and the 6 subgroups of the lambda chain variable region.

Variable Region	**Generic**	**Specific**
Light chain	V_L	$V_\kappa I$
Heavy chain	V_H	$V_\alpha III$
Constant Region	**Generic**	**Specific**
Light chain	C_L	C_λ
Heavy chain	C_H	$C_\gamma 1$

The surface membrane molecules that specifically bind immunoglobulins via the Fc portion are called Fc receptors, symbolized by FcR.[53] Receptors specific for a class of immunoglobulin are designated by a subscript Greek letter (corresponding to the class of immunoglobulin) between the c and the R. Subtypes are shown by unspaced roman numerals; species abbreviations are prefixes (e.g., hu for human, mo for mouse, rt for rat, and rb for rabbit).

$Fc_\alpha R$ $moFc_\gamma RII$

Fundamental Immunology,[54] among others, includes an overview of immunoglobulins.

23.5.6.2 Complement

Complement is a system of proteins involved in antigen–antibody reactions in all vertebrates. The components of the classical complement pathway are designated by the capital letter C, to which are attached arabic numerals[55]; the numerals do not follow the order of the reaction sequence of the classical pathway, which is C1, C4, C2, C3, C5, C6, C7, C8, C9.

Component C1 has subcomponents C1q, C1r, and C1s. Other complement factors, which take part in the alternative complement pathway, are represented by the letters B, D, H, I, and P.[56]

Fragments of complement are designated by added lowercase letters, generally with the larger fragment designated by the letter "b" and the smaller by "a"—for example, C3a and C3b. Inactive fragments are designated by the letter "i" as a suffix: C4bi. The loss of activity from a protein through peptide hydrolysis without fragmentation is indicated by a prefixed letter "i"—for example, "iC3b".

An overview of the complement pathways can be found in *Fundamental Immunology*.[54]

23.5.6.3 Lymphocytes and Surface Antigens of Immune Cells

The use of monoclonal antibodies to identify lymphocyte cell-surface antigens has resulted in development of the CD (cluster of differentiation) system of nomenclature.[57] A cluster of antibodies that display the same cellular reactivity are assigned numbers prefixed by "CD".

CD4 CD11a CD11b CD45 RB CDw186

A lowercase "w" indicates a provisional designation ("w" for the workshops at which decisions about designations are made). The CD system now encompasses about 350 surface molecules.[58]

23.5.6.4 Interleukins and Interferons

The nomenclature for interleukins is established by the WHO–International Union of Immunological Societies (IUIS) Nomenclature Subcommittee on Interleukin Designation.[59] There are now more than 20 interleukins, designated numerically.

interleukin-1α	interleukin-1β	interleukin-5	interleukin-12
IL-1α	IL-1β	IL-5	IL-12

The symbol for interferon is IFN.[60] The prefixes "Hu" and "Mu" indicate "human" and "murine", respectively, and a Greek letter, preceded by a hyphen, specifies the type of interferon. Nonallelic variants can be indicated by an arabic numeral or a capital letter immediately following the Greek letter.

IFN-β HuIFN-α3 MuIFN-γB

23.5.6.5 Allergens

A formal nomenclature system for allergens has been recommended by the International Union of Immunological Societies (IUIS).[61] Highly purified allergens are now designated by the first 3 letters of the genus, a space, the first letter of the species name, a space, and an arabic numeral, all in roman type. For example, pollen from perennial rye grass, *Lolium perenne*, would be designated Lol p 1. An allergen from the honey bee, *Apis mellifera*, would be Api m 4. If this system produces identical designations, either the genus or the species designation is extended by one or more letters (e.g., Can d for allergens from *Canis domesticus*, Cand a for allergens from *Candida albicans*). The use of roman type for allergens, rather than italics, conforms with accepted practices for genetic nomenclature, whereby italics is used to represent genes.

23.5.7 Renin–Angiotensin System

Systematic names and abbreviations for the renin–angiotensin system were established in 1979 by a nomenclature committee of the International Society of Hypertension.[62] The conventions include abbreviations for components of the system and superscript numbers indicating amino acid substitutions and their position.

ANG II, ANG-(1–8) for angiotensin-(1–8)octapeptide
$[Sar^1, Val^5, Ala^8]$ANG II *for* $[Sar^1,Val^5,Ala^8]$angiotensin-(1–8)octapeptide

23.5.8 Kidney and Renal Function

A detailed, standardized nomenclature for all components of renal structure was published by the International Union of Physiological Sciences in 1988.[63] The nomenclature of renal structures in nonhuman mammals is essentially the same as that for humans. In that report, terms are presented in an anatomically logical sequence, and a table summarizes the sequence and relationships among structures and gives abbreviations for the names of segments of the renal tubule and cell types.

OMCD	outer medullary collecting duct
DST cells	distal straight tubule cells

Symbols representing renal functions have not been well standardized, but many in use are generally accepted. Italicize single-letter symbols representing quantities (see

Sections 4.2 and 10.1.1.2) but not modifiers that do not represent quantities, as in the following example:

C_{cr} *for* creatinine clearance

23.5.9 Respiration

The symbols and abbreviations used in respiratory physiology that have been developed to describe human pulmonary function are also applied in reporting studies of other animals.

Use italic for single capital letters representing the major symbols in respiratory physiology (see Section 4.2, Table 4.5, and Section 10.1.1.2), but do not italicize multiletter symbols representing such variables. Use roman small capital or lowercase letters for modifiers (specifications) of the main symbols; these are set on the line with the main symbol or are subscripted, depending on the modifier. Separate multiple subscripts by commas. If there are several modifiers for a main symbol, place them in the following order: anatomic location (where), time (when), and condition or quality (what or how). The same letter may be used for different meanings (e.g., "*C*" for both "concentration" and "compliance"); similarly, the same letter may appear as a small capital letter or a lowercase letter, each carrying a different meaning (e.g., "A" for "alveolar" and "a" for "arterial"). More detailed descriptions of nomenclature and symbols are available.[22,64]

V_{CO_2} *for* CO_2 production per unit time
$FEF_{200-1,200}$ *for* forced expiratory flow between 200 and 1,200 mL of the forced vital capacity
*P*aw *for* pressure at any point along the airways

See Table 23.4 for the main symbols and examples of modified symbols.[64]

23.5.10 Thermal Physiology

The symbols for thermal physiology[65] represent physical quantities and may include modifying subscripts defining physical and physiological specificities. A subsequent publication[66] included a glossary to improve the precision of meaning and the uniformity

Table 23.4 Examples of symbols for respiratory function[a]

Variable	Symbol[b]	Variable	Symbol[b]
Main Symbols		Forced vital capacity	FVC
Concentration, compliance	*C*	Residual volume	RV
Fractional concentration of a gas	*F*	Total lung capacity	TLC
Pressure, partial pressure	*P*	Modifiers (On the Line)	
Gas volume	*V*	Alveolar (small capital)	A
Gas flow	$\dot{V}$	Expired (small capital)	E
Blood volume	*Q*	Inspired (small capital)	I
Blood flow	$\dot{Q}$	Lung (small capital)	L
Abbreviations		Arterial (lowercase)	a
Inspiratory capacity	IC	Blood (lowercase)	b
Forced expiratory volume	FEV	Capillary (lowercase)	c

[a] See Fishman's *Manual of Pulmonary Diseases and Disorders*[64] for a complete list of symbols.
[b] Main symbols are combined with modifiers and gas abbreviations (subscript) as appropriate (e.g., PA_{CO_2} for alveolar pressure of carbon dioxide, and Ca_{O_2} for arterial concentration of oxygen).

Table 23.5 Symbols for thermal physiology[a]

Quantity	Symbol	SI unit
Area, total body	A_b	m^2
Metabolic rate, basal	BMR	W, W m^{-2}, W kg^{-1}
Metabolic heat production	H	W m^{-2}
Radiant intensity, spectral	I_λ	W sr^{-1} nm^{-1}
Humidity, absolute	γ	kg m^{-3}
Pressure, vapor (saturated) at temperature T	$P_{s,T}$	Pa
Pressure, water vapor	P_w	Pa
Temperature, ambient	T_a	°C

[a] See Bligh and Johnson[66] for a complete list of symbols and modifiers.

Table 23.6 Symbols for the designation of generations

Symbol	Meaning	Example
F_1, F_2, . . .	Filial generations	$P_1 \times P_1 \rightarrow F_1$
		$F_1 \times F_1 \rightarrow F_2$
P_1, P_2, . . .	Parental generations	P_1 = parents of F_1
		P_2 = grandparents of F_1
B_1, B_2, . . .	Backcross generations	$F_1 \times P_1 \rightarrow B_1$
		$B_1 \times P_1 \rightarrow B_2$
S_1, S_2, . . .	Self-fertilized generations (only for plants)	Parental self-fertilization → S_1
		S_1 self-fertilization → S_2
I_1, I_2, . . .	Inbred generations	
E_1, E_2, . . .	Generations after experimental manipulation	
X_2	Offspring of F1 testcross	

of usage of technical terms in the field, along with a list of symbols, abbreviations, and SI units; however, some of these symbols have been given more than one meaning. Table 23.5 illustrates many of the symbols defined with illustrative modifiers and gives the proper SI units.[66] These 2 papers do not specify italicization of symbols for quantities as required by the relevant ISO standard (see Section 4.2, Table 4.5, Section 10.1.1.2, and ISO 80000[67]).

23.5.11 Reproduction: Inheritance and Pedigrees

The symbols used in genetics to designate generations are listed in Table 23.6. Each consists of a single roman capital letter and a subscript indicating the number of the generation.

The symbols recommended by the National Society of Genetic Counselors[68] for use in pedigree diagrams are illustrated in Figure 23.1. Squares represent males, circles females, and diamonds individuals of unknown sex. Generations are assigned roman numerals in consecutive order, beginning with the uppermost (first) generation in the diagram. Individuals on the same horizontal generation line are identified by arabic numerals assigned from left to right. Members of a generation may be grouped in a single symbol, enclosing a numeral to represent the number of individuals. Symbols

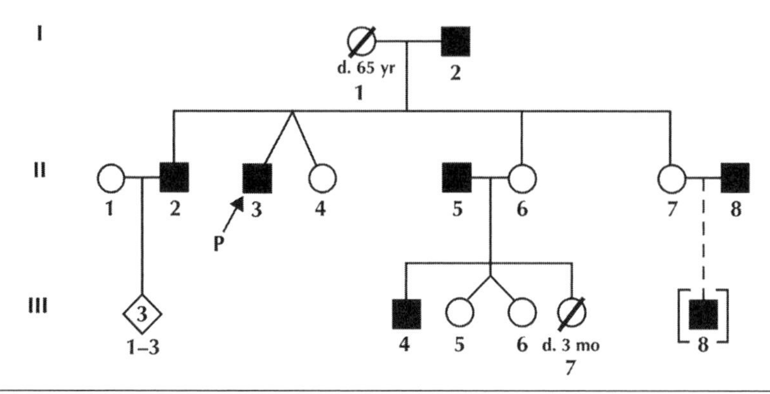

Figure 23.1 Symbols for pedigree diagrams.

for individuals carrying the trait of concern are solid; all others are open. Symbols may be subdivided to indicate multiple traits. A symbol for an individual identified by the letter P and an arrow represents a proband, "an affected individual coming to medical attention independent of other family members."[69]

Parents are connected by a horizontal line. Offspring are arrayed below on a parallel line in birth order from left to right. Identical (monozygotic) twins are connected to the horizontal sibling line by a single line. Fraternal (dizygotic) twins have separate lines that join the horizontal sibling line at the same point. For individuals with no offspring, a vertical line exits the symbol downward, terminating in a short horizontal line (no offspring by choice or reason unknown) or a double line (infertility). A slash through the symbol indicates that the individual has died. Additional symbols can be found in the recommendations of the National Society of Genetic Counselors.[68]

For nomenclature and symbols for chromosomes and genes, see Chapter 21.

CITED REFERENCES

1. Cann AJ. Principles of molecular virology. 5th ed. San Diego (CA): Academic Press; 2011.

2. Hull R. Mathews' plant virology. 4th ed. New York (NY): Academic Press; 2002.

3. Murphy FA, Gibbs EPJ, Horzinek MC, Studdert MJ. Veterinary virology. 3rd ed. San Diego (CA): Academic Press; 1999.

4. Lee JJ, Leedale GF, Bradbury PC. An illustrated guide to the protozoa. 2nd ed. Lawrence (KS): Society of Protozoologists; 2000.

5. Pommerville JC. Alcamo's fundamentals of microbiology. 9th ed. Boston (MA): Jones & Bartlett Publishers; 2010.

6. Hickey LJ. Classification of the architecture of dicotyledonous leaves. Am J Bot. 1973;60(1):17–33. http://dx.doi.org/10.2307/2441319.

7. Hopkins WG, Huner NPA. Introduction to plant physiology. 4th ed. New York (NY): John Wiley; 2008.

8. Fishman AP, editor. Handbook of physiology. Bethesda (MD): American Physiological Society; 1986.

9. Bailey J, editor. The Penguin dictionary of plant sciences. 2nd ed. London (UK): Penguin Books; 1999.

10. Gordh G, Headrick DH. A dictionary of entomology. 2nd ed. Wallingford (UK): CABI; 2010.

11. Holliday P. A dictionary of plant pathology. 2nd ed. Cambridge (UK): Cambridge University Press; 2001.

12. Kirk PM, Cannon PF, David JC, Stalpers JA, editors. Ainsworth and Bisby's dictionary of the fungi. 9th ed. New York (NY): CABI Publishing; 2001.

13. Lackie JM, Dow JAT, editors. The dictionary of cell and molecular biology. 3rd ed. San Diego (CA): Academic Press; 1999.

14. Salisbury FB, editor. Units, symbols, and terminology for plant physiology. A reference for presentation of research results in the plant sciences. New York (NY): Oxford University Press; 1996.

15. Tibbitts TW, Sager JC, Krizek DT. Guidelines for measuring and reporting environmental parameters in growth chambers. Biotronics. 2000;29(1):9–16. http://www.controlledenvironments.org/Guidelines/guidelines_biotronics.pdf.

16. Shibles R. Terminology pertaining to photosynthesis. Crop Sci. 1976;16(3):437–439. Report by the Crop Science Society of America Committee on Crop Terminology. http://dx.doi.org/10.2135/cropsci1976.0011183X001600030033x.

17. American Society of Agronomy; Crop Science Society of America; Soil Science Society of America. Publications handbook and style manual. Madison (WI): American Society of Agronomy; 2011 [accessed 2011 Nov 15]. http://www.agronomy.org/publications/style/.

18. Meier U, editor. Growth stages of mono- and dicotyledonous plants. 2nd ed. Berlin (Germany): Federal Biological Research Centre for Agriculture and Forestry; 2001.

19. Alaska Department of Fish and Game Interdivisional Publications Committee. ADF&G writer's guide. 3rd ed. Juneau (AK): Alaska Department of Fish and Game; 2010. Also available at http://www.adfg.alaska.gov/static/home/library/PDFs/writersguide_full.pdf.

20. Transactions of the American Fisheries Society: guide for authors. Trans Am Fish Soc. 2011;140:201–206. Also available at http://fisheries.org/docs/pub_tafs.pdf.

21. Koo TSY. Age designation in salmon. In: Koo TSY, editor. Studies of Alaska red salmon. Seattle (WA): University of Washington Press; 1962. p. 41–48. (University of Washington publications in fisheries, new series, vol. 1).

22. Iverson C, Flanagin A, Fontanarosa PB, Glass RM, Glitman P, Lantz JC, Meyer HS, Smith JM, Winker MA, Young RK. American Medical Association manual of style: a guide for authors and editors. 10th ed. Baltimore (MD): Williams & Wilkins; 2010.

23. World Association of Veterinary Anatomists, International Committee on Veterinary Gross Anatomical Nomenclature. Nomina anatomica veterinaria. 5th ed. Hannover (Germany): Editorial Committee; 2005 [accessed 2011 Nov 15]. 166 p. http://www.wava-amav.org/Downloads/nav_2005.pdf.

24. Baumel JJ, King AS, Breazile JE, Evans HE, Van den Berge JC, editors. Handbook of avian anatomy: nomina anatomica avium. 2nd ed. Cambridge (MA): Nuttall Ornithological Society; 1993. (Nuttall Ornithological Club publ.; 23). Prepared by the World Association of Veterinary Anatomists, International Committee on Avian Anatomical Nomenclature.

25. Chamberlain FW. Atlas of avian anatomy: osteology, arthrology, myology. East Lansing (MI): Michigan State College, Agricultural Experiment Station; 1943.

26. Kurtz DM, Prescott JS, Travlos GS, editors. Loeb and Quimby's clinical chemistry of laboratory animals. 3rd ed. London (UK): CRC Press; 2011.

27. Kaneko JJ, Harvey JW, Bruss ML, editors. Clinical biochemistry of domestic animals. 6th ed. San Diego (CA): Academic Press; 2008.

28. Andrews GA. Red blood cell antigens and blood groups in the dog and cat. In: Feldman BV, Schalm OW, Jain NC, Zinkl JG, editors. Schalm's veterinary hematology. 5th ed. Baltimore (MD): Lippincott Williams & Wilkins; 2000. p. 767–773.

29. Bowling AT. Red blood cell antigens and blood groups in the horse. In: Feldman BV, Schalm OW, Jain NC, Zinkl JG, editors. Schalm's veterinary hematology. 5th ed. Baltimore (MD): Lippincott Williams & Wilkins; 2000. p. 774–777.

30. Parfitt AM, Drezner MK, Glorieux FH, Kanis JA, Malluche H, Meunier PJ, Ott SM, Recker RR. Bone histomorphometry: standardization of nomenclature, symbols, and units. Report of the ASBMR Histomorphometry Nomenclature Committee. J Bone Miner Res. 1987;2(6):595–610. http://dx.doi.org/10.1002/jbmr.5650020617.

31. Hale AS, Gerlach JA. Major histocompatibility complex antigens. In: Feldman BV, Schalm OW, Jain NC, Zinkl JG, editors. Schalm's veterinary hematology. 5th ed. Baltimore (MD): Lippincott Williams & Wilkins; 2000. p. 789–794.

32. Penedo MCT. Red blood cell antigens and blood groups in the cow, pig, sheep, goat, and llama. In: Feldman BV, Schalm OW, Jain NC, Zinkl JG, editors. Schalm's veterinary hematology. 5th ed. Baltimore (MD): Lippincott Williams & Wilkins; 2000. p. 778–782.

33. Alaverdi N. Monoclonal antibodies to mouse cell-surface antigens. In: Coligan JE, Kruisbeek AM, Margulies DH, Shevach EM, Strober W, editors. Current protocols in immunology. New York (NY): John Wiley & Sons; c2003. Appendix 4B.

34. Puklavec JM, Barclay AN. Monoclonal antibodies to rat leukocyte surface antigens, MHC antigens, and immunoglobulins. In: Coligan JE, Kruisbeek AM, Margulies DH, Shevach EM, Strober W, editors. Current protocols in immunology. New York (NY): John Wiley & Sons; 2003. Appendix 4C.

35. Federative International Programme on Anatomical Terminologies. Terminologia anatomica. 2nd ed. New York (NY): Thieme Medical Publishers; 2011.

36. Klein H, Anstee D. Mollison's blood transfusion in clinical medicine. 11th ed. Oxford (UK): Blackwell Publishing; 2006.

37. Daniels GL, Fletcher A, Garratty G, Henry S, Jorgensen J, Judd WJ, Levene C, Lomas-Francis C, Moulds JJ, Moulds JM, et al. Blood group terminology 2004: from the International Society of Blood Transfusion committee on terminology for red cell surface antigens. Vox Sang. 2004;87(4):304–316. http://dx.doi.org/10.1111/j.1423-0410.2004.00564.x.

38. von dem Borne AEG, Decary F. ICSH/ISBT Working Party on Platelet Serology. Nomenclature of platelet-specific antigens. Vox Sang. 1990;58(2):176. http://dx.doi.org/10.1111/j.1423-0410.1990.tb02085.x.

39. Marder VJ, Mannucci PM, Firkin BG, Hoyer LW, Meyer D. Standard nomenclature for factor VIII and von Willebrand factor: a recommendation by the International Committee on Thrombosis and Haemostasis. Thromb Haemost. 1985;54(4):871–872.

40. American Heart Association, New York City Affiliate, Criteria Committee. Nomenclature and criteria for diagnosis of diseases of the heart and great vessels. 9th ed. Boston (MA): Little, Brown; 1994.

41. Young DS. Implementation of SI units for clinical laboratory data. Style specifications and conversion tables. Ann Intern Med. 1987;106(1):114–129. Errata in Ann Intern Med. 1987;107(2):265; Ann Intern Med. 1989;110(4):328; Ann Intern Med. 1991;114(2):172.

42. Marsh SG, Albert ED, Bodmer WF, Bontrop RE, Dupont B, Erlich HA, Geraghty DE, Hansen JA, Hurley CK, Mach B, et al. Nomenclature for factors of the HLA system, 2004. Tissue Antigens. 2005;65(4):301–369. http://dx.doi.org/10.1111/j.1399-0039.2005.00379.x.

43. Tissue antigens. Oxford (UK): Blackwell Publishing. Vol.1. 1971.

44. Human immunology. New York (NY): Elsevier/North-Holland. Vol. 1. 1980.

45. International Journal of Immunogenetics. Oxford (UK): Blackwell Publishing, Inc. Vol. 32. 2005. Continues: European Journal of Immunogenetics.

46. European Journal of Immunogenetics. Oxford (UK): Blackwell Scientific Publications. Vol. 18–31. 1991–2004. Continued by: International Journal of Immunogenetics.

47. National Marrow Donor Program. HLA educational resources. Minneapolis (MN): The Program; 2006–2011 [accessed 2011 Nov 15]. http://bioinformatics.nmdp.org/Education/HLA_Educational_Resources.aspx.

48. Gunther E, Walter L. The major histocompatibility complex of the rat (*Rattus norvegicus*). Immunogenetics. 2001;53(7):520–542. http://dx.doi.org/10.1007/s002510100361.

49. Wagner JL, Burnett RC, Storb R. Organization of the canine major histocompatibility complex: current perspectives. J Hered. 1999;90(1):35–38. http://dx.doi.org/10.1093/jhered/90.1.35.

50. Klein J, Benoist C, David CS, Demant P, Lindahl KF, Flaherty L, Flavell RA, Hammerling U, Hood LE, Hunt SE III, et al. Revised nomenclature of mouse H-2 genes. Immunogenetics. 1990;32(3):147–149. http://dx.doi.org/10.1007/BF02114967.

51. Klein J, Bontrop RE, Dawkins RL, Erlich HA, Gyllensten UB, Heise ER, Jones PP, Parham P, Wake-

land EK, Watkins DI. Nomenclature for the major histocompatibility complexes of different species: a proposal. Immunogenetics. 1990;31(4):217–219. Also available at http://www.springerlink.com/content/q825887606748722/fulltext.pdf.

52. Ellis SA. MHC studies in domestic animals. Eur J Immunogenet. 1994;21(3):209–215. http://dx.doi.org/10.1111/j.1744-313X.1994.tb00194.x.

53. Conrad D, Cooper M, Fridman WH, Kinet JP, Ravetch J. Nomenclature of Fc receptors. IUIS/WHO Subcommittee on Nomenclature of Fc receptors. Bull World Health Organ. 1994;72(5):809. Also available at http://www.ncbi.nlm.nih.gov/pmc/articles/PMC2486555/.

54. Paul WE, editor. Fundamental immunology. 6th ed. Philadelphia (PA): Lippincott Williams & Wilkins; 2008.

55. Austen KF, Becker EL, Borsos T, Lachmann PJ, Lepow IH, Mayer MM, Muller-Eberhard HJ, Nelson RA, Rapp HJ, Rosen FS, Trnka Z. Nomenclature of complement. Bull World Health Organ. 1968;39:935–938. Also available at http://www.ncbi.nlm.nih.gov/pmc/articles/PMC2554587/.

56. Alper CA, Austen KF, Cooper NR, Fearon DT, Gigli I, Hadding U, Lachmann PJ, Lambert PH, Lepow LH, Mayer MM, et al. Nomenclature of the alternative activating pathway of complement. Bull World Health Organ. 1981;59(3):489–491. Also available at http://www.ncbi.nlm.nih.gov/pmc/articles/PMC1536369/.

57. Beare A, Stockinger H, Zola H, Nicholson I. Monoclonal antibodies to human cell surface antigens. Current protocols in immunology. New York (NY): John Wiley & Sons; 2008. Appendix 4A.

58. Human cell differentiation molecules. North Adelaide (Australia): HCDM; 2011 [accessed 2011 Nov 15]. http://hcdm.org/.

59. Paul WE, Kashimoto T, Melchers F, Metcalf D, Mossman T, Oppenheim J, Ruddle N, Van Snick J. Nomenclature for secreted regulatory proteins of the immune system (interleukins). WHO-IUIS Nomenclature Subcommittee on Interleukin Designation. Bull World Health Organ. 1991;69(4)483–484. Also available at http://www.ncbi.nlm.nih.gov/pmc/articles/PMC1554309/.

60. Announcement on interferon nomenclature. Arch Virol. 1983;77(2–4):283–285.

61. King TP, Hoffman D, Lowenstein H, Marsh DG, Platts-Mills TAE, Thomas W. Allergen nomenclature. IUIS/WHO Allergen Nomenclature Subcommittee. Bull World Health Organ. 1994;72(5):797–800. Also available at http://www.allergen.org/.

62. International Society of Hypertension, Nomenclature Committee. Nomenclature of the renin-angiotensin system. Hypertension. 1979;1(6):654–656. http://dx.doi.org/10.1161/01.HYP.1.6.654.

63. Kriz W, Bankir L. A standard nomenclature for structures of the kidney. The Renal Commission of the International Union of Physiological Sciences. Kidney Int. 1988;33(1):1–7. http://dx.doi.org/10.1038/ki.1988.1.

64. Fishman AP, Elias JA, Fishman JA, Grippi MA, Kaiser LR, Senior RM, editors. Fishman's manual of pulmonary diseases and disorders. 3rd ed. New York (NY): McGraw-Hill; 2002.

65. Proposed standard system of symbols for thermal physiology. J Appl Physiol. 1969:27(3):439–446.

66. Bligh J, Johnson KG. Glossary of terms for thermal physiology. J Appl Physiol. 1973;35(6):941–961.

67. International Organization for Standardization. Quantities and units - Part 5: Thermodynamics. Geneva (Switzerland): ISO 80000-5:2007.

68. Bennett RL, Steinhaus KA, Uhrich SB, O'Sullivan CK, Resta RG, Lochner-Doyle D, Markel DS, Vincent V, Hamanishi J. Recommendations for standardized human pedigree nomenclature. Pedigree Standardization Task Force of the National Society of Genetic Counselors. Am J Hum Genet. 1995;56(3):745–752.

69. Bennett RL, Steinhaus KA, Uhrich SB, O'Sullivan CK, Resta RG, Lochner-Doyle D, Markel DS, Vincent V, Hamanishi J. Reply to Mazarita and Curtis [letter]. Am J Hum Genet. 1995;57(4):983–984.

ADDITIONAL REFERENCES

Blood DC, Studdert VP, compilers. Saunders comprehensive veterinary dictionary. 2nd ed. New York (NY): WB Saunders; 1999.

Boden E, editor. Black's veterinary dictionary. 20th ed. Lanham (MD): Barnes and Noble Books; 2001.

Floyd MR. The modified triadan system: nomenclature for veterinary dentistry. In: DeForge DH, Colmery

BH, editors. An atlas of veterinary dental radiology. Ames (IA): Iowa State University Press; 2000. p. 265–266.

Gibb M. Keyguide to information sources in veterinary medicine. London (UK): Mansell Publishers; 1990.

Hopkins WG, Huner NPA. Introduction to plant physiology. 3rd ed. New York (NY): John Wiley; 2004.

Stephens G. Animal health and veterinary science: a guide to the literature. In: Hutchinson B, Greider AP, editors. Using the agricultural, environmental and food literature. New York (NY): Marcel Dekker; 2002.

24 Disease Names

24.1 PLANT DISEASES

Plant disease names are not governed by formal rules of taxonomy and nomenclature. Names are usually based on a major symptom or sign or on the pathogen responsible; a few names refer to a geographic location, but eponyms are rare in plant pathology.

anther smut [refers to the plant tissue affected]
black root rot [refers to a symptom and the tissue affected]
powdery mildew [refers to the appearance of the fungal pathogen on the host plant surface]
bacterial soft rot [refers to the general pathogen group and a symptom]
Gibberella ear rot [refers to the pathogen by genus name and the part of the host affected]
Karnal bunt [refers to a region in India where this disease of wheat was first found]
northern corn leaf blight [refers to a general geographic area where the disease occurs, the host, the tissue affected, and the major symptom]

The same name may be used for diseases of several host species, even though each disease is caused by a different pathogen. For example, scab of wheat and barley is caused by any of several species of *Fusarium*, whereas common scab of potato is caused by *Streptomyces scabies*. Therefore, depending on the context, it may be important to include the host name as part of the disease name.

scab of wheat	*Phytophthora* rot of soybean
potato scab	southern blight of red clover

Journals that publish papers dealing with plant diseases differ in their use of italics when the scientific (Latin) name of a pathogen is part of the disease name. Some journals use both an initial capital and italics for the scientific name of the pathogen, others use neither an initial capital nor italics, and yet others use an initial capital but no italics. For consistency and accuracy, CSE recommends that any scientific name of a plant pathogenic organism forming part of a disease name be written in italics, with an initial capital if it is a genus name (see Section 22.2.3). Do not use an adjectival ending.

Fusarium head blight
not Fusarium head blight *or* fusarium head blight *or* fusarial head blight

A pathogen species may be moved to another genus for taxonomic reasons. In this situation, the original disease name is often retained because of its familiarity to lay people; it is preferable, however, to use the name that reflects the current nomenclatural status of the pathogen.

Stenocarpella stalk and ear rot of maize, caused by *Stenocarpella maydis*
[formerly *Diplodia* stalk and ear rot of maize, caused by *Diplodia maydis*]

Holliday[1] suggested 2 principles for naming plant diseases, designed to avoid the situations described above, but they are not always followed.

1) The name should be readily usable by plant growers because growers are concerned with the disease, not the pathogen.
2) The name should mention the host and the most conspicuous abnormality, instead of the name of the pathogen.

cotton wilt *for Verticillium* wilt
eucalyptus canker *for Cytospora* canker
soybean root rot *for Phytophthora* root rot
Dutch elm wilt *for* Dutch elm disease [wilt is the primary symptom, and the word "disease" should not appear in a plant disease name]

This system presents its own difficulties. If 2 or more pathogens cause similar symptoms, it may be difficult to coin distinct disease names based only on the host and the symptoms. For example, *Stagonospora nodorum* and *Septoria tritici* both cause a leaf blotch on wheat, with symptoms that can be difficult to distinguish. Therefore, the names *Stagonospora* blotch and *Septoria* blotch are used (rather than wheat blotch). Similarly, various stalk rots of corn are difficult to distinguish symptomatically, and pathogen names are therefore included as part of the disease names (e.g., *Fusarium* stalk rot, *Gibberella* stalk rot).

Diseases and injuries caused by nematodes are generally referred to by the common name of the nematode pathogen.

root knot of tomato [caused by root knot nematode]
lesion nematode on corn
soybean cyst nematode

The American Phytopathological Society maintains a list of recommended common names for plant diseases.[2] These do not carry the force of nomenclatural codes for organisms, but their use is encouraged for the publications of this scientific society. Plant disease names are listed also in *A Dictionary of Plant Pathology*[1] and *A Thesaurus of Agricultural Organisms: Pests, Weeds, and Diseases*.[3]

24.2 ANIMAL AND HUMAN DISEASES

24.2.1 Standard Disease Nomenclature

Various standard nomenclatures for diseases and histologic classifications of tumors, both national and international, are available, but many authors do not follow them. Generally accepted names and synonyms can be found in standard medical, dental, and veterinary dictionaries; authors who are unfamiliar with English terms should consult a multilingual thesaurus for the correct English form.

One authoritative source for verifying biomedical terminologies, including disease names, is the Unified Medical Language System (UMLS) metathesaurus,[4] a large, multipurpose, multilingual vocabulary database that contains information about biomedical and health-related concepts, their various names, and the relations among them. It is built from the electronic versions of more than 100 thesauri, classifications, code sets, and lists of controlled terms used in patient care, health services billing, public health statistics, indexing and cataloging of the biomedical literature, and basic, clinical, and health services research. The Medical Subject Headings (MeSH) vocabulary,[5] the National Library of Medicine's (NLM) controlled vocabulary thesaurus, is another source for verifying disease terminology for human and veterinary medicine. MeSH terms are used by NLM for indexing its own databases and by many other medical indexers to ensure that indexes in biomedical books and journals are compatible with the NLM databases.

The widespread adoption of electronic medical and health records has increased the need for standard code sets and terminologies for clinical data. Standards designated for use within the United States are identified in the UMLS metathesaurus. One comprehensive compilation that is gaining international recognition, and that is particularly relevant to usage in the United States and the United Kingdom, is the Systematized Nomenclature of Medicine–Clinical Terms[6] (SNOMED CT), originally created by the College of American Pathologists and now available from the International Health Terminology Standards Development Organisation or through the UMLS metathesaurus.[4] This nomenclature system covers diseases, disorders, syndromes, and their manifestations in humans and animals, as well as other topics, such as clinical procedures, devices, and pharmaceutical products.

The World Health Organization maintains the International Statistical Classification of Diseases and Related Health Problems[7] and adaptations, including the International Classification of Diseases for Oncology,[8] which are used for epidemiologic and health management purposes.

Standard nomenclatures for narrower fields have been published and may appear in the UMLS metathesaurus; journals in those fields may specify that terminology must follow the relevant standard nomenclature.

Psychiatric diagnoses may be specified with classification numbers and terms from the *Diagnostic and Statistical Manual of Mental Disorders*.[9]

The International Society for Human and Animal Mycology has made recommendations on naming fungal diseases.[10] The World Association for the Advancement of Veterinary Parasitology proposed a standardized nomenclature of parasitic diseases (SNOPAD[11]; formerly the standardized nomenclature of animal parasitic diseases, SNOAPAD[12]), although other approaches have been suggested.[13]

The World Health Organization has published a comprehensive series on the classification of human tumors.[14–23] The Union for International Cancer Control publishes the TNM (tumor, node, metastasis) classification system, which is widely used for describing tumors[24] (summarized in the *American Medical Association Manual of Style*[25]). Standardized nomenclatures for animal tumors are also available.[26–28]

24.2.2 Eponymic Disease Names

In general, use descriptive disease names rather than eponymic names. Medical dictionaries and specialized dictionaries[29,30] provide synonyms for eponymic names.

If an eponymic form is used, use the nonpossessive form rather than the possessive form (see Section 6.5.6). Do not capitalize adjectival and derivative forms of proper names used in disease names (see Section 9.7.2).

Crohn disease *not* Crohn's disease
addisonian anemia *from* Thomas Addison
parkinsonism *from* James Parkinson

CITED REFERENCES

1. Holliday P. A dictionary of plant pathology. 2nd ed. Cambridge (UK): Cambridge University Press; 2001.

2. American Phytopathological Society, Committee on Standardization of Common Names for Plant Diseases. Common names of plant diseases. St Paul (MN): The Society; 1978–2005 [accessed 2011 Nov 10]. http://www.apsnet.org/publications/commonnames/Pages/default.aspx.

3. Derwent Publications. A thesaurus of agricultural organisms: pests, weeds, and diseases. London (UK): Chapman and Hall; 1990. 2 vols.

4. Unified Medical Language System. Bethesda (MD): National Library of Medicine (US); 1999 [updated 2005 Nov 29; accessed 2011 Nov 10]. http://www.nlm.nih.gov/research/umls/.

5. Medical subject headings. Bethesda (MD): National Library of Medicine (US); 1999 [updated 2005 Dec 7; accessed 2011 Nov 10]. http://www.nlm.nih.gov/mesh/.

6. Systematized Nomenclature of Medicine–Clinical Terms (SNOMED CT). Copenhagen (Denmark): International Health Terminology Standards Development Organisation; 2008 [accessed 2011 Nov 10]. http://ihtsdo.org/snomed-ct/.

7. World Health Organization. International statistical classification of diseases and related health problems. 10th rev. Geneva (Switzerland): The Organization; 1994–2003 [accessed 2011 Nov 10]. http://www.who.int/classifications/icd/en/.

8. Fritz A, Percy C, Jack A, Shanmugarathan S, Sobin L, Parkin DM, Whelan S, editors. International classification of diseases for oncology (ICD-O). 3rd ed. Geneva (Switzerland): World Health Organization; 2000.

9. American Psychiatric Association. Diagnostic and statistical manual of mental disorders. 5th ed. Washington (DC): The Association; 2013.

10. Odds FC, Arai T, Disalvo AF, Evans EG, Hay RJ, Randhawa HS, Rinaldi MG, Walsh TJ. Nomenclature of fungal diseases: a report and recommendations from a Sub-Committee of the International Society for Human and Animal Mycology (ISHAM). J Med Vet Mycol. 1992;30(1):1–10.

11. Kassai T, Burt MD. A plea for consistency. Parasitol Today. 1994;10:127–128.

12. Kassai T, Cordero del Campillo M, Euzeby J, Gaafar S, Hiepe T, Himonas CA. Standardized nomenclature of animal parasitic diseases (SNOAPAD). Vet Parasitol. 1988;29(4):299–326.

13. Ashford RW. Current usage of nomenclature for parasitic diseases, with special reference to those involving arthropods. Med Vet Entomol. 2001;15(2):121–125.

14. Barnes L, Eveson JW, Reichart P, Sidransky D. Pathology and genetics of head and neck tumours. Geneva (Switzerland): World Health Organization; 2005. (WHO classification of tumours).

15. Bosman FT, Carneiro F, Hruban RH, Theise ND, editors. Pathology and genetics of tumours of digestive system. Geneva (Switzerland): World Health Organization; 2010. (WHO classification of tumours).

16. DeLellis RA, Lloyd RV, Heitz PU, Eng C. Pathology and genetics of tumours of endocrine organs. Geneva (Switzerland): World Health Organization; 2004. (WHO classification of tumours).

17. Eble JN, Sauter G, Epstein JI, Sesterhenn IA, editors. Pathology and genetics of tumours of the urinary system and male genital organs. Geneva (Switzerland): World Health Organization; 2004. (WHO classification of tumours).

18. Fletcher CDM, Unni KK, Mertens F, editors. Pathology and genetics of tumours of soft tissue and bone. Geneva (Switzerland): World Health Organization; 2002. (WHO classification of tumours).

19. LeBoit P, Burg G, Weedon D, Sarasin A, editors. Pathology and genetics of tumours of the skin. Geneva (Switzerland): World Health Organization; 2005. (WHO classification of tumours).

20. Louis DN, Ohgaki H, Wiestler OD, Cavenee WK, editors. Pathology and genetics of tumours of nervous system. Geneva (Switzerland): World Health Organization; 2007. (WHO classification of tumours).

21. Swerdlow SH, Campo E, Harris NL, Jaffe ES, Pileri SA, Stein H, Thiele J, Vardiman JW, editors. Pathology and genetics of tumours of haematopoietic and lymphoid tissues. Geneva (Switzerland): World Health Organization; 2008. (WHO classification of tumours).

22. Tavassoli FA, Devilee P, editors. Pathology and genetics of tumours of the breast and female genital organs. Geneva (Switzerland): World Health Organization; 2003. (WHO classification of tumours).

23. Travis WD, Brambilla E, Muller-Hermelink HK, Harris CC, editors. Pathology and genetics of tumours of the lung, pleura, thymus and heart. Geneva (Switzerland): World Health Organization; 2004. (WHO classification of tumours).

24. Sobin LH, Gospodarowicz MK, Wittekind C, editors. TNM classification of malignant tumours. 7th ed. New York (NY): Wiley-Blackwell; 2009.

25. Iverson C, Christiansen S, Flanagin A, Fontanarosa PB, Glass RM, Gregoline B, Lurie SJ, Meyer HS, Winker MA, Young RK. American Medical Association manual of style: a guide for authors and editors. 10th ed. New York (NY): Oxford University Press; 2007.

26. International histological classification of tumours of domestic animals. Bull World Health Organ. 1974;50(1–2):1–142.

27. International histological classification of tumours of domestic animals. Part 2. Bull World Health Organ. 1976;53(2–3):137–304.

28. Meuten DJ, editor. Tumors in domestic animals. 4th ed. Ames (IA): Blackwell; 2002.

29. Jablonski S. Jablonski's dictionary of syndromes and eponymic diseases. 2nd ed. Melbourne (FL): Krieger Publishing Company; 1991.

30. Magalini SI, Magalini SC, editors. Dictionary of medical syndromes. 4th ed. Philadelphia (PA): Lippincott; 1997.

25 The Earth

25.1 SCOPE

The atmosphere, rocks, soils, and bodies of water have been the subjects of intensive scientific study in the disciplines of meteorology, geology, soil science, limnology, oceanography, and related fields. This chapter covers the conventions used to report and describe research on these subjects.

25.2 GEOLOGIC TIME UNITS

The 3 most common classification systems used for geologic time units are chronostratigraphy, geochronology, and geochronometry. In chronostratigraphy (the time-rock or time-stratigraphic system) the units of time are based on all rocks formed during a particular span of time. In geochronology (geologic time) the unit is the interval of time when the chronostratigraphic unit was deposited. Chronostratigraphic units delineate the relative positions of the earth's strata in geologic time; geochronologic units delineate the ages of the earth's strata. For example, rocks in the Jurassic System (chronostratigraphic unit) were formed during the Jurassic Period (geochronologic unit). In geochronometry the unit boundaries are based on absolute time measured in years.

Chronostratigraphic units are, in order of decreasing rank, eonothem, erathem, system, series, stage, and chronozone. Geochronologic units are, in order of decreasing rank, eon, era, period, epoch, age, and chron. Chronozones and chrons, though nonhierarchical, are usually smaller than the other units.[1]

Terms for geochronologic rank may be used for geochronometric units when such terms have been formally accepted (e.g., Archean Eon and Proterozoic Eon). Geochronometric units may also correspond to chronostratigraphic units, though they are not defined by them.[1] The most common geochronometric units that have agreed-upon boundaries are divided into eons and eras. The International Commission on Stratigraphy (ICS) subdivides the Proterozoic Eon into the Paleoproterozoic, Mesoproterozoic, and Neoproterozoic eras and the Archean Eon into Eoarchean, Paleoarchean, Mesoarchean, and Neoarchean eras. The ICS also subdivides the eras of the Proterozoic into periods.[2] The North American Commission on Stratigraphic Nomenclature uses a similar but not identical naming system for eras within the Proterozoic and Archean: the Early, Middle, and Late Proterozoic eras are equivalent to the Paleoproterozoic, Mesoproterozoic, and Neoproterozoic; and the Archean is subdivided into the Early, Middle, and Late Archean eras.[1]

Chronostratigraphic and geochronologic units are correlated worldwide. Formal definitions for boundaries of geologic time divisions have been accepted and ratified through the International Union of Geological Sciences (IUGS). Terms for international geologic divisions of time are presented in the *International Stratigraphic Guide*[3] and the *International Stratigraphic Guide–Abridged Version*[4] and are listed in order of increasing geologic age in the *International Stratigraphic Chart*.[2] Different typography is used in the *International Stratigraphic Chart* to distinguish formal and informal names.

The capitalization style for geologic time terms is different for formal and informal names. Because the same words (e.g., "early" and "late") can modify both formal and informal names, ambiguity can result if style conventions adopted by authoritative bodies within the discipline are not followed.

Capitalize the initial letters of formal names, including modifiers such as "Early" or "Late" when they are part of the formal name.[1,5,6]

Phanerozoic Eon
Mesozoic Era
Cretaceous Period
Maastrichtian Stage
Late Jurassic

Series (rocks) are often termed "Upper" and "Lower" for the same epoch (time) that is termed "Late" or "Early". The paired terms referred to in the following sentence are "Upper/Lower" and "Late/Early". These paired terms are not interchangeable. "Middle" may be used as part of formal series or epoch names, in which case the initial letter is capitalized, or as an informal subdivision, in which case it is in lowercase letters.

Upper Triassic strata [position] were deposited in Late Triassic time [age].
Lower Cambrian rocks [position] contain Early Cambrian fossils [age].
Middle Devonian strata [position] formed before the middle Mississippian but after the Middle Ordovician [age].

Capitalize generic rank terms such as "system", "series", "period", and "epoch" when they are used with their specific modifiers. Do not use generic rank terms in map explanations or in illustrations or tables, except in a column heading in a table.

Phanerozoic Eonothem
Mesozoic Erathem
Cretaceous System
Upper Cretaceous Series
Maastrichtian Stage
Jurassic System

The specific name of a term may be used alone to imply the generic element.

Most of their work was concerned with the Jurassic.

Do not capitalize informal terms (e.g., "early" or "late") used as modifiers of formal names.

early Mesozoic Era [here, "early" is an informal modifier of the formal name "Mesozoic Era"]

For clarity, always use formal, full terms instead of shortened terms.

Permian and Pennsylvanian *not* Permo-Penn
Cambrian and Ordovician *not* Cambro-Ordovician
Middle Cambrian *not* Mid-Cambrian
Westwater Canyon Member *not* Westwater Member

The apparent inconsistency in style that may result when these capitalization guidelines are used properly in a document containing both formal and informal names is an acceptable consequence when the style convention conveys scientific meaning. However, the effectiveness of this convention in improving clarity depends on proper identification of geologic time terms as formal or informal. Authors and editors should consult an authoritative source to determine whether a term is formal or informal and should not assume that common usage is correct.

A question mark immediately after a tentative or provisional name can be used to show doubt about accuracy. The following examples using question marks to indicate doubt are more informative than a less specific term such as "probably Late Devonian".

Late? Devonian [accuracy of epoch is in doubt]
Late Devonian? [accuracy of period is in doubt]

To avoid confusion about the meaning of the question mark, avoid ending a sentence with a tentative or provisional name that is followed by a question mark.

Terms for geologic time may be abbreviated on maps (e.g., "K" for Cretaceous) (see Section 25.3.2), but do not abbreviate geologic time terms in running text. Express absolute dates in geologic time with the units giga-annum (Ga, 10^9 years), mega-annum (Ma, 10^6 years), or kiloannum (ka, 10^3 years). "Mega-annum" literally means millions of years ago. Express duration of time in years (Gy, My, or ky).[1] Spell out the units of time in running text when they are not preceded by a numeric value.

The Cretaceous Period lasted 80 My [duration] from 144 Ma to 65 Ma [absolute dates].

Use geochronologic units to designate the age of material units formed since 540 Ma. However, rocks older than 540 Ma are divided into geochronometric units with defined time boundaries. Because geochronometric units are essentially arbitrary time boundaries that have been internationally agreed upon, no type localities exist.

Archean Eon [older than 2,500 Ma]   Neoarchean [2,800 to 2,500 Ma]

Throughout geologic time, the earth's polarity has repeatedly reversed. The following polarity-chronostratigraphic units distinguish the primary magnetic-polarity record imposed when the rock was deposited, from Tertiary through Quaternary (Neogene) time, in order of increasing age: Brunhes, Matuyama, Gauss, and Gilbert.

the Matuyama Reversed-Polarity Chronozone

25.3 STRATIGRAPHIC AND MAP UNITS

A stratigraphic unit (includes lithostratigraphic and lithodemic units) name is the name of a formally defined mappable unit of rock or the local informal name applied to a rock unit. The map unit is recognized by its lithologic content and its boundaries. It is assigned a place within the geologic age sequence, has a stratigraphic rank, and is mappable.

25.3.1 Capitalization of Stratigraphic Names

Rock units are usually discussed chronologically in text, oldest (bottommost) first, youngest (topmost) last. Formally named units have compound names consisting of a geographic name and a rank term (e.g., group, formation, member, bed) or descriptive lithic term (e.g., shale, sandstone, limestone). Begin each word in a formally named stratigraphic unit with a capital letter except for the species name in a biostratigraphic unit.

Stony Point Tuff   Codell Sandstone Member   *Cordylodus intermedius* Zone
Baltimore Gneiss   St Louis Limestone

Refer to the *North American Stratigraphic Code*[1] and the *Lexicon of Canadian Stratigraphy*[7–9] for the recommended procedures for classifying and naming formal stratigraphic and related units.

For all informal stratigraphic units, capitalize only the place name; do not capitalize the descriptive lithic or rank terms.

tuff of Stony Point   St Louis coal
formation of Madeira Canyon   J sandstone

On first use, define and describe informal terms, including local or commercial names (as in mining or oil industries). A modifier such as "the informal" may precede the first mention of the name to alert the reader that the name represents an informal stratigraphic unit.

> the informal St Louis coal [on first use] *but* St Louis coal [for subsequent uses]

Do not capitalize a stratigraphic rank term when it is used alone in text as a generic reference to a stratigraphic unit.

> the formation *for* Nevada Formation group in the West *for* Uinta Mountain Group

Stratigraphic rank terms may be abbreviated (e.g., Gp for group, Fm for family, Mbr for member) in illustrations or tables but not in text. In text, the full name need not be repeated after the first usage. For example, after first usage "Cozzette Sandstone Member" may be referred to as "Cozzette Member" or "Cozzette" but not "Cozzette Sandstone". Follow the recommendations in the *International Stratigraphic Guide* for shortened forms of stratigraphic names.[3]

The North American Commission on Stratigraphic Nomenclature provides guidance on the capitalization style of formal and informal stratigraphic names.[1]

Biozones are often named after the predominant plant or animal species present in the zone (e.g., *Benueites benuensis* Zone). The taxonomy rules and formats for species names in biozones are the same as those for plant and animal names (see Chapter 22 in this manual).

25.3.2 Map Units

Map units on geologic maps and in stratigraphic sections are explained in a map legend. A map unit in a legend consists of a map symbol and a description of the unit. Letter symbols for map units consist of uppercase and lowercase letters, and sometimes subscripts, and convey information about the chronostratigraphic, geochronologic, or geochronometric units and the name of the map unit. A capitalized first letter in a map symbol represents the geologic period, for example, J for Jurassic, K for Cretaceous, T for Tertiary, and Q for Quaternary. Two capital letters (e.g., QT or TJ) indicate a range of 2 or more systems; the younger system is listed first. The lowercase letters that follow indicate the name of the map unit or the type of rock. Special characters may be used on geologic maps to distinguish the names of systems or periods that begin with the same letter. For example, the letter P is modified in the symbols for Paleogene, Paleozoic, Pennsylvanian, Precambrian, and Proterozoic to distinguish those symbols from the symbol for Permian, P.

Do not use map symbols in text as stand-alone abbreviations for map units. However, a map symbol may be presented in parentheses after the unit name in text, especially when the reader is referred to a map that uses the symbol.

> Locations of Jurassic Entrada Sandstone (Je) are shown on the map.
> *not* Locations of Je are shown on the map.

Letter symbols for map units are unique to each geologic map, but some parts of symbols are standardized for all maps. For more information on map symbols, see Hansen.[6]

In the description of a map unit the name of the stratigraphic unit is followed by its position (usually the series term) or age in parentheses. Describe units in order of increasing age (i.e., upper members of a formation before middle or lower members). The description of the unit is usually a brief account of the lithology, color, thickness, and other distinguishing characteristics of the rocks in each unit. The order in which these characteristics are described should be consistent within each legend. Some map units are simply combinations of units that are described elsewhere in the legend.

Tb	Bishop Conglomerate (Oligocene) — Light gray to pinkish gray very poorly sorted loosely cemented pebbly, cobbly to bouldery conglomerate and sandstone. Thickness highly varied; may locally exceed 60 m

25.3.3 Correlation Charts, Stratigraphic Tables, and Columnar Sections

Correlation charts are used to show the author's interpretation of rock units and their ages as they relate to units that other workers have recognized elsewhere. The relative or radiometric time is usually designated at the left. Units are listed from youngest (topmost) to oldest (bottommost). Identify rock units by name, whether formal or informal, without abbreviations if space permits.

Stratigraphic columnar sections are graphic illustrations that are used to describe and show in a vertical column the sequence and relations of rock or soil in a defined area. Colors or symbols used to distinguish units are unique to the particular illustration. Color terms used to describe rocks should be as specific as possible. Refer to the Geological Society of America's *Rock-Color Chart*.[10] Standard patterns are used to indicate rock types (see pages 376–377 in Compton[11]). For details on the construction of these charts and illustrations, consult Hansen.[6]

25.4 PHYSICAL DIVISIONS OF THE UNITED STATES

Physiographic divisions, provinces, and sections of the United States are specific physiographic entities and are capitalized as proper nouns.

Laurentian Upland	Appalachian Plateaus	Hudson Valley
New England Province	Floridian Section	Osage Plains

See page 344 of the *US Government Printing Office Style Manual*[12] or a reputable atlas such as the *National Geographic Atlas of the World*[13] for a comprehensive listing of physiographic division, province, and section names.

25.5 ROCKS AND MINERALS

The International Union of Geological Sciences Commission on Systematics in Petrology is working toward a comprehensive and internationally accepted nomenclature for igneous, metamorphic, and sedimentary rocks. Although some proposed classifications are widely used, the nomenclature has not been standardized. Detailed comments on nomenclature for igneous rocks and on application of modifiers to existing rock names can be found in Hansen[6] and Le Maitre.[14]

Mandarino and Back[15] list the correct spellings and chemical formulas for about 4,000 mineral species; this list is periodically updated and corrected. Avoid using colloquial, outdated, nonspecific, and varietal names unless the parent mineral species is referenced. Do not abbreviate names of minerals in text. Nickel and Grice[16] summarize the procedure and criteria for proposing a new mineral species and guidelines for mineral nomenclature according to recommendations of the Commission on New Minerals and Mineral Names of the International Mineralogical Association. Journals such as the *American Mineralogist* and *The Canadian Mineralogist* periodically publish new mineral names with characterizations of the newly designated minerals.

Hansen[6] recommends 4 guidelines for mineral-phase and mineral-component abbreviations to be used as symbols or as subscripts or superscripts.

1) Abbreviations should consist of 2 or 3 letters; capitalize only the first letter of the abbreviation if it is a mineral-phase symbol. Use all lowercase letters for a mineral-component symbol to distinguish it from a mineral-phase symbol.
2) The first letter should be the first letter of the mineral name; the other letters come from the rest of the name but preferably from the consonants.
3) Symbols should not be identical with those of the chemical elements (see Table 16.4 in this manual).
4) A symbol should not form a word likely to be used in scientific writing in any language.

Below are examples of abbreviations formed in accordance with these recommendations. Compilations can be found in Table 3 of Hansen[6] and in Kretz.[17]

Di [the mineral phase "diopside"]
di [the diopside component]
Agt aegirine-augite
Ak åkermanite
Hl halite
Rds rhodochrosite
Ft ferrotschermakite
Usp ulvöspinel

25.6 CRYSTALS

Crystals are distinguished by class, family, and system. A crystal class designates 1 of 32 categories determined by operations such as inversion and rotation (see Table 25.1) or combinations that leave a crystal invariant. A crystal family refers to 1 of 17 plane groups and 230 space groups. A crystal system refers to 1 of 7 categories classified by the unit-cell shape of the crystal's Bravais lattice or by the main symmetry elements of the class (see Table 25.2).

25.6.1 Symmetry, Planes, and Axes

A crystal is a periodic repetition of atoms or a group of atoms at equal intervals throughout the volume of a specimen. A lattice is an arrangement of points in space, where each point has identical surroundings of the same orientation. There are 14 types of 3-dimensional (or space) lattices, which are called Bravais lattices.

Table 25.1 Symbols for basic symmetry operations of a space group in crystallography[a]

Symbol	Description
$\{E/0\}$	Identity operation; no rotation, no translation
$\{C_n/0\}$	n-fold rotation
$\{\sigma/0\}$	Reflection
$\{I/0\}$	Inversion
$\{S_n/0\}$	n-fold rotation followed by a fractional translation in a direction parallel to the plane
$\{E/\tau\}$	Translation
$\{\sigma/\tau/m\}$	Reflection followed by a fractional translation in a direction parallel to the plane
$\{C_n/\tau/m\}$	Rotation followed by a fractional translation parallel to rotation axis

[a] Based on Bennett et al.[18]

Table 25.2 Symbols in crystallography

Symbol	Description
′	Prime applied to the symbol for any movement: the movement is accompanied by change of color
′	Prime applied to the prefix: change of color with translation
1	Not a diad
$\bar{1}$	Pure inversion
2	Diad
$\bar{2}$	Reflection in a plane perpendicular to the axis of rotation
A	Centered in yz plane
a	Glide reflection in the x direction
B	Centered in zx plane
b	Glide reflection in the y direction
C	Centered in xy plane
c	Centered net; glide rotation in the z direction
d	Diamond glide rotation
F	Centered in all 3 planes (xy, yz, zx)
g	Glide reflection; for plane groups (layer patterns) glide reflection in both x and y directions
I	Body-centered
i or I	Inversion
l'	Reflection of line l
l''	Reflection line of l'
m	Reflection lines; in parallel mirror lines, reflections, and glide reflections
n	Glide reflection in a diagonal direction for plane groups (layer patterns)
$\bar{n}$	Rotation through $360°/n$ ($2\pi/n$) combined with inversion
n_r	Rotation of $2\pi/n$ combined with an axial movement of r/n units
(n)	Number in parentheses as a superscript: number of colors in a polychromatic crystal
O	Center of inversion
P	Primitive plane
p	Primitive net
r	Row
σ	Reflection
T	Translation

The Miller indices of a plane are numbers derived from the reciprocals of the intercepts of a plane with the coordinate axes multiplied by the smallest number that will cause all numbers in the set of reciprocals to be integers. There are 3 principal axes. Principal axes are chosen parallel to translation directions, such that the axes are aligned with symmetry elements, and are characteristically labeled a, b, and c. If the intercepts are all equal, the reciprocals of the intercepts are in the ratio 1:1:1. Write the Miller indices of this plane within parentheses and with no spaces between the numbers, (111). When each of the 3 indices is less than 10, do not separate the numbers with commas; when even one of them is greater than 10, include commas to prevent confusion, (11,2,1). Use a bar above the number to indicate a negative intercept, $(11\bar{1})$. This example is read "1, 1, bar 1".

To refer to a group of planes, replace the parentheses with braces (curly brackets); "+" means included within the set.

$\{abc\} = (abc) + (acb) + (bac) + (bca) + (cab) + (cba)$
$\{100\} = (001) + (010) + (100)$

Denote the Miller indices for direction with square brackets, [*abc*]. Denote the set of directions that is equivalent by symmetry with angle brackets, ‹*abc*›.

A crystal is said to be right-handed or left-handed, depending on the direction in which it rotates polarized light. Some crystals are twinned. Japan, Dauphiné, and Brazil are 3 kinds of twinning. References to the faces of one twin are in lowercase italic letters; references to the faces of the other twin also are in lowercase italic letters but with a bar underneath.

$a, \underline{a}$ $\qquad z, \underline{z}$

25.6.2 Rotation, Reflection, Inversion, and Translation

Rotation leaves the coordinate system (either right-hand or left-hand) unchanged, whereas inversion changes a right-hand coordinate system to a left-hand one and vice versa. An integer *n* subscript denotes the number of types of rotation around a fixed point 0 (C_3, C_4, C_2 . . .), but more often the numbers are used alone (3, 4, 2 . . .). For example, D_5 indicates a dihedral crystal with pentad rotation and 5 mirror (reflection) lines.

$D_1 \quad D_2 \quad D_3 \quad D_4 \quad D_5 \quad D_6 \quad \ldots$

In international notation a number indicating rotation is followed by the letter *m* for the mirror lines (e.g., 3*m*). The mirror lines pass through the rotation center. For even-numbered types the lines are divided into 2 sets, those along the arms of the cross and those bisecting the angles so formed, whereas for odd-numbered types, the 2 sets are the same. Therefore, the mirror lines for even-numbered types are denoted by *mm*.

1*m* 2*mm* 3*m* 4*mm* 5*m* 6*mm* . . .

Translation means moving from one point in an object to some other point in the same object so that the environment around the 2 points is exactly the same. (For example, a point on the lower right side of a square in a cyclone fence is identical to all the other such points on the lower right sides of all the other squares in a cyclone fence.)

25.6.3 International (Hermann–Mauguin) Symbols

The Hermann–Mauguin symbol,[19] both the short and the full versions, consists of 2 parts: a letter indicating the centering type of the conventional cell and a set of characters indicating symmetry elements of the space group (modified point-group symbol). Use lowercase italic letters for 2 dimensions (nets) and capital letters for 3 dimensions (lattices).

Full	Short
$C1m1$ or $C11m$	Cm
$P2_1/n2_1/m2_1/a$	$Pnma$
$P6_3/m\ 2/m\ 2/c$	$P6_3/mmc$

25.6.4 Patterson Symmetry

Patterson symmetry gives the space group of the Patterson function $P(x,y,z)$. The Patterson function represents the convolution of a structure with its inverse or the pair-correlation function of a structure.

25.6.5 Laue Class and Cell

The space-group determination starts with the assignment of the Laue class to the weighted reciprocal lattice and the determination of the cell geometry. The Laue class determines the crystal system.

The axial system should be thought of as right-handed. For crystal systems with symmetry higher than orthorhombic, the symmetry directions and the convention that the cell should be taken as small as possible determine the axes uniquely. Three directions are fixed by symmetry for orthorhombic crystals, but any of the 3 may be called "a", "b", or "c". The convention is for $c < a < b$. Monoclinic crystals have one unique direction. If there are no special reasons to decide otherwise, the standard choice "b" is preferred. For triclinic crystals usually the reduced cell is taken, but the labeling of the axes remains a matter of choice.

25.7 SOILS

Soil information presented in technical publications may include pedon (profile) descriptions, laboratory data, soil classifications, and information from soil surveys. This section explains each of these kinds of information, how each relates to the others, and how each is presented.

25.7.1 Pedon Descriptions

The starting point for many soil investigations is a pedon description. A pedon is a 3-dimensional soil body about 1 m^2 at the surface and extending to the bottom of the soil, usually 1.5 to 2 m deep. (In contrast, a soil profile is 2-dimensional and cannot be sampled.) To prepare a pedon description, the pedon is divided into horizons that appear homogeneous. Typically, about 6 to 8 horizons are identified, fewer for shallow soils, more for deeper soils and detailed studies. The depth, horizon designation, and other morphological properties are recorded for each horizon. Pedon descriptions usually are presented in narrative form in soil surveys and in tabular form in journal articles, as illustrated in the following example for part of a pedon. Narrative pedon descriptions are generally written as a series of short descriptive phrases separated by semicolons. A narrative description includes more information than a table. In either form, authors present what they believe is most important. Guidelines for preparing pedon descriptions are available in the *Soil Survey Manual*[20] and the *Field Book for Describing and Sampling Soils*.[21] (See Sections 25.7.1.1 and 25.7.1.2 for descriptions of horizons and morphological properties and the abbreviations used in the following examples.)

Narrative description for part of pedon:

Ap—0 to 25 cm; dark grayish brown (10YR 4/2) silt loam, pale brown (10YR 6/3) dry; moderate fine granular structure; friable; many fine and very fine roots; neutral; abrupt smooth boundary. (15 to 25 cm thick)

E—25 to 33 cm; grayish brown (10YR 5/2) silt loam; weak fine subangular blocky structure; friable; common fine and very fine roots; common medium prominent yellowish brown (10YR 5/6) masses of iron accumulation in the matrix; moderately acid; clear smooth boundary. (0 to 15 cm thick)

Bt1—33 to 53 cm; yellowish brown (10YR 5/4) silty clay loam; moderate medium subangular blocky structure; firm; few fine and common very fine roots; common distinct dark grayish brown (10YR 4/2) clay films on faces of peds; many medium distinct light brownish gray (10YR 6/2) iron depletions in the matrix; moderately acid; clear wavy boundary. (10 to 40 cm thick)

Tabular pedon description:

Horizon	Depth (cm)	Color[a] (moist)	Texture[b]	Structure[c]	Consistence[d]	Ped coats
Ap	0–25	10YR 4/2	SiL	2 f gr	fr	
E	25–33	10YR 5/2 c 10YR 5/6	SiL	1 f sbk	fr	
Bt1	33–53	10YR 5/4 m 10YR 6/2	SiCL	2 m sbk	fi	clay, 10YR 4/2

[a] Munsell designation; abundance of mottles: c = common, m = many.
[b] C = clay, L = loam, Si = silt(y).
[c] Grade: 1 = weak, 2 = moderate; size: f = fine, m = medium; shape: gr = granular, sbk = subangular blocky.
[d] Consistence: fi = firm, fr = friable.

25.7.1.1 Horizon Designation

Horizon designations are shorthand labels that indicate the major properties and relations of the horizons. The sequence of horizons, from the surface down to 1.5 to 2 m, might be written as follows:

A, E, EB, 2Bt1, 2Bt2, 2Cd

The designation for an individual horizon designation may have up to 4 main parts:

1) A numeric prefix indicates the kind of parent material, the geological material from which the soil formed. The numeral 1, representing the upper parent material, is implied, whereas subsequent horizons are represented by numbers. Thus, in the horizon sequence above, the A, E, and EB horizons formed in one parent material, perhaps loess, and the 2Bt1 and lower horizons formed in a second parent material, perhaps glacial till. This sample has horizons from only 2 parent materials.
2) Master horizons, the most obvious horizons in the sample, are represented by single let-

Table 25.3 Brief definitions of master horizons

Horizon type	Description
O	Horizons that consist mainly of organic material, not mineral material like other horizons
A	Mineral horizons at the surface or below an O horizon that are relatively high in organic matter
E	Mineral horizons from which silicate clay, iron, aluminum, or some combination of these have been leached, usually to lower horizons
B	Horizons below an A, E, or O horizon that show one or more of the following:
	1) accumulation (moved in from above) of silicate clay, iron, aluminum, humus, or silica, alone or in combination
	2) removal of carbonates
	3) concentration of sesquioxides (iron and aluminum) because other material, such as silica, weathered out faster
	4) coatings of sesquioxides that make the horizon conspicuously darker, higher in chroma, or redder in hue than overlying and underlying horizons
	5) formation of granular, blocky, or prismatic structure
	6) brittleness
C	Horizons or layers, excluding hard bedrock, that are little affected by soil-forming processes and lack properties of O, A, E, or B horizons; most are mineral layers
L	Limnic soil materials: organic and inorganic materials that were deposited in water, including coprogenous earth (sedimentary peat), diatomaceous earth, and marl
R	Hard bedrock
W	A layer of ice within a soil, thus permafrost

ters. The letters A, E, B, and C in the example above denote master horizons (see Table 25.3). Less obvious horizons that are transitional between master horizons are assigned 2 letters (e.g., EB). Use uppercase letters for master horizon designations.

3) A horizon designation may have a suffix indicating major processes or properties. In the 2Bt1 and 2Bt2 horizons, the suffix t (for the German "ton", meaning clay) indicates that clay has moved downward in the pedon and has become immobilized in the Bt horizon. Use lowercase letters for these subordinate distinctions.
4) A numeric suffix indicates a minor subdivision. In the example above, the thick 2Bt horizon was split into 2 subhorizons (2Bt1 and 2Bt2) for sampling the soil.

25.7.1.2 Morphological Properties

The main morphological properties of soils are color, texture, structure, consistence, and ped coats. They are illustrated in the examples of pedon descriptions in Section 25.7.1.

Color is represented using designators described in the Munsell system.[22] In the designation 10YR 5/2, 10YR is the hue, 5 is the value, and 2 is the chroma. Descriptive words may also be used such as "grayish brown". In soil surveys, soil color is traditionally written with the word description of the color appearing first followed by the Munsell designation in parentheses.[23]

grayish brown (10YR 5/2) (10YR 5/6 c)

In many scientific journals only the Munsell designation is given, without parentheses. A description in parentheses may follow the Munsell designation the first time it is used in the abstract or text.[23]

10YR 5/2 (grayish brown) 10YR 5/6 c

Often, several colors are described. The dominant color is listed first (10YR 5/2), followed by minor colors (10YR 5/6). The abundance of mottles follows the color designation and is usually abbreviated as a lowercase letter (e.g., "c" for common, "f" for few, or "m" for many).

grayish brown (10YR 5/2) (10YR 5/6 c) *or* 10YR 5/2 (grayish brown) 10YR 5/6 c

Texture describes the amount and size of mineral particles in the soil. Texture classes are based on defined upper and lower limits on the percentages by weight of each component particle (sand, silt, or loam) in the soil that is less than 2 mm in effective diameter, as represented on a texture triangle. Examples of texture triangles can be found in the *Soil Survey Manual*[20] and the *Field Book for Describing and Sampling Soils*.[21] These texture classes are also used to describe sediment composition (see Section 25.10.2).

clay sand loam silty clay loam loamy sand

Texture modifiers are terms used to denote the presence of a condition or component other than sand, silt, or clay. A texture modifier can be used when the content of the component exceeds 15% by volume.

gravelly loam mucky loam

Structure refers to how soil particles are grouped together to form larger units, or peds, or, conversely, how the soil breaks into pieces. Three features of structure are described in a pedon description: grade or distinctness, size, and type or shape. Grade is represented in a pedon description by a number, size is represented by a lowercase letter, and type or shape is represented by lowercase letters. The following example is a description of a soil structure that has a moderate grade (2), fine-sized peds (granules) (f), and a granular shape (gr).

2 f gr

Consistence is a measure of soil strength and is described using terms such as loose, soft, friable, firm, or hard. Abbreviations for consistence are usually formed from the first letter or letters of the descriptive term and are written in lowercase letters (e.g., "l" for loose, "fr" for friable).

Coatings on peds, such as clay, organic matter, or iron oxides, reflect processes of soil formation. The Bt1 horizon in the example in Section 25.7.1 has dark grayish-brown clay coats on subangular blocky peds. Clay coats or films formed when clay moved down the profile and accumulated on ped surfaces in the Bt horizon.

Other properties that may be described include roots, pores, pH, effervescence (tested using hydrochloric acid, an indicator of carbonate minerals), cracks, rock fragments, and concentrations of iron or other materials.

25.7.1.3 Use of Pedon Descriptions

Pedon descriptions, the starting point for many soil investigations, have several applications:

1) as the basis for classifying the soil in a natural system
2) to divide pedons into units that can be sampled for laboratory analysis

3) as the basis for defining map units for soil surveys
4) to group individual soils into classes for specific uses of the soils such as crop production and engineering properties

25.7.2 Laboratory Data

Generally 2 kinds of soil samples are collected from soil horizons: bulk samples that will be dried, crushed, and sieved and natural-fabric (undisturbed) samples that will be used to measure a variety of physical properties and to study soil morphology under a microscope. Standard practice is to conduct laboratory analyses on bulk samples that have passed through a 2-mm sieve (material of less than 2 mm) and to report on samples that have been dried at 105 °C until the sample reaches a constant mass (oven dry). Laboratory data are used to classify a soil, to study soil formation processes, and to help decide how best to use a soil.

Pedon descriptions and laboratory data for many soils are available online through the National Cooperative Soil Survey Soil Characterization Data site.[24]

25.7.3 Soil Classification

Soils are classified, or placed into groups with similar properties, to organize knowledge about soils and to enable understanding of the relations among soils. Natural classification systems group soils according to natural relations, such as how soils formed. Technical classification systems group soils according to a specific purpose (crop production, engineering applications, erodibility, etc.). Generally, several technical classifications can be derived from a natural classification. Many scientific journals require that the classification of soils be given, just as for plants and animals.

Many countries have soil classification systems for soils within their boundaries—for example, the *Canadian System of Soil Classification*.[25] Two natural systems are designed to include all soils of the world, *Soil Taxonomy*[26] and the *World Reference Base for Soil Resources*.[27]

25.7.3.1 Soil Taxonomy

Soil Taxonomy[26] is a comprehensive soil classification system for the United States that has the potential to include all soils of the world. It is supported by the US National Cooperative Soil Survey, a coalition of the US Natural Resources Conservation Service, state universities, and other federal and state agencies involved with soil resources.

To use *Soil Taxonomy*, certain diagnostic horizons and materials must be identified. Most diagnostic horizons are subsoil horizons, but some are surface horizons. Diagnostic horizons are formed as a result of broad-scale soil-forming processes. For example, the argillic horizon formed as a result of clay migrating from upper horizons to lower horizons in the more humid, originally forested areas of the world. Diagnostic materials have distinctive properties that are often related to the origin of the material, such as volcanic activity.

25.7.3.1.1 CATEGORIES

Soil Taxonomy[26] has 6 classification categories or levels within the system. From highest to lowest, they are order, suborder, great group, subgroup, family, and series.

Most of the 12 soil orders are defined on the basis of diagnostic horizons or features that reflect soil-forming processes (see Table 25.4). In several of the soil orders, suborders are based on soil moisture regime, whereas in other orders, different criteria are used. For example, suborders in the Entisol order are based on the reason distinctive soil horizons have not formed, such as the soil is too sandy, too stony, or too steep. Great groups are based on the presence or absence of certain kinds of soil horizons and other soil properties. Subgroups indicate if the soil is in the central range of its great group or if it has marginal properties that tend toward those of another great group. Families have practical significance; their definitions are based on soil texture, mineralogy, temperature regime, and a few other properties. Soil series are defined by many properties and are named for the communities where they were first identified. The map units (see Section 25.3.2) in most published soil surveys are named after soil series; thus, series names are familiar at the local level. Series names are used to identify the entities reported in most scientific journals. Typical pedon descriptions for all series, their classification at the family level, and other details are given in the official series descriptions.[28]

25.7.3.1.2 CLASSIFICATION OF SOILS

To determine the taxonomic group of a soil, a morphological description and often laboratory data are required. Classification involves 6 steps:

1) Identify diagnostic soil horizons and diagnostic properties
2) Determine the soil order following a key to soil orders (e.g., Chapter 8 of *Soil Taxonomy*[26])
3) Identify the soil moisture regime

Table 25.4 Brief descriptions of the orders of *Soil Taxonomy*[a]

Soil orders	Description
Alfisols	Soils with a subsoil accumulation of silicate clay that are moderately weathered (have a high base saturation)
Andisols	Soils formed from volcanic materials
Aridisols	Soils of arid environments
Entisols	Very weakly developed soils, including many sandy soils
Gelisols	Soils with permafrost
Histosols	Soils formed from organic materials
Inceptisols	Weakly developed soils, excluding sandy soils
Mollisols	Soils with thick, dark surface horizons that are high in organic matter content
Oxisols	Very highly weathered soils of tropical areas that are high in iron- and aluminum-oxide minerals
Spodosols	Soils with a subsoil accumulation of aluminum, organic matter, and usually iron
Ultisols	Soils with a subsoil accumulation of clay that are highly weathered (have a low base saturation)
Vertisols	Soils that undergo much shrinking and swelling

[a] Based on Departmentof Agriculture (US), Natural Resources Conservation Service, Soil Survey Staff.[26]

4) Determine the subgroup following a key for one of the orders (e.g., Chapters 9 to 20 of *Soil Taxonomy*[26] or *Keys to Soil Taxonomy*[29])
5) Determine the family (Chapter 21 of *Soil Taxonomy*[26])
6) Match the description and data with official series descriptions to determine the series (this usually requires input from local experts)

25.7.3.1.3 NOMENCLATURE

Soil Taxonomy[26] uses nomenclature based largely on Greek and Latin formative elements. For example, names for wet soils contain the formative element "aqu" (from the Latin "aqua" for water). Soils with high base saturation may contain the element "eutr" (from the Greek "eutrophos" for fertile), also used in "eutrophication" to describe lakes when they become enriched with plant nutrients. Because of these derivations, the name of a class suggests some of the soil's properties. The form of the name also indicates the category within *Soil Taxonomy*.

The main rules for nomenclature are as follows:

1) Orders (see Table 25.4) have names with the suffix "sol" (e.g., Alfisol, Mollisol). In the example that follows, "alf" is the formative element of the Alfisol order. The soil has a subsoil clay accumulation and high base saturation. Capitalize the initial letter of order names.
2) Suborder names have 2 syllables, the last being the formative element of the order to which it belongs. For example, Aqualfs are wet Alfisols, and Udolls are Mollisols of humid climates. In the example that follows, the infix "ud" refers to a humid climate, and the soil is in the udic (moist but not waterlogged) soil moisture regime. Capitalize the initial letter of suborder names.
3) Great group names consist of a prefix added to a suborder name (e.g., Ochraqualfs, Argiudolls). In the example that follows, the prefix "Fragi" (for "fragilis") in Fragiudalfs means brittle. The soil has a fragipan, a subsurface layer that resists root penetration and water movement. Capitalize the initial letter of great group names.
4) Subgroup names are formed by adding an adjective to a great group name (e.g., Aeric Ochraqualfs, Typic Argiudolls). In the example that follows, the prefix "aqu" (for water) in Aquic Fragiudalfs means the soil is on the moist side of the udic moisture regime. Capitalize the initial letter of the subgroup adjective.
5) Family names are designated by modifiers that describe particle-size distribution, mineralogy, clay activity, and temperature. A fine-silty subsoil is high in silt and moderately high in clay. In a mixed soil, no single mineral dominates the mineralogy of the whole soil. A superactive soil has highly active clay (high ratio of cation exchange capacity to clay). Mesic soils are from a mesic, or medium, temperature regime. Family name modifiers are not capitalized except at the beginning of a sentence.
6) Series are defined by a typical pedon (central concept) and a range of characteristics (boundary conditions) and are named for a locality (e.g., Cincinnati).

These rules are illustrated in the following example of a name for the family level. For some scientific publications, a name at a higher taxonomic level may be appropriate, such as Fragiudalfs, a great group class.

fine-silty, mixed, superactive, mesic Aquic Fragiudalfs

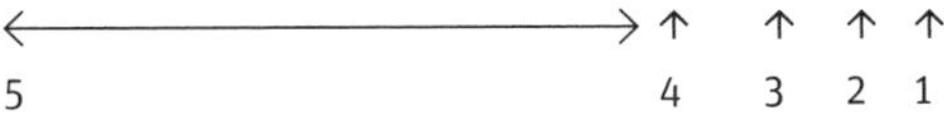

5 4 3 2 1

25.7.3.2 World Reference Base

The *World Reference Base for Soil Resources*[27] is an international system that grew from the legend of the Soil Map of the World, completed in 1974. The World Reference Base is not a national soil classification system like *Soil Taxonomy*[26] but rather a means of correlating many national systems. It also defines diagnostic horizons and materials, many of them similar to those of *Soil Taxonomy*. Two main categories in the World Reference Base are an upper level of 30 reference soil groups, similar to the order or suborder level of *Soil Taxonomy*, and a lower level of soil units defined and named by adding an adjective to the reference soil group name. For example, the Luvisols is a reference soil group similar to the Alfisols of *Soil Taxonomy*. The adjective Fragic may be added to that name (e.g., Fragic Luvisols) to describe the Luvisols with fragic horizons. They are essentially the Fragiudalfs of *Soil Taxonomy*. The World Reference Base defines 121 formative elements, but only a subset of them is used with each soil group. The subset varies from about 10 to 30 for the various soil groups.

25.7.4 Soil Surveys

Soil surveys comprise maps, text, and tables about soils and how to use them. They include pedon descriptions and soil names, as described above. Detailed soil surveys are usually published at a scale of about 1:12,000 to 1:24,000 on aerial photo backgrounds. The delineated units or map units are generally named for soil series. Map units always include soil series other than the one used in the map unit name. For example, the soil map unit below is named for the major soil series in it, Miami, but it may include areas of the Crosby series, the Russell series, and other series that are too small to be mapped.

Miami silt loam, 6% to 12% slopes

For some papers, the location, in a coordinate system used in the published soil survey, and the name of a soil mapping unit may be given.

Before about 2000, most soil surveys were published for legal entities such as counties. Copies of these publications are available in county agriculture offices, in many libraries, or through state universities. Since then, soil surveys in the United States have been published for major land resource areas, natural geographic areas characterized by similar soils and other similar characteristics. Most other countries have similar resource maps.

Traditionally soil surveys were published in print form. Now electronic versions of the maps, text, and tables for many areas are available and can be used with geographic information systems. In the United States the detailed soil surveys mentioned above constitute the Soil Survey Geographic Database (SSURGO).[30] The US General Soil Map (STATSGO2) consists of more general "soil association units"; it supersedes the State Soil Geographic (STATSGO) Database published in 1994.[31]

25.7.5 Soil Interpretations

Soil interpretations, predictions of how a soil might respond to various uses, are presented in soil surveys and other documents. They are based on research and experience with a few soils and are extended to similar soils through soil descriptions, laboratory data, soil classification systems, and soil surveys. The soil uses may be for agricultural production (e.g., land use capability classes, crop yield predictions, tillage systems, conservation practices); construction of buildings, highways, dams, and other structures (e.g., geotechnical soil classifications such as the Unified or the American Association of State Highway and Transportation Officials [AASHTO] systems); forestry (e.g., productivity, trees to plant); onsite sewage disposal (e.g., septic systems, lagoons); recreation (e.g., playgrounds, paths, camp areas); wildlife habitat; and other uses.

25.7.6 Locations

Give the locations of sites in a commonly used coordinate system such as latitude–longitude, Universal Transverse Mercator, state plane coordinate, or US rectangular survey (township-range-section) (see Section 14.2). If the US rectangular survey is used, give locations of specific points relative to their distances from 1 of the 4 corners or from the center of a section (1 mile × 1 mile).

100 m north and 56 m east of the southwest corner of section 4, T12N, R9E, Boise Meridian

The site location in the example above is more specific than an area description. (See Section 25.9 for area descriptions of tracts of land.)

25.7.7 Terminology and Additional Information

Definitions of soil science terms can be found in glossaries.[32,33] More information about soil science and many references are included in encyclopedias and handbooks of soil science.[34,35] Use of soil parent material (geologic material) and landform names should follow Bates and Jackson.[32]

25.7.8 Classification of Land Capability

The US Natural Resources Conservation Service recognizes 8 classes of land capability, numbered I to VIII.[26] Soils with the greatest capabilities for response to management and the fewest limitations on how they can be used (e.g., intensive crop farming, pasture, range, wildlife preserve, well drained, level, fertile) are in Class I. Those with the fewest capabilities and greatest limitations (e.g., not suitable for commercial plant production but restricted to recreation, wildlife, water supply, or aesthetic uses) are in Class VIII. The land classification system in Canada is similar.[36]

25.8 AQUIFERS

An aquifer may encompass thousands of square hectares. The term is difficult to define; Bates and Jackson[32] define it as "a body of rock that is sufficiently permeable to conduct ground water and to yield economically significant quantities of water to wells and springs". The following terms are not capitalized even when named: "aquifer", "aquifer system", "zone", and "confining unit". Terms such as "sand and gravel aquifer" and "limestone aquifer" are neither capitalized nor hyphenated. Do not capitalize adjectival modifiers and relative-position terms unless they are part of the formal geographic name.

Mississippi River alluvial aquifer   Upper Klamath aquifer   Upper Canada aquifer

Do not use quotation marks for aquifer names unless the term is a misnomer. Although the term "aquifer" may be imprecise, it is widely accepted and used; coining new terms will only add to the confusion. Do not use terms intended to be synonymous with "aquifer" or "aquifer system". Do not use "aquigroup" in place of "aquifer system". Distinguish hydrologic and geologic terms.

water from the Madison aquifer *not* Madison water
wells completed in Madison Limestone *or* wells completed in Madison aquifer *but not* Madison wells

25.9 TRACTS OF LAND

Locations of tracts of US public land are designated by rows of townships (each 6 miles square) and columns of ranges that make up the grid system used to survey all land west of the Ohio River, except for Texas, in relation to various east–west baselines and named north–south meridians. Note the following formats.

SE1/4 NW1/4 sec 4, T 12 S, R 15 E, of the Boise Meridian
[designates the southeast quarter of the northwest quarter of section 4, which is 12 townships south and 15 ranges east of the Boise Meridian]
N1/2 sec 20, T 7 N, R 2 W, sixth principal meridian
[designates the northern half of section 20, which is 7 townships north and 2 ranges west of the sixth principal meridian]

Omit periods from abbreviations. Do not use a space between the compass direction and the fraction. Use "half" or "quarter", not "one-half" or "one-quarter", if fractions are spelled out in land descriptions.

south half of T 47 N, R 64 E

The plural of "T" (for "township") is "Tps", and the plural of "R" (for "range") is "Rs".

Tps 9, 10, 11, and 12 S, Rs 12 and 13 W

If breaking a land-description symbol group at the end of a line is unavoidable, break it after a fraction and without a hyphen. For example, break the land description NE1/4 SE1/4 sec 4 at the end of a line as follows:

NE1/4 SE1/4   sec 4

For more information on writing public land descriptions, see *Specifications for Descriptions of Tracts of Land for Use in Land Orders and Proclamations*.[37] Land in Canada is divided using a similar system called the Canadian Lands Survey System; the Dominion Land Survey System is used for western Canada.[38]

25.10 SEDIMENT

Sediment is material deposited by water, wind, or glaciers. Sediment consists of rock and mineral fragments, organic material (shell, bone, or other plant and animal material), loess, desert sand, volcanic ash, precipitates from water, and/or settled dust and small atmospheric or cosmic particles. The study of sediments can provide information about the origin, transport, and age of material in river and lake beds and ocean basins. Sediments are classified according to their source, texture, and color.

25.10.1 Source

Lithogenous and terrigenous sediments consist of materials originating from weathered rocks; cosmogenous sediments consist of materials deposited from the atmosphere or space; hydrogenous sediments result from chemical reactions in water; and biogenous sediments consist largely of plant or animal remains. Sediment composed of more than 30% biogenous material is called an ooze.

25.10.2 Texture

Sediment texture is classified by grain size and sorting. The Wentworth and phi scales (see Table 25.5) are used to standardize grade classifications for particles based on maximum particle diameter.[39] Modifications of the Wentworth scale are also commonly used. In the phi scale, particle size, d (mm), is converted from the geometric Wentworth scale to an arithmetic scale: $\Phi = -\log_2 d$, in which Φ is dimensionless.

Sphericity is a measure of particle shape. It is the ratio of the shape of the particle to the shape of a perfect sphere. The lower the sphericity, the less round the particle is. Sphericity is a dimensionless measurement because it is a ratio.

Sorting is a measure of the similarity of particle sizes within sediment. The degree of sorting is an indicator of how sediment was transported. Uniformly sorted sediments (i.e., a mix of similarly sized particles) are generally transported by wind or waves, whereas

Table 25.5 Sediment classification according to grain size[a]

Size class	Diameter (mm)	Phi (Φ)
Boulder	>256	less than -8
Cobble	64 to 256	-8 to -6
Pebble	4 to 64	-6 to -2
Gravel	2 to 4	-2 to -1
Sand		
Very coarse sand	1 to 2	-1 to 0
Coarse sand	0.5 to 1	0 to 1
Medium sand	0.25 to 0.5	1 to 2
Fine sand	0.125 to 0.25	2 to 3
Very fine sand	0.0625 to 0.125	3 to 4
Silt		
Coarse silt	0.0310 to 0.0625	4 to 5
Medium silt	0.0156 to 0.0310	5
Fine silt	0.0078 to 0.0156	6
Very fine silt	0.0039 to 0.0078	7
Clay	<0.0039	≥8

[a] Adapted from Wentworth.[39]

poorly sorted sediments (i.e., a random mix of different-sized particles) may have been transported by sea ice, floods, or strong currents.

Sediments can be classified according to the relative composition of different-sized particles in the sediment using the nomenclature in Shepard.[40] A triangle similar to that used for categorizing soil texture (see Section 25.7.1.2) is used for categorizing sediments according to the relative percentages of sand, silt, and clay particles in the sediment.

25.11 WATER

25.11.1 Descriptive Terms

Descriptive terms for water usually contain the word "water" and a modifier. The same term is often used as both a noun and adjective form, but the construction of noun and adjective forms may vary. Water types generally are represented as 1- or 2-word constructions, and the modifier in 2-word forms is separated from the word "water" by either a hyphen or a space. Because the rules for forming these words are not consistent, consult a dictionary for recommended spelling and punctuation.

freshwater [noun or adjective]
deep water [noun] *but* deepwater [adjective]
open water [noun] *but* open-water [adjective]

25.11.2 Physical Properties of Seawater

This section includes only recommendations adopted by the International Association for the Physical Sciences of the Ocean (IAPSO).[41]

25.11.2.1 Salinity

The interpretation of the concept of salinity varies among oceanographers and limnologists. Salinity is defined, measured, and reported differently depending on the chemical composition of the water being studied.

Salinity, S, is a measure of the total amount of dissolved material in water. Salinity is also defined as the total ion concentration in water.[42,43] Absolute salinity, S_A, is the total mass of solid material dissolved in a sample of water divided by the mass of the sample when all of the carbonate has been converted into oxide, all of the bromine and iodine have been replaced by chlorine, and all of the organic matter has been completely oxidized.[44] Absolute salinity is difficult to measure directly. A variety of indirect measures of salinity are commonly used, and the measures vary according to the chemical composition of the water being examined.

Oceans have stable ionic compositions, but coastal and inland waters and salt lakes can be highly variable. Ocean salinity has been measured indirectly by determining chlorinity, the amount of chlorine ion (plus the chlorine equivalent of the bromine and iodine), by titration with silver nitrate.[44] The chlorinity of seawater, Cl, represents the mass of pure silver necessary to precipitate the halides contained in 0.328 523 4 kg of seawater. The following equation describes the relation between salinity and chlorinity:

$$S = 1.806\ 55 \times \mathrm{Cl}$$

Chlorinity is not an accurate measure of salinity in dilute seawater, inland waters, or salt lakes because the ratio of the constituent ions in those waters varies. Other measures of the colligative properties of water, such as osmotic pressure, boiling point, freezing point, density, conductivity, or total dissolved solids, are effective measures of the number of molecules in solution and generally produce salinity values that are precise enough for most limnological purposes.

Oceanographic disciplines are changing to a more accurate method of salinity measurement. Instead of using chlorinity, the electrical conductivity of seawater is being used to determine salinity.

25.11.2.2 Practical Salinity

Improved technology enables salinity to be determined more quickly and accurately through measurement of the electrical conductivity of seawater. Both salinity and chlorinity have been replaced by a single unit called practical salinity, which is based on the electrical conductivity of seawater. The symbol for salinity, *S*, is also used for practical salinity. Practical salinity is "the ratio K_{15} of the electrical conductivity of the seawater sample at the temperature of 15 °C and the pressure of one standard atmosphere, to that of a potassium chloride (KCl) solution, in which the mass fraction of KCl is 32.4356 × 10^{-3}, at the same temperature and pressure".[41] IAPSO recommends the use of practical salinity for the measurement of salinity of seawater.[41]

The term for the quantity should be written in full, "practical salinity", and not shortened to "salinity", to avoid confusion with absolute salinity or salinity measured by means other than electrical conductivity.

25.11.2.3 Salinity Units

Salinity and chlorinity have been expressed in units of parts per thousand (ppt or ‰), parts per million (ppm or ppM), or parts per billion (ppb). IAPSO recommends that these symbols be abandoned and replaced by a factor of 10 raised to the corresponding power (10^{-3} for ppt and ‰, 10^{-6} for ppm and ppM, and 10^{-9} for ppb).[41]

Within some disciplines salinity is reported as either a mass-to-mass ratio (g/kg) or a mass-to-volume ratio (g/L). If the use of a dimensionless unit (e.g., 10^{-3}) creates confusion as to which ratio is being reported, use SI units (g/kg or g/L) for clarification instead of non-SI units, such as parts per thousand, parts per million, or parts per billion.

Because practical salinity is a ratio of 2 electrical conductivities, it is a dimensionless unit. The practice of using the abbreviation PSU or psu to represent salinity units measured on the practical salinity scale is discouraged.[45]

> Salinity decreased to a practical salinity of 26 near the mouth of the estuary.
> *not* Salinity decreased to 26 PSU near the mouth of the estuary.

25.11.2.4 Density

Density of water represents the quantity mass divided by volume (kg/m³). Density varies with salinity, temperature, and pressure. Density is represented with the Greek letter rho (ρ) followed by the 3 parameters in parentheses (*S*, *t*, *p*) in the compulsory order:

salinity, temperature, and pressure. When values are used, the corresponding units may be indicated. Symbols and values are separated by commas. Avoid using subscripts.

$\rho(S, t, p)$ *not* $\rho S, t, p$ ρ(34.85, 3.17 °C, 17.20 MPa) *not* $\rho_{(34.85,\ 3.17\ °C,\ 17.20\ MPa)}$

Parameters may also be represented by a mixture of symbols and values.

$\rho(S$, 3.17 °C, 17.20 MPa)

Recommended units are degrees Celsius for temperature and megapascals for pressure. Practical salinity is dimensionless (see Section 25.11.2.3). When the values are written in the compulsory order and expressed in the recommended units, the unit symbol may be omitted within the parenthetic statement.

ρ(34.85, 3.17, 17.20)

The notations in the examples in this section are not for pure water, for which salinity would be zero. These rules also apply to quantities other than density that are functions of the same 3 parameters.

Density in the open ocean ranges from about 1,021 kg/m^3 at the surface to about 1,070 kg/m^3 at 10,000 m.[44] Density measurements in oceanography are represented by the Greek letter gamma (γ), and usually only the last 4 digits of the density measurement are written, expressed to 2 or 3 decimal places.

$\gamma = \rho - 1{,}000$

Thus, if ρ = 1,035.25 kg/m^3, then γ = 35.25 kg/m^3. Another convention is to use the Greek letter sigma (σ) for the dimensionless unit, as $\sigma(S, t, p)$, for reporting measurements of density. However, γ is preferred over $\sigma(S, t, p)$.[41]

The pressure effect on density can be ignored in many applications (e.g., for water at the sea surface). When the pressure is zero, seawater density is commonly represented by σ_t, pronounced "sigma tee". The use of this non-SI unit is strongly discouraged by IAPSO.[41] However, it is still widely used in oceanography; see Section 25.13 on use of non-SI units.

25.11.3 Currents and Stream Flow

Currents are generally horizontal movements of water. Currents may be temporary, small in scale, and variable, or permanent, large in scale, and continuously circulating. Currents are described by the velocity and direction of water flow, and current flow is expressed as a rate.

Currents are often depicted in vector diagrams or stick plots. In vector diagrams, currents are represented by arrows that point in the direction of the current flow (see Figure 25.1). The length of the arrow is proportional to the current speed.

In stick plots, lines originate from a horizontal axis (see Figure 25.2). The orientation of the line in relation to the axis indicates current direction. By convention, a line pointing to the top of the plot is pointing north and a line pointing to the right side of the plot is pointing east. The position of the line along the horizontal axis indicates the time when the current measurement was made. The length of the line is proportional to current speed. Accurate determination of current speed is made from the height of the stick as if it were parallel to the vertical or *y*-axis, not by drawing a horizontal line from the end of the angled stick over to the *y*-axis.

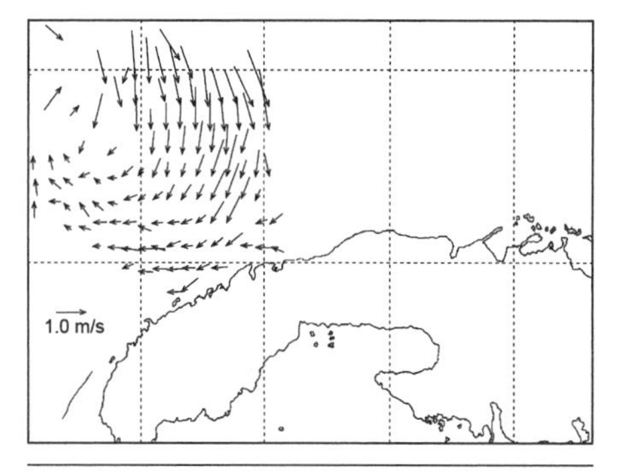

Figure 25.1 Example of a vector diagram.

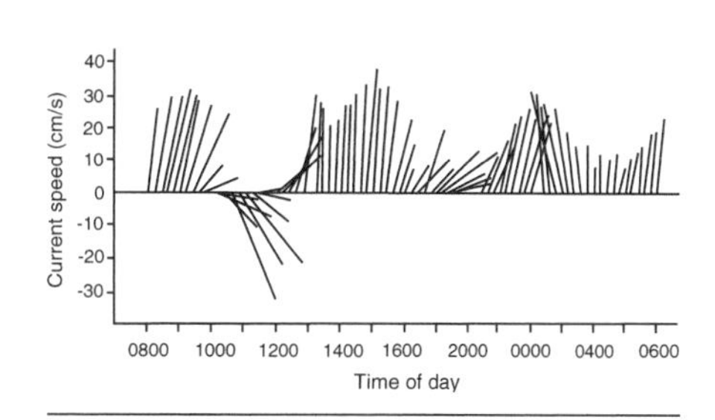

Figure 25.2 Example of a stick plot.

Vector diagrams and stick plots are also used to depict the direction and speed of wind, but a different convention is used for describing wind direction (see Section 25.12.2).

25.11.3.1 Names of Currents

Large-scale, permanent oceanic or coastal currents are generally named for the areas in which they occur. Capitalize current names.

South Equatorial Current   Benguela Current   Alaska Gyre

Currents off the coast of Japan are commonly referred to by their Japanese names, with or without the word "current" in the name. Both forms are acceptable.

Kuroshio *or* Kuroshio Current *or* Japan Current   Oyashio *or* Oyashio Current

The names of other well-known currents are commonly used without the word "Current" in the name. Both forms are acceptable.

Gulf Stream *or* Gulf Stream Current

25.11.3.2 Current Direction

The direction of a current is the direction in which the water is flowing. The convention for naming current direction is the opposite of the convention for naming wind direction described in Section 25.12.2.

A northerly ocean current flows to the north. *but* A northerly wind blows from the north.

Currents may circulate in clockwise or counterclockwise (anticlockwise) directions. These circulations may also be referred to as cyclonic or anticyclonic. Cyclonic currents are counterclockwise in the northern hemisphere and clockwise in the southern hemisphere, whereas anticyclonic currents are clockwise in the northern hemisphere and counterclockwise in the southern hemisphere.

25.11.3.3 Stream Flow

Stream flow is the volume of water moving past a location in a given period or the volume of water a stream or river discharges over a given period. Stream flow is also expressed as a rate, commonly in cubic meters per second (m^3/s) or megaliters per day (ML/d).

25.11.4 Tides

25.11.4.1 Datum

Tides are the periodic and alternating rise and fall of the surface of coastal ocean and marine waters resulting from gravitational forces of the sun and the moon on the earth. The height of the ocean surface at various stages of a tide is measured relative to a vertical datum. In this context, a datum is a base elevation from which to measure heights and depths. The datum is called a tidal datum when it is defined in terms of a phase of tide. The plural form of datum, when used to describe more than one base elevation, is "datums" and not the more common plural "data" used to describe multiple units of information.

The National Tidal Datum Convention of 1980[46] established a continuous tidal datum system for all marine waters of the United States, its territories, the Commonwealth of Puerto Rico, and the Trust Territory of the Pacific Islands. It provided a uniform tidal datum system independent of whether tides were diurnal or semidiurnal and established mean lower low water as the datum for nautical charts for both the Pacific and Atlantic coasts.

Other tidal datums are in common use. They are typically abbreviated to the first letter of each word in the datum name, in uppercase letters.

mean lower low water (MLLW)
mean higher high water (MHHW)
mean low water (MLW)
mean sea level (MSL)

Tidal datums are local, and no international standard exists. Canadian coastal charts use the tidal datum lower low water, large tide (LLWLT), which is the average of the lowest low waters, one from each year of predictions during the tidal datum epoch, a period of 19 years. The United Kingdom and Australia use the lowest astronomical tide (LAT) for the chart or sounding datum. The LAT is the lowest water level predicted in a 19-year period. Different datums may be used in other countries.

Nontidal datums are established as references for water height in upland rivers, lakes, inland waterways, and reservoirs. The International Great Lakes Datum for the Great Lakes and the Low Water Reference Plan for the middle Mississippi River are 2 examples of nontidal datums.

25.11.4.2 Tidal Height

Tidal height for a particular time is represented in meters relative to the datum for an area. Report a tide as a negative number if the tidal height is less than the datum and as a positive number if the height is above the datum. Although a plus sign is sometimes used with the numeric tidal height for tides above the datum, its use is not recommended unless required for clarity.

This morning's low tide of −13 m was the lowest of the spring tide series.
Mussels are exposed in the intertidal zone during tides of 1.5 m and lower.

In scientific writing, report tides in meters. In some nontechnical writing in the United States, tides may be reported in feet instead of meters.

25.11.5 Depth

Depth soundings on nautical charts represent the depth of the water at mean sea level, the average level measured over the tidal datum epoch. Nautical chart soundings in the United States are traditionally recorded in fathoms. (See Table 12.8 in this manual.) Nautical charts in other countries record depth in meters.

25.11.6 Seafloor Features

Capitalize the names of ocean floor features and the descriptive terms that follow, such as “ridge”, “trench”, “bank”, or “seamount”.

Marianas Trench Patton Seamount Great Barrier Reef
Mid-Ocean Ridge Grand Banks

General features of the seafloor are not capitalized.

continental shelf deep-sea trench fringing reef

25.12 AIR

25.12.1 Scales of Motion

Meteorology encompasses the study of atmospheric motion, in particular how such motion translates into the weather patterns experienced on the earth. These motions are classified according to the scale at which they occur. The largest is the planetary scale (order of thousands of kilometers), which covers the large waves in the jet stream that influence weather patterns. The synoptic scale (order of hundreds of kilometers) refers to the sorts of weather patterns that are typically presented on weather maps (also referred to as synoptic maps), including highs, lows, and fronts. Mesoscale phenomena (order of a few to tens of kilometers) are essentially regional in scope, with circulations as large as hurricanes often included in this classification but also those ranging down to the size of thunderstorms. Microscale phenomena (order of a kilometer or smaller) range from the size of clouds down to very small circulations, such as the flow of air past a building or the air flow within a few micrometers of the surface of a leaf.

The divisions among these scales are intentionally imprecise because different aspects of many phenomena span adjacent scales. Despite their imprecision, the divisions provide a useful framework for discussing the phenomena. Computer simulation models are often labeled according to the scales they are intended to reproduce; thus, computer models used for weather prediction might be referred to as synoptic or mesoscale numeric models.

25.12.2 Wind

In technical writing, report winds in terms of the direction from which they are blowing. Use the suffix “erly” to convey the same sense.

north wind *or* northerly wind [a wind blowing from the north]
south wind *or* southerly wind [a wind blowing from the south]

Note that the convention for reporting the direction of water currents (see Section 25.11.3.2) is the opposite of the convention for reporting the direction of winds.

In some nontechnical writing, wind may be expressed in terms of the direction toward which it is blowing; for this purpose use the suffix "ward":

> northward [a wind blowing toward the north]
> southward [a wind blowing toward the south]

Wind direction can also be expressed in degrees, with reference to the direction from which the wind is blowing. According to this convention, 0° and 360° winds are from the north, a 90° wind is from the east, a 180° wind is from the south, and a 270° wind is from the west.

25.12.3 Specialized Terminology

In addition to terms specific to meteorology, several common terms, such as those described in the following sections, are used in specialized ways. The American Meteorological Society publishes a glossary of meteorological terms.[47]

25.12.3.1 Oscillations and Teleconnections

Complicated interactions between the atmosphere and the oceans can cause mesoscale and synoptic scale wind and current circulations that may shift between 2 basic patterns (sometimes referred to as phases or states) in a more or less periodic manner over long time scales (months or years). Several of these oscillations affect seasonal weather patterns (as well as ocean currents) for specific regions of the globe. The most well known is the El Niño phenomenon. The term "El Niño" was originally used to refer to the period of warm surface ocean waters off the coast of Peru that occurs at irregular intervals ranging from 2 to 7 years. This event is now known to be coincident with seasonal weather anomalies in other parts of the world (i.e., a location might be warmer or colder or wetter or drier than normal), and these correlations are referred to as "teleconnections". El Niño is considered a part of an oscillation in the atmosphere–ocean system referred to as the El Niño/Southern Oscillation (ENSO). The phase of the oscillation opposite the El Niño period is called La Niña (i.e., when the ocean currents and wind and pressure patterns are reversed from the El Niño phase). Distinct but similar oscillations that are also potentially useful for seasonal forecasting include the Arctic Oscillation, the North Atlantic Oscillation, and the Pacific Decadal Oscillation.

25.12.3.2 Skill

In meteorology the term "skill" has a precise meaning that differs from its common usage. In meteorological usage, skill is a statistical evaluation of a forecast or class of forecasts in comparison with some other forecast or with a climatological average. When used in this manner, the word "skill" should not be replaced by a synonym (such as "aptitude"). The following are typical examples of correct usage in the meteorological literature:

> The skill of the 5-day forecast has improved dramatically, and such forecasts are now as skillful as the 3-day forecast was 15 years ago.
> During an active El Niño period, seasonal forecasts of wetter- or drier-than-normal conditions that show skill compared with a climatological average can be made for many regions of the country.

25.12.4 Satellite Names

Environmental satellites are used for tracking weather systems; for measuring winds, temperature, and the humidity structure of the atmosphere; for measuring chemical constituents, such as ozone; and for observations related to oceans and land surfaces. Most satellites are named according to the series to which they belong. During design and construction, a letter designating the specific satellite is added; after successful launch and operational implementation, the satellite is christened and the letter is changed to a number. In referring to a satellite that has not been christened, set the name in roman type; after christening, italicize the name (as for a ship's name). For example, in the Geostationary Orbiting Environmental Satellite (GOES) series, a satellite developed for launch in 1997 was named as follows:

GOES-K [during design and construction] *GOES-10* [after christening]

Use all uppercase letters if the name of a satellite or satellite series is an acronym (e.g., GOES or *GOES*). Spell out the full name of an acronym the first time it is used in the text. Capitalize only the initial letter of a satellite name or series that is not an acronym (e.g., Landsat).

25.13 NON-SI UNITS

Some non-SI units are still in common usage in the earth, oceanographic, and meteorological sciences (see Section 12.2.1 for SI and other non-SI units). Some terms convenient for crop and soil scientists are not converted to SI units. For example, land area is reported in acres instead of the metric equivalent hectares when area refers to the 160-acre parcels of land that the United States and Canadian governments surveyed in the 19th century. Geologic time is expressed in decimal multiples of the units annum and years (see Section 25.2). The unit board foot (bd ft) is still used in measurements of lumber (see Table 12.8).

A few units in meteorology and oceanography are not part of the standard SI list but are nonetheless pervasive in the technical literature. The most common non-SI unit found in meteorological literature is the millibar (mb), which has a long tradition of use in the field. This practice is generally discouraged, however, and millibar should be replaced by the pascal (or the appropriate decimal multiple such as the hectopascal [1 hPa = 1 mb]). Nautical mile (n mi) and knot (kt) sometimes appear in the literature, but these units are strongly discouraged (see Table 12.5). Use the metric equivalents kilometer instead of nautical mile and meters per second instead of knot. The sverdrup unit (Sv = 10^6 m^3/s) for volume transport in physical oceanography is not approved by the International Association for the Physical Sciences of the Ocean (IAPSO). Sigma-t (σ_t) is commonly used as a unit for seawater density (see Section 25.11.2.4), but IAPSO strongly discourages its use. This manual recommends using, whenever possible, the international system of units adopted by IAPSO.[41]

CITED REFERENCES

1. North American Commission on Stratigraphic Nomenclature. North American stratigraphic code. Am Assoc Pet Geol Bull. 2005;89(11):1547–1591. Also available at http://www.agiweb.org/nacsn/code2.html#anchor514748.

2. International Commission on Stratigraphy. International chronostratigraphic chart. [modified 2013 Jan; accessed 2013 Nov 19]. http://www.stratigraphy.org/index.php/ics-chart-timescale.

3. International Union of Geological Sciences. International stratigraphic guide: a guide to stratigraphic classification, terminology, and procedure. 2nd ed. Salvador A, editor. Boulder (CO): Geological Society of America Publications; 1994. 214 p.

4. Murphy MA, Salvador S, editors. International stratigraphic guide—abridged version. International Subcommission on Stratigraphic Classification of IUGS International Commission on Stratigraphy. Episodes. 1994;22(4):255–271.

5. Gradstein FM, Ogg JG. Geologic time scale 2004—why, how, and where next! Lethaia. 2004;(37):175–181.

6. Hansen WR, editor. Suggestions to authors of the reports of the United States Geological Survey. 7th ed. Reston (VA): Geological Survey (US); 1991 [modified 2004 Jun 8; accessed 2012 Mar 21]. 289 p. http://www.nwrc.usgs.gov/lib/lib_sta.htm.

7. Glass DJ, editor. Lexicon of Canadian stratigraphy [CD-ROM]. Vol. 4, Western Canada. Calgary (AB): Canadian Society of Petroleum Geologists; 1990. 772 p.

8. Lexicon of Canadian stratigraphy [CD-ROM]. Calgary (AB): Canadian Society of Petroleum Geologists; [date unknown]. Vol. 1, Arctic Archipelago; Vol. 2, Yukon–MacKenzie; Vol. 6, Atlantic Region.

9. Lexique stratigraphique canadien (Lexicon of Canadian stratigraphy). Québec (QC): Service Géologique du Québec; 1993. Vol. 5B, Région des Appalaches, des Basses-Terres du Saint-Laurent et des îles de la Madeleine.

10. Geological Society of America. Rock-color chart. Boulder (CO): The Society; 1995. 16 p. Rock-color chart committee: Goddard EN, Trask PD, DeFord RK, Rove ON, Singewald JT Jr, Overbeck RM.

11. Compton RR. Geology in the field. New York (NY): John Wiley & Sons, Inc.; 1985. 416 p.

12. Government Printing Office (US). Style manual. 30th ed. Washington (DC): The Office; 2008 [accessed 2012 Mar 21]. 344 p. http://www.gpo.gov/fdsys/pkg/GPO-STYLEMANUAL-2008/content-detail.html.

13. National Geographic atlas of the world. 8th ed. Washington (DC): National Geographic Society; 2005. 271 p.

14. Le Maitre RW, editor. Igneous rocks: a classification of igneous rocks and glossary of terms: recommendations of the International Union of Geological Sciences Subcommission on the Systematics of Igneous Rocks. 2nd ed. New York (NY): Cambridge University Press; 2002. 252 p. Le Maitre RW, Streckeisen A, Zanettin B, Le Bas MJ, Bonin B, Bateman P, Bellieni G, Dudek A, Efremova S, Keller J, Lameyre J, Sabine PA, Schmid R, Sorensen H, Woolley AR, contributors.

15. Mandarino JA, Back ME. Fleischer's glossary of mineral species. Tucson (AZ): Mineralogical Record; 2004. 309 p.

16. Nickel EH, Grice JD. The IMA Commission on New Minerals and Mineral Names: procedures and guidelines on mineral nomenclature. Can Mineral. 1998;(36):913–926.

17. Kretz R. Symbols for rock-forming minerals. Am Mineralog. 1983;(68):277–279.

18. Bennett A, Hamilton D, Maradudin A, Miller R, Murphy J. Crystals perfect and imperfect, by scientists of the Westinghouse Research Laboratories. Banigan S, editor. New York (NY): Walker; 1965. 237 p.

19. Hahn T, editor. International tables for crystallography. Vol. A, Space-group symmetry. 5th ed. Dordrecht (The Netherlands): Kluwer Academic Publishers; 1999. 912 p. Published for the International Union of Crystallography.

20. Department of Agriculture (US), Soil Conservation Service, Soil Survey Division Staff. 1993. Soil survey manual. Washington (DC): The Service; 1993 [modified 2011 March 28; accessed 2011 Nov 16]. http://soils.usda.gov/technical/manual/.

21. Schoeneberger PJ, Wysocki DA, Benham EC, Broderson WD, editors. Field book for describing and sampling soils. Ver. 2.0. Lincoln (NE): Department of Agriculture (US), Natural Resources Conservation Service, National Soil Survey Center; 2002 [accessed 2011 Nov 16]. 228 p. http://soils.usda.gov/technical/fieldbook/.

22. Department of Agriculture (US), Natural Resources Conservation Service. The color of soil. Washington (DC): The Service; [accessed 2011 Nov 15]. http://soils.usda.gov/education/resources/lessons/color/. Adapted from Lynn WC, Pearson MJ. The color of soil. Sci Teach. 2000 May;67(5):20–23.

23. American Society of Agronomy; Crop Science Society of America; Soil Science Society of America. Publications handbook and style manual. Madison (WI): American Society of Agronomy; 2011 [accessed 2011 Nov 16]. https://www.agronomy.org/publications/style.

24. National Cooperative Soil Survey soil characterization database. Lincoln (NE): Department of Agriculture (US), National Resources Conservation Service; [accessed 2011 Nov 16]. http://soils.usda.gov/survey/nscd/.

25. National Research Council Canada, Canadian Agricultural Services Coordinating Committee, Soil Classification Working Group. The Canadian system of soil classification. 3rd ed. Ottawa (ON): NRC Research Press; 1998. 188 p.

26. Department of Agriculture (US), Natural Resources Conservation Service, Soil Survey Staff. Soil taxonomy: a basic system of soil classification for making and interpreting soil surveys. 2nd ed. Washington (DC): The Service; 1999 [accessed 2011 Nov 16]. http://soils.usda.gov/technical/classification/taxonomy/.

27. International Society of Soil Science; International Soil Reference and Information Centre; Food and Agriculture Organization of the United Nations. World reference base for soil resources. Rome (Italy): Food and Agriculture Organization of the United Nations; 1998 [accessed 2011 Nov 16]. (World soil resources reports; 84). http://www.fao.org/docrep/w8594e/w8594e00.htm.

28. Soil series classification database. Washington (DC): Department of Agriculture (US), Natural Resources Conservation Service, Soil Survey Staff; [modified 2010 Sep 29; accessed 2011 Nov 16]. http://soils.usda.gov/technical/classification/scfile/index.html.

29. Department of Agriculture (US), National Resources Conservation Service, Soil Survey Staff. Keys to soil taxonomy. 11th ed. Washington (DC): The Service; 2010 [accessed 2011 Nov 16]. http://soils.usda.gov/technical/classification/tax_keys/index.html.

30. Soil Survey Geographic (SSURGO) Database. Lincoln (NE): Department of Agriculture (US), National Resources Conservation Service, Soil Survey Staff; [accessed 2011 Nov 16]. http://soils.usda.gov/survey/geography/ssurgo/.

31. US General Soil Map (STATSGO2). Lincoln (NE): Department of Agriculture (US), Natural Resources Conservation Service, Soil Survey Staff; [accessed 2011 Nov 16]. http://websoilsurvey.sc.egov.usda.gov/App/HomePage.htm.

32. Bates R, Jackson JA, editors. Glossary of geology [CD-ROM]. Alexandria (VA): American Geological Institute; 2000. DOS and Mac compatible.

33. Glossary of soil science terms. Madison (WI): Soil Science Society of America; 2011 [accessed 2011 Nov 16]. https://www.soils.org/publications/soils-glossary.

34. Lal R, editor. Encyclopedia of soil science. New York (NY): Marcel Dekker; 2002. 1450 p.

35. Sumner ME, editor. Handbook of soil science. Boca Raton (FL): CRC Press; 1999. 2148 p.

36. A national ecological framework for Canada. Ottawa (ON): Agriculture and Agri-Food Canada, Ecological Stratification Working Group; 1996 [modified 2008 Nov 27; accessed 2011 Nov 16]. http://sis.agr.gc.ca/cansis/publications/ecostrat/intro.html.

37. Department of the Interior (US), Bureau of Land Management. Specifications for descriptions of tracts of land for use in land orders and proclamations. Washington (DC): The Bureau; 1979 [accessed 2011 Nov 16]. http://www.blm.gov/or/gis/geoscience/files/Land_Descriptions.pdf.

38. Natural Resources Canada. Manual of instructions for the survey of Canada lands. 3rd ed. Vol. 2, Administrative requirements, general instructions and appendices. Ottawa (ON): Geomatics Canada Legal Surveys Division; 1996 [accessed 2011 Nov 16]. 135 p. http://clss.nrcan.gc.ca/standards-normes/data/Manual%20of%20Instructions-3rd%20Edition-Vol%202E.pdf.

39. Wentworth CK. A scale of grade and class terms for clastic sediments. J Geol. 1922;(30):377–392.

40. Shepard FP. Nomenclature based on sand-silt-clay ratios. J Sed Petrol. 1954;(24):151–158.

41. The international system of units (SI) in oceanography: report of IAPSO Working Group on Symbols, Units and Nomenclature in Physical Oceanography (SUN). Paris (France): UNESCO; 1985. 125 p. (UNESCO technical papers in marine science; 45); (IAPSO publication scientifique; no. 32). Also available at http://unesdoc.unesco.org/images/0006/000650/065031eb.pdf.

42. Hutchinson GE. A treatise on limnology. New York (NY): John Wiley & Sons; 1957. 1015 p.

43. Williams WD, Sherwood JE. Definition and measurement of salinity in salt lakes. Int J Salt Lake Res. 1994;(3):53–63.

44. Pickard GL, Emery WJ. Descriptive physical oceanography: an introduction. 5th enl. ed. (in SI units). Elmsford (NY): Pergamon Press, Inc.; 1990. 320 p.

45. Millero FJ. What is PSU? Oceanography. 1993;(6 Pt 3):67.

46. Hicks SD. National Tidal Datum Convention of 1980. Rockville (MD): National Oceanic and Atmospheric Administration (US), National Ocean Survey; 1980. 44 p.

47. Glickman TS, editor. Glossary of meteorology. 2nd ed. Boston (MA): American Meteorological Society; 2000. 855 p.

ADDITIONAL REFERENCES

Ashworth W, Little CE. Encyclopedia of environmental studies. New York (NY): Facts on File; 2001. 600 p.

Bates RL, Adkins-Heljeson MD, Buchanan RC, editors. Geowriting: a guide to writing, editing, and printing in earth science. 5th rev. ed. Alexandria (VA): American Geological Institute; 2004. 138 p.

Brady NC, Weil RR. Elements of the nature and properties of soils. 3rd ed. Upper Saddle River (NJ): Pearson Prentice Hall; 2010. 614 p.

Canarache A, Vintila I, Munteanu I. Elsevier's dictionary of soil science. Boston (MA): Elsevier; 2006. 1339 p.

Clark AM. Hey's mineral index: mineral species, varieties, and synonyms. 3rd ed. London (UK): Chapman & Hall; 1993. 848 p.

Fanning DS, Fanning MCB. Soil: morphology, genesis, and classification. New York (NY): Wiley; 1989. 395 p.

GID Editorial Board. Guide to authors: a guide for the preparation of Geological Survey of Canada maps and reports. Ottawa (ON): Geological Survey of Canada; 1998. 194 p. Also available at http://dx.doi.org/10.4095/209973.

Gradstein FM, Ogg JG, Smith AG, editors. A geologic time scale 2004. New York (NY): Cambridge University Press; 2005. 610 p.

Grant B. Geoscience reporting guidelines. Victoria (BC): [publisher unknown]; 2003. 356 p. Available for purchase from Prospectors and Developers Association of Canada, http://members.shaw.ca/geomanual/, and Geological Society of Canada, http://www.gac.ca/. Includes geologic time scale.

Haq BU, van Eynsinga FWB. Geological time table [chart]. 6th rev. ed. New York (NY): Elsevier Science; 2007.

Klein C, Dutrow B. Manual of mineral science. 23rd ed. New York (NY): John Wiley & Sons, Inc.; 2008. 675 p.

Ogg JG. Status of divisions of the international geologic time scale. Lethaia. 2004;(37):138–139.

Porteous A. Dictionary of environmental science and technology. 4th ed. New York (NY): John Wiley & Sons, Inc.; 2008. 797 p.

Texas A&M University, College of Geosciences. How to read a stick plot. College Station (TX): The University; [modified 2003 Apr 21; accessed 2013 Jan 28]. http://tabs.gerg.tamu.edu/Tglo/stick.html.

Thomas DS, Goudie A, editors. Dictionary of physical geography. 3rd ed. Oxford (UK): Blackwell; 2000. 672 p. Also available at http://dx.doi.org/10.1002/9781444313178.

Walsh SL, Gradstein FM, Ogg JG. History, philosophy, and application of the global stratotype section and point (GSSP). Lethaia. 2004;(37):201–218.

Wilson AJC. International tables for crystallography. Vol. C, Mathematical, physical, and chemical tables. Boston (MA): Kluwer Academic; 1992. Published for the International Union of Crystallography.

Witty JE, Arnold RW. Soil taxonomy: an overview. Outlook Agric. 1987;(16):8–13.

Wood DN, Hardy JE, Hardy AP, editors. Information sources in the earth sciences. 2nd ed. London (UK): Bowker-Saur; 1989. 518 p.

26 Astronomical Objects and Time Systems

26.1 NAMES AND ABBREVIATIONS

Through the past several decades, new instruments and methods for seeing what could not be seen as well, or at all, through optical telescopes have rapidly expanded our knowledge of what lies beyond the earth's surface and atmosphere. This change has led to increasing needs to accurately represent a still-growing array of astronomical objects and phenomena. The leading authority on this specialized and rapidly growing nomenclature is the International Astronomical Union.[1,2]

An astronomical object may be designated by a formal or proper name, a catalog number, or a composite name that indicates the type of object and its position; in general, the third of these systems is used for new designations.

Formal names may be traditional names established long before telescopic observation or names applied to objects of a class that is not too numerous for ready naming. Some examples are the names of constellations (see Table 26.1), stars prominent to the naked eye, the planets and dwarf planets, and certain small Solar System bodies.

Table 26.1 Constellation names and their abbreviations

Name of constellation	Abbreviation	Name of constellation	Abbreviation
Andromeda	And	Lacerta	Lac
Antlia	Ant	Leo	Leo
Apus	Aps	Leo Minor	LMi
Aquarius	Aqr	Lepus	Lep
Aquila	Aql	Libra	Lib
Ara	Ara	Lupus	Lup
Aries	Ari	Lynx	Lyn
Auriga	Aur	Lyra	Lyr
Boötes	Boo	Mensa	Men
Caelum	Cae	Microscopium	Mic
Camelopardus	Cam	Monoceros	Mon
Cancer	Cnc	Musca	Mus
Canes Venatici	CVn	Norma	Nor
Canis Major	CMa	Octans	Oct
Canis Minor	CMi	Ophiuchus	Oph
Capricornus	Cap	Orion	Ori
Carina	Car	Pavo	Pav
Cassiopeia	Cas	Pegasus	Peg
Centaurus	Cen	Perseus	Per
Cepheus	Cep	Phoenix	Phe
Cetus	Cet	Pictor	Pic
Chamaeleon	Cha	Pisces	Psc
Circinus	Cir	Piscis Austrinus[a]	PsA
Columba	Col	Puppis	Pup
Coma Berenices	Com	Pyxis	Pyx
Corona Austrina[a]	CrA	Reticulum	Ret
Corona Borealis	CrB	Sagitta	Sge
Corvus	Crv	Sagittarius	Sgr
Crater	Crt	Scorpius	Sco
Crux	Cru	Sculptor	Scl
Cygnus	Cyg	Scutum	Sct
Delphinus	Del	Serpens[b]	Ser
Dorado	Dor	Sextans	Sex
Draco	Dra	Taurus	Tau
Equuleus	Equ	Telescopium	Tel
Eridanus	Eri	Triangulum	Tri
Fornax	For	Triangulum Australe	TrA
Gemini	Gem	Tucana	Tuc
Grus	Gru	Ursa Major	UMa
Hercules	Her	Ursa Minor	UMi
Horologium	Hor	Vela	Vel
Hydra	Hya	Virgo	Vir
Hydrus	Hyi	Volans	Vol
Indus	Ind	Vulpecula	Vul

[a] "Australis" is sometimes used as the second term.
[b] Serpens may be divided into Serpens Caput and Serpens Cauda.

Table 26.2 Abbreviations for the titles of selected astronomical catalogs[a]

Abbreviation	Catalog
AGK *n*	*Astronomische Gesellschaft Katalog Nummer n*
BD	*Bonner Durchmusterung*
BS, BSC	*Bright Star Catalogue*
CD	*Cordoba Durchmusterung*
CPD	*Cape Photographic Durchmusterung*
FK *n*	*Fundamental Katalog Nummer n*
GCVS	*General Catalogue of Variable Stars*
HD	*Henry Draper Catalogue*
HDE	*Henry Draper (Catalogue) Extension*
HR	*Revised Harvard Photometry Catalogue*
IC	*Index Catalog*
NGC	*New General Catalog of Nebulae and Clusters of Stars*
SAO	*Smithsonian Astrophysical Observatory Star Catalog*
3C	*Third Cambridge Catalogue*

[a] For additional titles and their abbreviations, see Ridpath.[3]

Catalog numbers are the numbers assigned by the compiler of a catalog, and they are preceded by an abbreviation of the catalog name. (See Section 26.5 and its subsections for examples. For a list of some catalog names and their abbreviations, see Table 26.2. For a more comprehensive list, see *Norton's Star Atlas and Reference Handbook*.[3])

Some objects have 2 or more names of equal standing.

Ring Nebula = M57 [Messier catalog number] = NGC 6720 [New General Catalog number]
the Pleiades = 7 Sisters = M45

A temporary name may be given to an object at the time of first observation, with a new permanent name given when the object's properties have been established. Proper names and catalog numbers may be inadequate to unambiguously identify an object. A satisfactory composite name gives the essential characteristics of the object (by alphanumeric code-name) and its position. Do not introduce abbreviations unnecessarily; when necessary, an abbreviation should be unambiguous and should have at least 2 letters (for examples, see Sections 26.5.3 and 26.5.6.1).

26.2 CAPITALIZATION

Capitalize the proper names of constellations (see Table 26.1); the names of planets and their satellites, asteroids and comets, and stars; the names of other unique celestial objects; and the names of space programs.

the Crab Nebula	Project Apollo	Comet Biela	the North Star
the Coalsack	Solar System	Milky Way	Orion

For designations by catalog number, capitalize the catalog name (either its abbreviation or the full name).

NGC 6165 Bond 619 M81

Capitalize the word "earth" in an astronomical context, particularly when used with the names of other planets; in this situation, the definite article "the" is usually omitted. Similarly, capitalize "sun" and "moon" when used in an astronomical context. Use lowercase for these terms when they appear in more general contexts. (Note that some astronomy publications always capitalize "earth".)

The first 3 planets from the Sun are Mercury, Venus, and Earth.
Jupiter has many satellites, but Earth has only one.
Human exploration of the surface of the Moon has been followed by robotic exploration of the surface of Mars.
The sun provides energy for photosynthesis.
The moon appears to rise in the east.
Geology is the study of the earth's structure and composition.

Use lowercase for 3 main classes of terms:

1) descriptive terms applied to celestial phenomena (except for any proper nouns)

the gegenschein   the Cassini division
the rings of Saturn   the laplacian plane of Saturn's rings

2) terms applied to meteorologic and other atmospheric phenomena (except for any proper nouns)

aurora borealis   sun dog   meteor shower   Tunguska fireball

3) adjectival forms of proper nouns

Jupiter, jovian   Moon, lunar   Sun, solar   Mercury, mercurial

Capitalize the names of spacecraft and artificial satellites. Once a spacecraft is launched into orbit, it becomes a satellite. Satellite names are set in roman type during design and construction phases but in italics after christening (see Section 25.12.4).

Luna   *Columbia*   *Vega*   *Landsat*

26.3 CELESTIAL COORDINATES

Right ascension (designated α) is given in hours, minutes, and seconds of sidereal time, with no spaces between numbers and units. In less precise designations, the general convention for units may be acceptable (see Section 12.2.1).

$14^{h}6^{m}7^{s}$ ["14 h 40 min" in less precise designations]

Declination (designated δ) is given in degrees, minutes, and seconds of arc north (marked + or unmarked) or south (marked −) of the celestial equator.

49°8′11″ *or* +49°8′11″ *or* 48°08′11.00″   −87°41′08″

If decimal fractions of the basic units are used, the decimal point is placed between the unit symbol and the decimal value.

$26^{h}6^{m}7^{s}.2$ +34°.26

The mean equator and equinox serve to define a coordinate system by ignoring small, short-period variations in the motion of the celestial equator, but this system is affected by precession (the slow continuous westward motion of the equinoxes around the ecliptic that results from precession of the earth's axis). The positions of the mean equator and equinox at particular epochs are used to define standard reference systems, such as those for B1950.0 and J2000.0 (see Section 26.6.1.2 on the besselian and julian epochs).

See Table 26.3 for a list of abbreviations and symbols used frequently in the astronomical literature, including those related to celestial coordinates.

26.4 OBJECTS IN THE SOLAR SYSTEM

26.4.1 Planets, Satellites, and Rings

The Working Group for Planetary System Nomenclature of the International Astronomical Union is responsible for the adoption of names for the surface features of planets and satellites and for newly discovered members of the planetary system other than small Solar System bodies. Satellites are also designated by numbers assigned in the chronologic sequence of discovery. The decisions of the working group are reported in the transactions of the International Astronomical Union and are made available through the *Gazetteer of Planetary Nomenclature*.[4] The Solar System Dynamics Group of the Jet Propulsion Laboratory also maintains a list of natural satellite numbers and names.[5] When satellites of planets are listed, give them in order of mean distance from the planet.

In scientific publications, do not use the symbols for planets that are common in astrologic and other informal literature.

26.4.2 Dwarf Planets and Small Solar System Bodies (Asteroids and Comets)

In August 2006 the International Astronomical Union defined a planet as "a celestial body that (a) is in orbit around the Sun, (b) has sufficient mass for its self-gravity to overcome rigid body forces so that it assumes a hydrostatic equilibrium (nearly round) shape, and (c) has cleared the neighborhood around its orbit."[6]

At the same time, a new class of objects called dwarf planets was created. Like a planet, a dwarf planet must be in orbit around the Sun and be sufficiently massive to have become rounded by its own gravity. Unlike a planet, a dwarf planet may share its orbit with other objects such as asteroids. Dwarf planets are generally smaller than Mercury. Currently there are 5 objects accepted as dwarf planets: Ceres, Pluto, Eris, Makemake, and Haumea. The International Astronomical Union expects more to be added in coming years.

Pluto also belongs to a new class of Trans-Neptunian Objects that includes Eris, Makemake, and Haumea (but not Ceres). The International Astronomical Union has designated these objects "plutoids". Plutoids are dwarf planets that orbit the Sun at a semimajor axis greater than that of Neptune.

Objects orbiting the Sun that are insufficiently massive to be called either a planet or a dwarf planet are now officially referred to by the International Astronomical Union as "small Solar System bodies". (The term "minor planet" is still in use, but the new term is preferred in most contexts.) This class includes asteroids and comets.[6]

Asteroids are given numbers serially and proper names when reliable orbital elements have been defined.

878 Mildred   719 Albert   1627 Ivar   288 Glauke

Since 1995, newly discovered comets are assigned a designation reflecting the date of discovery.[7] The designation consists of the year of observation, an uppercase code letter

Table 26.3 Abbreviations and symbols frequently used in the astronomical literature

Abbreviation or symbol	Meaning
α	Right ascension *or* the brightest star in a constellation
AU	Astronomical unit (mean Earth–Sun distance); UA in French
B	Besselian
β	Second brightest star in a constellation
c	Speed of light
CMBR	Cosmic microwave background radiation
Δ T	Increment to be added to universal time to give terrestrial dynamical time
δ	Declination or fourth brightest star in a constellation
E	Color excess
ET	Ephemeris time (a measure of time for which a constant rate was defined); used from 1958 to 1983
G	Gravitation constant
γ	Third brightest star in a constellation
H_0	Hubble constant
H II region	Volume of hydrogen photoionized (into protons and electrons) by the ultraviolet radiation from a central, hot object
HA	Hour angle
k	Curvature index of space *or* Gaussian gravitational constant
kpc	kiloparsec
L	Luminosity, stellar
$L_{\odot}$	Luminosity, solar
λ	Celestial longitude
LAST	Local apparent sidereal time
LMST	Local mean sidereal time
LST	Local sidereal time
M	Absolute magnitude
$M_{\oplus}$	Mass of Earth
$M_{\odot}$	Mass, solar
MHD	Magnetohydrodynamics (also known as hydromagnetics)
mJy	milli-Jansky [unit of luminous flux]
Mpc	megaparsec, meaning a million parsecs or 10^6 pc
μg	microgauss
pc	parsec
PZT	Photographic zenith telescope [out of date]
Q	Aphelion (the point in solar orbit farthest from the Sun)
QSO	Quasistellar object, quasar
R	Cosmic scale factor (a measure of the size of the Universe as a function of time)
RA	Right ascension
$R_{\oplus}$	Radius of Earth
rv	Radial velocity
SRS	Southern reference system
t	Hour angle
TAI	International atomic time
TDB	Barycentric dynamical time
TDT	Terrestrial dynamical time
UT	Universal time

Table 26.3 Abbreviations and symbols frequently used in the astronomical literature (*continued*)

Abbreviation or symbol	Meaning
V	Visual magnitude
VLBI	Very-long-baseline radio interferometry
z	Red-shift parameter
ZAMS	Zero-age main sequence
ZHR	Zenithal hourly rate

identifying the half-month of the year in which the comet was observed, and a consecutive numeral indicating order of discovery during that half-month. An additional prefix (P/ for a periodic comet, C/ for a comet that is not periodic, X/ for a comet whose orbit cannot be computed, or D/ for a periodic comet that no longer exists) is used to indicate the nature of the object. Comets discovered before 1995 have multiple designations.

C/2005 A1, the first comet (nonperiodic) observed in the first half-month of 2005
C/2005 E2, the second comet (nonperiodic) observed in the fifth half-month of 2005, subsequently named C/McNaught

Before 1995, confirmed comets were designated by the name(s) of the discoverer(s) and a permanent number (in capital roman numerals); the number indicated for each year the chronologic sequence of the passages of all of the observed comets through perihelion (the point in its path at which an astronomical body is closest to the Sun).

Comet Arend–Roland 1957 III   Comet Kohoutek 1973 XII
Comet Tago–Sato–Kosaka 1969 IX

26.4.3 Meteor Showers, Meteors, and Meteorites

Meteor showers are usually named for the constellations in which their radiant points appear or according to the name of the comet with which they are associated.

the Arietids   the Lyrids   the Eta Aquarids   the Quadrantid

A large meteorite is usually identified by the name of the place near where it was found; a meteor crater, by its fixed geographic location. A fireball is usually identified by the date on which it was seen.

Seymour meteorite (Indiana)   Brent crater (Canada)

A comprehensive catalog of meteorite names and their synonyms is available.[8]

26.5 OBJECTS OUTSIDE THE SOLAR SYSTEM

26.5.1 Bright Stars

The Bayer designation of a bright star consists of a Greek letter followed by the name of (or standard 3-letter abbreviation for) the constellation in which it appears (see Table 26.4). The Greek letters are usually assigned in order of brightness (α, β, γ, δ, etc.). If Greek characters are not available, spell out the English name of the Greek letter. Present a bright star designation in roman type.

α Cen A *or* alpha Centauri A

In referring to any of the approximately 1,000 stars that have proper names derived from early Arabic names or peculiar characteristics, identify the star more precisely as well, especially if its name differs in other languages. *The Bright Star Catalogue*[9] lists about 900 such names and gives information on more than 9,110 objects, of which 9,096 are stars, brighter than magnitude 6.5. It also gives names based on the Bayer system (Greek letter followed by the star's constellation name) and the Flamsteed system (number and the star's constellation) (see Table 26.4).

About 1,500 bright stars are listed in the annual *Astronomical Almanac*[10]; the tabulation gives the corresponding Bayer, Flamsteed, and bright star designations, as well as the mean position and other information about each star.

Stars may be classified into 1 of 9 groups according to spectral temperature: O (for the hottest stars), B, A, F, G, K, M, L, and T (for the coolest stars, which are very red). In turn these groups can be subdivided into divisions designated by the numerals 0 to 9 appended to the alphabetic group designation (e.g., A4, F8, G0); there are no K8 or K9 stars.

Stars may also be classified by luminosity class with roman numerals, which may be subdivided by lowercase letters (see Table 26.5).

Use roman type for the spectral and luminosity classifications of stars and other similar symbols and abbreviations, but use italics for any symbols for quantities (see Section 4.2.2 and Table 4.5).

HD 190406	λ And	70 Oph A	AU Mic
Algol (β Per)	K2 III	F8	

26.5.2 Faint Stars

For stars not listed in catalogs and for which designations have not been given, assign designations in the form of an acronym and position in the standard form (see Section 26.5.7 on radiation sources). The acronym should refer in some way to the discoverer or literature citation in which the star is reported.

26.5.3 Variable Stars

The names of variable stars consist of the name of the constellation (see Table 26.1) preceded by a 1-letter or 2-letter code (capitalized) or the capital letter V followed by a number. When a classical name exists, use it in titles; for these names, consult the *General Catalogue of Variable Stars*[11] or the SIMBAD database[12] of the Centre de Données Astronomiques de Strasbourg.

W Vir	RV Tau	V356 Sgr	V827 Her 1987

Many binary (double) stars have variable-star names (e.g., RY Sct); alternatively, they may take the name of the catalog star with which they are associated (e.g., β Cet, α Her, Sirius A).

26.5.4 Novae or Cataclysmic Variables

The designation for a nova is the word "Nova" followed by the constellation name or its abbreviation (see Table 26.1) and the year; the elements are spaced without punctuation. The year is followed by a number if more than one nova has been discovered

Table 26.4 Examples of alternative designations for bright stars[a]

Traditional name	Bayer system	Flamsteed system
Betelgeuse	α Ori	58 Ori
Sirius	α CMa	9 CMa
Ras-Alhague	α Oph	55 Oph
Polaris	α UMi	14 UMi

[a] For other designations in the Bayer and Flamsteed systems, see the *Astronomical Almanac.*[10]

Table 26.5 Luminosity classes of stars

Class	Luminosity type	Class	Luminosity type
Ia and Ib	Supergiants	IV	Subgiants
II	Bright giants	V	Main sequence (dwarfs)
III	Giants		

in a constellation in the same year. Subsequently, a nova may be assigned a standard designation as a variable star.

Nova Sge 1987 Nova Sgr 1987 No. 1 Nova Per 1901 Nova Sge 1783

26.5.5 Supernovae

The designation for a supernova is "SN" followed by a space, then the year and an alphabetic designation for order of discovery: single capital letters from A to Z, then paired lowercase letters (aa, ab, ac, . . . ba, bb, bc, . . .). No punctuation is used.

SN 1985A SN 1985aa SN 1985az SN 1985ba

Newly discovered supernova remnants are designated SNR, followed by the position with respect to the system of J2000.0 (see Section 26.6.1.2 for besselian and julian epochs).

26.5.6 Nebulae, Galaxies, Clusters, Pulsars, and Quasars

Two catalogs of nebulae, galaxies, and clusters are in wide use. The one initially compiled by Charles Messier between 1771 and 1784 now comprises 110 entries.[13] The objects in the catalog (the brightest nebulae, galaxies, and star clusters visible from France) are referred to by Messier's initial and his serial number, unspaced.

M1 = Crab Nebula M45 = Pleiades (the Seven Sisters)

The second catalog, originally compiled by JLE Dreyer in 1888, is called the *New General Catalog* (NGC).[14,15] The objects in this catalog and its supplements (the *Index Catalog* or IC) are designated by the abbreviation of the catalog name, a space, and the catalog number of the object.

NGC 1952 = Crab Nebula h Persei = NGC 869

For the many catalogs that have been compiled since, the objects are usually referred to by an abbreviation (set in roman type) for the name of the catalog or of its compiler or by its compiler's initial and a serial number. Spell out abbreviations for the names of catalogs on first use (see Table 26.2 for the names and abbreviations of a few catalogs).

Another means of designating objects is by reference to the constellation, or large-scale star pattern, in which they reside (see Table 26.1 for constellations and their official abbreviations). The word "constellation" now refers to the whole of a specific region of the sky in which the pattern formed by the bright stars is seen; the ancient constellation patterns are still used, however, as a convenient guide to parts of the sky. Thus, Centaurus A was the first radio source to be discovered in the southern hemisphere constellation of Centaurus. Similarly, many especially famous galaxies, nebulae, and clusters are still known by individual names, often derived from their appearance and their more or less fanciful resemblance to terrestrial objects.

the Coalsack the Whirlpool galaxy the Eskimo

26.5.6.1 Planetary Nebulae

A few planetary nebulae have proper names, some have Messier numbers, many have *New General Catalog* (NGC, or IC, *Index Catalog*) numbers, many of them are given catalog numbers based on their discoverers' names, and all of them have catalog numbers based on their galactic coordinates.

Owl Nebula = M97 = NGC 3587 = PN G148.4+57.0 [a bright planetary nebula]
LoTr 5 = PN G339.9+88.4 [a faint planetary nebula discovered by Longmore and Tritton]

26.5.6.2 Galaxies

Some galaxies have proper names; these should also be identified by their Messier, NGC, or supplementary IC number.

26.5.6.3 Radio Sources

The International Astronomical Union–recommended designation (as of 1985) for radio sources is a catalog acronym followed by the celestial coordinates (right ascension and declination) (see Section 26.3) with respect to the reference system of J2000.0 (see Section 26.6.1.2 for besselian and julian epochs).

26.5.6.4 Star Clusters

Some clusters have proper names, but many are identified by their numbers in the Messier catalog or by their numbers in the NGC and supplementary IC.

NGC 4755 (Jewel Box) NGC 104 (47 Tuc) Terzan 2
NGC 6656 (M22) NGC 5139 (ω Cen) Grindlay 1

26.5.6.5 Pulsars

The position, in condensed notation, of the pulsar follows the abbreviation PSR. Older names, such as CP1919 (for the first pulsar ever discovered), are now obsolete.

PSR 1937 + 21 [the numeric designation following the abbreviation refers to right ascension of $19^{h}37^{m}$ and declination of $21^{\circ}.4$]

26.5.6.6 Quasistellar Objects or Quasars

No standard form has been assigned to the designation of quasars, but catalog numbers are used.

26.5.7 Radiation Sources

Designations of radiation sources are now taken from the catalogs of surveys made by particular satellites. When existing designations are used in listings, they should never be altered. To avoid ambiguity, source listings should also contain positional information or a second designation or both. The designation of an astronomical source consists of the following parts: acronymΔsequenceΔ(specifier), where "Δ" denotes a blank space. The designation of origin and sequence is necessary, but the specifier is optional and must appear in parentheses.

NGC 205 PKS 1817–43 CO J0326.0+3041.0 H20 G123.4+57.6 (VSLR=-185)

The acronym (formerly called origin) specifies the catalog or collection of sources. It may refer to catalog names (e.g., NGC, BD), the names of authors (RCW), instruments, observatories (PKS, VLA, IRAS, 3C, 51W), and so forth. It should consist of at least 2 letters or numerals but no special characters. The acronym element should never be further abbreviated.

The alphanumeric sequence (also called numbering) uniquely determines the source within a catalog or collection. It may consist of a sequence number within a catalog (e.g., HD 224801) or a combination of fields, or it may be based on coordinates. If coordinates are used, they are preceded by a code for the reference frame (also called a flag letter): G for galactic coordinates, B for besselian 1950, and J for julian 2000 equatorial coordinates. Coordinates used in designations are considered names; therefore, they are not changed even if the positions become more accurately known (e.g., at a different epoch, "BD -25 765" remains the designation, even though the source's declination is now -26). Subcomponents or multiples of objects are designated with letters or numerals added to the sequence with a colon: for example, "NGC 1818:B12".

The optional specifier allows an indication of association with larger radiating sources [for example, (M 31), (W 3)] or other object parameters. Because they are not required, they are enclosed within parentheses.[16]

Designation	**Position**	
[AcronymΔSequenceΔ(Specifier)]	[RA (2000)]	[Dec (2000)]
BD -3 5750	00 02 02.4	-02 45 59
H20 B0446.6+7253.7	04 46 37.3	+72 53 47
AC 211 (= 1E2127+119; in M15)	21 30 15.54	+11 43 39.0
PN G001.2–00.3	17 49 36.9	-28 03 59
R 136:a3	05 38 42.4	-69 06 03

The following examples are incorrect:

BD 4°14 [degree symbol should not be used; the declination sign is missing]
N221 [no space; the source is unclear: NGC or N in LMC?]
IRAS 5404–220 [the leading zero is missing; poor position]
P 43578 [the 1-letter origin is ambiguous]

26.6 DATES AND TIME

26.6.1 Dates

Definitions of astronomical dates include a specification of universal time (UT). For information about UT and coordinated universal time (UTC), see Sections 26.6.2.7 and 26.6.2.2, respectively.

26.6.1.1 Julian Date

The date on which an event occurred is usually best represented by the julian date (JD) accompanied by an appropriate conventional form. The julian date corresponding to any instant is the interval since noon on 1 January in the year 4713 BC. Accordingly, midnight on 1 January 1961 = 0000 UT 1 January 1961 = JD 2 437 500.5. In an astronomical context, the calendar date is expressed as year-month-day whether the month is spelled out or represented by a number. For the numeric form, represent months and days by 2-digit numbers from 01 to 12 for January to December and from 01 to 31 for the days of the month; use arabic, not roman, numerals. Separate the elements of a numeric date by hyphens.

1998 December 31 1998-12-31

The basis of the date and time system used must be clearly stated, because no system is free from ambiguity. The calendar year 1 BC is followed by the year AD 1; for astronomical purposes, it is convenient to denote the year 1 BC as year 0 and the year n BC as year $-(n - 1)$; for example, for astronomical purposes, 350 BC is year −349.

In converting from calendar date and time (see Sections 13.3 and 13.2, respectively) to julian date, remember that the julian day begins at 1200 UT (noon on the Greenwich meridian), whereas current calendar days begin at 0000 UT. Before 1925, Greenwich mean time (GMT) often, but not always, referred to the day that began at noon (rather than at midnight) on the Greenwich meridian. There is still some ambiguity about the precise meaning of GMT, so it is preferable to use coordinated universal time (see Section 26.6.2.2) instead of GMT for reporting astronomical observations.

The julian system may be used in conjunction with other time systems, such as ephemeris time and international atomic time (TAI). The name "julian ephemeris day" (JED) was introduced for the former system, but it is now more appropriate to use the abbreviation JD combined with the abbreviation for the time scale (see Section 26.6.2 and its subsections). For example, a column heading in a table could be JD (TAI), whereas for individual values the time scale might be indicated after the numeric value.

The modified julian date (MJD) is sometimes used for current dates. The MJD provides a shorter number for which the decimal part is zero at 0000 UT (or TAI or UTC, as specified); MJD = JD − 2 400 000.5.

26.6.1.2 Besselian and Julian Epochs

Beginning in 1984, astronomical epochs have been expressed in julian days (JD), the julian century being composed of 36,525 julian days. Julian epochs are denoted by the letter J. The date J2000.0 is JD 2 451 545.0, and it occurs at noon on 1 January 2000 (expressed as 2000 January 1.5). Julian days may be used for 3 different time scales:

coordinated universal time (UTC), international atomic time (TAI), and ephemeris time or dynamical time (T_{eph}), each of which is discussed below. The choice of time scale should be denoted as follows: J2000.0 for UTC, J2000.0 (TAI) for TAI, and JED 2000.0 for T_{eph}.

Before 1984, besselian years were used, these being the period of the earth's orbit measured between 2 successive passages of the apparent right ascension of the mean sun through the value of $18^{h}40^{min}$. Besselian epochs are denoted by the letter B (e.g., B1950.0). The relation between the julian and besselian time scales may be obtained from the following identities:

B1900.0 = JED 2 415 020.313 52
B1950.0 = JED 2 433 282.423 459 05
B2000.0 = JED 2 451 544.533 398 1

26.6.2 Time Systems

Seven major time scales are currently used in astronomy, although in reality 2 of them are more properly considered angles.

26.6.2.1 International Atomic Time

International atomic time (TAI) is a statistical time scale based on an ensemble of atomic clocks around the world measuring the SI second (the duration of 9,192,631,770 periods of the radiation corresponding to the transition between the 2 hyperfine levels of the ground state of the cesium-133 atom). TAI is broadcast by radio for use by scientists throughout the world. The specific duration of the SI second was chosen to match 1/86,400 the length of Newcomb's value for the mean solar day.

26.6.2.2 Coordinated Universal Time

Coordinated universal time (UTC) is used for civil time-keeping. It is the time kept by someone's wristwatch in the same time zone as Greenwich, England. UTC runs at the same rate as international atomic time (TAI); however, UTC is occasionally adjusted so that it keeps approximate pace with the rotation of the earth, which is generally slowing down. The adjustments to UTC take the form of insertions (or, possibly, deletions) of "leap seconds" into UTC, either at the end of June or at the end of December. Thus, UTC departs from TAI slowly, by a series of 1-s steps. (In the 1970s, there were also slight rate adjustments in addition to the forward or backward jumps in time.) A table of the leap seconds and rate adjustments (i.e., the difference between TAI and UTC) is available from the Earth Orientation Center at the Observatoire de Paris.[17]

26.6.2.3 Terrestrial Time

For historical continuity with previous uses of ephemeris time, terrestrial time (TT, formerly known as terrestrial dynamical time or TDT) is defined as international atomic time + 32.184 s. It is used for the printed listings of crude astronomical ephemerides and as a rough approximation to T_{eph} (see Section 26.6.2.4).

26.6.2.4 Ephemeris Time or Dynamical Time

The independent variable in the equations of motion for the bodies in the Solar System was once called "ephemeris time" (ET). In 1976, barycentric dynamical time (abbreviated TDB, for "temps dynamique barycentrique") was defined. However, neither the definition of ET nor that of TDB properly expressed the "free-flowing time" used in the ephemerides. That quantity is rigorously a relativistic coordinate time and has more recently been referred to as T_{eph}. It is what ET and TDB were intended to be.

T_{eph} is related to clocks on earth (terrestrial time [TT] and therefore international atomic time [TAI] and coordinated universal time [UTC]) through a formula based in relativity. The form of the formula is chosen so that T_{eph} never departs from TT by more than 0.002 s of time:

$$|T_{eph} - TT| < 0.002 \text{ s}$$

26.6.2.5 Coordinated Barycentric Time

Coordinated barycentric time (TCB) is the dynamical time recently defined by the International Astronomical Union; it is strictly based upon the SI second. Like T_{eph}, TCB is a relativistic coordinate time. TCB has the advantage that all quantities in a TCB-based system are expressed in SI units; however, there is a disadvantage: the form of the equation relating TCB to terrestrial time [TT] is such that TCB runs at a different rate than TT and therefore departs from it at the rate of about 0.5 s per year. Thus, TT cannot be used to approximate TCB. The 2 times were defined as equal in 1977.0 and have been departing from each other ever since.

26.6.2.6 Greenwich Mean Sidereal Time

Sidereal time should be thought of as an angle; it is, effectively, the spin of the earth in space. One orients the earth in inertial space by first aligning the earth's pole at the desired epoch, using the formulae of precession and nutation, and then by rotating the earth on its polar axis through the sidereal time.

Before the advent of atomic time, the rotation of the earth was used for time-keeping. This involved the determination of the earth's orientation in space and, thus, the determination of sidereal time. Now, the accuracy of atomic clocks allows the process to be reversed, and the measurement of sidereal time is used to show the variation of the earth's rotation.

Sidereal time is given by the hour angle of the of-date equinox. At a particular location, the hour angle gives the local sidereal time; at Greenwich, it gives Greenwich mean sidereal time (GMST).

26.6.2.7 Universal Time: Rotation of the Earth with Respect to Mean Sun

Universal time (UT) is based on the earth's rotation with respect to mean sun. The more refined UT1 is now used; it accounts for the wandering of the earth's rotational pole with respect to its surface. Like Greenwich mean sidereal time (GMST), UT1 should be thought of as an angle. It is related to GMST strictly by a given formula and is therefore determined directly by measurements of sidereal time. It is UT1 to which coordinated universal time (UTC) is adjusted by the use of leap seconds. If the difference between UTC and UT1 is more than 0.9 s, then UTC is adjusted. Thus, the mean sun at Greenwich is overhead at UT1 noontime and not exactly at UTC noontime.

CITED REFERENCES

1. International Astronomical Union. Paris (France): IAU Secretariat; [accessed 2011 Nov 13]. http://www.iau.org/.

2. Wilkins GA. IAU style manual. Norwell (MA): Kluwer Academic Publishers; 1989 [accessed 2011 Nov 13]. Also published in Trans Int Astron Union. 1988;20B:S1–S50. Also available at http://www.iau.org/static/publications/stylemanual1989.pdf.

3. Ridpath I, editor. Norton's star atlas and reference handbook. 20th ed. New York (NY): Pi Press; 2004.

4. United States Geological Survey, Astrogeology Science Center. Gazetteer of planetary nomenclature. Flagstaff (AZ): The Center; [accessed 2011 Nov 13]. http://planetarynames.wr.usgs.gov/.

5. Solar System dynamics. Pasadena (CA): National Aeronautics and Space Administration (US), Jet Propulsion Laboratory, Solar System Dynamics Group, Planetary satellite discovery circumstances; [modified 2011 Nov 9; accessed 2011 Nov 13]. http://ssd.jpl.nasa.gov/?sat_discovery.

6. International Astronomical Union. Pluto and the developing landscape of our Solar System. Paris (France): IAU Secretariat; [accessed 2011 Nov 13]. http://www.iau.org/public/pluto/.

7. International Astronomical Union, Committee on Small Body Nomenclature. Cometary names. Paris (France): The Union; 1997 [updated 2011 Feb 8; accessed 2011 Nov 13]. http://www.ss.astro.umd.edu/IAU/csbn/comet.shtml.

8. Grady MM. Catalogue of meteorites: with special reference to those represented in the collection of the Natural History Museum, London. 5th ed. Cambridge (UK): Cambridge University Press; 2000.

9. Hoffleit D, Warren WH Jr. The Bright Star Catalogue. 5th ed. New Haven (CT): Yale University Observatory; 1991 [modified 2004 Nov 8]. Also available at http://heasarc.gsfc.nasa.gov/W3Browse/star-catalog/bsc5p.html.

10. United States Naval Observatory, Nautical Almanac Office. Astronomical almanac for the year 2012 and its companion, the astronomical almanac online. Washington (DC): The Observatory; 2010 [accessed 2011 Nov 13]. http://asa.usno.navy.mil.

11. GCVS: General catalogue of variable stars. Moscow (Russia): Sternberg Astronomical Institute; 2011 Jan [accessed 2011 Nov 13]. http://www.sai.msu.su/gcvs/gcvs/index.htm.

12. SIMBAD Astronomical Database. Release 4. Strasbourg (France): Centre de Données Astronomiques de Strasbourg; [modified 2011 Oct 26; accessed 2011 Nov 13]. http://simbad.u-strasbg.fr/simbad.

13. The Messier catalog. Tucson (AZ): Students for the Development and Exploration of Space; [modified 2008 Feb 25; accessed 2011 Nov 13]. http://www.seds.org/messier/.

14. Sinnott RW, editor. NGC 2000.0. The complete new general catalogue and index catalogues of nebulae and star clusters by J.L.E. Dreyer. Cambridge (MA): Sky Publishing; 1988.

15. The interactive NGC catalog online. Tucson (AZ): Students for the Development and Exploration of Space; [accessed 2011 Nov 13]. Online interactive version based on Sinnott (reference 14). http://www.seds.org/~spider/ngc/ngc.html.

16. International Astronomical Union. Specifications concerning designations for astronomical radiation sources outside the Solar System, IAU Recommendations. Paris (France): The Union; 2008 Nov 26 [accessed 2011 Nov 13]. http://cdsweb.u-strasbg.fr/Dic/iau-spec.htx.

17. Earth Orientation Center. When should we introduce leap second in UTC? Paris (France): Observatoire de Paris. Table 1, Relationship between TAI and UTC [accessed 2011 Nov 13]. http://hpiers.obspm.fr/eop-pc/earthor/utc/leapsecond.html.

ADDITIONAL REFERENCES

Cox AN, editor. Allen's astrophysical quantities. 4th ed. New York (NY): AIP Press; 2000.

Dictionary of nomenclature of celestial objects. Strasbourg (France): Centre de Données Astronomiques de

Strasbourg; [updated 2011 Nov 4; accessed 2011 Nov 14]. Up-to-date electronic version of Lortet et al. (below). http://cdsweb.u-strasbg.fr/cgi-bin/Dic.

Hoffleit D. Supplement to the Bright Star Catalogue. New Haven (CT): Yale University, Department of Astronomy; 1984.

International Astronomical Union. Designations and nomenclature of celestial objects. Paris (France): IAU Secretariat; [updated 2011 Nov 4; accessed 2011 Nov 14]. http://cdsweb.u-strasbg.fr/Dic/iau-spec.html.

Lang KR. The Cambridge guide to the Solar System. 2nd ed. Cambridge (UK): Cambridge University Press; 2011.

Lortet MC, Borde S, Ochsenbein F. Second reference dictionary of the nomenclature of celestial objects. Astron Astrophys Suppl Ser. 1994;107(2):193–218.

Maran SP, editor. The astronomy and astrophysics encyclopedia. New York (NY): Van Nostrand Reinhold; 1992.

Mason BD, Wycoff GL, Hartkopf WI. The Washington double star catalog. Washington (DC): US Naval Observatory, Astrometry Department; [updated 2008 Oct 2; accessed 2011 Nov 14]. http://ad.usno.navy.mil/wds/.

Schmadel LD. Dictionary of minor planet names. 6th ed. Berlin (Germany): Springer; 2012.

Seidelmann PK, editor. Explanatory supplement to the astronomical almanac. Sausalito (CA): University Science Books; 2005.

4

Technical Elements of Publications

27 Journal Style and Format

27.1 DEFINITIONS

Journals, also known as serials to librarians and bibliographers, are "publications appearing or intended to appear at regular or stated intervals, generally more frequently than annually, each issue of which normally contains separate articles or papers."[1]

The usefulness and ease of use of a journal are determined not only by its content but also by its style and format. Decisions on these characteristics should aim to serve the needs of readers, librarians, and bibliographers. Priority should be given to clarity, accuracy, and adequacy of information; these can be served by carefully attending to the many details of style and format described here. The recommendations in this chapter are derived mainly from the relevant standards of the National Information Standards Organization (NISO)[1–8] and the International Organization for Standardization (ISO)[9–15]; see their websites (http://www.niso.org/ and http://www.iso.org/).

This chapter primarily presents style and format issues concerning the print journal. Standards for electronic journals and other nonprint formats have not been established by the NISO and ISO. Many of the recommendations in this chapter are, however, relevant in principle to elements of the presentation of nonprint journals. (See Section 27.3.1.2 for some special considerations regarding the publication of electronic journals and Sections 29.3.7.12 and 29.3.7.13 for information on their citation.)

27.2 FORMAT

Although the print journal is still currently considered the predominant format, the last decade has seen a dramatic shift to online publication as a complement to the traditional print journal. In addition, many new and established journals have begun to forego print editions completely, and this trend will likely continue throughout the next decade. Journals published in print are generally static in nature; their content is fixed at the time of publication and does not change. Online journal articles may also be static, but they can be updated to correct errors or make similar changes. Likewise, most journals now publish early online versions of many articles, allowing articles to be available sooner than the traditional print production process would permit.

Because many journals are issued in more than one format, publishers should designate one of these as the "version of record"—that is, the authoritative version in which errata and other changes are recorded, and inform the reader which version that is. Although it may no longer be necessary to differentiate which format is being cited in a reference list, if multiple versions of an article exist, it is important that readers know which version is being referenced. For example, if an early online version of an article was used, an author should not cite the final version of that article without consulting it first to ensure that no significant changes were made. (See Chapter 29 for details on citation.)

27.3 PUBLICATION

27.3.1 Publishers and Publishing

27.3.1.1 Overall Process

A journal publisher provides the following services: editorial, production, sales and marketing, subscription fulfillment, distribution, and finance. Managing editors, also called acquisitions editors, represent the publisher to journal editors and authors. In addition to financial and production management, managing editors provide oversight to manuscript editors and editorial assistants and serve as liaison to the journal editors.[16]

Journals vary widely in the methods by which their editors choose to accept or reject manuscripts. Most editors apply the following criteria[17]:

1) relevance to the journal's scope and audience
2) importance of the message to most of the journal's audience
3) newness and innovative nature of the message
4) scientific validity of the evidence supporting the conclusions presented
5) usefulness to the journal in maintaining a suitable range of topics

Most journal editors are assisted in their decisions on manuscripts by a system of external review. Individuals termed "peer reviewers", "referees", or "assessors" are invited to provide expert opinion. Such reviewers, usually a minimum of 2, are provided with a form or other list containing the type of information sought on the originality of the work, accuracy of the methods, and soundness of the conclusions.[16] Some journals mask the reviewers to the identity of the authors, although this practice is falling out of favor. Although most journals maintain the anonymity of the peer reviewers, some journals allow the peer reviewers' identity to be known to the authors, either generally or with the permission of the peer reviewer. This is termed "open peer review" and is finding increasing favor. Some journals also ask authors to suggest suitable peer reviewers for their manuscript, who may or may not be used by the journal. Regardless of the particular practice followed, journals should publish the names of their editors and their policy on peer review.

Once a manuscript has been accepted, manuscript editors provide the link between publisher and author, adapting the manuscript to the journal's house style, checking for accuracy and consistency, and looking for discrepancies and omissions. Copy editors also read the proofs and work with authors and editors on corrections needed.

Many books are available to provide details of the publishing and editing process. Day's *How to Write and Publish a Scientific Paper*,[18] Hall's *How to Write a Paper*,[16] Huth's *Writing and Publishing in Medicine*,[17] and Zeiger's *Essentials of Writing Biomedical Research Papers*[19] are a few examples. Others may be found in the Bibliography in this manual.

27.3.1.2 Online Publication

27.3.1.2.1 ADVANTAGES AND DISADVANTAGES

It is the standard today for journals to be published online, whether or not they also appear in print. Online publication provides many advantages over print[20]:

1) The size limitations of print no longer apply. Some journals that publish in both print and online formats are placing longer versions of their print articles on the Internet. The *BMJ*'s ELPS program—Electronic Long, Print Short—is one example. Other journals are publishing additional articles online only.
2) A variety of types of supplemental material that would be impossible or impractical to have in print are being made available online, including data files, complex tables and graphs, film and video clips, and animation.
3) Hyperlinks (electronic links to another place in the same document or to another document) make articles more usable. For example, a hyperlink can join an in-text reference to the reference list or to a full-text copy of the document being referenced.
4) Color can be added without the appreciably higher costs incurred for it in print.
5) Overall production is faster. Manuscripts are submitted and edited online, and editorial changes can be made easily. Many journals also are making some articles or entire issues available online ahead of the print publication to reach their constituents faster.
6) Online distribution is easier than mail service and has a lower cost because such processes as printing and binding are eliminated. Online distribution also provides publishers a much wider audience for their journals and enables them to make available large archives of previous volumes.
7) For online-only journals, production and inventory costs are greatly reduced, allowing for new distribution possibilities, such as open access publication.

For all of these reasons, online publication has become extremely desirable. However, some unresolved issues in online journal publishing include the following:

1) Access. Currently online journals offer a bewildering array of options for access. A few publishers offer open access, permitting free entry to all articles. Some embargo the current issue; others embargo all but the current issue. Still others password-protect their journals to subscribers only and offer pay-per-view for nonsubscribers. Various organizations have also entered the access fray. Some associations and societies offer free online access to all of their journals as a benefit of membership; other institutions pay for licenses to enable their members or employees to gain access to relevant titles. The National Institutes of Health (NIH) have established guidelines for access to articles based on NIH-funded research (see http://publicaccess.nih.gov/), but otherwise the availability of articles can vary widely.
2) Copyright. Because journal articles published online can be easily copied and altered, many publishers have justifiable concerns about their intellectual property rights.
3) Archiving. Because of the high cost of maintaining archives, a number of publishers have reached agreements with archiving organizations such as HighWire (http://highwire.org/), ScienceDirect (http://sciencedirect.com), and PubMed Central (http://www.pubmedcentral.gov/). However, no standards currently exist for the archiving of electronic material, and the problems of data and media degradation, as well as migration to newer formats, are real concerns.
4) Numbering. Although journals with print counterparts continue to use the traditional practice of numbering volumes and issues, online-only journals display a wide variety of schemes, ranging from a document number with no volume to no numbering at all,

only a date of publication. Online documents without print versions may not carry page numbers, and some journals that publish online before print use their web-based article numbering system as their print pagination system. These numbering issues make bibliographic citation more difficult (see Section 29.3.7.13).

For a discussion of costs and related business issues, see *Electronic Scientific, Technical, and Medical Journal Publishing and Its Implications*,[21] a report of a National Academy of Sciences symposium.

27.3.1.2.2 PRODUCTION ISSUES

Many publishers choose to use PDF, Adobe's Portable Document Format, to make their journals available electronically because it is the easiest form to implement. Numerous conversion programs[22] exist to take documents from such commonly used formats as Microsoft Word into PDF form. These programs range from costly to free of charge, depending on the complexity of the text to be converted. An example of freeware is the National Library of Medicine's DocMorph.[23]

Although easily produced, a PDF file is simply an electronic copy of the pages created for print. To take advantage of the full features and capabilities that the electronic world provides, producing journals in a markup language such as HTML (Hypertext Markup Language), XHTML (Extensible Hypertext Markup Language), SGML (Standard Generalized Markup Language), or XML (Extensible Markup Language) is required. For example, interrupting the text with a definition or other explanation, although helpful for the novice, can be annoying to a knowledgeable reader. With HTML, the definition can be provided by means of a hyperlink, accessible by a mouse click or mouseover, if needed. However, because text produced by HTML and the other markup languages is nonlinear, users can easily get lost in the links. Such devices as sidebars for an expandable table of contents are therefore essential. This nonlinear format also makes producing a version of the journal in a printable format a challenge. Many books about design are available, and representative ones are included in the Bibliography of this manual.

In addition to the usual elements required for print publication, 3 newer ones—DTDs (Document Type Definitions), metadata, and DOIs (Digital Object Identifiers)—must be considered for online journals. A DTD is a scheme for organizing the contents of a document. It defines the parts of a document called "elements", provides ways to describe the elements by means of attributes, and enables units of code to be created that can be referred to by names called "entities".[20] In a journal article, for example, an article title is one element and an author name is another. A DTD defines the structure of a document, enabling use of the newer features of markup languages and therefore leading to improved production of the finished document.

Metadata is standardized, descriptive information about a publication presented in a machine-readable, structured form. Two major metadata vocabularies in use for publishing are the Dublin Core Metadata Initiative (DCMI) (http://dublincore.org/) and the Online Information Exchange (ONIX) (http://www.editeur.org/8/ONIX/). DCMI describes 15 metadata elements: title, creator, subject, description, publisher, contributor, date, type, format, identifier, source, language, relation, coverage, and rights. Both the ISO and NISO have developed standards for the use of Dublin Core.[8,14] ONIX uses more than 200 metadata elements.[20]

The purpose of metadata is to assist in 1) identifying, locating, and organizing resources; 2) promoting interoperability across systems; 3) providing for digital identification using such identifiers as ISSNs (International Standard Serial Numbers) and URLs (Uniform Resource Locators); and 4) archiving and preservation.[24] Metadata, for example, is used by online search engines such as Google Scholar to locate documents satisfying search criteria; therefore, the presence of metadata in an electronic journal article enhances its retrieval. Two publications provided by the NISO are useful introductions: *Understanding Metadata*[24] and *Metadata Demystified: A Guide for Publishers*.[25]

The DOI was developed by the International DOI Foundation (http://www.doi.org/) to provide for persistent identification of documents across networks. It does this by means of a naming convention consisting of a prefix and a suffix. The prefix contains the name of the particular DOI directory and the content owner's identifier; the suffix is a numeric or alphanumeric string supplied by the publisher. Publishers register with the International DOI Foundation to obtain their owner's identifier, and then submit their suffix to a DOI registration agency along with the URL and appropriate metadata for the particular document being registered.[8] See NISO Z39.84 *Syntax for the Digital Object Identifier*[8] for further information. Although DOIs can be applied to any order or level of web-based objects, most publishers are currently assigning DOIs at the article level.

The Columbia Guide to Digital Publishing[20] provides a comprehensive discussion of online and electronic documents. For a detailed listing of publications on the topic, see Charles Bailey's *Scholarly Electronic Publishing Bibliography 2010*.[26] (See also the Bibliography at the end of this manual for further reading and Chapter 31, "Typography and Manuscript Preparation". For information on citing online journals, see Sections 29.3.7.12 and 29.3.7.13.)

27.3.2 Copyright

Many journals require authors to assign copyright to the journal, but the copyright status of a given article may vary. For example, articles produced by employees of the US government cannot be copyrighted. A newer trend, particularly with open access journals, is to permit authors to retain copyright. When this occurs, journals usually require authors to grant them the exclusive right to publish their article, both in print and in electronic/online form.[16]

Journal editors should make their policy on copyright clear[27] and include this information in the instructions to authors (see Section 27.6.3.5). (See Chapter 3, "The Basics of Copyright", for details.)

27.3.3 Changes and Irregularities

In general, changes should not be made in the characteristics of a journal (title, trim size, frequency, design, and so on) without careful thought as to the possible consequences and the expectation that the changes will apply for an indefinitely long period. Frequent and hastily adopted changes can injure a journal's reputation among authors, readers, and librarians.

27.3.3.1 Change in Trim Size

A change in trim size should be made only with the first issue of a volume. The former trim size (and other aspects of format) should be applied, however, to any parts of the preceding volume produced after the first issue of the volume with the new size, such as a volume index or a supplement that is part of the preceding volume.

27.3.3.2 Change in Frequency of Issues

A new frequency of issues (e.g., weekly instead of the previous monthly) should begin with a new volume, with the change prominently announced (on the cover and the table of contents) in the last issue of the previous frequency and the first issue with the change. Readers, librarians, bibliographers, and subscription agencies should also be alerted to the change, well in advance of its occurrence.

27.3.3.3 Irregularities

Any changes other than the kinds considered in this section should be called to the attention of readers in the first affected issue, on the cover, at the head of the table of contents, and on the masthead page, with a description of the change and its expected duration.

When issues are combined or separated into parts, the issue number should carry in its various locations (cover, table of contents, masthead) clear identification representing the change. For example, 2 combined issues could be represented by "Volume 17, Issues 1 and 2", an issue split into 2 parts by "Volume 17, Issue 5, Part 1" and "Volume 17, Issue 5, Part 2".

27.4 JOURNAL TITLES

27.4.1 Function and Length

A new title should be chosen with great care. The title should represent the best compromise between a title long enough to adequately convey the journal's scope and content and short enough to not annoy persons having to record it in bibliographic records.

> *Current Research* [Does this journal cover all current research in all fields? Some fields within a broad discipline such as biology? Research only by members of the sponsoring society?]
>
> *The International Archives of Botany, Plant Physiology, Zoology, and Related Fields* [The scope appears to be all of biology. Why try to specify some components and not others? Why not devise the shortest title that represents the scope reasonably well—e.g., *The International Archives of Biology*?]

A single-word title can be amplified in meaning with a compound title or subtitle that details the journal's scope.

> *Medusa: A Multidisciplinary Journal of Mythology*

Note, however, that such a title might be represented in databases and indexes by the initial term only.

Acronymic titles, such as "PABA" for *Proceedings of the American Bariatrics Associa-*

tion, can cause confusion. They may be readily understood by readers in the journal's field but are cryptic to others and fail to indicate the scope and content of the journal. If they are used along with the full title they represent, users of the journal may not know which version is preferred for references.

If the scope of a journal changes, a title change may be desirable (see Section 27.4.5).

27.4.2 Uniqueness

Care should be taken to ensure that the title does not duplicate another title in use or owned by another publisher. Duplication might represent infringement on the legal right of a publisher to a title and, at the very least, could confuse readers, librarians, and bibliographers. Even choosing a title very similar to an existing one can cause confusion and lead to conflict with the first title's owner.

Existing titles can be identified in many different sources, both print catalogs and database listings of serial titles; examples are the *List of Serials Indexed for Online Users* of the National Library of Medicine (http://www.nlm.nih.gov/tsd/serials/lsiou.html), *BIOSIS Serial Sources*[28] for the BIOSIS Previews Database, and *Ulrich's Periodicals Directory*.[29] (See Appendix 29.1 for a list of sources for scientific journal titles.) These and other sources will be searched by the National Serials Data Program in response to requests for information on the uniqueness of a proposed title or to a request for an ISSN (see Section 27.4.4). Information on additional sources is usually available from librarians in science and other academic libraries.

27.4.3 Locations and Form

The identical and complete form of the title should be used wherever it appears: cover, spine, table of contents, masthead, officers' page, running heads and footers, and text. If the title must be abbreviated to save space, the abbreviation should have the form specified by *Information and Documentation: Rules for the Abbreviation of Title Words and Titles of Publications* (ISO 4).[15] The database for this standard, maintained by the ISSN International Centre in Paris (http://www.issn.org/), is available by subscription. Sources of title-word abbreviations and abbreviated titles prepared in accordance with this standard can be found in Appendix 29.1.

The location and typography of the title should be such as to avoid any ambiguity with other text near it. An acronymic designation of a society or publisher's name should not be placed near a title if it might be thought to be an acronymic form of a title.

27.4.4 ISSN and CODEN

Two widely used numbering systems for the identification of serials are the ISSN[5,10] and the CODEN. The ISSN is an 8-digit code number that uniquely identifies a journal title.

Journal of Cyclic Nucleotide Research ISSN 0095-1544
[continued as]
Journal of Cyclic Nucleotide and Protein Phosphorylation Research ISSN 0746-3898

Note that the eighth character is a check digit, which may be the letter X (when the check-digit number would be 10) rather than a numeral. In the United States, a request for an ISSN for the title of a new journal or a new title for an existing journal should be made to the National Serials Data Program, Library of Congress, Washington (http://www.loc.gov/issn/). See the website of the ISSN International Centre in Paris (http://www.issn.org) for information about obtaining ISSNs in other countries. The request should be made before publication of the new title so that the ISSN can appear on the first issue under this title.

A CODEN is a unique 6-character identifier of a serial.

Journal of Cyclic Nucleotide Research	JCNRDU
Journal of Cyclic Nucleotide and Protein Phosphorylation Research	JCNREV

CODENs are assigned by the International CODEN Service, maintained by the Chemical Abstracts Service, Columbus, Ohio (http://www.cas.org/products/other-cas-products). CODENs are assigned to journals in all subject areas, not just in chemistry.

27.4.5 Change of Title

A change in the title of a journal should be made only for important, carefully considered reasons, such as a change in topical scope or sponsorship or the need for a clearer, more specific title. The change should be made at the beginning of a volume, preferably the first volume and issue of a calendar year. The precautions that should be taken in selecting a new title are the same as for choosing the title of a new journal (see Sections 27.4.1 and 27.4.2). The title will have to be changed in all of its locations (see Section 27.4.3). In these locations, the former title should be in positions and a size subordinate to the new title. The former title (preceded by "Formerly") should appear in these locations for 1 year after the change.

When 2 journals are combined and 1 of the previous titles is continued, the recommendations in the paragraph above apply; the volume-number sequence of the retained title should continue. When a new title is applied to 2 combined journals, apply the recommendations for a new journal. For 1 year, the new title should be accompanied in the stipulated locations by a clarifying statement, "[New Title] represents the combined [Previous Title 1] and [Previous Title 2]."

If a journal must be divided into 2 or more new journals and its title is continued for 1 of the journals, that journal should continue its ISSN and volume numbering; the new journals should apply the recommendations for a new journal, but the cover, table of contents, and masthead should carry for 1 year the statement "Continues in part [Original Title, ISSN]".

Any changes in title should be called to the attention of all subscribers with a notice mailed separately from issues of the journal or journals making the change. Librarians, database producers, and subscription agencies should also be alerted to the change well in advance.

27.5 JOURNAL VOLUMES AND THEIR PARTS

The division of journals into units and subunits, usually designated as "volumes" and "issues", is a long-standing convention that brings readers the journal in readily handled dimensions, facilitates the journal's display and storage in libraries, and provides conventions for bibliographic identification. Volumes and issues should be organized and formatted with a view to efficiently serving the needs of readers for many kinds of information.

27.5.1 Period of Publication

Each volume should represent no more than 1 full calendar year of issues. If a single volume carrying all issues in a calendar year would be too bulky for reliable binding and easy handling, a year's issues can be grouped into 2, 3, or more volumes, but all volumes should include the same number of issues.

Because the recommended formats for bibliographic references to journals and their articles include the year of publication (see Section 29.3.7.1.4), planning for a new journal ought to avoid having any volume include issues from more than 1 calendar year. Thus, for a journal with monthly issues to be gathered in 2 volumes, the starting date would be January or July; for a semimonthly (twice-a-month) journal, the appropriate starting date would be 1 January or 1 July; for a weekly journal, the date of the first Monday in January or July. For semimonthly journals, the successive dates are the arbitrary choices of the 1st and 15th of each month.

27.5.2 Numbering

Volumes should be numbered sequentially with arabic numerals starting with 1, the numeral to follow "Volume". If the journal's title is changed, common practice is to continue the volume numbering sequence with the new title, but numbering may begin again with 1. Roman numeral sequences should not be used.

27.5.3 Pagination

Page numbers of the text pages must begin at 1 and run sequentially through all text pages of the issues making up the volume, rather than restarting sequential numbers at 1 in individual issues, as is usually done in popular magazines. That practice can make finding articles in bound volumes more difficult.

The preliminary pages of a volume (title page, table of contents, information pages; see Sections 27.5.4.1 to 27.5.4.3) are generally numbered to distinguish them from text pages. Lowercase roman numerals (i, ii, iii, iv, v, and so on) have been conventionally used in books and journals, but an alternative system of alphanumeric designators (such P1, P2, P3, and so on) may be used. Pages not to be bound into the volume, such as advertising pages, can also be numbered alphanumerically to distinguish them from text pages, such as A1, A2, A3, and so on.

The pages of the index to the volume should continue with the numeric sequence of the text pages if the index is placed at the end of the bound volume. If the journal places the index just before the main text of the bound volume, the pagination should continue that of the preliminary pages opening the volume. Alternatively, a system of alphanumeric designators, such as I1, I2, and so on, may be used, especially if the index is printed and bound separately.

Note that publishers and printers often refer to page numbers as "folios".

Online journals with print counterparts usually carry the same pagination scheme. However, online-only journals may have no page numbers (see Section 27.3.1.2.1).

27.5.4 Parts of Volumes

Traditionally, each volume included at least 4 parts: 1) title page; 2) volume table of contents; 3) text of the volume; 4) index.

Although online archives are increasingly replacing bound volumes, some journals and publishers still prefer this method for collecting a volume's worth of content. In these cases, information pages for the entire volume can be helpful to many readers and will be included in the bound volume in a library even if the usual information pages are cut out of the individual issues in the bound volume. These pages can provide the information given on the masthead page of each issue (see Section 27.6.3.3.4 and Table 27.5) and in the Information for Authors section (especially if they are published within the advertising pages of one or more issues rather than within the text pages; see Section 27.6.3.5 and Table 27.6).

To facilitate prompt binding of the issues of a volume, the title page, the volume table of contents, the information pages, and the index are usually published as the end pages of the last issue of the volume. If these pages are, instead, routinely sent to subscribers later in a separate form or are sent only on request, this practice should be noted on the masthead of each issue. In separate form these pages must have the same trim size as issues.

27.5.4.1 Title Page

The title page for the volume should carry information needed to identify fully the journal, the volume, the issues included, the first and last dates of issues, the ISSN and CODEN, and the publisher's name and location (see Table 27.1); it should be a right-hand (recto) page but need not carry a page number. The editor's name may be included here or on the masthead or both.

The reverse (verso) of the title page is usually blank but can be used to carry additional information about the journal, such as that usually appearing in the masthead of issues (see Section 27.6.3.3.4).

27.5.4.2 Table of Contents

The volume table of contents should provide all the information carried on the issue tables of contents (see Section 27.6.3.3) and in the issue sequence. It is conveniently formed from the tables of contents of individual issues; maintaining separation of the contents listing into issues (and their dates) rather than providing an uninterrupted sequence for the entire volume is probably preferred by most readers. If additional preliminary pages such as a masthead page or Information for Authors pages follow the table of contents, these should be listed before the listing of the contents of the first issue.

The table should begin on a right-hand (recto) page (see Section 27.5.3).

27.5.4.3 Information Pages

These pages can include information carried in each issue that would not be available in bound volumes—for example, an Information for Authors section published within the advertising pages of issues. For recommendations on the content of a masthead page and Information for Authors, see Sections 27.6.3.3.4 and 27.6.3.5 and Tables 27.5 and 27.6.

Table 27.1 Volume title page: recommended information

Information	Notes
Complete title (Section 27.4)	The title should be the dominant element and preferably near the top. If the complete title is a compound title or includes a subtitle, all elements should be included. If an acronymic version of the title is the title proper, there should be no ambiguity with the complete title; if the acronymic version is not the title proper, the acronym should be inferior to the formal title both in location and font size.
ISSN and CODEN (Section 27.4.4)	These can be in a subordinate location, such as the foot of the page.
Publisher's name and location (Section 27.5.4.1)	If the journal has a sponsoring organization (such as a professional society) that is not the publisher, the organization name and location should also appear.
Editor's name (Section 27.5.4.1)	This information need not appear here if it is on the masthead page or other information pages included among the preliminary pages.
Volume number and issue numbers	First and last issue numbers of its issues (in arabic numerals) separated by a hyphen or en dash; if volumes are divided into parts, a part title page should carry only numbers of the issues included.
Dates of the first and last issues	Including the year, as in "January–December 2014".

27.5.4.4 Index

The text pages should be indexed for subjects of articles and other separate units (such as editorials and letters) and their authors. Separate subject and author indexes are easier to consult than single indexes carrying both categories. Index pages should not be interrupted by advertising or news pages, and their pagination should be continuous with that of the preceding text pages. Alternatively, an alphanumeric page numbering system, such as I1, I2, and so on, may be used, particularly if the index is bound separately. The index is preferably the last element of the text pages of the last issue in a volume (see Section 30.6).

27.6 JOURNAL ISSUES AND THEIR PARTS

Issues are collections of papers and other documents assembled and published with dimensions and a frequency that are economical for the publisher and convenient for the reader. For consistency, most journals that publish both a print and online version continue to follow these conventions, as do many online-only journals. Although there are notable exceptions, such as *PLOS ONE*, that publish articles online individually and do not collect them in an "issue", the PDF versions of these articles still comply with many of the text guidelines described in Sections 27.6.3.4 and 27.7.

27.6.1 Trim Size

All issues should have the same trim size (height and width). A large number of scientific journals have the trim size of 8.5 by 11.0 inches (21.5 by 28.0 cm or a similar metric standard). This size is desirable for several reasons: it accommodates more text per weight of paper than smaller trim sizes, which may reduce both paper and postal costs; it facilitates the sale of advertising space; it allows conveniently for 2- or 3-column page formats; and bound volumes will fit on most library shelving. Smaller trim sizes make up a bulkier issue for a given amount of text.

27.6.2 Frequency, Numbering, and Dates

The choice of frequency (quarterly, monthly, or weekly for most scholarly journals) is usually determined by the amount of text to be published and the relative merits of getting papers to readers without unnecessary delay and in a conveniently handled format. These factors are judged against the lower costs of printing, binding, and distributing with a lower frequency of publication.

Issues are identified sequentially with arabic numerals beginning with 1 within each volume; for example, the January issue of a monthly journal is "Issue 1". Roman numerals should not be used.

The needs of publishers, readers, and librarians are best served with a constant frequency and issuance on fixed dates (see Table 27.2 for recommendations on dates of issues and see Sections 27.6.3.1 to 27.6.3.3 and Tables 27.1 and 27.3 on locations for issue dates).

27.6.3 Parts Of Issues

27.6.3.1 Cover

The front cover must carry the information needed to identify the journal and the issue unequivocally (see Table 27.3). The covers of particular issues may need additional information (see Table 27.4).

Bar codes (Code 128) identifying a journal's ISSN, its title, and the particular issue

Table 27.2 Issue dates

Frequency of issues	Recommended dating
Weekly	The same day of each week.
Quasiweekly (for example, 50 issues per year)	The same day of each week in which an issue is published.
Semimonthly (twice a month)	Dates representing approximately equal intervals between issues. Convenient dates are the 1st and15th days of each month (such as 1 January 2014 and 15 January 2014), even if the mailing date does not coincide with the issue date.
Monthly	The month of issuance; a day of the month need not be specified. The date of mailing should be as constant as possible from month to month.
Bimonthly (2-month intervals)	The pair of months issued, even if the mailing date (which should be as constant as possible) is not in the first month; for example, January–February 2014.
Quarterly or thirdly	The span of months represented by each issue as with bimonthly issuance, such as January–March 2014 (quarterly) and January–April 2014 (thirdly). Designators such as "Winter 2013", which could mean mailing in December 2013 or in February 2014, should be avoided.

Table 27.3 Issue front cover: needed information and location

Element	Recommendations
Complete title	With subtitle if used. In a dominant position on the upper portion of the cover.
Volume number and issue number	In arabic numerals, such as "Volume 68, Number 9", placed near the title. If issues of a journal are combined, the numbers of the combined issues and their dates should appear, such as "Volume 68, Numbers 9–10, September–October 2014".
Issue date	Placed near the title.
Weekly and semimonthly	A specific date, not a specific day, in ascending or descending sequence of elements, for example, "15 January 2014" or "2014 January 15", not "January 15, 2014" (see Section 13.3.3).
Monthly	Month and year.
Bimonthly	Single issue number, paired months, year, as in "Number 2, March–April 2014".
Quarterly or lower frequency	Issue number, months represented, and year; see "Bimonthly" above. Seasonal designators such as "Winter 2013" should be avoided.
Name of publisher and of sponsor	Not duplicated if they are the same. Sponsor names are also often placed with the masthead or elsewhere, at the discretion of the publisher.
URL for journal's website	Complete URL, including "http://"
Location of the table of contents (page numbers)	Only if the table of contents is not on the front or back cover or within the first 3 pages.
Issue contents	If length of contents information and space available on the cover permit.
Bar code	See Section 27.6.3.1.

Table 27.4 Issue front cover: additional information for particular issues

Element	Recommendations
Volume index	Location given in page numbers.
Supplements, special issues	Identified as such on the front cover, including topic or title and correct citation (i.e., "Volume 32, Supplement 1", "Volume 32, Number 3, Supplement B", or "Number 9, Part 2", with the usual part identified as "Number 9, Part 1". See Section 27.6.3.9).
Change of title	Previous title within parentheses or after "Formerly" under the new title on the covers of all issues in the first year of the volume with the new title. See Section 27.4.5.
Changes in trim size, frequency; irregularities	See Sections 27.3.3.2 to 27.3.3.3.

are widely used by popular magazines to speed accession of issues in libraries. The recommended location for the code is the lower left-hand corner of the front cover in the horizontal orientation. Information can be obtained from the Serials Industry Systems Advisory Committee (SISAC) (http://www.barcodeisland.com/sisac.phtml).

The information recommended in Table 27.3 should appear on the front cover of every issue in the same location, typefaces, and fonts. The title should be the most prominent element on the cover and preferably be placed in the upper half of the cover. If titles are presented in more than one language, the primarily known title should be emphasized by its size and location. Formatting that might produce ambiguity or obscurity of this information (e.g., from advertising) should be avoided. The cover should not be included in the pagination of the text.

27.6.3.2 Spine

If the spine is wide enough to accommodate legible type, it should carry the most important information identifying an issue:

1) the title of the journal or its standard abbreviation (see Section 27.4)
2) the volume and issue numbers
3) the issue date
4) the page numbers of the issue if the volume is paginated continuously

If the spine is wide enough, this information should be printed so that it can be read when the issue stands upright on a shelf. On a narrower spine, the information should be positioned so that it can be read from left to right when the issue is lying with the front cover up.

27.6.3.3 Table of Contents Page

The table of contents of each issue must identify fully all of the articles and the sections into which they are grouped. Material of minor and ephemeral value might be identified solely by a section title, such as "News Notes".

The front cover was the preferred location in scientific journals for the contents listing, but this is no longer the most common practice. The next preference is to place it on a right-hand (recto) page as close to the front of the journal as possible. Care should be taken not to bury the contents list among advertisements and other introductory material. If the table of contents is on the front cover, one should also appear within the first 5 pages as a precaution against loss or damage of the front cover.

Tables of contents of past or coming issues should be clearly identified and placed to eliminate confusion with the table for the present issue.

27.6.3.3.1 BIBLIOGRAPHIC IDENTIFICATION

All information needed to identify the journal and the issue should be included at the head of the table of contents page:

1) complete title and subtitle
2) abbreviated title in standard form (see Section 27.4 and Appendix 29.1) and the ISSN (see Section 27.4.4), both of which need not be adjacent to the complete title but should be in an obvious place and not buried in text
3) volume and issue numbers
4) date of issue (exact for a weekly, biweekly, or twice-monthly [semimonthly] journal); only month or month range for a monthly, bimonthly, or quarterly journal (see Table 27.2)
5) notice of a change in title or frequency

27.6.3.3.2 INFORMATION ON ARTICLES AND SECTIONS

All articles except those of minor and ephemeral value must be fully identified. Titles must have the same complete form as on the title pages of articles. Complete titles must be provided for articles of more than ephemeral value that appear under section headings, such as "Editorials".

Author names should have the same form as on the articles if space permits; if it does not, given names should be represented by initials as in bibliographic references (see Section 29.3.6.1).

Only initial page numbers need be given for articles paged continuously. Section-grouped articles and other content should be completely identified under the headings of the sections if space allows. Sections with many short articles—for example, book reviews, letters to the editor, and news notices—may be represented only by the section heading and the initial page of the section.

Special features (such as an Information for Authors section, an erratum or retraction notice, and a volume index) that do not appear in every issue must be represented by title and initial page number. If an Information for Authors section is published only at infrequent intervals, the issues in which it does appear should be identified on the contents page of all other issues. A feature appearing in every issue and serving important needs of readers and authors, such as a journal-policy page or an Information for Authors section, should be represented with a page number, even if carried in an advertising section.

27.6.3.3.3 FORMAT AND TYPOGRAPHY

Bibliographic data, including the journal title, should precede the title "Table of Contents" (see Section 27.6.3.3.1).

Section and article titles should generally appear in the same sequence in the table of contents as in the issue. For those journals whose articles are published directly online, however, articles may appear in date order or a subject-oriented sequence in the table of contents. The most legible arrangement for article titles is probably that of 3 columns: article titles in the first, author names in the second; page numbers in the third. If space does not allow this format, article titles and author names should be listed on the left and page numbers in a right-hand column. Different typefaces for titles and author names may help with legibility, such as a roman face for titles and an italic face for names.

Black ink is preferred to allow for clear reproduction of the page in an alerting service publication such as *Current Contents*[30] or in photocopying. The type should be preferably 10-point or larger if the text page is more than 7 inches (17.75 cm) wide.

27.6.3.3.4 THE MASTHEAD

Each issue should include a section describing the journal, its availability, its ownership, and any additional information often needed by authors, readers, subscribers, librarians, dealers, indexers, bibliographers, archivists, and advertisers. (See Table 27.5 for details on the information to be provided on the masthead page; note that some of the recommendations are relevant only to publication in the United States.) The content

Table 27.5 The masthead

Type of information	Notes
Minimally Needed Information	
Title, subtitle, International Standard Serial Number (ISSN), CODEN	If the journal is published with titles in more than one language or with an acronymic title, this information should be included; the complete (not the acronymic) title most widely used should be given first place.
Publisher	Name and complete address (postal [including postal code], email, facsimile-transmission ["fax"] addresses); telephone numbers (including toll-free number), with area (or regional) code.
Volume number per year, issue number per volume, and issue frequency	Frequency as "weekly", "biweekly" (not "fortnightly"), "semimonthly" (or "twice a month"), "monthly", "bimonthly", "quarterly", "semiannually" (or "twice a year").
Postal data required	Information required by the postal service of the countries in which the journal is published and distributed.
Copyright information	The legally required notice of copyright with the copyright sign (©), the copyright year, and the name of the copyright holder. Also, the publisher's policy on fair-use copying and mechanism for copying royalties (Copyright Clearance Center or other means) should be stated.
URL for the journal's website	Complete URL, including "http://"
Subscription rates, single-issue price, ordering information	The address to which orders should be sent, even if it is the same as that of the publisher. Ordering information: terms of payment (currency, payment methods [check, money order, cabled payment, credit-card payment], including telephone number for ordering), availability of back issues and other media (such as microforms, online services).
Address-change procedure	Address to which notice of changes should be sent.
Sponsorship	Name and address of the sponsor if not also the publisher.
Advertising management	Names, addresses, and telephone numbers of staff persons or agencies able to give information on display advertising, classified advertising, or both kinds.
Logo and name of circulation auditor	
Statement and logo indicating publication on acid-free paper	The statement recommended by the NISO standard: "This paper meets the requirements of ANSI / NISO Z39.48-1992 (Permanence of Paper)". The logo is an infinity symbol ∞ within a circle.
Additional Useful Information	
Editors and other staff persons	Names, addresses, and telephone numbers of persons receiving manuscripts and other mail intended for the editor.
Editorial board or committee	Additional editors or associate editors who assist the editor with peer review and acceptance decisions.
Publication committee	Representatives of the sponsoring or publishing organization.
Peer-review policy	Details useful to authors.
Indexing, abstracting	Identification of the print and online services providing indexing data and abstracts from the journal.
Statement on use of recycled paper	

specified by "Additional useful information" should be considered for inclusion if it is not placed in an Information for Authors section.

The masthead should appear within the first 5 pages, preferably on the same page as or immediately adjacent to the table of contents. Its location should be constant from issue to issue. If it appears on the inside front cover or within pages not likely to be bound in the volume, it should be included with the preliminary pages (title page, table of contents) that are provided with the conclusion of each volume (see Section 27.5.4). If the masthead information is on the table of contents page, the layout and typography should make clear the information most likely to be needed by most users of the journal.

27.6.3.4 Text Pages

27.6.3.4.1 FORMAT

The text can be carried as a solid text block on each page or divided into 2 or 3 columns, separated from each other by a gutter (white space). Text in columns must begin at the upper left-hand corner of the text block, proceed down the first column, and continue at the top of the next column.

The choice of format from among the single-, double-, and triple-column possibilities will hinge on various considerations. The single-column format used in a journal with a relatively large trim size (such as the 8.5 by 11.0 inch [21.5 by 28.0 cm] size widely used in the United States) may have lines of type too long for convenient reading. The double-column format produces shorter and more readily read lines, which, in addition, can have less space between lines (called "leading") and permit variation in placement of figures and tables, with saving in space needed for a given amount of text. A 3-column format may produce lines that are too short for efficient reading unless the margins around the text block are narrow or the trim size is larger. The choice among the possible formats should be made in consultation with a typographic designer.

The margins must be wide enough to allow for some trimming in the binding process and for the binding itself. Suggested minimums for the inner gutter are 1.0 inch (2.5 cm); for the top, outer, and bottom margins, 0.75 inch (1.9 cm).

Figures and tables should be aligned within the margins of the text block and be readable in the same orientation as the text. If a table is too large for this orientation, it can be rotated counter-clockwise 90° so that its bottom is against the inner margin if it is on a left-hand page and against the outer margin if it is on a right-hand page. A large table can be split between 2 facing pages, but the row headings must be repeated on the right-hand half of the table. Larger tables may be run across more than 2 pages, but again, headings must be repeated. Rules for the most pleasing placement of figures and tables on pages can be developed by the journal's designer, but, in general, tables and figures are best placed in the upper part of a page and the text in the lower part (see also Chapter 30, "Accessories to Text").

27.6.3.4.2 INFORMATION CARRIED ON PAGES

Facing pages of text (adjacent left-hand and right-hand pages) should together carry the information that identifies the journal and issue: the journal title, abbreviated if necessary; the volume number and issue numbers; and issue date. If space permits, facing pages should also carry a short version of the article title or the name of the journal section (such as "Book Reviews"). These elements can be in the top margin as a "running head" or in the bottom margin as a "running foot" or "footline". A widely used convention is the top margin for article or section information and the bottom margin for journal title, volume and issue numbers, and date.

All pages must be numbered sequentially: right-hand pages odd numbered and left-hand pages even numbered. The recommended position for page numbers ("folios") is in the bottom margin below the text block, aligned with the left-hand edge of the text block on left-hand pages and with the right-hand edge on right-hand pages. Placement of page numbers in the top margin above the text block, aligned with the left-hand edge for left hand pages and the right hand edge for right hand pages, is also commonly used. (See Section 27.5.2 on sequential numbering through each volume.)

27.6.3.5 Information for Authors

Journals can help themselves and potential authors who are preparing and submitting papers by maintaining online and by publishing at regular intervals one or more pages describing journal policies and manuscript requirements (for recommended content, see Table 27.6). These pages, also often called "Instructions for Authors", should appear in the same place in each issue, preferably within the issue rather than on the cover, with the page listed in the table of contents (see Section 27.5.4.2). If they are not published in each issue, the URL for the online Instructions for Authors should be provided on the table of contents, masthead page, or cover.

For assistance in preparing author information and instructions, please refer to examples available online from journals and in compilations such as *Instructions to Authors in the Health Sciences* (http://mulford.meduohio.edu/instr/) prepared by the Mulford Health Science Library of the University of Toledo.

27.6.3.6 Advertisements

The pagination for advertising pages should be in the same sequence as that of other pages also likely to be discarded when the journal is bound for library use or prepared for online archiving. A system of pagination distinguishing these pages from the text pages can help librarians and binderies identify pages to be excluded from bound volumes (see Section 27.5.3).

Advertising that might be confused with the scientific text of the journal should be refused, especially if it is formatted to resemble text pages (an "advertorial"). In addition, the placement of advertisements within an issue can raise ethical issues.

Advertisements placed immediately adjacent to an article discussing the same topic, such as an advertisement for an antidepressant next to an article on drug treatment for depression, can be problematic.

27.6.3.7 Corrections and Retractions

Statements of authors' errors, the editor's or publisher's errors, omissions, retractions, or other matters that merit being called to the attention of readers should be published in the same place in the journal in any issue carrying such statements. The location

Table 27.6 Information for authors: potential content

Potential content	Specific items
Distribution	Circulation: domestic, foreign. Categories of subscribers: personal, institutional, student.
Other availability	Online sources (URL); CD-ROM formats; microforms.
Indexing and abstracting services	Names of services covering the journal, addresses, and sources of additional information.
Content	Categories of papers considered; specifications for their specific content, expected format, allowed length. Descriptions of appropriate content for sections of the journal.
Submission of manuscripts	Content of submission letter: must include name of author responsible for further communication, address, and communications numbers (telephone, fax, email); information on prior or duplicate publication of a paper's content (whether identical or not).
	URL for online submission and peer review, if available.
	For mail submission, number of copies; spacing; kind of paper. Requirements for figures.
	Mailing specifications. Alternative formats, such as word-processing diskettes and acceptable word-processing programs.
	Copyright-transfer requirements; other required information, such as competing interests of authors, research funding, authorship statements.
	Page charges.
Ethical considerations	Criteria for authorship; basis for sequence of author names.
	Conflict-of-interest identification.
	Definition of prior and duplicate publication.
Editorial and review process	Procedure followed by the editors and time constraints.
	Information provided to authors.
	Policy on peer reviewing (e.g., whether blinded or anonymous).
Manuscript style and format	Details on manuscript format: title page, limit on title length, abstract structure and length, text and headings. Details of scientific style, including units of measurement, nomenclature.
	Format for references. Details of requirements for figures and tables. Charges for color figures. Recommended style manual and dictionary.
Publication	Scheduling; handling of proofs; reprint orders; prepublication release of information (author preprints).

should be prominent, such as on a page with an editorial or with letters to the editor or in a section titled "Corrections". If the statement is not on the table of contents page, its location should be identified on that page, with page number. The statement should make clear its function—for example, with a title beginning "Correction" or "Notice of Error" or "Errata" and including a brief term or phrase identifying the subject or title of the previously published article; the statement should fully identify the corrected item (if it was published) with bibliographic data. The material corrected and the correction should be made clear in the notice.

Retraction notices should be published with attention to the same considerations of format, but they must also identify the person or organization responsible for the retraction.

Both correction and retraction notices must be indexed in volume indexes. In addition to the index entries for the notices themselves, the index entries for the articles that are the subject of a notice should include a parenthetic statement cross-referencing the related notice. In fields with indexing databases that flag indexed articles with notices of retraction or correction (such as MEDLINE/PubMed of the National Library of Medicine), these notices should be called to the attention of the relevant database producer for possible action. For further information about the Council of Science Editors' position on corrections and retractions, see CSE's White Paper on Promoting Integrity in Scientific Journal Publications, 2012 Update.[31]

27.6.3.8 Articles in Installments or in Series

Some articles may be too long for a single issue and must be published in 2 or more installments. The successive sections should carry the article's title with the additional designation of part, such as "Part 1", "Part 2", and so on. The concluding part should include notice of conclusion, such as "Part 2 (Conclusion)", preferably in the title or in a statement that the article concludes the series. Arabic, not roman, numerals should be used.

If possible, such articles should be scheduled so that all parts appear in consecutive issues of a single volume, with notices on their title pages of the titles and pages of the preceding parts. If the subjects of the divisions of such an article and space in the table of contents permit, each part can be helpfully titled to represent its content:

Recent Developments in the Geomorphology of the United States: Part 1, The East
Recent Developments in the Geomorphology of the United States: Part 2, The Midwest
Recent Developments in the Geomorphology of the United States: Part 3, The West (Conclusion)

Individual articles representing a series should also be published within a single volume if possible, with titles relating the articles in the series:

The Economics of Health Services: Canada
The Economics of Health Services: France
The Economics of Health Services: Great Britain

27.6.3.9 Supplements and Special Issues

To serve special editorial objectives, journals may occasionally publish supplements or special issues. These are usually collections of papers on a specific topic or the papers or abstracts presented at a conference. Supplements and other special issues should have the same trim size, design, and format as regular issues to avoid confusing readers or producing difficulties for library binding (see Table 27.7 for details). To determine pagination, if the supplement is designated as Part 2 of a specific issue and is to be bound immediately after Part 1 of the issue, the pagination should continue from the preceding text (Part 1). If it is a volume supplement, it can be paginated at the end of the usual set of issues (but before the volume index). However, supplements are often published in the middle of a volume or with an issue as yet unpaginated; in these cases, separate pagination is needed, and the page numbers should include an alphabetic modifier (such as "S" for "supplement"): S1, S2, S3, and so on.

Table 27.7 Supplements and special issues

Elements	Notes
Supplement	
Trim size	Same as regular issues.
Cover	Separate cover; design, format, and identification same as for regular issues; indication of supplement, including title or topic; volume and issue number should indicate the number or other designator of the supplement, such as "Volume 31, Supplement 2", "Volume 31, Number 3, Supplement B", or "Number 3, Part 2".
Title page, table of contents	Specific to the supplement.
Bibliographic identification	The supplement: "Part 2" or "Supplement B" of the regular issue it supplements. In a bibliographic reference the identifier follows the issue number, as in "(6 Pt 2)" or "31(3 Suppl B)".
Title	Supplement title; if the supplement carries a single article (such as a very long review article or a dissertation), the supplement title is the article title. A multiarticle supplement (such as conference proceedings) must carry individual article titles and preferably also carry a supplement title.
Pagination	Preferably continuous with regular issues; if paginated separately, an alphabetic designation such as "S" usually precedes the page number, such as "S324".
Indexing	Indexed with issues and other supplements in the volume.
Special issue	Same trim size, design, format, and pagination as regular issues; cover and table of contents may carry a title identifying the theme or title of the special issue.

The content of the supplement should be indexed in the volume's index, whether the supplement has the volume's pagination or its own.

27.7 JOURNAL ARTICLES AND THEIR PARTS

This section applies to aspects of format and style common to the principal kind of articles in scientific journals, both in their print and PDF versions (articles in HTML or other dynamic formats typically contain the elements described in this section but their arrangement can vary widely). Special additional considerations are needed for research papers (see Section 27.7.5.1) and for other kinds of articles (see Section 27.7.5), such as reviews, editorials, letters to the editor, and book reviews.

27.7.1 Title Page

The first page of an article should carry all the information needed to enable a reader to rapidly identify the contents, author(s), and origin of the article and its location in the journal. These elements are summarized in Table 27.8.

27.7.1.1 Article Title

The title should be as informative as possible within the length limit stipulated by the journal in its Information for Authors section. A title should be straightforwardly descriptive and eschew hyperbolic rhetoric. A title may be a declaratory statement if it does not imply a generalizability of the reported findings beyond what is supported by the described evidence. An interrogatory title can indicate the question considered,

Table 27.8 Title page of a scientific article: necessary elements (see Section 27.7.1)

Elements	Notes
Title	Should describe the subject and include terms used in recognized thesauri or other vocabularies.
Author statement (byline)	All author names, in the sequence preferred by the authors or stipulated by the journal; surnames accompanied by given names (preferable) or all of their initials. See Section 27.7.1.2.
Author affiliation	The institution (or institutions) that was the site of the reported research or study. Current affiliations and addresses can be provided in a note at the end of the article.
Abstract of the article	Abstract in the format and within the length specified by the journal.
Bibliographic reference	The reference by which the article should be identified in bibliographies published in the subject field of the journal.
Beginning text	Depending on availability of space.
Footnotes	May be needed for extensive group authorship or affiliation lists or to call attention to current affiliations and addresses.
Footline elements	Journal title, volume and issue numbers, copyright notice, DOI if applicable.
Initial page number	

such as the subject of the research reported or the matter considered in an article, such as an editorial.

> Large-scale Eradication of Rabies in Southern Belgium with Recombinant Vaccinia-Rabies Vaccine
>
> *not* Large-scale Eradication of Rabies with Recombinant Vaccine [readers may want to know that the campaign was regional, not national or international, and the specific kind of vaccine]
>
> Does Astronomy Really Need the Hubble Space Telescope?

The title should start with a word or term representing the most important aspect of the article, with the following terms in descending order of importance if possible. Terms subordinate to the subject of the research—a description of the research design, for example—can be carried in a subtitle.

> Hypertension in Young Adult Black Males Treated with Hydrochlorothiazide or Weight Reduction: A Randomized Controlled Trial
>
> *not* A Randomized Controlled Trial of Treatment of Hypertension in Young-Adult Black Males with Hydrochlorothiazide or Weight Reduction [move the study-design descriptor to a subtitle]

Another factor to consider in writing article titles is whether they are easily retrievable by a search of online databases. Individuals depend on searching online databases to locate articles because 1) science has become so interdisciplinary that relevant articles may appear in a variety of journals in several specialties; 2) the number of published articles has increased each year; 3) it is becoming more difficult to find the time to scan issues of journals; and 4) students and researchers who grew up with the Internet demand instant access to information. Databases have moved from being available only by complex command language searching performed by librarians and other trained searchers to more user-friendly, online products accessible to a wide audience.

With regard to the results of an online search, for most people the title is the most important factor in deciding whether to pursue an article. This is particularly true for those databases and articles that do not have abstracts. Searchers also often restrict a search to words in the title to increase the relevance of their results.

The following guidelines apply to the writing of titles in general as well as their searchability in online databases:

1) Do not write erudite or "cutesy" titles that appear clever but do not convey useful information to the user. For example, an article with the title "Whither Oncology?" discusses the future of surgical oncology as a recognized surgical specialty, but the title certainly does not convey this and would not be retrieved on a search for the topic.
2) Be as specific as possible. Do not use broad terms when only narrower concepts are discussed. For example, the article with the title "Tranquilizers for Gastrointestinal Distress" in reality only discusses the use of fluoxetine to treat ulcers. A title search for fluoxetine and ulcers would not retrieve this article.
3) Do not leave the audience in doubt about what is being discussed. Titles beginning with such wording as "A New Use for . . ." are not helpful. What use? Diagnostic? Therapeutic? Titles that present concepts but do not state the relation discussed are also not useful, such as "Growth hormone therapy and thyrotropin-releasing hormone".
4) Use standard terms from formal scientific nomenclatures, such as the BIOSIS vocabulary guides[32] and the Medical Subject Headings[33] of the National Library of Medicine, instead of common or nonstandard terms.
5) Avoid using acronyms and initialisms whenever possible. For example, MCV could stand for mean corpuscular volume or motor conduction velocity. Also, database searching is not case sensitive. A search cannot distinguish between Ca for calcium and CA for cancer.
6) Use common or generic names whenever possible for drugs and chemicals instead of formulas and numbers. For example, the drug 2-chloro-10-(3-dimethylaminopropyl)-phen othiazine can only be searched in many databases by combining the constituents: chloro AND dimethylaminopropyl AND phenothiazine. Such a search could retrieve a number of irrelevant articles; using "chlorpromazine" instead avoids the confusion. Numbers for chemicals and drugs also have the potential for confusion. Many online systems replace a hyphen with a space or eliminate the hyphen character altogether. Thus, STI-571 becomes either STI 571 or STI571. Use "imatinib" instead.
7) Write out scientific names. Many online databases do not permit searching by a single letter. Thus, *Salmonella typhi* appearing as *S. typhi* in a title could be retrieved only by a search for "typhi" alone, causing false retrieval of *Rickettsia typhi.* In those databases that do permit single letter searching, the possibilities for irrelevant retrieval (e.g., S for sulfur) are limitless.
8) Be aware of guidelines for article titles issued by authoritative bodies. For example, the CONSORT (Consolidated Standards of Reporting Trials) statement[34] contains a checklist of what must be included in articles reporting on clinical trials.
9) Add alternative or synonymous terms in the title where appropriate. For example, titles with common names for plants or animals might include the proper taxonomic name, and titles with generic drug names might include the trade name if the research were concerned with specific products.

   The North American Distribution of the House Fly (*Musca domestica*)
   The Incidence of Adverse Effects from Phenytoin (Dilantin)

10) Use simple word order and common word combinations. Online searchers often use adjacency—a search for one word directly in front of another—to retrieve more relevant results. Thus, for example, use "juvenile delinquency" rather than "delinquency in juveniles".
11) Do not use roman numerals for articles published in parts. For example, a searcher looking for blood coagulation factor III would not wish to retrieve Part III of an article.
12) Do not assume that journal title words can assist in a search. Most databases do not

make the individual words in a journal title searchable. For example, the title "Natural History and Prevention of Radiation Injury" does not indicate that the injury discussed occurs in dentistry, although the article appears in *Advances in Dental Research*. A better title is "Natural History and Prevention of Radiation Injury in Dentistry".

13) Become familiar with database "stopwords". Every database has a list of words that cannot be used in searching. Stopwords are commonly used words that would overburden the system if they were searchable. Examples are "a", "the", "really", "very". However, some meaningful words often used in titles appear on stopword lists. "Being" is one example; many articles discuss the well-being of laboratory animals. In this example, use "welfare" or another synonym.

The conventions for typefaces and other standards appropriate to scientific nomenclature should also be applied in titles, such as *Musca domestica* in the example above (see Chapters 15 to 26 of this manual). Punctuation in titles should be kept to a minimum. Commas are acceptable for separating the elements of a series of related terms (see Section 5.3.3.1); colons are acceptable for coupling main and subordinate elements in a title (see Section 5.3.1.1). Semicolons and dashes should generally be avoided. (For capitalization in titles, see Section 9.3.)

27.7.1.2 Author Statement (Byline)

Only those persons qualifying for authorship by standards accepted in the scientific community[27,35–38] should be listed as authors. In general, these standards ensure that persons credited as authors can take public responsibility for the report through having participated in designing the research, carrying it out, and writing or revising drafts of the article. Journals can help ensure that author statements in their articles represent responsible authorship by requiring putative authors to sign statements before acceptance of their articles, thereby vouching for the authenticity and validity of their articles through adequate participation in the article's research and reporting.

Some journals require a statement that clearly indicates the role of each author in the research that led to an article as well as his or her role in the preparation of the manuscript. This statement, under the heading of "contributors" or "author contributions", usually appears at the end of the article before the reference list. This practice has its roots in a 1998 article[39] by Drummond Rennie, then the deputy editor of *JAMA*, in which he urged disclosure of such information to readers. (See Chapter 2 for further information on the responsibilities of authors.)

Authors should be identified by at least 1 name (not just initials) in addition to the surname (family name) to reduce ambiguity in author names in indexing services. If space for names must be limited, 2 or more initials for names preceding the surname should be used if available. Note that there are exceptions for names from some cultures in which only 1 name is used.

If a publication allows for use of an ORCID identifier (iD), authors can register for an iD, which is a unique identification number assigned to researchers that ties their name and nonsensitive identification information to their published works, thus eliminating the chance for confusion in identity of authors with similar names. For more information about ORCID, go to www.orcid.org.

The basis of the sequence of names should be stipulated by the journal, such as alphabetic by surname (see Section 29.3.3) or in order of decreasing degree of responsibility for the research and reporting it. The basis should be consistently applied within the journal and should be stated in its Information for Authors section. In addition, the correct author surname should be made evident in the author byline. Identification of the correct surname is extremely important because references, indexes, and databases all use surnames for their entries, and users of such tools depend upon the surname for retrieval. Note that surnames are the first element of a name in some cultures, and some compound surnames are not hyphenated (e.g., Elizabeth Scott Parker's surname might be Parker or Scott Parker); in parts of the world where the surname is the last element of a personal name, the sequence in name elements should be consistent for all names. Journal publishers are urged to use typography, such as all-capital letters or some other scheme, to readily identify authors' surnames (see Section 8.1).

A journal must establish its own policy on inclusion or exclusion of academic degrees and honorific designations with author names. The scientific standard that the validity and importance of research should stand on its evidence and not on the "authority" of authors suggests that academic degrees and honorific designations should not appear in scientific articles.

For large-scale research involving many investigators, authorship can be indicated by a collective (corporate) title.

The Stanford–CERN Collaboration
The French Fullerene Development Group
The Chinese Ecologic Research Coalition
The Scottish Oncologic Study Network

For such articles, further information on participants is usually desirable; a minimum is the name and address of the investigator who will respond to inquiries about, or criticism of, the reported research. A note or footnote can list persons (and their affiliations) who were responsible for writing the paper and who signed the voucher described above (perhaps as a "Writing Committee" or "Principal Investigators"); additional listings can identify participants by function, such as "Detector Design", "Statistical Analysis", and so on. Alternatively, the journal can maintain in its editorial office the names and addresses of participants in the reported research and be prepared to require them to respond publicly in the journal to inquiries or criticisms.

27.7.1.3 Author Affiliation and Site of Research

The institution(s) in which the author(s) carried out the reported research (or for other kinds of articles, in which the intellectual work forming the basis for the article was completed) should be identified under the author statement. Many journals have a special area for this on the title page, in a separate box or in a margin or footer. For a large group of authors and multiple institutions, this information may have to be given in a footnote or on a sidebar on the article's title page. Affiliations not directly related to reported research should not be included. In addition, the responsible institution(s) is often identified in a line beginning with "From the . . ." when the authors' affiliations do not make that clear.

To enable interested persons in the scientific community to communicate with the authors, their affiliations, addresses, and other information (such as telephone numbers and email addresses) should be provided on the title page (see Section 14.3); if this information could consume too much space on the title page, it may be more conveniently provided in a note at the end of the article. If it is thus placed, a footnote on the title page should indicate the location. A journal may choose to limit such information to that for a single author, usually termed the "corresponding author", who agrees to be responsible for responding to all communications.

27.7.1.4 Abstract

NISO[4] and ISO[3] mandate the inclusion of an abstract with every journal article, essay, and discussion, and most journals require one. The abstract should come after the author statement (and the author affiliation statement if it follows the authors) and before the text. It should be no longer than the length limit stipulated in the journal's Information for Authors section. In some disciplines, the major indexing services provide a maximum length by stipulating the length it will allow in its database or the length at which a longer abstract will be truncated. Because abstracts also appear in abstract journals and online databases, separated from the articles they describe, abstracts should be complete and understandable unto themselves, without reference to the full article.

Both NISO and ISO state that an abstract should contain the following elements: purpose, methodology, results, and conclusions. In general, abstracts should be single paragraphs and carry no subheadings. An exception is the structured abstract developed by Haynes and associates[40,41] and required by many journals for articles of clinical interest. In a structured abstract, each heading has its own paragraph, labeled with the appropriate title: objective, design, setting, patients or participants, interventions, measurements and results, and conclusions.

Abstracts of research papers should be informative, giving specific summaries of all elements of content. For reviews and other similarly long and wide-scope articles, abstracts may have to be indicative, simply sketching out the topics of the article and not summarizing evidence and conclusions. In either type of abstract, abbreviations should not be used unless they are understood when standing alone (such as DNA, pH, USA). Abstracts should not include bibliographic references or tabulated data.

As an aid to database searchers, abstracts may be used to provide synonyms for terms provided in the article's title. For example, if the term "renal" appears in the title, the word "kidney" may be used in the abstract.

See NISO Z39.14[4] and ISO 214[3] for additional information.

27.7.1.5 Bibliographic Reference

The title page of an article should include the bibliographic reference by which the article should be represented in bibliographies. The format should be that specified by the journal for references in its articles (see Chapter 29 for recommended formats). The reference is conveniently placed at the end of the abstract. In an electronic journal, it may be placed more conveniently for most readers at the top of the first display screen, above the title of the article.

27.7.1.6 Keywords

Some journals ask authors to provide a brief list of words (called "keywords") that characterize the main topics of the article. Such terms are used to assist the journal's indexers and the readers of the article. When selecting keywords, use recognized vocabularies related to the discipline discussed, such as the BIOSIS Search Guide[32] and the MeSH thesaurus[33] of the National Library of Medicine. When used, keywords are placed below the abstract.

27.7.1.7 Financial and Other Support

Information on financial support, such as funding from grants and contracts and supply of equipment and drugs, is often placed with acknowledgments (see Section 27.7.3.3) at the end of the article, but can also be placed in a footnote on the title page. The International Committee of Medical Journal Editors (ICMJE), for example, mandates placement of such information on the title page.[27] If grants and contracts are mentioned, provide the relevant grant and contract numbers and the name of the supporting institution. Many journals are also publishing information on competing interests, such as honoraria and stock holdings, which could be construed as affecting the authors' view of a company's products.

27.7.1.8 Text on the Title Page

The text usually begins on the title page unless the title, author and affiliation statements, and abstract are very long. The page design, layout, and typefaces should enable readers to distinguish readily between the abstract and the beginning of the text. Generally, figures and tables should not be placed on the title page lest they occupy too much space for clear presentation of the other elements.

27.7.1.9 The Footline

The title page should carry in its running foot or footline(s) the title of the journal, the volume and issue numbers, the issue date, and the copyright notice for the article. Many journals also include the URL for the journal's website here, as well as the DOI for the article.

27.7.2 Text

27.7.2.1 Text Subheadings

Readers scanning articles are helped by subheadings for parts of the text that indicate their specific content. Subheadings can indicate what structural element of the article the section represents (e.g., "Methods") or inform on the subject of the section (e.g., "Antisepsis Antibodies"). Additionally, subheadings break up what would otherwise be long uninterrupted blocks of type that give pages a "gray look", which is intimidating to readers.

Subheadings can represent different hierarchical levels of sections. First-level subheadings indicate the main divisions of text within the article. Second-level subheadings indicate divisions of subject matter within those main divisions, and so on. In general, a heading should be followed by at least 2 subheadings at the next level down, not by a single subheading.

Subheadings representing different hierarchical levels must be readily distinguished by readers. This need must govern the choices of typefaces and their styles, sizes, and weights. A designer or typographer should be consulted when setting a style for subheadings. In general, the subheadings for all levels should be in, or closely related to, the typeface of the text or in a contrasting, but consistent, typeface (sans serif heading typeface with serif text typeface, for example). The hierarchy of subheadings can also be suggested by their placement: centered in the type column above the following text, flush with the left edge of a column and spaced above the following text, or flush with the left edge but running into the paragraph. A typical hierarchy is illustrated below:

First level: bold capitals and lowercase	**Methods** Xxx xxxxx xxx xxx xxxx xxxxx xxx x xxxx.
Second: capitals and lowercase	Chemical Analyses
Third: italic small capitals	*CALCIUM*
Fourth: italic capitals and lowercase	*Ionized Calcium* Xx xxxx xxxxx x xx xxxxxxx x xxx xxxx . . .

Decisions on how many levels are needed should hinge on the length of sections. Relatively short sections of text may be structured clearly enough with paragraphing that represents topical divisions almost as well as low-level subheadings. Many journals will find 2 levels adequate.

27.7.2.2 In-Text References

References are presented in 2 ways in journal articles. At the end of the article, all of the references that contributed to the work are presented in a list variously called "References", "End References", "Reference List", "Literature Cited", or "Bibliography". Within the text of a document, these references are presented in an abbreviated format that refers to the end references. Such abbreviated references within the text of a work are called "in-text references". (For recommendations on the location and spacing of in-text references within the text of an article and in tables, see Chapter 29.)

27.7.2.3 Footnotes and Endnotes

Footnotes are occasionally used for references, for additional information about manufacturers and suppliers, and for other parenthetic information. Although this practice may be convenient for readers, it raises composition costs and is generally not used in scientific journals; it is still used in some books. (For recommendations on footnotes in tables and footnote signs, see Section 30.1.1.8.)

The functions of footnotes can often be served in journal text with parenthetic statements where extended comment is not needed.

> The patients in Group 1 were treated with a sodium salt of panmalignomycin (supplied by Northeast Pharmaceuticals; Boston, Massachusetts) and with prednisone; Group 2 patients were treated with a microsomal form of the potassium salt (supplied by Occidental Drugs, Incorporated; Seattle, Washington) and the same dosage form of prednisone.

Endnotes serve the same function as footnotes and are useful when extended parenthetic comment is needed but is too long or too frequent and would interrupt the text. They can be placed at the end of the text under a heading such as "Notes". References in the text to the notes can be superscript lowercase letters, a device that avoids confusion with superscript numbers for in-text references (see Section 29.2.4). The corresponding notes are identified in the "Notes" section with preceding online letters:

> The patients were treated with a sodium salt of panmalignomycin[a] and with prednisone; Group 2 patients were treated with a microsomal form of the potassium salt[b] and the same dosage form of prednisone.
>
> **Notes**
>
> a. Northeast Pharmaceuticals; Boston, Massachusetts. Note that . . .
>
> b. Occidental Drugs, Incorporated; Seattle, Washington. This form is also . . .

This style is likely to be applied more frequently with the development of online journals that provide rapid hypertext links between in-text references and the cited text (references, notes).

Some journals combine references and notes in a single section and cite both in the text with a single sequence of numbers. The journal *Science*, published by the American Association for the Advancement of Science, is a widely available example of this style.

27.7.2.4 Tables and Figures

Tables and figures in journal articles are generally governed by the same considerations of style and format applicable to those in books (see Chapter 28). Note, however, that the larger dimensions of journal pages may tempt the person responsible for page layout to place too many tables or figures on individual pages, thus producing pages that may visually confuse the reader as to the sequence of text. A useful general principle is placing tables or figures at the top of pages and the text at the bottom. Most journals locate tables and figures where convenient after their first mention in the text, ensuring a mix of tables, figures, and text. When an article has many figures, some may be combined with appropriate cropping into multipart figures that can be treated in layout as single figures. (For detailed information on tables and figures, see Chapter 30, "Accessories to Text".)

27.7.3 Notes, Appendixes, Addendums, and Acknowledgments

Various sections for functions subordinate to the text proper of an article are usually placed between the text and the reference section. A "Notes" section is described above in Section 27.7.2.3.

27.7.3.1 Appendixes

Appendix sections can be used for aspects of an article's subject that may be needed by some readers but are too long and detailed to be put in the text and would interrupt the flow of information for many readers. Appendixes should be self-contained and capable of standing alone, without reference to the article. Content of this kind includes long descriptions of unusual methods, long quotations from cited documents, questionnaires, guidelines, and patient handouts. An appendix should be headed with a specific explanatory title, such as "Appendix 1: Ethical Standards of Some Major Scientific Societies", not simply "Appendix 1". Arabic rather than roman numerals should be used.

27.7.3.2 Addendums

An addendum section is usually used for information an author needs to add late in the publication sequence, such as evidence from a just-published paper by another author, additional validating evidence from the author's own work, an overlooked reference, and so on.

27.7.3.3 Acknowledgments

An acknowledgments section can carry notices of permission to cite unpublished work, identification of grants and other kinds of financial support, and credits for contributions to the reported work that did not justify authorship. If grants and contracts are mentioned, provide the relevant grant and contract numbers as well as the name of the supporting institution. Note that the ICMJE[27] requires financial information on the title page.

27.7.4 References and Bibliography

References are presented in 2 ways in scientific publications. At the end of a document, such as a journal article, book, or book chapter, all of the references that contributed to the work are presented in a list variously called "References", "End References", "Reference List", "Literature Cited", or "Bibliography". The term "end references" is primarily used in this manual. Within the text of a document, these references are presented in an abbreviated format that refers to the end references. Such abbreviated references within the text of a work are called "in-text references". (The 6th edition of *Scientific Style and Format*[42] referred to in-text references as citations, but this term is ambiguous because librarians and many others use "citation" interchangeably with "reference".)

If the list includes both works used to write the article and other works that might be of interest to readers, the list is often called a "Bibliography". If the list consists only of the works that the author has cited in the text, the citations are usually designated "References". Another option is that used in this manual: a "References" section divided into 2 subsections, "Cited References" and "Additional References". A variety of other wording, such as "Literature Cited", may be found. (See Chapter 29 for forms of citations and the arrangement of the components of references.)

27.7.5 Special Types of Articles

27.7.5.1 Research Papers

A wide variety of formats may be needed for formal reports of research, and no single format can serve all possible needs. A journal covering a clearly defined field of science should, if possible, stipulate one or more specific formats most useful for that field. These formats should be summarized on the journal's Information for Authors section (see Section 27.6.3.5 and Table 27.6). Nevertheless, some general principles can govern the structuring and formatting of reports in many fields.

Most scientific articles must have the structure of critical argument[17]: a question or hypothesis is posed; the evidence, pro and con, bearing on the answer is presented and assessed; and an answer is reached. Reports of scientific research generally follow this structure but include additional elements. First, the means by which the investigator gathered his or her evidence bearing on the question or hypothesis is described in detail sufficient to enable another investigator to replicate the research. Second, the evidence gathered by the investigator is presented separate from that available in the scientific literature.

These considerations have led to a format widely used in many scientific fields: division and sequencing of a report so that it explicitly appears with sections that can be headed sequentially Introduction, Methods, Results, and Discussion. These subheadings need not appear or may appear but not in this form; for example, a report can open with a paragraph serving the functions of an introduction but not headed with "Introduction". If the materials used in the research must be described in detail, the methods section can be titled "Materials and Methods". Further, a research paper is more likely to be clearly understood if the research is described in the sequence in which it was conceived, designed, and carried out, its results analyzed, and its conclusions reached; this chronological sequence corresponds to the sequence of critical argument. Therefore, whichever headings are assigned sequentially to sections within a report, they should have been selected with these principles applied. Table 27.9 indicates a typical sequence of headings and the functions served by the sections they head.

27.7.5.2 Review Articles

Properly conceived review articles have the same basic structure of critical argument as a research paper. Their subheadings should make clear the subtopics considered sequentially. Explicit description of the methods and standards applied in selecting the references cited helps review articles meet the same intellectual standards[43] as research

Table 27.9 Sections of a research report: typical headings and functions

Headings for sections	Function of section; comments
Introduction	Describes the state of knowledge that gave rise to the question examined by, or the hypothesis posed for, the research. States the question (not necessarily as an explicit question) or hypothesis.
	A review of existing literature also may be found in the introduction.
Materials and Methods	Describes the research design, the materials and methods used in the research (subjects, their selection, equipment, laboratory or field procedures), and how the findings were analyzed.
	Various disciplines have highly specific needs for such descriptions, and journals should specify what they expect to find in a methods section.
Results	Findings in the described research. Tables and figures supporting the text.
Discussion	Brief summary of the decisive findings and tentative conclusions. Examination of other evidence supporting or contradicting the tentative conclusions. Final answer. Consideration of generalizability of the answer. Implications for further research.
References	Sources of documents relevant to elements of the argument and descriptions of materials and methods used.

papers; such a description can be carried in a methods section near the beginning of the review or in an appendix.

27.7.5.3 Editorials

Because of their brevity, editorials usually do not need text subheadings, although their intellectual structure should have the elements of critical argument. Many journals do not indicate authorship of editorials. Those that do can choose whether to place the authorship statement (see Section 27.7.1.2) under the title, to emphasize the authority of the author, or at the end of the text, to lessen the emphasis on the author.

27.7.5.4 Letters to the Editor

Letters are usually published largely as submitted, hence with the typical structure of a letter. Each letter or group of letters on a single topic should be headed with a short title to clarify for scanners of a letters section what topics are covered. Many journals edit the salutations of letters to a standard form such as "To the Editor:". An editor should decide how much identification should be given the authors of letters: name only, name and academic degrees, name and postal address, or some other combination. In some journals, scientific comment on articles submitted by readers is called "Discussion".

27.7.5.5 Book Reviews

The structures of a book review section and its contents should be designed for the convenience of the reader. A section with many reviews might group them by topic with appropriate headings for groups. Bibliographic information, the price, and other source information about the book such as an ISBN (International Standard Book Number) should head a review. For book reviews, it is usually more useful to begin the bibliographic information with the title, then the author, rather than the author-first format found in end references (see Chapter 29).

27.8 REPRINTS AND OFFPRINTS

Reprints and offprints should retain the trim size, format, and pagination of the original. Two or more articles combined in a single reprint may carry new, additional pagination that begins on the first right-hand text page with arabic numeral 1, but the original page numbers, within parentheses (round brackets), should accompany the new page numbers.

The usual bibliographic identifiers on the title pages of articles should also appear on the reprints. If a reprint of one or more articles is issued under a new cover, the cover should also carry full titles, author designations, and bibliographic identifiers.

27.9 POSTAL REQUIREMENTS

National postal services may have requirements for periodic publication of notices concerning the ownership, circulation, and other characteristics of journals; the frequency and required location in the journal of such notices may also be specified. The postal service in the country of publication or distribution should be consulted for detailed information on such requirements.

CITED REFERENCES

1. National Information Standards Organization. Bibliographic references. Bethesda (MD): NISO Press; 2005. (ANSI/NISO Z39.29-2005). Also available at http://www.niso.org/.

2. National Information Standards Organization. The Dublin Core metadata element set. Bethesda (MD): NISO Press; 2007. (ANSI/NISO Z39.85-2007). Also available at http://www.niso.org/.

3. International Organization for Standardization. Documentation - Abstracts for publications and documentation. Geneva (Switzerland): ISO; 1976. (ISO 214).

4. National Information Standards Organization. Guidelines for abstracts. Bethesda (MD): NISO Press; 1997. (ANSI/NISO Z39.14-1997 (R2009)). Also available at http://www.niso.org/.

5. National Information Standards Organization. International standard serial numbering (ISSN). Bethesda (MD): NISO Press; 2001. (ANSI/NISO Z39.9-1992 (R2001)). Also available at http://www.niso.org/.

6. National Information Standards Organization. Permanence of paper for publications and documents in libraries and archives. Bethesda (MD): NISO Press; 2009. (ANSI/NISO Z39.48-1992 (R2009)). Also available at http://www.niso.org/.

7. National Information Standards Organization. Printed information on spines. Bethesda (MD): NISO Press; 2009. (ANSI/NISO Z39.41-1997 (R2009)). Also available at http://www.niso.org/.

8. National Information Standards Organization. Syntax for the digital object identifier. Bethesda (MD): NISO Press; 2005. (ANSI/NISO Z39.84-2005). Also available at http://www.niso.org/.

9. International Organization for Standardization. Information and Documentation - Electronic manuscript preparation and markup. Geneva (Switzerland): ISO; 1994. (ISO 12083).

10. International Organization for Standardization. Documentation - International standard serial numbering (ISSN). Geneva (Switzerland): ISO; 2007. (ISO 3297).

11. International Organization for Standardization. Documentation - Presentation of contributions to periodicals and other serials. Geneva (Switzerland): ISO; 1986. (ISO 215).

12. International Organization for Standardization. Documentation - Presentation of periodicals. Geneva (Switzerland): ISO; 1977. (ISO 8).

13. International Organization for Standardization. Documentation - Presentation of title information of series. Geneva (Switzerland): ISO; 1985. (ISO 7275).

14. International Organization for Standardization. Information and documentation - The Dublin Core metadata element set. Geneva (Switzerland): ISO; 2009. (ISO 15836).

15. International Organization for Standardization. Information and documentation: Rules for the abbreviation of title words and titles of publications. Geneva (Switzerland): ISO; 1997. (ISO 4).

16. Hall GM, editor. How to write a paper. 5th ed. Hoboken (NJ): John Wiley & Sons; 2013.

17. Huth EJ. Writing and publishing in medicine. 3rd ed. Baltimore (MD): Williams & Wilkins; c1999.

18. Day RA, Gastel B. How to write and publish a scientific paper. 6th ed. Westport (CT): Greenwood; 2006.

19. Zeiger M. Essentials of writing biomedical research papers. 2nd ed. New York (NY): McGraw-Hill, Health Professions Division; c2000.

20. Kasdorf WE, editor. The Columbia guide to digital publishing. New York (NY): Columbia University Press; c2003.

21. National Academy of Sciences, Committee on Electronic Scientific, Technical, and Medical Journal Publishing. Electronic scientific, technical, and medical journal publishing and its implications: report of a symposium. Washington (DC): National Academies Press; c2004. Also available at http://www.nap.edu/books/0309091616/html.

22. Goldstein H. Acrobat/PDF resources. Highland (MD): Goldray Consulting Group; c2010 [accessed 2012 Nov 12]. http://goldray.com/webdesign/acrobat_resources.htm.

23. DocMorph. Bethesda (MD): National Library of Medicine (US); [updated 2010 Dec 13; accessed 2012 Nov 12]. http://docmorph.nlm.nih.gov/docmorph/docmorph.htm.

24. National Information Standards Organization. Understanding metadata. Bethesda (MD): NISO Press; c2004 [accessed 2012 Nov 12]. http://www.niso.org/standards/resources/UnderstandingMetadata.pdf.

25. Brand A, Daly F, Meyers B. Metadata demystified: a guide for publishers. Hanover (PA): Sheridan Press; c2003. Also available at http://www.niso.org/standards/resources/Metadata_Demystified.pdf.

26. Bailey CW Jr. Scholarly electronic publishing bibliography 2010. Houston (TX): Digital Scholarship; 2010 [accessed 2013 Nov 19]. http://www.digital-scholarship.org/sepb/annual/sepb2010.pdf.

27. International Committee of Medical Journal Editors. Uniform requirements for manuscripts submitted to biomedical journals: writing and editing for biomedical publication. Philadelphia (PA): ICMJE; [updated 2010; accessed 2012 Nov 12]. http://www.icmje.org/.

28. BIOSIS serial sources. Philadelphia (PA): Thomson; 2001.

29. Ulrich's periodicals directory. 50th ed. New Providence (NJ): Bowker; 2011.

30. Current Contents: Life Sciences. Philadelphia (PA): Institute for Scientific Information. Vol. 10, Jan 3, 1967. Weekly. Also published in other sections, such as: Agriculture, Biology & Environmental Sciences; Clinical Medicine; Physical, Chemical & Earth Sciences; Engineering, Computing & Technology.

31. Scott-Lichter D and the Editorial Policy Committee, Council of Science Editors. CSE's white paper on promoting integrity in scientific journal publications, 2012 update. 3rd ed. Wheat Ridge (CO): CSE; 2012.

32. BIOSIS search guide 2001–2002. Philadelphia (PA): Biological Abstracts, Inc.; c2001.

33. MeSH Browser. Bethesda (MD): National Library of Medicine (US); [updated 2012 Sep 4; accessed 2012 Nov 12]. http://www.nlm.nih.gov/mesh/MBrowser.html.

34. Begg C, Cho M, Eastwood S, Horton R, Moher D, Olkin I, Pitkin R, Rennie D, Schulz KF, Simel D, Stroup DF. Improving the quality of reporting of randomized controlled trials. The CONSORT statement. JAMA. 1996;276(8):637–639.

35. Dodd JS, editor. The ACS style guide: a manual for authors and editors. 2nd ed. Washington (DC): American Chemical Society; 1997.

36. Huth EJ. Guidelines on authorship of medical papers. Ann Intern Med. 1986;104(2):269–274.

37. Huth EJ. Scientific authorship and publication: process, standards, problems, suggestions. Washington (DC): Institute of Medicine; 1988.

38. Publication manual of the American Psychological Association. 6th ed. Washington (DC): American Psychological Association; 2009.

39. Rennie D. Freedom and responsibility in medical publication: setting the balance right. JAMA. 1998;280(3):300–302.

40. Ad Hoc Working Group for Critical Appraisal of the Medical Literature. A proposal for more informative abstracts of clinical articles. Ann Intern Med. 1987;106(4):598–604.

41. Haynes RB, Mulrow CD, Huth EJ, Altman DG, Gardner MJ. More informative abstracts revisited. Ann Intern Med. 1990;113(1):69–76.

42. Council of Biology Editors, Style Manual Committee. Scientific style and format: the CBE manual for authors, editors, and publishers. 6th ed. New York (NY): Cambridge University Press; 1994.

43. Mulrow C, Thacker S, Pugh J. A proposal for more informative abstracts of review articles. Ann Intern Med. 1988;108(4):613–615.

28 Published Media

28.1 DEFINITIONS

The definition of "published media" has drastically changed in recent years. With the increased use, functionality, and availability of virtual and electronic media, print is no longer the primary vehicle with which to publish or disseminate scholarly information. Many academic journals are increasing their readership by adding online versions of their publication to the available platforms, and some are even moving to online-only formats with mobile-optimized options for readers with e-readers, tablets, or smart phones. Any form of writing or audio or video recording is now considered to be published simply by being made available online, whether it is in the form of an e-publication, database or archive entry, blog post, podcast, video, or any of the other myriad platforms that are constantly becoming available for market consumption. For the purposes of this manual, we will focus on peer-reviewed materials published by and/or for academic and scholarly institutions, keeping in mind that virtually anything that is printed can also be made available electronically.

The term "monograph" is often understood to mean a scholarly treatise on a specific subject, usually limited in scope and usually written by a specialist in the field. In this manual, the term is used in its broader sense to mean a publication that is complete in a single volume or in a limited number of volumes, such as a book that is published in 3 physical pieces. Books, textbooks, technical reports, conference proceedings, master's theses and dissertations, bibliographies, festschrifts, and patents, as well as the scholarly treatises described above, are all examples of monographs. Most of this chapter focuses on the standard print book and textbook. Because technical reports and other specific types of monographs have additional special features, they are treated in their own section (see Section 28.6). See Section 28.6.3 for further specifics relating to electronic publications.

Some books are published in what is known as a monographic series. Each book in the series can stand on its own as a separate publication, but the series brings together titles on a specific subject or by a specific organization. For example, the book *Differentiation of Embryonic Stem Cells*, published by Elsevier Academic Press, is also volume 365 of the series *Methods in Enzymology*, and *Bisphosphonates and Metastatic Bone Disease*, published by the Parthenon Publishing Group, is volume 10 of a series of conferences sponsored by Ciba Pharmaceuticals called the *International Congress, Symposium, and Seminar Series*.

Journals and newspapers, also known as serials, are publications appearing or intended to appear indefinitely at regular or stated intervals, generally more frequently than annually, each issue of which normally contains separate articles or papers[1] (see Chapter 27 for journal style and format).

Book publishers and printers use many technical terms, some of which should be known by authors as well as by editors. Among the most widely used are "recto", meaning a right-hand page, and "verso", meaning a left-hand page. Verso is also used to refer to the back or reverse side of a page, as in the verso of the title page. Recto pages are always odd-numbered and verso pages always even-numbered. A printed page number may also be called a folio (which has, however, additional meanings) and a page number at the bottom of the page is a drop folio. Other more specific terms are defined in their relevant sections below. Definitions of additional technical terms may be found in glossaries in standards of the National Information Standards Organization (NISO),[1-10] *The Chicago Manual of Style*,[11] and *Butcher's Copy-editing: The Cambridge Handbook for Editors, Authors, and Publishers*.[12]

28.2 FORMAT

Monographs may appear in a variety of physical formats. In addition to the traditional paper form, they may be published as electronic documents, which include CD-ROMs (compact disc read-only memory), DVDs (digital versatile disc or digital video disc), and anything available online. Regardless of their physical format, in scientific publishing all types of monographs generally share the same main characteristics. They typically have 3 sections: the pages or screens of the opening section, usually called "front matter" or "preliminary pages" or "prelims"; the main body of the publication or "text", also known as "subject matter" or "main text" or "body text"; and accessory closing pages or screens, usually called "back matter" or "endmatter". See Table 28.1 for an alphabetic list of components and their location.

Monographs published in print are generally static in nature, meaning that their content is fixed at the time of publication and does not change until a new or revised edition is published. The content is presented in a linear form, from front to back. Electronic books may also be static but are increasingly more dynamic in their functionality; e-books and other online publications are frequently updated and take advantage of such capabilities as hyperlinking, the ability to create links to another place in the same document or to an entirely different document. For example, links may be created from an in-text reference to the citation in the reference list and from the citation to the complete article. Large data sets and other types of ancillary material may also be included in an electronic monograph, either within the text or as supplemental material through a link.

It is increasingly common for books to be published in more than one format, usually in print and online. If a book is issued in more than one format, the publisher should designate one of these as the "version of record"—that is, the authoritative version in which errata and other changes are recorded, and inform the reader which version that is. When deciding on the version of record, it may be helpful to keep in mind that errata can most easily be updated online. It is also important that readers of a book published in more than one format cite the specific version seen and used in a reference list. For example, an author should not cite the print version if the online version was used (see Chapter 29 for details).

There is no single universally agreed-on standard for the structure and format of books. In the United States and Canada, *The Chicago Manual of Style*[11] has become the de facto standard because of its detailed description of conventional practice in book layout and formats. The legalities and customs of publishing in Canada are described

Table 28.1 Components of books

Component	Location[a]	Section reference
Abbreviations list	Front matter	28.4.2.11.4
Abstract	Front matter	28.4.2.4.9
	Text	28.4.3.4
Acknowledgments, permissions, and credits	Front matter	28.4.2.4.8
		28.4.2.11.1
Addendums	Back matter (recto, preferably)	28.4.4.1
Appendixes	Back matter (recto, preferably)	28.4.4.2
Bar code	Cover and jacket	28.4.1
Bibliography	Back matter (recto, preferably)	28.4.4.5
Cataloging- in-Publication	Front matter	28.4.2.4.4
Chapter 1	Text (recto, 1 or 3)	28.4.3.4
Colophon	Cover and jacket	28.4.1
	Front matter	28.4.2.3
Colophon page	Back matter (verso)	28.4.4.7
Contents list	Front matter (recto for initial page, v or vii)	28.4.2.6
Contributors	Front matter	28.4.2.11.3
Copyright notice	Front matter (verso, iv)	28.4.2.4.1
Copyright page	Front matter (verso, iv)	28.4.2.4
Country in which printed	Front matter (verso, iv)	28.4.2.4.6
Cover	Cover and jacket	28.4.1
Dedication	Front matter (recto, v)	28.4.2.5
Endpapers	Cover and jacket	28.4.1
Epigraph	Text	28.4.3.2
Errata	Front matter	28.4.2.7
Footnotes	Text	28.4.3.4.3
	Back matter	28.4.4.3
Foreword	Front matter (recto)	28.4.2.9
Frontispiece	Front matter	28.4.2.2
Glossary	Back matter (recto, preferably)	28.4.4.4
Half title	Front matter (recto, i)	28.4.2.1
Half title, facing page	Front matter	28.4.2.1
Half title, second	Text	28.4.3.3
	Back matter (recto, preferably)	28.4.4.2
Half title, verso	Front matter (verso, ii)	28.4.2.1
History of publication	Front matter	28.4.2.4.2
Illustrations list	Front matter	28.4.2.8
Indexes	Back matter (recto, preferably)	28.4.4.6
International Standard Book Number	Front matter	28.4.2.4.5
Introduction	Front matter	28.4.2.11.2
	Text	28.4.3.1
Jacket	Cover and jacket	28.4.1
Notes	Text	28.4.3.4.3
	Back matter (recto, preferably)	28.4.4.3
Parts	Text	28.4.3.4
Permanence of paper statement	Front matter	28.4.2.4.7

Table 28.1 Components of books (*continued*)

Component	Location[a]	Section reference
Preface	Front matter (recto)	28.4.2.10
Printer	Front matter	28.4.2.4.10
	Back matter	28.4.4.7
Publisher's address	Front matter	28.4.2.4.3
Reference list	Back matter (recto, preferably)	28.4.4.5
Running heads and feet	Text	28.5.2
Second half title	Text	28.4.3.3
	Back matter (recto, preferably)	28.4.4.2
Section numbering	Text	28.4.3.4.2
Sections	Text	28.4.3.4
Series	Cover and jacket	28.4.1
	Front matter	28.4.2.1
Spine	Cover and jacket	28.4.1
Subheadings of chapters	Text	28.4.3.4.1
Tables list	Front matter	28.4.2.8
Title page of book	Front matter (recto, iii)	28.4.2.3
Title page, verso	Front matter (verso, iv)	28.4.2.4
Title pages of chapters	Text (recto, preferably)	28.4.3.4

[a] Where a specific page (recto or verso) or a specific page number is preferred, this indication follows the location.

on the website of the Library and Archives Canada.[13] British practices are well reflected in Chapters 7–9 of *Butcher's Copy-editing: The Cambridge Handbook for Editors, Authors, and Publishers.*[12]

Many individual standards published by the International Organization for Standardization (ISO) and the National Information Standards Organization (NISO) in the United States are relevant to aspects of book formatting and are cited in this chapter. Governments also issue publishing standards through agencies such as the US Government Printing Office[14] and, in the United Kingdom, the Office of Public Sector Information (formerly Her Majesty's Stationery Office)[15]; these should be consulted when books are produced as official government documents.

28.3 PUBLICATION

Publication is "the reproduction in tangible form and the general distribution to the public of copies of a work from which it can be read or otherwise visually perceived" (UNESCO)[16] and "covers any general distribution to the public of works in print and of non-print works, including visual, audiovisual, and electronic works".[1]

28.3.1 Publishers

The individual or organization responsible for making a publication available to the public is called a publisher. The advent of the Internet and other online media has stretched the definition of both "publication" and "publisher". However, in electronic terms a publisher is defined as the individual or organization that produces or sponsors the website.

Publishers of scientific monographs fall into 4 major categories: commercial scholarly publishers, professional organizations, university presses, and trade publishers. For-profit firms specializing in publishing books for a specific discipline, such as the physical or life sciences, are the largest class of scientific publisher. They market their publications extensively and are widely known to their specific audiences. The publishing arms of professional organizations, such as the British Medical Association and the American Chemical Society, are the next largest group of scholarly publishers. A society's membership is a built-in audience, and these organizations often publish monographs that commercial publishers would reject because of the small market potential. In the third category, university presses produce high-quality publications, but their output is more likely to be limited to monographs of regional or limited interest. Finally, trade publishers, large commercial organizations that produce publications of mass appeal, publish scientific monographs if they feel there would be enough interest among the general reading public.[17] Professional organizations and university presses both have editorial boards composed of association members or faculty that make final publication decisions. In contrast, such decisions are usually made by the staff editors of commercial and trade publishers.[18]

A publisher's tasks include negotiating contracts with authors and their agents, editing the author's manuscript, designing the physical book (typography, layout, etc.) and/or the electronic version, producing the finished product (printing, binding, etc.), marketing the work, and making arrangements for its distribution through various market channels[19] (see Section 28.3.2).

28.3.2 Publication Process

The publication process for monographs begins when an acquisitions editor solicits a manuscript from an author on behalf of a publisher or when an author submits a book proposal to a publisher. The latter is the most common avenue. Before submitting a proposal, also called a prospectus, the prospective author should always contact the potential publisher regarding any specific instructions for such technical details as permissible word processing and illustration formats as well as other submission requirements, such as text length. The proposal usually contains the following information[12]:

- the rationale for the book, i.e., its purpose
- its intended audience
- information on competing books and why the proposed one is superior
- the proposed content and structure of the book
- a completed chapter
- estimated completion date
- some marketing suggestions

Once a proposal has been accepted, the acquisitions editor or commissioning editor usually works with the author to develop a contract that defines the relationship between the author and the publisher. See Chapter 2 of this manual for authors' and editors' responsibilities and ethical considerations.

After the author submits the manuscript, which is typically done electronically via email or through a submission website, it is the copy editor's role to ensure that the text and illustrations are clear, correct, and consistent; that the spelling, grammar, and punctuation are in accord with the publisher's house style; and that agreement exists among all of the parts, such as headings to text and in-text references to the reference list. The technical or substantive editor looks for such items as errors in fact, incoherent or ambiguous sentences, misleading language, and agreement of the text with the stated level of audience. With some publishers, the copy editor and the technical or substantive editor are the same person. The manuscript is marked up, usually using tracked changes in a word-processing program or in the form of annotations in a PDF, and is then returned to the author for comment once substantive editing is complete, and any needed negotiation on changes is carried out. Following this process, the copy editor prepares the manuscript for the designer and typesetter by marking or coding those portions of the text requiring their attention, such as chapter and section headings, illustrations, and typefaces and fonts. Finally, the typeset copy, often called the galley proof, is returned to the author and the publisher for a last review and check against the original copy. Details of this entire process may be found in *The Chicago Manual of Style*[11] and other sources.[12,20] See Section 28.5 for a discussion of design elements.

28.4 PARTS OF MONOGRAPHS

28.4.1 Cover, Jacket, and Endpapers

A book cover is composed of a front, a back, and a spine. The front cover usually contains the names of the authors (or editors), the title, and any statement of edition. It may also include the subtitle, a title under which the book was previously published, a volume number of the publication if issued in multiple parts, the name of the publisher or its colophon (publisher's trademark or graphic device used as an embellishment), the date of publication, and any indication of series. Names of authors are often abbreviated to the last name (surname) and of publishers to a single word or colophon. For a book with a jacket, the cover may be left blank.

The back cover is usually blank except for the ISBN (International Standard Book Number) and the bar code, which must be included[11] unless it appears on the jacket. The ISBN, a number assigned to a book by its publisher, should appear in human-readable font above the bar code preceded by the letters ISBN. See Section 28.4.2.4.5 for further discussion of the ISBN. The bar code, technically called the Bookland EAN (European Article Number) code, consists of 13 digits: the first 3 digits are always "978" (meaning a book); the next 9 digits are the first 9 digits of the ISBN, with the hyphens deleted; and the final number is a check digit. A separate 5-digit add-on to the right of the bar code is often used to indicate the retail price.[21] When printed, the Bookland EAN symbol is to be positioned at the bottom of the back cover or jacket and must be printed in a color that can be read by a scanner.[11] See *Bar Codes for the Book Industry*[21] for details.

The recently developed technology of QR (quick response) Codes and other 2-dimensional matrix-style codes is leading to the increasing use of such codes on book covers, jackets, and inside pages.[22] Two-dimensional codes are becoming more commonly used than standard barcodes because of their ability to contain more data, the speed at which the data can be accessed, and the wide variety of types of data they can store.[23]

Both NISO (NISO Z39.41)[6] and the International Organization for Standardization (ISO 6357)[23] have standards relating to the content and location of spine information. NISO reserves the top two-thirds of a spine for the author(s) or editor(s), title, edition statement, and any "numerical, alphabetical, or chronological" data. Such data include the number or name of a volume, part, or series and the year. The bottom third is divided into 2 parts: a library information area above an area for publisher information. The library information area is blank, providing space for individual libraries to add a local call number or bar code. ISO reserves the top two-thirds of a spine for the title, the authors whenever possible, and the publisher if sufficient space exists. The bottom third (at least 30 mm from the bottom) is reserved for library identification.

Book jackets are used to protect the cover and also contain a front, back, and spine. Although the front and spine usually contain the same information as their corresponding parts on the cover, the back is often used for promotional information from the publisher. The back also contains the bar code or two-dimensional code if it is not included on the cover. The left inside flap of the jacket (the inside front of the jacket) may contain a summary of the book, and the right flap (the inside back of the jacket) may contain a brief biography of the author, perhaps with his or her photograph.

It is important that the title appearing on the cover and the jacket is the same and that it agrees with the title page.

See also Section 28.5 for design elements.

The endpapers (a folded paper, half of which is pasted to the inside cover of a hardcover book and the other half of which is adjacent to the first or last leaf of the book) are convenient places for maps, chronologies, pronunciation tables, and other information that will be consulted frequently by the reader. Bear in mind that a paperback book has no endpapers.

Be cautious regarding the value of material placed as endpapers, however. Libraries and other archives often put bookplates and other accessioning information inside the covers, thus obscuring any text on the endpapers.

28.4.2 Front Matter

The usual components of the front matter (the "preliminary" pages of the book) and their location are listed in Table 28.1. Other components may be incorporated and other sequences may be suitable. The components highly likely to be used by most readers should start on right-hand pages (rectos), but this rule can be ignored for some components if economical use of pages is a necessity. The pages of the front matter are usually given lowercase roman numeral page numbers. If the front matter begins with a half title page, the numeral for that page is "i"; if the first page is the title page, it instead is designated "i". However, the roman numeral is usually not printed on the title page or its verso, the half title pages, or blank pages. Every page of the front matter is counted for pagination purposes, even if no page number is printed. For example, the title page and its verso have no page numbers printed on them, but the page following them would be numbered "iii".

The various components of front matter that follow are given in the order in which they usually appear in a monograph.

28.4.2.1 Half Title and Its Verso

The right-hand page (recto) preceding the title page (usually the first right-hand page of the book) is generally used for the half title (also called "bastard title" or "false title"). The half title page, an artifact from when books were printed without covers, may be eliminated for economy of pages. Usually this page carries only the main title with no subtitle and no author's name. It sometimes carries a series name or an epigraph (a brief quotation or motto illustrating the theme of the book). The epigraph may also be included as part of the text (see Section 28.4.3.2). The half title should be the same as that on a half title page (called a second half title) placed to immediately precede the text. If the book is divided into parts or units, each may be preceded by a half title page bearing the title of the part.

The left-hand page (verso) of the half title page faces the title page and is usually blank but sometimes is used to carry an advertisement. This advertisement (also known as an "ad card", "book card", "card page", or "face title") lists other books by the author, usually with phrasing such as "John Jones is also the author of . . ." or "Other books by John Jones". Alternatively, if a book is part of a series, this page can carry the series title, its editor, and a list of the other titles in the series.

28.4.2.2 Frontispiece

A frontispiece, an illustration setting the tone of the book by representing an important aspect of the book (e.g., a portrait of the subject of a biography), faces the title page. A book without such an illustration may use this half title verso page for information on other books by the author, series title and number, or other books in the series represented by this book.

28.4.2.3 Title Page

The title page presents the full title of the book, including any subtitle. The book title as it appears here is the authoritative title—that is, the title that is used in a reference list and by library catalogers. The title that appears on other pages, including the half title, cover, and jacket, should agree with this title. Note that the contents of a book may be copyrighted but not its title (see Table 3.1).

The title page should be designed before the rest of the front matter, and it and all other preliminary pages should harmonize with the rest of the book. The full title should follow conventions of title capitalization and scientific conventions such as those for symbolization and italicization (see Chapters 4, 9, and 15–26). The title and the subtitle (if there is one) should be clearly delineated by consistent type style and size, any needed punctuation, and the layout of the page. For example, if the title page reads:

> ASCA
> Style Manual for Authors and Editors

it is unclear whether the correct title is "ASCA Style Manual for Authors and Editors" or just "Style Manual for Authors and Editors".

Using capital and lowercase letters will make proper capitalization of the title clear to readers not familiar with the conventions of capitalization. A subtitle should appear below the main title and in a smaller type size; a superior position can suggest that the subtitle is the main title or part of it.

The title page should include not only the full title of the book but also the names of all authors and the publisher's name and location (city or cities). ISO 1066[24] also states that when organizations serve as authors, the organizational name and its constituent parts should be presented in hierarchical order. The title page may also carry additional elements, such as the name of a translator, editor, illustrator, or author of the foreword or introduction. The roles of each should be clearly stated, as "John Jones, editor" or "Edited by John Jones". If each chapter or part of a book has its own author, only the names of the editor or editors of the book as a whole appear on the title page; the chapter authors appear on a separate recto page under the heading "Contributors" (see Section 28.4.2.11). A colophon, a symbol or emblem representing the publisher's imprint, may also be present here or on the verso of the title page.

If the book has been revised or represents a new edition, the title page should carry the number of the edition.

CSE Style Manual
6th edition, revised

Similarly, if the book is published in more than one physical volume, the volume number, its title (if any), and the name of the volume editor (if any) for each volume should be placed on the title page. Alternatively, this information may be placed in each volume on a separate title page in addition to the title page of the publication as a whole.

Although it is becoming less common thanks to online capabilities, some books may be accompanied by a CD-ROM, DVD, or other nonprint media format to permit the publisher to include material such as data sets and illustrative material that would not be practical to have in print. If such additional formats are present, they should be acknowledged on the title page, as "Includes one CD-ROM" or "See [URL] for additional materials".

There is some controversy regarding placement of the date of publication. ISO 1066[24] mandates that this date appear on the title page and Butcher et al.[12] permits it there; however, *The Chicago Manual of Style*[11] states that the date is "best avoided on the title page, particularly if it conflicts with copyright information." It is not uncommon for a book to be copyrighted a year or more in advance of publication. If the date of publication is not given on the title page, it must appear on its verso.

The arrangement of elements on the title page depends on the book. Usually the title comes first, but if the author is famous, his or her name may be placed above the title. In general, the proper sequence of elements from the top down is as follows: title, subtitle, the author and/or editor statement, the edition identifier, any information on accompanying materials, the publisher's name, the publisher's location, and the year of publication. In scholarly monographs, the author (or editor) information may include a short statement of the author's academic or research affiliation.

Occasionally a book designer may want to add graphic interest and specificity to a book's title page by designing an expansive title "spread" that spans both pages ii and

iii. If so, the design should carry the "imprint" information (publisher, place of publication, and date of publication) on page iii.

28.4.2.4 Verso of Title Page (Copyright Page)

The verso or back of the title page is often called "the copyright page" because it is usually the location of the copyright notice. It is termed the "biblio page" in the United Kingdom because this page is also used to carry additional detailed information that identifies the book more fully than its title page.

28.4.2.4.1 COPYRIGHT NOTICE

Although the copyright notice is customarily placed on the verso of the title page, it may also appear on the title page itself. Until 1976, US copyright law specified that the notice appear on the title page or the page immediately following. This is no longer required, but custom prevails. The copyright notice must consist of the word "Copyright" or the abbreviation "Copr" or the symbol ©, accompanied by the name of the copyright holder and the year of copyright. The order of the 3 elements is not important.

For details on copyright, including information on regulations in various countries, consult Chapter 3, "The Basics of Copyright".

28.4.2.4.2 HISTORY OF PUBLICATION

The copyright notice is often followed by statements presenting the year of first publication, the years of subsequent editions, and the years of each impression. For further details consult Chapter 1 of *The Chicago Manual of Style*.[11]

28.4.2.4.3 PUBLISHER'S ADDRESS

The complete address (or abbreviated address consisting only of the city and/or URL, for example) of the publisher, and that of its agents in other countries, is often provided on the verso of the title page. *The Chicago Manual of Style*[11] recommends placing this information on the copyright page.

28.4.2.4.4 CATALOGING-IN-PUBLICATION (CIP)

Cataloging-in-Publication (CIP) is a prepublication cataloging program established in 1971 by the US Library of Congress to enable libraries to catalog books immediately on their arrival. The US program is available only for books carrying a US place of publication. The Library creates a bibliographic record (known as CIP data) for each publication and transmits it to the publisher, who in turn prints the record on the verso of the title page. CIP data are distributed weekly "in machine readable form to large libraries, bibliographic services, and book vendors around the world. Many of these organizations redistribute these records in products and services designed to alert the library community to forthcoming publications and to facilitate acquisition."[25] See the CIP program website (http://www.loc.gov/publish/cip/about/) for details.

The CIP Programme is administered in the United Kingdom and Ireland by the British Library (http://www.bl.uk/bibliographic/cip.html), in Canada by Library and Archives Canada (http://www.collectionscanada.gc.ca/cip/index-e.html), and in Australia by the

National Library of Australia (http://www.nla.gov.au/cip). Other countries, including China and Malaysia, have also established similar CIP programs; contact the appropriate national library in your own country for details.

28.4.2.4.5 INTERNATIONAL STANDARD BOOK NUMBER (ISBN)

The International Standard Book Number (ISBN) uniquely identifies a book and thus facilitates handling orders and keeping track of inventory by computer. The identifying series of 13 numbers (10 in the years preceding 2005) should appear on the jacket and at the foot of the outside back cover as well as on the verso of the title page.

An ISBN identifies one title, or edition of a title, from one publisher and is unique to that title or edition. If the book is published in paperback and in clothbound editions, or in paper and online versions, a separate ISBN is assigned to each. When a work is published in 2 or more volumes, a separate ISBN is assigned to each volume unless the work is sold only as a set; nevertheless, the ISBN should be printed on the verso of the title page of each volume.

In the United States, the ISBN Agency is the RR Bowker Company. For more information, contact them at

U.S. ISBN Agency
630 Central Avenue
New Providence, NJ 07974-1154
Fax: 908-219-0188
Email: isbn-san@bowker.com
Website: http://isbn.org

The International ISBN Agency is located in London under the umbrella of EDItEUR (http://www.isbn-international.org/).

28.4.2.4.6 COUNTRY IN WHICH PRINTED

A statement naming the country in which the book was printed is conventionally placed on the verso of the title page.

Printed in the United States of America   Printed in Canada

Such a statement must appear on the cover or jacket if a book is printed in other than the country of publication.[11]

28.4.2.4.7 PERMANENCE OF PAPER STATEMENT

When the paper used in a book meets the standards for the permanence of paper set forth in *Permanence of Paper for Publications and Documents in Libraries and Archives*,[5] this fact should be represented by the statement "♾ This paper meets the requirements of ANSI/NISO Z39.48-1992 (Permanence of Paper)." on the verso of the title page.

28.4.2.4.8 ACKNOWLEDGMENTS, PERMISSIONS, AND CREDITS

An author's recognition of persons, groups, and contributions required to produce the book is stated in the acknowledgment. An acknowledgment that involves copyright may appear here on the verso of the title page, as is necessary for anthologies. It may also be included in the preface (see Section 28.4.2.10) or appear as a separate recto page if length is a consideration (see Section 28.4.2.11).

Permission is clearance from a copyright owner for the author to quote passages or reproduce illustrations from another publication. Photo credits may also appear here instead of with the illustrations. When permissions and credits are extensive, it is sometimes desirable to include them on a separate recto page (see Section 28.4.2.11).

28.4.2.4.9 ABSTRACT

Some nonfiction books on narrow topics not accurately reflected by the book's title can usefully carry an abstract on the verso of the title page or on the right-hand page following it. If separate abstracts are supplied for chapters, they should appear on the page preceding the start of a chapter or on its opening page. See NISO Z39.14[4] and ISO 214[26] for standards for writing abstracts as well as the style manuals of various organizations, such as the American Medical Association and American Chemical Society.

28.4.2.4.10 PRINTER

The name of the compositor, printer, or binder, or any combination thereof, may be included on the verso of the title page. This information may also be placed on the colophon page (see Section 28.4.4.7).

28.4.2.4.11 ADDITIONAL INFORMATION ABOUT THE VERSO OF THE TITLE PAGE

Additional information on appropriate contents of the verso of the title page, such as credit for cover artwork, may be found in *The Chicago Manual of Style*.[11] Information on contents for the United Kingdom may be found in *Butcher's Copy-editing: The Cambridge Handbook for Editors, Authors, and Publishers*.[12]

28.4.2.5 Dedication

Whether a book will carry a dedication, an inscription to honor or compliment a patron, relative, or friend, and what it will say are the author's choices; brevity should be encouraged. The dedication usually faces the verso of the title page, that is, on the first right-hand page following the title page, but sometimes it is on the verso of the title page. If it is on a recto page, it should be followed by a blank verso page.

28.4.2.6 Contents List (Table of Contents)

The list of contents sets forth the titles (usually only the chapter titles) representing the content of the book. It simplifies finding particular materials in the text and back matter and clarifies the organization of the book. If the number of subheadings for chapters is not too great and their listing would assist readers in finding needed content rapidly (as in a reference book), they can be given under the chapter titles; the design of the contents list should then be such as to enable readers to distinguish readily chapter titles from subheadings.

The numbers of beginning pages of chapters (and sections of chapters if these are also listed) are conveniently indicated immediately after the chapter (and section) titles. However, titles of parts of the book (chapter groups) are not followed by beginning-page numbers. Formerly, titles were always placed flush left and the page numbers flush right, but now that leaders (dots leading from the title to the page number) have gone out of vogue, a closer placement of titles and page numbers allows the reader's eye to quickly pick out the appropriate page number. Traditionally, the contents list has followed all front matter except an introduction, but if the book contains a large amount of front matter, the list should appear as early as possible and list everything that follows. The contents list contains no reference to items that precede it. This section can be simply titled "Contents". Its ease of use will depend greatly on the qualities of typography and layout established by the designer of the book.

Note that European practice often places the contents list at the end of the book.

28.4.2.7 Errata

If a list of errors in the book is needed, placing it immediately after the contents list will increase the probability that it will be seen by readers. If the book has already been bound when the errors are discovered, a separate leaf may be tipped in there. Each erratum notice should give the page number, line number, error, and correction.

> page 85, line 8: For "Einstein's theory" read "Einstein's wife's theory".

This section reports only substantive, never simply typographic, errors.

28.4.2.8 Lists of Illustrations (Figures) and Tables

Decisions on whether to list illustrations and tables and their page numbers will depend on the importance of directing readers to them from the front matter rather than simply from the text for which they are relevant. The front matter of a heavily illustrated book might be burdened by long lists, but a reference book in which tables are major sources of information could helpfully list its tables. If the illustrations are on unnumbered inserts, their location should be identified with phrasing such as "preceding page 11" or "following page 120". If tables are listed, the list comes after the list of illustrations; table titles may be shortened for the listing.

Both kinds of lists should either be a continuation of the contents list and in harmony with its design or start on right-hand pages immediately following the contents list.

See Chapter 30, "Accessories to Text", for information on the structure and format of illustrative material.

28.4.2.9 Foreword

The foreword (not "foreward" or "forward") is often of a scope similar to that of an author's preface but is by someone other than the author. It often serves to place the book in the context of the related literature and to point out merits of the book as seen by an authority other than the author (or editor). It typically runs to no more than 2 pages and ends with its author's name and, if desired, title and affiliation.

28.4.2.10 Preface

The preface is the author's comment about the book as a whole and describes its purpose, sources, and extent. It could be a background note on why the author wrote the book or on the process of writing it. It differs from the introduction in that the introduction discusses the text and prepares the reader for the contents or explains it. Customarily the author ends by acknowledging indebtedness to other authorities or to persons who helped write or proof the book or who contributed information. The preface is not part of the text and is numbered with the appropriate sequence of lowercase roman numerals. An editor's preface, if used, usually precedes the author's preface. The current preface is placed before prefaces to any previous editions.

28.4.2.11 Miscellaneous Front Matter

28.4.2.11.1 ACKNOWLEDGMENTS

An author's recognition of persons, groups, and contributions important to the production of the book appears under this heading. These may be placed on a separate recto page if not included in the preface (see Section 28.4.2.10) or on the verso of the title page (see Section 28.4.2.4).

28.4.2.11.2 INTRODUCTION

The introduction defines the limits and organization of the work and states the purpose, scope, and author's approach. This may be included on a separate recto page if not a part of the text, which is its more usual placement (see Section 28.4.3.1).

28.4.2.11.3 CONTRIBUTORS

For multiauthor books in which only the editor's name appears on the title page, it is appropriate to list the contributors with statements of their affiliations and other relevant information following the contents list. If each author is given a biographical note, the notes are placed at the end of the book. An alternative is to list the contributors' names only in the contents and give a note about each author at the beginning of the appropriate chapter.

28.4.2.11.4 LIST OF ABBREVIATIONS

A separate list of abbreviations and the terms or names they stand for is needed only when many are used throughout the text. Alternatively, abbreviations may be incorporated within a glossary.

28.4.3 Text

The first page of the text or body of a book begins the sequence of pagination with arabic numerals. If this first page carries a part or chapter designation but no substantive text, the page number is omitted, even though the page is counted as page 1. The first page number appears on the first substantive text page: "2" if the first substantive text page is on the verso of the first page, "3" if it is on the following recto. In general, components of the text (parts and chapters) should begin on recto pages unless space is a prime consideration.

All text pages are numbered consecutively whether the page number is expressed or not. For example, page numbers are not expressed on pages containing only illustrations or tables.

Although the text section of many books simply starts with the first chapter, which should begin on a recto page, some books may carry several types of content preceding the actual text but on text-numbered pages; see "Introduction" and other content below.

28.4.3.1 Introduction

The introduction, part of the text, is usually set like the text. It defines the limits and organization of the work and states the purpose, scope, and author's approach. Sometimes it is signed by its author, but if the author of the book also wrote the introduction, the signature is not necessary. Note that some style manuals treat the introduction as a part of front matter (see Section 28.4.2.11).

28.4.3.2 Epigraph

The epigraph, a quotation intended to convey a theme or tone developed in the text that follows, may appear on the title page or on the back of the dedication (see Section 28.4.2.5). It may replace the second half title (see Section 28.4.3.3) or be on its verso and face the first page of text. Its source need not be documented beyond the name of the quoted author and, perhaps, the title of the quoted work.

> When a man knows he's to be hanged in a fortnight's time, it concentrates his mind wonderfully.
>
> —Samuel Johnson

Epigraphs may be used for the same purpose for individual chapters; these should appear on chapter title pages between the title and the chapter's text.

28.4.3.3 Second Half Title

If a second half title page is used, which usually occurs when there is a large amount of front matter, the first page is customarily designated with arabic numerals and the preceding pages with lowercase roman numerals. The number for the second half title page is usually not expressed and its verso (back) is normally blank.

The second half title page marks the end of the front matter and the beginning of the text. If the text of the book is divided into parts, a second half title page is not needed; its function is performed by the first part-title page.

28.4.3.4 Parts, Sections, Chapters

Chapters of a book may be logically grouped into numbered parts (or sections). The part number and title appear on a right-hand page preceding the text of the part; its verso is usually blank. Although the title pages of parts are counted in the pagination, their page numbers are not expressed. Each part may have an introduction. This introduction may appear on a new right-hand page following the title page of the part, on the verso of the title page, or on the title page itself. Part numbers should be in arabic numerals. Chapters within parts are numbered consecutively throughout the book and do not begin again with chapter 1 in each part.

Most prose books are divided into chapters, not necessarily of similar length. Each usually starts on a new page, preferably recto (right-hand page), but the verso (left-hand) location can be used for economical use of pages. Note that beginning a chapter on a recto page reduces costs for offprints because the printer does not need to recompose pages to print the offprints of each chapter.

A chapter should begin with a chapter number and title (chapter head), which are placed above the beginning of the text. The opening page carries a page number at the bottom of the page (the drop folio) and no running head (a short title printed at the top of the page). In a multiauthor book, each author's affiliation or other identification is usually given in a footnote on the first page of the chapter, if not placed in a list of contributors in the front matter.[11] This information may appear immediately below the names in some texts.[12]

Abstracts are not commonly used for book chapters. If an abstract is needed, it should appear at the head of the chapter beneath its title (and author statement, if there is one).

28.4.3.4.1 SUBHEADINGS

As an aid to the reader, chapters may be subdivided with subheadings, secondary subheadings, and tertiary subheadings. These should be succinct and meaningful, accurately representing the sequence and, with secondary and lower subheadings, hierarchies of content. When a chapter is subdivided with subheadings, there should be at least 2 subheadings in most cases. Subheadings are usually set on a separate line from the first line of the section, but if space is a consideration, the lowest level subheading may be placed on the first line of the text of the section.

The levels of subheadings should be made clear by appropriate choices of type styles; the design should make them clearly distinguishable from the text proper. Subheadings should not contain citations to references or notes and should not be used to define abbreviations or terms.

The first sentence after a subheading should be complete unto itself and not assume that the reader has read the subheading.

> Subheadings
> Subheadings should not contain references
> *not*
> Subheadings
> They should not contain references

28.4.3.4.2 SECTION NUMBERING

Sections and subsections of chapters may be numbered to help guide readers quickly to parts of the text. This device is especially useful in a reference book, like this manual, unlikely to be read sequentially but used mainly to obtain answers to specific questions. A common method is double numeration: the number of a section consists of the number of the chapter, a demarcation point (in this manual a period), and the number of the section within the chapter. For example, 4.8 is the eighth section of chapter 4.

For books such as instruction manuals, another method is to use the chapter number followed by a demarcation point and a paragraph number.

28.4.3.4.3 FOOTNOTES

Footnotes on text pages should generally be avoided (see "Notes", Section 28.4.4.3). One exception occurs in a multiauthor book, when they may be used to provide the authors' affiliation or other author identification. If footnotes must be used, consult sources such as Chapter 14 of *The Chicago Manual of Style*[11] and Chapter 9 of *Butcher's Copy-editing: The Cambridge Handbook for Editors, Authors, and Publishers*.[12]

28.4.4 Back Matter

The back matter (also "reference matter" or "end matter") constitutes the pages following the main text of the book. Pagination of this section continues with arabic numerals and includes any of the components described below; the sequence is generally the one shown here.

28.4.4.1 Addendums

Brief supplemental data that became available too late for inclusion in the text may be added at the end of the book in a section placed immediately following the text and titled "Addendum". Such a section should be listed in the book's contents list. In general, use of such a section should be avoided, because its contents are likely to be overlooked by readers as they go through the text.

Addendums can also be added as online-only material; however, this is more valuable to the reader if the addendum is associated with an online or mobile-optimized version of the book, in which the reader can click on a hyperlink that links to the website containing the addendum.

28.4.4.2 Appendixes

Appendixes carry supplemental material that illustrates, enlarges on, or otherwise supports the text without distracting readers from the main line of the text's exposition: letters, lists, large tables, forms, speeches, detailed protocols, questionnaires, and other supporting documents. The first appendix begins on a recto page and is preceded by a half title ("Appendix" or "Appendixes" as appropriate) if space permits.

An appendix may be placed at the end of its respective chapter, especially in a multiauthor book, if it is needed for that chapter and not for the others. This also applies where offprinting is anticipated and each author's appendix will appear with the chapter to which it relates. When an appendix follows its chapter, it may begin on either a left- or right-hand page, or it may run on at the end of the chapter's text.

If the work carries only 1 appendix, it is not numbered and the word Appendix appears by itself. If 2 or more appendixes are carried, they are numbered or lettered consecutively.

Appendixes can also be housed online in an archived database. Just like an online addendum, an online appendix is more valuable to the reader if linked to an online or mobile-optimized version of the book, because the reader can view the appendix effortlessly through a hyperlink.

28.4.4.3 Notes

Notes consist of information of interest to scholars and researchers. They usually follow the chapter to which they refer, but they may be placed at the end of the book (in which case they should be grouped by chapter). Notes are usually set 1 or 2 points smaller than the text. The numbering of notes usually begins with "1" for each chapter.

Footnotes on text pages should generally be avoided. They increase the difficulty and cost of page make-up, reduce text areas on facing pages to uneven depths, disarrange the orderly appearance of the page, and may interrupt the flow of reading.

28.4.4.4 Glossary

A glossary is a dictionary of terms and concepts used in the book. It usually precedes the bibliography. Entries should not be capitalized except for proper nouns.

28.4.4.5 Reference List (Bibliography)

The reference list provides citations to works on the subject of the book. If the list includes both works used to write the book and other works that might be of interest to readers, the list is often called a "bibliography". If the list consists only of the works that the author has cited in the text, the citations are usually designated "references". Alternatively, the reference list may be divided into "cited references" and "additional references". A variety of other wording, such as "literature cited", may be found. See Chapter 29 for forms of citations and the arrangement of the components of references.

References should preferably appear at the end of the chapter in which they are cited. This is particularly desirable when large numbers of references are included; the reader will find them more readily at the end of the relevant chapter than in a long compilation in the back matter.

28.4.4.6 Indexes

An index is an alphabetic list following the text that cites names, places, and topics discussed in the book and gives the page numbers on which discussion of these topics occurs. If section numbers are used in chapters, they may be used rather than page numbers.

In long books citing many authors, it may preferable to have separate author and subject indexes. With this division, the author index precedes the subject index.

An index can be set in smaller type than the text, in 2 or more columns. Main headings should be clearly distinguished from subheadings. It is important to capitalize only those entry terms that are properly capitalized as proper nouns (see Section 30.6.1.2; see Section 30.6 and its subsections for additional details of format and style for indexes).

28.4.4.7 Colophon Page

The colophon ("finishing touch") appears on the last page: a recto or verso page following the last page of the book, or in scholarly books, the last page of the index. It consists of details on the design and manufacture of the book, including such items as the name of the printer, typeface used, paper and other materials used, and style of binding. The names of those responsible for the design and production are also sometimes included. Lately, the title page and the copyright page have taken the place of the colophon in recording this information.

The word colophon is also used for the trade emblem or device of a printer or publisher (required by British law, but not American, to appear in a book). Such trade emblems usually are placed on the cover and title page. The publisher's name and location, called the imprint, should appear on the title page (see Section 28.4.2.3) and the publisher's address on the verso (back) of the title page (see Section 28.4.2.4).

28.5 DESIGN ELEMENTS

Good book design should not demonstrate the skill of the designer but, rather, complement the text, presenting the author's words "noiselessly" to enhance readability.[27] Book design is a complex task for which many reference texts are available; a selection of these is included in the Bibliography of this manual. The sections below cover 2 areas of design that will be of particular interest to authors and editors.

28.5.1 Pagination

All text pages are numbered consecutively whether the page number is expressed or not. The most common location for the page number is at the top of the page and flush with the outside margin, the left margin for verso (left-hand) pages and the right margin for recto (right-hand) pages. If the number appears at the bottom of the page, it is called a drop folio. The opening page of each chapter carries a drop folio. The outer location (top or the "drop folio" location) is preferred to centered locations because it is more readily seen by readers skimming through a book.

The front matter (everything preceding the second half title page) is numbered with lowercase roman numerals. Here the drop folio location is preferred.

Page numbers are not expressed on pages that only contain illustrations or tables.

In multivolume works, pagination may be consecutive throughout the volumes or begin anew with each volume. The former is preferred.

28.5.2 Running Heads and Running Feet

Running heads are headings that appear across the tops of pages; running feet appear across pages below the text. Both are intended to indicate content. Often the title of the book appears on the left-hand pages and a short title for the chapter on the right-hand pages. An alternative scheme is a running head for the chapter at the top of verso pages and running heads for sections of chapters at the top of recto pages.

Both running heads and running feet are discretionary. Running heads are usually omitted on display pages such as chapter title-pages and pages of illustrations. However, both types of heads can be useful to readers in skimming the text and in identifying the source when individual chapters or other pages have been photocopied and become separated from the book.

28.6 SPECIFIC TYPES OF MONOGRAPHS

28.6.1 Technical Reports

A technical report (also called a scientific and technical report) is defined by NISO as "a separately issued record of research results, research in progress, or other technical studies".[1] The purpose of a report is to disseminate the results of that research.[7] Most technical reports are issued by government agencies, usually at the federal, state, or provincial level, but reports also originate from universities and other types of research institutions.

Most of the components of a technical report are akin to those of books, but some special needs are detailed below. The preceding sections on books are relevant to the components that reports and books have in common. The major differences between the standard book and a technical report lie in their authorship and in the provision of information about sponsorship, including any report, contract, and grant numbers. In preparing a technical report, it is important to identify both the sponsoring organization (the organization that funded the research) and the performing organization (the organization that conducted the research) and to indicate which of them actually published the report. In some cases, the same organization performs both functions. For example, the National Institute on Aging has intramural scientists and may publish their work in report format. Often, however, the sponsoring organization provides funds to another organization that actually performs the research. These funds are disbursed through grants and contracts. When this occurs, either the sponsoring organization or the performing organization may publish the report. Thus, there are 3 possible scenarios for publication of a technical report:

- written and published by the sponsoring organization
- written by the performing organization and published by the sponsoring organization
- written and published by the performing organization

The following recommendations are based on ANSI/NISO Z39.18 *Scientific and Technical Reports: Preparation, Presentation, and Preservation*,[7] with particular attention to the elements that differ from those in a standard book. Consult this publication for further details (see also Section 29.3.7.4 for information on citing technical reports).

28.6.1.1 Front Matter

28.6.1.1.1 COVER

The cover, an optional component of a report, identifies its subject and indicates whether it contains classified or proprietary information. The relevant information includes the following elements:

1) report number (alternatively on the back cover or on both)
2) report title and subtitle
3) title and numbering of series, if relevant
4) author(s), principal investigator(s), editor(s), and/or compiler(s)
5) publisher (the organization assuming responsibility for the publication; this may or may not be the sponsoring organization)
6) publication date
7) any limitations on distribution of the report
8) sponsoring organization
9) International Standard Book Number (ISBN) or International Standard Serial Number (ISSN) (also included on the back cover)

10) technical requirements relating to the format of the report, such as audio or video recording, CD-ROM, e-publication, etc.
11) subject of the report

Legal or other requirements may mandate additional information.

28.6.1.1.2 TITLE SECTION

The title section is required. It indicates the subject and content of the report and provides the information needed for bibliographic description and access. The title section also clearly identifies the responsibilities—that is, performance and sponsorship, of the organizations involved with the report. The relevant information includes the following elements:

1) report number (see ANSI/NISO Z39.23[8] and ISO 10444[28])
2) report title and subtitle
3) title and numbering of series, if relevant
4) author(s), principal investigator(s), editor(s), and/or compiler(s)
5) performing organization (affiliation of the author or authors)
6) publication data (place of publication, publisher, and date)
7) type of report and period covered by the report, if applicable
8) contract and grant number, as appropriate
9) sponsoring or issuing organization (if it differs from the performing organization)
10) subject descriptors

Legal or other requirements may mandate additional information.

28.6.1.1.3 NOTICES

Special notices on the verso of the title page or inside the front cover call attention to particular conditions for the report. A copyright section includes the copyright symbol, year, and copyright holder; full names of the sponsoring organization and any other organization providing funding for the report; authority to copy the contents or require permission to copy; the complete name, address, and telephone number of the producer; and how additional copies of the report may be obtained. A distribution limitations section provides information on security classification, restricted distribution, or proprietary status; states that findings are preliminary or that the report is a draft or working paper intended to elicit comments and ideas; indicates if the report is a preprint of a paper to be presented at a professional meeting; or presents legal conditions, such as the use of brand or trade names. Disclaimers should be avoided, but if necessary, they go here.

28.6.1.1.4 REPORT DOCUMENTATION PAGE

Forms such as the National Technical Information Service (NTIS) bibliographic data sheet or Standard Form 298 are used by agencies within the US government. Some agencies require documentation pages and specify their location. A report documentation page is optional for reports produced by academic and industrial institutions, which typically place them as back matter.

28.6.1.1.5 ABSTRACT

The abstract, a required element, states in approximately 200 words the purpose, scope, and major findings of the report. It contains no references or illustrations and should be understood when seen without the text. The abstract usually appears in its own section between the title section and table of contents, but agencies that require a report documentation page include the abstract there. See ANSI/NISO Z39.14 (R2002) *Guidelines for Abstracts*[4] and ISO 214 *Documentation: Abstracts for Publications and Documentation.*[26]

28.6.1.2 Text

28.6.1.2.1 SUMMARY

The actual technical report begins with a brief (500 to 1,000 words) summary that states the problem being investigated and the principal results, conclusions, and recommendations made by the report. A summary introduces no new material, but defines all symbols, abbreviations, acronyms, and unusual terms used. Some organizations prefer the term "executive summary", and some also place the summary at the end of front matter rather than at the beginning of the text.

28.6.1.2.2 REPORT TEXT

The text of the report closely resembles a journal article in format and style, including an introduction, description of methods and procedures, presentation of results, discussion, conclusions, recommendations, and references.

28.6.1.3 Back Matter

The back matter clarifies and supplements the text of the report. It contains appendixes; a bibliography; lists of symbols, abbreviations, and acronyms; a glossary; an index; and the distribution list. The latter is a complete list of the names and addresses of all persons and organizations that will receive a copy of the report, thus providing a permanent record of the initial distribution.

28.6.2 Conference Proceedings

Conference proceedings consist primarily of the full papers or abstracts of papers presented at a conference or other type of meeting. They also may include such material as information on posters presented, sponsors of the conference, lists of sessions and other information for the organizers and attendees of the conference, advertisements for supporting organizations, and information on forthcoming conferences of the sponsor.

Proceedings share many characteristics with books, and the preceding sections on books provide useful information on their preparation. The major differences lie in their titles and the inclusion of information about the dates, location, and sponsorship of the conference or meeting represented.

Conference proceedings often have 3 titles: the title of the book of proceedings, the title of the conference, and the title of the conference series, as in the following example:

Book title:	Building high performance healthcare organizations
Conference title:	MEDINFO 2004
Conference series title:	11th World Congress on Medical Informatics

Alternatively, without a separate book title, they may have 2 titles:

Conference title: CIDU 2011
Conference series title: Proceedings of the 2011 Conference on Intelligent Data Understanding

or only 1 title:

Conference series title: 2008 Fifth International Conference BioMedical Visualization: Information Visualization in Medical and Biomedical Informatics

Include all titles on the title page of a conference proceeding. Also provide information on the place and inclusive dates of the conference or other meeting. The place should include the city, state or Canadian province, and country if not readily discernible from the city name. Include also the name of the specific location, such as the university or other venue, if relevant. Finally, give the names of any sponsoring organizations.

See ANSI/NISO Z39.82 *Title Pages for Conference Publications* for further information[10] (see also Section 29.3.7.3 for information on citing conference proceedings and papers).

28.6.3 Electronic Monographs

Books may be published in a variety of electronic media: in CD-ROM and DVD formats; online; as print and graphics file downloads to computers or mobile-optimized for e-readers, taablets, and mobile smart phones; as audio CDs; as single audio or video file downloads; and as audio and video podcasts. Electronic formats provide many advantages over print[29]:

- Size is not always as important as it is with printed matter. Although editorial and production costs will increase with the size of the book, the restrictions on size imposed in the print world no longer apply. It is important to note, however, that excessively large electronic files may take long times to download or require large amounts of bandwidth to stream.
- A range of supplementary content, such as large data files and multimedia, which would be prohibited by cost and physical limitations in print, may be added to broaden the scope and usability of the text. Large amounts of supplementary content can be located online; these can be reached via hyperlinks from the digital text (online or on mobile-optimized platforms). For example, a video showing a surgical or laboratory procedure greatly enhances a print description, as does animation of complex molecular activity. Hyperlinks to another place in the same document or to an entirely different document or website can also provide immediate access to numerous additional resources.
- Searchability is greatly enhanced by the ability to look for words and phrases anywhere within the text, and navigability is improved by hyperlinks.
- Specialized dictionaries can be built in, allowing the reader to check the meaning of a term without having to leave his or her place in the text.
- Color and the range of available type fonts are dependent only on the instrument used for reading; online, this is almost limitless. Colors and fonts can be used not only by the publisher to indicate levels or sections of a book but also by the user to mark text or add notes in a saved copy.
- Editorial changes and updates can be made easily before publication without costly page changes, and errors can be corrected or text updated with the same ease after initial publication. Multiple versions can also be produced from the same text base.
- Maintaining a physical inventory is not necessary. There are minimal storage and distribution costs, and the danger of producing more copies than can be sold is eliminated.
- An electronic book need never go out of print, and printed copies can be generated in numbers as little as one book at a time (print on demand).

Since about 2000, electronic publication has moved from a novelty to a force of nature, with electronic monographs and serials now standing in active competition with print publications. Sales of electronic books and periodicals are on a steady increase and ever-improving instruments used to view these documents—e-readers, tablets, smart phones—hit the market on a regular basis. The PDF, once the standard of electronic publishing, has had to make room for a variety of electronic book formats. These formats generally have arisen from markup languages, such as HTML (HyperText Markup Language), XHTML (Extensible HyperText Markup Language), and XML (Extensible Markup Language), and provide the reader with the capabilities and versatility that only the electronic world could provide.

A variety of digital formats have been developed to deliver reading matter to mobile-optimized platforms, such as e-readers, tablets, and smart phones, whose screens may be too small to view a traditional PDF file. In some cases these electronic formats are proprietary, and documents created in these formats cannot be read using platforms that support competing formats. Amazon's AZW, for example, is a proprietary digital format that is the only means by which Amazon's Kindle books can be read. To make these books available to a larger market, however, Amazon has created a Kindle "app" (application), which can be used on various brands of e-readers, tablets, or smart phones.

Other digital formats are more generic. The EPUB format is an open standard format developed by the International Digital Publishing Forum (http://idpf.org/epub). EPUB was designed to be a universal format and is presently the fastest growing digital format for e-readers. With the exception of the Kindle e-reader, most major platforms can accept EPUB. It is expected that Amazon will make adjustments to the Kindle in the future that will allow EPUB-generated books to be read on that instrument.

Although electronic books differ radically in physical form from the standard print book, their content should include all of the elements discussed in Section 28.4. In addition to the usual elements required for print publication, 2 other items—metadata and DOIs—must be considered for electronic books. Metadata is standardized, descriptive information about a publication presented in a machine-readable, structured form. Two major metadata vocabularies in use for publishing are the Dublin Core Metadata Initiative (DCMI) (http://dublincore.org/) and the Online Information Exchange (ONIX) for Books (http://www.editeur.org/83/Overview/). Simple DCMI describes 15 metadata elements: title, subject, description, type, source, relation, coverage, creator, publisher, contributor, rights, date, format, identifier, and language; qualified DCMI adds 3 more elements: audience, provenance, and rights holder.[30] ONIX, with over 200 metadata elements, describes itself as the "international standard for representing and communicating book industry product information in electronic form".[31]

The purpose of metadata is to assist in 1) identifying, locating, and organizing resources; 2) promoting interoperability across systems; 3) providing for digital identification using such identifiers as ISBNs and URLs (Uniform Resource Locators); and 4) archiving and preservation.[32] Metadata, for example, is used by online search engines, such as Google, to locate documents satisfying search criteria; therefore, the presence of metadata in an electronic book enhances its retrieval. Two publications provided by

NISO are useful introductions: *Understanding Metadata*[32] and *Metadata Demystified: A Guide for Publishers*.[33]

The DOI (Digital Object Identifier) was developed by the International DOI Foundation (http://www.doi.org/) to provide for persistent identification of documents across networks. It does this by means of a naming convention consisting of a prefix and a suffix. The prefix contains the name of the particular DOI directory and the content owner's identifier; the suffix is a numeric or alphanumeric string supplied by the publisher. Publishers register with the International DOI Foundation to obtain their owner's identifier and then submit their suffix to a DOI registration agency along with the URL and appropriate metadata for the particular document being registered.[9] For additional information see Paskin's "Digital Object Identifier (DOI) System".[34] Publishers currently assign DOIs at both the book and chapter level. Some assign one DOI to a book regardless of changes in edition or format, whereas others give a unique DOI to each version. Many reference citation styles now require inclusion of a DOI if available.

CITED REFERENCES

1. National Information Standards Organization. Bibliographic references. Bethesda (MD): NISO Press; 2005. (ANSI/NISO Z39.29-2005). Also available at http://www.niso.org/.

2. National Information Standards Organization. The Dublin Core metadata element set. Bethesda (MD): NISO Press; 2001. (ANSI/NISO Z39.85-2001). Also available at http://www.niso.org/.

3. National Information Standards Organization. Electronic manuscript preparation and markup. Bethesda (MD): NISO Press; 2002. (NISO/ANSI/ISO 12083-1995 (R2002)). Also available at http://www.niso.org/.

4. National Information Standards Organization. Guidelines for abstracts. Bethesda (MD): NISO Press; 2002. (ANSI/NISO Z39.14-1997 (R2002)). Also available at http://www.niso.org/.

5. National Information Standards Organization. Permanence of paper for publications and documents in libraries and archives. Bethesda (MD): NISO Press; 2002. (ANSI/NISO Z39.48-1992 (R2002)). Also available at http://www.niso.org/.

6. National Information Standards Organization. Printed information on spines. Bethesda (MD): NISO Press; 2002. (ANSI/NISO Z39.41-1997 (R2002)). Also available at http://www.niso.org/.

7. National Information Standards Organization. Scientific and technical reports: preparation, presentation, and preservation. Bethesda (MD): NISO Press; 2005. (ANSI/NISO Z39.18-2005). Also available at http://www.niso.org/.

8. National Information Standards Organization. Standard technical report number format and creation. Bethesda (MD): NISO Press; 2002. (ANSI/NISO Z39.23-1997 (R2002)). Also available at http://www.niso.org/.

9. National Information Standards Organization. Syntax for the digital object identifier. Bethesda (MD): NISO Press; 2000. (ANSI/NISO Z39.84-2000). Also available at http://www.niso.org/.

10. National Information Standards Organization. Title pages for conference publications. Bethesda (MD): NISO Press; 2001. (ANSI/NISO Z39.82-2001). Also available at http://www.niso.org/.

11. The Chicago manual of style: the essential guide for writers, editors, and publishers. 16th ed. Chicago (IL): University of Chicago Press; 2010. Also available at http://www.chicagomanualofstyle.org.

12. Butcher J, Drake C, Leach M. Butcher's copy-editing: the Cambridge handbook for editors, copy-editors, and proofreaders. 4th ed. Cambridge (UK): Cambridge University Press; 2006.

13. Library and Archives Canada. Ottawa (ON): Government of Canada; [accessed 2013 Jan 27]. http://www.collectionscanada.gc.ca/.

14. United States Government Printing Office style manual. Washington (DC): US Government Printing Office; 2008. http://www.gpo.gov/fdsys/pkg/GPO-STYLEMANUAL-2008/pdf/GPO-STYLEMANUAL-2008.pdf.

15. Office of Public Sector Information. London (UK): The Office; [updated 2010 Oct 5; accessed 2013 Jan 27]. http://www.opsi.gov.uk/psi/.

16. UNESCO Universal Copyright Convention. Washington (DC), Geneva (Switzerland): United Nations Educational, Scientific and Cultural Organization; 1952 [accessed 5 Feb 2012]. (Universal Copyright Convention, Article VI. Copyright Office (US)); 1974 [accessed 2013 Jul 28]. (International copyright conventions circular; 38c). http://portal.unesco.org/en/ev.php-URL_ID=15381&URL_DO=DO_TOPIC&URL_SECTION=201.html.

17. Huth EJ. Writing and publishing in medicine. 3rd ed. Baltimore (MD): Williams & Wilkins; c1999.

18. Gibaldi J. MLA style manual and guide to scholarly publishing. 3rd ed. New York (NY): Modern Language Association; 2008.

19. Reitz JM. ODLIS: online dictionary of library and information science. Santa Barbara (CA): ABC-CLIO; c2004–2012 [accessed 2013 Jan 27]. http://www.abc-clio.com/ODLIS/odlis_A.aspx.

20. Clark G, Phillips A. Inside book publishing. 4th ed. London (UK): Routledge; 2008. http://www.insidebookpublishing.com/.

21. Bar codes for the book industry. Chicago (IL): Bar Code Graphics, Inc.; [accessed 2013 Jan 27]. http://www.barcode-us.com/isbn/isbnSymbols.html.

22. Borko F, editor. Handbook of augmented reality. New York (NY): Springer; 2011. http://dx.doi.org/10.1007/978-1-4614-0064-6_16.

23. International Organization for Standardization. Documentation - Spine titles on books and other publications. Geneva (Switzerland): ISO; 1985. (ISO 6357).

24. International Organization for Standardization. Documentation - Title leaves of books. 2nd ed. Geneva (Switzerland): ISO; 1991. (ISO 1066).

25. Library of Congress (US), Cataloging in Publication Division. The Cataloging in Publication Program, About CIP. Washington (DC): The Library of Congress; [accessed 2013 Jan 27]. http://www.loc.gov/publish/cip/about/index.html.

26. International Organization for Standardization. Documentation: abstracts for publications and documentation. Geneva (Switzerland): ISO; 1976. (ISO 214).

27. Peacock J. Book production. 2nd ed. London (UK): Blueprint; 1995.

28. International Organization for Standardization. Information and documentation - International standard technical report number (ISRN). Geneva (Switzerland): ISO; 1994. (ISO 10444).

29. Kasdorf WE, editor. The Columbia guide to digital publishing. New York (NY): Columbia University Press; c2003.

30. Hillman D. Using Dublin Core: the elements. Singapore: Dublin Core Metadata Initiative; [accessed 2013 Jan 27]. http://dublincore.org/documents/usageguide/elements.shtml.

31. ONIX for books. London (UK): EDItEUR; 2012 [accessed 2013 Jan 27]. http://www.editeur.org/11/Books/.

32. National Information Standards Organization. Understanding metadata. Bethesda (MD): NISO Press; c2004 [accessed 2013 Jan 27]. http://www.niso.org/publications/press/UnderstandingMetadata.pdf.

33. Brand A, Daly F, Meyers B. Metadata demystified: a guide for publishers. Hanover (PA): Sheridan Press; c2003 [accessed 2013 Jan 27]. http://www.niso.org/standards/resources/Metadata_Demystified.pdf.

34. Paskin N. Digital Object Identifier (DOI) System. In Encyclopedia of Library and Information Sciences. New York (NY): Taylor & Francis, 2008.

29 References

29.1 DEFINITIONS

The bibliographic description of a journal article, book, or other published work is called a reference. References fulfill 2 essential roles in the research and publishing process, ensuring intellectual integrity by[1]:

- Giving credit to those individuals and organizations whose published works have contributed to the research being reported
- Providing users of references with sufficient information to uniquely identify and locate a published work

This chapter presents information on the components of a reference, describes how these components should be determined and formatted, and finally indicates how they should be combined to form references to the various types of works found in the scientific literature.

References are presented in 2 ways in scientific publications. At the end of any document, such as a journal article, book, or book chapter, all of the references that contributed to the work are presented in a list variously called references, end references, reference list, literature cited, or bibliography. The term "end references" is primarily used within this manual. Within the text of a document, these references are presented in an abbreviated format that refers to the end references. Such abbreviated references within the text of a work are called "in-text references". (An earlier edition of *Scientific Style and Format*[2] referred to in-text references as citations, but this term is ambiguous because librarians and many others use "citation" interchangeably with "reference".) This chapter presents information on the 3 main systems for creating in-text references that are widely used in the scientific literature. The system of in-text references chosen determines the order of references in the reference list.

There are many methods of formatting end references. A representative list of those in common use, with an emphasis on the ones used in science, is presented at the end of this manual in the Bibliography. The Council of Science Editors (CSE) Style Manual Committee for the previous edition of *Scientific Style and Format* selected the *National Library of Medicine Recommended Formats for Bibliographic Citation*[3] as the preferred method, and its use has continued with this publication. The National Library of Medicine (NLM) method was chosen because it is based on principles set forth by the National Information Standards Organization (NISO)[1] and the International Organization for Standardization (ISO)[4,5] and because it has adapted the NISO and ISO standards to scientific material. (See Section 29.3.2 for a discussion of these standards.)

The need to cite non-English publications arises frequently in science. Throughout this chapter are recommendations to romanize words and names appearing in Cyrillic, Greek, Arabic, or Hebrew and to romanize or translate those appearing in character-based languages such as Chinese, Japanese, and Korean. Romanization, a form of transliteration, means simply to represent in the Roman (Latin) alphabet the letters or characters of another alphabet. Many systems of romanization are available. The ISO has produced a series of 11 standards[6–15] covering various languages. A frequently used authority in the United States is the *ALA-LC Romanization Tables*[16] produced by a joint effort of the Library of Congress and the American Library Association.

29.2 REFERENCES WITHIN THE TEXT (IN-TEXT REFERENCES)

Although many variations exist, there are 3 major systems for referring to a reference within the text of a published work: citation–sequence, name–year, and citation–name. Each of these systems is described below along with its respective advantages and disadvantages. The various methods for formatting in-text references (e.g., in italics) are also discussed. Lastly, the rationale for the selection of the citation–sequence system for use within this manual is presented.

29.2.1 Systems of In-Text References

29.2.1.1 Citation–Sequence

In the citation–sequence system, use numbers within the text to refer to the end references. Number the references and order them within the list in the sequence in which they first appear within the text. For example, if a reference by Smith is the first one mentioned in the text, then the complete reference to the Smith work will be number 1 in the end references. Use the same number for subsequent in-text references to the same document.

In-text reference:

Traumatic life events and posttraumatic stress disorder (PTSD) are endemic among American civilians.[1] Each year . . .

End reference:

1. Kessler RC, Sonnega A, Bromet E, Hughes M, Nelson CB. Posttraumatic stress disorder in the National Comorbidity Survey. Arch Gen Psychiatry. 1995 Dec;52(12):1048–1060.

In-text reference numbers that are not in a continuous numeric sequence should be separated using commas with no spaces. For more than 2 numbers in a continuous sequence, connect the first and last numbers with an en dash (or a hyphen if preferred); if there are only 2 consecutive numbers, separate them with a comma. If possible, place the numbers in superscript at each point of citation to eliminate confusion between in-text references and parenthetic numbers.

Modern scientific nomenclature really began with Linnaeus in botany,[1] but other disciplines[2,3] were not many years behind in developing various systems[4–7] for nomenclature and symbolization.

. . . have been shown[1,2,5,7,11–15] to abrogate the requirements of T cells . . .

In publications whose style requires placing in-text references on the line and within parentheses instead of in superscript, ensure that numbers other than in-text references that appear within parentheses are accompanied by a term, unit, or symbol that unambiguously distinguishes them from in-text references.

. . . and in our case series (12 patients) we found (37) that . . .

. . . the high value (12 g/L) previously reported (17) was later shown to have been . . .

29.2.1.2 Name–Year

In the name–year system, in-text references consist of the surname of the author or authors and the year of publication of the document. Enclose the name and year in parentheses.

In-text reference:

The NIH has called for a change in smallpox vaccination policy (Fauci 2002) that . . .

End reference:

Fauci AS. 2002. Smallpox vaccination policy—the need for dialogue. N Engl J Med. 346(17):1319–1320.

This system is widely known as the "Harvard system" because it was developed by Edward Mark of Harvard's zoological laboratory,[17] but this appellation is ambiguous. For example, the system used to compile references to the legal literature[4] is also known as "Harvard" because the Harvard Law Review Association has long been the publisher.

Variations in the name–year format are needed to accommodate multiple works written by the same author, authors with identical surnames, multiple authors, corporations and other organizations serving as authors, works without identifiable authorship, and works whose citations contain multiple dates.

29.2.1.2.1 MULTIPLE WORKS PUBLISHED BY THE SAME AUTHOR IN DIFFERENT YEARS

Distinguish works by the same author published in different years by placing the years after the author name in chronological sequence.

In-text reference:

Smith's studies of arbovirus infections (Smith 1970, 1975) have shown that . . .

End references:

Smith CE. 1970. Studies on arbovirus epidemiology associated with established and developing rice culture. Introduction. Trans R Soc Trop Med Hyg. 64(4):481–482.

Smith CE. 1975. The significance of mosquito longevity and blood-feeding behaviour in the dynamics of arbovirus infections. Med Biol. 53(5):288–294.

29.2.1.2.2 MULTIPLE WORKS PUBLISHED BY THE SAME AUTHOR IN THE SAME YEAR

For 2 or more works published by the same author in the same year, add an alphabetic designator to the year in both the in-text reference and the end reference.

In-text reference:

Cold hardiness in cereals (Andrews 1960a, 1960b) is affected by . . .

End references:

Andrews JE. 1960a. Cold hardiness of sprouting wheat as affected by duration of hardening and hardening temperature. Can J Plant Sci. 40(1):93–102.

Andrews JE. 1960b. Cold hardening and cold hardiness of young winter rye seedlings as affected by stage of development and temperature. Can J Bot. 38(3):353–363.

The sequence of the alphabetic designations is preferably determined by the sequence of publication of the works, earliest to latest. For example, an article published in January 2002 would be designated 2002a and one published in February 2002 would be designated 2002b. If this sequence cannot be determined, put the references in alphabetic order by article title.

29.2.1.2.3 AUTHORS WITH IDENTICAL SURNAMES

When the authors of 2 works published in the same year have identical surnames, include their initials in the in-text reference and separate the 2 in-text references by a semicolon and a space.

In-text reference:

Earlier commentary on animal experimentation (Dawson J 1986; Dawson M 1986) showed . . .

End references:

Dawson J. 1986. Animal experiments: conference report. BMJ. 292(6536):1654–1655.

Dawson M. 1986. Some examples of necessary continuation and possible discontinuation of animal experiments. Acta Physiol Scand Suppl. 554:194–197.

29.2.1.2.4 TWO OR MORE AUTHORS

If a work has 2 authors, give both names in the in-text reference, separated by "and".

In-text reference:

. . . and the most recent work on albuterol administration (Mazan and Hoffman 2001) is . . .

End reference:

Mazan MR, Hoffman AM. 2001. Effects of aerosolized albuterol on physiologic responses to exercise in standardbreds. Am J Vet Res. 62(11):1812–1817.

If both authors have the same surname, add their initials.

(Smith TL and Smith UV 1990)

Note that in the case of 2 authors, the separator is different for the in-text reference ("and") and the end reference (comma).

If a reference has 3 or more authors, give only the first author's name followed by "et al." and the publication year.

. . . but later studies (Ito et al. 1999) established that . . .

If the first author's names and the years of publication are identical for several references, include enough coauthor names in the in-text references to eliminate ambiguity. For example:

(Martinez, Fuentes, et al. 1990) or (Martinez, Fuentes, Ortiz, et al. 1990)

depending on what other Martinez et al. 1990 references are cited.

29.2.1.2.5 CORPORATIONS OR ORGANIZATIONS AS AUTHORS

If the author of a reference is a corporation, university, committee, or other organization, a shortened form may be created for the in-text reference to avoid interrupting the text with a long string of words. Use the initial letter of each part of the name or a readily recognizable abbreviation. For clarity, the abbreviation appears as the initial element in the end reference, within square brackets.

In-text reference:
The landmark report on legalized abortion (IOM 1975) was . . .

End reference:
[IOM] Institute of Medicine (US). 1975. Legalized abortion and the public health; report of a study by a committee of the Institute of Medicine. Washington (DC): National Academy of Sciences.

However, if the document has few in-text references, full organizational names are acceptable.

The landmark report on legalized abortion (Institute of Medicine 1975) was . . .

29.2.1.2.6 WORKS WITHOUT IDENTIFIABLE AUTHORSHIP

For works whose authorship cannot be determined (see Section 29.3.6.1 for where to find author information), do not use "anonymous".[1] Begin the in-text reference with the first word or first few words of the title, followed by an ellipsis. Use only as many words of the title as are needed to distinguish it from other titles being used as references.

In-text reference:
Drug dosage recommendations for elderly patients (Handbook . . . c2000) depart from . . .

End reference:
Handbook of geriatric drug therapy. c2000. Springhouse (PA): Springhouse.

29.2.1.2.7 WORKS WHOSE CITATIONS CONTAIN MULTIPLE DATES

It is possible for a work to have more than one date. This occurs with journals whose volumes span calendar years, with books published in several volumes over time, and with electronic documents for which a date of publication, a date of copyright, a date of modification, and a date of citation may all be available. For in-text references referring to publications with a range of dates, give the first and last years of publication, separated by an en dash (or a hyphen if preferred), as "1998–2000".

In electronic publications for which a date of publication, a date of copyright, a date of modification, and a date of citation may all be available, include only one of these dates in the in-text reference in the following order of preference: 1) date of publication; 2) date of copyright; 3) date of modification, update, or revision; and 4) accessed date.

In-text references:
(Johnson and Becker 1995–1999)
(Allen c2000)
(Morris [mod 1999])
(Handel et al. [accessed 2002])

End references:

Johnson KA, Becker JA. 1995–1999. The whole brain atlas. Boston (MA): Harvard Medical School; [modified 1999 Jan 12; accessed 2001 Mar 7]. http://www.med.harvard.edu/AANLIB/home.html.

Allen TF. c2000. The encyclopedia of pure materia medica. France: Homeopathe International; [accessed 2001 Apr 5]. http://homeoint.org/allen/.

Morris C, editor. [modified 1999 Mar 25]. Academic Press dictionary of science and technology. Orlando (FL): Harcourt, Inc.; [accessed 2001 Mar 28]. http://www.harcourt.com/dictionary/.

Handel A, Cross GH, Little CD. [accessed 2002 Jun 5]. Imaging of the spine. New York (NY): Mazel. http://www.mazel.com/.

Because both dates of modification or revision and dates of citation are within square brackets in the reference list (see Section 29.3.7.13), they also appear within square brackets in the in-text reference.

29.2.1.2.8 WORKS WHOSE DATES CANNOT BE DETERMINED

On rare occasions, it is not possible to determine any date associated with a publication. This occurs most frequently with older publications and manuscripts. For in-text references referring to publications with no date, place the words "date unknown" within square brackets.

In-text reference:

An early Belgian study (Lederer [date unknown]) on nutrition showed . . .

End reference:

Lederer J. [date unknown]. Alimentation et cancer [Diet and cancer]. 3rd ed. Brussels (Belgium): Nauwelaerts.

29.2.1.2.9 ACCOMMODATIONS FOR LOCATION OF IN-TEXT REFERENCES

Additional variations in the name–year system are needed to accommodate the location of the in-text reference, such as a work cited near the author's name and multiple in-text references at the same point. If works by the same author (or author group) are cited close to a mention of the author's name in the text and there is no uncertainty as to author identification, the in-text reference may be limited to the publication year.

When Chen's studies (1990, 1992a, 1992b, 1994) are examined closely . . .

When several in-text references occur at the same point, give them in chronologic sequence from earliest to latest, separated by semicolons. Sequence by month those published in the same year or, if this information is not available, alphabetically by author names.

. . . and the main contributors (Dawson and Briggs 1974; Dawson and Jones 1974; Smith AL 1978; Smith GT 1978; Smith et al. 1978; Tyndall et al. 1978; Zymgomoski 1978; Brown 1980) established beyond a doubt that . . .

29.2.1.3 Citation–Name

In the citation–name system, complete the list of end references for the work and then sequence them first alphabetically by author, then by title; see Section 29.3.3 for details. Number the references in that sequence, such that a work authored by Adam is number 1, by Brown is number 2, and so on. These numbers assigned to the end references are used for the in-text references regardless of the sequence in which they appear in the text of the work. For example, if a work by Zielinski is the first in-text reference appearing in a document and the end reference for Zielinski is number 56 in the list, the in-text reference will be number 56 also. Use the same number for subsequent in-text references to the same document.

In-text reference:

Traumatic life events and posttraumatic stress disorder (PTSD) are endemic among American civilians.[22] Each year . . .

End reference:

22. Kessler RC, Sonnega A, Bromet E, Hughes M, Nelson CB. Posttraumatic stress disorder in the National Comorbidity Survey. Arch Gen Psychiatry. 1995 Dec;52(12):1048–1060.

When several in-text references occur at the same point, place their corresponding reference list numbers in numeric order, as in the example below, to make it easy for readers to find them in the reference list. Separate in-text reference numbers that are not in a continuous numeric sequence by commas with no spaces. For more than 2 numbers in a continuous sequence, connect the first and last numbers by an en dash (or a hyphen if preferred); if there are only 2 consecutive numbers, separate them by a comma. If possible, place the numbers in superscript at each point of citation to eliminate confusion between in-text references and parenthetic numbers.

. . . have been shown[3,8–10,17,33,34] to abrogate the requirements of T-cells . . .

In publications whose style requires placing in-text references on the line and within parentheses instead of in superscript, ensure that numbers other than in-text references that appear within parentheses are accompanied by a term, unit, or symbol that unambiguously distinguishes them from in-text references.

. . . and in our case series (12 patients) we found (37) that . . .
. . . the high value (12 g/L) previously reported (17) was later shown to have been . . .

29.2.2 Advantages and Disadvantages of the Systems

There are 2 main advantages to the citation–sequence system. Because in-text references consist of numbers only, they interrupt the reading of the work only minimally. This advantage is significant when long sequences of in-text references occur, such as in review articles. Using numbers for in-text references also saves space and paper, and therefore cost, over the name–year system.

Another advantage of the citation–sequence system is that little decision making is needed by the compiler to format either an in-text reference or an end reference. Unlike the name–year system, where a series of rules must be applied to accommodate such frequent occurrences as multiple authors and corporate authorship, the in-text reference simply consists of a number. End references are formatted according to standard bibliographic rules without the need to move the date from its usual position and place it after the author name(s), as is necessary in the name–year system. This factor is important in the current online environment, when many authors use references that have been downloaded from databases.

However, the citation–sequence system has 3 disadvantages. First, readers who might be familiar with the subject and author of the work cited cannot identify the author from the in-text reference; they must turn to the list of end references to get this information. Some people also feel that the visibility and therefore the reputations of authors suffer if their names appear only in the end references and not within the text. Second, because in-text references are numbered in the order in which they appear within the work, the numbers change as the manuscript changes, which results in some clerical burden. Finally, the reference list in the citation–sequence system has little use on its own. Because references are ordered by their appearance within the text, there is some difficulty in locating the works of a specific author, particularly in long lists, and works by the same author are usually not listed together.

The name–year system has certain advantages. It is easier to add and delete references from the manuscript, since doing so does not necessitate any renumbering of the list, as would be the case in the citation–sequence system. The name–year system also recognizes the authors in the text without referral to the end references. In addition, the date provided with the author name in the text may provide useful information to the reader. Finally, since the reference list is arranged alphabetically by author, it is easy to locate works by specific authors, as works by the same first author are grouped together.

The most significant disadvantage of the name–year system relates to the numerous rules that must be followed to properly format an in-text reference. Using and applying rules for multiple authors, works published by the same author in the same year, and works authored by organizations can become burdensome to the user. These rules also conflict with recommended bibliographic standards, such as the use of "and" between 2 author names in the in-text reference; newer formats such as the Internet make these rules even more complex. Finally, long strings of in-text references in the name–year format interrupt the text and may therefore be irritating to the reader.

The third system for managing in-text references, the citation–name system, also has advantages and disadvantages. It shares with the citation–sequence system the advantages that in-text references only minimally interrupt the text and that end references are presented in standard bibliographic format. It shares with the name–year system the advantages that end references are listed in alphabetical order by author, making identification of works by specific authors easier and grouping works by the same first author together.

There are 2 disadvantages of the citation–name system. It shares with the citation–sequence system the disadvantage of the reader not being able to identify authors from the in-text references and the problem of reference numbers changing as the manuscript is revised.

29.2.3 Rationale for Using the Citation–Sequence System within This Manual

Although all 3 of the systems for generating in-text references discussed previously are widely used and accepted, the compilers have chosen to use the citation–sequence system within this manual. In this context, the advantages of using numbers for in-text references so that the text is not severely interrupted and showing end references in standard bibliographic format outweigh the disadvantages. Having end references in standard format is particularly important because the information for many references is now downloaded from online systems. The name–year advantage of author recognition within the text is not as important in this manual, because sources cover a wide range of scientific disciplines. Using citation–sequence also makes it easier for readers to find the appropriate end reference as they read along, unlike the citation–name system, where end references are arranged alphabetically by author. Thus, the citation–sequence system combines the best advantages of both of the other systems and does not have their major disadvantages.

In the sections that follow, the examples show how references would be formatted in the citation–sequence system. This should prevent the confusion expressed by some users of previous editions of this manual, which presented examples for both the citation–name system and the name–year system. However, these examples can be converted easily to either of the other systems.

- To convert in-text references from the citation–sequence system to the name–year system, follow the rules described in Section 29.2.1.2.
- To transform end references from the citation–sequence system into the name–year system, move the date information to immediately follow the author list and change the semicolon following the publisher to a period; see Section 29.2.1.2 for details.
- To convert to the citation–name system, reorder the end references alphabetically by author rather than in the sequence of their appearance in the text; the format for end references remains the same.

The examples used throughout are taken from actual publications and have been selected because they illustrate the particular point being made. Note that, because it is important not to change authors' actual words when creating a reference, the various titles within the examples may not follow CSE recommendations for style and format. For example, this manual recommends that "T cells" should be written without a hyphen or en dash, but if a title uses "T-cells", it has not been changed.

29.2.4 Punctuation, Typography, and Placement of In-Text References

Previous sections have presented information on the composition of an in-text reference: a name–year in-text reference contains the author name and publication year, and a citation–sequence or citation–name in-text reference consists of a citation number. This section discusses the location of in-text references and how they should be formatted. The recommendations in this chapter do not specify type styles for in-text references but simply indicate what information they should include. Some publications may choose to use italic or bold type to help distinguish in-text references for the reader; however, too many variations in type styles may confuse more than help.

29.2.4.1 General Format and Punctuation

In the name–year system, place in-text references on the same line as the text and within parentheses. Follow the last name of the author by the year of publication without intervening punctuation.

(Chang 1999)

Some other style manuals recommend separating the name and the year with a comma, as (Rizzo, 1998), placing the name and year in superscript either with or without parentheses, as [Rizzo 1998] or [(Rizzo 1998)], or placing the name and year in italics, but these practices are not recommended because they all entail extra work at the keyboard.

In both the citation–name and citation–year systems, place the reference number in superscript, usually in a slightly smaller type size.

> . . . has been shown[5] to replace IL-3 for the transient growth . . .

Other manuals recommend placing the in-text reference numbers in superscript within parentheses or in italics or both, as [(9)], [*9*], or [*(9)*], or placing them in parentheses on the line, with or without italics, as (10) or (*10*). However, the latter practice may create confusion with numerals other than reference numbers that appear in the text.

> In our series of patients (45) we found (37) that . . .

If in-text references are placed on the line, accompany any other numbers by a term, unit, or symbol that unambiguously distinguishes them from reference numbers.

> In our case series (45 patients) we found (37) that . . .

29.2.4.2 Placement

To avoid ambiguity about what is being referenced, an in-text reference should immediately follow the title, word, or phrase to which it is directly relevant, rather than appearing at the end of long clauses or sentences. Place in-text references adjacent to text if located mid-sentence. With regard to nearby punctuation, place any superscript in-text reference numbers immediately after commas and ending punctuation but before colons and semicolons.

> The most recent report (Slack 2002) on the use of . . .
> Only one study[3] relating the adverse effects of . . .
> In Jablonski's landmark work,[23]. . .

For additional information about handling in-text references, such as multiple in-text references at one point and multiple in-text references by the same author, see Sections 29.2.1.1, 29.2.1.2, and 29.2.1.3.

29.2.4.3 Use in Tables and Figures

In-text references in tables are usually most appropriately put in footnotes (see Section 30.1.1.8). If in-text references must appear within the field of a table, use a separate column or row for them and supply an appropriate heading to identify them.

If either the citation–sequence or citation–name system is used in a document, do not attach superscript in-text reference numbers to numbers in the field of a table, because they could be interpreted as exponents. Superscript alphabetic symbols can represent footnotes to the table, and the in-text references can be cited in the footnotes.

Do not use in-text references within figures, charts, graphs, or illustrations. If such references are needed to support the data or methods, put them in the legend.

If a reference is cited only in a table or figure caption and not again in the text, and if the journal style is to use citation–sequence, the reference should be numbered according to the location in the text where the table or figure is called out.

29.3 END REFERENCES AND REFERENCE LISTS

At the end of a document, list the references to sources that have been cited within the text, including those found in tables and figures, under the heading "References", "Cited References", "Literature Cited", or "Bibliography". To distinguish references for additional reading or other purposes that are not cited within the text, list them alphabetically by author under a separate heading (e.g., "Additional References", "Additional Reading", or "Supplemental References").

29.3.1 General Principles

29.3.1.1 What to Include in a Reference List

An author should never place in a reference list a document that he or she has not seen. The practice of citing documents only on the basis of information from other documents has led to the perpetuation of many erroneous references in the literature. If an author feels that the information in a document is important enough to be cited within the text, it is equally important to verify this information by examining the original. Accompanying this basic rule is the assumption that the author has in his or her possession a copy of all documents cited or can point to a public archive such as a library where a copy may be found. When it is not possible to see an original document, cite the source of your information on it; do not cite the original assuming that the secondary source is correct.

References to documents such as journal articles, books and book chapters, conference proceedings and conference papers, technical reports, dissertations and theses, audiovisual material, and websites are accepted by virtually all publishers. For other types of publications, such as those listed below, accessibility may be less reliable, and the decision of whether or not they can be included in a reference list is left to individual publishers and their editors:

- documents that have been accepted for publication but not yet published
- papers or abstracts of papers presented at meetings for which full articles were never published
- written personal communication, such as letters and email
- documented personal conversations
- in-house documents such as memoranda
- "trade" documents, such as manufacturers' catalogs and instruction manuals

The proliferation of "citable" items on the Internet has brought forth the new task of citing the version of the document seen. Many publishers are producing documents in multiple versions, in print and online. These various versions may appear identical in content, but because errors may be introduced in converting a document from one format to another, and because corrections and other changes are easily incorporated into an electronic version, they may in fact differ in significant ways (see Section 29.3.7.13 for a further discussion of this situation). When producing a reference for a document, the individual doing the citing must always cite the specific version seen; therefore, for example, do not cite the print version when only the electronic version was seen.

29.3.1.2 What to Include in a Reference

Section 29.3.6 provides detailed information on the components of an end reference and the order in which they should be given, primarily for the standard reference to a book or journal article. Section 29.3.7 gives information on the components of references to other types of documents, such as technical reports and patents. Not all documents will have all the possible components listed. For example, a book may not have an editor or may not be part of a series, yet information is provided on how to cite these elements if they are present.

Some components of a reference (e.g., the title) are required, whereas other components (e.g., the total number of pages for a book and the month of publication of a journal article) are considered optional. Optional components provide additional useful information to a reader but are not necessary to identify or find the document. Each component of a reference discussed in this manual is clearly labeled as required or optional.

Square brackets are a convention used in references to indicate that the material enclosed within them has not been found in the document itself but rather has been added by the person doing the citing. The following are examples of the type of information to be enclosed within square brackets:

- English-language translation of a non-English title
- physical format of a document, as "[DVD]"
- date of citation for an Internet document, as "[accessed 2004 Sep 15]"

Even if not provided in a document, some types of information must be accounted for in a reference. For example, the title is a required component of all end references. If no discernible title can be identified, "construct" a title from the first few words of the document and place it within square brackets. Similarly, the place of publication, publisher, and date of publication are all required for a reference to a book. If they do not appear in a document, use the words "place unknown", "publisher unknown", and "date unknown" in the position these components would normally occupy in the reference. If the place of publication can be reasonably inferred (e.g., Chicago for a document published by the American Medical Association), name the place but enclose it within square brackets.

[Chicago (IL)]: American Medical Association; 2001.
[place unknown]: Association of Refinery Workers; 1961.
Rome (Italy): [publisher unknown]; 1935.
Ames (Iowa): Iowa Department of Agriculture; [date unknown].
[place unknown: publisher unknown]; 1999.
[place, publisher, date unknown].

29.3.2 Standards for References

Many style manuals provide information about bibliographic references. The American Medical Association,[18] the American Chemical Society,[19] and the American Psychological Association,[20] among others, produce subject-oriented manuals, whereas such publications as *The Chicago Manual of Style*[21] and Kate Turabian's *A Manual for Writers of Term Papers, Theses, and Dissertations*[22] enjoy wide, general popularity. Less widely known are the reference standards produced by the ISO[4,5] and the National Information Standards Organization (NISO).[1]

Why, then, is the CSE including a chapter on reference style in its own manual? The CSE is committed to furthering the use of national and international standards. However, although the ISO and NISO reference standards mentioned above certainly meet these criteria, they have been produced for a general audience, not a scientific one, and often lack the detail needed for citing scientific materials. The examples they provide are also limited, failing to cover the breadth of material needed for scientific specialties. For these reasons the NLM has produced its own guides, *The National Library of Medicine Recommended Formats for Bibliographic Citation*[3] and *The National Library of Medicine Recommended Formats for Bibliographic Citation: Supplement: Internet Formats*.[23] One of the stated purposes of these publications is to adapt ISO and NISO principles to the medical literature. Because in many ways the literature of medicine is among the most complex of the scientific disciplines, CSE has chosen to use the NLM publications as the basis for its reference style. Because NLM permits a variety of options in its recommendations, CSE has identified those most relevant for the scientific literature and incorporated them into this chapter.

29.3.3 Sequence of References within a Reference List

In the citation–sequence system (see Section 29.2.1.1), list and number end references in the order in which each referenced document is first cited in the text. In the citation–name system (see Section 29.2.1.3), place references in alphabetic order by author, and then number the references in the same sequence. These numbers are used for the in-text references. Finally, for the name–year system (see Section 29.2.1.2), place references in alphabetic order by author.

When using the citation–name and name–year systems, alphabetic sequencing is determined by the first author's surname (family name) and then, if necessary, by letter-by-letter alphabetic sequencing determined by the initials of the first author and the beginning letters of any following surnames. The following additional principles apply to the alphabetization of author names:

1) Treat particles such as "de", "la", "van", "van de", and "von" as part of the surname. Alphabetize according to the particle, regardless of nationality, unless contradictory author preference is stated in the publication.

   Carter A
   de la Salle KL
   Gatlin BG
   Harris BN
   ten Asbroek AH
   van de Kamp J

2) Enter initial elements of surnames as they are written, not in terms of the full name that they represent. For example, do not alphabetize "St Louis" as if it were "Saint Louis" or "McGinnis" as if it were "MacGinnis".
3) Ignore apostrophes within surnames. For example, order "M'Veigh" as if it were "Mveigh" and "A'Amar" as if it were "Aamar".

4) Ignore diacritics, accents, and special characters within names. Treat a marked letter as if it were not marked; treat ligated letters as if they were not ligated. This rule ignores some conventions used in non-English languages to simplify rules for English-language publications.

Å treated as A	æ treated as ae	ü treated as u	Ĝ treated as g
Ø treated as O	Ç treated as C	à treated as a	Ł treated as L

5) When organizations serve as authors, drop "The" in an organizational name for the purposes of alphabetizing, as "American Chemical Society" not "The American Chemical Society".
6) When a component of an organization is an author, place components in descending hierarchical order.

   University of North Carolina, School of Dentistry, Department of Dental Ecology.

7) In the name–year system, when an acronym or initialism has been used for an organization as the in-text reference, order the reference by the full name, not the acronym or initialism.

   [ACS] American Chemical Society   [ABU] Australian Biochemical Union

If no authors are present, order items by title. The following principles apply to titles:

1) Ignore "a", "an", and "the" at the beginning of a title when alphabetizing. Thus, "The American Experience" is written with "The" at the beginning of the title, but is filed under "A".
2) Order initialisms and acronyms in document titles as if they were words, not by the full name for which they stand. Thus, "The CAS Registry" as a title is ordered before "Chemical Abstracts Service" as an author.
3) Order a title beginning with a number as if it were written out. Thus, the title "10 rules for healthy living" is filed in the Ts as if it were "Ten rules . . ."

Sequence references beginning with the same author list or the same title words according to the following rules:

1) For multiple items by the same author, order the items by title in the citation–name system, but by year in ascending order in the name–year system:

   Citation–name:
   Smith A. New approaches to staining. New York (NY): Putnam; 1995.
   Smith A. Three new approaches. New York (NY): Lippincott; 1990.

   Name–year:
   Smith A. 1990. Three new approaches. New York (NY): Lippincott.
   Smith A. 1995. New approaches to staining. New York (NY): Putnam.

2) If there are several items with the same first author, alone and/or with coauthors, present items written by a single author before items with coauthors, regardless of title. In such a grouping, give the multiauthor publications in alphabetical order by the second authors' surname, regardless of the number of authors.

   Citation–name:
   Smith A. History repeats itself.
   Smith A. New approaches to staining.
   Smith A, Jones B. Downs syndrome.
   Smith A, Jones B, Carson C. Alzheimer disease.
   Smith A, Martin G. Cutting corners.

Name–year:
Smith A. 1999. New approaches to staining.
Smith A. 2001. History repeats itself.
Smith A, Jones B. 2004. Downs syndrome.
Smith A, Jones B, Carson C. 2000. Alzheimer disease.
Smith A, Martin G. 2001. Cutting corners.

3) For organizations as authors, bring together all items written by the same organization. An overall organization alone, without any indication of subsidiary components, is considered one author; a component of an organization is another author. Therefore, start with all those references written by the overall organization alone, then just the overall organization with another organization, then the organization with parts, then the organization with parts with another organization. Use a semicolon to separate 2 organizations as authors and a comma to separate an organization and its parts.

International Union of Pure and Applied Chemistry.
International Union of Pure and Applied Chemistry; International Union of Biochemistry.
International Union of Pure and Applied Chemistry, Commission on the Nomenclature of Organic Chemistry.
International Union of Pure and Applied Chemistry, Macromolecular Division, Commission on Macromolecular Nomenclature.
International Union of Pure and Applied Chemistry, Macromolecular Division, Commission on Macromolecular Nomenclature; International Union of Biochemistry.
International Union of Pure and Applied Chemistry, Macromolecular Division, Commission on Macromolecular Nomenclature; International Union of Biochemistry, Nomenclature Commission.

4) Apply the following hierarchy of delimiters when sequencing references: period, colon, comma, space.

Titles beginning with the same word:
Aging.
Aging: three ways to avoid it.
Aging, getting bald, and other indignities.
Aging at fifty.

Authors:
St Louis AM
Stickley PB

29.3.4 Typography

As noted in Section 29.2.4, typography is not covered by the recommendations in this manual. Editors and publishers may chose to underline, make bold, place in italics, or capitalize some components of a reference such as a title or volume to make them more visible to the reader. However, because too many variations in type styles may actually make the reference harder to read, CSE urges caution in applying these conventions and has not used any of them within this manual.

29.3.5 Punctuation

The system of punctuation for references recommended in this manual has been adapted from the NLM recommendations.[3] Punctuation is used as a delimiter, indicating the end of a component of a reference or separating individual items within a component. It also serves to indicate certain components of a reference by surrounding them. This system uses 8 punctuation marks and a space. Their general uses are described in the list below; see Section 29.3.6 and its component subsections for greater detail.

Punctuation Mark	Use
Period	To indicate the end of a component of a reference, such as the author list or a book or journal title
	To indicate the end of a reference
	To indicate abbreviation of words (except in references to journal articles, where they are omitted within the abbreviation for a journal title and for parts of a volume or issue)
Comma	To separate related items within a component of a reference, such as author names; levels of an organization serving as author or publisher, as "American Medical Association, Committee on Ethics"; items within an author address
	To separate an author from the author role, as "Smith J, editor"
Semicolon	Between multiple occurrences of components of a reference, such as 2 organizations serving as authors
	Between logically related groups of items, such as the name, date, and place of a conference
	After the publisher name except in name–year references
	Before a volume of a journal (or issue if there is no volume), except in name–year references
Colon	To separate a title from a subordinate title
	After a connective phrase, such as "In:"
	Before any identifying number, as "Report No.:"
	Before the pages of a journal article
En dash or hyphen	To separate the first and last element of a range (en dash or hyphen), such as page numbers for a journal article or years of publication
	To indicate indefinite continuation of publication (en dash or hyphen), such as a journal title or database
Parentheses	To surround a component of a reference, such as an author address, an issue number, the name of a state or country in a place of publication, a series name
Square brackets	To indicate information not provided within the document being referenced, such as a translation of a title
	To indicate information provided for clarity, such as the type of journal article, (e.g., editorial and letter) or the medium of publication (e.g., audiovisual)
Equals sign	To separate parallel titles (when a title is published in 2 or more languages)
Space	After periods and all other punctuation marks (except the semicolon that separates the date from the volume number in journal references), unless they are immediately followed by another mark, as "Contract No.:"
	Between components of a reference when there is no punctuation, such as between an author's last name and initials

29.3.6 Components of End References and Their Sequence

The term "monograph" is often understood to mean a scholarly treatise on a specific subject, usually limited in scope and usually written by a specialist in the field. In this manual, the term is used in its broader sense to mean a publication that is complete in a single volume or in a limited number of volumes, such as a book that is published in 3 physical pieces. Books, textbooks, technical reports, conference proceedings, master's theses and dissertations, bibliographies, festschrift, and patents, as well as the scholarly treatises described above, are all examples of monographs.

Journals and newspapers, also known as serials, are publications "appearing or intended to appear indefinitely at regular or stated intervals, generally more frequently than annually, each issue of which normally contains separate articles or papers".[1] A major difference in the way references to monographs and serials are constructed relates to publisher information. References to monographs contain the place of publication and the publisher; references to journal articles do not. This difference probably relates to the fact that the procedures for citing monographs arose from formal library cataloging, whereas the system for citing journals grew out of printed bibliographies, then online databases. Common usage thus permits a great deal of abbreviation within journal references.

Consistent ordering of the components within a reference is necessary to avoid confusion and to enable the reader to locate needed information easily. The following sequence is prescribed by NISO[1]:

Books and other monographs	Journal and newspaper articles
Author(s)	Author(s)
Title	Article title
Content designator	Content designator
Medium designator	Journal or newspaper title
Edition	Edition
Secondary author	Medium designator
Place of publication	Date
Publisher	Volume
Date	Issue
Extent (pagination)	Location (pagination)
Physical description	Physical description
Series	Notes
Notes	

Any single component may be taken out of this sequence to meet a particular need as long as all the other components remain in order. An entry in a reference list for the name–year system of in-text references, in which the year of publication follows authorship, is one example.

A particular publication being cited may not possess all the components given in the table above. For example, not all monographs have editions or are part of a series. Also, not all of these components are required for an end reference. For example, the physical description is usually optional. Information included there, such as the total number of pages of a book or the size of a map, may be useful to the reader but is not essential. For further explanation, see the descriptions of the components below, as well as Section 29.3.7 and its subsections for the various types of publications.

The information on authors, titles, and other components of a reference provided in this section pertains primarily to standard journal articles and books. Conference publications, technical reports, patents, audiovisual materials, Internet documents, and other specialized types of publications have more complex rules; see Section 29.3.7 and its subsections for details.

29.3.6.1 Author

In all of the 3 systems discussed in this chapter—citation–sequence, name–year, and citation–name—an end reference begins with author information. An author is defined by NISO as "a person, committee, organization, or other party responsible for the creation of the intellectual or artistic content of a work".[1] Author information is usually found on the title page of a book or other monograph, at the beginning (top) of a journal article (although occasionally at the end of the article), on the opening screens or accompanying booklet of an audiovisual, or on the opening screen or at the end of a website.

Individuals who are authors (personal authors), organizations that serve as authors, and editors and others who serve as secondary authors are discussed below. If no author can be determined for a document, omit authorship from the reference. The use of "anonymous" is not permitted.[1] When the name–year system of in-text references is being used and no author can be identified, begin the in-text reference with the first word or first few words of the title, followed by an ellipsis. Use only as many words as are necessary to distinguish the document being cited from other documents without authors (see also Section 29.2.1.2).

In-text reference:
Drug dosage recommendations in elderly patients (Handbook . . . c2000) depart from . . .

End reference:
Handbook of geriatric drug therapy. c2000. Springhouse (PA): Springhouse.

29.3.6.1.1 PERSONAL AUTHORS

In a reference, list the names of the authors in the order in which they appear in the original text. Begin with the surname (family name or last name) of the first author, followed by the initial letters of the given (first) name and middle name. When there are 2 to 10 authors, list all of them, including the 10th author; if there are more than 10 authors, list the first 10 followed by "et al." (which is preferred) or "and others". This is the style followed within this manual. Some publishers depart from this recommendation and require fewer author names, most often 3 or 6, whereas others, such as the NLM in its MEDLINE database, use all author names.

Separate the surname and initials of an author by a space; separate successive author names from each other by a comma and a space. Follow the last item in the author list with a period unless 1) there are more than 10 authors or 2) an editor or other type of secondary author (see Section 29.3.6.1.3) is serving as author. In these cases, follow the last author with a comma and either "et al." or the type of secondary author.

Takagi Y, Harada J, Chiarugi A, Moskowitz MA.
Jones AR, Smith KR, Williams AB, Carter F, White RY, Little RT, Kane TR, Larosa J, Mann FD, Swartz MN, et al. Matthews DD, Bellenir K, editors.

As an option, full first names of authors may be given. In this case separate the last name from the first name or initials by a comma and separate successive names by a semicolon.

Takagi, Yasushi; Harada, Jun; Chiarugi, Alberto; Moskowitz, Michael A.
Jones, Arthur R.; Smith, Keith R.; Williams, Alice B.; Carter, Frank; White, Ralph Y.; Little, Richard T.; Kane, T.R.; Larosa, John; Mann, Frederick D.; Swartz, Mary N., et al.
McKee, Phillip H.; Calonje, Eduardo; Granter, Scott R., editors.

The preference for initials in this manual is to reduce differences between formats of references to journal articles, in which author initials predominate, and to books, in which full given names usually appear. Consistency in style within a reference list reduces work for typists and editors and also decreases chances for errors. Saving space is also a consideration.

Format of personal author names: The conventions for designating surname vary by country. If the list of authors does not indicate which part of the name is the surname and/or provides names in all capital letters, the table of contents or author index to a publication, if available, may provide assistance. If the document itself is not helpful, see Section 8.1 of this manual for guidance on the format to use.

- Omit degrees, titles, and honors that may follow a personal name, such as MD; omit rank and honors that may precede a name, such as Colonel or Sir.
- Place designations of rank within a family after the initials, without punctuation; convert designations in roman numerals to arabic ordinals.

  Vincent T. DeVita Jr *becomes* DeVita VT Jr
  John A. Adams III *becomes* Adams JA 3rd
  James G. Jones II *becomes* Jones JG 2nd
  Henry B. Cooper IV *becomes* Cooper HB 4th

- Capitalize names and enter spaces within names as they appear in the document being cited on the assumption that the author has approved the form used. The following names, for example, could appear in various forms:

  Van Der Horn KH *or* van der Horn KH
  Le Sage R *or* LeSage R *or* Lesage R
  De Wolf F *or* de Wolf F *or* DeWolf F *or* Dewolf F

- With the exception of hyphenated surnames and surnames with apostrophes after particles such as O', D', and L', omit punctuation within a name.

  Estelle Palmer-Canton *becomes* Palmer-Canton E
  Charles A. St. James *becomes* St James CA
  Alan D. O'Brien *becomes* O'Brien AD

- Convert given or first names and middle names to initials. The following rules apply:

  1) Enter given names and middle names containing a prefix, a preposition, or other particle by the first letter only.

  D'Arcy Hart *becomes* Hart D
  De la Broquerie Fortier *becomes* Fortier D
  W. St. John Patterson *becomes* Patterson WS
  Craig McC. Brooks *becomes* Brooks CM

  2) Disregard hyphens or en dashes joining given names.

  Jean-Louis Lagrot *becomes* Lagrot JL

  3) Disregard traditional abbreviations of given names. Some non-US publications use abbreviations of conventional given names rather than single initials, such as "St." for "Stefan". Use only the first letter of the abbreviation.

  Ch. Wunderly *becomes* Wunderly C
  C. Fr. Erdman *becomes* Erdman CF

  4) For transliterated names in which the original initial is represented by more than one letter, give the required letters, but capitalize only the first one.

  Iu. A. Iakontov *becomes* Iakontov IuA
  G. Th. Tsakalos *becomes* Tsakalos GTh

Author roles: Follow the authors of particular types of monographs with a comma and the particular author role. For example, authors of bibliographies are termed "compilers", of patents "inventors", and of maps "cartographers" (see Sections 29.3.7.6, 29.3.7.7, and 29.3.7.9).

Selden CR, Kelliher R, compilers. Distance education in public health [bibliography].
McClellan E, cartographer. Cholera epidemic of 1893 [map].

Authors of parts of monographs: When citing a part of a book or other monograph, such as a chapter that has separate authors, begin the reference with the author(s) of the part and the title of the part, followed by the connective phrase "in" and a colon (see Section 29.3.6.12), and then the editors and other citation information for the monograph.

Weinstein L, Swartz MN. Pathologic properties of invading organisms. In: Sodeman WA, editor. Pathologic physiology: mechanisms of disease. 5th ed. Philadelphia (PA): Saunders; 1974. p. 135–140.

29.3.6.1.2 ORGANIZATIONS AS AUTHORS

An organization such as a university, society, association, corporation, or governmental body may also serve as an author. However, if both a personal author and an organizational author appear on the title page of a document, use the personal author only and add the organization name as a note. An exception may be made to this rule when the organization name has specific significance to the intended audience.

If a division or other part of the organization is provided on the publication, give the parts of the name in descending hierarchical order, separated by commas. This requirement may call for breaking and reordering the elements of an organizational author name as it appears on the referenced document. For example:

Committee on Enzymes of the Scandinavian Society for Clinical Chemistry and Clinical Physiology
becomes
Scandinavian Society for Clinical Chemistry and Clinical Physiology, Committee on Enzymes

The hierarchy need not include all possible elements presented in a publication; the highest used should be the one most likely to be known by the readership. For example, although the National Cancer Institute (NCI) is hierarchically under the National Institutes of Health, which is under the Department of Health and Human Services, the NCI is well known in its own right and a book identified as coming from the Division of Cancer Prevention of the National Cancer Institute would be given the authorship designation "National Cancer Institute (US), Division of Cancer Prevention", not "Department of Health and Human Services (US), National Institutes of Health, National Cancer Institute, Division of Cancer Prevention". The general rule is to ignore the word "The" when it begins an organizational name. For example, "The American Chemical Society" becomes simply "American Chemical Society". However, in some cases such as "The Ohio State University" or "The Joint Commission", the "The" is part of the official name of the organization and should be retained; check the organization's website if unsure.

In citing organizations that are national bodies, such as government agencies, if a nationality is not included with the name, place the country after the name, within parentheses. Use the 2-letter ISO country code[24] to indicate the nationality (see Appendix 29.3 for a list of the codes for selected countries).

National Academy of Sciences (US)
National Fire Service (GB)
Royal Institute of Technology (SE)

If the organizational name is identical with that of one or more other organizations, add a geographic modifier for clarification; for example, because there are many Beth Israel hospitals in the United States, specify the city as "Beth Israel Hospital (Boston)".

For names of organizations in non-Roman alphabets, romanize the names if in Cyrillic, Greek, Arabic, or Hebrew; romanize or translate them if in character-based languages (e.g., Chinese, Japanese). Whenever possible, follow a non-English name with a translation. Place all translations within square brackets.

> [Chinese Ornithological Society].
> Universitatis Helsingiensis, Instituti Medico-Legalis [Helsinki University, Medical–Legal Institute].

If 2 or more organizations appear as authors, give them in the order listed in the publication and separate them by a semicolon and a space.

> American Medical Association; American Heart Association.

When using the name–year system of in-text references, use the initial letter of each part of the name or a readily recognizable abbreviation to create a shortened form for the in-text reference. These letters or the acronym, within square brackets, become the initial element in the end reference (see Section 29.2.1.2). Alphabetize these letters or acronyms within the end references as if they were spelled out; for example, treat "[ACS]" as "American Chemical Society".

> **In-text reference:**
> The landmark report on legalized abortion (IOM 1975) . . .
>
> **End reference:**
> [IOM] Institute of Medicine (US). 1975. Legalized abortion and the public health: report of a study by a committee of the Institute of Medicine. Washington (DC): National Academy of Sciences (US).

29.3.6.1.3 SECONDARY AUTHORS

A secondary author is "the person, committee, organization, or other party responsible for adopting, interpreting, or otherwise modifying the intellectual content of a preexisting work".[1] Editors, translators, illustrators, and producers are examples of secondary authors. If a work has secondary authors in addition to a personal or organizational author, the names of such secondary authors are an optional component of a reference. Place the secondary author after the title of the work or any edition statement that may follow the title, followed by a comma and the role or roles of the secondary author. Separate multiple secondary authors by commas; separate secondary authors occupying different roles by semicolons. Give secondary authors in the order in which they appear in the publication.

> Martin EW. Hazards of medication. Ruskin A, Napke E, editors. Richer PM. Artistic anatomy. Hale RB, translator and editor.
> Luzikow VN. Mitochondrial biogenesis and breakdown. Galkin AV, translator; Roodyn DB, editor.

If a work has no personal or organizational author but has an editor or a translator, the name is a required component of a reference. Place the editor(s) or translator(s) at the beginning of the reference in the author position.

> Diener HC, Wilkinson M, editors. Drug-induced headache.

If no primary author, editor, or translator can be determined for a work, omit any indication of authorship and begin the reference with the title. The use of "anonymous" is not permitted.[1]

29.3.6.2 Title and Subtitle

Journals have both a title for each of the individual articles within them and a title or name for the journal. Specific rules for journal titles are discussed in detail in Section 29.3.7.1.3. Books and other monographs have a title for the entire document and may have individual volume, chapter, or section titles within them. Some books also may be part of a series with its own title (see Section 29.3.6.11 for series information). In both the citation–sequence and citation–name reference systems, the title for a journal article or a book follows the author list; if no authors are present, the title begins the reference. In the name–year system, the title follows the year of publication.

A journal article or monograph title may be followed by a subtitle, the "term or phrase following the chief, distinguishing title that is subordinate to and can be grammatically and intellectually separated from it without loss of meaning to the title. It completes and qualifies the title or makes it more explicit".[1] Journal titles also often have subtitles, but these are not included in a reference unless the entire journal title is being cited (see Section 29.3.7.1.9).

29.3.6.2.1 FORMAT OF TITLES

In general, record the words of a title as they appear in the original document. For example, this manual recommends against using periods within an acronym. However, if an acronym appearing within a title has periods, retain them in the reference. It is important not to change the actual words of the author when creating a reference. There are 6 major exceptions to this general rule.

1) For a title in a non-Roman alphabet, romanize the title if it is in Cyrillic, Greek, Arabic, or Hebrew; romanize or translate it if in a character-based language (e.g., Chinese, Japanese). Whenever possible, follow a non-English title with a translation. Place all translated titles within square brackets.
2) If a title contains a Greek letter or some other symbol that cannot be reproduced with the type fonts available, substitute the name for the symbol, (e.g., Ω becomes omega).
3) If a title is presented in 2 languages, representing the bilingual nature of the text, as often occurs in Canadian publications, give both titles in the order in which they are presented in the text, with an equals sign between them.
4) Abbreviate titles for journals using ISO rules[25] (see Section 29.3.7.1.3).
5) Apply a specific order and punctuation to titles of conference proceedings (see Section 29.3.7.3).
6) Consider such terms as "final report" or "annual report" following titles of technical reports as edition statements, not part of the title (see Section 29.3.7.4).

Occasionally a publication does not appear to have any title. An article or other short document may simply begin with the text. In this unusual circumstance, "construct" a title from the first few words of the text, using enough words to make the constructed title meaningful, and place them within square brackets.

Content and medium designators, used to provide information on the nature and physical format of a publication, may follow a title (see Section 29.3.6.4).

For monographs and articles in journals and newspapers, capitalize only the first word of a title, proper nouns, proper adjectives, acronyms, and initialisms. If the title begins with a Greek letter, chemical formula, or another special character that might lose its meaning if capitalized, retain the lowercase. For titles not in English, follow the conventions of the particular language regarding capitalization. Journal and newspaper titles differ in that every word is capitalized (see Sections 29.3.7.1.3 and 29.3.7.8.3 for details).

29.3.6.2.2 PUNCTUATION

Retain whatever punctuation appears within the original title. If no punctuation separates the title from a subtitle, use a colon followed by a space. End the title with a period unless another form of punctuation such as a question mark or an exclamation point is already present. An exception is titles of conference proceedings, for which a specific order and punctuation are applied (see Section 29.3.7.3).

Pediatric cancer: methods and protocols.
Rational diuretic management in congestive heart failure: a case-based review.
Which role for radiation therapy in ethmoid cancer? A retrospective analysis of 84 cases from a single institution.
Let's defeat cancer! The biological effect of deuterium depletion.
Whither surgical oncology?

29.3.6.3 Edition

For books and other monographs, edition information "identifies a different form or version of a previously published work, such as 2nd edition or version 4.0".[1] Give edition statements in their original language, but romanize names presented in Cyrillic, Greek, Arabic, or Hebrew; romanize or translate them if in a character-based language (e.g., Chinese, Japanese). If desired, follow a non-English edition statement with a translation, surrounded by square brackets. Express numbers representing editions in arabic ordinals:

first *becomes* 1st
second *becomes* 2nd
third *becomes* 3rd
IV *becomes* 4th
quatrième *becomes* 4th

Place edition information after the title. Abbreviate the word "edition" and other descriptive words for print publications according to ISO 832[26] and follow them with a period. First editions not known to be followed by new editions usually need not be identified as "1st ed.", but if later editions have been published, identify the first edition as such when cited. For English-language edition statements, capitalize only the first word of the edition statement unless it contains proper nouns; for non-English statements, abide by the conventions of the particular language regarding capitalization. Follow the complete edition statement with a period.

Eighth Edition *becomes* 8th ed.
New revised edition *becomes* New rev. ed.
Third American Edition *becomes* 3rd Am. ed.
Edizione Italiana *becomes* Ed. Ital.

Some words expressing edition for electronic publications such as "version" are becoming more common and also can be abbreviated (see Sections 29.3.7.12 and 29.3.7.13), but less common words such as "update" or "release" should not be abbreviated.

Ver. 4.0   Update 1.2

Titles of technical reports are often followed by such words as "final report" or "annual report". These are considered edition statements (see Section 29.3.7.4).

Editions for journal titles are abbreviated according to ISO 4.[25] Unlike editions for books and other monographs, they are usually published simultaneously and are considered a part of the title. Separate them from the title proper by a space, surrounded them by parentheses, and capitalize each word. Do not follow abbreviated words with a period, but conclude all the title information with a period.

Hosp Pract (Off Ed). Tierarztl Prax (Ausg Klient Heimtiere).
BMJ (Int Ed). Indice Espec Farm Intercon (Ed Farm).

These differences reflect the long-held practice of treating journal references differently from books and other monographs (see Section 29.3.7.1.3).

Newspapers may be published in different editions; edition designations here are also considered a part of the title (see Section 29.3.7.8.3).

29.3.6.4 Content and Medium Designators

Content and medium designators are used to provide information on the nature and physical format of a document. For journal articles, content designators indicate forms such as editorials, letters to the editor, news, or meeting abstracts; for books and other monographs, they indicate forms such as dissertations and bibliographies; for electronic documents, they identify forms such as databases and computer programs. Medium designators inform the user that the item is in a nonprint format and that special equipment is needed to read or view it. Examples are "videocassette" or "microfiche". A medium designator is a required part of a reference; a content designator is optional. Because inclusion of a URL in a reference to an online source makes the medium obvious, it is not necessary to include "Internet" as a medium designator for online sources.

Example of an online reference:

GDB: The Human Genome Database. Washington (DC): US Department of Energy; 2005 [accessed 2011 Nov 19]. http://www.ornl.gov/sci/techresources/Human_Genome/home.shtml.

Because both content and medium designators provide clarification to title information, place them within the period that ends the title information; because they provide information not included in the title proper, place them within square brackets. Neither type of designator is capitalized, with the exception of CD-ROM and DVD because they are initialisms.

Examples of a content designator following a title:

Smith KL. New dangers in our field [editorial]. Am J Nucl Eng. 1991;13(1):15–16.

Ritzmann RE. The snapping mechanism of Alpheid shrimp [dissertation]. Charlottesville (VA): University of Virginia; 1974.

Example of a medium designator following a title:

Goldie S. A calendar of letters of Florence Nightingale [microfiche]. Oxford (England): O.M.P.; 1977. 40 microfiche: 4 x 6 in.

It is possible for a document to carry both a content designator and a medium designator, for example, "[dissertation on microfilm]"and "[bibliography on CD-ROM]".

29.3.6.5 Place of Publication

Place of publication is defined as the "name of the city where the publisher is located".[1] For books and other monographs, the place of publication is required in a reference. For journal articles, this information is omitted, in accordance with long practice.

The title page or verso (back) of the title page is the source for place of publication information. If more than one city of publication is provided, use the first one encountered or the one in the largest or bold type font. An alternative is to use the city likely to be most familiar to the audience of the reference list (e.g., New York for a publication intended for an American audience and London for a British one); do not list multiple cities. End place of publication information with a colon.

If no place of publication can be found on the title page or its verso, but one can be found elsewhere within the publication or can be reasonably inferred (e.g., Chicago as the place for a publication of the American Medical Association), put the city within square brackets, as "[Chicago (IL)]". When no city can be located or inferred, use the words "place unknown" within square brackets.

Mueller FO, Schindler RD. Annual survey of football injury research 1931–1985. [place unknown]: American Football Coaches Association; 1986.

Additional information may be included with the city name to provide any necessary clarification. Although many city names such as New York or Stockholm are almost universally known and can stand alone, others are not so well known or may be confused with cities of the same name in other countries. For example, if a book has been published in Paris, Texas, the state name may be supplied to avoid confusion with Paris, France. Follow US and Canadian cities with the 2-letter postal abbreviation for the state or province. Follow cities in other countries with the name of the country, either written out or as the 2-letter ISO country code[24] (see Appendix 29.3 for a list of the codes for selected countries).

Use the name of the country as it is found on the publication, such as England, Great Britain, or the United Kingdom. If the audience for the reference is an English-speaking one, the anglicized form for a city is preferred (e.g., "Rome" for "Roma"). However, the name as it appears on the publication is always correct. Place the state, province, country, or ISO country code in parentheses after the city.

New York, NY *becomes* New York: *or* New York (NY):
Boca Raton, Florida *becomes* Boca Raton (FL):
London, England *becomes* London: *or* London (England): *or* London (GB):
London, Ontario, Canada *becomes* London (ON):

For consistency within a reference list, treat each occurrence of a city the same (e.g., always "New York:" or always "New York (NY):"). To avoid any ambiguity, the state, province, or country name has been included for all cities of publication in the references within this manual.

Place of publication is handled differently in citing newspapers. If the newspaper title does not indicate the location where it is published, add the location within parentheses after or within the title, as appropriate, unless it is a national paper (see Section 29.3.7.8).

El Dia (Mexico City) or Bergen County (NJ) Record *but* Christian Science Monitor

29.3.6.6 Publisher

A publisher is defined by NISO as the "person, firm, or corporate body responsible for making a work available (issuing it) to the public".[1] Publisher information is usually found on the title page, the verso (back) of the title page, or the cover of monographs and on the title page, masthead, or cover of journal issues. If the publisher name cannot be found in these locations but can be found elsewhere within the publication, such as the preface, use square brackets around the name. Publisher name is omitted from references to journal and newspaper articles but is required for all other types of publications. If no publisher can be determined, use the words "publisher unknown" surrounded by square brackets.

Record the name as it appears in the publication, using whatever capitalization and punctuation is found there, omitting only an initial "The" of a name, such that "The Entomological Society of America" becomes "Entomological Society of America". Give names of publishers in their original language, but romanize names presented in Cyrillic, Greek, Arabic, or Hebrew; romanize or translate them if in a character-based language (e.g., Chinese, Japanese). If desired, follow a non-English name with a translation. Place all translated publisher names within square brackets. If the name of a division of other part of an organization is included in the publisher information, give the names in hierarchical order from highest to lowest.

[Chinese Medical Society]
University of Pennsylvania, Department of Psychology
Elsevier Science, Biomedical Division

The hierarchy need not include all possible elements presented in a publication; the highest used should be the one most likely to be known by the readership. For example, although the National Weather Service (NWS) is hierarchically under the National Oceanic and Atmospheric Administration, which is under the US Department of Commerce, the NWS is well known in its own right, and a publication identified as coming from the Climate Prediction Center of the National Weather Service would be given the publisher designation "National Weather Service (US), Climate Prediction Center" not "Department of Commerce (US), National Oceanic and Atmospheric Administration, National Weather Service, Climate Prediction Center". In citing organizations such as government agencies that are national bodies, if a nationality is not included with the name, place the ISO 2-letter country code[24] after the name, within parentheses (see Appendix 29.3 for a list of the codes for selected countries).

National Academy of Sciences (US)
National Fire Service (GB)
Royal Institute of Technology (SE)

If more than one publisher is found in a document, use the first one given or the one set in the largest type or set in bold. An alternative is to use the publisher likely to be most familiar to the audience of the reference list (e.g., an American publisher for a US audience and a London publisher for a British one). Do not list multiple publishers. For those publications with joint or co-publishers, use the name provided first as the publisher and include the name of the second as a note, if desired, as "Jointly published by the Canadian Pharmacists Association". End publisher information with a semicolon when using the citation–name and citation–sequence systems and a period when using the name–year system.

Do not confuse the publisher with the distributor, who disseminates documents for the publisher. For example, the most common distributors of US government publications are the US Government Printing Office (GPO) and the National Technical Information Service (NTIS). Designate the agency making the publication available as the publisher and include distributor information as a note (see Section 29.3.6.13).

Levedahl JW, Matsumoto M. U.S. domestic food assistance programs: lessons from the past. Washington (DC): Department of Agriculture (US), Economic Research Service; 1990 Jan. 17 p. Available from: NTIS, Springfield, VA; PB90-182932.

29.3.6.7 Date

The date of publication is required for all references. Date information can be found on the title page or the verso of the title page for books and other monographs and on the title page, cover, or masthead for serial publications such as a journal issue or newspaper. Some audiovisuals and many electronic publications lack a traditional title page, but date information can be found in a variety of other locations. For audiovisuals, look for date information on the opening screen, on the label of a CD-ROM or DVD, on the carrying case, or in accompanying written documentation. Internet date information can also be found in a number of places, including the opening screen, the bottom or end of the document, and the HTML, XML, or other source code viewable through your Web browser.

If a date of publication cannot be found in any of these locations, the following rules apply:

1) For print publications, if no date can be found on the title page or the verso of the title page for books and other monographs or on the title page, cover, or masthead of a journal issue, and the date can be determined from elsewhere within the publication, place the date within square brackets.
2) If no year of publication can be determined, use the year of copyright, preceded by "c", as "c2000".
3) If dates for both publication and copyright appear, use only the date of publication, unless at least 3 years separate the 2 dates. In the latter situation, use both dates, beginning with the publication date, and separate them by a comma and a space, as "2000, c1997". This convention alerts the user that the information contained in the publication is older than the publication date implies.
4) If neither a year of publication nor the copyright date can be found, use the words "date unknown" within square brackets.

Lederer J. Alimentation et cancer [Diet and cancer]. Brussels (Belgium): Nauwelaerts; [date unknown].

An exception to rule 4 occurs when citing electronic publications. Because dates of publication or copyright are often absent from electronic documents, the dates of update/revision or citation (or both) are used instead (see Section 29.3.7.13 for details).

The year of publication is required for all types of references. The month and day of the month or the season must be included in 4 situations:

1) when citing a journal that has no volume or issue number;
2) when citing patents;
3) with newspaper articles; and
4) for dates of update/revision and citation when citing electronic publications. If desired, the month, day, and season may be used in other situations.

For multiple years of publication, separate the first and last years by an en dash (or a hyphen if preferred), as "2004–2005". When the month is used, abbreviate it to the first 3 letters of the English name for the month, without a period, and place it after the year, such that "March 2002" becomes "2002 Mar". Separate multiple months of publication by an en dash (or a hyphen if preferred), as "Jan–Feb" and "1999 Dec–2000 Jan". Follow the month abbreviation by the day of the month, as "1992 Apr 3", and separate multiple days of publication by an en dash, as "2005 Mar 1–15". Capitalize names of seasons and give them in English, but do not abbreviate them, such that Summer, Fall, Winter, and Spring are written out. Separate multiple seasons by an en dash, as "Fall–Winter".

In references to journal articles using the citation–sequence and citation–name systems, follow date information with a semicolon unless there is no volume or issue, in which case use a colon. For monographs, follow date information with a period. When using the name–year system, follow date information with a period in both journal and monograph references.

Because of the volatile nature of Internet publications, there are 3 dates of importance in citing them:

1) the date the document was posted on the Internet (the date of publication) or, alternatively, was copyrighted
2) the latest date of any update or revision
3) the date the person doing the citing actually saw the publication

For instructions on including these dates when citing Internet documents, follow the guidelines given in Section 29.3.7.13.

Databases also have special rules for date information when citing them (see Sections 29.3.7.12.5 and 29.3.7.13.7 for instructions).

29.3.6.8 Volume and Issue

Volume and issue information is found in both journals and monographs. It can be located on the title page or the verso of the title page for books and other monographs and on the title page, cover, or masthead for serial publications such as a journal issue or newspaper. According to NISO,[1] a volume is "usually the primary or most inclusive level of enumeration supplied by the publisher to identify the bibliographic units of a work. Such designation may be alphabetic, numeric, or alphanumeric" and an issue is "an alphabetic, numeric, or alphanumeric designation supplied by the publisher on the work that usually divides a larger designation, such as a volume number, into logical subunits". Both volume and issue information, if present, is required in references for monographs and journals. However, this information is usually omitted for newspaper articles (see Section 29.3.7.8).

29.3.6.8.1 JOURNAL VOLUMES AND ISSUES

Both volume and issue numbers are required in references to journal articles. Write all volume numbers in arabic numerals, such that "LX" becomes "60", and ignore words such as "Volume", "Vol.", or "Band" that precede them. Separate multiple volumes by an en dash (or a hyphen if preferred), as "42–43". If a volume has a supplement, part, special number, or other division, place it after the volume number in abbreviated form, as "Suppl", "Pt", "Spec No". For example:

30 Suppl	30 Spec No 2	30(Pt A)	30 Suppl A	30 Suppl 1 Pt 2
30(Pt 1)	30 Suppl 1	30 Spec Issue 3	30(Pt 2 Suppl)	

Sample full citation:

Ochi K, Sugiura N, Komatsuzaki Y, Nishino H, Ohashi T. Patency of inferior meatal antrostomy. Auris Nasus Larynx. 2003 Feb;30 Suppl:S57–S60.

Parentheses surrounding a part of a volume and en dashes or hyphens separating multiple volume numbers are the only punctuation permitted within volume information. Follow volume information with a colon unless there is also issue information.

An issue number must be included in a journal reference if it is present in the document. The source for issue information is (in order of preference) the title page of the issue, the issue cover, and the masthead. Journal running heads often lack the issue number, even when it is provided in these other locations, and publishers have been known to forget to change running heads between issues. Write all issue numbers in arabic numerals, such that "VI" becomes "6", and ignore words such as "Number", "No.", or "Heft" that precede them. Separate multiple issues by an en dash (or a hyphen if preferred), as "1–2". If an issue has a supplement, part, special number, or other division, place it after the issue number in abbreviated form, as "Suppl", "Pt", or "Spec No". Place all issue information within parentheses. For example:

A supplement to issue number 1 becomes (1 Suppl)
Supplement A to issue number 4 becomes (4 Suppl A)
Part 1 of issue number 5 becomes (5 Pt 1)
Special number 2 to issue number 6 becomes (6 Spec No 2)

Sample full citation:

Menter A. Pharmacokinetics and safety of tazarotene. J Am Acad Dermatol. 2000 Aug;43(2 Pt 3):S31–S35.

Parentheses surrounding issue information and en dashes or hyphens separating multiple issues are the only punctuation permitted, even when there is no volume information. Follow issue information with a colon.

Soins Chir. 1996;(179):6–9.

See Section 29.3.7.1.5 for additional information and examples.

29.3.6.8.2 BOOK AND OTHER MONOGRAPH VOLUMES

Volume numbers are used for books and other monographs when citing one volume of a multivolume set. Write these numbers in arabic numerals, such that "III" becomes "3". Unlike references to volume numbers in journal articles, which use just the number, a volume number for a monograph is preceded by "Vol.", as "Vol. 3". Use English names for volume (e.g., "Band" becomes "Vol."). The volume number follows the title of the set and precedes the title of the individual volume.

Parenteral nutrition. Vol. 2, Clinical nutrition.

Monograph volumes that are part of a series may themselves have volumes and issues (see Section 29.3.6.11).

29.3.6.9 Location within a Work and Extent of a Work (Pagination)

Location within a work "indicates the specific point in a publication at which the item being referenced is located"[1] and is a required part of an end reference. The pages on which a journal article resides or on which the specific chapter of a book is found are examples of location. In this manual the word "location" is used rather than the more familiar "pagination" to reflect the increasing citation of online and other nonprint publications that often do not have traditional page numbers (see Sections 29.3.7.12 and 29.3.7.13).

Extent of a work is "a number that indicates the total physical extent or size of the work"[1] and is an optional part of a reference. "Extent" is used rather than "pagination" because many newer types of items being cited, such as websites, lack pagination. The total number of pages of a book, the number of reels of a motion picture, and the number of bytes contained in an online document are examples of extent.

29.3.6.9.1 LOCATION WITHIN A WORK (PAGINATION)

For a journal article, location consists of the first page and the last page of the article, separated by an en dash (or by a hyphen if preferred) if the article has continuous pagination. If the pagination is discontinuous, as often occurs when an article is interrupted by advertisements, separate the groupings of page numbers by a comma and a space.

345–346, 348–349, 351, 355–357, 360, 362

If space is a consideration, use up to 3 groupings of page numbers and the word "cont" or "passim" in place of the fourth and later occurrences.

345–346, 348–349, 351, 355–357, 360, 362 *becomes* 345–346, 348–349, 351 cont

Place location information after the volume and/or issue (or the year of publication if there is no volume or issue) and precede it with a colon. Conclude the location information with a period.

Cho YW, Cecka KM. Organ procurement organization and transplant center effects on cadaver renal transplant outcomes. Clin Transplant. 1996:427–441.

Robinson BE. Dementia: a three-pronged strategy for primary care. Geriatrics. 1986;41(2):75–77, 81, 84 passim.

Retain letters that precede or follow page numbers, as often occurs with appendices or supplements. Give them in uppercase or lowercase, whichever appears in the publication.

Bodi M, Ardanuy C, Rello J. Impact of gram-positive resistance on outcome of nosocomial pneumonia. Crit Care Med. 2001;29(4 Suppl):N82–N86.

Pariza MW, Loretz LJ, Storkson JM, Holland NC. Mutagens and modulator of mutagenesis in fried ground beef. Cancer Res. 1983 May;43(5 Suppl):2444s–2446s.

Unlike the practice with volume and issue numbers, retain roman numerals expressing location. Give them in uppercase or lowercase, whichever appears in the publication.

Nagpal S. An orphan meets family members in skin. J Invest Dermatol. 2003;120(2):viii–x.

Gothefors L. Nyheter i vaccinations programmet för barn [Current aspects in vaccination program for children]. Vardfacket. 1991;15(16):X–XI. Swedish.

Location information for a book chapter or other part of a monograph consists of the beginning and concluding pages of the chapter or part, separated by an en dash (or a hyphen if preferred). If the chapter or part has a separate author, place location information after the date, preceded by "p." and a space, as "p. 456–463". Follow location information with a period.

Todd JK. Streptococcus pneumoniae (*Pneumococcus*). In: Behrman RE, Kliegman RM, Jenson HB, editors. Nelson textbook of pediatrics. 16th ed. Philadelphia (PA): W.B. Saunders Company; c2000. p. 799–801.

If the part being cited has no separate author, place the location after the title of the part and separate location information from the title by a semicolon.

Marsden CD, Fowler TJ. Clinical neurology. London (England): Edward Arnold; 1989. Table 3.4, Differences between upper and lower motor neuron lesions; p. 79.

It may occasionally be necessary to cite a journal article or a part of a book that has no page numbers. If only the particular item to be cited has no page numbers, it may be possible to identify the location in relation to those pages that are numbered, as "preceding p. 1" or "following p. 503". Surround such phrases with square brackets. If this is not possible or if the entire publication has no page numbers, give the total number of pages of the item being cited, within square brackets, as "[5 p.]".

Katz LC, Weliky M, Dalva M. Relationships between local synaptic connections and orientation domains in primary visual cortex. Neuron. 1998;20(4):[following p. 819].

The humane care and use of animals: statement of journal policy. Int J Neurosci. 1988;42(3–4):[3 p.].

Location information for newspaper articles, audiovisuals, and Internet publications is more complex (see Sections 29.3.7.8, 29.3.7.11, and 29.3.7.13, respectively, for details).

29.3.6.9.2 EXTENT OF A WORK (PAGINATION)

Although the extent of an item is considered an optional part of an end reference, this information may be useful to the reader. A 10-page journal article, for example, would be considered substantive, whereas a 10-page book probably would not. For books and other print monographs, express extent as the total number of pages of the publication, followed by a space and "p.", as "346 p." If a publication has no page numbers, count the number of physical pages and express the extent in terms of leaves, as "16 leaves". If a book is published in more than one physical volume, give the total number of volumes in place of the number of pages, as "3 vol."

Dye DL. Space proton doses at points within the human body. Seattle (WA): Boeing Company; 1962. 29 leaves.

Haubrich WS, Schaffner F, Berk JE, editors. Bockus gastroenterology. 5th ed. Philadelphia (PA): Saunders; c1995. 4 vol.

Because audiovisual materials such as CDs, DVDs, motion pictures, and slides have no pagination, express their extent in terms of the total number of physical pieces and, if possible, run time (length).

123 slides.   3 CDs: 6 hrs.   1 DVD: 91 min.

Descriptive terms, such as hours and minutes, may be spelled out or abbreviated according to ISO 832[26]; do not use SI symbols (International System of Units) in this context (see Section 29.3.7.11 for details).

Many online publications lack traditional page numbers and are often nonlinear, providing numerous hyperlinks. For that reason, the extent of online publications is usually omitted. If this information is included, express extent for these publications in terms of the total number of screens, paragraphs, lines, or bytes. Print fonts and screen sizes vary with computer hardware and software; therefore, unless the length is supplied by the publisher, calculate the extent in the best manner possible and place it within square brackets, as "[about 5 screens]", "[about 21 p.]" (see Section 29.3.7.13 for details).

29.3.6.10 Physical Description

As its name implies, this part of a reference describes the physical attributes of a publication, including size, color, reproduction rate or ratio if a microform, and scale if a map. Physical description information is always an optional part of a reference.

Physical description is most useful in describing audiovisuals and microforms. Give the total number of physical pieces, followed by a colon and the description, separating the individual parts of the description with commas. Abbreviate words used in the description according to ISO 832,[26] rather than using SI symbols (International System of Units). To reduce work at the keyboard, use the letter "x" rather than the mathematical times symbol.

3 videocassettes: 285 min., sound, color, 1/2 in.
124 slides: color, 2 x 2 in.
31 microfiche: black & white, 4 x 6 in.
5 reels: color, 35 mm.
1 reel: 16 mm.

See Appendix 29.2 for a list of English words commonly used in physical description.

29.3.6.11 Series

When a book or other monograph contains a collective title in addition to its own title, it is part of a series, and the series title (the collective title) may be included in a reference. For example, a book published by Karger in 2001 has the title *Mechanisms of DNA Tumor Virus Transformation* and a series or collective title *Monographs in Virology*. Although series information is optional, it may be useful, particularly if a series is numbered, to aid in identification and retrieval. When series information is included, it follows the date of publication (and pagination, if included) in the citation–sequence and citation–name systems and follows the publisher in the name–year system. In all systems, the series is enclosed within parentheses. Give the name of the series first, capitalizing only the first word and proper nouns, and follow it with any numeration provided, such as a volume or issue number. Separate the title and the numeration by a semicolon. Conclude series information with a period, which falls outside the closing parenthesis.

Rosenthal LJ, editor. Mechanisms of DNA tumor virus transformation. New York (NY): Karger; 2001. 163 p. (Monographs in virology; vol. 23).

Crepaldi G, Tiengo A, Del Prato S, editors. Insulin resistance, metabolic diseases, and diabetic complications. Proceedings of the 7th European Symposium on Metabolism; 1998 Sep 30–Oct 3; Padua, Italy. New York (NY): Elsevier; 1999. 276 p. (Excerpta Medica international congress series; 1117).

Some series contain an editor's name. Include this information if desired.

Wassarman PM, Keller GM, editors. Differentiation of embryonic cells. Boston (MA): Elsevier Academic Press; c2003. (Abelson JN, Simon MI, editors. Methods in enzymology; vol. 365).

29.3.6.12 Connective Phrase

As the name implies, a connective phrase joins one component of a reference to another. "In" is the most common example. When citing a chapter that has authors separate from the authors or editors of the book as a whole, "In" is used to connect information about the chapter to information about the book.

Goldhagan JL. Child health in the developing world. In: Behrman RE, Kliegman RM, Jenson HB, editors. Nelson textbook of pediatrics. 16th ed. Philadelphia (PA): W.B. Saunders Company; 2000. p. 11–14.

"In" is also used to connect information about a paper in a conference proceedings to information about the proceedings itself. It would seem logical that "In" would also be used in journal references to join information about the article to information about the journal, but long practice omits it.

"Available from:" and "Located at:" are also examples of connective phrases. These phrases precede information about the source or location of a publication, such as a clearinghouse or library collection.

Available from: NTIS, Springfield, VA; PB91-182030.
Located at: National Library of Medicine, Bethesda, MD; WZ 345 P314n 1991.

29.3.6.13 Notes

"Notes" is a collective name for information about a publication that enhances the reference or provides clarification. Most of the information contained in notes is of interest to the reader but not essential to the reference. Examples of nonessential notes include availability information, such as a library or other collection where an unusual book may be found; the language of the publication if other than English; the name and location of the organization that sponsored the publication when it differs from the publisher; numbers that would facilitate acquisition, such as an International Standard Serial Number (ISSN) or DOI (Digital Object Identifier); and system requirements for electronic documents, including the computer operating system on which it is designed to run and the names and version of any software required.

There are 3 circumstances in which notes information is required. The report or contract number of a technical report, the Uniform Resource Locator (URL) of an online document, and the location of a manuscript are all essential information that must be included (see Sections 29.3.7.4, 29.3.7.13, and 29.3.7.15.2, respectively, for details).

29.3.7 References for Specific Types of Publications, with Examples

The sections below discuss the major issues in citing specific types of publications. For details about any specific component of a reference, consult the appropriate subsection of Section 29.3.6.

The examples used throughout are taken from actual publications and have been selected because they illustrate the particular point being made. Note that, because it is important not to change authors' actual words when creating a reference, the various titles within the examples may not follow CSE recommendations for style and format.

For example, this manual recommends that "T cells" should be written without an en dash or hyphen, but if a title uses "T-cells", it has not been changed.

29.3.7.1 Journals

This section on journals covers the components of a reference to a journal article and their sequence, including punctuation. It shows by example how to handle many common variations. For more information on punctuation, see Section 29.3.5, and for details on the individual components of a reference, see the appropriate subsections of Section 29.3.6. Examples of references to journals appearing in print or in microform (microfilm, microfiche) are included. For references to journal articles in audiovisual format, see Section 29.3.7.11, and in electronic form, see Sections 29.3.7.12 and 29.3.7.13.

By tradition, the rules for formatting references to journal articles permit substantial abbreviation. References to articles omit information on place of publication and publisher, whereas monograph references carry these details. The words for "volume" and "number" (or their abbreviations) are omitted when citing journal articles but are included when citing monographs. Journal titles are abbreviated; monograph titles are not. This brevity in citing journal articles stems from the need to conserve space in printed bibliographies and early databases.

The general format for a reference to a journal article, including punctuation, is as follows:

Citation–sequence and citation–name:
Author(s). Article title. Journal title. Date;volume(issue):location.

Example:
Smart N, Fang ZY, Marwick TH. A practical guide to exercise training for heart failure patients. J Card Fail. 2003;9(1):49–58.

Name–year:
Author(s). Date. Article title. Journal title. Volume(issue):location.

Example:
Smart N, Fang ZY, Marwick TH. 2003. A practical guide to exercise training for heart failure patients. J Card Fail. 9(1):49–58.

The placement of the year of publication is the major difference between the systems. For a complete discussion of these systems, see Section 29.2.1 and its subsections. To conserve space, most of the examples throughout this section show only the citation–sequence format. To convert to the name–year format, move the year to follow the author names and insert a period after it.

To cite a part of a journal article rather than the entire article, see Section 29.3.7.1.8; to cite an entire journal, see Section 29.3.7.1.9.

29.3.7.1.1 AUTHORS OF JOURNAL ARTICLES

List the names of the authors in the order in which they appear on the title page of an article. Begin with the surname (family name or last name) of the first author, followed by the initial letters of the given (first) names and middle names. Continue with the surname and initials of the second and subsequent authors. For details on the format of author names, such as the way names from various nationalities are handled, see Section 29.3.6.1.1.

When there are 2 to 10 authors, list them all, including the 10th author; if there are more than 10 authors, list the first 10 followed by "et al." or "and others". Separate the surname and initials of an author by a space; separate successive author names from each other by a comma and a space. Follow the last item in the author list with a period.

Jansen NW, Kant IJ, van den Brandt PA. Need for recovery in the working population: description and associations with fatigue and psychological distress. Int J Behav Med. 2002;9(4):322–340.

Kent DD Jr, Arnold DF Sr, Nydegger RV. Effect of selected psychological characteristics upon choice-shift patterns found within hierarchical groups of public accountants. Psychol Rep. 2002;91(1):85–104.

Pizzi C, Caraglia M, Cianciulli M, Fabbrocini A, Libroia A, Matano E, Contegiacomo A, Del Prete S, Abbruzzese A, Martignetti A, et al. Low-dose recombinant IL-2 induces psychological changes: monitoring by Minnesota Multiphasic Personality Inventory (MMPI). Anticancer Res. 2002;22(2A):727–732.

An organization such as a university, society, association, corporation, or governmental body may also serve as an author. However, if both a personal author and an organizational author appear on the title page of an article, use the personal author only. An exception may be made to this rule if the organization name has long been used within the discipline when referring to the publication. If a division or other part of the organization is mentioned on the publication, give the parts of the name in descending hierarchical order, separated by commas. Separate 2 or more different organizations by a semicolon. In citing organizations that are national bodies such as government agencies, if a nationality is not included with the name, place the country after the name, within parentheses, using the 2-letter ISO country code[24] (see Appendix 29.3 for a list of the codes for selected countries and see Section 29.3.6.1.2 for details on organizations as authors).

American College of Surgeons, Committee on Trauma, Ad Hoc Subcommittee on Outcomes, Working Group. Practice management guidelines for emergency department thoracotomy. J Am Coll Surg. 2001;193(3):303–309.

United Nations, Department of International Economic and Social Affairs, Population Division; United Nations Fund for Population Activities. Algeria. Popul Policy Compend. 1985 Jun:1–8.

Royal Marsden Hospital Bone-Marrow Transplantation Team (GB). Failure of Syngeneic bone-marrow graft without preconditioning in post-hepatitis marrow aplasia. Lancet. 1977;2(8041):742–744.

When using the name–year system of in-text references, use the initial letter of each part of the name or a readily recognizable abbreviation to create a shortened form for the in-text reference. This shortened form, within square brackets, becomes the initial element in the end reference (see Section 29.2.1.2 for details).

In-text reference:

Emergency department guidelines on thoracotomy (ACS 2001) . . .

End reference:

[ACS] American College of Surgeons, Committee on Trauma, Ad Hoc Subcommittee on Outcomes, Working Group. 2001. Practice management guidelines for emergency department thoracotomy. J Am Coll Surg. 193(3):303–309.

If neither a personal nor an organizational author can be found, begin the end reference with the article title. The use of "anonymous" is not permitted. In the name–year system, begin the end reference with the title, followed by the year; begin the in-text reference for such publications with the first word or first few words of the title, followed by an ellipsis.

End reference:

Changes made to policy, standards for disease-specific care certification. 2003. Jt Comm Perspect. 23(1):9–10.

In-text reference:

Revisions for disease-specific care certification (Changes . . . 2003) have been . . .

29.3.7.1.2 TITLES OF JOURNAL ARTICLES

In general, record the words of an article title exactly as they appear on the title page. It is important not to change the words of the author(s). There are 3 major exceptions to this rule.

1) For a title in a non-Roman alphabet, romanize the title if it is in Cyrillic, Greek, Arabic, or Hebrew; romanize or translate it if in a character-based language (e.g., Chinese, Japanese). Whenever possible, follow a non-English title with a translation. Place all translated titles within square brackets.

    Euvrard S, Kanitakis J, Claudy A. Tumeurs cutanées chez les greffes d'organe [Cutaneous tumors in organ transplant recipients]. Presse Med. 2002;31(40):1895–1903. French.
    Jin XH, Huang WQ. [Research progress in immune regulation of gonadorelin]. Sheng Li Ke Xue Jin Zhan. 2000;31(2):169–172. Chinese.

2) If a title contains a Greek letter or some other symbol that cannot be reproduced with the type fonts available, substitute the word for the symbol (e.g., Ω becomes omega).
3) If a title is presented in 2 languages, representing the bilingual nature of the text, as often occurs in Canadian publications, give both titles in the order in which they are presented in the text, with an equals sign between them.

    Aouizerate P, Guizard M. Prise en charge des thrombocytopenies induites par l'heparine = Management of heparin-induced thrombcytopenia. Therapie. 2002;57(6):577–588. French, English.

Two situations require special mention. Journal articles sometimes contain a header at the top of the page, such as "news", "case report", or "clinical study", to indicate a section of the issue. Do not include such a header as part of the article title, unless the table of contents for the journal issue indicates that it is. Occasionally a journal article does not appear to have any title; the article or other short document simply begins with the text. In this unusual circumstance, "construct" a title from the first few words of the text, using enough words to make the constructed title meaningful, and place them within square brackets.

A content designator may be placed at the end of an article title. For journal articles, content designators indicate such various forms as editorials, letters to the editor, news, or meeting abstracts. Such content designators are optional. However, the indication that the cited document is an abstract rather than a complete article may be particularly useful to a reader. If a content designator is used, place it within square brackets immediately following the article title, separated by a space.

Weil MH, Tang W, Bisera J. Cardiopulmonary resuscitation: one size does not fit all [editorial]. Circulation. 2003;107(6):794.
Brzezinska AK, Chilian WM. Vascular endothelial cells conduct superoxide via CIC-3 chloride channel [abstract]. Circulation. 2002;106(19 Suppl 2):II212.

Separate the article title from the closing period of the author list by a space. Capitalize only the first word of the title, proper nouns, proper adjectives, acronyms, and initialisms. If the title begins with a Greek letter, chemical formula, or another special character that would lose its meaning if capitalized, retain the lowercase. For titles not in English, follow the conventions of the particular language regarding capitalization.

Gorge G, Kunz T, Kirstein M. Die nicht-chirurgische Therapie des iatrogenen Aneurysma spurium [Nonsurgical therapy of iatrogenic false aneurysms]. Dtsch Med Wochenschr. 2003;128(1–2):36–40. German.

Retain whatever punctuation appears within the original title. If no punctuation separates the title from a subtitle, use a colon followed by a space unless some other form of punctuation such as a question mark or an exclamation point is already present. End the title with a period unless another form of punctuation already exists.

29.3.7.1.3 TITLES OF JOURNALS

Abbreviate the "significant" words in a journal title according to ISO 4,[25] and omit other words, such as articles, conjunctions, and prepositions; capitalize all abbreviated words. Sources for title-word abbreviations such as the *List of Journals Indexed for MEDLINE*[27] published by the National Library of Medicine and *BIOSIS Serial Sources*[28] by Biological Abstracts, and other sources that follow ISO, may be used (see Appendix 29.1 for a list of authoritative sources for journal title abbreviations). ISO 4 does not permit abbreviation of journal titles that consist of a single word and does not cover titles written in a character-based language (e.g., Chinese, Japanese). Thus, by common practice, character-based language titles are never abbreviated; capitalize words in them according to the practice of the particular language. As an option, journal titles in character-based languages may be translated. If translated, follow the standard abbreviation rules.

Akita Igaku. *may become* Akita J Med.

Journal subtitles are not considered a part of the title proper and therefore are not included in the abbreviation.

JAMA: the Journal of the American Medical Association *becomes* JAMA.

It is important to cite the journal name as it was published at the time. For example, the *British Medical Journal* officially changed its title to *BMJ* in 1985. Cite articles from 1984 and earlier as Br Med J, not BMJ.

Separate the journal title from the closing period of the article title by a space. Capitalize all title-word abbreviations. Do not use a period after each abbreviation, but place a period at the end of the entire title unless an edition statement or medium designator is included (see below). Omit any punctuation found within a title.

In 2 situations, punctuation marks are added to a title: parentheses for an edition statement and square brackets for a medium designator. If an edition statement appears in the journal, include the edition information after the title, separated from it by a space and surrounded by parentheses. Abbreviate the words in the edition statement according to ISO 4.[25]

Hospital Practice. Office Edition. *becomes* Hosp Pract (Off Ed).
BMJ. International Edition. *becomes* BMJ (Int Ed).
Indice de Especialidades Farmacéuticas Intercon. Edición para Farmacias. *becomes* Indice Espec Farm Intercon (Ed Farm).
Tierärztliche Praxis. Ausgabe Kleintiere Heimtiere. *becomes* Tierarztl Prax (Ausg Klient Heimtiere).

A medium designator tells the user that the journal is published in other than print form, such as microfilm, microfiche, CD-ROM, DVD, or videocassette. Do not capitalize medium designators except for formats known more commonly by the acronym such as CD-ROM or DVD. Separate a medium designator from the title by a space and surround it with square brackets.

Aesthetic Reconstr Plast Surg [microfiche].

For more information on citing journals in videocassette or other audiovisual format, see Section 29.3.7.11; for electronic formats such as CD-ROM and websites, see Sections 29.3.7.12 and 29.3.7.13.

Some bibliographies and online databases show a place of publication after the journal title, such as "Oncology (Huntingt)". This is a library convention to indicate that the library or other organization indexing the journal has 2 or more journal titles with the same name in its collection; the name of the city where the journal is published is used to distinguish among the various titles. The city is usually shown in abbreviated format, as "Huntingt" for "Huntington" in the example above. If you are using a bibliography or database to verify your reference and a place name is included, you may retain it if you wish.

Journal editors may choose to underline journal titles, make them bold, or set them in italics. Because too many variations in type styles may actually make the reference harder to read, CSE urges caution in adopting these conventions and has not used any of them in the references within this manual.

29.3.7.1.4 DATES OF JOURNAL ARTICLES

The year of publication is required for all journal article references. In the name–year system, the year follows the author list; in the citation–sequence and citation–name systems, it follows the journal title. The month and day of the month or the season must be included in only 2 situations: 1) when citing a journal article that has no volume or issue number and 2) for the dates of update/revision and citation when citing electronic journal articles. However, they may be included in other situations if desired (see Sections 29.3.7.12 and 29.3.7.13 for more information about dates for electronic journals).

For multiple years of publication, separate the first and last year of publication by an en dash (or a hyphen if preferred), as "2003–2004". When the month is used, place it after the year, abbreviated to the first 3 letters of the English name for the month, without a period, such that "March 2002" becomes "2002 Mar". Separate multiple months of publication by an en dash (or a hyphen if preferred), as "2005 Jan–Feb" and "1999

Dec–2000 Jan". Follow the month abbreviation by the day of the month, as "1992 Apr 3". Capitalize names of seasons and give them in English, but do not abbreviate them, such that Summer, Fall, Winter, and Spring are written out. Separate multiple seasons by an en dash, as "Fall–Winter".

Separate date information from the journal title by a space in the citation–name and citation–sequence systems, and from the author list by a space in the name–year system. Follow the date information in the citation–name and citation–sequence systems with a semicolon unless there is no volume or issue, in which case use a colon; follow the date with a period in the name–year system.

Citation–name and citation–sequence:

Shibbal CS, Lipinska SD. Alzheimer's educational/support group: considerations for success. J Gerontol Soc Work. 1985–1986;9(2):41–48.

Sampat P. Groundwater shock: the polluting of the world's major freshwater stores. World Watch. 2000 Jan–Feb:10–22.

South Africa says it can't afford anti-AIDS drug AZT. AIDS Wkly. 1999 Nov 29–Dec 6:20–21.

Menting A. The village and the children. Harv AIDS Rev. 2000 Spring–Summer:15–18.

Danoek K. Skiing in and through the history of medicine. Nord Medicinhist Arsb. 1982:86–100.

Name–year:

Shibbal CS, Lipinska SD. 1985–1986. Alzheimer's educational/support group: considerations for success. J Gerontol Soc Work. 9(2):41–48.

Sampat P. 2000 Jan–Feb. Groundwater shock: the polluting of the world's major freshwater stores. World Watch. 10–22.

South Africa says it can't afford anti-AIDS drug AZT. 1999 Nov 29–Dec 6. AIDS Wkly. 20–21.

Menting A. 2000 Spring–Summer. The village and the children. Harv AIDS Rev. 15–18.

Danoek K. 1982. Skiing in and through the history of medicine. Nord Medicinhist Arsb. 86–100.

Occasionally a journal will publish a supplement or other part to a year, instead of to a particular volume or issue. Place such supplements and parts after the date (citation–name and citation–sequence systems) or after the journal title (name–year system).

Citation–name and citation–sequence:

Kaufert JM, O'Neil JD, Koolage WW. The cultural and political context of informed consent for Native Canadians. Arctic Med Res. 1991;Suppl:181–184.

Name–year:

Kaufert JM, O'Neil JD, Koolage WW. 1991. The cultural and political context of informed consent for Native Canadians. Arctic Med Res. Suppl:181–184.

29.3.7.1.5 VOLUMES AND ISSUES OF JOURNAL ARTICLES

The volume number must be included in a reference to a journal article when it is present in the document. Write all volume numbers in arabic numerals. Convert numeric terms and roman numerals to arabic, such that "sixtieth" becomes "60" and "IX" becomes "9". Separate multiple volume numbers by an en dash (or a hyphen if preferred).

In addition to issue numbers, discussed below, volumes may be further subdivided by supplements, parts, and special numbers. Abbreviate these, without periods, as "Suppl", "Pt", and "Spec No".

30 Suppl	30 Spec No 2	30(Pt A)	30 Suppl A	30 Suppl 1 Pt 2
30(Pt 1)	30 Suppl 1	30 Spec Issue 3	30(Pt 2 Suppl)	

Parentheses surrounding a part number and an en dash (or a hyphen if preferred) separating multiple volume numbers are the only punctuation permitted within volume information. Follow volume information with a colon unless there is also issue information.

Volume with no issue or other subdivision:

Laskowski DA. Physical and chemical properties of pyrethroids. Rev Environ Contam Toxicol. 2002;174:49–170.

Multiple volume numbers:

Duan J, Gregory J. Coagulation by hydrolysing metal salts. Adv Colloid Interface Sci. 2003;100–102:475–502.

Volume with supplement:

Heemskerk J, Tobin AJ, Ravina B. From chemical to drug: neurodegeneration drug screening and the ethics of clinical trials. Nat Neurosci. 2002;5 Suppl:1027–1029.

Volume with numbered supplement:

Greene GJ, Kipen HM. The vomeronasal organ and chemical sensitivity: a hypothesis. Environ Health Perspect. 2002;110 Suppl 4:655–661.

Black HR. Evaluation of the elderly hypertensive. Geriatrics. 1989;44 Suppl B:15–19.

Volume with part:

Betz A, Hopkins PC, Le Bonniec BF, Stone SR. Contribution of interactions with the core domain of hirudin to the stability of its complex with thrombin. Biochem J. 1994;298(Pt 2):507–510.

Volume with a supplement and a part:

Catalan J, Ho B. HIV-1-associated panic disorders. Clin Neuropharmacol. 1992;15 Suppl 1 Pt 1:368A–369A.

Volume with special number:

Berger A, Pereira D, Baker K, O'Mara A, Bolle J. A commentary: social and cultural determinants of end-of-life care for elderly persons. Gerontologist. 2002 Oct;42 Spec No 3:49–53.

An issue number must be included in a reference to a journal article if it is present in the document. Write all issue numbers in arabic numerals. Convert numeric terms and roman numerals to arabic, such that "fifth" becomes "5" and "III" becomes "3". Separate multiple issue numbers by an en dash (or a hyphen if preferred), as "(10–12)". Issue numbers usually follow volume numbers in a reference but may stand on their own. If the only numeration on a journal is "number x", treat this number as an issue. Journal running heads or running footers often lack an issue number; refer to the title page of the issue, the issue cover, or the masthead when possible for authoritative information on issue numbering.

Issues may be further subdivided by supplements, parts, and special numbers. Abbreviate these, without periods, as "Suppl", "Pt", and "Spec No".

A supplement to issue number 1 becomes (1 Suppl)
Supplement A to issue number 4 becomes (4 Suppl A)
Part 1 of issue number 5 becomes (5 Pt 1)
Special number 2 to issue number 6 becomes (6 Spec No 2)

Parentheses surrounding issue information are the only punctuation permitted (except for the en dash or hyphen separating multiple issues) and are used even when there is no volume information. Follow issue information with a colon.

Issue with no volume:

Sabatier R. Reorienting health and social services. AIDS STD Health Promot Exch. 1995;(4):1–3.

Issue with volume:

Lawlor-Klean P, Swanson SC. Network's pathway provides smoother neonatal transport. Hosp Case Manag. 1995;3(2):27–30.

Multiple issue numbers with volume:

Ramstrom O, Bunyapaiboonsri T, Lohmann S, Lehn JM. Chemical biology of dynamic combinatorial libraries. Biochim Biophys Acta. 2002;1572(2–3):178–186.

Multiple issue numbers without volume:

Tamari S, Scott A. Fertility of Palestinian women between national perspective and social reality. Popul Bull ESCWA. 1989;(35–37):5–42.

Issue with volume and supplement:

Gardos G, Cole JO, Haskell D, Marby D, Paine SS, Moore P. The natural history of tardive dyskinesia. J Clin Pharmacol. 1988;8(4 Suppl):31S–37S.

Issue with volume and supplement with further subdivision:

Koski CL. Therapy of CIDP and related immune-mediated neuropathies. Neurology. 2002;59(12 Suppl 6):S22–S27.

Issue with a part:

Clark TW, Soulen MC. Chemical ablation of hepatocellular carcinoma. J Vasc Interv Radiol. 2002;13(9 Pt 2):S245–S252.

Issue with volume plus a part and a supplement:

Baxter CR. Foreword: comments on critical pathways. J Burn Care Rehabil. 1995;16(2 Pt 2 Suppl):189.

Issue with volume and special number:

Muller H, Scott R, Weber W, Meier R. Colorectal cancer: lessons for genetic counseling and care for families. Clin Genet. 1994;46(1 Spec No):106–114.

Occasionally, a journal will publish a supplement or other part to a year instead of to a particular volume or issue. Place such supplements and parts after the date in the citation–name and citation–sequence systems and following the journal title in name–year.

Citation–name and citation–sequence:

Kaufert JM, O'Neil JD, Koolage WW. The cultural and political context of informed consent for Native Canadians. Arctic Med Res. 1991;Suppl:181–184.

Martonyi CL. Photographic slit-lamp biomicroscopes. Ophthalmology. 1989 Sep;(Pt 2):6–19.

Name–year:

Kaufert JM, O'Neil JD, Koolage WW. 1991. The cultural and political context of informed consent for Native Canadians. Arctic Med Res. Suppl:181–184.

Martonyi CL. 1989 Sep. Photographic slit-lamp biomicroscopes. Ophthalmology. (Pt 2):6–19.

For journals that do not have a volume, issue, or other subdivision, follow the year of publication with a colon and the page numbers in the citation–name and citation–sequence systems and follow the journal title with the page numbers in the name–year system. Include a month of publication if present.

Citation–name and citation–sequence:

Thacker LR, McCune TR, Harland RC. Kidney sharing by centers of the South-Eastern Organ Procurement Foundation. Clin Transpl. 1996:129–137.

Name–year:

Thacker LR, McCune TR, Harland RC. 1996. Kidney sharing by centers of the South-Eastern Organ Procurement Foundation. Clin Transpl. 129–137.

29.3.7.1.6 LOCATION (PAGINATION) OF JOURNAL ARTICLES

The location where a journal article appears in the issue is required in all journal article references. For print journals, give the first and last page numbers of the article if it appears on continuous pages; do not omit any digits of the closing page number. Separate the first and last pages by an en dash (or a hyphen if preferred). If the pagination is discontinuous, as can occur when an article is interrupted by advertisements, separate the groupings of page numbers by a comma and a space.

345–346, 348–349, 351, 355–357, 360, 362

If space is a consideration, use up to 3 groupings of page numbers and the word "cont" or "passim" in place of the fourth and later occurrences.

345–346, 348–349, 351, 355–357, 360, 362 *becomes* 345–346, 348–349, 351 cont

Letters may appear as a part of page numbers, often associated with supplements or other special issues. Retain these letters, as "S34–S41" and "566A–569A". Retain also page numbers written in roman numerals and give them in either uppercase or lowercase, whichever appears in the original document, as "ix–xx" or "III–VI". Follow all page information with a period.

Beckman EJ. Using CO2 to produce chemical products sustainably. Environ Sci Technol. 2002;36(17):347A–353A.
Clark TW, Soulen MC. Chemical ablation of hepatocellular carcinoma. J Vasc Interv Radiol. 2002;13(9 Pt 2):S245–S252.
Buckley S. Explores the social nature of early development and the way in which social and cognitive development interacts. Downs Syndr Res Pract. 2002;8(2):v–vii.
Ronne Y. Ansvarsfall. Blodtransfusion till fel patient [Liability case. Blood transfusion to the wrong patient]. Vardfacket. 1989;13(2):XXVI–XXVII. Swedish.

Occasionally, a journal article will have no page numbers. If only the particular item to be cited has no page numbers, it may be possible to identify the location in relation to those pages that are numbered, as "preceding p. 1" or "following p. 503". Surround such phrases with square brackets. If the entire publication has no page numbers, give the total number of pages of the item you wish to cite, surrounded by square brackets, as "[5 p.]".

Kolloch RE. In der Praxis andert sich fur de Einsatz von Kalziumantagonisten eigentlich nichts [In practice nothing has changed in the indication for therapeutic use of calcium antagonists]. Fortschr Med. 1997;115(33):[preceding p. 55]. German.
Neue Nifedipin-Zubeitung ermoglicht tagliche Einmalgabe [New nifedipine preparation makes single daily dose possible]. Fortschr Med. 1997;115(33):[following p. 54]. German.
Glenwright HD, Martin MV. Infection control in dentistry. A practitioner's guide. Br Dent J. 1993; 175(1 Suppl):[8 p.].

Many online journal articles lack traditional page numbers (see Section 29.3.7.13 for guidance). Audiovisual journals, for example, those posted online as videos, also lack page numbers (see Section 29.3.7.11 for guidance).

29.3.7.1.7 NOTES IN JOURNAL ARTICLES

"Notes" is a collective name for information about a publication that enhances the reference or provides clarification. Most of the information contained in notes is of interest to the reader but not essential to the reference. Examples of notes for journal articles include the following:

- Language of publication may be added if the article is not in English.
- If the authors are part of a named group, such as a committee or task force, provide the name here.
- The name and dates of the meeting may be useful for a reference to a meeting abstract.
- The name of a library or other archive where the journal is held would be helpful for a title not widely available.
- The ISSN of the journal would aid in identification.

A note follows the location and generally has no specific format or punctuation; not even a complete sentence is required. Conclude a note with a period. Notes are usually optional. However, there is one circumstance in which notes information is required for a journal reference. For journal articles on the Internet, the Uniform Resource Locator (URL) must be included (see Section 29.3.7.13 for details).

29.3.7.1.8 PORTIONS OF JOURNAL ARTICLES

At times, it may be desirable to cite a separately identified portion of an article rather than the article as a whole. Sections, tables, figures, charts, graphs, photographs, appendixes, and the like are considered parts of articles when they are written or compiled by the authors of the article. To cite only a part of a journal article, begin the end reference with information on the article and follow that with information about the part. Part information includes the name of the part and any numeration, the title of the part, if present, and the pagination of the part. Capitalize only the first word of the title, proper nouns, proper adjectives, acronyms, and initialisms. Separate the numeration and title with a comma and a space; separate the title and the pagination with a semicolon and a space. Precede page information with "p."

Gards G, Vole JO, Haskell D, Marby D, Paine SS, Moore P. The natural history of tardive dyskinesia. J Clin Psychopharmacol. 1988;8(4 Suppl):31S–37S. Table 3, Occurrence in the United States; p. 32S.

Franklin PD, Panzer RJ, Brideau LP, Griner PF. Innovations in clinical practice through hospital-funded grants. Acad Med. 1990 Jun;65(6):355–360. Figure 1, Cost and net annual savings per study; p. 358.

29.3.7.1.9 COMPLETE JOURNAL TITLES

Rather than citing an individual journal article, it may be desirable to include information on a journal as a whole in a reference list. The general format for a reference to an entire journal, including punctuation, is as follows:

Citation–sequence and citation–name:
Title (Edition). Place of publication: publisher. Beginning Vol., date–Ending Vol., date. Notes.

Examples:
Annals of Internal Medicine. Philadelphia (PA): American College of Physicians. Vol. 1, 1927– .
Hospital Practice (Office Edition). New York (NY): McGraw-Hill. Vol. 16, 1981–Vol. 36, 2001. Continues in part: Hospital Practice.

Name–year:
Title (Edition). Date. Place of publication: publisher. Beginning Vol.–Ending Vol. Notes.

Examples:
Annals of Internal Medicine. 1927. Philadelphia (PA): American College of Physicians. Vol. 1– .
Hospital Practice (Office Edition). 1981–2001. New York (NY): McGraw-Hill. Vol. 16–Vol. 36. Continues in part: Hospital Practice.

To cite an entire journal, begin with the full name of the journal in unabbreviated format, using whatever capitalization and punctuation are found within the title. Follow the title with a colon and any subtitle. When citing a specific edition of a journal, place the edition statement following the title, surrounded by parentheses.

American Journal of Medical Quality: the Official Journal of the American College of Medical Quality.
BMJ (Clinical Research Edition).

Use the guidelines in Sections 29.3.6.5 and 29.3.6.6 for formatting place of publication and publisher, and conclude this information with a period. Enter the beginning volume number of the journal, preceded by "Vol.", then a comma and the beginning year of publication. If the journal is still being published, follow the year with an en dash (or a hyphen if preferred) and a space. End the reference with a period.

American Journal of Medical Quality: the Official Journal of the American College of Medical Quality. Lawrence (KS): Allen Press. Vol. 7, 1992– .
BMJ (Clinical Research Edition). London (England): British Medical Association. Vol. 297, 1998– .

29.3.7.2 Books

References to books appearing in print or in microform (microfilm, microfiche) are included in this section. For references to books in audiovisual format, see Section 29.3.7.11; in electronic form, see Section 29.3.7.12; and online, see Section 29.3.7.13. The standard book, such as a textbook, is covered here. Special types of books, such as conference proceedings, technical reports, dissertations and theses, bibliographies, and patents, are treated in their own sections below. For more information on punctuation, see Section 29.3.5, and for details on rules for the components of a reference, see Section 29.3.6 and its subsections.

The general format for a reference to a book, including punctuation, is as follows:

Citation–sequence and citation–name:
Author(s). Title. Edition. Place of publication: publisher; date. Extent. Notes.

Example:
Schott J, Priest J. Leading antenatal classes: a practical guide. 2nd ed. Boston (MA): Books for Midwives; 2002.

Name–year:
Author(s). Date. Title. Edition. Place of publication: publisher. Extent. Notes.

Example:
Schott J, Priest J. 2002. Leading antenatal classes: a practical guide. 2nd ed. Boston (MA): Books for Midwives.

The placement of the year of publication is the major difference between the systems. For a complete discussion of these systems, see Section 29.2.1 and its subsections. To conserve space, most of the examples throughout this section show only the first format. To convert to the name–year format, move the date to follow the author names.

29.3.7.2.1 AUTHORS AND SECONDARY AUTHORS OF BOOKS

List the names of the authors in the order in which they appear in the text. Begin with the surname (family name or last name) of the first author, followed by the initial letters of the given names. Continue with the surname and initials of the second and subsequent authors. For details on the format of author names, such as the way names from various nationalities are handled, see Section 29.3.6.1.1.

When there are 2 to 10 authors, list them all, including the 10th author; if there are more than 10 authors, list the first 10 followed by "et al." or "and others". Separate the surname and initials of an author by a space; separate successive author names from each other by a comma and a space. End the author list with a period.

> Ferrozzi F, Garlaschi G, Bova D. CT of metastases. New York (NY): Springer; 2000.
> Wenger NK, Sivarajan Froelicher E, Smith LK, Ades PA, Berra K, Blumenthal JA, Certo CME, Dattilo AM, Davis D, DeBusk RF, et al. Cardiac rehabilitation. Rockville (MD): Agency for Health Care Policy and Research (US); 1995.

An organization such as a university, society, association, corporation, or governmental body may also serve as an author. However, if both a personal author and an organizational author appear on the title page of a book, use the personal author only. An exception may be made to this rule when the organization name has specific significance to the intended audience, such as when a group or committee issues a report at regular intervals. When the decision has been made to use the organization as author, do not also use the personal authors in the reference except as a note (see Section 29.3.7.2.9).

If a division or other part of an organization is included on the publication, give the parts of the name in descending hierarchical order, separated by commas. Separate 2 or more different organizations by a semicolon. In citing organizations that are national bodies, such as government agencies, if a nationality is not included with the name, place the country after the name, within parentheses, using the 2-letter ISO country code[24] (see Appendix 29.3 for a list of the codes for selected countries).

> Advanced Life Support Group. Acute medical emergencies: the practical approach. London (England): BMJ Books; 2001.
> National Research Council (US), Subcommittee to Review the Hanford Thyroid Study. Final results and report. Review of the Hanford Thyroid Disease Study draft final report. Washington (DC): National Academy Press (US); 2000.
> National Alliance of State and Territorial AIDS Directors (US); Academy for Educational Development; Centers for Disease Control and Prevention (US). Bright ideas 2001: innovative or promising practices in HIV prevention and HIV prevention community planning. 2nd ed. Atlanta (GA): Centers for Disease Control and Prevention (US); 2001.

An editor assumes the place of an author in a reference if no author can be determined. List the names of the editors in the order in which they appear in the text. Follow the last-named editor by a comma and the word "editor" or "editors".

> Leeper FJ, Vederas JC, editors. Biosynthesis: polyketides and vitamins. New York (NY): Springer; c2000.

If both authors and editors are present, give the authors first and place the editors after the title and any edition statement.

Martin EW. Hazards of medication. 2nd ed. Ruskin A, Napke E, Alexander SF, Kelsey FO, Farage DJ, Mills DH, Elkas RW, editors. Philadelphia (PA): J. B. Lippincott Company; c1978.

Editors are considered secondary authors because they modify the work of authors. Translators and illustrators are other examples of secondary authors. If an author is named, inclusion of any secondary authors in the reference is optional. If there is more than one type of secondary author, such as an editor and an illustrator, give them in the order provided in the text, separated by a semicolon.

Stahl SM. Essential psychopharmacology of depression and bipolar disorder. Muntner N, illustrator. New York (NY): Cambridge University Press; 2000.

Luzikov VN. Mitochondrial biogenesis and breakdown. Galkin AV, translator; Roodyn DB, editor. New York (NY): Consultants Bureau; 1985.

If neither a personal nor an organizational author or editor can be found, begin the reference with the title of the book. The use of "anonymous" is not permitted.

Citation–sequence and citation–name:

Directory of AIDS resources in the metropolitan Washington area. Washington (DC): WTTG Television; 1988.

Name–year:

Directory of AIDS resources in the metropolitan Washington area. 1988. Washington (DC): WTTG Television.

29.3.7.2.2 TITLES OF BOOKS

It is important not to change the words in the title of a book. Therefore, in general, record the words of a book title exactly as they appear in the publication. There are 3 major exceptions to this rule.

1) For a title in a non-Roman alphabet, romanize the title if it is in Cyrillic, Greek, Arabic, or Hebrew; romanize or translate it if in a character-based language (e.g., Chinese, Japanese). Whenever possible follow a non-English title with a translation. Place all translated titles within square brackets.

   Gomez Lavin C, Zapata Garcia R. Psiquiatría, salud mental y trabajo social [Psychiatry, mental health and social work]. Pamplona (Spain): Ediciones Eunate; c2000. Spanish.

   Iudin SV, Kiku PF. Gigienicheskie aspekty rasprostranennosti onkologicheskikh zabolevanii [Public health aspects of the incidence of oncologic diseases]. Vladivostok (Russia): Dal'nauka; 2002. Russian.

   Sun Y, Liu C. [Practical remedies for diabetes]. Beijing (China): Ren min wei sheng chu ban she; 1999. Chinese.

2) If a title contains a Greek letter or some other symbol that cannot be reproduced with the type fonts available, substitute the word for the symbol (e.g., Ω becomes omega).
3) If a title is presented in 2 languages, representing the bilingual nature of the text, as often occurs in Canadian publications, give both titles in the order in which they are presented in the text, with an equals sign between them.

   Adoption of an opinion on ethical aspects of human stem cell research and use = Adoption d'un avis sur les aspects éthiques de la recherche sur les cellules souches humaines et leur utilisation. Rev. ed. Brussels (Belgium): European Group on Ethics in Science and New Technologies to the European Commission; 2001. English, French.

Titles to conference proceedings and titles of books on the Internet have special rules (see Sections 29.3.7.3 and 29.3.7.13, respectively).

Separate the title of the book from the period at the end of the author list by a space. Capitalize only the first word of the title, proper nouns, proper adjectives, acronyms, and initialisms. If the title begins with a Greek letter, chemical formula, or another special character that might lose its meaning if capitalized, retain the lowercase. For non-English titles, follow the conventions of the particular language regarding capitalization. Whenever possible, include an English translation, and place it within square brackets.

Lindner UK, Raftopoulo A. EKG in Notfallen [EKG in emergencies]. 2nd ed. Berlin (Germany): Springer-Verlag; c2001. German.

Retain whatever punctuation appears within the title. If no punctuation separates the title from a subtitle, use a colon followed by a space. End the title with a period unless another form of punctuation, such as a question mark or an exclamation point, is already present. The only punctuation to be added to a title is the square brackets for a medium designator. Medium designators tell the user that the book is published in other than print form, such as microfilm, microfiche, or CD-ROM. Separate a medium designator from the title by a space and surround it by square brackets.

Elder abuse and neglect: a synthesis of research [microfiche].

For more information on citing books in DVD or other audiovisual format, see Section 29.3.7.11; for electronic formats such as the CD-ROM and the Internet, see Section 29.3.7.13.

29.3.7.2.3 EDITIONS OF BOOKS

If a book has been published in more than one edition, an edition statement is a required part of an end reference. Give edition statements in their original language, but romanize names presented in Cyrillic, Greek, Arabic, or Hebrew; romanize or translate them if in a character-based language (e.g., Chinese, Japanese). If desired, follow a non-English edition statement with a translation, placed within square brackets. Express numbers representing editions in arabic ordinals:

second *becomes* 2nd
III *becomes* 3rd
quatrième *becomes* 4th

Place edition information after the title. Abbreviate the word "edition" and other descriptive words for print publications according to ISO 832[26] and follow them with a period. First editions not known to be followed by new editions usually need not be identified as "1st ed.", but if later editions have been published, the first edition should be identified as such when cited. For English-language edition statements, capitalize only the first word of the edition statement unless it contains proper nouns; for non-English statements, follow the conventions of the particular language. Follow the complete edition statement with a period.

Eighth Edition *becomes* 8th ed.
New revised edition *becomes* New rev. ed.
Fifth American Edition *becomes* 5th Am. ed.
Edizione Italiana *becomes* Ed. Ital.

29.3.7.2.4 PLACE OF PUBLICATION OF BOOKS

Place of publication is defined as the city with its state, province, and/or country where the book was published and is a required component of a book reference. Additional information may be included with the city name to provide any necessary clarification. Although many city names such as London or New York are almost universally known and can stand alone, others are not so well known or may be confused with cities of the same name in other countries. For example, if a book has been published in Paris, Texas, the state name may be supplied to avoid confusion with Paris, France. Follow US and Canadian cities with the 2-letter abbreviation for the state or province. Follow cities in other countries with the name of the country, either written out or as the 2-letter ISO country code[24] (see Appendix 29.3 for a list of the codes for selected countries). Use the name of the country as it is found on the publication, such as England, Great Britain, or the United Kingdom. If the audience for the reference is an English-speaking one, the anglicized form for a city is preferred (e.g., "Rome" for "Roma"). However, the name as it appears on the publication is always correct. Place the state, province, country, or ISO code in parentheses after the city. End place of publication information with a colon.

> New York, NY *becomes* New York: *or* New York (NY):
> Boca Raton, Florida *becomes* Boca Raton (FL):
> London, England *becomes* London: *or* London (England): *or* London (GB):
> London, Ontario, Canada *becomes* London (ON):

If no place of publication can be found on the title page or its verso, but one can be found elsewhere within the publication or can be reasonably inferred (e.g., Chicago as the place for a publication of the American Medical Association), place the city within square brackets, as "[Chicago]". When no city can be located or inferred, supply the words "place unknown" within square brackets.

> Oakes J, Riewe R, editors. Issues in the North. [place unknown]: Canadian Circumpolar Institute; 1996.

For consistency within a reference list, treat each occurrence of a city the same way (e.g., always "New York:" or always "New York (NY):"). To avoid any ambiguity, the state, province, or country name has been added to all cities of publication in the references within this manual.

29.3.7.2.5 PUBLISHERS OF BOOKS

The name of the publisher is a required component of a reference to a book. Record the name as it appears in the publication, using whatever capitalization and punctuation is found there. If more than one publisher is listed on a document, use the first one given or the one set in the largest type or set in bold. An alternative is to use the publisher likely to be most familiar to the audience of the reference list (e.g., an American publisher for a US audience and a London publisher for a British one). Do not list multiple publishers. If a publication has been issued jointly by 2 publishers, use the name provided first as the publisher and include the name of the second as a note, if desired, as "Jointly published by the Canadian Pharmacists Association".

Give names of publishers in their original language, but romanize names presented in Cyrillic, Greek, Arabic, or Hebrew; romanize or translate them if in a character-based language (e.g., Chinese, Japanese). If desired, follow a non-English name with a translation. Place all translated publisher names within square brackets. If the name of a division of other part of an organization is included in the publisher information, give the names in descending hierarchical order.

Tartuskii Universitet [Tartu University]
Institut d'Etudes du Massif Central, Centre d'Histoire des Enterprises et des Communautes
University of Pennsylvania, Department of Psychology
Elsevier Science, Biomedical Division

In citing organizations such as government agencies that are national bodies, if a nationality is not included with the name, place the ISO 2-letter country code[24] after the name, within parentheses (see Appendix 29.3 for a list of the codes for selected countries).

National Academy of Sciences (US)
National Fire Service (GB)
Royal Institute of Technology (SE), Medical Division

End publisher information with a semicolon when using the citation–name and citation–sequence systems and a period when using the name–year system. If no publisher can be determined, use the words "publisher unknown" surrounded by square brackets (see Section 29.3.6.6 for additional information about publishers).

29.3.7.2.6 DATES OF BOOKS

A date is a required component of a reference to a book. Always give the year of publication. If the publication date includes a month, as often occurs in technical reports and government publications, you may include it. Place the month after the year, abbreviated to the first 3 letters of the English name, such that September becomes "Sep". For multiple years of publication, separate the first and last years by an en dash (or a hyphen if preferred), as "1998–1999". If date information cannot be found on the title page of a book, the verso of the title page, or its cover, but can be determined from elsewhere within the book, place the year within square brackets, as "[2001]". If no year of publication can be determined, use the year of copyright, preceded by "c", as "c2002".

Some publications have both a year of publication and a year of copyright. Use only the year of publication unless at least 3 years separate the 2 dates. In the latter situation, use both dates, beginning with the year of publication, and separate them by a comma and a space, as "2002, c1997". This convention alerts the user that the information contained in the publication is older than the year of publication implies. If neither a year of publication nor a year of copyright can be found, use the words "date unknown" within square brackets. End date information with a period.

Chang TMS. Blood substitutes: principles, methods, products, and clinical trials. New York (NY): Karger Landes Systems; 1997–1998.

Drummond MF. Disease management: who needs it and why? York (England): University of York, Centre for Health Economics; [1997].

Kruger L, editor. Pain and touch. San Diego (CA): Academic Press; c1996.

Mark BS, Incorvaia J, editors. The handbook of infant, child, and adolescent psychotherapy. Northvale (NJ): Jason Aronson Inc.; c1995–1997.

Karlson RG. Essays on the history of psychology. New York (NY): Modern History Press; 1999, c1995.

Lederer J. Alimentation et cancer [Diet and cancer]. Brussels (Belgium): Nauwelaerts; [date unknown]. French.

Date information for audiovisuals and Internet publications is more complex (see Sections 29.3.7.11 and 29.3.7.13, respectively).

29.3.7.2.7 EXTENT (PAGINATION) OF BOOKS

Extent or pagination is an optional component of a book reference, but it can provide useful information to the reader. When citing an entire book (see Section 29.3.7.2.10 for citing parts of books), use for extent the total number of pages on which the text of the publication appears. Do not count pages for such items as introductory material, appendixes, and indexes unless they are included in the pagination of the text. For example, if a book has introductory pages i–xii, text on pages 1–340, and an unpaginated appendix, the extent is "340 p." If a book has no page numbers, count the number of pages and express extent in terms of leaves, as "10 leaves". Some books are published in more than one physical volume. When this occurs, cite the total number of volumes instead of the number of pages, as "4 vol."

For books in audiovisual format, such as a CD-ROM or a DVD, express extent as the number of physical pieces, as "1 DVD", "3 CD-ROMs", or "56 slides" (see Section 29.3.7.11 for details). For books on the Internet, page numbers are not usually present unless the book is in PDF format. When citing such PDF files, give extent as you would for a print publication. Express extent for other types of Internet publications in terms of the total number of screens, paragraphs, lines, or bytes (see Section 29.3.7.13 for details).

End the information on extent with a period unless a period is already present at the end of an abbreviated word. Abbreviate pages as "p." and volumes as "vol."; other words for extent, such as "leaves", "CD-ROMs", and "screens", are not abbreviated, to avoid confusion. Separate extent from the date information that precedes it by a space.

29.3.7.2.8 SERIES OF BOOKS

When a book has a collective title in addition to its own title, it is part of a series; the series title (the collective title) may be included in a reference but is not required. When used, series information follows the date of publication (and pagination, if included) in the citation–sequence and citation–name systems and follows the publisher in the name–year system. Series information is enclosed within parentheses in all systems. Give the name of the series first, capitalizing only the first word and proper nouns, and follow it with any numeration provided, such as a volume or issue number. Separate the title and the numeration by a semicolon. Conclude series information with a period, which falls outside the closing parenthesis.

> Ambudkar SV, Gottesman MM, editors. ABC transporters: biochemical, cellular, and molecular aspects. San Diego (CA): Academic Press; c1998. (Methods in enzymology; vol. 292).

29.3.7.2.9 NOTES IN BOOKS

"Notes" is a collective name for information about a publication that provides clarification or assists in location. Examples of notes for books include the language of a publication if other than English, the International Standard Book Number (ISBN), and the name of a library or other collection where an unusual book may be found.

Escobar-Jiménez F, Herrera Pombo JL. Actualizaciones clínicas y terapéuticas en la lesión nerviosa del diabético [Clinical and therapeutic update on nervous injury of the diabetic]. Barcelona (Spain): Masson; c2000. Spanish.

Singer C. TB in South Africa: the people's plague. McKenzie P, photographer. [place unknown]: [South Africa Department of Health]; [1997]. ISBN: 0-621-17646-x. Located at: National Library of Medicine, Bethesda, MD; 1997 I797.

There are 3 circumstances in which notes information is required. The report and contract numbers of a technical report, the URL of an online document, and the location of a manuscript are all essential information that must be included (see Sections 29.3.7.4, 29.3.7.13, and 29.3.7.15.2, respectively, for details).

A note follows the date of publication (and the pagination and series, if included) in the citation–sequence and citation–name systems and follows the publisher in the name–year system. A note generally has no specific format or punctuation; not even a complete sentence is required. Conclude a note with a period.

29.3.7.2.10 PARTS OF BOOKS AND CONTRIBUTIONS TO BOOKS

At times it may be desirable to cite a separately identified portion of a book rather than the book as a whole. Chapters, sections, tables, charts, graphs, photographs, appendixes, and the like are considered parts of books when they are written or composed by the author or one of the authors of a book; they are treated as contributions when written by someone other than the author or one of the authors of a book. Because a reference should start with the individual or organization with responsibility for the intellectual content of the publication, begin a reference to a part of a book with the book itself, and follow it with the information on the part; begin a reference to a contribution with information on the contribution, followed by the word "In:" and information about the book itself.

Separate information on a part of a book from information on the book as a whole by a space. Begin part information with the name of the part and any numeration, followed by a comma and the title of the part. Separate the title of the part from the pagination of the part by a semicolon and a space.

Examples of parts:

Shakelford RT. Surgery of the alimentary tract. Philadelphia (PA): W.B. Saunders; 1978. Chapter 2, Esophagoscopy; p. 29–40.

Magalini SL, Magalini SC, de Francisci G, editors. Dictionary of medical syndromes. 3rd ed. Philadelphia (PA): Lippincott; 1990. Dean-Barnes; p. 230–231.

Sissons HA, Murray RO, Kemp HBS. Orthopaedic diagnosis: clinical, radiological, and pathological coordinates. Berlin (Germany): Springer-Verlag; 1984. Figure 1.3, Stress fractures of the spine; p. 236.

Follow the author(s) and title of a contribution with a space and the word "In:". Conclude the reference with a space and the pagination of the contribution.

Examples of contributions:

Anderson RJ, Schrier RW. Acute renal failure. In: Braunwald E, Isselbacher KJ, Petersdorf RG, editors. Harrison's principles of internal medicine. 15th ed. New York (NY): McGraw-Hill; c2001. p. 1149–1155.

Hazeltine WA. AIDS. In: The encyclopedia Americana. International ed. Danbury (CT): Grolier Incorporated; 1990. p. 365–366.

29.3.7.3 Conference Proceedings and Conference Papers

Although not all conference proceedings are published in book format anymore (many are now published in journal issues or supplemental issues online), they do share many characteristics with books (see Section 29.3.7.2 and its subsections). The major difference in citing them lies in their titles and in the provision of information about the dates and places of the conference. Conference papers are cited in the same way as contributions to books (see Section 29.3.7.2.10).

The general format for a reference to a conference proceedings, including punctuation, is as follows:

Citation–sequence and citation–name:

Editor(s). Title of book. Number and name of conference; date of conference; place of conference. Place of publication: publisher; date. Extent. Notes.

Examples:

Callaos N, Margenstern M, Zhang J, Castillo O, Doberkat EE, editors. SCI 2003. Proceedings of the 7th World Multiconference on Systemics, Cybernetics and Informatics; 2003 Jul 27–30; Orlando, FL. Orlando (FL): International Institute of Informatics and Systematics.

Antonioli GE, editor. Pacemaker leads 1997. Proceedings of the 3rd International Symposium on Pacemaker Leads; 1997 Sep 11–13; Ferrara, Italy. Bologna (Italy): Monducci Editore; c1997.

Name–year:

Editor(s). Date. Title of book. Number and name of conference; date of conference; place of conference. Place of publication: publisher. Extent. Notes.

Examples:

Callaos N, Margenstern M, Zhang J, Castillo O, Doberkat EE, editors. c2003. SCI 2003. Proceedings of the 7th World Multiconference on Systemics, Cybernetics and Informatics; Orlando, FL. Orlando (FL): International Institute of Informatics and Systematics.

Antonioli GE, editor. c1997. Pacemaker leads 1997. Proceedings of the 3rd International Symposium on Pacemaker Leads; 1997 Sep 11–13; Ferrara, Italy. Bologna (Italy): Monducci Editore.

The general format for a reference to a conference paper, including punctuation, is as follows:

Citation–sequence and citation–name:

Author(s) of paper. Title of paper. In: Editor(s). Title of book. Number and name of conference; date of conference; place of conference. Place of publication: publisher; date. Location. Notes.

Example:

Lee DJ, Bates D, Dromey C, Xu X, Antani S. An imaging system correlating lip shapes with tongue contact patterns for speech pathology research. In: Krol M, Mitra S, Lee DJ, editors. CBMS 2003. Proceedings of the 16th IEEE Symposium on Computer-Based Medical Systems; 2003 Jun 26–27; New York. Los Alamitos (CA): IEEE Computer Society; c2003. p. 307–313.

Name–year:

Author(s) of paper. Date. Title of paper. In: Editor(s). Title of book. Number and name of conference; date of conference; place of conference. Place of publication: publisher. Location. Notes.

Example:

Lee DJ, Bates D, Dromey C, Xu X, Antani S. c2003. An imaging system correlating lip shapes with tongue contact patterns for speech pathology research. In: Krol M, Mitra S, Lee DJ, editors. CBMS 2003. Proceedings of the 16th IEEE Symposium on Computer-Based Medical Systems; New York. Los Alamitos (CA): IEEE Computer Society. p. 307–313.

The placement of the year of publication is the major difference between the systems. For a complete discussion of these systems, see Section 29.2.1 and its subsections. To conserve space, most of the examples throughout this section show only the first format. To convert to the name–year format, move the date to follow the author names.

For information on citing papers presented but never published, see Section 29.3.7.15.1.

29.3.7.3.1 AUTHORS AND SECONDARY AUTHORS OF CONFERENCE PUBLICATIONS

Because conference proceedings are a collection of the papers presented at a conference, most of them have editors instead of authors. Although editors are usually considered secondary authors, when there are no authors place the names of editors in the author position at the head of the reference. List the names in the order in which they appear in the text. Begin with the surname (family name or last name) of the first editor, followed by the initial letters of the given names. Continue with the surname and initials of the second and subsequent editors. For details on the format of names, such as the way names from various nationalities are handled, see Section 29.3.6.1.1.

When there are 2 to 10 editors, list them all, including the 10th editor; if there are more than 10 editors, list the first 10 followed by "et al." or "and others". Separate the surname and initials of an editor by a space; separate successive editor names from each other by a comma and a space. Follow the last item in the editor list with a comma, the word "editor" or "editors" and a period.

Conference publications may have other secondary authors, such as illustrators and translators, in addition to editors. Inclusion of these other types of secondary authors in a reference is optional. Place the secondary author after the title of the work or any edition statement that may follow the title, followed by a comma and the role or roles of the secondary author. Separate multiple secondary authors by commas; separate secondary authors occupying different roles by semicolons. Give secondary authors in the order in which they appear in the publication.

> Doring R, editor. The German health reporting system and current approaches in Europe: a comparative view on differences, parallels, and trends. Proceedings of the International Conference; 2001 Nov; Berlin. Carney A, Frank M, translators. Berlin (Germany): Robert Koch-Institut; 2002.

If no authors or editors can be found for a conference publication, begin the reference with the title. The use of "anonymous" is not permitted.[1]

Follow the same rules for the author(s) of conference papers, omitting of course the word "editor". End the author list with a period.

29.3.7.3.2 TITLES OF CONFERENCE PROCEEDINGS

Conference proceedings often have 2 titles: the title of the book of proceedings and the name of the conference. When both titles are present, begin with the title of the book. In general, record the words of the book title exactly as they appear in the publication. For the 3 major exceptions to this rule, see Section 29.3.7.2.2. Separate the title of the book from the period at the end of the list of editors by a space. Capitalize only the first word of the title, proper nouns, proper adjectives, acronyms, and initialisms. For non-English titles, abide by the conventions of the particular language regarding capitalization; follow the original language with a translation if possible, surrounded by square brackets.

Llombart-Bosch A, editor. Patologia del cancer de higado. 1st Reunion Germano-Espanola de Anatomia Patologica [Pathology of liver cancer. 1st German-Spanish Meeting on Pathologic Anatomy]; 1976 Sep 23–25; Lloret de Mar, Spain. Valencia (Spain): Fundación Garcia Munoz; 1977. Spanish.

Retain whatever punctuation is found within the title. If no punctuation separates the title from a subtitle, use a colon followed by a space. End the title with a period unless another form of punctuation such as a question mark or an exclamation point is already present. The only punctuation to be added to a title is the square brackets for a medium designator. Medium designators tell the user that the book is published in other than print form, such as microfilm, microfiche, CD-ROM, or DVD. Separate a medium designator from the title by a space and surround it by square brackets.

Bhattathiri VN, editor. Cervical cancer in developing countries [microfiche].

Follow the book title with information about the conference. Unlike book titles, which follow whatever appears on the publication itself, conference information has a specific order and punctuation. Begin with the name of the conference, capitalizing all significant words and converting all numbers to arabic ordinals. Follow the name with the dates of the conference, in year month day order, abbreviating names of months to the first 3 letters of the English name, as "Feb" for "February". Conclude conference information with the city in which the conference was held, followed by the 2-letter abbreviation for the US state or Canadian province or the country (if not the United States or Canada). Insert a semicolon and a space after the conference name and after the date; close the conference information with a period.

Bhattathiri VN, editor. Cervical cancer in developing countries. Proceedings of the 11th Meeting of the International Working Party for Treatment of Cancer of the Cervix in Developing Areas; 1992 Oct 12–14; Trivandrum, India. Trivandrum (India): Regional Cancer Centre; c1993.

If the only title available is the name of the conference, place this information after the editor(s).

Sasaki Y, Nomura Y, editors. Symposium on Nasal Polyp; 1984 Oct 5–6; Tokyo, Japan. Stockholm (Sweden): Almqvist & Wiksell; 1986.

29.3.7.3.3 TITLES OF CONFERENCE PAPERS

In general, record the words of a conference paper title exactly as they appear in the publication. It is important not to change the words of the author(s). For the 3 major exceptions to this rule, see Section 29.3.7.1.2. The content designator "[abstract]" may be placed at the end of the title when the reference refers to a meeting abstract rather than a complete paper. If a content designator is used, place it within square brackets immediately following the title, preceded by a space.

Follow the title of the conference paper (and the content designator if appropriate) with a period, a space, the word "In:", and the conference proceedings information.

Church JA, Marshall G, Lang W. Thrombotic thrombocytopenic purpura (TTP) in an HIV-infected child [abstract]. In: Morisset RA, editor. V International Conference on AIDS: the Scientific and Social Challenge; 1989 Jun 4–9, Montreal. Ottawa (ON): International Development Research Centre; 1989. p. 494.

29.3.7.3.4 PLACE OF PUBLICATION, PUBLISHER, AND DATE OF PUBLICATION OF CONFERENCE PUBLICATIONS

The rules for place of publication, publisher, and date of publication are the same for conference proceedings as they are for books (see Sections 29.3.7.2.4 through 29.3.7.2.6).

29.3.7.3.5 EXTENT AND LOCATION (PAGINATION) OF CONFERENCE PUBLICATIONS

The extent of a conference proceedings is an optional component of a reference, but it can provide useful information to the reader. Use for extent the total number of pages on which the text appears, as "235 p." Do not count pages for such items as introductory material, appendixes, and indexes unless they are included in the pagination of the text. Some proceedings are published in more than one physical volume. When this occurs, cite the total number of volumes instead of the number of pages, as "2 vol." Abbreviate pages as "p." and volumes as "vol." Separate extent from date information by a space.

The location of a conference paper within the proceedings is a required component of a reference. Begin with "p." and follow it with the first and last page numbers of the paper, separated by an en dash (or a hyphen if preferred) (see Section 29.3.7.1.6 for further information on pagination).

> Thorneycroft IH. Oral contraceptives and myocardial infarction. In: Toward a new standard in oral contraception. Proceedings of a symposium held at the 12th World Congress on Fertility and Sterility; 1989 Oct 3; Marrakesh, Morocco. St Louis (MO): Mosby-Year Book; 1990. p. 1393–1397.

29.3.7.3.6 NOTES IN CONFERENCE PUBLICATIONS

"Notes" is a collective name for information about a publication that enhances the reference or provides clarification. Most of the information contained in notes is of interest to the reader but not essential to the reference. Examples of useful notes for conference proceedings and papers include the language of publication if other than English, information on the sponsorship of the conference, and the name and call number of a library where the proceedings might be viewed.

29.3.7.4 Technical Reports

A technical report is defined by NISO as "a separately issued record of research results, research in progress, or other technical studies".[1] Most technical reports are issued by governmental agencies, usually at the federal or state level, but reports also originate from universities and other types of research institutions. Technical reports share many characteristics with the books described in Section 29.3.7.2 and its subsections. The major difference in citing them lies in their authorship and in the provision of information about sponsorship, including any report, contract, and grant numbers.

In citing a technical report, it is important to identify both the sponsoring organization (i.e., the organization that funded the research) and the performing organization (i.e., the organization that conducted the research) and to determine which of them actually published the report. In some cases, the same organization performs both functions.

For example, the National Cancer Institute has intramural scientists and may publish their work in report format. Often, however, the sponsoring organization provides funds to another organization that actually performs the research. These funds are disbursed through grants and contracts. When this occurs, either the sponsoring organization or the performing organization may publish the report. Thus, there are 3 possible scenarios for publication of a technical report. It may be:

- written and published by the sponsoring organization.
- written by the performing organization and published by the sponsoring organization.
- written and published by the performing organization.

The general format for a reference to a technical report that is written and published by the sponsoring organization, including punctuation, is as follows:

Citation–sequence and citation–name:

Author(s). Title of report. Edition. Place of publication: publisher; date. Extent. Report No.: Notes.

Example:

Feller BA. Health characteristics of persons with chronic activity limitation, United States, 1979. Hyattsville (MD): National Center for Health Statistics (US); 1981. Report No.: VHS-SER-10/137. Available from: NTIS, Springfield, VA; PB88-228622.

Name–year:

Author(s). Date. Title of report. Edition. Place of publication: Publisher. Extent. Report No.: Notes.

Example:

Feller BA. 1981. Health characteristics of persons with chronic activity limitation, United States, 1979. Hyattsville (MD): National Center for Health Statistics (US). Report No.: VHS-SER-10/137. Available from: NTIS, Springfield, VA; PB88-228622.

The placement of the year of publication is the major difference between the systems. For a complete discussion of these systems, see Section 29.2.1 and its subsections. To conserve space, most of the examples throughout this section show only the first format. To convert to the name–year format, move the year to follow the author names.

The general format for a reference to a technical report that is written by the performing organization and published by the sponsoring organization, including punctuation, is as follows:

Citation–sequence and citation–name:

Author(s) (performing organization name and address). Title of report. Edition. Place of publication: publisher; date. Extent. Report No.: Contract No.: Grant No.: Notes.

Example:

Cooper LN (Department of Physics, Brown University, Providence, RI). Theoretical and experimental research into biological mechanisms underlying learning and memory. Final progress report 1 Aug 88–31 Jul 89. Washington (DC): Air Force Office of Scientific Research (US); 1990. Report No.: AFOSR-TR-90-0672. Contract No.: AFOSR-88-0228;2305;B4. Available from: NTIS, Springfield, VA; AD-A223615.

Name–year:

Author(s) (performing organization name and address). Date. Title of report. Edition. Place of publication: publisher. Extent. Report No.: Contract No.: Grant No.: Notes.

Example:

Cooper LN (Department of Physics, Brown University, Providence, RI). 1990. Theoretical and experimental research into biological mechanisms underlying learning and memory. Final progress report 1 Aug 88–31 Jul 89. Washington (DC): Air Force Office of Scientific Research (US). Report No.: AFOSR-TR-90-0672. Contract No.: AFOSR-88-0228;2305;B4. Available from: NTIS, Springfield, VA; AD-A223615.

The general format for a reference to a technical report that is written and published by the performing organization, including punctuation, is as follows:

Citation–sequence and citation–name:

Author(s). Title of report. Edition. Place of publication: publisher; date. Extent. Report No.: Contract No.: Grant No.: Notes.

Example:

Moray NP, Huey BM. Human factors research and nuclear safety. Washington (DC): National Academy Press; 1988. Contract No.: NRC-04-86-301. Available from: NTIS, Springfield, VA; PB89-175517. Sponsored by the Nuclear Regulatory Commission.

Name–year:

Author(s). Date. Title of report. Edition. Place of publication: publisher. Extent. Report No.: Contract No.: Grant No.: Notes.

Example:

Moray NP, Huey BM. 1988. Human factors research and nuclear safety. Washington (DC): National Academy Press. Contract No.: NRC-04-86-301. Available from: NTIS, Springfield, VA; PB89-175517. Sponsored by the Nuclear Regulatory Commission.

29.3.7.4.1 AUTHORS AND SECONDARY AUTHORS OF TECHNICAL REPORTS

List the names of the authors in the order in which they appear in the text. Begin with the surname (family name or last name) of the first author, followed by the initial letters of the given names. Continue with the surname and initials of the second and subsequent authors. Separate the surname and initials of an author by a space; separate successive author names from each other by a comma and a space. End author information with a period. For details on the format of author names, such as the way names from various nationalities are handled, see Section 29.3.6.1.1.

In a reference to a technical report that is written by the performing organization and published by the sponsoring organization, place the name and address of the performing organization within parentheses after the last-named author. Address information consists of the department or division, organizational name, city, the 2-letter abbreviation for the US state or Canadian province, and the country (if not the United States or Canada). Write country names in full or use the 2-letter ISO country code[24] (see Appendix 29.3 for a list of the codes for selected countries).

Carlson BL, Lother Y, Johnson HG (Department of Psychology, University of Pittsburgh, Pittsburgh, PA).

Editors, translators, and illustrators are examples of secondary authors. If a technical report has secondary authors in addition to a personal or organizational author, it is optional to include them in a reference. Place the secondary author after the title of the report or any edition statement that may follow the title, followed by a comma and the role or roles of the secondary author (see Section 29.3.6.1.3 for details of secondary authors).

Carlson BL, Lother Y, Johnson HG (Department of Psychology, University of Pittsburgh, Pittsburgh, PA). Effects of task learning under long-term isolation. Final report. Roehr E, illustrator.

29.3.7.4.2 TITLES AND EDITIONS OF TECHNICAL REPORTS

Titles of technical reports follow the same format as that used for books (see Section 29.3.7.2.2). Edition statements in technical reports follow the title and often differ from edition statements for books, being expressed by such terms as "Annual report", "Interim report", or "Final report". They often contain date information, as "Annual report 2012" or "Interim report 1 Aug 2011–31 Jul 2012". Use whatever wording is provided, capitalizing only the first word and proper nouns of the edition statement and ending it with a period.

Thomas DL (Ohio Aerospace Institute, Brook Park, Ohio). Design and analysis of UHTC leading edge attachment. Final report. Washington (DC): National Aeronautics and Space Administration (US); 2002. Report No.: NAS126211505. Grant No.: NCC3756.

29.3.7.4.3 PLACE OF PUBLICATION, PUBLISHER, DATE, EXTENT, AND NOTES OF TECHNICAL REPORTS

The rules for place of publication, publisher, date, extent, and notes are the same for technical reports as they are for books (see Sections 29.3.7.2.4 through 29.3.7.2.9). There is, however, one type of additional information that is required in a technical report. Place any report numbers, contract numbers, or grant numbers after the date of publication (and pagination, if provided). Precede each number with its identifying words and end each number with a period. List report numbers first.

Report No.: ODAM-52-86. Contract No.: HRSA-240-84-0123.
Report No.: MPS-8701-02. Grant No.: EY2456.

Other types of information are optional when citing technical reports but useful to the reader. If the report has been written and published by the performing organization, the name of the sponsoring organization may be included as a note, as "Sponsored by the American Cancer Society". Also, because most technical reports cannot be obtained from the publishing or sponsoring organizations, a clearinghouse or other source address may be provided after the report and other numbers. To assist the reader, precede source information with the words "Available from:" and include any acquisition numbers provided. Separate the acquisition number from the address by a semicolon. End availability information with a period.

Available from: NTIS, Springfield, VA; PB90-155268.
Available from: US GPO, Washington, DC; HE 20.9002:Ac 7.
Available from: ERIC, Arlington, VA; ED297746.

29.3.7.4.4 PARTS OF TECHNICAL REPORTS

To cite only a part of a technical report, such a chapter, table, or graph, rather than the entire publication, begin the reference with information on the report and follow that with information about the part; see the rules provided in Section 29.3.7.2.10.

29.3.7.5 Dissertations and Theses

Dissertations and theses are written in support of academic degrees above the baccalaureate level. Although some European and other countries use the term "thesis" to refer to material written for a doctorate, the term is reserved in this manual for work at the master's level, whereas "dissertation" is used for the doctorate.

The general format for a reference to a dissertation or thesis, including punctuation, is as follows:

Citation–sequence and citation–name:

Author(s). Title of dissertation or thesis [content designator]. Place of publication: publisher; date. Extent. Notes.

Examples:

Lutz M. 1903: American nervousness and the economy of cultural change [dissertation]. [Stanford (CA)]: Stanford University; 1989.

Oviedo S. Adolescent pregnancy: voices heard in the everyday lives of pregnant teenagers [master's thesis]. [Denton (TX)]: University of North Texas; 1995.

Name–year:

Author(s). Date. Title of dissertation or thesis [content designator]. Place of publication: publisher. Extent. Notes.

Examples:

Lutz M. 1989. 1903: American nervousness and the economy of cultural change [dissertation]. [Stanford (CA)]: Stanford University.

Oviedo S. 1995. Adolescent pregnancy: voices heard in the everyday lives of pregnant teenagers [master's thesis]. [Denton (TX)]: University of North Texas.

The placement of the year of publication is the major difference between the systems. For a complete discussion of these systems, see Section 29.2.1 and its subsections. To conserve space, most of the examples throughout this section show only the first format. To convert to the name–year format, move the year to follow the author names.

29.3.7.5.1 AUTHORS OF DISSERTATIONS AND THESES

With rare exceptions, dissertations have only 1 author. Most master's theses also have only a single author, but occasionally a thesis will have 2 authors. Begin with the surname (family name or last name) of the author, followed by the initial letters of the given names. Separate multiple authors by a comma and a space. See Section 29.3.6.1.1 for details on the format of author names, such as the way names from various nationalities are handled.

29.3.7.5.2 TITLES OF DISSERTATIONS AND THESES

In general, record the words of the title of a dissertation or thesis exactly as they appear in the publication. For the 3 major exceptions to this rule, see Section 29.3.7.2.2. Separate the title of the book from the period at the end of the author list by a space. Capitalize only the first word of the title, proper nouns, proper adjectives, acronyms, and initialisms. For non-English titles, use the conventions of the particular language regarding capitalization; follow the original language with a translation if possible, surrounded by square brackets.

Klein P. Familie und agrarisch-heimgewerbliche Verflechtung. Eine demographische Studie zu Spenge (Ravensberg), 1768–1868 [Family and proto-industrialization: a demographic study of Spenge (Ravensberg), 1768–1868] [dissertation]. Bielefeld (Germany): Universitat Bielefeld, Institut fur Bevolkerungsforschung und Sozialpolitik; 1993. German.

Dissertations and theses are rigorous reports of original research. To inform the user that the particular document being referenced is a dissertation or thesis, follow the title with a space and the appropriate content designator—that is, dissertation or master's thesis, placed within square brackets.

29.3.7.5.3 PLACE OF PUBLICATION AND PUBLISHER OF DISSERTATIONS AND THESES

The place of publication for a thesis or dissertation is the city where the university or other institution granting the degree is located. Additional information may be included with the city name to provide any necessary clarification. Place the state, province, and country name or ISO country code[24] within parentheses after the city (see Appendix 29.3 for a list of the codes for selected countries).

Many dissertations, particularly those of US universities, do not state the place of publication. When this occurs, obtain the city name from some other source and place it within square brackets. Conclude place of publication information with a colon.

Lerner PF. Hysterical men: war, neurosis and German mental medicine, 1914–1921 [dissertation]. [New York (NY)]: Columbia University; 1997.

Dwivedi G. The association of breast feeding and nutritional status of children 13–36 months of age [master's thesis]. [Montreal (QC)]: McGill University; 1992.

If the place cannot be determined, as may occur with some older dissertations and theses, give the words "place unknown" within square brackets.

The publisher is the university or other institution granting the degree. Names of publishers should be given in their original language, but romanize names presented in Cyrillic, Greek, Arabic, or Hebrew; romanize or translate them if in a character-based language (e.g., Chinese, Japanese). If desired, follow a non-English name with a translation. Surround all translated titles by square brackets. Begin publisher information with the name of the institution and follow it with a comma and any subsidiary division.

Miller LE, Sperry BM. Central American women's experience of prenatal care [master's thesis]. [Boston (MA)]: Massachusetts General Hospital, Institute of Health Professions; 1992.

Cisse A. Connaissances et comportements sexuels des jeunes de 15–29 ans sur les M.T.S. et le SIDA a Bamako [Sexual knowledge and behavior of young people 15–29 years of age concerning sexually transmitted diseases (STDs) and AIDS in Bamako] [master's thesis]. [Quebec City (QC)]: Laval University; 1993. French.

29.3.7.5.4 DATES OF DISSERTATIONS AND THESES

A date is a required component of a reference to a dissertation or thesis. Always give the year of publication. If the publication date includes a month, as sometimes occurs, you may include it. Place the month after the year, abbreviated to the first 3 letters of the English name, such that September becomes "Sep". End date information with a period.

29.3.7.5.5 EXTENT AND NOTES OF DISSERTATIONS AND THESES

Extent (pagination) and notes are optional components of a reference, but both can provide useful information to the reader. Use for extent the total number of pages on which the text appears, as "235 p." Do not count pages for such items as introductory material, appendixes, and indexes unless they are included in the pagination of the text. Separate extent from date information by a space.

Notes for a dissertation or thesis provide clarification, such as grant or other sponsorship information, or assist in location. The latter is important because these documents may not be easily obtainable from the university or other institution granting the degree. A clearinghouse or other source address may be provided after the date (or pagination, if included). To aid the user, precede source information with the words "Available from:" or "Located at:" and include any acquisition numbers provided. Separate the acquisition number from the address by a semicolon. End availability information with a period.

St Hill PF. Acceptability and use of family planning services by refugee Haitian women in Miami [dissertation]. [San Francisco (CA)]: University of California, San Francisco; 1992. Available from: UMI, Ann Arbor, MI; AAD13-39383.

Lerner PF. Hysterical men: war, neurosis and German mental medicine, 1914–1921 [dissertation]. [New York (NY)]: Columbia University; 1996. Located at: National Library of Medicine, Bethesda, MD; 1997 F-178.

29.3.7.5.6 PARTS OF DISSERTATIONS AND THESES

To cite only a part of a dissertation or thesis, such as a chapter, table, or graph, rather than the entire publication, begin the reference with information on the dissertation or thesis and follow that with information about the part; see the rules provided in Section 29.3.7.2.10.

Lerner PF. Hysterical men: war, neurosis and German mental medicine, 1914–1921 [dissertation]. [New York (NY)]: Columbia University; 1996. Chapter 7, The pension issue after the war; p. 377–410. Located at: National Library of Medicine, Bethesda, MD; 1997 F-178.

29.3.7.6 Bibliographies

Bibliographies are collections of references made for a specific purpose, such as to bring together references on a specific subject or by a particular author. Their citation format is identical to that of the standard book (see Section 29.3.7.2) with 3 exceptions: authors are termed "compilers"; the content designator "[bibliography]" is added after the title if the word is not included there; and descriptive information, such as the number of citations included and the period covered, may be added as notes.

The general format for a reference to a bibliography, including punctuation, is as follows:

Citation-sequence and citation-name:

Author(s), compilers. Title of bibliography [content designator]. Place of publication: publisher; date. Extent. Notes.

Example:

Cabirac D, Warmbrodt R, compilers. Biotechnology and bioethics [bibliography]. Beltsville (MD): National Agricultural Library (US); 1993. 97 citations from the AGRICOLA database from January 1985–December 1992.

Name–year:

Author(s), compilers. Date. Title of bibliography [content designator]. Place of publication: publisher. Extent. Notes.

Example:

Cabirac D, Warmbrodt R, compilers. 1993. Biotechnology and bioethics [bibliography]. Beltsville (MD): National Agricultural Library (US). 97 citations from the AGRICOLA database from January 1985–December 1992.

The placement of the year of publication is the major difference between the systems. For a complete discussion of these systems, see Section 29.2.1 and its subsections.

29.3.7.7 Patents

A patent is "a title of legal protection of an invention, issued, upon application and subject to meeting legal criteria, by a government office".[1] Patents are a special type of monograph requiring specific information on the legal aspects of the patent.

The general format for a reference to a patent, including punctuation, is as follows:

Citation–sequence and citation–name:

Author(s), inventors; patent holder, assignee. Title of patent. Country issuing the patent country code patent number. Publication date. Extent.

Example:

Blanco EE, Meade JC, Richards WD, inventors; Ophthalmic Ventures, assignee. Surgical stapling system. United States patent US 4,969,591. 1990 Nov 13.

Name–year:

Author(s), inventors; patent holder, assignee. Date. Title of patent. Country issuing the patent country code patent number. Extent.

Example:

Blanco EE, Meade JC, Richards WD, inventors; Ophthalmic Ventures, assignee. 1990 Nov 13. Surgical stapling system. United States patent US 4,969,591.

The placement of the year of publication is the major difference between the systems. For a complete discussion of these systems, see Section 29.2.1 and its subsections. To conserve space, most of the examples throughout this section show only the first format. To convert to the name–year format, move the year to follow the assignee names. For patents, unlike most other types of publications, always provide the full date of publication, including month and day.

29.3.7.7.1 AUTHORS (INVENTORS AND ASSIGNEES) OF PATENTS

Patents have 2 types of "authors": the inventor of the device, process, or other entity being patented, and the "assignee", the individual(s) or organization holding legal title to the patent. Begin with the inventors. List the names of the inventors in the order in which they appear in the text. Begin with the surname (family name or last name) of the first inventor, followed by the initial letters of the given names, a comma, and a space. Continue with the surname and initials of the second and subsequent inventors, each followed by a comma and a space. Follow the last-named inventor with a comma, a space, and the word "inventor" or "inventors". For details on the format of author names, such as the way names from various nationalities are handled, see Section 29.3.6.1.1. End inventor information with a semicolon.

Follow the list of inventors with the assignee. Assignees are often organizations but may also be individuals. Follow the last-named assignee with a comma and the word "assignee" or "assignees". End assignee information with a period.

Harred JF, Knight AR, McIntyre JS, inventors; Dow Chemical Company, assignee.
Leibensohn S, inventor; Leibensohn S, assignee.

29.3.7.7.2 TITLES OF PATENTS

In general, record the words of the title of a patent exactly as they appear in the publication. For the 3 major exceptions to this rule, see Section 29.3.7.2.2. Separate the title of the patent from the assignee by a space. Capitalize only the first word of the title, proper nouns, proper adjectives, acronyms, and initialisms. For non-English titles, use the conventions of the particular language regarding capitalization; follow the original language with a translation if possible, surrounded by square brackets.

Bottcher H, Juraszyk H, Hausberg HH, Greiner H, Seyfried C, Minck KO, Bergmann R, inventors; Merck Patent GMBH, assignee. Indolderivate [Indole derivatives]. German patent DE 3,907,974. 1990 Sep 13.

29.3.7.7.3 DESCRIPTIVE INFORMATION FOR PATENTS

To adequately describe the patent, begin with the name of the country granting it. Use the adjectival form of the English name for the country (e.g., "French" for France and "German" for Germany). Follow the country name with the word "patent" or whatever word for patent is used by the granting country, the 2-letter ISO country code[7] (see Appendix 29.3 for a list of the codes for selected countries), and the patent number. Conclude the descriptive information with a period.

United States patent US 3,654,317.
French patent FR 2,456,525.
German Offenlegungsschrift DE 1,966,199.

29.3.7.7.4 DATE AND EXTENT (PAGINATION) OF PATENTS

The date the patent was granted is a mandatory component of a reference. Begin with the year followed by the month and day. Give the month in English, abbreviated to the first 3 letters, such that "January" becomes "Jan". Unlike references to many other types of publications, the full date is required for patents.

As an option, you may include the number of pages of the patent, as "8 p."

29.3.7.7.5 NOTES IN PATENTS

"Notes" is a collective name for information about a publication that enhances the reference or provides clarification. In the note for a patent, place information on the country name and number granted to the patent in other countries and the international patent classification number, if known.

29.3.7.8 Newspapers

References to articles in newspapers are similar to those for journal articles (see Section 29.3.7.1 and its subsections). There are 3 major differences: 1) names of newspapers are never abbreviated (e.g., "The Washington Post" not "Wash Post"), although a leading "The" may be dropped if desired; 2) section information, if present, replaces volume and issue information; 3) and column location is added.

The general format for a reference to a newspaper, including punctuation, is as follows:

Citation–sequence and citation–name:
Author(s). Title of article. Title of newspaper (edition). Date;section:beginning page of article (column no.).

Example:
Weiss R. Study shows problems in cloning people: researchers find replicating primates will be harder than other mammals. Washington Post (Home Ed.). 2003 Apr 11;Sect. A:12 (col. 1).

Name–year:
Author(s). Date. Title of article. Title of newspaper (edition). Section:beginning page of article (column no.).

Example:
Weiss R. 2003 Apr 11. Study shows problems in cloning people: researchers find replicating primates will be harder than other mammals. Washington Post (Home Ed.). Sect. A:12 (col. 1).

The placement of the year of publication is the major difference between the systems. For a complete discussion of these systems, see Section 29.2.1 and its subsections. To conserve space, most of the examples throughout this section show only the first format. To convert to the name–year format, move the year to follow the author names. For newspapers, unlike most other types of publications, always provide the full date of publication, including month and day.

29.3.7.8.1 AUTHORS OF NEWSPAPER ARTICLES

Newspaper articles may be signed, either at the beginning of the article or at its conclusion, or unsigned. If the article is signed, begin the reference with the surname (family name or last name) of the author, followed by the initial letters of the given names. Separate multiple authors by a comma and a space. See Section 29.3.6.1.1 for details on the format of author names, such as the way names from various nationalities are handled. End author information with a period. If the article is unsigned, begin the reference with the title of the article; the use of "anonymous" is not permitted.[1] When using the name–year system for unsigned articles, the date follows the title.

29.3.7.8.2 TITLES OF NEWSPAPER ARTICLES

In general, record the words of the title of a newspaper article exactly as they appear in the publication. For the 3 major exceptions to this rule, see Section 29.3.7.1.2. Separate the title of the article from the author(s) by a space. Capitalize only the first word of the title, proper nouns, proper adjectives, acronyms, and initialisms. For non-English titles, use the conventions of the particular language regarding capitalization; follow the original language with a translation if possible, surrounded by square brackets.

Abatir accidentes de trabajo, una via de aumentar la productividad [Reducing accidents in the workplace, a way of increasing productivity]. El Dia (Mexico City). 1990 Nov 21;29(10231):5 (col. 1). Spanish.

Newspaper articles sometimes appear on a page with a header, such as "National News". Do not include headers as part of the article title.

29.3.7.8.3 TITLES AND EDITIONS OF NEWSPAPERS

Unlike the titles of journals in references to journal articles, the titles of newspapers are not abbreviated in references to newspaper articles. Record the words of the newspaper title exactly as they appear in the publication, using whatever capitalization and punctuation are found. As an option, a leading "The" may be dropped from the name.

The Wall Street Journal *or* Wall Street Journal
L'Action Francaise
Deutsche Volks-Zeitung

If a newspaper title does not indicate the location where it is published, add the location either within or after the title, as appropriate. For most newspapers, the location is the city. If the city is not well known, follow US and Canadian cities with the 2-letter postal abbreviation for the state or province; follow cities in other countries with the name of the country, either written out or as the 2-letter ISO country code[24] (see Appendix 29.3 for a list of the codes for selected countries). Surround location information with parentheses.

The Times Dispatch (Richmond, VA) *or* Times Dispatch (Richmond, VA)
Bergen County (NJ) Record
El Dia (Mexico City)
Les Echos (Paris)
Le Journal du Jura (Bienne, Switzerland)

An exception to the rule regarding addition of location information occurs when the newspaper is a national one without a particular geographic base, such as the *Christian Science Monitor* or *Le Monde*.

Newspapers may be published in different editions. Because the content and pagination may differ among editions, always include the edition after the title within parentheses. Abbreviate words used in the edition statement according to ISO 832.[26] End abbreviated words with a period; end title information with a period.

The New York Times (Long Island Ed.). *or* New York Times (Long Island Ed.).
Pravda (English Ed.).
The Washington Post (Final Ed.). *or* Washington Post (Final Ed.).

29.3.7.8.4 DATES OF NEWSPAPER ARTICLES

Because volume and issue information is usually omitted from references to newspapers, always provide the complete date of publication. Begin with the year followed by the month and day. Give the month in English, abbreviated to the first 3 letters, such that "January" becomes "Jan". Conclude date information with a semicolon.

2003 Mar 10;Sect. A:34 (col. 2).

29.3.7.8.5 VOLUMES AND ISSUES OF NEWSPAPER ARTICLES

Long tradition omits volume and issue numbers from references to newspapers. The section name, number, or letter substitutes for volume information. Abbreviate "Section" to "Sect." and follow it by the letter or number, as "Sect. A" and "Sect. 3". Give arabic numerals for sections, such that "II" becomes "2". Sections may be named, as "Metro Section" or "Weekend Section". Use these names in a reference only when the section lacks a number or letter. Abbreviate names of sections according to ISO 832.[26]

Kunkle F. Grievance panel seeks reprimand for Gansler. Washington Post (Home Ed.). 2003 Jul 3;Montgomery Extra:10 (col. 1).

As an option, however, volume and issue numbers may be included the same way they are for journal articles (see Section 29.3.7.1.5 for details of formats). Note that section letters are incorporated into the page number when volume and issues are used.

Weiss R. Study shows problems in cloning people: researchers find replicating primates will be harder than other mammals. Washington Post (Home Ed.). 2003 Apr 1;126(127):A12 (col. 1).

29.3.7.8.6 LOCATION (PAGINATION) OF NEWSPAPER ARTICLES

Pagination in a reference to a newspaper article differs from pagination in all other types of publications in that only the first page of the article is given. Follow the first page by the column number in which the article begins. Abbreviate column as "col." Surround column information with parentheses and conclude it with a period.

Sect. 1:42 (col. 4). Sect. A:2 (col. 5).

If a newspaper section is lettered, the letter is omitted before the beginning page number. For example, an article in Section C on page C13 is given as "Sect. C:13".

29.3.7.8.7 NOTES IN NEWSPAPER ARTICLES

A note for a newspaper usually consists of a dateline, which informs the reader of the place and date where a news item was created, if these differ from the date and place of the newspaper as a whole. Follow the name of the city by a comma and the month and day. Abbreviate months to the first 3 letters of the English name.

Health workers win suit over smoking ban. New York Times (Final Ed.). 1990 Dec 2;Sect. 1:33 (col. 1). Washington, Dec 1.

29.3.7.9 Maps

Maps in this section refer to those that are published as independent sheets. For citing maps appearing in books and atlases, see Section 29.3.7.2.10. References to maps are similar to books, with 2 major exceptions: authors of maps are termed "cartographers" and information on the type of map is provided after the title.

The general format for a reference to a map, including punctuation, is as follows:

Citation-sequence and citation-name:

Author, cartographer. Title of map [map type]. Place of publication: publisher; date. Physical description. Notes.

Example:

Doyon R, Donovan T, cartographers. AIDS in Massachusetts, 1985–1991 [demographic map]. Amherst (MA): University of Massachusetts, Department of Geology & Geography; 1992.

Name–year:

Author, cartographer. Date. Title of map [map type]. Place of publication: publisher. Physical description. Notes.

Example:

Doyon R, Donovan T, cartographers. 1992. AIDS in Massachusetts, 1985–1991 [demographic map]. Amherst (MA): University of Massachusetts, Department of Geology & Geography.

The placement of the year of publication is the major difference between the systems. For a complete discussion of these systems, see Section 29.2.1 and its subsections. To conserve space, most of the examples throughout this section show only the first format. To convert to the name–year system, move the year to follow the word cartographer(s).

29.3.7.9.1 AUTHORS (CARTOGRAPHERS) OF MAPS

List the names of the cartographers in the order in which they appear on the map. Begin with the surname (family name or last name) of the first cartographer, followed by the initial letters of the given names. Continue with the surname and initials of the second and subsequent cartographers. Separate multiple cartographers by a comma and a space. End cartographer information with a period. For details on the format of author names, such as the way names from various nationalities are handled, see Section 29.3.6.1.1.

An organization such as a university, society, association, corporation, or governmental body may also serve as cartographer. If a division or other part of an organization is included on the map, give the parts of the name in descending hierarchical order, separated by commas. In citing organizations that are national bodies such as government agencies, if a nationality is not included with the name, place the country after the name, within parentheses, using the 2-letter ISO country code[24] (see Appendix 29.3 for a list of the codes for selected countries). See Section 29.3.6.1.2 for more details on organizations as authors.

State of California, Division of Mines and Geology, cartographer.
Bureau of the Census (US), Geography Division, cartographer.

If a map has no cartographer, begin the reference with the title. The use of "anonymous" is prohibited.[1]

29.3.7.9.2 TITLES OF MAPS

In general, record the words of map titles exactly as they appear on the map. For the 3 major exceptions to this rule, see Section 29.3.7.2.2. Separate the title of the map from the end of the list of cartographers by a space. Capitalize only the first word of the title, proper nouns, proper adjectives, acronyms, and initialisms. For non-English titles, use the conventions of the particular language regarding capitalization; follow the original language with a translation if possible, surrounded by square brackets.

If the map has no title, use the geographic area it covers as the title and place the area name within square brackets, as "[United States]", "[Iowa]", or "[Philadelphia]". Follow the title of a map with the type of map, within square brackets, as "[demographic map]", "[topographic map]", or "[political map]". If in doubt about the particular type, simply use the word "[map]".

US Geological Survey, cartographer. [Alaska] [computer map]. Reston (VA): USGS; 1999.

29.3.7.9.3 PLACE OF PUBLICATION, PUBLISHER, AND DATE OF PUBLICATION OF MAPS

Publication information for a map is the same as for a standard book (see Sections 29.3.7.2.4 through 29.3.7.2.6).

29.3.7.9.4 PHYSICAL DESCRIPTION OF MAPS

As an option, information about the physical characteristics of a map may be provided. This could include the number of sheets, the map's physical size, whether the map is in color or black and white, and its scale or projection. Note that the words and abbreviations used in the description are taken from ISO 832[26] and reflect standard library practice; they are not SI symbols. To reduce work at the keyboard, use the letter "x" rather than the mathematical times symbol.

Environmental health facilities, Cincinnati, Ohio [demographic map]. [Washington (DC)]: Department of Health, Education, and Welfare (US), Public Health Service; [1964]. 1 sheet: black & white, 12 × 15 in.

Other examples:
1 sheet: color, 53 × 65 cm.
9 maps on 1 sheet: color, each map 21 × 13 cm. or smaller.
1 sheet: color, 88 × 66 cm., scale 1:1,600,000.
3 sheets.

See Appendix 29.2 for a list of English words commonly used in physical description.

29.3.7.9.5 NOTES FOR MAPS

"Notes" is a collective name for optional information about a reference that provides clarification or assists in location. Examples of notes for maps include the language of the map if other than English, the area covered by the map if it is not included in the title, and the name of a library or other collection where an unusual map may be found. Precede library information with the words "Located at:" and include the library's call number or other finding aid number:

Environmental health facilities, Cincinnati, Ohio [demographic map]. [Washington (DC)]: Department of Health, Education, and Welfare (US), Public Health Service; [1964]. 1 sheet: black & white, 12 × 15 in. Located at: National Library of Medicine, Bethesda, MD; WA11 AA1.C25 no. 17.

29.3.7.10 Legal Materials

References to legal documents differ greatly in style from those generally used in the sciences and the humanities. Their style is described in detail in *A Uniform System of Citation* published by the Harvard Law Review and others.[29] Even though these references may be barely comprehensible to most authors and readers in science, they identify legal documents accurately for retrieval from law and general libraries. The examples of common types of legal citations provided here represent the standards followed in the United States and may not be applicable to legal documents in other countries.

Public law:
Preventive Health Amendments of 1993, Pub. L. No. 103-183, 107 Stat. 2226 (Dec. 14, 1993).

Unenacted bill:
Medical Records Confidentiality Act of 1995, S. 1360, 104th Cong., 1st Sess. (1995).

Code of Federal Regulations:
Informed Consent, 42 C.F.R. Sect. 441.257 (1995).

Hearing:
Increased Drug Abuse: the Impact on the Nation's Emergency Rooms: Hearings Before the Subcomm. on Human Resources and Intergovernmental Relations of the House Comm. on Government Operations, 103rd Cong., 1st Sess. (May 26, 1993).

Because these formats are prescribed by the legal community, the format of end references is the same for the citation–sequence, citation–name, and name–year systems; do not take the date out of sequence for name–year. In-text references for the name–year system are handled the same way as are books without authors. Begin the in-text reference with the first word or first few words of the title, followed by an ellipsis and the year of publication. Use only as many words from the title as are necessary to distinguish it from other references being cited by title.

(Preventive . . . 1993)   (Informed . . . 1995)

An item from the *Federal Register* is cited the same way as a standard journal article. Abbreviate the title to *Fed Regist.* and see Section 29.3.7.1 for journal article citation information.

29.3.7.11 Audiovisuals

Audiovisuals include older media such as videocassettes, audiocassettes, videodiscs, motion pictures, and slides, but also more recent technologies such as CD-ROMs, DVDs, and Blu-ray discs (see Section 29.3.7.12), as well as online videos and podcasts (see Section 29.3.7.13). An audiovisual may be published either in monograph form, such as a book on DVD, or as a journal distributed on CD-ROM. Cite audiovisuals using the standard format for journals and books (see Sections 29.3.7.1 and 29.3.7.2), but because special equipment is needed to view these materials, add a medium designator after the title to alert the user.

The general format for a reference to an audiovisual, including punctuation, is as follows:

Citation–sequence and citation–name:

Books in audiovisual format:
Author(s). Title of audiovisual [medium designator]. Edition. Place of publication: publisher; date. Notes.

Example:
Johnson D, editor. Surgical techniques in orthopaedics: anterior cruciate ligament reconstruction [Blu-ray]. Rosemont (IL): American Academy of Orthopaedic Surgeons; c2002. 1 Blu-ray.

Journal articles in audiovisual format:
Author(s) of article. Title of article. Title of journal (edition) [medium designator]. Date;volume(issue): location. Notes.

Example:
Wilson JH, Singhoffer JH. Paradoxical embolus in evolution: report of a case. Dyn Cardiovasc Imaging [CD]. 1990;3(1):[presentation 2, 4:25 min.]. 1 CD.

Name–year:

Books in audiovisual format:
Author(s). Date. Title of audiovisual [medium designator]. Edition. Place of publication: publisher. Notes.

Example:

Johnson D, editor. c2002. Surgical techniques in orthopaedics: anterior cruciate ligament reconstruction [Blu-ray]. Rosemont (IL): American Academy of Orthopaedic Surgeons. 1 Blu-ray.

Journal articles in audiovisual format:

Author(s) of article. Date. Title of article. Title of journal (edition) [medium designator]. Volume(issue): location. Notes.

Example:

Wilson JH, Singhoffer JH. 1990. Paradoxical embolus in evolution: report of a case. Dyn Cardiovasc Imaging [videocassette]. 3(1):[presentation 2, 4 min.]. 1 videocassette.

The placement of the year of publication is the major difference between the systems. For a complete discussion of these systems, see Section 29.2.1 and its subsections. To conserve space, most of the examples throughout this section show only the first format. To convert to the name–year format, move the year to follow the author list.

29.3.7.11.1 AUTHORS AND SECONDARY AUTHORS OF AUDIOVISUALS

List the names of the authors in the order in which they appear on the opening screen(s) or in the accompanying text. Begin with the surname (family name or last name) of the first author, followed by the initial letters of the given names. Continue with the surname and initials of the second and subsequent authors. When there are 2 to 10 authors, list all of them, including the 10th author; if there are more than 10 authors, list the first 10 followed by "et al." or "and others". For details on the format of author names, such as the way names from various nationalities are handled, see Section 29.3.6.1.1. Separate the surname and initials of an author by a space; separate successive author names from each other by a comma and a space. Follow the last name in the author list with a period.

An organization such as a university, society, association, corporation, or governmental body may also serve as author. If a division or other part of an organization is included on the publication, give the parts of the name in descending hierarchical order, separated by commas. In citing organizations that are national bodies such as government agencies, if a nationality is not included with the name, place the country after the name, within parentheses, using the 2-letter ISO country code[24] (see Appendix 29.3 for a list of the codes for selected countries). See Section 29.3.6.1.2 for more details on organizations as authors.

American Psychological Association, Committee on Ethics.
Bureau of the Census (US), Geography Division.

Books and other monographs in audiovisual format may have secondary authors, such as editors and producers. If an audiovisual has secondary authors in addition to a personal or organizational author, it is optional to include them in a reference. Place the secondary author after the title of the work (or any edition statement that may follow the title), followed by a comma and the role or roles of the secondary author. Give the secondary authors in the order in which they appear in the publication. Separate multiple secondary authors by commas; separate secondary authors occupying different roles by a semicolon.

Mangold LA, Lindskog FO, Eaton B. IPPB therapy [slides]. Taylor JP, editor; University of Kansas, Respiratory Therapy Faculty, producer. Bowie (MD): Robert J. Brady; c1977. 387 slides: color, 2 × 2 in.

Bates B. Abdomen [videocassette]. 3rd ed. Blanchard-Healy, Inc., producer. Philadelphia (PA): Lippincott; c1995. 1 videocassette.

If an audiovisual has no personal or organizational author but has an editor or editors, place the editor(s) at the beginning of the reference.

Wood RM, editor. New horizons in esthetic dentistry [videocassette]. Visualeyes Productions, producer. Chicago (IL): Chicago Dental Society; 1989. 2 videocassettes.

If no primary author or editor can be determined for an audiovisual, omit any indication of authorship and begin the reference with the title. The use of "anonymous" is not permitted.[1]

29.3.7.11.2 TITLES OF BOOKS AND JOURNAL ARTICLES IN AUDIOVISUAL FORMAT

In general, record the words of titles of books and other monographs as well as journal article titles in audiovisual format exactly as they appear on the opening screen(s) or in the accompanying documentation. For the 3 major exceptions to this rule, see Sections 29.3.7.2.2 and 29.3.7.1.2. Separate the title of the audiovisual from the period at the end of the author list by a space. Capitalize only the first word of the title, proper nouns, proper adjectives, acronyms, and initialisms. For non-English titles, use the conventions of the particular language regarding capitalization; follow the original language with a translation if possible, surrounded by square brackets.

Saint-Anne Dargassies S. Apparition et évolution de l'infirmité motrice cérébrale chez le nourrisson [Appearance and evolution of cerebral palsy in the infant]

Follow the title of a book in audiovisual format with a medium designator within square brackets to indicate the type of audiovisual, as "[DVD]" or "[motion picture]". Conclude title information with a period.

Saint-Anne Dargassies S. Apparition et évolution de l'infirmité motrice cérébrale chez le nourrisson [Appearance and evolution of cerebral palsy in the infant] [DVD].

Medium designators for journals in audiovisual format follow the journal title, not the article title (see Section 29.3.7.11.3).

If a book has been published in more than one edition, follow the medium designator with the edition statement. Editions for books in audiovisual format are often termed "version" (abbreviated "ver.") or "release" (see Section 29.3.6.3 for details on edition statements).

O'Carroll MK. Advanced radiographic techniques [videocassette]. Rev. ed. Chapel Hill (NC): Health Sciences Consortium; 1993. 2 videocassettes: 39 min.

Anhydrous ammonia [DVD]. Ver. 3.0. Lawrence D, producer. Edgartown (MA): Emergency Film Group; c2001. 1 DVD: 29 min.

29.3.7.11.3 TITLES OF JOURNALS IN AUDIOVISUAL FORMAT

Locate journal titles for audiovisuals on the opening screen(s) or in the accompanying documentation and cite them in the same manner as print journal titles (see Section 29.3.7.1.3). Abbreviate the "significant" words in a journal title according to ISO 4,[25] and omit other words, such as articles, conjunctions, and prepositions. ISO 4 does not permit abbreviation of journal titles that consist of a single word and does not cover titles written in a character-based language (e.g., Chinese, Japanese). By common practice, character-based language titles are never abbreviated. As an option, however, they may be translated and abbreviated following the rules for English language titles.

Journal subtitles are not considered a part of the title proper and therefore are not included in the abbreviation.

> Video Journal of Obstetrics & Gynecology: VJOG *becomes* Video J Obstet Gynecol

Follow the abbreviated journal title with a medium designator within square brackets to indicate the type of audiovisual, as "[DVD]" or "[CD-ROM]". Conclude title information with a period.

> Netw Contin Med Educ [DVD].
> Dialogues Dermatol [CD-ROM].

29.3.7.11.4 PLACE OF PUBLICATION AND PUBLISHER OF AUDIOVISUALS

Publication information for an audiovisual in book format is the same as for a standard book (see Sections 29.3.7.2.4 and 29.3.7.2.5). Follow place of publication by a colon and a space; follow publisher information by a semicolon and a space unless the name–year system is used, when a period concludes it. Long practice omits such information for journal articles.

29.3.7.11.5 DATES OF AUDIOVISUALS

A date is a required component of a reference to an audiovisual. Always give the year of publication as found on the opening screen(s) or in the accompanying documentation. If a date includes a month, day of the month, or season, as often occurs in journal articles in audiovisual format, you may include them; these elements are required only when citing a journal that has no volume or issue number (see Section 29.3.7.11.6). Give months in English, abbreviated to the first 3 letters, such that September becomes "Sep". For multiple years of publication, separate the first and last year by an en dash (or a hyphen if preferred), as "2011–2012". If no year of publication can be determined, use the year of copyright, preceded by "c", as "c2012".

Some audiovisuals have both a year of publication and a year of copyright. Use only the year of publication unless at least 3 years separate the 2 dates. In the latter situation, use both dates, beginning with the year of publication, and separate them by a comma and a space, as "2012, c2007". If neither a year of publication nor a year of copyright can be found, use the words "date unknown" within square brackets. End date information with a period.

For books in audiovisual format using the citation–name and citation–sequence systems, place the date of publication after the publisher name and follow it with a period. When using the name–year system, place the date after the author list. For journal articles, place the date after the medium designator and follow it with a semicolon unless no volume or issue is found (see Section 29.3.7.11.6). In the latter case, end date information with a colon.

29.3.7.11.6 VOLUMES AND ISSUES FOR JOURNAL ARTICLES IN AUDIOVISUAL FORMATS

Volume numbers must be included in references to journal articles in audiovisual format when they are present in the document. Write all volume numbers in arabic numerals. Convert numeric terms and roman numerals to arabic, such that "sixtieth" becomes "60" and "IX" becomes "9". Separate multiple volume numbers by an en dash (or a hyphen if preferred); see Section 29.3.7.1.5 for further details on volume information. Follow volume information with a colon unless there is also issue information.

Issue numbers must also be included in references to journal articles in audiovisual format when such issue numbers are present. Write all issue numbers in arabic numerals. Convert numeric terms and roman numerals to arabic, such that "fifth" becomes "5" and "III" becomes "3". Issue numbers usually follow volume numbers in a reference but may stand on their own. If the only numeration on a journal is "number x", treat this number as an issue. Separate multiple issue numbers by an en dash (or a hyphen if preferred). For further details on issues, see Section 29.3.7.1.5. Surround all issue information by parentheses, even when there is no volume information. Follow issue information with a colon.

Gen Surg [audiocassette]. 1990;37(2):   Video J Neurosurg [videocassette]. 1991;5(3 Pt 2):

29.3.7.11.7 EXTENT AND LOCATION (PAGINATION) OF AUDIOVISUALS

The extent or length of a book in audiovisual format is an optional component of a reference, but may be included to provide useful information to the reader. Provide extent in terms of the total number of physical pieces and the run time if appropriate.

2 DVDs: 140 min.   24 CD-ROMs: 480 min.   387 slides.

Unlike extent for books in audiovisual formats, which is optional, the location of journal articles within an issue is a required component of a reference. Since no page numbers appear in journal articles in audio or video format, express location by 2 means: the order of the article in the journal issue, as "presentation 1" or "article 3", and the total length (run time) of the article, as "5 min." Place all location information within square brackets and conclude it with a period.

Ball WC Jr. Bronchiolitis obliterans organizing pneumonia. Johns Hopkins Med Grand Rounds [audiocassette]. 1990;16(5):[presentation 5, 13 min.].

Pohost GM. NMR spectroscopy and imaging of the heart. Video J Cardiol [videocassette]. 1990;5(3):[article 1, 17 min.].

29.3.7.11.8 NOTES IN AUDIOVISUALS

"Notes" is a collective name for a variety of types of optional information about a reference that provides clarification or assists in location. Examples of notes for audiovisuals include mention of any printed manual or other material accompanying the audiovisual, any sponsoring organization, an indication whether the audiovisual was produced for any special occasion, and a library or other archive where the audiovisual may be found. If actors, graphic artists, musicians, or other individuals associated with production of the audiovisual appear in acknowledgments, their names may be provided here in Notes.

29.3.7.12 Electronic Formats (Non-Internet)

Although it is increasingly rare for a publication to maintain a version of a document in electronic form that is not online, some documents may still be found in a variety of other electronic media, such as CD-ROM, DVD, or Blu-ray (see Section 29.3.7.13 for online formats). They also take a variety of forms, such as books, journals, databases, and computer software programs. Although electronic documents differ radically in physical form from the usual book or journal, the basic rules for citing them are the same.

The general format for a reference to an electronic document, including punctuation, is as follows:

Citation–sequence and citation–name:

Books in electronic format:

Author(s). Title of book [medium designator]. Edition. Place of publication: publisher; date. Notes.

Example:

Shlegel W, Mahr A. 3D conformal radiation therapy: multimedia introduction to methods and techniques [CD-ROM]. New York (NY): Springer Verlag; c2002. 1 CD-ROM. System Requirements: Pentium 200 or faster; 32 MB RAM, Windows 98/2000/Me/NT, operating system that supports the Joliet CD-ROM file system (long file names), screen resolution of 1024 × 768 (1280 × 1024 recommended), HTML browser that supports frames and Java script (recommended for Windows: IE 5 or higher; MPEG-1 player (recommended for Windows: Microsoft Media Player 6.4 or higher), 8x CD-ROM drive.

Journal articles in electronic format:

Author(s) of article. Title of article. Title of journal (edition) [medium designator]. Date;volume(issue): location. Notes.

Example:

Lam YWF, Shepherds AMM. Perspectives in rational management: clinical psychology I. Cyberlog (Release 1.1) [disk]. 1989;(12):[2 disks]. 2 computer disks. System Requirements: IBM PC, 192K RAM, DOS 2.0 or higher, color graphics adapter, composite or RGB Monitor.

Databases in electronic format:

Author(s). Title of database [medium designator]. Edition. Place of publication: publisher. Beginning date–ending date. Physical description. Notes.

Example:

Pharm-line: database for pharmacy practice and prescribing [CD-ROM]. London (UK): Guy's and St Thomas' Hospital. 1997–2002. 1 CD-ROM. System Requirements: Windows 3.1 or higher, QuickTime, Sound Blaster, CD-ROM player. Also available at http://www.pharm-line.nhs.uk/ by subscription.

Name–year:

Books in electronic format:

Author(s). Date. Title of book [medium designator]. Edition. Place of publication: publisher. Notes.

Example:

Shlegel W, Mahr A. c2002. 3D conformal radiation therapy: multimedia introduction to methods and techniques [CD-ROM]. New York (NY): Springer Verlag. 1 CD-ROM. System Requirements: Pentium 200 or Faster, 32 MB RAM, Windows 98/2000/Me /NT, operating system that supports the Joliet CD-ROM file system (long file names), screen resolution of 1024 × 768 (1280 × 1024 recommended), HTML browser that supports frames and Java script (recommended for Windows: IE 5 or higher, MPEG-1 player (recommended for Windows: Microsoft Media Player 6.4 or higher), 8x CD-ROM drive.

Journal articles in electronic format:

Author(s) of article. Date. Title of article. Title of journal (edition) [medium designator]. Volume(issue): location. Notes.

Example:

Lam YWF, Shepherds AMM. 1989. Perspectives in rational management: clinical psychology I. Cyberlog (Release 1.1) [disk]. (12):[2 disks]. 2 computer disks. System Requirements: IBM PC, 192K RAM, DOS 2.0 or higher, color graphics adapter, composite or RGB Monitor.

Databases in electronic format:

Author(s). Beginning date–ending date. Title of database [medium designator]. Edition. Place of publication: publisher. Notes.

Example:

Pharm-line: database for pharmacy practice and prescribing [CD-ROM]. 1997–2002. London (UK): Guy's and St Thomas' Hospital. 1 CD-ROM. System Requirements: Windows 3.1 or higher, QuickTime, Sound Blaster, CD-ROM player. Also available at http://www.pharm-line.nhs.uk/ by subscription.

The placement of the year of publication is the major difference between the systems. For a complete discussion of these systems, see Section 29.2.1 and its subsections. To conserve space, most of the examples throughout this section show only the first format. To convert to the name–year format, simply move the year to follow the author list.

29.3.7.12.1 AUTHORS AND SECONDARY AUTHORS IN ELECTRONIC FORMATS (NON-INTERNET)

List the names of authors in the order in which they appear on the opening screen(s) or in the accompanying text. Begin with the surname (family name or last name) of the first author, followed by the initial letters of the given names. Continue with the surname and initials of the second and subsequent authors. When there are 2 to 10 authors, list all of them, including the 10th author; if there are more than 10 authors, list the first 10 followed by "et al." or "and others". For details on the format of author names, such as the way names from various nationalities are handled, see Section 29.3.6.1.1. Separate the surname and initials of an author by a space; separate successive author names from each other by a comma and a space. Follow the last name in the author list with a period.

An organization such as a university, society, association, corporation, or governmental body may also serve as author. If a division or other part of an organization is included on the publication, give the parts of the name in descending hierarchical order, separated by commas. In citing organizations that are national bodies such as government agencies, if a nationality is not included with the name, place the country after the name, within parentheses, using the 2-letter ISO country code[24] (see Appendix 29.3 for a list of the codes for selected countries). See Section 29.3.6.1.2 for more information on organizations as authors.

Her Majesty's Government Communications Centre (UK).
National Cancer Institute (US), Division of Cancer Biology.

Books and other monographs in electronic format may have secondary authors, such as editors and producers. If an electronic document has secondary authors in addition to a personal or organizational author, it is optional to include the secondary author in a reference. Place the secondary author after the title of the work or any edition statement that may follow the title, followed by a comma and the role or roles of the secondary author. Give the secondary authors in the order in which they appear in the publication. Separate multiple secondary authors by commas; separate secondary authors occupying different roles by a semicolon.

Sierpina VS. Challenger integrative health care: complementary and alternative therapies for the whole person [CD-ROM]. Gooding L, editor. Memphis (TN): Challenger Corp.; 2002. 1 CD-ROM.
Lander ES, Schreiber S. Scanning life's matrix: genes, proteins, and small molecules [DVD]. Sutherland Media Productions, producer. Chevy Chase (MD): Howard Hughes Medical Institute; 2003. 2 DVDs: 240 min.

If an electronic document has no personal or organizational author but has an editor or editors, place the editor(s) at the beginning of the reference.

Hawkins RJ, Pearl ML, editors. The athlete's shoulder [DVD]. Rosemont (IL): American Academy of Orthopaedic Surgeons; c2002. 1 DVD.
Mildvan D, editor. AIDS [CD-ROM]. 2nd ed. Philadelphia (PA): Current Medicine; c1996. 1 CD-ROM.

If no primary author or editor can be determined for an electronic document, omit any indication of authorship and begin the reference with the title. The use of "anonymous" is not permitted.

29.3.7.12.2 TITLES OF BOOKS, JOURNAL ARTICLES, AND DATABASES IN ELECTRONIC FORMATS (NON-INTERNET)

In general, record the words of titles of books and other monographs, journal articles, and databases in electronic format as they appear on the opening screen(s) or in the accompanying documentation. For the 3 major exceptions to this rule, see Sections 29.3.7.2.2 and 29.3.7.1.2. Separate the title of the electronic document from the period at the end of the author list by a space. Capitalize only the first word of the title, proper nouns, proper adjectives, acronyms, and initialisms. For non-English titles, use the conventions of the particular language regarding capitalization; follow the original language with a translation if possible, surrounded by square brackets.

Fortin CA. Endométriectomie [Endometrial ablation]

Follow the title of books and databases in electronic format with a medium designator within square brackets to indicate the type of format, as "[CD-ROM]" or "[DVD]". Conclude title information with a period.

Fortin CA. Endométriectomie [Endometrial ablation] [CD-ROM].

Medium designators for journal articles in electronic format follow the journal title, not the article title.

If a book or database has been published in more than one edition, follow the medium designator with the edition statement. Editions for electronic books are often termed "version" ("ver.") or "release" (see Section 29.3.6.3 for more information on editions).

Advanced problems in cardiac arrhythmias [CD-ROM]. Ver. 4.0 for Windows. Massachusetts General Hospital, Laboratory of Computer Science, producer. Baltimore (MD): Williams & Wilkins; c1997. 1 CD-ROM. System Requirements: 80386 processor; 4 megabytes or RAM, 10 megabytes of disk space; Windows 3.1; CD-ROM drive; display adapter with 256 col. capabilities, but may be satisfactory with 16 colors.

29.3.7.12.3 TITLES OF JOURNALS IN ELECTRONIC FORMATS (NON-INTERNET)

Locate journal titles for electronic formats on the opening screen(s) or in the accompanying documentation and cite them in the same manner as print journal titles (see Section 29.3.7.1.3). Follow the abbreviated journal title with a medium designator within square brackets to indicate the type of electronic format, as "[CD-ROM]" or "[DVD]". Conclude title information with a period.

Electron J Pathol Histol [CD-ROM].

A journal title in electronic format will sometimes be produced in different editions. When this occurs, place the edition statement after the abbreviated journal title and follow it with the medium designator (see Section 29.3.7.1.3 for more information on editions in journals).

Cardiovasc Risk Factors (Madrid Ed) [CD-ROM].

29.3.7.12.4 PLACE OF PUBLICATION AND PUBLISHER IN ELECTRONIC FORMATS (NON-INTERNET)

Publication information for a book in electronic format is the same as for a standard book (see Sections 29.3.7.2.4 and 29.3.7.2.5). Long practice omits such information for journal articles.

29.3.7.12.5 DATES IN ELECTRONIC FORMATS (NON-INTERNET)

A date is a required component of a reference to an electronic publication. Always give the year of publication as found on the opening screen(s), on the face of the CD-ROM or DVD, on the jewel case (carrying case), or in the accompanying documentation. If a date includes a month, day of the month, or season, as often occurs in journal articles in electronic format, you may include them; these elements are required when citing a journal that has no volume or issue number (see Section 29.3.7.11.6). Give months in English, abbreviated to the first 3 letters, such that September becomes "Sep". For multiple years of publication, separate the first and last year by an en dash (or a hyphen if preferred), as "2012–2013". If no year of publication can be determined, use the year of copyright, preceded by "c", as "c2012".

Some electronic publications have both a year of publication and a year of copyright. Use only the year of publication unless at least 3 years separate the 2 dates. In the latter situation, use both dates, beginning with the year of publication, and separate them by a comma and a space, as "2012, c2007". If neither a year of publication nor a year of copyright can be found, use the words "date unknown" within square brackets. End date information with a period.

For books in electronic format, place the date of publication after the publisher name in the citation–sequence and citation–name systems and after the author list in the name–year system; follow it with a period in all systems. For journal articles, place the date after the medium designator in the citation–sequence and citation–name systems and after the author list in the name–year system. Follow the date with a semicolon in citation–sequence and citation–name unless no volume or issue is found (see Section 29.3.7.11.6). In the name–year system, end date information with period.

Databases have additional rules applied to them regarding dates, because most of them contain records or other entries that have been collected over time. If the database being cited is open—that is, it is continuing to be updated, provide the beginning date of the database followed by an en dash (or a hyphen if preferred), a space, and a period.

European veterinary dissertations [CD-ROM]. Bilthoven (The Netherlands): Euroscience. 2003– . 1 CD-ROM. Quarterly. System Requirements: AskSam Viewer (included on disc); CD-ROM drive.

If the database being cited is closed—that is, it is no longer being updated, provide the beginning and ending dates, separated by an en dash (or a hyphen if preferred). End the date information with a period.

Health and psychosocial instruments [CD-ROM]. Pittsburgh (PA): Behavioral Measurement Database Services. 1992–2003. 1 CD-ROM. System Requirements: IBM PC or compatible; 640 RAM; CD-ROM drive.

If no date can be determined, provide the words "date unknown" within square brackets. When the database is open, follow [date unknown] with an en dash (or a hyphen if preferred), a space, and a period.

29.3.7.12.6 VOLUMES AND ISSUES FOR JOURNAL ARTICLES IN ELECTRONIC FORMATS (NON-INTERNET)

Volume numbers must be included in references to journal articles in electronic format when they are present in the document. Write all volume numbers in arabic numerals. Convert numeric terms and roman numerals to arabic, such that "sixtieth" becomes "60" and "IX" becomes "9". Separate multiple volume numbers by an en dash (or a hyphen if preferred). See Section 29.3.7.1.5 for further details on volume information. Follow volume information with a colon unless there is also issue information.

Issue numbers must also be included in references to journal articles in electronic format when such issue numbers are present. Write all issue numbers in arabic numerals. Convert numeric terms and roman numerals to arabic, such that "fifth" becomes "5" and "III" becomes "3". Issue numbers usually follow volume numbers in a reference but may stand on their own. If the only numeration on a journal is "number x", treat this number as an issue. Separate multiple issue numbers by an en dash (or a hyphen if preferred). For further details on issues, see Section 29.3.7.1.5. Surround all issue information by parentheses, even when there is no volume information. Follow issue information with a colon.

J Venom Anim Toxins [CD-ROM]. 2003;9(1):

29.3.7.12.7 EXTENT AND LOCATION IN ELECTRONIC FORMATS (NON-INTERNET)

The extent or length of a book in electronic format is an optional component of a reference but may be included to provide useful information to the reader. Provide extent in terms of the total number of physical pieces.

2 CD-ROMs   1 DVD

Unlike extent for books in electronic formats, which is optional, the location of journal articles within an issue is a required component of a reference. If traditional page numbers are provided, enter them the same as you would for a print journal (see Section 29.3.7.1.6). When page numbers are absent, as often occurs with journal articles in electronic format, calculate the extent of the article by the best means possible and place it within square brackets, such as "[10 paragraphs]", "[about 6 screens]", and "[about 5 p.]". Of course screen size, font used, and printers vary greatly, but the purpose is to give the user of the citation an indication of the length of the item. Note that when the number is approximated, the word "about" appears before the length indicator. Conclude location information or extent with a period.

Whitehead RE, MacDonald CB, Melhem ER, McMahon L. Spontaneous labyrinthine hemorrhage in sickle cell disease. Am J Neuroradiol [CD-ROM]. 1998 Sep;19(8):1437–1440. 1 CD-ROM.
Giovanni Cannavo G, Adriana Favati A, Mule D. Mobbing: aspetti medico–legali [Mobbing: medical-legal aspects]. Tagete [CD-ROM]. 2002 Mar;8(1):[about 5 screens]. 1 CD-ROM. Italian.

29.3.7.12.8 NOTES IN ELECTRONIC FORMATS (NON-INTERNET)

"Notes" is a collective name for a variety of types of optional information about a reference that provide clarification or assist in location. Particularly useful for electronic documents are any special viewing requirements, such as the particular software or equipment needed to view the item. For databases, the date range covered would be helpful. Examples of other notes for electronic publications include mention of any printed manual or other material accompanying the item, any sponsoring organization, indication whether the item was produced for any special occasion, and a library or other archive where the item may be found. If actors, graphic artists, musicians, or other individuals associated with production of the item appear in acknowledgments, provide their names in notes if desired.

29.3.7.13 Websites and Other Online Formats

The basic rules for citing—that is, what constitutes a reference, do not differ markedly for websites from what is required for print. There is still an author or organization with responsibility for the item, a title, a place of "publication", a publisher, a date of publication, and the extent of the item (i.e., number of pages or the equivalent). Anyone preparing a citation to an online document should attempt to locate all of these elements. Simply adding a URL or other electronic address to a title is not sufficient. Websites disappear with great frequency, and users of a citation must be given some other identifying information if they are to locate it. It is true, however, that some elements are more difficult to locate when citing online items. For example, a number of poorly constructed websites do not contain dates, and authorship or publishing responsibility may be unclear or absent. It also may be difficult to discern the title from the collage of graphics presented. The person doing the citing must therefore work with the information provided.

Some elements require expansion for an online citation to provide useful information to the user. For example, the date of publication is a required element in any citation, but many online items are updated or otherwise modified several times after the date of publication. Any date of update/revision should therefore be included in a citation along with the date accessed—that is, the date (and time in some cases) when the person doing the citing saw the item online. This is necessary in the volatile online environment, where changes can be easily made and an item seen one day may not be the same in crucial ways when viewed the next day. For this reason, it is strongly recommended that, when possible, the person doing the citing produce a print or other copy for future reference.

It may be a useful construct to begin a citation to material found online by first locating all of the information needed to cite it as if it were a print document, then adding the Internet-specific items—that is, the date of any update, if known, and date accessed within square brackets after the date of publication, and the URL. For example, to cite a book found online:

Print information:
Lefebvre P. Molecular and genetic maps of the nuclear genome. Durham (NC): Duke University, Department of Biology; 2002.

With added website information:
Lefebvre P. Molecular and genetic maps of the nuclear genome. Durham (NC): Duke University, Department of Biology; 2002 [modified 2002 Dec 11; accessed 2003 Sep 5]. http://www.biology.duke.edu/chlamy_genome/nuclear_maps.html.

The general format for a reference to an online document, including punctuation, is as follows:

Citation–sequence and citation–name:

Websites and Homepages:
Title of Homepage. Edition. Place of publication: publisher; date of publication [date updated; date accessed]. Notes.

Example:
APSnet: plant pathology. St Paul (MN): American Phytopathological Association; c1994–2005 [accessed 2005 Jun 20]. http://www.apsnet.org/.

e-Books:
Author(s). Title of book. Edition. Place of publication: publisher; date of publication [date updated; date accessed]. Notes.

Example:
Griffiths AJF, Miller JH, Suzuki DT, Lewontin RC, Gelbart WM. Introduction to genetic analysis. 7th ed. New York (NY): W. H. Freeman & Co.; c2000 [accessed 2005 May 31]. http://www.ncbi.nlm.nih.gov/books/bv.fcgi?call=bv.View..ShowTOC&rid=iga. TOC.

Online Journal Articles:
Author(s) of article. Title of article. Title of journal (edition). Date of publication [date updated; date accessed];volume(issue):location. Notes.

Example:
Savage E, Ramsay M, White J, Beard S, Lawson H, Hunjan R, Brown D. Mumps outbreaks across England and Wales in 2004: observational study. BMJ. 2005 [accessed 2005 May 31];330(7500):1119–1120. http://bmj.bmjjournals.com/cgi/reprint/330/7500/1119. doi:10.1136/bmj.330.7500.1119.

Online Databases:
Title of Database. Edition. Place of publication: publisher. Beginning date–ending date [date updated; date accessed]. Notes.

Example:
IMGT/HLA Sequence Database. Release 2.9.0. Cambridge (England): European Bioinformatics Institute. 2003– [updated 2005 Jun 1; accessed 2005 Jun 22]. http://www.ebi.ac.uk/imgt/hla/.

Online Image or Infographic:
Artist's Name. Title [descriptive word]. City (ST): Publisher or Producer. [accessed date]. URL.

Example:
Jameson E. Mind on fire [image]. California: Elizabeth Jameson. [accessed 2012 Oct 23]. http://www.jamesonfineart.com/large-multi-view/Art%20of%20the%20Brain/1992765-1-171407/Art%20of%20the%20Brain.html.

Example:
Hiptype. The DNA of a successful book [infographic]. Nowsourcing. [accessed 2012 Oct 23]. http://www.hiptype.com/infographic.

Podcast or Webcast:
Narrator's Name. Title of podcast episode [descriptive word, episode number if available]. Name of podcast show. Producer. Date first aired, length. [accessed date]. URL.

Example:
Vuolo M. Our dying words [podcast, episode 16]. Lexicon Valley. Slate. 2012 July 9, 21:10 minutes. [accessed 2012 Oct 23]. http://www.slate.com/articles/podcasts/lexicon_valley/2012/07/lexicon_valley_why_should_we_care_if_a_language_goes_extinct_.html.

Example:
Curlin F. Of all the physicians is there a physician? Irony in the practice of medicine [webcast]. Library of Congress. 2012 June 28, 66 minutes. [accessed 2012 July 7]. http://www.loc.gov/today/cyberlc/feature_wdesc.php?rec=5579.

Online Video:
Title of video [descriptive word, episode number if available]. Title of program. Producer. Date first aired or posted, length. [accessed date]. URL.

Example:
How smart are animals? [video]. NOVA scienceNOW. PBS. 2011 Feb 9, 53:06 minutes. [accessed 2012 March 25]. http://video.pbs.org/video/1777525840.

Blog:
Author's name. Title of post [descriptive word]. Title of blog. [accessed date]. URL.

Example:
Fogarty M. Formatting titles on Twitter and Facebook [blog]. Grammar Girl: Quick and Dirty Tips for Better Writing. [accessed 2012 Oct 19]. http://grammar.quickanddirtytips.com/formatting-titles-on-twitter-and-facebook.aspx.

Social Networking Sites:
Username or group/page name. Network name [descriptive word for page type, post type]. Date and time posted, if available. [accessed date]. URL.

Example:
Neil deGrasse Tyson. Facebook [fan page, shared link]. 2012 Sept 10, 7:47 a.m. [2012 Oct 24]. https://www.facebook.com/neildegrassetyson.

Example:
Council of Science Editors. LinkedIn [group page, Katharine O'Moore-Klopf, ELS, user post]. 2012 Jan. [accessed 2012 Oct 24]. http://www.linkedin.com/groupItem?view=&gid=3103324&type=member&item=85278600&qid=3cd9f629-10ce-45eb-9fe4-860224d9fda2&trk=group_most_popular-0-b-ttl&goback=.gmp_3103324.

Name–year:

Websites and Homepages:
Title of Homepage. Date of publication. Edition. Place of publication: publisher; [date updated; date accessed]. Notes.

Example:

APSnet: plant pathology online. c1994–2005. St Paul (MN): American Phytopathological Association; [accessed 2005 Jun 20]. http://www.apsnet.org/.

e-Books:

Author(s). Date of publication. Title of book. Edition. Place of publication: publisher; [date updated; date accessed]. Notes.

Example:

Griffiths AJF, Miller JH, Suzuki DT, Lewontin RC, Gelbart WM. c2000. Introduction to genetic analysis. 7th ed. New York (NY): W. H. Freeman & Co.; [accessed 2005 May 31]. http://www.ncbi.nlm.nih.gov/books/bv.fcgi?call=bv.View..ShowTOC&rid=iga. TOC.

Online Journal Articles:

Author(s) of article. Date of publication. Title of article. Title of journal (edition). [date updated; date accessed];Volume(issue):location. Notes.

Example:

Savage E, Ramsay M, White J, Beard S, Lawson H, Hunjan R, Brown D. 2005. Mumps outbreaks across England and Wales in 2004: observational study. BMJ. [accessed 2005 May 31];330(7500):1119–1120. http://bmj.bmjjournals.com/cgi/reprint/330/7500/1119. doi:10.1136/bmj.330.7500.1119.

Online Databases:

Title of Database. Beginning date–ending date. Edition. Place of publication: publisher. [date updated; date accessed]. Notes.

Example:

IMGT/HLA Sequence Database. 2003– . Release 2.9.0. Cambridge (England): European Bioinformatics Institute. [updated 2005 Jun 1; accessed 2005 Jun 22]. http://www.ebi.ac.uk/imgt/hla/.

The placement of the year of publication is the major difference between the systems. For a complete discussion of these systems, see Section 29.2.1 and its subsections. To conserve space, most of the examples throughout this section show only the first format. To convert to the name–year format, simply move the year to follow the author list. For a more complete discussion of citing Internet documents, including many sample references, see the National Library of Medicine Internet Formats supplement.[23]

29.3.7.13.1 GENERAL RULES AND GUIDELINES FOR WEBSITES AND OTHER ONLINE FORMATS

A homepage is defined as the first or introductory page of a website and usually provides a table of contents or index to the contents of the site.[1] A citation to a website as a whole is made primarily from information found on its homepage. See the various sections below for instructions on citing homepages.

A journal article, a book or other type of monograph, or a database that simply resides on the Internet is cited the same way that such materials appearing in print are cited, with these exceptions:

- Include any date of update or revision and a date of citation within square brackets, along with the date of publication. Use the dates for the individual journal article, monograph, or database, not the dates of the website as a whole, unless no dates can be found for the individual item (see Section 29.3.7.13.7).

- When the extent of the work or the location within the work are absent, as often occurs, calculate them by the best means possible, for example, in terms of screens, paragraphs, or bytes (see Sections 29.3.7.13.8 and 29.3.7.13.9).
- Use the Notes element to provide the URL or other electronic address of the item (see Section 29.3.7.13.10).

Most difficulties in Internet citation arise when citing only a portion of a website that is not a journal article, book, or database (see Section 29.3.7.13.11 for instructions).

Overall guidelines for the specific components of a citation are presented in the sections that follow. The individual sections in this manual for the specific formats, for example, monographs (Section 29.3.7.2) and journals (Section 29.3.7.1), should also be consulted for further details of citation.

29.3.7.13.2 AUTHORS OF WEBSITES AND OTHER ONLINE FORMATS

Although there are monographs and journal articles on the Internet that clearly state the names of the authors, most sites—particularly homepages—do not. Do not assume that an individual named as webmaster or contact person is the author; he or she most probably is not, especially for homepages produced by large organizations. Some sites will give a name in association with a copyright statement, such as "copyright 2013 by John A. Smith". It is not safe to assume that this individual is the author, either. If the only personal name given on a site is associated with a copyright statement, use that individual's name as the publisher. Most sites will display an organization's name rather than a person's name. In such cases when the organization appears to be serving as both author and publisher, place the organization in the publisher position (see Section 29.3.7.13.6). Do not use the word "anonymous" in a citation if an author cannot be determined.[1]

> Public health response to biological and chemical weapons: WHO guidance. 2nd ed. Geneva (Switzerland): World Health Organization; 2004 [accessed 2005 May 31]. http://www.who.int /csr/delibepidemics/biochemguide/en/.

If a personal author is present, use the surname (family or last name) followed by the initials of the given names, such as "Smith JA". It is also correct to use the full first name, such as "Smith, John A." See Section 29.3.6.1 for details on handling author names, including organizations as authors.

29.3.7.13.3 TITLES OF WEBSITES AND OTHER ONLINE FORMATS

Online monographs will usually display clearly identifiable titles, and serials will have both the title of the article and the title of the journal. Homepages, on the other hand, may display only the name of the organization responsible for the site. If so, this name becomes the title. Here are some basic rules to follow for identifying wording as a title:

1) Look for what is the most prominent (usually the largest) wording on the screen.
2) Look for wording followed by a copyright, trademark, or registered trademark symbol (© or ™ or ®).
3) Look at the title bar of the web browser (generally in the top left corner).
4) Look for the title in the source code of the document.

If a title cannot be determined, construct a title by using the first series of words on the screen.

Once you have determined the title, its format depends on the type of document. For books, other types of monographs, journal article titles, and journal titles, follow the rules for print publications, capitalizing only the first word and proper nouns (or other conventions of the particular language for non-English works; see Sections 29.3.6.2, 29.3.7.2.2, 29.3.7.1.2, and 29.3.7.1.3). However, an exception is made when special characters or nonstandard typographic features are present, in which case the title should approximate the way it appears on the screen, such as the journal title "Psicologia .com". Abbreviate journal titles in the same manner that they would be for print (e.g., Biomedical Engineering Online becomes Biomed Eng Online). ISO 4[25] of the International Organization for Standardization is the source used for abbreviations (see Appendix 29.1 for a list of authoritative sources for journal title abbreviations).

For all other types of Internet material, such as homepages and databases, reproduce the title for citation purposes as closely as possible to the wording that appears on the screen, duplicating capitalization, spacing, and punctuation. This may include all capital letters or all lowercase letters, capital letters within words, and run-together words. Some examples are: netLibrary, medicinebydesign, and (Wĕbopēdia).

29.3.7.13.4 EDITIONS OF WEBSITES AND OTHER ONLINE FORMATS

Most Internet publications, with the exception of monographs and databases, will not have an edition statement. Some with print counterparts will say "online release". Other words used to express edition in the Internet world include "version" ("ver."), "level", and "update", such as "ver. 5.1" or "third update". Regardless of the particular wording, include any indication of edition in a citation. Occasionally both an edition and a version or release will be given; use both in the order found on the screen.

> Beers MH, editor. The Merck manual. 2nd home ed. Online release. Whitehouse Station (NJ): Merck Research Laboratories; c2004–2005 [accessed 2005 Jun 28]. http://www.merck.com/mmhe/index.html.

Capitalize the first word and any proper nouns in an edition statement. Abbreviate the word edition to "ed.", but do not abbreviate words like release and update, which are not as common. Convert numbers to arabic ordinals, such that "third" becomes "3rd" and "first" becomes "1st".

> Richardson ML. Approaches to differential diagnosis in musculoskeletal imaging. Ver. 2.0. Seattle (WA): University of Washington School of Medicine, Department of Radiology; c2000 [updated 2001 Oct 1; accessed 2005 May 31]. http://www.rad.washington.edu/mskbook/index.html.

29.3.7.13.5 PLACE OF PUBLICATION OF WEBSITES AND OTHER ONLINE FORMATS

Place of publication is defined as the city in which the individual or the organization issuing or sponsoring the publication resides. In the case of the Internet, the place would be the location of the Web or other site. This information is usually found at the bottom of a homepage but may also be at the top of the first screen or at the end of a document. If it is not in one of these locations, it may be obtained from a linkage within the site, usually under a "contact us" or similar link. There are 2 options if the place cannot be determined from the site itself:

1) If the city can be reasonably inferred, place the name within square brackets (e.g., Bethesda as the place of publication of a report issued by the National Cancer Institute).
2) If it is not possible to infer the city, put the words "place unknown" within square brackets.

Follow the name of the city with the 2-letter abbreviation for the US state or Canadian province in parentheses—for example, "Bethesda (MD)" or "[Bethesda (MD)]" if inferred. Follow cities in other countries with the name of the country, either written out or as the 2-letter ISO country code[24], as "Frankfurt (Germany)" or "Frankfurt (DE)" (see Appendix 29.3 for a list of the codes for selected countries). State or country information is generally omitted if the place is well known; thus, it is "New York" not "New York (NY)" and "Paris" not "Paris (FR)". For consistency within a reference list, treat each occurrence of a city the same way (e.g., always "Washington:" or always "Washington (DC):"). To avoid any ambiguity, the state, province, or country name has been included for all cities of publication in the references within this manual.

Hollenberg NK. Hypertension: mechanisms and therapy. [London (England)]: Current Medicine Group Ltd; c2005 [accessed 2005 May 31]. http://www.norvasc-braunwald.com/01/0100vpre.asp. Registration required.

29.3.7.13.6 PUBLISHERS OF WEBSITES AND OTHER ONLINE FORMATS

The advent of the Internet and other online sources has stretched the definition of "publication" and "publisher". However, in Internet terms a publisher is defined as the individual or organization that produces or sponsors the site. As with the place of publication, this information is usually found at the bottom of a homepage, at the top or on a sidebar of the first screen, or at the end of a document. The publisher may also be identified by looking for the organization named after a copyright statement (e.g., copyright 1997 by the American Chemical Society). If wording such as "this site is maintained by XYZ Corporation for ABC Organization" appears, ABC Organization is considered the publisher and XYZ Corporation the distributor. Publisher information is required in a citation; distributor information may be included as a note (see Section 29.3.7.13.10).

Give the publisher name as it appears on the screen, with whatever capitalization and punctuation is used there. If the title of a site such as a homepage and the name of the organization that sponsors it are the same, it is an option to give the name in an abbreviated form as publisher. For example, if the "University of Maryland" is the title of the homepage, it may be abbreviated to "The University" as publisher. If no publisher can be determined, use the words "publisher unknown" in the citation, within square brackets.

29.3.7.13.7 DATES ON WEBSITES AND OTHER ONLINE FORMATS

Because of the volatile nature of Internet publications, there are 3 dates of importance in citing them:

1) the date the publication was placed online or, alternatively, was copyrighted
2) the latest date any update or revision occurred
3) the date the person doing the citing accessed the publication online

The date of publication must always be included in a citation if available. The date of copyright should be included only if the date of publication is absent or it differs from the date of publication by at least 3 years (e.g., 2012, c2007). Note that a copyright date is always preceded by the letter "c".

Unfortunately, the dates of publication and the dates of any update or revision are often absent from a website, making the third date all the more important. One way to determine a date for an online document is to use the web browser to view the document page information or source code. Webmasters who do not put a date in the online display may put one in the source code, either at the top or at the bottom. Error correction or other changes to online publications may occur between scheduled or advertised updates or revisions, and these dates of update/revision may not be known. Therefore, the date the publication was actually seen online is a required part of an end reference.

Dates should be expressed in the format of "year month day". For the date of publication, an example would be "2006 Jan 3". For any dates of update or revision and the date of citation, the format may be the same, such as "[updated 1996 Feb 4; accessed 1997 Nov 4]". Various words may be found on a website to express the fact that a document has been updated or revised, such as amended, modified, or reviewed. Use whatever wording is provided by the site. Note that the dates of update or revision and citation are always placed within square brackets. As mentioned above, the date of citation must always be included in a reference.

To accommodate the name–year system of citation, date information must be split to avoid long strings of dates interrupting the text. Place the date of publication after the author list (or after the title if there is no author) and the dates of update or revision and citation following the publisher.

Citation–sequence and citation–name:

Lawrence RA. A review of the medical benefits and contraindications to breastfeeding in the United States. Arlington (VA): National Center for Education in Maternal and Child Health (US); 1997 [accessed 2005 Jun 28]. 40 p. http://www.ncemch.org/pubs/PDFs/breastfeedingTIB.pdf.

Alderson P, Green S, Higgins J, editors. Cochrane handbook for systematic reviews of interventions. Ver. 4.2.4. Oxford (UK): The Cochrane Collaboration; [modified 2005 Jun 6; accessed 2005 Jun 20]. http://www.cochrane.org/resources/handbook/.

Name–year:

Lawrence RA. 1997. A review of the medical benefits and contraindications to breastfeeding in the United States. Arlington (VA): National Center for Education in Maternal and Child Health (US); [accessed 2005 Jun 28]. http://www.ncemch.org/pubs/PDFs/breastfeedingTIB.pdf.

Alderson P, Green S, Higgins J, editors. [modified 2005 Jun 6]. Cochrane handbook for systematic reviews of interventions. Ver. 4.2.4. Oxford (UK): The Cochrane Collaboration; [accessed 2005 Jun 20]. http://www.cochrane.org/resources/handbook/.

Databases have additional rules applied to them regarding dates because most of them contain records or other entries that have been collected over time.

1) If the database being cited is open—that is, records are still being added to it, provide the beginning date of the database followed by an en dash (or a hyphen if preferred)

and a space. Place any dates of update/revision and the date of citation within square brackets. End the reference with a period and the availability statement.

Cattle Genome Database. Ver. 0.9. Brisbane (Australia): Queensland Biosciences Precinct. 1966– [revised 2002 Dec 10; accessed 2005 Jun 22]. http://www.cgd.csiro.au/.

2) If the database being cited is closed—that is, records are no longer being added, provide the beginning and ending dates, separated by an en dash (or a hyphen if preferred). Place any dates of update or revision and the date of citation within square brackets. End the reference with the URL and a period.

GDB: the Genome Database. Toronto (ON): Hospital for Sick Children. 1999–2002 [accessed 2003 Sep 5]. http://www.gdb.org/.

3) With either type of database, if the dates cannot be determined, use "date unknown" surrounded by square brackets. If the database is open, follow [date unknown] with an en dash (or a hyphen if preferred) and a space.

RGD: Rat Genome Database. Milwaukee (WI): Medical College of Wisconsin, Bioinformatics Research Center. [date unknown]– [updated 2005 Jun 20; accessed 2005 Jun 22]. http://rgd.mcw.edu/.

29.3.7.13.8 LOCATION OF WEBSITES AND OTHER ONLINE FORMATS

Location, called pagination in the print world, indicates the exact position of a document such as a journal article within a journal issue. It is also used to specify the position of a chapter in a book or of a chart or graph that is being cited. Express location in terms of a page number or number range, separating the first and last page number with an en dash (or a hyphen if preferred). For parts of journal articles and books, precede the first page number by the letter "p" and a period and a space, as "p."

Holden K, Hodzic E, Feng S, Freet KJ, Lefebvre RB, Barthold SW. Coinfection with *Anaplasma phagocytophilum* alters *Borrelia burgdorferi* population distribution in C3H/HeN mice. Infect Immun. 2005 [accessed 2005 Jun 20];73(6):3440–3444. http://iai.asm.org/cgi/content/full/73/6/3440?view=long&pmid=15908372.

Saxon LA. Sudden cardiac death: epidemiology and temporal trends. Rev Cardiovasc Med. 2005; 6 Suppl 2:S12–20. Figure 3, Schematic illustrating the complexity of multifaceted heart failure syndrome; [accessed 2005 Jun 26]; p. S15. http://www.medreviews.com/pdfs/articles/ RICM_6Suppl2_S12.pdf.

Institute for Laboratory Animal Research, Committee on Occupational Health and Safety in the Care and Use of Nonhuman Primates. Occupational health and safety in the care and use of nonhuman primates. Washington (DC): National Academies Press; c2005. Chapter 4, Identifying noninfectious hazards; [accessed 2005 Jun 20]; p. 59–67. http://www.nap.edu/books/030908914X/html.

Although traditional page numbers are often found online, the nonlinear nature of this medium has led some online journal publishers to adopt a document number scheme for journal articles, either in addition to a volume and/or issue number or as the only numeration. When a volume and/or issue number are present, place the document number in the location position. For example:

Pediatrics. 2000 Nov;106(5):e70. [e70 is the document number]

Use whatever wording for "document number" is supplied by the journal and abbreviate it according to standard practice. In the first example above, the journal has supplied no wording, simply using the "e" to indicate "electronic". As an option, an indication of the length of the article may be included after the document number in the extent position (see below). If this length is not stated within the document itself, it should be estimated by the best means possible in terms of paragraphs, screens, or pages and placed within square brackets. Because estimates of page counts and number of screens vary by font size and screen size, precede such counts by the word "about".

Pediatrics. 2000 Nov;106(5):e70. [about 2 p.].

When no volume or issue number is present, place the document number in the volume position, again using whatever wording for "document number" is supplied by the journal. Follow this number with a colon and the extent of the document.

Online J Curr Clin Trials. 1999;Doc No 134:[10 paragraphs].

If an online document displays neither page numbers nor document numbers, the extent or length of the item being cited should always be included in a citation (see Section 29.3.7.13.9).

29.3.7.13.9 EXTENT OF WEBSITES AND OTHER ONLINE FORMATS

Extent is the length of the item being cited, usually expressed by the total number of pages of a print item or the number of minutes of run time for an audiovisual. Although the extent of an item is an optional element in any citation, the length of a document provides useful information for the user. Many online monographs lack traditional page numbers and homepages are nonlinear, often having numerous hypertext links. For publications other than homepages, show extent in terms of the number of screens, lines, paragraphs, or bytes. If an online document is printed out or appears as a PDF (portable document format) document, express it in the traditional number of pages. Unless the length is supplied by the publisher, which sometimes occurs when a list of items with their size is presented to the user for assistance with downloading or when the item being cited is a PDF document, calculate the extent by the best means possible and place it within square brackets, such as "[about 5 screens]", "[10 paragraphs]", "[about 21 p.]", "[332K bytes]". Of course screen size, font used, and printers vary greatly, but the purpose is to give the user of the citation an indication of the length of an item. When the number is approximated, use the word "about" before the length indicator.

29.3.7.13.10 NOTES ON WEBSITES AND OTHER ONLINE FORMATS

The notes component of a reference usually has no specified format and is not required. Its purpose is to give the reader useful information not provided elsewhere in a citation. However, availability information is a required part of all citations to online sources. This is the location at which the document may be found, expressed in terms of an FTP or web address.

Available from: Telnet to dialog.com
http://www.nlm.nih.gov/pubs/cbm/dental_caries.html

For website addresses, the location displayed by the web browser is usually the one to use. Sometimes, however, a site found by a hyperlink may not be addressable directly. It is therefore good practice to verify the address before including it in a citation. If the web address is not directly addressable, provide instructions for navigating to it from the closest addressable site.

> http://www.projects.rodlin.ac.uk/bola/bolahome.html after clicking on the Nomenclature link.

If a URL must be broken to conserve space, break it before or after a slash or other punctuation that is already included in the URL. Do not insert a hyphen, a space, or a hard line break within a URL.

Examples of other types of information that may be included in notes are the language of the item if other than English; any special viewing requirements, such as a particular web browser, version of a browser, or software; the DOI; the name and email address of the webmaster or other contact individual; additional information about the publisher such as the street address; or information about the creation of a publication, such as creation for a particular conference or to commemorate an event.

> System Requirements: Java 6.0 or higher; Adobe Acrobat Reader 9.0; Real Player 7; color monitor; speakers.
> doi:10.1056/NEJM200505193522019

29.3.7.13.11 PARTS OF AND CONTRIBUTIONS TO WEBSITES AND OTHER ONLINE FORMATS

Most difficulties in online citation arise when citing only a portion of a website that is not a journal article, book, or database. If the portion being referenced has no author other than the author of the site itself:

1) Begin the reference by citing the homepage.
2) Give the title of the portion followed by a semicolon and space (see Section 29.3.7.13.3).
3) Provide the dates of the portion: the date of publication, if available, followed by a space and the date of citation within square brackets.
4) End date information with a semicolon and a space.
5) Enter the extent of the portion in terms of lines, screens, paragraphs (see Section 29.3.7.13.9).
6) End the reference with a period and a space, followed by the URL of the portion and, if desired, its DOI or other additional information.

> UT Southwestern Medical Center. Dallas (TX): University of Texas Southwestern Medical Center; c2005. Holographic movies show promise for medical, military applications; 2005 Jun 14 [accessed 2005 Jun 26]; [about 4 screens]. http://www8.utsouthwestern.edu/utsw/cda/dept37389/files/228328.html.

If the portion being referenced has authors (individual, organization, or corporation) apart from the author of the entire website, begin the citation with the author(s) and title of the portion. Follow the rules pertaining to citing a book or other monograph provided above.

29.3.7.14 Forthcoming Material

Forthcoming material consists of journal articles or books that have been accepted for publication but have not yet been published. Forthcoming has replaced the former "in press" because changes in the publishing industry have made the latter term obsolete.

Note that some publishers will not accept references to unpublished items in a reference list.

29.3.7.14.1 JOURNAL ARTICLES AS FORTHCOMING MATERIAL

See Section 29.3.7.1 for instructions on entering the author(s), article title, and journal title. Follow this information with the word "Forthcoming". When using the citation–name and citation–sequence systems, conclude the reference with the estimated date of publication, if known; in the name–year system, place the estimated date after the author list (see Section 29.3.7.1.4 for information on entering dates).

The general format for a reference to a forthcoming journal article, including punctuation, is as follows:

Citation–sequence and citation–name:
Author(s). Article title. Journal title. Forthcoming date.

Example:
Farley T, Galves A, Dickinson LM, Perez MJ. Stress, coping, and health: a comparison of Mexican immigrants, Mexican-Americans, and non-Hispanic whites. J Immigr Health. Forthcoming 2005 Jul.

Name–year:
Author(s). Forthcoming date. Article title. Journal title.

Example:
Farley T, Galves A, Dickinson LM, Perez MJ. Forthcoming 2005 Jul. Stress, coping, and health: a comparison of Mexican immigrants, Mexican-Americans, and non-Hispanic whites. J Immigr Health.

As an option, the affiliation of the first author of the paper may be added to facilitate retrieval in the event there may be some delay or change in final publication. Give affiliation information in hierarchical order, from lowest to highest, and surround it with parentheses. Use the 2-letter postal code for US states and Canadian provinces. For other countries, either provide the name of the country in full or use the 2-letter ISO country code[24] (see Appendix 29.3 for a list of the codes for selected countries).

Atwine B (Department of Community Medicine, University Hospital, Lund University, Malmo, Sweden), Cantor-Graae E, Bajunirwe F. Psychological distress among AIDS orphans in rural Uganda. Soc Sci Med. Forthcoming 2005 Aug.

Also as an option, for journal articles you may include the exact volume and issue number if known.

Atwine B (Department of Community Medicine, University Hospital, Lund University, Malmo, Sweden), Cantor-Graae E, Bajunirwe F. Psychological distress among AIDS orphans in rural Uganda. Soc Sci Med. 61(4). Forthcoming 2005 Aug.

29.3.7.14.2 BOOKS AS FORTHCOMING MATERIAL

See Section 29.3.7.2 for instructions on entering the author(s), title, edition, place of publication, and publisher. Follow this information with the word "Forthcoming". When using the citation–name and citation–sequence systems, conclude the reference with the estimated date of publication, if known; in the name–year system, place the estimated date after the author list (see Section 29.3.7.2.6 for information on entering dates).

The general format for a reference to a forthcoming book, including punctuation, is as follows:

Citation-sequence and citation-name:
Author(s). Title. Edition. Place of publication: publisher. Forthcoming date.

Example:
Goldstein DS. Adrenaline and the inner world: an introduction to scientific integrative medicine. Baltimore (MD): Johns Hopkins University Press. Forthcoming 2006.

Name-year:
Author(s). Forthcoming date. Title. Edition. Place of publication: publisher.

Example:
Goldstein DS. Forthcoming 2006. Adrenaline and the inner world: an introduction to scientific integrative medicine. Baltimore (MD): Johns Hopkins University Press.

As an option, the affiliation of the first author of the book may be added to facilitate retrieval in the event there may be some delay or change in final publication. Give affiliation information in hierarchical order, from lowest to highest, and surround it with parentheses. Use the 2-letter postal code for US states and Canadian provinces. For other countries, either provide the name of the country in full or use the 2-letter ISO country code[24] (see Appendix 29.3 for a list of the codes for selected countries).

Goldstein DS (Clinical Neurocardiology Section, National Institute of Neurological Disorders and Stroke, Bethesda, MD). Adrenaline and the inner world: an introduction to scientific integrative medicine. Baltimore (MD): Johns Hopkins University Press. Forthcoming 2006.

29.3.7.15 Unpublished Material

Unpublished material includes papers and poster sessions presented at meetings, manuscripts, and personal communication such as letters, email messages, and conversations. The general format is similar to that of references for other types of documents, but information on the availability of the item must be included whenever possible. Documents generally available to scholars in an archive or from a depository can usually be included in a reference list. Note, however, that many publishers do not permit placing any form of unpublished material in the end references.

29.3.7.15.1 PAPERS AND POSTER SESSIONS PRESENTED AT MEETINGS

Papers and poster sessions presented at meetings include both items that were presented but never published and items for which any subsequent publication is unknown. If subsequent publication is known, cite the published form rather than the item presented.

The general format for a reference to a paper presented at a meeting, including punctuation, is as follows:

Citation-sequence and citation-name:
Author(s). Title of paper. Paper presented at: Title of conference. Number and name of the conference; date of the conference; place of the conference.

Example:
Antani S, Long LR, Thoma GR, Lee DJ. Anatomical shape representation in spine x-ray images. Paper presented at: VIIP 2003. Proceedings of the 3rd IASTED International Conference on Visualization, Imaging and Image Processing; 2003 Sep 8–10; Benalmadena, Spain.

Charles L, Gordner R. Analysis of MedlinePlus en Español customer service requests. Poster session presented at: Futuro magnifico! Celebrating our diversity. MLA '05: Medical Library Association Annual Meeting; 2005 May 14–19; San Antonio, TX.

Name–year:
Author(s). Date of the conference. Title of paper. Paper presented at: Title of conference. Number and name of the conference; place of the conference.

Example:
Antani S, Long LR, Thoma GR, Lee DJ. 2003. Anatomical shape representation in spine x-ray images. Paper presented at: VIIP 2003. Proceedings of the 3rd IASTED International Conference on Visualization, Imaging and Image Processing; Benalmadena, Spain.
Charles L, Gordner R. 2005. Analysis of MedlinePlus en Español customer service requests. Poster session presented at: Futuro magnifico! Celebrating our diversity. MLA '05: Medical Library Association Annual Meeting; San Antonio, TX.

Begin the reference with information on the author(s) and title of the paper or poster session, following the same format as that used for journal articles (see Sections 29.3.7.1.1 and 29.3.7.1.2). Use the phrases "Paper presented at:" and "Poster session presented at:" to connect author and title information with information on the meeting. See Section 29.3.7.3 for details of the format for meeting information.

29.3.7.15.2 MANUSCRIPTS OTHER THAN JOURNAL ARTICLES

Manuscripts are unpublished books or other documents and may be either handwritten or typed. For inclusion in end references, cite only documents that reside in a public archive such as a library or other repository that permits public access. Both collections of documents and individual manuscripts may be cited. See Section 29.3.7.15.3 for information on citing letters and other personal communication not contained in a library or archive.

The general format for a reference to a manuscript collection, including punctuation, is as follows:

Citation–sequence and citation–name:
Author(s). Title. Date. Physical description. Notes.

Example:
Axelrod J. The Julius Axelrod papers. 1915–1998. 22 boxes. Located at: Modern Manuscripts Collection, History of Medicine Division, National Library of Medicine, Bethesda, MD; MS C 494.

Name–year:
Author(s). Date. Title. Physical description. Notes.

Example:
Axelrod J. 1915–1998. The Julius Axelrod papers. 22 boxes. Located at: Modern Manuscripts Collection, History of Medicine Division, National Library of Medicine, Bethesda, MD; MS C 494.

The general format for a reference to an individual manuscript, including punctuation, is as follows:

Citation–sequence and citation–name:
Author(s). Title. Date. Physical description. Notes.

Example:
Stearns AA. Armory Square Hospital nursing diary. 1864. 70 leaves. Located at: History of Medicine Division, National Library of Medicine, Bethesda, MD; MS B 372.

Name–year:
Author(s). Date. Title. Physical description. Notes.

Example:
Stearns AA. 1864. Armory Square Hospital nursing diary. 70 leaves. Located at: History of Medicine Division, National Library of Medicine, Bethesda, MD; MS B 372.

Authors of manuscripts: Begin with the surname (family name or last name) of the author, followed by the initial letters of the given names. For details on the format of author names, such as the way names from various nationalities are handled, see Section 29.3.6.1.1.

If a manuscript has no discernible author, begin the reference with the title. The use of "anonymous" is not permitted.[25] When the name–year system of citation is being used, begin the in-text reference with the first word or first few words of the title, followed by an ellipsis. Use only as many words of the title as are needed to distinguish it from other titles being used as references.

Titles of manuscripts: In general, record the words of a manuscript title exactly as they appear in the publication. Separate the title from the period at the end of the author list by a space. Capitalize only the first word of the title, proper nouns, proper adjectives, acronyms, and initialisms. If the title begins with a Greek letter, chemical formula, or another special character that might lose its meaning if capitalized, retain the lowercase. For non-English titles, follow the conventions of the particular language regarding capitalization. Whenever possible, place an English translation within square brackets after the original title.

If a manuscript does not appear to have a formal title, simply construct a title from the first few words of the text, using enough words to make the constructed title meaningful, and place them within square brackets. By convention, manuscript collections have constructed titles, such as "Collected papers of . . ." or "Papers and collection of . . ." Do not place manuscript collection titles within square brackets. End titles with a period.

Dates of manuscripts: For a manuscript collection, use the first and last years of the items in the collection, separated by an en dash (or a hyphen if preferred), as "1958–2003". For an individual manuscript, use either the single year or span of years provided. If there is no date on the manuscript or dates on a manuscript collection, but a date or dates can be estimated, place the year or years within square brackets, as "[1858]" or "[1920–1930]". If no year can be determined, use "date unknown" within square brackets.

Physical description of manuscripts: Because manuscripts and manuscript collections often require special arrangements with the library or other archive where they are housed to view them, it is highly recommended that the extent of the item be included in the reference. Describe manuscript collections in terms of the number of items within the collection, the numbers of boxes housing the collection, and the number of linear feet of shelf space occupied.

1,500 items: 10 boxes. 200 items: 3.5 feet. 5 boxes.

Describe individual manuscripts that are unpaginated in terms of the number of physical pages and express them in terms of leaves, as "5 leaves". If a manuscript is paginated, provide the total numbered pages, as "15 p."

Notes for manuscripts: "Notes" is a collective term for information about a publication that provides clarification or assists in location. Always include in notes the location where the manuscript or manuscript collection may be found. Begin location information with the words "Located at:". Include the department and name of the institution, the city of the institution, the US state or Canadian province, and the country if other than the United States or Canada. Either write out the name of the country or use the 2-letter ISO country code[24] (see Appendix 29.3 for a list of the codes for selected countries). Follow address information with the library call number or other finding aid of the institution. Use commas to separate segments of the address and a semicolon to separate institution information from the library call number or other finding number of the institution. End location information with a period.

Located at: Manuscript Division, Library of Congress, Washington, DC; MSS75867.
Located at: Edinburgh University Library, Edinburgh, Scotland; MS D.c 2-392.

Types of additional optional information that may be included in notes are the language of publication if other than English, further description of the material such as its history, and any restrictions on the use of the material by the library or other archive.

29.3.7.15.3 PERSONAL COMMUNICATIONS

CSE recommends placing references to personal communications such as letters and conversations within the running text, not as formal end references. The nature and source of the cited information should be identified by an appropriate statement. Place the source information within parentheses, using a term or terms to indicate clearly that the citation is not represented in the reference list.

. . . and most of these meningiomas proved to be inoperable (2003 letter from RS Grant to me; unreferenced, see "Notes") while a few were not.

The author must provide written permission to the publisher from the cited person (if living) or from the cited organization if it is carried in a document such as an internal memorandum that is not accessible to scholars. The permission should be acknowledged in an "Acknowledgments" or a "Notes" section that follows the text of an article or is placed at the end of a book's main text; such statements may include additional details, such as the reason for the communication.

APPENDIX 29.1 AUTHORITATIVE SOURCES FOR JOURNAL TITLE ABBREVIATIONS

ISO 4[a] of the International Organization for Standardization is the recognized authority for abbreviating words in journal titles. The database for this standard, maintained by the ISSN International Centre in Paris (http://www.issn.org), is available by subscription.

The source list below provides a variety of alternative sources for journal title abbreviations. Each is considered authoritative within its specific area or discipline. Be aware, however, that although many of them use the ISO standard as their basis, some do not. For example, ISO does not permit abbreviations of the romanized words from titles appearing in character-based language, (e.g., Chinese, Japanese, or Korean), but some of these sources do permit them.

Titles of journals not represented in the Source List may be abbreviated using the Rules for Abbreviating Journal Titles that follow the Source List.

Source List

AIP style manual. Appendix B: Journal title abbreviations. New York (NY): American Institute of Physics; c1990–1997. [accessed 2012 Jan 18]. p. 46–54. http://www.aip.org/pubservs/style/4thed/AIP_Style_4thed.pdf.

Alkire LG, editor. Periodical title abbreviations. 16th ed. Detroit (MI): Thompson Gale; c2006. 2 vol. Vol. 1, By abbreviation; vol. 2, By title.

American Mathematical Society. Abbreviations of names of serials. Providence (RI): The Society; c2005. [accessed 2012 Jan 18]. http://www.ams.org/msnhtml/serials.pdf.

A Brief Guide for Authors. Appendix C, Standard journal abbreviations. Boston (MA): American Meteorological Society; c2010. [accessed 2012 Jan 18]. http://www.ametsoc.org/PUBS/Authorsguide/pdf_vs/authguide.pdf.

BioABACUS: biotechnology abbreviation & acronym uncovering service. Las Cruces (NM): New Mexico State University; 2000 [accessed 2012 Nov 8]. http://www.acronymfinder.com/Biotechnology-Abbreviation-and-Acronym-Uncovering-Service-(BioABACUS).html.

BIOSIS serial sources. Philadelphia (PA): Thomson Reuters; 2012.

CAS: a division of the American Chemical Society. CAplus core journal coverage list. Columbus (OH): Chemical Abstracts Service; c2013. [updated 2013 Jun; accessed 2013 Nov 19]. http://www.cas.org/content/references/corejournals.

Coghill AM, Garson LR, editors. The ACS style guide: effective communication of scientific information. 3rd ed. Washington (DC): American Chemical Society; 2006.

CSA. CSA serials title lists. Ann Arbor (MI): ProQuest; c2013 [accessed 2013 Nov 19]. http://www.proquest.com/en-US/products/titlelists/tl-csa.shtml.

EMBASE list of journals indexed. St Louis (MO): Elsevier; 2006.

Frontiers in Bioscience. Abbreviated journal names. Irvine (CA): Frontiers in Bioscience; c2013 [accessed 2013 Nov 19]. http://www.bioscience.org/references.

Genamics JournalSeek: a searchable database of online scholarly journals. Dublin (OH): OCLC. [date unknown]– [accessed 2013 Nov 19]. http://journalseek.net/index.htm.

HCMR Library. List of journal titles abbreviations. Anavyssos and Heraklion (Greece): Hellenic Centre for Marine Research (GR), Library Services; [accessed 2013 Nov 13]. http://www.hcmr.gr/gr/listview3.php?id=703.

INSPEC: list of journals indexed. London (UK): The Institution of Engineering and Technology; 2004 [updated 2011 Nov 17; accessed 2013 Nov 19]. http://ovidsupport.custhelp.com/app/answers/detail/a_id/1658.

ISI journal abbreviations index. Pasadena (CA): Caltech Library System; c2013 [accessed 2013 Nov 19]. http://library.caltech.edu/reference/abbreviations.

Journal Citation Reports. Philadelphia (PA): Thomson Scientific Inc.; c2011 [accessed 2013 Nov 19]. http://thomsonreuters.com/products/ip-science/04_031/jcrwebfs.pdf.

Journals indexed in AGRICOLA. Beltsville (MD): National Agricultural Library (US). 1992– [modified 2013 Nov 14; accessed 2013 Nov 19]. http://www.nal.usda.gov/nal-catalog/journals-indexed-agricola.jia.

Journal titles and abbreviations (BIOSIS format). Gainesville (FL): University of Florida, Institute of Food and Agricultural Sciences, Entomology and Nemotology Department; [revised 1998 May; accessed 2013 Nov 19]. http://entnem.ifas.ufl.edu/choate/insectclass/all_journals.htm.

NLM catalog: journals referenced in the NCBI databases. Bethesda (MD): National Center for Biotechnology Information, National Library of Medicine (US); 2013 [accessed 2013 Nov 19]. http://www.ncbi.nlm.nih.gov/nlmcatalog/journals.

Rules for Abbreviating Journal Titles

The rules provided below have been adapted and updated from a CBE Views article[b] discussing the abbreviation rules of the NLM.

1) The title to be abbreviated is the title proper: the name of the journal in its fullest form. The names of sponsoring organizations are excluded unless the name is syntactically connected to a generic term for a serial publication, such as "Annals", "Journal", "Bulletin", "Archives".

   Bulletin of the University of Nebraska State Museum. *becomes* Bull Univ Nebr State Mus.

   If the full title is preceded by an abbreviated form serving as an acronym, only the acronym is selected. If a full title is followed by an acronym, only the full title is selected for abbreviation.

   JOP: Journal of the Pancreas. *becomes* JOP.
   The American Journal of Bioethics: AJOB. *becomes* Am J Bioeth.

2) The initial letter of each abbreviated word is capitalized; the abbreviated title ends with a period.
3) Single word titles and titles in character-based languages, such as Chinese, Japanese, and Korean, are not abbreviated.

   Science. *remains* Science.
   Virology. *remains* Virology.
   Kansenshogaku Zasshi. *remains* Kansenshogaku Zasshi.

4) Single-syllable words and words of 5 or fewer letters (in singular form) are not usually abbreviated.

   Blood Cells. *remains* Blood Cells.

   However, there are some exceptions. For example, "Human" is abbreviated "Hum".
5) Abbreviation is preferred by truncation: at least the 2 final letters of a word are dropped.

   Geological Survey of Finland Bulletin. *becomes* Geol Surv Finl Bull.
   Journal of Heterocyclic Chemistry. *becomes* J Heterocycl Chem.

6) Cognates and variant forms with the same stem in the same, or a related language, are represented by the same abbreviation.

   Entomologische Abhandlungen. *becomes* Entomol Abh.
   Entomologica Americana. *becomes* Entomol Am.
   Entomologist's Gazette. *becomes* Entomol Gaz.
   Entomologie et Phytopathologie Appliquees. *becomes* Entomol Phytopathol Appl.

7) Commonly used words with the same meaning are abbreviated in the same manner across languages.

   Journal *becomes* J
   Zeitschrift *becomes* Z

8) Words may also be abbreviated by contraction (omission of internal letters).

   Zeitung *becomes* Ztg
   Country *becomes* Ctry

9) Articles, conjunctions, and prepositions are dropped unless they are part of a personal or place name, are in a scientific or technical term, or are part of a standard phrase.

   La Pediatria Medica e Chirurgica. *becomes* Pediatr Med Chir.
   In Vitro Cellular and Developmental Biology. *becomes* In Vitro Cell Dev Biol.

10) All punctuation within titles is omitted, including apostrophes, ampersands, commas, dashes, and hyphens.

   Biomaterials, Artificial Cells, and Artificial Organs. *becomes* Biomater Artif Cells Artif Organs.
   Epilepsy & Behavior. *becomes* Epilepsy Behav.
   Women's Health. *becomes* Womens Health.
   L'Orthodontie Francaise. *becomes* Orthod Fr.

Hyphenated terms are treated as 2 words if they can stand alone or are collapsed to 1 word if they cannot.

Suicide & Life-Threatening Behavior. *becomes* Suicide Life Threat Behav.
Journal of Neuro-Oncology. *becomes* J Neurooncol.

11) All diacritical marks are omitted.

[a] Based on International Organization for Standardization. Information and documentation - Rules for the abbreviation of title words and titles of publications. Geneva (Switzerland): The Organization; 1997. (ISO 4: 1997). Updated online by subscription only.

[b] Based on Arenales D, Sinn S. How to amputate; rules for journal title abbreviations. CBE Views. 1989;12(6):106–108.

APPENDIX 29.2 COMMON ENGLISH WORDS USED IN BIBLIOGRAPHIC DESCRIPTION AND THEIR ABBREVIATIONS[a]

abridged = abr.
abstract = abstr.
academy = acad.
adaptation = adapt.
American = Amer.
and others = et al. or and others
annotation = annot.
annual = annu.
association = assoc.
augmented = augm.
authorized = authoriz.

biannual = biannu.
bibliography = bibliogr.
bimonthly = bimonth.
biography = biogr.
black and white = black & white
brochure = broch.
bulletin = bull.

catalog = cat.
centimeters = cm.
chapter = chap.
color = n.a.[b]
column = col.
commission = commiss.
company = co.
compiler = comp.
conference = conf.
continued = cont.
corporation = corp.

department = dept.
diagram = diagr.
dictionary = dict.
director = dir.
directory = dir.
dissertation = diss.
distribution = dist.
division = div.
document = doc.

edition = ed.
encyclopedia = encycl.
English = Engl.
enlarged = enl.
European = Europ.
executive = exec.
explanation = expl.
extract = extr.

facsimile = facs.
faculty = fac.
figure = fig.
foundation = found.
frontispiece = front.

gazette = gaz.
government = gov.

handbook = handb.

illustration = ill.
illustrator = ill.
impression = impr.
inch = in.
inches = in.
inches per second = ips
inclusive = incl.
incomplete = incompl.
index = ind.
information = inform.
institute = inst.
international = intern.
introduction = introd.
invariable = invar.

laboratory = lab.
library = libr.
literature = lit.

manual = man.
manuscript = ms.
meeting = meet.
microfiche = mfiche.
microfilm = mf.
millimeter = mm.
miscellaneous = misc.
modified = mod.
monograph = monogr.
monthly = month.

national = nat.
new series = n.s.
newspaper = newsp.
notice = not.
number = no.

observation = observ.
original = orig.

page = p.
pamphlet = pamph.
paperback = pbk.
part = pt.
periodical = period.
photography = phot.
picture = pict.
portrait = portr.
posthumous = posth.
preface = pref.
preliminary = prelim.
preparation = prep
preprint = prepr.
printed = print.
proceedings = proc.
professor = prof.
program = progr.
pseudonym = pseud.
publication = publ.
publisher = publ.

quarterly = quart.

reference = ref.
reprint = repr.
reproduction = reprod.
responsible = resp.
revised = rev.
revolutions per minute = rpm.

scientific = sci.
section = sect.
separate = sep.
series = ser.
session = sess.
silent = n.a.
society = soc.
sound = n.a.
special = spec.
successor = success.
summary = summ.
supplement = suppl.
symposium = symp.

table = tab.
times (in physical desc.) = x
translation = transl.
translator = transl.
transliteration = translit.

university = univ.

version = ver.
volume = vol.

year = y.
yearbook = yb.

[a] Based on International Organization for Standardization. Information and documentation - Bibliographic description and references - rules for the abbreviation of bibliographic terms. 2nd ed. Geneva (Switzerland): The Organization; 1994. (ISO 832: 1994).

[b] n.a. = not abbreviated.

APPENDIX 29.3 ISO COUNTRY CODES FOR SELECTED COUNTRIES[a]

Afghanistan	AF	Lesotho	LS
Argentina	AR	Liberia	LR
Australia	AU	Libya	LY
Austria	AT	Liechtenstein	LI
Belgium	BE	Lithuania	LT
Bolivia	BO	Mexico	MX
Brazil	BR	Morocco	MA
Bulgaria	BG	Netherlands	NL
Chile	CL	New Zealand	NZ
China	CN	Nicaragua	NI
Czech Republic	CZ	Nigeria	NG
Denmark	DK	Norway	NO
Dominican Republic	DO	Paraguay	PY
Ecuador	EC	Peru	PE
Egypt	EG	Philippines	PH
El Salvador	SV	Poland	PL
Estonia	EE	Portugal	PT
Ethiopia	ET	Romania	RO
Finland	FI	Russian Federation	RU
France	FR	Rwanda	RW
Georgia	GE	Saudi Arabia	SA
Germany	DE	Serbia	RS
Greece	GR	Singapore	SG
Grenada	GD	Slovakia	SK
Greenland	GL	Slovenia	SI
Guatemala	GT	South Africa	ZA
Honduras	HN	Spain	ES
Hong Kong	HK	Sudan	SD
Hungary	HU	Swaziland	SZ
Iceland	IS	Sweden	SE
India	IN	Switzerland	CH
Indonesia	ID	Taiwan	TW
Iran	IR	Thailand	TH
Iraq	IQ	Tunisia	TN
Ireland	IE	Turkey	TR
Israel	IL	Uganda	UG
Italy	IT	Ukraine	UA
Jamaica	JM	United Arab Emirates	AE
Japan	JP	United Kingdom	GB
Kenya	KE	United States	US
Korea, Democratic People's Republic	KP	Uruguay	UY
Korea, Republic	KR	Venezuela	VE
Latvia	LV	Zambia	ZM
Lebanon	LB	Zimbabwe	ZW

[a] Adapted from International Organization for Standardization. Country codes. Geneva (Switzerland): The Organization; 1997 [updated 2012 Aug 2]. http://www.iso.org/iso/country_names_and_code_elements. (ISO 3166-1-alpha-2: 1997).

CITED REFERENCES

1. National Information Standards Organization (US). Bibliographic references. Bethesda (MD): NISO Press; 2005. (ANSI/NISO Z39.29-2005).

2. Council of Biology Editors, Style Manual Committee. Scientific style and format: the CBE manual for authors, editors, and publishers. 6th ed. New York (NY): Cambridge University Press; 1994.

3. Patrias K. National Library of Medicine recommended formats for bibliographic citation. Bethesda (MD): The Library; 1991 [accessed 2013 Feb 1]. http://www.nlm.nih.gov/pubs/formats/recommendedformats.pdf. Second edition titled Citing Medicine available in 2006 from the NCBI Bookshelf at http://www.ncbi.nlm.nih.gov/entrez/query.fcgi?db=Books.

4. International Organization for Standardization. Documentation - Bibliographic references - content, form and structure. 2nd ed. Geneva (Switzerland): The Organization; 1987. (ISO 690: 1987).

5. International Organization for Standardization. Information and documentation - Bibliographic references. Part 2: Electronic documents or parts thereof. Geneva (Switzerland): The Organization; 1997. (ISO 690-2: 1997).

6. International Organization for Standardization. Information and documentation - Transliteration of Arabic characters into Latin characters. Geneva (Switzerland): The Organization; 1984. (ISO 233: 1984).

7. International Organization for Standardization. Information and documentation - Transliteration of Arabic characters into Latin characters. Part 2: Arabic language - simplified transliteration. Geneva (Switzerland): The Organization; 1993. (ISO 233-2: 1993).

8. International Organization for Standardization. Information and documentation - Transliteration of Armenian characters into Latin characters. Geneva (Switzerland): The Organization; 1996. (ISO 9985: 1996).

9. International Organization for Standardization. Information and documentation - Transliteration of Cyrillic characters into Latin characters - Slavic and non-Slavic languages. Geneva (Switzerland): The Organization; 1995. (ISO 9: 1995).

10. International Organization for Standardization. Information and documentation - Transliteration of Devanagari and related Indic scripts into Latin characters. Geneva (Switzerland): The Organization; 2001. (ISO 15919: 2001).

11. International Organization for Standardization. Information and documentation - Transliteration of Georgian characters into Latin characters. Geneva (Switzerland): The Organization; 1996. (ISO 9984: 1996).

12. International Organization for Standardization. Information and documentation - Transliteration of Hebrew characters into Latin characters. Geneva (Switzerland): The Organization; 1984. (ISO 259: 1984).

13. International Organization for Standardization. Information and documentation - Transliteration of Hebrew characters into Latin characters. Part 2: Simplified transliteration. Geneva (Switzerland): The Organization; 1994. (ISO 259-2: 1994).

14. International Organization for Standardization. Information and documentation - Transliteration of Korean script into Latin characters. Geneva (Switzerland): The Organization; 1996. (ISO/TR 11941: 1996).

15. International Organization for Standardization. Information and documentation - Transliteration of Thai. Geneva (Switzerland): The Organization; 1998. (ISO 11940: 1998).

16. Library of Congress (US), Cataloging Policy and Support Office. ALA-LC romanization tables. Washington (DC): The Library; 1997 [revised 2012; updated 2013 Jan 8; accessed 2013 Feb 1]. http://www.loc.gov/catdir/cpso/roman.html.

17. Chernin E. The "Harvard system": a mystery dispelled. BMJ. 1988;297(6655):1062–1063.

18. Iverson C, Flanagin A, Fontanarosa PB, Glass RM, Glitman P, Lantz JC, Meyer HS, Smith JM, Winker MA, Young RK. American Medical Association manual of style: a guide for authors and editors. 9th ed. Baltimore (MD): Williams & Wilkins; 1997.

19. Dodd JS, editor. The ACS style guide: a manual for authors and editors. 2nd ed. Washington (DC): American Chemical Society; 1997.

20. Publication manual of the American Psychological Association. 5th ed. Washington (DC): The Association; c2001.

21. The Chicago manual of style: the essential guide for authors, editors, and publishers. 16th ed. Chicago (IL): University of Chicago Press; 2010. Also available at http://www.chicagomanualofstyle.org.

22. Turabian KL. A manual for writers of term papers, theses, and dissertations. 7th ed. rev. Chicago (IL): University of Chicago Press; 2007.

23. Patrias K. National Library of Medicine recommended formats for bibliographic citation. Supplement: Internet formats. Bethesda (MD): The Library; 2001 Jul [accessed 2013 Feb 1]. http://www.nlm.nih.gov/pubs/formats/internet.pdf. Updated edition titled Citing Medicine available in 2007 from the NCBI Bookshelf at http://www.nlm.nih.gov/pubs/formats/recommendedformats.html.

24. International Organization for Standardization. Codes for the representation of names of countries and their subdivisions. Part 1: country codes. Geneva (Switzerland): The Organization; 1997 [updated 2012 Aug 2]. (ISO 3166-1-alpha-2: 1997). http://www.iso.org/iso/country_names_and_code_elements.

25. International Organization for Standardization. Information and documentation - Rules for the abbreviation of title words and titles of publications. Geneva (Switzerland): The Organization; 1997. (ISO 4: 1997). Updated online by subscription only.

26. International Organization for Standardization. Information and documentation - Bibliographic description and references - rules for the abbreviation of bibliographic terms. 2nd ed. Geneva (Switzerland): The Organization; 1994. (ISO 832: 1994).

27. NLM Catalog: Journals referenced in the NCBI Databases. Bethesda (MD): National Library of Medicine (US); 2005. Also available at http://www.ncbi.nlm.nih.gov/nlmcatalog/journals.

28. BIOSIS serial sources. Philadelphia (PA): Thomson; 2005.

29. Columbia Law Review, Harvard Law Review, University of Pennsylvania Law review, Yale Law Review, compilers. The bluebook: a uniform system of citation. 17th ed. Cambridge (MA): Harvard Law Review Association; 2000.

30 Accessories to Text: Tables, Figures, and Indexes

30.1 TABLES

In scientific publications, data need to be presented as accurately, as completely, as clearly, and as concisely as the accounts of how they were collected, analyzed, and interpreted. Thus, tables and graphs are as important as text and should be prepared with the same care and attention. As in writing, a good table, graph, and image should help readers understand, find, remember, and use information as quickly and as easily as possible.

Tables are to be distinguished from a series in running text and from short lists and tabulations, which have a heading and only 1 or 2 columns without a rule, number, or title. Lists and tabulations are embedded in the text and so rely on the preceding text to explain their meaning. In contrast, a table has a number, title, and both row and column headings, and it is set off from the text close to where it is first cited.

[A series in running text] The drug had unexpected side effects in several patients: nausea, vomiting, headache, fever, fatigue, and body aches. Treatment consisted of . . .

or

[A list] The drug had unexpected side effects in several patients:

Nausea
Vomiting
Headache
Fever
Fatigue
Body aches

Treatment consisted of . . .

or

[A tabulation] The drug had unexpected side effects in several patients:

Nausea 12%
Vomiting 10%
Headache 10%
Fever 6%
Fatigue 4%
Body aches 3%

Treatment consisted of . . .

Preparing effective tables often calls for creativity rather than rigidly applying rules. The following guidelines may be helpful, but they may not apply in every case. See Lang[1] for the principles of table construction.

1) Ensure that the table, including its title, headnotes, and footnotes, is complete enough to be understood without any (or undue) reference to the text.
2) Make the table orderly, logical, and as simple as possible.
3) Ensure that the units, symbols, and data of the table are consistent with those in the text.
4) Ensure that all the data in each row and column are consistent with the row and column headings.
5) Ensure that tables containing similar types of information have parallel formats.
6) Avoid duplicating tabular data in figures or text.
7) Do not use table parts (e.g., Table 1a, Table 1b).
8) Prepare all tables using the table function in a word-processing program. Put each entry into a unique cell; do not position the entries with line breaks and tab stops, which are easily lost in file transfers and conversions.

30.1.1 Parts of a Table

Tables have up to 9 parts: 1) a number, 2) a title (not a caption), 3) a headnote, 4) column headings, 5) row headings, 6) the data field (the individual data cells), 7) expanded abbreviations, 8) footnotes, and 9) a credit or source line. All tables should always have a title, column and row headings, and a data field.

30.1.1.1 Table Number

If a document has only one table, the table is not labeled or numbered and is cited in the text as "Table". This practice is necessary to avoid confusion, because citing a Table 1 implies that there is a Table 2.

If a document has 2 or more tables, label them with the word "Table" and number them all with consecutive arabic numerals in order of mention in the text. For long documents, reports, or books that are divided into sections or chapters, tables may be found more easily by numbering the section or chapter independently from the table and separating the numbers with a period: for example: Table 1.1, Table 1.2, and Table 3.4. Do not use table parts (e.g., Table 1a, Table 1b).

30.1.1.2 Table Title

The table title (and any headnotes or footnotes to the title) must be clear and sufficient to allow the reader to understand the fundamental aspects of the table without any (or undue) reference to the text. Abbreviations may be used in the title (see Section 30.1.3), although units of measure are best introduced in column (see Section 30.1.1.4) or row headings (see Section 30.1.1.5), unless the unit applies to every cell in the data field.

Begin the title one space after the table number.

The title consists of a single phrase in sentence case without a closing period. The title should identify the data in the data field (see Section 30.1.1.6) but without simply repeating the column and row headings. The title may include other information as well, such as 1) interventions and outcomes; 2) the source of the data (e.g., the subjects or species, the chemical or physical elements represented); 3) dates and other key details of data collection, provided these apply to all data presented in the table; and 4) sample size(s). Alternatively, this information may be placed in headnotes, footnotes, or headings.

Table 12 Values of water quality variables for samples from Lake Weir, Florida, taken between June and August, 2001

not

Table 12 Mean dissolved oxygen, pH, hardness, chlorophyll a, and temperature for 16 samples from Lake Weir, Florida, June through August 2001

In large documents with a series of similar tables, the titles should allow the reader to easily distinguish among individual tables. Related tables can be differentiated by generic initial phrasing in titles followed by a colon and subtitle: a phrase specific for the particular table placed after a colon.

Table 1 Infectious diseases in China: incidence by socioeconomic class
Table 2 Infectious diseases in China: incidence by region
Table 3 Infectious diseases in Japan: incidence by socioeconomic class
Table 4 Infectious diseases in Japan: incidence by region

In some disciplines, especially in basic science, a table title and its accompanying headnote can include a summary of the experimental conditions under which the data were collected and sometimes the interpretation of these data. Including this information can appropriately create a long title containing paragraphs of information. Information presented in the title need not be duplicated in the text.

30.1.1.3 Headnotes

A headnote consists of one or more sentences containing supplemental information that follows the title and that generally requires more emphasis than footnotes, which appear below the table. Headnotes are useful for expanding or qualifying the information in the title, such as indicating how or when the data were collected.

Table 5 Number of prescriptions written by house staff between January 1 and June 30, 2010, by medical specialty. Data are from 23 clinics responding to the 2011 survey of area health care providers with more than 5 physicians.

30.1.1.4 Column Headings

Column headings (also called "heads" or "boxheads") identify the data in the columns of the table. Each column of a table, including the column on the far left that contains the row heads, must have a heading. Headings that span two or more columns are called "spanner headings".

Numbers are usually more easily compared if they are presented side-by-side. Hence, the independent (explanatory) variables (e.g., group names) are often more usefully presented in the column headings and the dependent (response) variables in the row headings. Other factors may affect how a table is organized, however (see Section 30.1.3).

Use a word or short phrase with sentence-style capitalization for each column heading. A common heading format is to name the variable, give a descriptive statistic, and give the units of measurement. The sample size is also often given or repeated in column (or row) headings.

Television watching, mean (SD), h/d
Cancer diagnosis, n (%)
Bone mineral density, range, g/cm^2

Litter #1 (n=12)
Live births (n=34)

Give groups descriptive labels, such as low-dose and high-dose, not Group 1 and Group 2. Letters and numbers become arbitrary "codes" that readers have to learn and remember and so make the table (and the text) harder to follow.

Avoid a column heading consisting solely of a unit designation (such as n or %) unless the designation is identified in a spanner heading. It is generally better to provide additional descriptive information: "Percent positive" or "Positive, %" or "Mass, g". When symbols for variables are used alone as headings, the meaning of the symbol must be immediately clear from the table title and any accompanying text. In particular, avoid using "*n*" or "*N*" alone; instead, use an explicit heading, such as "Number of subjects" or "Patients, *n*", or explain the meaning in a footnote.

To conserve space, use abbreviations, symbols, and other short forms in column headings. Define these elements in a footnote if necessary. If the same abbreviations and symbols recur in later tables, they may or may not need to be redefined in subsequent tables, depending on the nature of the topic.

To save space and organize complex material, the common elements of 2 or more

Table *z* Table title (short, with no closing punctuation)[a]

	Spanner head 1[b]			Spanner head 2		
Column heading for stub	Column heading	Column heading	Column heading[c]	Column heading	Column heading	Column heading
Stub heading						
Row heading	x.x	x.x	x.x	x.x	x.x	x.x
Row heading	xxx	xxx	xxx	xxx	xxx	xxx
Row heading	x.x	x.x	x.x	x.x	x.x	x.x
Stub heading						
Row heading	x.x	x.x[d]	x.x	x.x	x.x	x.x
Row heading	xxx	xxx	xxx	xxx	xxx	xxx
Row heading	xx	xx	xx	xx	xx	xx

[a] Footnote crediting source of information if reproduced, adapted, or based on another published table.
[b] Footnote explaining spanner head 1.
[c] Footnote explaining the column heading.
[d] Footnote explaining a data cell nuance.

Figure 30.1 The major parts of a table.

Table *z* Summary of nutrients

Iron (mg)	Vitamin A (IU)	Vitamin E (IU)

becomes

	Vitamins (IU)	
Iron (mg)	A	E

Figure 30.2 Consolidation of column headings under spanner headings.

adjacent column headings can be gathered into a spanner (or "straddle heading" or "decked heading"; see Fig. 30.1). All information in the spanner head must apply to every column encompassed by the spanner. A horizontal rule, called a spanner rule, runs the width of all the columns to which the spanner applies (Fig. 30.2).

Whenever a spanner is used, every column under the spanner must have its own column heading. Nesting of spanners (i.e., a spanner heading above 2 or more spanner headings) is sometimes appropriate, but nesting that exceeds 3 levels (including the column headings) should be used with care because complex nesting can be confusing. Do not use a spanner that encompasses all the column headings; instead, put the information into the title of the table.

Position column headings flush left or centered over their respective columns. Headings can be appropriately broken and nested or stacked vertically as spanners to save horizontal space (see Section 30.1.5).

30.1.1.5 Row Headings

The left-most column of a table, the "stub", consists of a column heading at the top and row headings below. Like other column headings, the column heading for the stub should identify the row headings beneath it. A row heading identifies the data appearing in that row (the cells to the right of the row heading). Guidelines for column headings

Table *z* Measured and calculated values of KA series samples

Sample	Temperature (°C)	*L*	FC index (dyn/cm^2)	*W* [a]
KA-100	20	2.17	3.472	0.86
KA-100	40	3.53	4.774	0.86
KA-102	20	2.04	5.962	0.86
KA-102	40	3.46	4.627	0.86
KA-104	20	1.86	8.388	0.86
KA-104	40	3.29	5.981	0.86

[a] Calculated value.

Table *z* Measured and calculated values of KA series samples

Sample	*L*	FC index (dyn/cm^2)	*W* [a]
KA-100			
20 °C	2.17	3.472	0.86
40 °C	3.53	4.774	0.86
KA-102			
20 °C	2.04	5.962	0.86
40 °C	3.46	4.627	0.86
KA-104			
20 °C	1.86	8.388	0.86
40 °C	3.29	5.981	0.86

[a] Calculated value.

Figure 30.3 Use of stub headings.

apply to row headings (see Section 30.1.1.4). Unlike a column heading, however, a row heading may consist solely of a unit designation.

Information such as constants or experimental conditions should not appear in columns of the table but should be incorporated into the stub—for example, through indented subheadings or cut-in headings that span the table (Fig. 30.3).

Stub headings organize row headings into groups (just as spanner headings organize columns into groups), with all row headings in a group described by the stub heading immediately above. Stub headings are left-justified and capitalized sentence style. The only information in a row with a stub heading is the wording of the stub heading itself and perhaps the units for the data, if the units apply to all the row headings under the stub heading. No data are given for a stub heading.

An alternative to a stub heading is a "cut-in" heading. Cut-in headings essentially break a table into 2 or more sub-tables by grouping related rows under the same column headings. Whereas stub headings act more as qualifiers for row headings, cut-in headings often identify different groups, each of which has the same set of row headings.

Set off a cut-in heading with a full-width horizontal rule across the entire data field with spaces above and below the rule.

The decision to use a left-justified stub heading or a cut-in heading is a matter of style and judgment (see Fig. 30.4). In general, use left-justified stub headings when they are

Table *z* Mineral survey samples meeting minimum specifications

Location	Survey 1	Survey 2
Mineral-containing samples (%)		
Idaho	56.4	51.4
Wyoming	44.3	46.3
Mineral content (wt % ± SD)		
Idaho	35.1 ± 7.9	25.4 ± 6.7
Wyoming	32.6 ± 6.6	27.1 ± 8.5

Table *y* Mineral survey samples meeting minimum specifications

Location	Survey 1	Survey 2
	Mineral-containing samples (%)	
Idaho	56.4	51.4
Wyoming	44.3	46.3
	Mineral content (wt % ± SD)	
Idaho	35.1 ± 7.9	25.4 ± 6.7
Wyoming	32.6 ± 6.6	27.1 ± 8.5

Figure 30.4 Examples of left-justified (Table *z*) and centered (Table *y*) stub headings.

brief (e.g., up to about 4 words and numbers), the number of included row headings is limited, and the total number of stub headings is high. Use cut-in headings to distinguish different sets of row headings when the text is longer, the number of row headings below is more extensive, and the total number of headings is limited.

Indent all row heads and the second and subsequent lines of any row headings that carry over to another line. Entries opposite a multiple-line row head should be aligned on the first line of the heading.

30.1.1.6 Data Field

The body of the table, or "data field", may contain numbers, text, or symbols. The table title must clearly explain what information is being presented in the data field (see Section 30.1.1.2). Each entry appears in a cell formed by the intersection of a column and a row and must be consistent with the information in its respective column and row headings.

Related quantities or values in a table should have the same degree of precision. Similarly, quantities or values from a table that are repeated in the text should have the same precision in both places. Thus, if a number in the text is rounded, then it is likely that the value in the table is more exact than necessary and should be rounded to the same extent (see Sections 12.1.3.3 and 12.1.3.4). However, numbers should always be rounded to the nearest meaningful digit. Although a patient's weight can be measured accurately in hundredths of a kilogram, for many applications, rounding to a tenth of a kilogram may be sufficient (e.g., 42.81 kg vs. 42.8 kg).

The data field can be organized in 2 basic ways, depending on whether the intent is to help readers evaluate the data or to help them find specific values. If the purpose is to aid in evaluating the data, the data field should be ordered by some characteristic of the data. If the purpose of the table is to help readers find specific values, then the data

Table z Abundances of elements in meteorite samples A and B

Element	Sample A, mean (SD)	Sample B	Relative abundance (A/B)
Chromium, mg/kg	96.6 (1.2)	2,250	0.043
Iridium, mg/kg	13.1 (4.7)	514	0.025
Selenium, mg/kg	17.3 (8.0)	19.5	0.89
Zinc, wt%	3.46 (0.01)	0.030	114

Figure 30.5 Alignment of columns containing values in different units.

field should be ordered by some characteristic of the row or column headings. For example, in a table showing the prevalence of a disease in several countries, the data field could be organized by presenting the prevalence data from high to low; alternatively, the country names in the row headings could be presented in alphabetical order to make it easier to find the prevalence for a specific country.

30.1.1.6.1 NUMBERS

Align numbers within a column on the decimal point, actual or implied. When the units vary down a column (because the units are different for each row), alignment can be challenging (Fig. 30.5). In this situation, it may be preferable to left-justify or center all entries. If the entries in a column do not carry the same units, or if they are symbols (e.g., arrows, plus signs) they may be aligned as appropriate. If the columns have a common element, such as parentheses, a decimal, or a multiplier (e.g., ×10), align the entries on that element.

Values can, and often should, be summed across columns, down rows, or both, depending on the need to show sums. Explain totals that are not the exact sum of the numbers in a column or row in a footnote (e.g., "Totals may not sum to 100% because of rounding").

30.1.1.6.2 TEXT

Single words or short phrases may be either flush left or centered, but entries requiring several lines should be flush left. Place a blank line between rows in a table containing only text, if needed to aid comprehension, and use sentence-style capitalization in most such tables (but see Section 30.1.3). Entries consisting of complete sentences (with the exception of quotations) are seldom effective or necessary, but when they are needed, capitalize and punctuate in sentence style. Similar entries in a column or row should use the same or parallel phrasing (i.e., all nouns, all noun phrases, all verbs, all verb phrases).

> Treated species, Untreated species *not* Treated species, species not treated
> Cut, Sutured, Ablated *not* Cut, Suturing, Ablation

Conventional table design calls for an initial capital for each heading or cell entry; however, technical tables may contain words, abbreviations, or symbols in which capitalization, or the lack thereof, is an essential element of recognition (e.g., pH, cDNA, Fe, *c-jun*, AU, sin, log, mtDNA). If the reader will not be confused or misled (e.g., in tables that consist solely of text), capitalize headings and cell entries in sentence style; see, for example, Tables 30.1 and 30.2 in this chapter. If the reader might be misled by the use of capitals for words that should not be capitalized in running text, do not use sentence-style capitalization; instead, use uppercase or lowercase as dictated by the conventions for the terms used (see Tables 12.2 and 12.9). The variable capitalization style for the tables in this manual reflects variability in the types of terms being displayed.

To conserve space in the data field, use standard symbols and abbreviations (those allowed by the publisher without definition or expansion) as well as abbreviations already introduced in the text; other abbreviations may also be used and defined in a footnote. Abbreviations defined in the text should be redefined if used in a table to allow the table to stand alone.

30.1.1.6.3 SYMBOLS

Symbols should usually be centered in the cell. Variation in the horizontal placement of symbols can be confusing because such variation may be seen as meaningful when it is not. The meaning of each symbol should be explained in a footnote.

30.1.1.6.4 EMPTY CELLS

A table may have cells for which no information is available or possible. Empty cells (particularly if there is only one) can cause uncertainty about whether the cells are meant to be empty or are empty because data were unintentionally omitted. The cells may be left empty if it is obvious that they should be empty. However, when values (including zeros) could logically be expected in cells for which data are unavailable, the cells should not be left empty. Instead, to avoid ambiguity, place an ellipsis (. . .) or an appropriate abbreviation (e.g., ND for "no data") in the cell to indicate that the cell contains no data, and specify the meaning of the abbreviation in a footnote. Avoid using similar abbreviations with different meanings in the same table (e.g., ND = not detectable, nd = no data).

A cell should be left empty only if an entry would not make logical sense. For example, in a table of chemical compositions of pure compounds, the intersection of a column labeled "Sodium, wt %" with a row labeled "Ca(OH)2" should remain blank because there is no sodium in calcium hydroxide. If, however, the table presents trace impurities in chemical compounds, this cell should not remain blank because a preparation of calcium hydroxide could conceivably contain some sodium.

In a large table with many empty cells, empty cells are less likely to be interpreted as omissions, but they can make it difficult to follow rows from the stub to entries several columns across the page, particularly if the table must appear in landscape format (i.e., the table is turned counterclockwise 90°). When a table has many empty cells, consider redesigning the table.

30.1.1.7 Expanded Abbreviations

Place expanded abbreviations immediately below the data field, either stacked vertically or as a list in which the abbreviation is separated from its expansion by a comma and the units are separate by semicolons.

TEE, transesophageal echocardiography
CVD, cardiovascular disease
OS, overall survival
TEE, transesophageal echocardiography; CVD, cardiovascular disease; OS, overall survival

30.1.1.8 Footnotes

Footnotes are used to explain special aspects of the title, column headings, row headings, or the data. Use superscript lowercase letters to direct readers to the related footnote. Superscript letters are preferred to numbers or symbols for 3 reasons.

1) More letters are available to serve as footnote signs than devices such as asterisks and daggers.
2) Alphabetic order is readily recognizable.
3) Because most scientific tables contain numbers rather than words, the use of letters (rather than numbers) reduces the likelihood that the footnote symbol will be wrongly identified as an exponent, typographic error, or reference citation (in documents citing references by numbers).

Assign footnote letters in alphabetic order from left to right and from top to bottom starting with the table title and then progressing through the column headings and finally the rows of the table, including the row heads. If a table contains spanners (see Section 30.1.1.4), the assignment order for column headings is from left to right, as follows: 1) the stub column heading, superior spanners, and headings of any columns with unspanned headings; 2) inferior spanners; and 3) the column headings under spanners.

The footnotes themselves are placed in alphabetic order below any expanded abbreviations. Each superscript letter is placed flush with the left margin of the table and is separated from the footnote by a space. Carryover lines should be aligned with the first letter of the first line of the note, so that the footnote letters are clearly visible. If the footnotes are numerous but short and the table is wide, several columns of footnotes may be positioned across the bottom of the table to avoid a wasteful vertical series of short lines below the table.

30.1.1.9 Credit or Source Lines

A credit or source line identifying the source of the information in a table or the table itself may be placed as a headnote, a footnote, or the lowest element of the table below the data field. If a credit line is needed, the form depends on the style of reference citations used in the document and the specificity needed. The credit line should take 1 of 3 forms: "Reprinted from" (for an exact reproduction), "Adapted from" (where either the information or the form of the table has been changed from the original), or "Based on" (where information comes from a particular source, but the table is not reproduced or adapted). When a table is reproduced or adapted from a published source, it is not sufficient to cite the source and list it in the reference list. Permission to reuse the table must be obtained from the copyright holder (see Section 3.3.3). The credit line should have the form and content requested by the copyright holder.

When a few cells contain data from a different source, an overall source line can be used, and the data from the minor source can be credited by footnotes to the individual data. When the data in a table come equally from several sources, the sources may be credited in the column or row headings, either by footnotes to the headings or by listing the sources themselves (e.g., "Reference 10"; "Smith 1989") with the column or row headings.

30.1.2 Text References to Tables

In general, references to tables should be parenthetic to the text, although in some cases clarity and simplicity are achieved by referring to the table in the subject of the sentence. As a general rule, state the interpretation of the data in the text and cite the table with the supporting data in parentheses at the end of the sentence.

> The tensile strength of the new alloy was double that of the old one (Table 3).
> *not*
> Table 3 shows the results of tensile strength testing. [This information should be apparent in the title of the table and need not be mentioned in the text.]

30.1.3 Preparing and Designing Tables

Before preparing or editing a table, try to eliminate any tables that might be better presented in the text or more effectively presented as figures.[2] Data that would require only 1 or 2 columns and only 2 or 3 rows should almost always be converted to text (unless the numbers have many digits, which are more easily read and compared in a table), but the same may be true of even longer tables (Fig. 30.6). The text below is just as functional as the table and saves considerable space.

> Average length (standard deviation) varied among the 5 species: *P. borealis*, 11.1 mm (1.4); *P. platyceros*, 13.6 mm (2.0); *P. dispar*, 14.7 mm (1.5); *P. hypsinotis*, 14.8 mm (1.8); *P. goniuris*, 15.0 mm (0.9).

In this example, the information about test groups has also been eliminated, on the grounds that these details do not help interpret the findings. Day[3] presents an excellent discussion of data that are more appropriate for text than for tables (e.g., tables with a large number of zeros or empty cells).

The question of which data should be in rows and which in columns has no simple answer. As discussed in Section 30.1.1.6, numbers are more easily compared side by side, which suggests that the response (dependent) variables should be identified in row headings and the explanatory (independent) variables (e.g., group names) in column headings. The final choice should be based on the constraints of the page size, the width of cell entries, the number of rows and columns, and the number of properties being

Table x Average lengths and standard deviations of 5 shrimp species tested in experimental grow-out conditions

Variable	*P. borealis*	*P. platyceros*	*P. dispar*	*P. hypsinotis*	*P. goniuris*
Test group	4A	3C	5B	4B	1C
Mean length	11.1	13.6	14.7	14.8	15.0
SD	1.4	2.0	1.5	1.8	0.9

Figure 30.6 A table that could be easily converted to text.

compared. In considering the possibilities, note that the number of row headings is virtually unlimited (i.e., can extend down over several pages), whereas column headings are limited by page width (whether portrait or landscape page orientation); in addition, more space is required between column headings than between single-spaced row headings.

Several techniques can reduce table size.

- Remove columns and rows that do not contain enough variation to justify inclusion by moving the information to the text or a footnote (see Fig. 30.3).
- Combine columns or rows. For example, sex and age presented in two columns can often be combined into one column labeled "sex/age, y".
- Combine columns under spanner heads to avoid duplicating headings.
- Move information in a column to the stub as part of the row heading (Section 30.1.1.5) or as a stub heading (Section 30.1.1.4).
- Optimize the horizontal and vertical dimensions of the cells (see if widening one column will reduce the length of all columns).
- Widen the text margins in each column.
- Reduce the height of each row.
- Use a type font with the tightest spacing ("times new roman" in Times New Roman is shorter than "times new roman" in Arial, which is shorter than "`times new roman`" in Courier).
- Avoid bolding or italicizing words, which makes them longer.
- Use lower-case letters when possible.
- Edit as tightly as possible.
- Use abbreviations and symbols.

30.1.4 Designating Units of Measurement

The units in a table should correspond to those used in the text. If all the numbers in the data field are in the same units, designate those units in the table title. If the units vary across columns or rows, do not include unit designations in the title; instead, present the relevant units in the column or row headings. If a few scattered cells have units that differ from those specified in the title or column or row heads, the exceptions can be identified by individual footnotes (but see the next paragraph).

Section 12.2.1.2 (item 7) specifies that SI prefixes should be used to convert numbers less than 0.1 and greater than 1,000 to values within that range (e.g., 0.008 km becomes 8 m). However, tables are generally a good place to bend this rule. Therefore, do not adjust entries with values outside that range (e.g., changing 0.067 L to 0.67 dL); instead, keep the units consistent to avoid the need to use footnotes for the exceptions (and the risk that the reader will miss the footnote). This practice also aids the reader in assessing the magnitude of differences among the values being compared.

When the information in a column is not expressed in SI units (e.g., the number of cells in a culture or the number of occurrences of an event), very large or very small numbers are expressed with multipliers in the column heading, preferably using scientific notation (see Section 12.1.5) rather than alternative formats (words [e.g., thousands, millions] or numerals [e.g., 1,000s; 1,000,000s]). Such multipliers can easily cause confusion and uncertainty: Should the reader do the multiplication or has the author already performed the multiplication? The preferred presentation, although somewhat less likely to be misunderstood than the discouraged forms, can still be misconstrued; an explanatory footnote may therefore be the safest solution.

Preferred	Discouraged
Mass × 10^3, kg	10^3 × Mass, kg
	Mass (× 10^3, kg)

For the examples above, the number "23" in a table cell would represent a mass of 23,000 kg. Placing the multiplier before the unit indicates that the values in the cells are already in the multiplied form, whereas placing it after the unit may be interpreted to mean that the reader must do the multiplication. When a table presents only nonscientific information, such as budget data, the traditional form of heading, "$ (millions)", may be used.

When the range of values in a column is extremely wide, it may be necessary to use a heading that requires the entries in each cell to be in full scientific notation (e.g., 1.596×10^{-4}). Because of the space problems presented by the width of such entries, every effort should be made to avoid this option.

30.1.5 Spacing

Separate the table title and column headings by 1 line and a table-width horizontal rule (line). Separate the column headings from the field and the field from the footnotes with table-width rules. Vertical lines are unnecessary if columns are properly spaced, and they generally clutter a table without adding clarity.

Put space between adjacent columns to separate them visually. Observing minimum spacing is most important when the columns are numerous and the field values are close to each other.

In large tables of numbers, particularly those containing closely spaced columns, break the rows with a blank line for visual ease, usually about every 5 lines (range, 3 to 6). For full-page tables that would carry over to the following page if such blank lines were inserted, a light screen (shading) can be used over alternating sets of rows, providing a visual marker without the loss of vertical space.

30.2 GENERAL INFORMATION ON FIGURES

Figures are visual displays of data and information and include charts and graphs, photographs, biomedical images (e.g., radiographs, PET scans), maps, diagrams, drawings, and other types of illustrations. More detail on the characteristics of various kinds of images and their proper preparation can be found in Lang[1] and Peterson.[4]

When preparing figures, strive for the characteristics listed below.

- The figure should have a purpose; it should support claims made in the text.
- The data and information displayed should be accurate.
- The figure should be as simple as possible, focusing attention on important aspects and minimizing distracting elements.
- In general, data should be presented only once. An exception can be made for "selective redundancy", or duplication that confirms the accuracy of a key value.

- Similar elements within a figure and between related figures should be presented uniformly and consistently.
- Drawn images should be visually appealing.

For all figures in a given document, make the style and format (e.g., form of caption, style and capitalization in axis labels, typeface, type size) as consistent as possible. However, this consistency may be relaxed for review documents in which the figures are reproduced from a variety of published works.

Use the same terms, symbols, and abbreviations in figures as are used in the text. Abbreviations introduced in the text should usually be redefined in the figure, to allow the figure to stand alone.

30.2.1 Figure Number

If a document has only one figure, the figure is cited in the text as "Figure" and it is not numbered.

If a document has 2 or more figures, label each with the abbreviation "Fig." and number them all with consecutive arabic numerals in order of mention in the text. For long documents, reports, or books that are divided into sections or chapters, figures may be found more easily by numbering the section or chapter independently from the figure and separating the numbers with a period: for example: Fig. 2.1, Fig. 2.3, and Fig. 5.5.

Unlike tables, figures can have parts (e.g., Fig. 1a, Fig. 1b). Figure parts are 2 or more adjacent data displays or images that show information from the same source or that is to be compared with that in the adjacent parts. Common applications are figure parts showing several measurements from the same animal or sample, the same measurement from several animals or samples, "before and after" data, time series data or photos, anterior-posterior views, and close-up or expanded views.

Indicate figure parts with lowercase letters assigned in alphabetic order, placed immediately to the right of the figure number. If the separate parts of the figure call for so much information that individual captions are needed, the composite should be divided into separately numbered figures.

30.2.2 Figure Caption

The figure caption must be clear and sufficient to allow the reader to understand the fundamental aspects of the figure without having to refer unduly to the text.

The caption (not a title) consists of a single phrase in sentence case without a closing period (unless a headnote is included; see below). The caption should identify the data or image in the data field or image area. Captions for figures with parts should identify each part by its letter, followed by a closing parenthesis, and then identify the information shown in the figure part. Typically, information that applies to all parts is presented first and the distinguishing information in each part is presented after identifying the part as described.

> Fig. 3. Rainfall in the greater Seattle area, 2010. a) March, b) June, c) September, and d) December.

For graphs, the caption should not simply repeat the axis labels; instead, it should describe what the data show and the relations or comparisons intended.

Fig. 5 Association between air temperature and survival of Tanner crabs released as illegal bycatch in the Bering Sea bottom trawl fishery

not

Fig. 5 A plot of air temperature and survival of Tanner crabs released as illegal bycatch in the Bering Sea bottom trawl fishery

30.2.3 Headnotes

As with tables, qualifying information about the figure beyond that contained in the caption can be presented in a headnote, which follows the figure caption and the part captions, if any. In addition, the headnote may also contain expanded abbreviations, keys that identify symbols in the data field or image area, and credit or source information.

As with tables, some disciplines, especially in basic science, appropriately include in the caption or headnote a summary of the experimental conditions and sometimes the interpretation of the image. Information presented in the title need not be duplicated in the text.

30.2.4 Text References to Figures

In general, as with tables, references to figures should be parenthetic to the text, although in some cases clarity and simplicity are achieved by presenting the figure reference as an integral part of the sentence. If the reference to a figure begins a sentence, spell out the word "Figure"; elsewhere, the capitalized abbreviation "Fig." should be used without expansion.

Figure 5 is the result of this combination.
The incubator (Fig. 2) was custom-made.
The incubator shown in Fig. 2 was custom-made; that shown in Fig. 3 was a commercial model.

Use the phrasing of the third example only when the figure literally shows a drawing or photograph of something physical. Avoid rhetorical forms stating that a figure is itself evidence (rather, a figure contains evidence). Exceptions are diagnostic and laboratory images, which *are* the evidence.

Separation was complete after 20 h (Fig. 5).
not Fig. 5 shows that separation was complete after 20 h.
Additive X was the most effective (Fig. 13).
not Fig. 13 proves that additive X was the most effective.
Figure 3 [an echocardiogram] clearly shows mitral regurgitation.

30.3 GRAPHS, CHARTS, AND PARTS OF A GRAPH

Graphs and charts are visual presentations of data and the relations among them. We focus here on Cartesian or coordinate graphs, which have 2 or sometimes 3 axes that identify the values of the data. Many of the guidelines here are applicable to other types of graphs and charts (e.g., pie charts, box plots, bar and column charts), but addressing other presentations is beyond the scope of this manual.

Graphs should be of high quality, with legible symbols and type, appropriate resolution (pixels per inch or dots per inch) and line thicknesses, easily distinguished shadings (try to keep a 20% difference between adjacent shadings), and complete labeling.[5] Shading is preferred to colors because, even if the publication is printed in color, it may be later photocopied or printed from PDFs in black and white.

If the software package used to create a graphic image cannot generate a figure of adequate quality, consider having a professional artist prepare the figure. Figures prepared as slides for oral presentations and those prepared for print or electronic publication have different visual objectives and thus are seldom interchangeable.

Graphs may have several parts, depending on their format: 1) a figure number, 2) a caption (not a title), 3) a headnote, 4) a data field, 5) axes and scales, 6) symbols, 7) legends, and 8) a credit or source line. Graphs should always have at least a caption, axes and scales, symbols, and a data field. Guidelines for assigning figure numbers, captions, and headnotes are described above; the other parts are described below.

30.3.1 Data Field

Data fields should be well defined and as free from distracting information as possible.[6–8] A square encompassing a margin beyond the zero-zero point of the axes defines the data field and provides boundaries at the top and right that can help identify the positions of data in the field. Dotted lines can be added to indicate the value of zero on both axes.

30.3.2 Axes and Scales

For most purposes, design a graph so that the vertical axis (ordinate, *y* axis) represents the dependent variable and the horizontal axis (abscissa, *x* axis) represents the independent variable. Hence, time is always on the *x* axis.

Make the scale on the axes slightly larger than the range of values being plotted, so that the data occupy most of the range of each axis. If an extremely large range must be covered and cannot be practically shown with a continuous scale, indicate a discontinuity in the scale *and the data field* with paired diagonal lines (—//—) indicating a missing extent of the range. Most readers assume that the graphs begin at the zero-zero point. If beginning either scale at zero is not practical because the range of data is large and adding a discontinuity would be distracting, you may begin the scale with a value slightly smaller than the minimum value of the graphed data.

For the numbers used to mark axes, choose simple, appropriate multiples of the quantity graphed, typically multiples of 2, 5, 10, or 25. Divide the scales with interval (tick) marks, rather than showing a grid over the entire graph, and place tick marks and scale marks outside the data field. Place the scale numbers outside of the data field as well, just left of the *y* axis and just below the *x* axis and centered on their respective tick marks. For numbers less than 1, include the digit zero before the decimal.

Some graphs use a second scale outside the right vertical axis. The right-side vertical axis should be scaled, numbered (with numbers to the right of the axis), and labeled in the same manner as the left-side scale. Be cautious when adding a right-hand scale that is not mathematically related to the left-hand scale for the purpose of graphing an additional variable. The ratio of one vertical scale to the other of different units can be varied to manipulate the visual relationship between the respective curves. Each curve in such a graph must be clearly identified, preferably by arrows in the data field that point to each curve or in a legend or in the figure caption.

An axis label, consisting of a word or short phrase, identifies the variable plotted on each axis. Use sentence-style capitalization for the labels, and include the appropriate unit after a comma at the end of the label. Center each label on the length of the corresponding axis. The *y* axis label should be rotated 90 degrees from normal so it will face and be parallel to the y axis. When a multiplier is needed for the axis values, the quantity, not the unit of measurement, is multiplied by the appropriate factor.

Median recovery time, y
10^{-3} × Radioactivity, counts/min *not* Radioactivity, counts/min × 10^{-3}

Make axis labels as unambiguous and descriptive as possible (e.g., "Number of stem cells" rather than "Cell numbers"). Occasionally, an axis shows a complex quantity for which a suitable label may consist only of a unit; in such a case, the unit symbol should be spelled out, especially if both SI and non-SI units are involved (see Section 12.2.1.2, item 4).

Nanograms per cell per day *not* ng/cell·d *or* ng·cell^{-1}·d^{-1}
but Calcium uptake, ng/cell·d

If an axis is labeled with only a simple unit, the appropriate quantity should be added; for example, a label of "km" for "kilometers" should usually be "Distance, km".

30.3.3 Plotting Symbols

Plotting symbols need to be distinct and legible and provide good contrast between the figure in the foreground and the background.[7] Open and closed circles provide the best contrast and are more effective than the combination of open circles and open squares. A good symbol system will also be able to indicate overlapping values. The symbols ○, ●, ◉, ⊙, ⊗, and ⃠ work well, but there are many other options.

Graphs should be simple and should contain no more information than is needed to make the point. Several curves can be displayed on a single graph, provided they are not close enough to obscure any of the individual curves. When the number of lines is great or when overlap prevents following each line, consider using "small multiples" in which each line is graphed separately in its own figure part.[9–11] Figures presented this way should have identical *x* and *y* axes.

In a series of graphs within a document, make symbols, shadings, and line patterns consistent among all figures. That is, the symbols representing control and treatment groups in the first figure should be the same for these groups in later figures. If the early and later figures represent 2 sets of experiments, however, the later set might logically be represented by different symbols.

30.3.4 Legends

Legends identify the meaning of plotting symbols in a graph. Although the data field should be as uncluttered as possible, it is usually better to label lines and symbols directly in the data field than to include a legend. When direct labeling is not feasible, labels can be grouped into a legend also placed in the data field. However, placing this information into the caption is sometimes preferable, even though it requires readers to look back and forth between the graph and the caption.

30.3.5 Credit or Source Lines

Credit and source lines for figures should be formatted as they are for tables (see Section 30.1.1.9) and placed after the caption or headnote.

30.4 PHOTOGRAPHS AND CLINICAL AND LABORATORY IMAGES

Image contents vary widely, depending on the format of the image (e.g., an illustration, echocardiogram, intraoperative photo, blots and gels) as well as the nature of the image itself, so specific guidelines are not possible. As a general guideline, image areas should be well defined, be as free from distracting information as possible, and provide good contrast between the figure in the foreground and the background.

Photographs and diagnostic images should show the detail of interest without undue distraction; in most cases, photographs need to be cropped to ensure that the reader's attention is directed correctly.

30.4.1 Scale

Indicators of scale are often needed to make photos meaningful. Such indicators may be included in the photograph itself (e.g., a geology hammer showing the scale of a rock formation) or added later. Scales are most useful when they are visual, such as a bar denoting a given length in a micrograph, rather than mathematical (a statement of magnification such as "×1,000"). Visual indicators also eliminate a need to recalculate the magnification if the original photo must be reduced or enlarged for printing.

30.4.2 Callouts

Type for callouts identifying parts or structures of an image should be consistent with the type in other document figures. Reduction of a photograph or drawing before printing should be taken into account so that the final size of the type is suitable for a printed page.

30.4.3 Supportive Information

For each image (especially micrographs and biomedical images), certain information may need to be provided:

- Identify the subject of the image; explain what is being shown.
- Tell how the image was acquired (e.g., equipment and settings; position of the subject; stains used).
- Explain why the image was selected for publication over other images of the same subject (i.e., is the image typical, atypical, from the current research or from another source?).

- If the image was modified, explain the modifications. Undisclosed image modification, especially if the modifications appear to intentionally misrepresent or falsify the data, is considered to be scientific misconduct.
- Draw the viewer's attention to important features by using circles, lines, arrows, or other graphic devices superimposed on the image.
- Tell what the image means: what it represents and its implication for the research.

More detailed recommendations are given in Lang et al.[12] and Rossner and Yamada.[13]

30.5 MAPS AND PARTS OF A MAP

Maps are an integral part of publishing in geoscience and other disciplines. Maps combine elements of cartography, statistics, graphic design, and fine art, so guidelines for preparing them must be general and flexible. The guidelines here are for printed maps, although many guidelines also apply to maps viewed online.

Maps are primarily visual, so legibility is essential. Care must to taken to balance the size, number, and density of elements; the contrast between different elements; the number and use of colors; the degree of abstraction (e.g., an arial photo, a pictorial map, a line drawing); the relationship between background and foreground images; and the quality of the rendering to create a map with a clear message.

The most common parts of a map are described below. Not all maps will have all the parts.

30.5.1 Title

Identify the locations or data depicted on the map. The purpose of showing the map should be made clear in the text. The title may appear anywhere on the map, as long as it is a dominant element and does not obscure important details, or above or below the map.

30.5.2 Source

Identify the source or sources of the data. This information is typically placed in small type at the bottom of the map and often, but not necessarily, outside the border.

30.5.3 Preparer

The person or group who created the map are usually acknowledged with the source of the data as described above.

30.5.4 Publication Date and Copyright

Credit the copyright holder and the author, company, or publisher of the map. Again, this information is typically placed in small type at the bottom of the map and often, but not necessarily, outside the border.

30.5.5 Projection

Identify the type of projection (e.g., Mercator, Goode homolosine, sinusoidal), also in small type at the bottom on the map. Because all projections have strengths and weaknesses, choose the projection that best illustrates the points to be made.

30.5.6 Compass Rose

Indicate the directions on the map. By convention, the top of the map is north. If the layout of the map on a page prevents this orientation, the directional should clearly indicate north. The term "compass rose" takes its name from the artistic tradition of indicating 8 or even 16 directions with points on the compass, but a single arrow pointing north may be sufficient. The compass rose can be placed anywhere on the map as long as it does not obscure important information.

30.5.7 Scale

The scale of the map is the ratio of the map dimensions to the actual geographic distances. It is useful to present the scale both as a ratio (e.g., 1:50) and as a conversion unit (1 cm = 750 km). The scale is often placed close to the projection at the bottom of the map.

30.5.8 Coordinate System

Sites on the map are usually located with a coordinate system, such as lines of latitude and longitude (parallels and meridians); a grid pattern with alphabetized or numbered rows and columns; or numerical scales along the sides of the map (Cartesian coordinates). The details of the coordinate system should be obvious and clear and consistent with any associated index. Inset maps (see below) may have different coordinate systems.

30.5.9 Identification Symbol System

Maps use a variety of shading, cross-hatching and stippling, symbols, pictographs, type fonts and sizes, and color to identify various features of geography, demography, economic activities, and so on. The symbols need to be distinct, defined, and legible. Standard symbol systems have been proposed for several applications and should be used when available.

30.5.10 Quantitative Symbol System

Many maps overlay statistical graphics on locations to indicate measures of quantity and intensity, density, or gradations. The symbols used should be distinct enough to make meaningful comparisons. Values on a linear scale are more easily understood and compared than those expressed by gradation in area (e.g., circles or images of different sizes). When indicating quantity with differences in area, it is preferable to graph the data with a few ordinal categories, each associated with a symbol large enough to be distinct from those used to indicate larger quantities.

Especially problematic is how to indicate both positive and negative values, such as elevations above and below sea level or increases or decreases in populations.

30.5.11 Legend

The legend is the section of the map that defines the meaning of graphic conventions and identification and quantitative symbols (e.g., dotted lines for national boundaries, orange for high temperatures, stippling for forests, and so on). Every convention and symbol on the map should be defined in the legend. The legend is usually placed on the edge of the map in a position that does not obscure details of the map itself.

30.5.12 Inset Map

An inset map is a smaller map drawn at a larger scale to expand a section of the primary map (e.g., small maps of major cities that appear on a larger state highway map). It may be placed anywhere and often has its own coordinate system for locating places.

30.5.13 Locator Map

A locator map is a smaller map drawn at a smaller scale to indicate the location and orientation of the primary map in relation to a larger geographical area (e.g., a drawing of a country with a shaded square indicating the region shown on the primary map). It may be placed anywhere.

By their very nature, maps distort spatial relationships: areas, angles, gross shapes, distances, and directions. Further, printing can radically alter the appearance of a map. In particular, be careful of distortions created by 1) emphasizing certain elements, 2) deemphasizing contradictory elements, 3) confusing the image with details, and 4) using provocative symbols.

The editor, author, or data specialist should check to make sure that names of attributes used in metadata files correspond with names of attributes used on the plotted map. More information on the production and editing of maps is included in *Suggestions to Authors of the Reports of the United States Geological Survey*.[14]

30.6 INDEXES

An index is compiled to enable readers to find information easily; therefore, the needs of the intended or likely readers should guide what terms to select for the index, what cross-references should be inserted, and what depth of indexing will be most appropriate. This manual does not discuss how to deal with these questions; guidance can be found in texts on the entire process of indexing[15-17]; another helpful source is *The Chicago Manual of Style*.[18] After these initial decisions come some decisions on the structure of the index (see Table 30.1), which must be made at the outset.

30.6.1 Entry and Subentry Terms

Indexes can have different levels for entry terms: 1 level (only single-level entries), 2 levels (top-level terms and subterms), or, rarely, more levels. The next possible variation is subentry terms that are stacked (indented under the main entry terms, with each subentry term flush left) or run-in (continued on the same line as their main entry and in paragraph form, a space-saving device). In both formats, the subheadings are alphabetized.

eponymic terms
  capitalization 90
  clinical genetic syndromes 237
  rules for use 132–133

or

eponymic terms: capitalization 90; clinical genetic syndromes 237; rules for use 132–133

Table 30.1 Checklist for decisions on content and formatting of indexes[a]

Peculiarities of topic and particular audiences: Is the document political, with a need to index each person's name at any mention in the text; geographic, with a need to index each place name at first mention; historical, with a need to incorporate dates?
Is only a subject index needed? Or should an index of cited authors also be prepared? For journals and multiauthor documents, is an author index needed?
Typical number of index entry terms selected per page of text (depth of indexing)?
Levels of index entries: main headings only, main headings with subheadings, main headings with subheadings and sub-subheadings?
Alphabetization of entries letter-by-letter or word-by-word?
Inclusive or initial page numbers? Or section numbers? Or both?
Separate inclusive page numbers with an en dash or a hyphen?
Subheadings stacked or run-in?
Degree of indentation for subheadings?
Degree of indentation for second line of entries that extend longer than one line (turnovers)?
Punctuation (comma) or spaces between the entry and the page or section number? How many spaces?
Need to observe scientific conventions, such as italicization of scientific names for organisms, small capitals for some chemical-name prefixes?
Italicize "see" and "see also"?
Commas between page or section numbers?

[a] These decisions should be made by the indexer and discussed with the designer to ensure the layout of the index follows the indexer's intention.

For main entries and subentries, only nouns with no preceding articles and modifiers should serve as index terms.

"Albright syndrome" *not* "the Albright syndrome"

Retain compound nouns as presented. Reducing such entries to the root noun could render the term unhelpful or meaningless.

red blood cell	ground truth	polar bear
intensive care	North Pole	dead weight

Because the reader will infer that a subentry is subordinate to the next higher level (usually a main entry at the top level), a preposition indicating the relation of the subentry to the entry above is seldom needed. If a preposition is needed for clarity, follow logical grammatical order and alphabetization.

metastases
  from liver 45–92
  to liver 1–5

Whether to use common abbreviations and acronyms as main headings also depends on their comprehensibility in the prospective audience.

AIDS *or* immunodeficiency syndrome, acquired *or* acquired immunodeficiency syndrome
DADDS *or* N,N′-(sulfonyldi-4,1-phenylene)bisacetamide
DNA *or* deoxyribonucleic acid

30.6.1.1 Alphabetization

Entries in an index are alphabetized by 1 of 2 systems. In the letter-by-letter system, all words in a term comprised of 2 or more words are considered as a whole (as if run together). In the word-by-word system, terms comprised of 2 or more words beginning with the same word or abbreviation are grouped and then alphabetized by the sequence of letters in the second word. That is, in the letter-by-letter system, ignore the space when alphabetizing, and in word-by-word system, start the alphabetic sequence again after the space. In both systems, begin alphabetizing again after a comma or paren.

30.6.1.1.1 LETTER-BY-LETTER OR WORD-BY-WORD

The letter-by-letter system is generally preferred.

Letter-by-Letter	**Word-by-Word**
Saint, Eva Marie	Saint, Eva Marie
Saint Augustine	Saint Augustine
Sanhedrin	San Jacinto
sanitarium	San Xavier
San Jacinto	Sanhedrin
Sanskrit	sanitarium
Santa Claus	Sanskrit
Santa Isabela	Santa Claus
San Xavier	Santa Isabela

Computer programs designed for purposes other than indexing but that have an indexing component (such as word-processing and desk-top publishing programs) may not automatically produce an accurate letter-by-letter sequence, and the sequence may have to be corrected. Computer programs designed specifically for indexes can automatically produce accurate letter-by-letter sorts.

30.6.1.1.2 PREFIXES

For chemical names beginning with a numeral, the entries are alphabetized by the letters as usual, but the numeric order is used to sort entries within the alphabetic sort (see example below). Other descriptive prefixes (e.g., chloro-, α-, β-, D-, L-) are ignored. Alphabetization carried out by computer sorting may place terms with such prefixes at the end of the alphabetic sequence along with other nonstandard spellings and hyphenated words. If this occurs, rearrange the terms in the proper alphabetic sequence.

butanol
1-butanol
1-butanol , 1-cyclohexyl-
1-butanol, 2-cyclohexyl-
2-butanol
tert-butanolysis
16,17-butanomorphinan
1-butanone
2-butanone
2-butanone, 3,3-dimethyl-
2-butanone, 3-hydroxy-
8,9-butano-9-nonanolide
Butanox
butanoyl chloride
butanoyl disulfide
butanoyl nitrite

30.6.1.1.3 "MAC", "MC", AND "M'"

In British practice, names beginning with "Mac" or "Mc" or "M'" may be alphabetized as though they are all spelled "Mac". CSE recommends a letter-by-letter sequence for such names so that "Mac" would precede "Mc", although this sequence assumes that readers searching for a name will know how the name is spelled.

Recommended Sequence	British Sequence
Mably, G	Mably, G
Macalister, D	McAdam, J
MacArthur, A	McAdook, W
Macbeth, R	Macalister, D
MacBride, E	McAllister, A
MacWhirter, J	McAlpine, W
Macy, J	McAneny, G
Madach, I	MacArthur, A
McAdam, J	Macbeth, R
McAdook, W	McBey, J
McAlpine, W	MacBride, E
McAllister, A	McClintock, F
McAneny, G	M'Clintock, J
McBey, J	MacWhirter, J
McClintock, F	Macy, J
M'Clintock, J	Madach, I

30.6.1.1.4 ABBREVIATIONS

Abbreviations are sometimes alphabetized separately at the beginning of an index of their respective alphabetic sections. This manual recommends placing them in a normal letter-by-letter sequence.

30.6.1.1.5 NON-ENGLISH ROMAN CHARACTERS

In some European languages, words beginning with a ligature or a letter with a diacritic may be placed in a sequence not expected by Anglophones. For example, in a Danish dictionary words beginning with the ligature "æ" follow words beginning with "z"; words beginning with "ø" follow those beginning with "æ"; words beginning with "å" follow those beginning with "ø". As in the recommendations for alphabetization of author names in reference lists (see item 4 in Section 29.3.3), diacritics should be ignored in alphabetizing non-English words for English-language indexes; the terms should be treated as beginning with a letter (or letters) without regard to diacritical marks or ligations.

30.6.1.2 Capitalization

Some indexers always capitalize main-entry terms and set subentries in lowercase. In the sciences, capitalization or noncapitalization of terms may be needed to differentiate unequivocally among classes of what the terms represent. For example, capitalizing proprietary names of drugs distinguishes them from the noncapitalized nonproprietary names. Therefore, indexes of scientific publications should not have arbitrary capitalization of all entry terms; instead, the rules for use of capital or lowercase initial letters specified elsewhere in this manual should be applied, such as those for drug names and for trade names of commercial products (see Section 9.7.7).

30.6.2 Punctuation

Punctuation is generally not needed with single-level or stacked multilevel indexes and single initial-page numbers or section numbers following entry terms; spacing suffices.

anatomy 57
bacteriology 62

Commas are needed to separate multiple initial-page or inclusive-page numbers.

anatomy 57, 69, 101
bacteriology 62, 64–67, 89

If the index will contain entries that incorporate numbers, it may be visually helpful to separate the entry term from the location number by a comma. In indexes that contain entries where punctuation is meaningful to the concept, as in an index with chemical names, extra space may be preferred to the extra comma.

history
  14th century, 12–13
  1799–1825, 40, 44–47
  1850–1890, 52–57, 66, 70
  Freemasons, 99, 102

$C_2H_4O_2$ 199–200, 204
$C_5H_{10}O_2$ 272–274, 410
dextrin-1,6-glucosidase 5, 7, 10
vitamin A 11
vitamin 7–9, 11

30.6.3 Page Numbers

Use inclusive page numbers when indexing, not initial page numbers. Separate the initial and final page numbers with an en dash.

30.6.4 Section Numbers

Some texts are organized by section number. They can be indexed by section number, page number, or both. Again, use inclusive page ranges.

age 203.1 (4)	*or*	age 203.1
AIDS		AIDS
certification 203.5 (17–20)		certification 203.5
confidentiality 203.6 (21)		confidentiality 203.6
precautions 203.7 (27)		precautions 203.7
anemia 400.3 (23)		anemia 400.3

30.6.5 Cross-Referencing

A helpful index includes references to related entries or to the term under which the information has been indexed, indicated by "see" and "see also". "See" and "see also" are often italicized to make clear that they are not entry terms. "See also" entries appear either before or after the other subheadings; this manual recommends the latter. Placement and typeface, as well as spacing, should be decided at the outset. It is inconsiderate to refer the reader to a second entry and then give only one page number; it would be better to cite the page number in both places. Some information should be indexed as a subentry under a main entry and again as a main heading.

Pennsylvania
  Amish 25–29
  barns 25, 31, 33–35
  Lake Erie 20–22
  rivers 11–15

BIA *see* Bureau of Indian Affairs

barns
  Amish 25
  New England 5–10
  Pennsylvania 25, 31, 33–35
  tertiary use 40–45
  see also houses

30.6.6 Clarifying Homonyms

Words or names with the same spelling but with different referents—for example, the names of 2 persons spelled exactly the same—may have to be indexed. When 2 or more such homonyms are found, some means should be found to differentiate them so as to not misdirect the user of the index to an undesired part of the text. Such means might include identifying position, birthplace or birth date, or type of object, even if the indexed text does not carry such information.

Jones, John [American chemist]
Jones, John [English explorer]
Jones, John [New Zealand geologist]

condenser [electrical]
condenser [chemical]

CITED REFERENCES

1. Lang T. How to write, publish, and present in the health sciences: a guide for clinicians and laboratory researchers. Philadelphia: American College of Physicians; 2010. Chapter 4, How to display data in tables and graphs.

2. Gelman A, Pasarica C, Dodhia R. Let's practice what we preach: turning tables into graphs. Am Stat. 2002;56(1):121–130.

3. Day RA, Gastel B. How to write and publish a scientific paper. 6th ed. Westport (CT): Greenwood Press; 2006.

4. Peterson SM. CSE GuideLines: editing science graphs. Reston (VA): Council of Science Editors; 2000.

5. Loos EM. Evaluating scientific illustrations: basics for editors. Sci Ed. 2000;23(4):124–125.

6. Cleveland WS, McGill R. Graphical perception and graphical methods for analyzing scientific data. Science. 1985;229:828–833.

7. Cleveland WS. The elements of graphing data. 2nd ed. Summit (NJ): Hobart Press; 1994.

8. Cleveland WS. Visualizing data. Summit (NJ): Hobart Press; 1993.

9. Tufte ER. Envisioning information. Cheshire (CT): Graphics Press; 1990.

10. Tufte ER. Visual explanations: images and quantities, evidence and narrative. Cheshire (CT): Graphics Press; 1997.

11. Tufte ER. The cognitive style of PowerPoint. Cheshire (CT): Graphics Press; 2004.

12. Lang TA, Talerico C, Siontis GCM. Documenting clinical and laboratory images in publications: the CLIP principles. CHEST. 2012;141(6):1626–1632.

13. Rossner M, Yamada KM. What's in a picture? The temptation of image manipulation. J Cellular Bio. 2004;66:11–15. doi:10.1083/jcb.200406019.

14. Hansen WR, editor. Suggestions to authors of the reports of the United States Geological Survey. 7th ed. Reston (VA): United States Geological Survey; 1991 [modified 2012 Dec 31; accessed 2013 Jan 31]. http://www.nwrc.usgs.gov/lib/lib_sta.htm.

15. Lancaster FW. Indexing and abstracting in theory and practice. 3rd ed. Champaign (IL): University of Illinois, Graduate School of Library and Information Science; 2003.

16. Mulvaney NC. Indexing books. 2nd ed. Chicago (IL): University of Chicago Press; 2005.

17. Wellisch HH. Indexing from A to Z. 2nd ed. New York: HW Wilson; 1995.

18. The Chicago manual of style: the essential guide for writers, editors, and publishers. 16th ed. Chicago (IL): University of Chicago Press; 2010. Also available at http://www.chicagomanualofstyle.org.

ADDITIONAL REFERENCES

American Society for Indexing. Indexing evaluation checklist. Wheat Ridge (CO): ASI; [accessed 2012 June 20]. http://www.asindexing.org/about-indexing/index-evaluation-checklist/.

Briscoe MH. Preparing scientific illustrations: a guide to better posters, presentations, and publications. 2nd ed. New York (NY): Springer-Verlag; 1996.

Effective cartography: mapping with quantitave data. [Cambridge (MA): Harvard University, Graduate School of Design]; [date unknown]. http://www.gsd.harvard.edu/gis/manual/normalize/.

Frankel F. Envisioning science: the design and craft of the science image. Cambridge (MA): MIT Press; 2002.

Hodges ERS, editor. The Guild handbook of scientific illustration. 2nd ed. New York (NY): Wiley; 2003.

International Standards Organization. Information and documentation—guidelines for the content, organization and presentation of indexes. 2nd ed. Geneva (Switzerland): ISO; 1996. (ISO 999).

Kasdorf WE. The Columbia guide to digital publishing. New York (NY): Columbia University Press; 2003.

Monmonier M. How to lie with maps. 2nd ed. Chicago (IL): University of Chicago Press; 1996.

Schofield EK. Quality of graphs in scientific journals: an exploratory study. Sci Ed. 2002;25(2):39–41.

Tufte ER. The visual display of quantitative information. 2nd ed. Cheshire (CT): Graphics Press; 2001.

The UGS publications warehouse. http://pubs.usgs.gov.

Wainer H. How to display data badly. Am Stat. 1984;38(2):137–147.

Wainer H. Graphic discovery. A trout in the milk and other visual adventures. Princeton (NJ): Princeton University Press; 2005.

Wright P. Presenting technical information: a survey of research findings. Instruct Sci 1977;6(1):93–134.

31 Typography and Manuscript Preparation

31.1 TYPE SPECIFICATIONS

Text attributes (such as roman, italic, and boldface), case (such as capital letters and small capitals), and position of characters (such as superscript and subscript letters and numerals) convey special meanings in scientific publication. These conventions are summarized in Chapter 10 and presented in detail in the chapters of Part 3 of this manual. Authors and copy editors should be familiar with these conventions. In traditional paper-based publication processes, copy editors mark up manuscripts to indicate these type specifications to the typesetter or graphic designer. However, in electronic processes, authors and editors usually need to ensure only that the correct type specifications are used in the word-processing file.

In traditional processes, copy editors may mark up manuscripts for other details in publication style and format, such as the size of type to be used in titles, headings, text, and tables; the leading (spacing) between type lines; and other particulars that specify the format of a journal, book, report, or article. These specifications and how they are indicated usually are established by the publisher and typesetter or graphic designer, especially for journals. In electronic processes, such specifications established in advance usually obviate the need for mark-up but may necessitate use of style functions in word-processing software or a coding program. Copy editors should note that additional specifications may have to be indicated or established for particular manuscripts.

This manual does not attempt to describe all of the decisions, steps, and means that must be applied in preparing a book or journal manuscript for publication. Detailed guidance on these aspects of manuscript preparation can be found in such publications as *The Chicago Manual of Style*[1] and *The Copyeditor's Handbook*.[2]

31.2 CHARACTERISTICS OF TYPE

Editors should be familiar with the chief characteristics of type: typeface styles and weights, type sizes, and the spacing of characters and lines. These elements must be specified to define the appearance of both printed matter and electronic publication (e.g., fonts used on a journal's website). Generally authors need not be concerned with this information, but those with some knowledge of typographic characteristics may better understand how their manuscripts have been prepared for the printer.

The Chicago Manual of Style[1] is an excellent, convenient source of detail on elements of typography. *The Elements of Typographic Style*[3] has clear and concise descriptions of practical and aesthetic aspects of typographic design and of typefaces.

31.2.1 Typeface Styles and Weights

A typeface (sometimes called simply a "face") is a particular coherently designed group of sets of letters, numerals, and, often, additional characters such as punctuation marks and other symbols used together as needed in printing text. A font is a complete set of the characters of a typeface in a given size. The meanings of "typeface" and "font" have been blurred in recent years in electronic-based publication processes, where "font" is often applied where "typeface" has traditionally been used in the printing industry and among typographers.

The classifications of typefaces are complex. Typefaces used in printing English-language texts are generally based on the style of capital (uppercase) letters used in the classical Roman culture and related lowercase letters developed later; hence, the faces are generally called "roman typefaces". Typefaces based on the style of lowercase letters developed in Italy early in the 16th century are generally known as "italic typefaces". A second classification is that of serif and sans serif faces. Serif typefaces are those in which letters of the alphabet have cross-elements terminating the main strokes of letters; serifs were a characteristic of classical Roman letters and define present roman typefaces; sans serif typefaces have basic characteristics of roman faces but lack serifs terminating letter strokes.

Council of Science Editors	[a roman serif typeface]
European Association of Science Editors	[a roman sans serif typeface]

A third classification is that of upright and slanted (oblique) faces. Typical roman typefaces are upright (not slanted). In typography, a typeface created by slanting the vertical axis of a roman typeface is called an oblique typeface. By contrast, italic typefaces are also slanted but are designed specifically with the slanted vertical axis rather than simply slanted from a roman face. However, the similarity between oblique and italic typefaces has led to the term "italic" being also applied to slanted (oblique) letters of a roman typeface, such as those produced in many word-processing programs. Section 10.1.1 summarizes general and scientific uses of italic type. This manual follows the convention of using "roman" to mean "upright" and "italic" to mean "slanted".

Scientific Style and Format	[an italic version of a serif typeface]
Scientific Style and Format	[an italic version of a sans serif typeface]

Each particular typeface is given an identifying name, which may be that of the designer, may refer to a geographic or historical origin, or may have been coined to be evocative of the design's aesthetic character.

> the study of human physiology
> [set in a roman typeface named Times; it is a serif face related in its design to a face designed for *The Times* of London]

There are many other varieties of typefaces for the Roman alphabet, some with scientific applications, such as Fraktur faces (sometimes called "black letter script") based on Germanic origins, faces for phonetic symbols, and others not used in scientific publication, such as script typefaces designed to simulate handwritten letters. Greek letters, both uppercase and lowercase, are widely used in scientific and mathematic publications and are available in most typefaces and fonts, but they are generally represented as symbols rather than an alphabetic font.

Additional variations available within many typefaces are useful in scientific publishing; an example is small capitals, letters with the basic structure of capital letters but of the same height as lowercase letters such as the letter x or m; see Section 10.1.4 for uses of small capitals.

> COUNCIL OF SCIENCE EDITORS [in capital and small-capital letters]

A particular typeface may give rise to closely related faces differing in their weight, a characteristic determined mainly by the width of letter strokes; such a derived face with heavy or wide strokes is known as a boldface version.

> ***A clone is a type of plant cultivar.*** [in boldface italic type]

Uses of boldface type in scientific publication are summarized in Section 10.1.3.

31.2.2 Type Sizes and Spacing

Linear dimensions in the printing industry are based on a system of measurement that has the point as its basic unit; the larger unit in the system is the pica. Points and picas are defined as fractions of an inch.

> 1 point = 1/12 pica = 1/72 inch [approximately 0.35 mm]
> 1 pica = 12 points = 1/6 inch [approximately 4.2 mm]
> 6 picas = 72 points = 1 inch [approximately 25.4 mm]

Points are used to describe the size of characters in a font; the spacing between characters, between words, and between lines of type; and the thickness of ruled lines.

Picas are used to describe the dimensions of a type page (the block of text on a page), the spacing between columns of type (sometimes given in ems; see the next paragraph), and other elements in the design of pages. However, the outside dimensions of an entire page (type page and its margins) may be stated in inches (in those parts of the publishing and printing industries still working with nonmetric units).

The size of a font is defined in points by the maximal vertical dimension of its characters measured from the lowest point of "descenders" of characters (such as the part of the letter j extending below the bottom of lowercase characters such as the letters x and m in a line of type) to the highest point of "ascenders" (such as the upward stroke

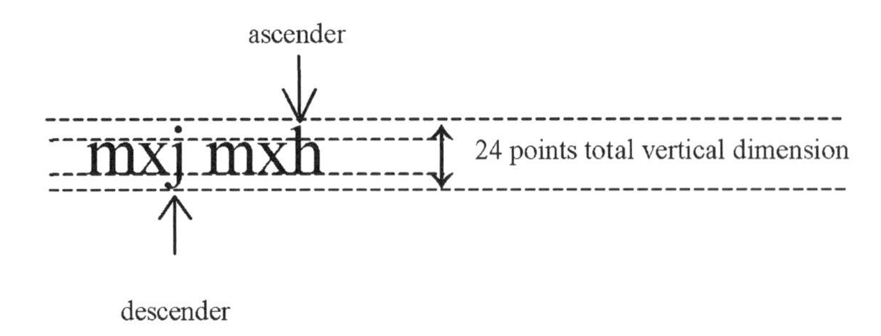

Enzymes are proteins.	[in 8-point Times]
Enzymes are proteins.	[in 10-point Times]
Enzymes are proteins.	[in 12-point Times]
Enzymes are proteins.	[in 14-point Times]
Enzymes are proteins.	[in 18-point Times]

Figure 31.1 Visual representation of ascender, descender, and sample font sizes.

of the letter h that goes above the upper line of lowercase characters such as x and m; see Figure 31.1). The height of the lowercase characters without ascenders or descenders such as x and m is called the x-height.

The basic unit of horizontal linear dimensions is the em, a name derived from the width of the capital letter M. The en has traditionally been defined as half the length of the em; however, these 2 units are less rigidly defined in today's electronic composition systems. Ems and ens are used to specify horizontal distances for such design elements as indentions, word and letter spacing, and dashes (see Sections 5.3.5.1 and 5.3.5.3 for the em dash and the en dash).The term "em quad" is used for the symbol designating a horizontal space in a type line equal to a whole-number multiple of an em: "1-em quad" for a space in a line equal to 1 em; "2-em quad" for a space equal to 2 ems.

31.3 MANUSCRIPT MARKING FOR THE TYPESETTER OR GRAPHIC DESIGNER

In traditional publication, manuscripts are generally sent to the printer with 2 kinds of marks. Copyediting marks are used to correct typing errors, misspellings, grammatical errors, and elements of publication style (such as units of measurement) not in accordance with that of the publisher. In addition, mark-up is used to specify type styles and sizes, spacing, and other typographic details. Recent advances in computer programs and software used for the preparation of manuscripts and typesetting of pages has modified, and often simplified or obviated, traditional copyediting marks and typographic mark-up.

31.3.1 Copyediting

Before a manuscript is sent to the typesetter or graphic designer for typesetting, the manuscript editor (also called "copy editor", or "subeditor" in British parlance) checks the manuscript for adherence to the journal's style specifications (such as the style details set forth in this manual), for correctness and proper formatting of references, for correctness of spelling and punctuation, for adherence to scientific conventions, and for other aspects of copy editing. Software functions that aid in editing, such as the Spelling & Grammar function, the macro function of Microsoft Word, or specialized programs for common error correction or the formatting of references, are frequently used to automate some aspects of this process. Today, most documents undergo electronic editing, and some or all of the changes may be marked with a tracking function of the software.

If the author's intent with any detail of the manuscript is not clear, the copy editor should discuss the question with the author, preferably before the manuscript is sent to the typesetter or graphic designer. Alternatively, the copy editor can decide on the apparently correct interpretation and put a query for the author within the electronic file of the manuscript (to be sent to the author with the proof). If substantive alterations may change the meaning of the text, the manuscript must be returned to the author, typically via email, for approval of changes. Publishers, working with the typesetter or graphic designer, establish the specifics of author approval (at what stage it occurs, how and in what form the document is sent to the author, and the turnaround time) to fit with other aspects of the production process and schedule.

Manuscripts should be prepared either on paper (in traditional publishing processes) or electronically and should indicate use of italics, bold-face, small capitals, special characters, and other type characteristics using either mark-up or word-processing; a requirement of this kind should be made clear in the journal's information for authors in a section on manuscript style and format (see Section 27.6.3.5 and Table 27.6).

Most manuscripts today are prepared electronically, but for those prepared on hard copy, the mark-up needed to specify type characteristics is presented in Table 31.1.

Table 31.1 Typographic conventions and markings[a]

Convention	Mark	Example	Printed appearance
Italic type[b]	Single underline	Escherichia coli	*Escherichia coli*
Boldface type[c]	Wavy underline	the Paramyxoviridae	**the Paramyxoviridae**
Capital letter[d]	Triple underline	professor of geology	Professor of Geology
Small capital[e]	Double underline	d-cystathionine	D-cystathionine

[a] Additional marks frequently needed by manuscript editors for giving instructions to printers are represented in Table 32.1.

[b] For additional discussion of the use of italic type, see Section 10.1.1.

[c] For additional discussion of the use of boldface type, see Section 10.1.2.

[d] For additional discussion of the use of capital letters, see Section 10.1.3.

[e] For additional discussion of the use of small-capital letters, see Section 10.1.4.

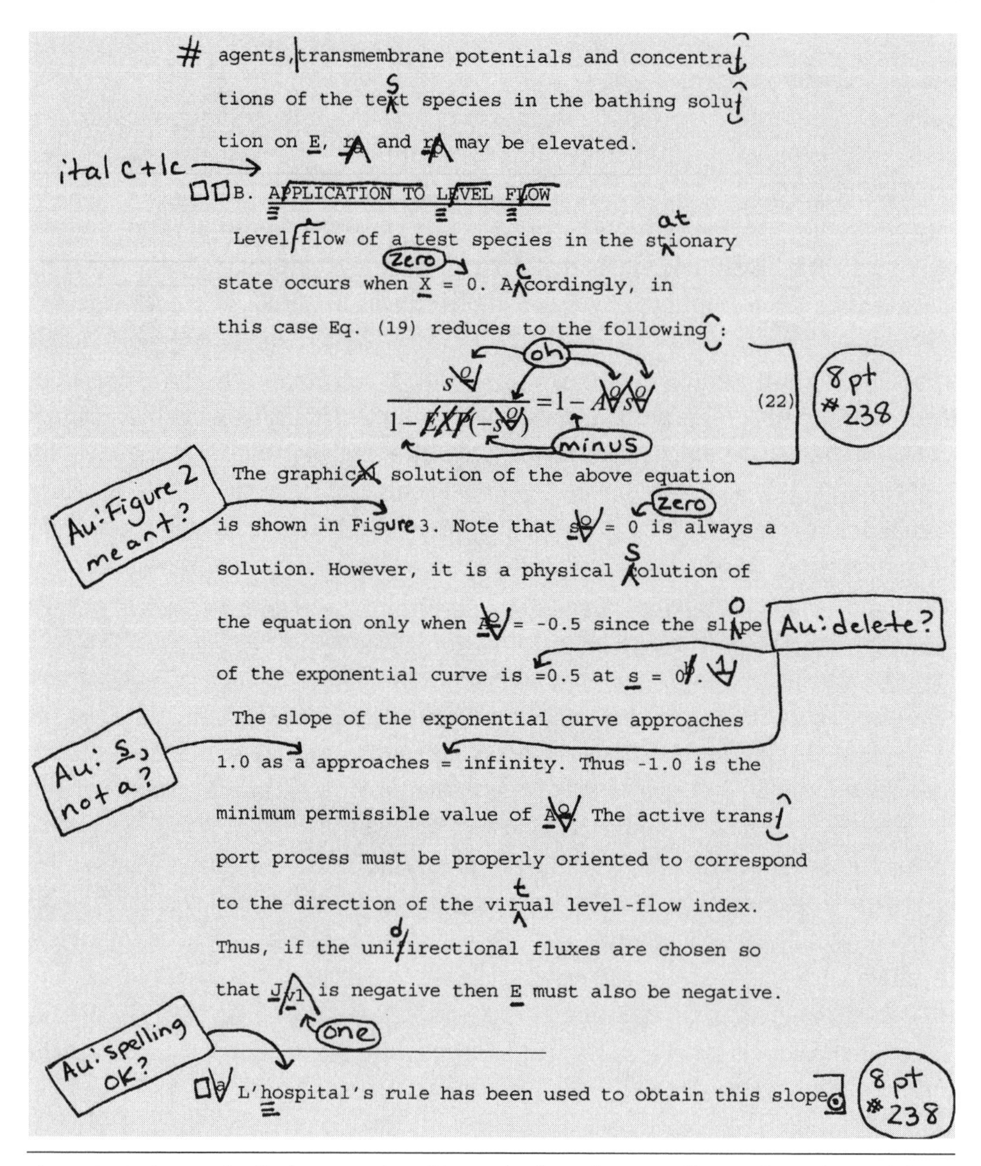
agents, transmembrane potentials and concentra-
tions of the text species in the bathing solution on E, rA and rA may be elevated.

B. APPLICATION TO LEVEL FLOW

Level flow of a test species in the stionary
state occurs when X = 0. Accordingly, in
this case Eq. (19) reduces to the following:

$$\frac{s\theta}{1-EXP(-s\theta)} = 1 - A\theta s\theta \qquad (22)$$

The graphical solution of the above equation
is shown in Figure 3. Note that s = 0 is always a
solution. However, it is a physical tolution of
the equation only when A = -0.5 since the slape
of the exponential curve is -0.5 at s = 0.

The slope of the exponential curve approaches
1.0 as a approaches infinity. Thus -1.0 is the
minimum permissible value of A. The active transport process must be properly oriented to correspond
to the direction of the virual level-flow index.
Thus, if the unitirectional fluxes are chosen so
that J_{v1} is negative then E must also be negative.

L'hospital's rule has been used to obtain this slope.

Figure 31.2 A page of edited manuscript with queries from the copy editor to the author and directions to the printer.

Figure 31.2 is an example of a manuscript showing both copyediting and mark-up. See *The Chicago Manual of Style*[1] for detailed instructions on copyediting.

31.3.2 Traditional and Electronic Mark-Up

The specifications for a journal's paper, design, and detailed typographic characteristics are usually established by the journal's publisher or editor in consultation with a graphic designer, who can propose ways to convey the publisher's intentions using graphical elements. Specifications may be developed when a new journal is about to begin publication or when the design of an existing journal is changed.

Table 31.2 Specifications for a journal: dimensions, paper, design, and typographic details

Characteristic	Notes
Trim size	Overall dimensions of the cover and pages: width and height[a]
Paper: kind and weight	Selection determined by kind and quality of paper needed and cost; the kind influences the quality of color and halftone illustrations
Layout of pages[b]	
Type-page margins	Space around the type page (area for text)
Columns	1-column, 2-column, or 3-column layout of the type page
Typefaces	Determined in part by aesthetic and readability considerations, but needs for fonts of special characters, as in chemistry, mathematics, and physics, may limit the choice
Text sizes	For titles; body; references; figure captions; and table titles, headings, bodies, and footnotes
Text colors[c]	For titles, text heads, and in-text call-outs for tables, figures, appendixes
Leading	Space between lines, specified for each size of type
Title page for articles	Layout of necessary elements[d]
Text format	Indentations; spacing between paragraphs and around figures and tables
Text heads and subheads	Typefaces and sizes; decision on number and type characteristics of different levels[e]
Table characteristics	Rules, spacing, column and row headings, footnote signs[f]
Design of special pages	Contents page, information for authors page[g]
Cover design	Layout, title elements, and other details[h]

[a] See Section 27.6.1.
[b] See Section 27.6.3.4.
[c] Color may be an expensive and impractical aspect for a printed publication, but for online-only publications, it may be an option. Be sure to choose colors that are aesthetically pleasing on all computer monitors, regardless of individual monitor color calibration.
[d] See Section 27.7.1.
[e] See Section 27.7.2.1.
[f] See Section 30.1 and its subsections.
[g] See Sections 27.6.3.3 and 27.6.3.5.
[h] See Section 27.6.3.1.

Detailed specifications help the journal's manuscript editors decide what needs to be marked in manuscripts as instructions to the typesetter or graphic designer and what will be understood from the agreed-on specifications. The details most often established for a journal's specifications are summarized in Table 31.2.

In traditional publishing processes, elements in the manuscript for which specifications have been defined and that are readily identified by the typesetter need not be marked (e.g., running text in Roman-alphabet letters with normal capitalization, single-level headings, article titles, and abstracts). Elements must be marked if there may be ambiguity (such as superscripts, subscripts, special characters, and different levels of headings) or if there are no specifications. In hard-copy mark-up, place instructions to the printer immediately adjacent to the element in question and circle the instructions to indicate that they are not editing revisions to the manuscript.

Queries to the author that arise during copyediting must be clearly distinguished from directions to the typesetter or graphic designer (mark-up) and from copyediting marks. For example, author queries can be placed within a square enclosure and should be marked specifically for the author's attention (e.g., "?A:" or "Au: . . . ?"). For manuscripts submitted in electronic format, the copy editor generally returns the copyedited manuscript with queries to the author embedded in the text or in electronic comments, or presented separately at the beginning or end of the manuscript, or in an accompanying letter (email message).

Development of electronic typesetting and layout has allowed automatic application of typefaces, styles, and formats to components of books and journals. This may be accomplished in various ways. Some systems convert styles applied in word-processing programs into type characteristics in layout software or into mark-up languages. Alternatively, tags or codes that designate the functional elements of text, equations, and tables are applied manually or in an automated way to electronic manuscripts. These may take the form of generic codes, style tags, or mark-up languages.[1] The system that processes the document for print layout or electronic publication is programmed to identify the tags in the electronic manuscript and to automatically apply stipulated specifications (such as for typeface, type weight, and type size) to the corresponding tagged or coded components.

Although some tagging systems are proprietary, standards intended to lead to uniformity and interchangeability in electronic manuscripts have been developed and are in general use. The National Information Standards Organization standard for preparing and tagging electronic manuscripts[4] draws on the international standard that defines the Standard Generalized Markup Language (SGML) (ISO 8879[5]). The Extensible Markup Language (XML),[6] now widely used for tagging documents intended for electronic delivery, is derived from SGML but is less complex. Both XML and SGML are metalanguages, "languages with which to make up and define mark-up systems".[7] As such, they specify the structure of a document, rather than its appearance; in contrast, the HyperText Markup Language (HTML; also derived from SGML), eXtensible HyperText Markup Language (XHTML), HTML5 (an improved and more advanced version of HTML, anticipated for release in 2014[8]), and Adobe Acrobat's Portable Document Format (PDF), which are used for electronic display, define appearance but not structure.[7,9] XML and SGML do not specify tags as such; rather, tags are defined by the user according to need (e.g., <heading> for a heading). A variety of processing systems are then used to prepare the tagged document for presentation through any of several media (e.g., print, CD-ROM, Internet, e-books, or mobile devices). Each system has a style sheet to convert the tags to its own coding system, which determines the final appearance of each element.[7] These tags are also valuable for allowing the content of a document to be further manipulated to extract additional utilities from the tagged material (e.g., increasing reusability and searchability in online abstracting or indexing databases).

The document type definition (DTD) defines the overall structure of a document to be coded in a formal mark-up language. It specifies the tags to be used and the rules for using them. The DTD is a required component of documents marked in SGML. DTDs are optional with XML, but their use helps to establish consistency in a set of documents.[7]

For further discussion of document structure and XML, see Chapter 1.

CITED REFERENCES

1. The Chicago manual of style: the essential guide for writers, editors, and publishers. 16th ed. Chicago (IL): University of Chicago Press; 2010. Also available at http://www.chicagomanualofstyle.org.

2. Einsohn A. The copyeditor's handbook: a guide for book publishing and corporate communications. 3rd ed. Berkeley (CA): University of California Press; 2011.

3. Bringhurst R. The elements of typographic style. 2nd ed. Vancouver (BC): Hartley & Marks; 2002.

4. National Information Standards Organization. Electronic manuscript preparation and markup. Bethesda (MD): NISO Press; 1995. (NISO/ANSI/ISO 12083 - 1995 (R2002)).

5. International Organization for Standardization. Information processing—text and office systems—Standard Generalized Markup Language (SGML). Geneva (Switzerland): ISO; 1989. (ISO 8879). ISO 8879: 1986/Amd 1:1988; ISO 8879:1986/Cor 1:1996; ISO 8879:1986/Cor 2:1999.

6. W3C XML Core Working Group. Extensible Markup Language (XML) 1.0. 5th ed. W3C Recommendation. [place unknown]: W3C Consortium; 2008 [accessed 2013 Jan 27]. http://www.w3.org/TR/xml/.

7. Kasdorf B. XML and PDF: why we need both. An introduction to the two core technologies for publishing. JP News J Publ. 2003;(2):1,3–14. Also available at http://www1.allenpress.com/newsletters/pdf/JP-2003-02.pdf.

8. W3C HTML Working Group. W3C confirms May 2011 for HTML5 last call, targets 2014 for HTML5 Standard. [accessed 2012 June 18]. http://www.w3.org/2011/02/htmlwg-pr.html.

9. Beebe L, Meyers B. Digital workflow: managing the process electronically. Hanover (PA): Sheridan Press; 2000.

ADDITIONAL REFERENCES

Butcher J, Drake C, Leach M. Butcher's copy-editing: the Cambridge handbook for editors, copy-editors, and proofreaders. 4th ed. Cambridge (UK): Cambridge University Press; 2006.

Copy Editor: Language News for the Publishing Profession. Vol. 1. Phoenix (AZ): McMurry; 1990– .

The Editorial Eye. Vol. 1. Alexandria (VA): EEI Press; 1978– .

Judd K. Copyediting: a practical guide. 3rd ed. Menlo Park (CA): Crisp Publications; 2001.

Kasdorf WE, editor. The Columbia guide to digital publishing. New York (NY): Columbia University Press; 2003.

New Hart's rules: the handbook of style for writers and editors. Oxford (UK): Oxford University Press; 2005.

32 Proof Correction

32.1 PROOFS

Most publishers provide authors with the printer's proofs of the text (article, chapter, book, or other document) being published, including proofs of any illustrations. A proof used to be defined as an impression on paper of text in typeset form and illustrations as prepared for publication; however, in the increasingly common "paperless" offices of today, a proof is now more commonly identified as a Portable Document Format (PDF) representation of the formatted and typeset text with tables and figures incorporated and set in their final locations.

Proofs, which are usually sent as PDF attachments to an email (or may be sent via regular mail, although this is no longer recommended), are sometimes accompanied by a copy of the original manuscript showing the changes made by the editor or copy editor. The author is responsible for checking the proof of the typeset text and illustrations against the originals and for making any corrections on the proof. Usually this is the last time the author sees the text and illustrations before publication, so thoroughness is critical. Proofs must be returned to the printer, editor, or publisher by a specific deadline; response by email is often requested, either in the form of an attached annotated PDF or some other electronic method. See Section 32.5 for details about annotating a PDF.

Editors should inform corresponding authors when to expect the arrival of proofs so authors can take appropriate steps if they will not be available at that time, such as asking a coauthor or colleague to check the proofs. Manuscripts prepared in camera-ready form for offset printing must be proofread carefully before being submitted to the publisher (authors will likely not see proofs of papers prepared for this type of publication).

Some publishers do not supply proofs of any kind; instead, they send the author the copyedited manuscript before typesetting. The author checks the correctness of any changes made by the copy editor and responds to any queries raised. The corrections are made before the manuscripts are sent for typesetting, which minimizes changes at later stages. This system works well for manuscripts that are copyedited electronically.

32.2 KINDS OF PROOF

The text of the manuscript and any accompanying tabular material may be set in type by various systems. In hot-metal composition, the standard system for many years, type was cast from molten metal with a Linotype or Monotype machine. The metal lines or pieces of type so produced were placed in long trays called galleys, and proofs were produced by direct impression of the inked type onto paper (galley proofs). Changes were marked on the paper proofs, corrections were made in the type, and then text, figures, and tables were assembled in page format. In recent years, however, electronic and automated composition, either by an external supplier or an in-house layout department, has effectively replaced hot-metal composition. In electronic composition, proofs are generally prepared in the format of pages, showing all components of the document, and are supplied to authors as PDF page proofs rather than paper galley proofs.

The kinds of proof provided for line drawings and continuous-tone (halftone) illustrations will depend on the printing or electronic distribution process to be used for the publication. For example, authors may receive a PDF version of line art, but hard-copy proofs, with quality similar to that of the final publication, may be needed for halftone or color illustrations.

32.3 CHECKING PROOFS

Proofreading encompasses a wide variety of tasks, such as ensuring that all components of the document are present, verifying that document specifications have been followed, checking end-of-line hyphenation, and checking page numbers and running heads. Several important aspects of proof correction are described briefly here; detailed guidelines for editors and proofreaders are available in *The Chicago Manual of Style*[1] and *Mark My Words*.[2]

32.3.1 Text and Composed Tables

32.3.1.1 Author's Responsibilities

When authors are asked to check proofs, the following instructions should accompany the proofs.

1) The author should not mark the original manuscript if it is returned with proofs. It is the record of what the printer was asked to typeset and is the basis for billing for author's and editor's alterations.
2) The author should only comment on or respond to, not delete or change, any of the queries, suggestions for changes, and corrections made by the editor or a proofreader at an earlier proofing step.

3) The author should not delete any identifying or other information on proofs that will be needed by the printer; these marks will be deleted when the pages are made up.
4) The author should plan to read proofs at least twice. In the first reading the author should follow the text of the proof while another person reads the manuscript aloud, to ensure that all components of the manuscript have been included and set correctly. In the second reading the author should read the proof alone, for sense.
5) Authors are expected to limit changes at this stage to the correction of errors; the proof stage is not the time to make trivial changes, improve prose style, add new material (unless absolutely essential to the scientific or scholarly outcome), or delete existing material. Corrections cost more than initial composition, and errors may be introduced.

Printer's errors (PEs) should be corrected at no cost to the journal or author. Author's alterations (AAs) are usually charged to the publisher and ultimately may be charged to the author. Authors and editors should indicate printer's errors by marking "PE" near the correction. For text revisions in older composition systems, it was advisable to balance the number of any characters and spaces added by deletion of an equal number of characters and spaces (and vice versa), to minimize the effect of changes on other lines of text. In electronic systems, however, character-for-character replacement is less critical.

Authors may want to add at the proof stage some text on important observations made since submitting the manuscript. The decision to allow such additions must be that of the editor, rather than the copy editor. Adding new content to a peer-reviewed article under an old "received" date is generally considered unethical if that content has not been judged for its acceptability by the peer reviewers or the scientific editor. The editor may suggest including a dated addendum or "note added in proof" containing the new material, which will obviate the need for changes in the text.

32.3.1.2 Accuracy

Carefully check proofs against the original text for accuracy of equations and numeric data (especially in tables) and for proper spelling, punctuation, separation of paragraphs, order of headings, and citation of references, figures, and tables. Pay attention to the appropriate location of tables and figures in relation to their first mention in the text.

32.3.1.3 Typography

Check letters and paragraphs closely for possible repetition or transposition. Copy editors at the publisher's office and at the printing plant are responsible for correcting poor typography and alignment, but authors should verify the proper alignment of statistical data, chemical formulas and equations, and mathematical expressions. Check symbols for agreement with recognized conventions.

32.3.1.4 Word Division

Check end-of-line word breaks on proofs carefully. Most computer systems for composition have their own rules for hyphenating words at the ends of lines. The extent to which such systems are programmed to handle exceptions to general rules differs, so take special care in checking proofs of text prepared in this way. In the American system, word division is based on pronunciation, whereas in the British system it is based on the etymologic derivation of words. In checking divided words, pay particular attention to Latin words and specialized scientific terms. Some principles are outlined here.

Each Latin word is divided into as many syllables as it has separate vowels and diphthongs. Use the following guidelines for dividing such words, including the scientific names of organisms.

1) Join the consonant "h" between 2 vowels with the second vowel (rni-hi, co-hors).
2) Join an "x" between 2 vowels with the preceding vowel (sax-urn, ax-il·la).
3) Do not break the combinations "ch", "ph", "th", "gu", and "qu", which are treated as single consonants (Te-thys, lin-gua).
4) Separate "gl", "tl", and "thl" (Ag-la·o·ne·ma, At-las, ath-let·i·cus).
5) Join a single consonant between the last 2 vowels of the word or between any 2 unaccented vowels with the final or second vowel (pa-ter), but join a single consonant before or after an accented vowel with the accented syllable (i-tin´·e·ra, dom´-i·nus).
6) Separate 2 consonants between vowels (cor-pus, for-ma).
7) When 3 consonants occur between vowels, join the last consonant with the vowel that follows (emp-tor).
8) When 4 consonants occur between vowels, join 2 consonants to each vowel (trans-trum).

As in other aspects of Latin grammar, there are exceptions to these rules, but they are too technical or unusual to merit treatment here.

Take special care with scientific terms that have hyphens as part of their proper structure; chemical names are the most frequently used terms of this kind (see Chapter 17). When such words must be broken at the end of a line, the break should, if possible, use the word's existing hyphen as the end-of-the-line hyphen; an additional hyphen should not appear at the beginning of the following line.

L-glutamine amido- ligase	*not*	L-glutamine amido -ligase

For an equation or other mathematical expression in running text that must be broken at the end of the line, try to have an operator symbol (such as the equals symbol) at the end of the line, not at the beginning of the following line (see also Section 12.4).

. . . and if x = 3.51*y* + 75.89, then . . .	*not*	. . . and if x = 3.51*y* + 75.89, then . . .

If possible, avoid a break that produces part of a word that can be read as a word in its own right with a meaning that might be confusing, amusing, or offensive in the context. If a break cannot be avoided, break the word so as to avoid the possibly disturbing effect even if a standard rule for breaks must be violated; often an inappropriate break can be avoided by rearranging the sentence.

More detailed guidance on word breaks can be found in *New Hart's Rules*.[3] *The Chicago Manual of Style*[1] includes recommendations for word-breaks in several non-English languages.

32.3.2 Illustrations

Ensure that all figures are present and that they have been numbered and oriented correctly. If figure proofs are supplied separately from the text and there is a potential for incorrect orientation in the layout, mark the top of each figure proof.

Table 32.1 Proofreaders' marks and symbols

Instruction	ANSI–NISO Standard marks[a]	
	Marginal mark	In-line mark
Delete	ℓ	the red book
Close up	⁀	the bo ok
Delete and close up	ℓ	the bfook
Restore deletion	stet	the red book
Insert in line	red	the book
Substitute in line	red	the black book
	e	tha book
Insert space in line	#	thebook
Equalize spacing	eq #	the yellow book
Lead (space between lines)	# or ld	The red book was lost.
Remove leads (space) between lines	ℓ# or ℓld	The red book was found.
Insert hair space or thin space	hr# or thin #	100/000
Begin new paragraph	¶ or L	The red book was lost.¶The black book was found.
Run paragraphs together	no ¶	The black book was lost. The red book was found.
Insert 1-em quad (indent)	□	The red book
Insert 2-em quad (indent)	□□ or 2	was found
Insert 3-em quad (indent)	□□□ or 3	at night.
Move to left	[	the book
Move to right	]	the book
Center	ctr	]the book[
Move up	⌐¬	the book
Move down	└┘	the book

Examine the proof of each line drawing to ensure that all elements (lettering, symbols, lines) are present and legible. Indicate clearly any corrections that are needed. Depending on the process used for illustrations, the printer may be able to make the corrections or a corrected original from the author may be required.

Check halftone proofs to ensure that any necessary labels are present and that areas of interest are visible (contrast not too light or dark).

Review each figure in conjunction with its caption to ensure that the numbering matches and to confirm that the caption adequately describes the figure.

Table 32.1 Proofreaders' marks and symbols (*continued*)

British Standards Institution marks[b]		
Marginal mark	In-line mark	Corrected text
	the red book	the book
	the bo ok	the book
	the bbook	the book
	the red book	the red book
red	the book	the red book
red	the black book	the red book
e	thd book	the book
	thebook	the book
	the yellow book	the yellow book
extend text mark [specify the space if necessary]	The red book was lost.	The red book was lost.
extend text mark [specify the space if necessary]	The red was found.	The red book was found.
thin	100000	100 000
	The red book was lost. The black book was found.	The red book was lost. The black book was found.
	The black book was lost. The red book was found.	The black book was lost. The red book was found.
1	The red book	The red book
2	was found	was found
3	at night.	at night.
	the book	the book
	the book	the book
	the book	the book
	the book	the book
	the book	the book

32.3.3 Documents for Online Publication

For material that is intended for online publication, either instead of or in addition to print publication, the same considerations of accuracy and completeness apply, but additional aspects must be checked. In particular, the proofreader or author must confirm the appearance and functionality of the electronic document in the intended medium (e.g., special characters must appear as intended on screen, and links among parts of the document and to external resources must work).

Table 32.1 Proofreaders' marks and symbols (*continued*)

	ANSI–NISO Standard marks[a]	
Instruction	**Marginal mark**	**In-line mark**
Align vertically	‖ *or* align	The book / was lost / in the fog.
Align horizontally	= *or* straighten	The book was found.
Transpose	tr.	The book was found. The found book was .
Spell out	(sp)	He arrived 1st.
Push down quad (spacing material)	⊥	the book
Reset broken letter	×	the book
Turn right side up	9	the green book
Lowercase letter	lc	the Green book
Capitalize as marked	cap	The good Book
Set as small capital	sc	D-glucose
Set in italic type	ital	The Good Book
Set in roman type (upright type)	rom	the *book*
Set in boldface type	bf	The Good Book
Set in lightface type	lf	The **book**
Set in capitals and small capitals	c+sc	Dong geng
Set in boldface italics, capitals and lowercase	bf ital c+lc	a style manual
Wrong font; reset	wf	body type
Reset as superscript (superior)	2	*e* = *mc*2
Reset as subscript (inferior)	2	H2S
Insert as superscript (superior)	b	1203
Insert as subscript (inferior)	2	HO
Period (full stop)	⊙	Read the book
Comma	,	leaves, buds and branches
Semicolon	;	Think then decide.
Colon	:	Read these books
Hyphen	=/=	graft versus host disease
Apostrophe	'	*Donovans Demise*
Double quotation marks	"/"	He said book.

Table 32.1 Proofreaders' marks and symbols (*continued*)

British Standards Institution marks[b]		
Marginal mark	**In-line mark**	**Corrected text**
‖	The book was lost in the fog.	The book was lost in the fog.
=	The book was found.	The book was found.
	The book was found. The found book was.	The book was found. The book was found.
first	He arrived 1st.	He arrived first.
⊥	the book	the book
×	the book	the book
	the green book	the green book
≠	the green book	the green book
≡	The good Book	The Good Book
=	D-glucose	D-glucose
	The Good Book	*The Good Book*
	the *book*	the book
	The Good Book	**The Good Book**
	The **book**	The book
	Dong geng	Dong Geng
	a style manual	***A Style Manual***
⊗	body type	body type
	$e = mc2$	$e = mc^2$
	H2S	H_2S
	1203b	1203^b
	H2O	H_2O
⊙	Read the book	Read the book.
,	leaves, buds and branches	leaves, buds, and branches
;	Think then decide.	Think; then decide.
:	Read these books	Read these books:
	graft versus host disease	graft-versus-host disease
'	*Donovans Demise*	*Donovan's Demise*
	He said book.	He said "book".

Table 32.1 Proofreaders' marks and symbols (*continued*)

	ANSI–NISO Standard marks[a]	
Instruction	**Marginal mark**	**In-line mark**
Single quotation marks	‘/’	"Don't cry Fire!"
Question mark	?	Is this your book/
En-dash (rule)	1/N	pages 10 15
Em-dash	1/M	His book find it!
3-em dash	3/M	His reply, "nuts"!
Parenthesis marks	(/)	report Smith 1992 was
Brackets (square brackets)	[/]	the book a manual
Slash (slant line, oblique)	/	5 ms

[a] National Information Standards Organization.[4]
[b] British Standards Institution.[5]

32.4 MARKING PAPER PROOFS

Although hard-copy proofs are decreasing significantly in use, some journals continue to use them or allow authors to return hand-marked printed copies of their PDF proofs. For marking corrections on paper or printed-out proofs, use standard proofreaders' marks. These differ slightly in the American[4] and British[5,6] systems. No standardized system of European proofreaders' marks exists, many of which differ from those of the American and British systems; most European printers recognize American and British marks or provide authors with lists of marks they prefer. See Table 32.1 for the American and British marks.

Make proof corrections in the margins of the galley or page proofs (not above lines of type as in correcting a manuscript; see Section 31.3.2 and Figure 31.2). In examining proofs for needed changes, the typesetter or compositor scans the margins of proofs and makes only the alterations that are indicated there.

If the printer has supplied hard-copy proofs, make marks for corrections in colors different from any already on the proof, or use the colors specified on instructions accompanying the proofs. For example, if the printer's proofreader has marked corrections in green, printer's errors can be indicated in red and author's alterations in blue. If proofs are to be returned by fax, use a tool that will generate markings suitable for transmission (i.e., not pencil). Make the corrections neatly so that they will be clearly legible. If a blanket order is called for (e.g., "set *Rosa* in italics throughout"), mark each occurrence of the change on the proof.

Queries to the author from the editor or the printer may appear in the margins of the proof. The author should respond to all such queries, as follows. To have the line remain as set, the author should draw a line through the query but not erase it. If the editor or a proofreader has indicated a change that is not acceptable, the author should

Table 32.1 Proofreaders' marks and symbols (*continued*)

	British Standards Institution marks[b]	
Marginal mark	**In-line mark**	**Corrected text**
‘ ’ ?	"Don't cry/Fire!"	"Don't cry 'Fire'!"
1 en	Is this your book/	Is this your book?
1 em	pages 10/15	pages 10–15
3 em	His book/find it!	His book—find it!
( )	His reply, /####/!	His reply, ———!
[ ]	report/Smith 1992/was	report (Smith 1992) was
⊘	the book/a manual/	the book [a manual]
	5 ms	5 m/s

draw a line through the suggested change, add a marginal note ("OK as set" or "stet"), and place dots under the material that must be retained. Notations that are not to be set in type, such as a note to the printer, must be circled.

32.4.1 Correction Marks Needed

For each correction, at least 2 marks are needed on the proof: one or more in the text (in-line) and one or more in the margin (marginal) nearest the in-line mark; see Figure 32.1. In-line marks include the caret (^) to show where an addition is to be made and a line drawn through a character or a word to be deleted. A marginal mark specifies the change. This may take the form of one or more characters or words to be inserted, or it may be a proofreader's symbol, such as a space or deletion sign. Make each mark on an imaginary extension (preferably to the right) of the line to which it applies. To indicate several corrections for a single line, arrange them in sequence from left to right in the nearer margin, and separate adjacent corrections by a slash (/).

32.4.2 Long Corrections

If a correction or insertion consists of more than 1 or 2 lines, print it on a separate sheet and attach the sheet to the proof at the appropriate place with tape or paper clip, never with a pin or staple. Indicate clearly where the new copy is to be inserted. Mark each insert with a letter and the number of the proof sheet to which it belongs—for example, "A for proof sheet 2".

32.5 ANNOTATING PDF PROOFS

As publishers, printers, and journal offices move toward paperless workflows, it is becoming standard practice to send PDF proofs via email instead of mailing hard-copy galley

Proof Marked for Corrections

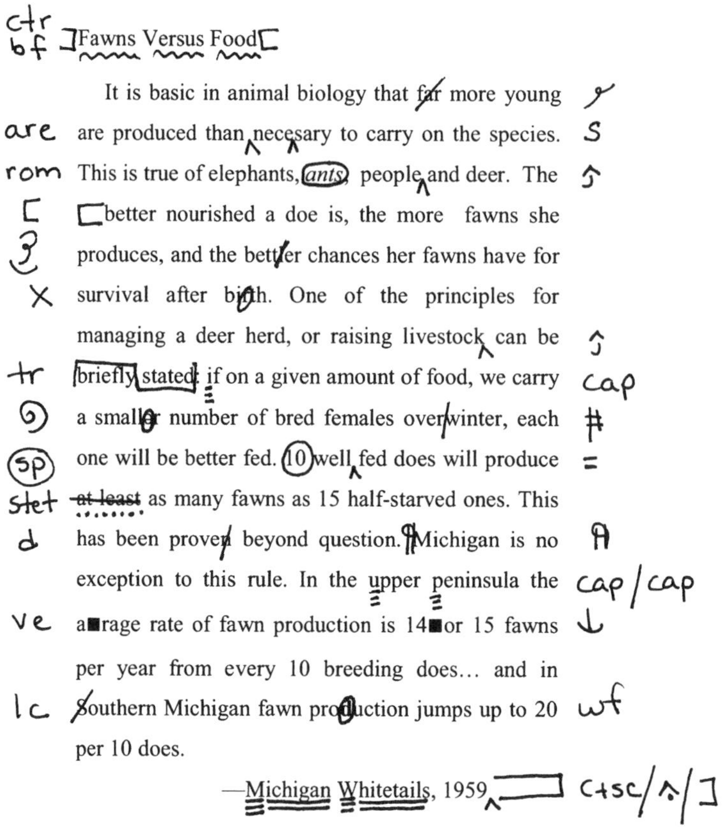

Fawns Versus Food

It is basic in animal biology that far more young are produced than necsary to carry on the species. This is true of elephants, ants, people and deer. The better nourished a doe is, the more fawns she produces, and the bettter chances her fawns have for survival after birth. One of the principles for managing a deer herd, or raising livestock can be briefly stated: if on a given amount of food, we carry a smallr number of bred females overwinter, each one will be better fed. 10 well fed does will produce ~~at least~~ as many fawns as 15 half-starved ones. This has been prove beyond question. Michigan is no exception to this rule. In the upper peninsula the a■rage rate of fawn production is 14■or 15 fawns per year from every 10 breeding does... and in Southern Michigan fawn prouction jumps up to 20 per 10 does.

—Michigan Whitetails, 1959

Proof after Corrections

Fawns Versus Food

It is basic in animal biology that more young are produced than are necessary to carry on the species. This is true of elephants, ants, people, and deer. The better nourished a doe is, the more fawns she produces, and the better chances her fawns have for survival after birth. One of the principles for managing a deer herd, or raising livestock, can be stated briefly: If on a given amount of food, we carry a smaller number of bred females over winter, each one will be better fed. Ten well-fed does will produce at least as many fawns as 15 half-starved ones. This has been proved beyond question.

Michigan is no exception to this rule. In the Upper Peninsula the average rate of fawn production is 14 or 15 fawns per year from every 10 breeding does … and in southern Michigan fawn production jumps up to 20 per 10 does.

— MICHIGAN WHITETAILS, 1959.

Figure 32.1 Marked and corrected paper proof.

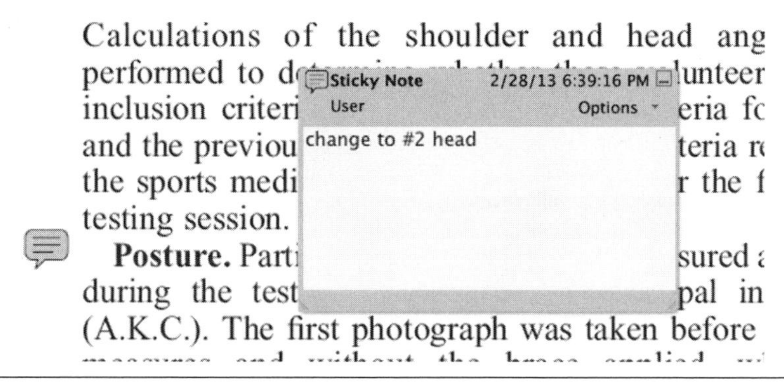

Figure 32.2 Sample of a PDF annotated with a Sticky Note containing an author's correction.

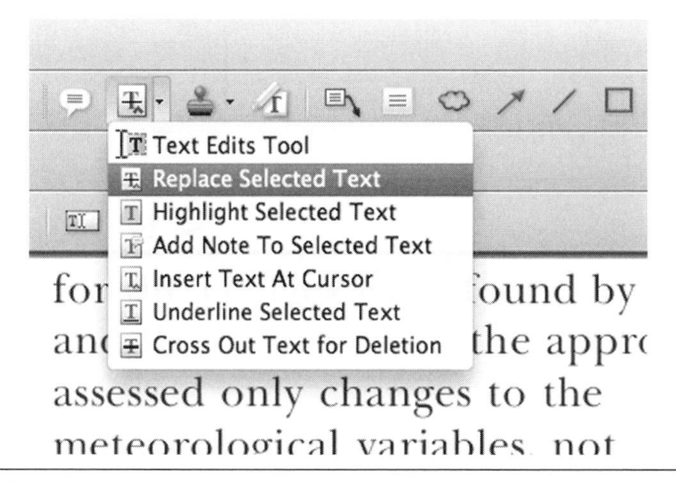

Figure 32.3 Adobe Acrobat Professional offers several tools for marking corrections.

proofs. There are many "best practices" available for how to annotate (mark) PDFs, but much of it depends on which version of Adobe Acrobat you have on your computer. If you do not already have Adobe Reader on your computer, you may download the most recent version for free at http://get.adobe.com/reader/. Adobe Reader will allow you to mark corrections on PDF proofs using the Sticky Note function (see Figure 32.2). If you have access to a version of Adobe Acrobat Professional, you will be able to use other editing tools such as the Replace Text, Insert Text, and Cross Out functions (see Figures 32.3 and 32.4).

32.6 RETURNING PROOFS

Authors must return proofs to whomever the publisher stipulates in the instructions sent with the proofs. If illustrations are to be remade, the printer must have the original artwork.

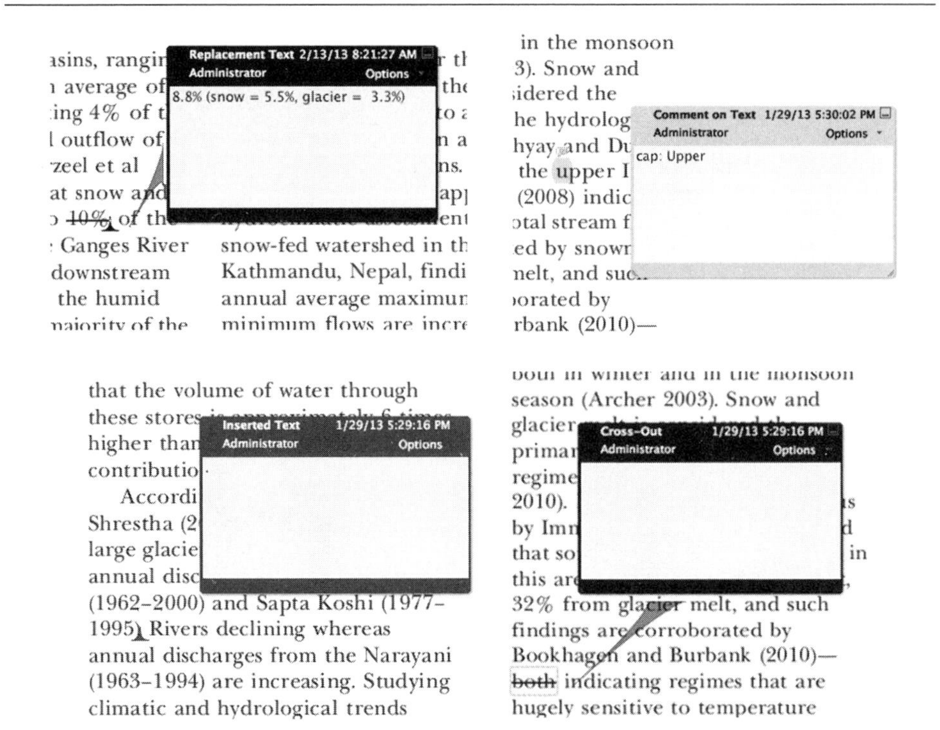

Figure 32.4 Annotated PDF proof corrections.

CITED REFERENCES

1. The Chicago manual of style: the essential guide for writers, editors, and publishers. 16th ed. Chicago (IL): University of Chicago Press; 2010. Also available at http:///www.chicagomanualofstyle.org.

2. Smith P. Mark my words: instruction and practice in proofreading. 3rd ed. Alexandria (VA): EEI Press; 1997.

3. New Hart's rules: the handbook of style for writers and editors. New York (NY): Oxford University Press; 2005.

4. National Information Standards Organization. Proof corrections: American National Standard proof corrections. Bethesda (MD): NISO; 1991. (ANSI/NISO Z39.22-1989.) No longer in print.

5. British Standards Institution. Copy preparation and proof correction: specification for typographic requirements, marks for copy preparation and proof correction, proofing procedure. London (UK): British Standards Institution; 2005. (BS 5261-2:2005).

6. Butcher J, Drake C, Leach M. Butcher's copy-editing: the Cambridge handbook for editors, copy-editors, and proofreaders. 4th ed. Cambridge (UK): Cambridge University Press; 2006.

ADDITIONAL REFERENCE

Lomangino K, Kaufman CS, Wills AJ. Digital art in scholarly periodical publishing. Hanover (NH): Sheridan Press; 2002. Also available at http://www.kwfco.com/sites/default/files/resources/Digital%20Art%20White%20Paper.pdf.

BIBLIOGRAPHY

The works listed here are books and other sources useful in scientific writing and publishing. Additional sources with narrower scope are listed under "References" or "Additional References" at the end of each chapter.

STYLE MANUALS AND OTHER WRITING GUIDES

General

Batschelet M. Web writing/Web designing. Boston (MA): Allyn and Bacon; c2001.

The Chicago manual of style: the essential guide for writers, editors, and publishers. 16th ed. Chicago (IL): University of Chicago Press; 2010.

Dorner J. Writing for the Internet. Oxford (UK): Oxford University Press; 2002.

Editing Canadian English. 2nd ed. Toronto (ON): Macfarlane Walter & Ross; 2000. Prepared for the Editors' Association of Canada.

Electronic publishing: guide to best practices for Canadian publishers. Ver. 1.0. Ottawa (ON): National Library of Canada; 2001 [accessed 2013 Feb 4]. http://www.collectionscanada.ca/obj/p13/f2/01-e.pdf.

Garrand TP. Writing for multimedia and the Web: a practical guide to content development for interactive media. 3rd ed. Boston (MA): Focal Press; c2006.

Hacker D, Sommers N. A pocket style manual. 6th ed. Boston (MA): Bedford/St. Martin's; c2011.

Hammerich I, Harrison C. Developing online content: the principles of writing and editing for the Web. Hoboken (NJ): John Wiley & Sons, Inc.; 2002.

Lipson C, Day M, editors. Technical communication and the World Wide Web. Mahwah (NJ): Lawrence Erlbaum Associates; 2005.

Lunsford AA. Easy writer: a pocket guide. 3rd ed. Boston (MA): Bedford Books; c2005.

Maciuba-Koppel D. The Web writer's guide. New York (NY): Focal Press; c2002.

Merriam-Webster's manual for writers and editors. Completely rev. ed. Springfield (MA): Merriam-Webster; 1998.

Modern Language Association. The MLA style manual and guide to scholarly publishing. New York (NY): Modern Language Association of America; 2008.

Modern Language Association. MLA handbook for writers of research papers. 7th ed. New York (NY): Modern Language Association of America; 2009.
New Hart's rules: the handbook of style for writers and editors. Oxford (UK): Oxford University Press; 2005.
Newman J, Cusick E, La Tourette A, editors. The writer's workbook. London (UK): Arnold; c2004.
Public Works and Government Services Canada, Translation Bureau. The Canadian style: a guide to writing and editing. 2nd ed. Toronto (ON): Dundern Press; 1997.
Ritter RM. The Oxford guide to style. Oxford (UK): Oxford University Press; 2002.
Ross-Larson B. Writing for the Information Age: elements of style for the 21st century. New York (NY): W.W. Norton & Company, Inc.; 2002.
Ruszkiewicz JJ, Friend CE, Seward DE, Hairston Emerita ME. The Scott, Foresman handbook for writers. 9th ed. Essex (UK): Longman; c2010.
Thibault D. Bibliographic style manual. Ottawa (ON): National Library of Canada; 1990.
Thibault D. Bibliographic style manual. Section 3, Electronic documents. Ottawa (ON): National Library of Canada; 1998.
Turabian KL. A manual for writers of term papers, theses and dissertations. 7th ed. Chicago (IL): University of Chicago Press; 2007.
Sabin WA. The Gregg reference manual: a manual of style, grammar, usage, and formatting. 11th ed. Boston (MA): McGraw-Hill/Irwin; 2010.
Skillin ME, Gay RM. Words into type. 3rd ed. Englewood Cliffs (NJ): Prentice-Hall; 1974.
Strunk W, White EB. The elements of style. 4th ed. New York (NY): Simon and Schuster Trade; 1999.
Style manual for authors, editors and printers. 6th ed. Brisbane (Australia): John Wiley & Sons, Australia Ltd; 2002.
United States Government Printing Office style manual. Washington (DC): The Office; 2008. Also available at http://www.gpoaccess.gov/stylemanual/.
Weiss EH. The elements of international English style: a guide to writing correspondence, reports, technical documents, and internet pages for a global audience. Armonk (NY): M.E. Sharpe; c2005.

Scientific

AIP style manual. 4th ed. New York (NY): American Institute of Physics; c1990–1997 [accessed 2013 Feb 4]. http://www.aip.org/pubservs/style/4thed/toc.html.
Albert T. The A–Z of medical writing. London (UK): BMJ Books; 2000.
ASA quick tips for ASA style. Washington (DC): American Sociological Association; 2010 [accessed 2013 Nov 19]. http://www.asanet.org/documents/teaching/pdfs/Quick_Tips_for_ASA_Style.pdf.
ASA style guide. 4th ed. Washington (DC): American Sociological Association; 2010.
ASM style manual for journals and books. Washington (DC): American Society for Microbiology; 1991.
Author packages for publishing with the AMS. Providence (RI): American Mathematical Society; c2013 [accessed 2013 Feb 4]. http://www.ams.org/publications/authors/tex/tex-hold-author-info.
Author resources. Washington (DC): American Geophysical Union; [updated 2013; accessed 2013 Feb 5]. http://publications.agu.org/author-resource-center/.
AMS journals authors guide. Boston (MA): American Meteorological Society; 2012 [accessed 2013 Feb 4]. http://www.ametsoc.org/pubs/authorsguide/pdf_vs/authguide.pdf.
Bates RL, Adkins-Heljeson MD, Buchanan RC, editors. Geowriting: a guide to writing, editing, and printing in earth science. 5th rev. ed. Alexandria (VA): American Geological Institute; 1995.
Berglund C, Saltman D, editors. Communication for health care. Melbourne (Australia): Oxford University Press; 2002.
Browner WS. Publishing and presenting clinical research. 3rd ed. Philadelphia (PA): Lippincott Williams & Wilkins; c2012.
Clare J, Hamilton H, editors. Writing research: transforming data into text. New York (NY): Churchill Livingstone; 2003.
The Columbia Law Review, Harvard Law Review, University of Pennsylvania Law Review, Yale Law Journal, compilers. The bluebook: a uniform system of citation. 19th ed. Cambridge (MA): The Harvard Law Review Association; 2010.
Day RA, Gastel B. How to write and publish a scientific paper. 6th ed. Westport (CT): Greenwood; 2006.
Day RA, Sakaduski N. Scientific English: a guide for scientists and other professionals. 3rd ed. Westport (CT): Greenwood; 2011.
Coghill AM, Garson LR, editors. The ACS style guide: effective communication of scientific information. 3rd ed. Washington (DC): American Chemical Society; 2006.

Frequently asked questions: authors. Piscataway (NJ): The Institute of Electrical and Electronics Engineers, Inc.; 2012 [accessed 2013 Feb 5]. http://www.ieee.org/documents/publications_faq.pdf.
Gastel B. Health writer's handbook. 2nd ed. Ames (IA): Blackwell Publishing; c2004.
Gustavii B. How to write & illustrate a scientific paper. 2nd ed. New York (NY): Cambridge University Press; 2008.
Hall GM, editor. How to write a paper. 4th ed. London (UK): BMJ Books; 2008.
Hancock E. Ideas into words: mastering the craft of science writing. Baltimore (MD): Johns Hopkins University Press; 2003.
Hansen WR, editor. Suggestions to authors of the reports of the United States Geological Survey. 7th ed. Reston (VA): United States Geological Survey; 1991 [modified 2012 Dec 31; accessed 2013 Feb 5]. http://www.nwrc.usgs.gov/lib/lib_sta.htm.
Health Care Communications Group. Writing, speaking & communication skills for health professionals. New Haven (CT): Yale University Press; c2001.
Heifferon B. Writing in the health professions. New York (NY): Pearson Longman; 2005.
Huth EJ. Writing and publishing in medicine. 3rd ed. Baltimore (MD): Williams & Wilkins; c1999.
International Committee of Medical Journal Editors. Uniform requirements for manuscripts submitted to biomedical journals. Philadelphia (PA): ICMJE; [updated 2010 Apr; accessed 2013 Feb 5]. http://www.icmje.org/urm_main.html.
JAMA and Archives Journals. American Medical Association manual of style: a guide for authors and editors. 10th ed. New York: Oxford Press; 2007.
Johnstone MJ. Effective writing for health professionals: a practical guide to getting published. Crows Nest (Australia): Allen & Unwin; 2004.
Knatterud ME. First do no harm: empathy and the writing of medical journal articles. New York (NY): Routledge; 2002.
Lang TA, Secic M. How to report statistics in medicine: annotated guidelines for authors, editors, and reviewers. 2nd ed. Philadelphia (PA): American College of Physicians; 2006.
Maisonneuve H, Enckell PH, Polderman A, Thapa R, Vekony M, editors. The science editors' handbook. Tampere (Finland): EASE; 2003. EASE is the European Association of Science Editors.
Matthews JR, Matthews RW. Successful scientific writing: a step-by-step guide for the biological and medical sciences. 3rd ed. Cambridge (UK): Cambridge University Press; 2007.
McMillan VE. Writing papers in the biological sciences. 4th ed. New York (NY): Bedford/St. Martins; 2006.
Mill D. Content is king: writing and editing online. San Diego (CA): Elsevier Science & Technology Books; 2005.
Montgomery SL. The Chicago guide to communicating science. Chicago (IL): University of Chicago Press; 2003.
Paradis JG, Zimmerman ML. The MIT guide to science and engineering communication. 2nd ed. Cambridge (MA): MIT Press; c2002.
Patrias K. Citing medicine: the NLM style guide for authors, editors, and publishers. 2nd ed. Bethesda (MD): National Library of Medicine; 2007 [accessed 2013 Feb 5]. http://www.nlm.nih.gov/pubs/formats/recommendedformats.html.
Patrias K. Citing medicine: the NLM style guide for authors, editors, and publishers. Supplement: Internet formats. Bethesda (MD): National Library of Medicine; 2001 [accessed 2013 Feb 5]. http://www.nlm.nih.gov/pubs/formats/internet2001.pdf.
Peat J, Elliott E, Baur L, Keena V. Scientific writing: easy when you know how. London (UK): BMJ Books; 2002.
Publication manual of the American Psychological Association. 6th ed. Washington (DC): The Association; c2009.
Publications handbook and style manual. Madison (WI): American Society of Agronomy; c2005 [accessed 2013 Feb 5]. Jointly published with the Crop Science Society of America and the Soil Science Society of America. https://www.agronomy.org/publications/style.
Rubens P, editor. Science and technical writing: a manual of style. 2nd ed. New York (NY): Routledge; 2001.
Swanson E, O'Sean A, Schleyer A. Mathematics into type. Updated ed. Providence (RI): American Mathematical Society; 1999.
Terryberry KJ. Writing for the health professions. Clifton Park (NY): Thomson/Delmar Learning; c2005.
Thomas SA. How to write health sciences papers, dissertations, and theses. New York (NY): Churchill Livingstone; 2000.
Waldron A, Judd P, Miller V, editors. Physical Review style and notation guide. Woodbury (NY): American

Physical Society; 1983 [revised 1993 Feb; minor revision 2005 Jun; minor revision 2011 Jun; accessed 2013 Feb 5]. http://forms.aps.org/author/styleguide.pdf.
Williams D. Writing skills in practice: a practical guide for health professionals. Philadelphia (PA): Jessica Kingsley Publishers; 2002.
Wilkins GA. IAU style manual. Paris (France): International Agronomical Union. 20B:S1-S50; 1989 [accessed 2013 Feb 5]. http://www.iau.org/static/publications/stylemanual1989.pdf.
Wilkinson AM. The scientist's handbook for writing papers and dissertations. Englewood Cliffs (NJ): Prentice-Hall; 1991.
WHO style guide. Geneva (Switzerland): World Health Organization; 2004.
Woodford FP. How to teach scientific communication. Reston (VA): Council of Biology Editors; 1999.
Zeiger M. Essentials of writing biomedical research papers. 2nd ed. New York (NY): McGraw-Hill, Health Professions Division; c2000.

DICTIONARIES

General

Abbreviations. New York (NY): Oxford University Press; 2013. http://public.oed.com/how-to-use-the-oed/abbreviations/.
The American heritage abbreviations dictionary. 3rd ed., updated. Boston (MA): Houghton Mifflin; c2007.
The American heritage dictionary. 5th ed. Boston (MA): Houghton Mifflin Harcourt; c2011. http://ahdictionary.com/.
Cambridge essential English dictionary. 2nd ed. Cambridge (UK): Cambridge University Press; 2004. http://dictionary.cambridge.org/dictionary/essential-british-english/.
The Chambers dictionary. 11th ed. Edinburgh (Scotland): Chambers; c2008.
Cohen SB, editor. The Columbia gazetteer of the world. New York (NY): Columbia University Press; 1998. http://www.columbiagazetteer.org.
Cowie AP, Mackin R. Oxford dictionary of current idiomatic English. Vol. 1, Verbs with prepositions and particles. Oxford (UK): Oxford University Press; 1975.
Cowie AP, Mackin R, McCaig IR. Oxford dictionary of current idiomatic English. Vol. 2, Phrase, clause and sentence idioms. Oxford (UK): Oxford University Press; 1983.
Lindberg CA, compiler. The Oxford American writer's thesaurus. 3rd ed. Oxford (UK): Oxford University Press; 2012.
Longman dictionary of contemporary English. New York (NY): Pearson ELT; 2003. http://www.ldoceonline.com/.
Merriam-Webster's collegiate dictionary. 11th ed. Springfield (MA): Merriam-Webster, Inc.; 2003.
Merriam-Webster's geographical dictionary. 3rd ed., revised and updated. Springfield (MA): Merriam-Webster; c2007.
Pearsall J, Trumble B, editors. The Oxford English reference dictionary. 2nd ed, revised and updated. New York (NY): Oxford University Press; 2003.
Random House unabridged dictionary. 2nd ed. New York (NY): Random House; 1993.
Soanes C, Stevenson A, editors. Concise Oxford English dictionary. 11th ed., revised. New York (NY): Oxford University Press; 2008.
Stahl D, Kerchelich K. Abbreviations dictionary. 10th ed. Boca Raton (FL): CRC Press; 2001. Originated by Ralph De Sola.
Trumble WR, Stevenson A, editors. Shorter Oxford English dictionary on historical principles. 5th ed. Oxford (UK): Oxford University Press; 2002.
Ritter RM, compiler and editor. The Oxford dictionary for writers and editors. 2nd ed. Oxford (UK): Oxford University Press; 2000.

General Science

The American heritage science dictionary. Boston (MA): Houghton Mifflin Harcourt; c2011.
Clugston MJ, editor. The new Penguin dictionary of science. 3rd ed. New York (NY): Penguin Books; 2009.
Collin SMH. Dictionary of science and technology. London (UK): Bloomsbury; 2003.
Collins dictionary of science. Glasgow (Scotland): HarperCollins Publishers; 2005.
Fenna D. A dictionary of weights, measures, and units. Oxford (UK): Oxford University Press; 2002.
Jerrard HG, McNeil DB. A dictionary of scientific units: including dimensionless numbers and scales. 6th ed. New York (NY): Chapman & Hall; 1992.
McGraw-Hill dictionary of scientific and technical terms. 6th ed. New York (NY): McGraw-Hill; c2003.
Random House concise dictionary of science & computers. New York (NY): Random House Reference; c2004.

Wildi T. Metric units and conversion charts: a metrication handbook for engineers, technologists, and scientists. 2nd ed. New York (NY): IEEE Press; c1996.

Biology and Medicine

Dorland's illustrated medical dictionary. 29th ed. London (UK): W.B. Saunders; 2000.
Hale WG, Saunders VA, Margham JP. Dictionary of biology. New ed. Glasgow (Scotland): HarperCollins; 2003.
Hine R, editor. The Facts On File dictionary of biology. 4th ed. New York (NY): Facts On File; 2005.
Hine R, Martin E, eds. A dictionary of biology. 5th ed. Oxford (UK): Oxford University Press; 2004.
Konstantinidis G, compiler. Elsevier's dictionary of medicine and biology: in English, Greek, German, Italian, and Latin. Boston (MA): Elsevier; 2005.
Lawrence E, editor. Henderson's dictionary of biology. 15th ed. New York (NY): Pearson; 2011.
Thain M, Hickman M. The Penguin dictionary of biology. 11th ed. New York (NY): Penguin Books; 2004.
Stedman's abbrev: abbreviations, acronyms & symbols. 3rd ed. Baltimore (MD): Lippincott Williams & Wilkins; 2003.
Stedman's medical dictionary. 28th ed. Philadelphia (PA): Lippincott Williams & Wilkins; c2005.
Stedman's medical dictionary for the health professions and nursing. Illus. 7th ed. Philadelphia (PA): Lippincott Williams & Wilkins; 2011.
Tsur SA, compiler. Elsevier's dictionary of abbreviations, acronyms, synonyms, and symbols used in medicine. 2nd enl. ed. Amsterdam (Netherlands): Elsevier; 2004.

Chemistry

Daintith J, editor. A dictionary of chemistry. 6th ed. Oxford (UK): Oxford University Press; 2008.
Daintith J, editor. The Facts On File dictionary of chemistry. 4th ed. New York (NY): Facts On File; 2005.
Lewis RJ Sr, reviser. Hawley's condensed chemical dictionary. 15th ed. New York (NY): Wiley; c2007.
McGraw-Hill dictionary of chemistry. 2nd ed. New York (NY): McGraw-Hill; c2003.

Engineering

ASTM dictionary of engineering science & technology. 10th ed. West Conshohocken (PA): ASTM International; 2005. Sponsored by ASTM Committee E02 on Terminology.
Bignami M, compiler. Elsevier's dictionary of engineering: in English, German, French, Italian, Spanish and Portuguese/Brazilian. Amsterdam (Netherlands): Elsevier; 2004.
Laplante PA, editor. Dictionary of computer science, engineering, and technology. Boca Raton (FL): CRC Press; c2001.
McGraw-Hill dictionary of engineering. 2nd ed. New York (NY): McGraw-Hill; c2003.
Timings RL, Twigg P. The pocket illustrated dictionary of engineering terms. Boston (MA): Butterworth Heinemann; 2001.

Geology

Lapidus DF. Collins dictionary geology. New ed. rev. MacDonald J, Burton C, revisers. Glasgow (Scotland): HarperCollins; 2003.
McGraw-Hill dictionary of geology and mineralogy. 2nd ed. New York (NY): McGraw-Hill; c2003.

Mathematics

Clapham C. The concise Oxford dictionary of mathematics. 4th ed. Oxford (UK): Oxford University Press; 2009.
Daintith J, Clark J, editors. The Facts On File dictionary of mathematics. 4th ed. New York (NY): Facts on File; 2005.
McGraw-Hill dictionary of mathematics. 2nd ed. New York (NY): McGraw-Hill; c2003.
Nelson D, editor. The Penguin dictionary of mathematics. 4th ed. New York (NY): Penguin Books; 2008.

Physics

Daintith J, Rennie R, editors. The Facts On File dictionary of physics. 4th ed. New York (NY): Facts On File; c2005.
Dictionary of physics. Basingstoke (UK): Palgrave Macmillan; 2004. An updated translation of the Lexicon der Physik, published in German.
A dictionary of physics. 6th ed. Oxford (UK): Oxford University Press; 2012.
McGraw-Hill dictionary of physics. 3rd ed. New York (NY): McGraw-Hill; c2003.

GUIDES TO USAGE AND PROSE STYLE

Many books offer guidance on clear writing and appropriate usage. This list includes primarily guides with broad scope.

Bernstein TM. The careful writer: a modern guide to English usage. New York (NY): Atheneum; 1973.
Brown RW. Composition of scientific words. Rev. ed. Washington (DC): Smithsonian Institution Press; 1956.
Chalker S. Oxford dictionary of English grammar. New York (NY): Oxford University Press; 1998.
Fogarty M. Grammar Girl's quick and dirty tips for better writing. New York (NY): Holt Paperbacks; 2008.
Fowler HW. A dictionary of modern English usage. 2nd ed. Gower E, reviser. Oxford (UK): Oxford University Press; 1965.
Fulwiler T, Hayakawa AR. The Blair handbook. 5th ed. New York (NY): Longman; c2009.
Good CE. Who's (oops whose) grammar book is this anyway? All the grammar you need to succeed in life. New York (NY): Barnes & Noble Books; 2002.
Gorrell D. Style and difference: a guide for writers. Boston (MA): Houghton Mifflin Co.; c2005.
Gower E. The complete plain words. Greenbaum S, Whitcut J, revisers. Boston (MA): DR Godine; 2002.
Graves R, Hodge A. The use and abuse of the English language. 2nd ed. Washington (DC): Marlowe & Co.; 1995. Formerly titled The Reader over Your Shoulder.
Greenbaum S. The Oxford English grammar. London (UK): Oxford University Press; 1996.
Halliday MAK. An introduction to functional grammar. 3rd ed. Matthiessen CMIM, reviser. London (UK): Arnold; 2004.
Lanham RA. Revising prose. 5th ed. New York (NY): Longman; c2006.
Lauther H. Lauther's complete punctuation thesaurus of the English language. Boston (MA): Branden; 1991.
Law J, editor. The language toolkit: practical advice on English grammar and usage. Oxford (UK): Oxford University Press; 2002.
Mager NH, Mager SK. Prentice Hall encyclopedic dictionary of English usage. 2nd ed. Englewood Cliffs (NJ): Prentice Hall; c1993.
Merriam-Webster's dictionary of English usage. Springfield (MA): Merriam-Webster, Inc.; c1994.
Merriam-Webster's guide to punctuation and style. 2nd ed. Springfield (MA): Merriam-Webster; c2001.
Peters P. The Cambridge guide to English usage. Cambridge (UK): Cambridge University Press; 2004.
Seely J. Oxford A–Z of grammar and punctuation. 2nd ed. Oxford (UK): Oxford University Press; 2010.
Truss L. Eats, shoots & leaves: the zero tolerance approach to punctuation. New York (NY): Gotham Books; c2006.
Villemaire L, Villemaire D. Grammar and writing skills for the health professional. Albany (NY): Delmar/Thomson Learning; c2001.
Walsh B. The elephants of style: a trunkload of tips on the big issues and gray areas of contemporary American English. New York (NY): McGraw-Hill; c2004.
Walsh B. Lapsing into a comma: a curmudgeon's guide to the many things that can go wrong in print—and how to avoid them. Lincolnwood (IL): Contemporary Books; c2000.
Williams JM, Colomb GG. Style: the basics of clarity and grace. 4th ed. New York (NY): Longman; c2010.
Wilson KG. The Columbia guide to standard American English. New York (NY): MJI Books; 1998.
Witte F. Basic grammar and usage for biomedical communicators. Dubuqe (IA): Kendall/Hunt Publishing; 2003.

EDITING

Angelbeck JD, Hughes S. Editing on screen: effective working practice, a teach-yourself course. South Woodchester (UK): EPE Books; 1999.
Bowles DA, Borden DL. Creative editing. 6th ed. Belmont (CA): Wadsworth Publishing; 2010.
Brooks BS, Pinson JL. The art of editing: in the age of convergence. 9th ed. Boston (MA): Pearson; c2008.
Burnard L, O Keeffe KO, Unsworth J, editors. Electronic textual editing. New York (NY): Modern Language Association of America; c2006.
Butcher J, Drake C, Leach M. Butcher's copy-editing: the Cambridge handbook for editors, copy-editors, and proofreaders. 4th ed. Cambridge (UK): Cambridge University Press; 2006.
Cain JS. Eye on editing 1: developing writing skills through grammar. Boston (MA): Pearson ESL; 2001.
Cain JS. Eye on editing 2: developing editing skills for writing. Boston (MA): Pearson ESL; 2002.
Camp SC. Developing proofreading and editing skills. 5th ed. Columbus (OH): McGraw-Hill School Education Group; 2004.
Campbell S. The editing book: a guide to clear and forceful writing. Victoria (BC): Trafford Publishing; 2001.

Cheney TA. Getting the words right: how to revise, edit and rewrite. Cincinnati (OH): F & W Publications, Inc.; 2005.
Council of Biology Editors, Journal Procedures and Practice Committee. Editorial forms: a guide to journal management. Bethesda (MD): The Council; 1987.
Dale C, Pilgrim T. Fearless editing: crafting words and images for print, Web, and public relations. Boston (MA): Allyn & Bacon, Incorporated; 2004.
Dodds J. The ready reference handbook: writing, revising, editing. 3rd ed. White Plains (NY): Longman Publishing Group; 2002.
Editors' Association of Canada. Meeting editorial standards. Rev. ed. Concord (ON): Captus Press; 2000.
Einsohn A. The copyeditor's handbook: a guide for book publishing and corporate communications, with exercises and answer keys. 3rd ed. Berkeley (CA): University of California Press; 2011.
Ellis BG. The copy-editing and headline handbook. Cambridge (MA): Perseus Pub.; c2001.
Flann E, Hill B. The Australian editing handbook. 2nd ed., rev. and updated. Brisbane (Australia): John Wiley & Sons Australia, Ltd.; 2004.
Josephson D, Hidden L. Write it right: the ground rules for self-editing like the pros. Ridgeland (SC): Cameo Publications, LLC; 2005.
Judd K. Copyediting: a practical guide. 3rd ed. Menlo Park (CA): Crisp Learning; c2001.
LaRocque P. The concise guide to copy editing: preparing written work for readers. Oak Park (IL): Marion Street Press, Inc.; 2003.
Lieb T. Editing for clear communication. Burr Ridge (IL): McGraw-Hill Higher Education; 2001.
O'Connor M. How to copyedit scientific books and journals. Baltimore (MD): Williams & Wilkins; 1986.
Professional editorial standards. Toronto (ON): Editors' Association of Canada; 2009 [accessed 2013 Feb 6]. http://www.editors.ca/resources/eac_publications/pes/index.html.
Quinn S. Digital sub-editing and design. San Diego (CA): Elsevier Science & Technology Books; 2001.
Rew LJ. Editing for writers. Upper Saddle River (NJ): Prentice Hall; c1999.
Rooney EJ, Witte OR. Copy editing for professionals. Champaign (IL): Stipes Publishing L.L.C.; 2000.
Ruby J. Electronic editing with Microsoft Word and Adobe Acrobat: skills and drills workbook. Riva (MD): IconLogic, Inc.; 2003.
Rude C, Eaton A. Technical editing. 5th ed. White Plains (NY): Longman; 2010.
Russial J. Strategic copy editing. New York (NY): Guilford Publications, Inc.; 2003.
Speck BW. Editing: an annotated bibliography. Westport (CT): Greenwood Press; 1991.
Speck BW, Hinnen DA, Hinnen K, compilers. Teaching revising and editing: an annotated bibliography. Westport (CT): Greenwood Press; 2003.
Stainton EM. The fine art of copyediting: including advice to editors on how to get along with authors, and tips on style for both. 2nd ed., rev. and expanded. New York (NY): Columbia University Press; c2002.
Tarutz J. Technical editing: the practical guide for editors and writers. Boston (MA): Addison-Wesley Longman, Inc.; 2000.
Trinkle DA, Mullins K. The elements of e-style. Armonk (NY): M.E. Sharpe Inc.; 2002.
Weil BH. Technical editing. Temecula (CA): Textbook Publishers; 2003.
White JV. Editing by design: for designers, art directors, and editors. New York (NY): Allworth Press; 2003.
Wilber R. The writer's handbook for editing & revision. Lincolnwood (IL): NTC Pub. Group; c1997.

EDITORS' NEWSLETTERS

Editing Matters: the Magazine for Editors and Proofreaders. London (UK): Society for Editors and Proofreaders. 2004– .
Editor & Publisher. New York (NY): Editor & Publisher Co. Vol. 1, 1901– . http://www.editorandpublisher.com/Newsletter/.
The Editorial Eye. Alexandria (VA): EEI Press. Vol. 1, 1978– .
The Scholarly Kitchen. Wheat Ridge (CO): Society for Scholarly Publishing. http://scholarlykitchen.sspnet.org/.
Science Editor: a publication of the Council of Science Editors. Vol. 23. Wheat Ridge (CO): The Council; 2000– . Continues: CBE Views.

PUBLISHING

Abel R, Newlin L, editors. Scholarly publishing: books, journals, publishers, and libraries in the twentieth century. New York (NY): Wiley; c2002.
Bailey CW Jr. Scholarly electronic publishing bibliography. Ver. 64. Houston (TX): University of Houston Libraries; c1996–2006 [accessed 2013 Feb 6]. http://repositories.tdl.org/uh-ir/handle/10657/162.
Balkin R. A writer's guide to book publishing. 3rd ed., rev. and expanded. New York (NY): Plume; c1994.

Bergsland D. Introduction to digital publishing. Clifton Park (NY): Thomson/Delmar Learning; c2002.
Bolter J. Writing space: computers, hypertext, and the remediation of print. 2nd ed. Mahwah (NJ): Lawrence Erlbaum Associates; c2001.
Clark GN. Inside book publishing. 4th ed. London (UK): Routledge; 2008.
Cope B, Kalantzis D, editors. Print and electronic text convergence. Altona (Australia): Common Ground Pub.; 2001.
Cope B, Mason D, editors. Digital book production and supply chain management. Altona (Australia): Common Ground Pub.; 2001.
Cox J, Cox L. Scholarly publishing practice: the ALPSP report on academic journal publishers' policies and practices in online publishing. Worthing (UK): Association of Learned & Professional Society Publishers; 2003.
Crawford SY, Hurd JM, Weller AC. From print to electronic: the transformation of scientific communication. Medford (NJ): Information Today; 1996.
Dennis EE, LaMay CL, Pease EC, editors. Publishing books. New Brunswick (NJ): Transaction Publishers; 1997.
Dolin PA. Exploring digital workflow. Clifton Park (NY): Thomson Delmar Learning; 2006.
Epublishing. Dallas (TX): Writers Write, Inc.; c1997–2012 [accessed 2013 Feb 6]. http://www.writerswrite.com/epublishing/.
Feather J. Communicating knowledge: publishing in the 21st century. Munich (Germany): K.G. Saur; 2003.
Friedlander A, Bessette RS. The implications of information technology for scientific journal publishing: a literature review. Arlington (VA): National Science Foundation (US); 2003 [accessed 2013 Feb 6]. http://www.nsf.gov/statistics/nsf03323/pdf/nsf03323.pdf.
Gans JS, editor. Publishing economics: analyses of the academic journal market in economics. Northampton (MA): E. Elgar; c2000.
Gardiner J. The publishing handbook. London (UK): Routledge; 2002.
Henke H. Electronic books and epublishing: a practical guide for authors. New York (NY): Springer; c2001.
Interactive electronic publishing. Luxembourg (Luxembourg): European Union, Office for Official Publications of the European Communities; 2000.
Kasdorf WE, editor. The Columbia guide to digital publishing. New York (NY): Columbia University Press; c2003.
Lee M. Bookmaking: the illustrated guide to editing, design and production. 3rd ed. New York (NY): W.W. Norton & Company, Incorporated; 2009.
National Academy of Sciences (US), Committee on Electronic Scientific, Technical, and Medical Journal Publishing. Electronic scientific, technical, and medical journal publishing and its implications: report of a symposium. Washington (DC): National Academies Press; c2004. Also available at http://www.nap.edu/openbook.php?isbn=0309091616.
Page G, Campbell R, Meadows J. Journal publishing. Rev., updated and expanded ed. New York (NY): Cambridge University Press; 1997.
Peacock J. Book production. London (UK): Blueprint; 1995.
Potter CN. Who does what and why in book publishing. Secaucus (NJ): Carol Pub. Group; c1990.
Poynter D. Book production: composition, layout, editing and design; getting it ready for printing. Santa Barbara (CA): Para Publishing; 2000.
Smith GM. The peer-reviewed journal: a comprehensive guide through the editorial process. 3rd ed. New Orleans (LA): Chatgris Press; 2000.
Tenopir C, King DW. Towards electronic journals: realities for scientists, librarians, and publishers. Washington (DC): Special Libraries Association; c2000.

GRAPHICS AND DESIGN

Bringhurst R. The elements of typographic style. 3rd ed. Point Roberts (WA): Hartley & Marks; 2004.
Cleveland WS. Visualizing data. Summit (NJ): Hobart Press; 1993.
Council of Biology Editors, Scientific Illustration Committee. Illustrating science: standards for publication. Bethesda (MD): The Council; 1988.
Dillon A. Designing usable electronic text. 2nd ed. New York (NY): Taylor & Francis; 2003.
Felici J. The complete manual of typography: a guide to setting perfect type. 2nd ed. Berkeley (CA): Adobe Press; 2011.
Heller S, Meggs PB, editors. Texts on type: critical writings on typography. New York (NY): Allworth Press; 2001.
Hill W. The complete typographer: a manual for designing with type. 3rd ed. Upper Saddle River (NJ): Prentice Hall; 2010.

Hodges ERS, ed. The Guild handbook of scientific illustration. 2nd ed., rev. Hoboken (NJ): John Wiley & Sons, Inc.; 2003.
Holleley D. Digital book design and publishing. Elmira Heights (NY): Clarellen; 2001.
Lupton E. Thinking with type: a primer for designers, writers and editors. New York (NY): Princeton Architectural Press; 2004.
Manual of style: being a compilation of the typographical rules in force at the University of Chicago Press, to which are appended specimens of types in use. Chicago (IL): University of Chicago Press; 2003.
Skopec D. Digital layout for the Internet and other media. Heimann M, illustrator; Luhman NJ, translator. New York (NY): Sterling Pub. Co., Inc.; c2003.
Tufte E. Envisioning information. Cheshire (CT): Graphics Press; 1990.
Tufte ER. The visual display of quantitative information. 2nd ed. Cheshire (CT): Graphics Press; 2001.
Spilker B, Schoenfelder J. Presentation of clinical data. New York (NY): Raven Press; c1990.
Swanson G, editor. Graphic design and reading: explorations of an uneasy relationship. New York (NY): Allworth Press; c2000.

STANDARDS FOR EDITING AND PUBLISHING

In the United States, standards issued by the National Information Standards Organization (NISO) can be purchased from NISO Press (http://www.niso.org/). A NISO-developed and approved standard becomes an American National Standard after the American National Standards Institute (ANSI) verifies that the process for approval has met the ANSI criteria.

Standards of the International Organization for Standardization (ISO) can be purchased from ISO (http://www.iso.org/) or the American National Standards Institute (http://www.ansi.org/).

In Canada, the standards issued by the Canadian Standards Association and by ANSI, ISO, and other standards organization can be purchased from the Standards Council of Canada (http://www.scc.ca/).

British standards can be purchased from BSI British Standards (http://www.bsigroup.com/); in the United States, they can be purchased from the American National Standards Institute.

See the individual chapters of this manual for specific standards.

INDEX

The letter "A" preceding a number denotes an appendix; the letter "F" preceding a number denotes a figure; the letter "T" preceding a number denotes a table. Entries in quotation marks refer to words as such.